Differential Equations

A Modeling Perspective

To use or not to use: that is the question:
Whether 'tis nobler in the mind to suffer
The stings and errors of outrageous computation,
Or to apply technology against a sea of questions,
And by its power resolve them?
To model: perchance to visualize: ay, there's the key:
For in that model what insights may come. . .

Differential Equations

A Modeling Perspective

Robert L. Borrelli
Courtney S. Coleman

•

Harvey Mudd College

John Wiley & Sons, Inc.

New York • Chichester • Weinheim • Brisbane • Singapore • Toronto

MATHEMATICS EDITOR	Barbara Holland
MARKETING MANAGER	Leslie Hines
SENIOR PRODUCTION EDITOR	Elizabeth Swain
DESIGNER	Kevin Murphy
PHOTO RESEARCHER	Kim Khatchatourian
COVER ART	Wides & Holl
ILLUSTRATIONS	Will Suckow

Cover Photo: © Richard Megna / Fundamental Photographs

This book was set in Times Roman using LATEX and macros written
especially for this project by David Richards. This book was printed
and bound by Donnelley/Willard. The cover was printed by Lehigh Press.

Library of Congress Cataloging in Publication Data:
Borrelli, Robert L.
    Differential equations : a modeling perspective / Robert L.
Borrelli, Courtney S. Coleman.
        p.   cm.
    Includes index.
    ISBN 0-471-04230-7 (cloth : alk. paper)
        1. Differential Equations.   I. Coleman, Courtney S.
    II. Title.
    QA371.B743   1998
    515'.35--dc21
                                                                97-28427
                                                                CIP

Printed in the United States of America

10   9   8   7   6   5   4   3   2   1

Preface

Differential equations are a powerful tool in constructing mathematical models for the physical world. Their use in industry and engineering is so widespread and they perform their task so well that they are clearly one of the most successful of modeling tools. This is an exciting time to study differential equations because interactive computer solvers can quickly and easily generate striking graphical displays that can provide amazing insights into the properties of dynamical systems.

Course and Text Objectives

This is an introductory textbook for students of science, mathematics, and engineering that features modeling and graphical visualization as central themes. Differential systems and numerical methods are introduced early and students are encouraged to use numerical solvers from the start. Our goal is to present this material in a way that's clear and understandable to students at all levels, that motivates them to ask "why" and that communicates to them our enthusiasm and excitement for the study of ODEs.

While we adopt the modern view of differential systems as evolving dynamical systems, we retain the topics and objectives of a traditional course. We introduce modern topics such as sensitivity, long-term behavior, bifurcation, and chaos, but we also present the solution formulas and theory expected in a first course.

Prerequisites

This text presumes a familiarity with single variable calculus. A few sections (mostly in the latter chapters) require some basic knowledge of partial differentiation. A course in linear algebra is not assumed; linear concepts are developed as needed. This text works best if students have access to numerical solvers, but it can be used successfully even without any computers at all.

Available Solvers

No programming knowledge is required to use this text. There are many differential equations solvers available today which do not require the user to be computer literate. Solvers can be found for nearly any platform, be it a hand-held calculator, personal computer, workstation, or large computer system. We presume no specific solver.

Highlights

•

- *Dynamical Systems Approach.* Throughout the text we adopt a dynamical systems approach that models natural processes that evolve in time. We treat the basic questions of existence, uniqueness, long-term behavior and sensitivity to the data as recurring themes.

- *Mathematical Modeling.* Every picture tells a story: building a model is like drawing a picture of the system, and interpreting the solution of the model equations is like telling a story. There are a great number of models in the text from which to choose. Some sections are completely devoted to a single model, but for the most part models comprise only a portion of a section, and the text allows flexibility in the treatment of modeling.

- *Graphical Visualization Emphasized.* The solutions of an ordinary differential equation are functions whose graphs are curves. These curves may be computer-generated and provide compelling visual evidence of mathematical deductions and a clear understanding of complicated solution formulas. Every graph in this text is accompanied by the data necessary to reproduce it. These graphs are the actual output of a numerical solver, not artist renderings. The text and the hundreds of graphs of solutions emphasize this visual connection with the theory.

- *Numerical Solvers Introduced Early.* With the ready availability of excellent and inexpensive numerical solvers, it makes a great deal of sense to introduce a numerical solution approach very early so that students can begin to examine the geometry of solutions and the way solutions change when the elements of a differential equation are perturbed. The introduction of computers into the course leads to heightened interest in understanding dynamical systems. The basic properties of dynamical systems serve as a valuable tool for interpreting visual displays of solutions of differential equations.

- *Systems Introduced Early.* From the start, simple systems of differential equations are treated matter-of-factly in the modeling process, because it is natural to do so. This does not present a problem since computer solvers can handle a system of first-order differential equations as easily as a single differential equation.

- *Appendices.* Appendix A contains proofs of the mathematical underpinnings of first-order differential equations. Appendix B contains useful background material.

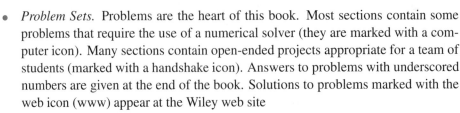

- *Problem Sets.* Problems are the heart of this book. Most sections contain some problems that require the use of a numerical solver (they are marked with a computer icon). Many sections contain open-ended projects appropriate for a team of students (marked with a handshake icon). Answers to problems with underscored numbers are given at the end of the book. Solutions to problems marked with the web icon (www) appear at the Wiley web site

WWW

http://www.wiley.com/college/borrelli

A Course Based on this Text
•

Many courses based on this text are possible; here is a semester course that does not assume any prior exposure to differential equations:

    Chapter 1:   Sections 1.1–1.7
    Chapter 2:   Sections 2.1–2.6
    Chapter 3:   Sections 3.1–3.6
    Chapter 4:   Section 4.1, 4.2, or 4.3
    Chapter 5:   Sections 5.1, 5.2, and 5.3 or 5.4
    Chapter 7:   Sections 7.1–7.7
    Chapter 8:   Sections 8.1, 8.2
    Chapter 9:   Sections 9.1, 9.2

The sections from Chapters 8 and 9 can be replaced by material from Chapter 10 or 11, if it is more appropriate for the course.

Supplements
•

A Student Resource Manual (SRM) gives complete solutions (along with graphs) to every other part of every odd-numbered problem (group problems not included). Also included are longer proofs of theorems in the text where appropriate. A sampling of solutions in the SRM is found at the Wiley web site for problems in the text marked with www.

The authors, together with William Boyce, have produced a collection of computer graphics experiments and modeling projects (also published by Wiley) with the title *Differential Equations Laboratory Workbook*. The Workbook supplements a course on ordinary differential equations. An appendix in the Workbook contains a telegraphic overview of three modeling environments: rate processes, electrical circuits, and mechanics. The Workbook also contains many graphs of solutions of differential equations which serve as a handy visual reference. This text was written to be independent of the Workbook, although the two would obviously fit well together.

Supplements for the instructor are available from the publisher.

Dedication
•

To my wife Ursula Marie whose patience and understanding were at times stretched to the limit by the outrageous work schedule required to produce this text. Petrarca said it well when he wrote: *Tu che dentro mi vedi e'l mio mal senti \\ et sola puoi finir tanto dolore \\ con la tua ombra acqueta i miei lamenti.*   (RLB)

This book is dedicated to my wonderful and patient wife Julia and our children, their spouses, and our grandchildren: David, Sally, Elizabeth, Brittany, Rebecca, Timothy, Margaret, Chuck, Erica, Katie, Diane, David, and Christopher.   (CSC)

Acknowledgements

We indebted to three wonderful people who have worked with us on this book. Tony Leneis is the principal author of ODEToolkit, a robust interactive command script interface over the ODE solver DEQSolve that was used to produce all the graphs. Dave Richards has a remarkable eye for composition and artwork, and his LaTeX typesetting ability is evident on every page. Jenny Switkes helped us with all aspects of the text. Her good humor, patience, and remarkable abilities are greatly appreciated. It was a real pleasure to work with these three talented colleagues.

Our appreciation also goes out to many others who contributed much to this book. Professor Beverly West gave us many insightful suggestions. Will Suckow designed the delightful little computer guys that appear throughout, and Sally Arroyo gave us excellent advice on design matters. Kevin Carosso, Ned Freed, and Dan Newman authored DEQSolve, which is based on LSODA, a descendent of C. W. Gear's DIFFSUB, part of ODEPACK, developed by Alan Hindmarsh at Lawrence Livermore National Laboratories. We also want to thank the mathematics department and the administration of Harvey Mudd College for their support and encouragement.

We especially extend our gratitude to Tiffany Arnal, Claire Launay, and Joel Miller who contributed in many important ways to the final phases of the project. We also want to thank the following students who have done much to improve the text and the problems: Aaron Archer, Patri Forwalter-Friedman, Motoya Kohtani, Aaron Lamb, Christie Lee, Dan Lopez, Susan McMains, Robert Prestegard, Justin Radick, Marie Snipes, Kal Wong, Xuemei Wu, Kaiqi Xiong, and Bob Zirpoli.

The authors are indebted to the reviewers of the early versions of this text; their comments and suggestions were extremely valuable. We particularly want to thank Professors David Arnold, Ulrich Daepp, Steven R. Dunbar, Rahim Eighanmi, Richard Elderkin, Mark Farris, Roland di Franco, Mark Fuller, Ben Fusaro, Matthias Kawski, David Kraines, David Lerner, Zhongyuan Li, Michael Montano, Michael Moody, Mike Pepe, Karl E. Petersen, Bhagat Singh, Ed Spitznagel, Kenneth Stolarsky, David Voss, Rich West, and Christina M. Yuengling. In addition we want to thank the following student reviewers: Matthew Anderson, Shannon Holland, Kevin Huffenberger, Itai Seggev, and Treasa Sweek. The responsibility for any errors, of course, is ours.

And finally, we owe a debt of gratitude to Barbara Holland, our Wiley Mathematics Editor, for her generous support, advice, and encouragement during the writing of this text. Our warmest thanks, Barbara!

R. L. Borrelli   borrelli@hmc.edu
C. S. Coleman   coleman@hmc.edu
Claremont, August 1997

Student Perspective

We are the students who helped refine this textbook. Our experience with this text began when we took Differential Equations where we used a preliminary edition of the book. Then somehow (we don't quite know how) we were sucked into this project. When we started, we really had no idea what we were getting into. We have discovered one thing: writing a good, comprehensive text is hard! If we had known how much work was involved, especially in the final stages of development, we all might have chosen to have a summer vacation instead. But we really do believe that the final result was worth the effort.

We had a mission: to use our own experiences with the text to make it better for the student. We critically read every chapter, and made suggestions on how explanations and examples could be made easier to understand. We were familiar with some of the chapters from our own class, so we already knew what changes we felt would improve the text. Other chapters were completely new to us, and we were truly students learning the material for the first time (without a professor's guidance).

By the time we had gotten through each chapter, the authors had stacks of suggestions to sift through. They considered all of the comments and used them to come to a consensus not only between themselves, but also with us, on how the book could be improved. Some sections of the book needed only minor changes; other parts were nearly rewritten. We can each see our comments and suggestions not only in the minor changes but the major ones as well. We believe that these changes have made the book better.

This text started out a bit different from the rest, and it still is. Its emphasis on modeling gives differential equations a purpose. This book not only teaches the student how to solve the differential equations that model a situation, but how to create that model as well. The emphasis on graphical analysis and visualization makes the concepts more intuitive. The numerous examples help drive the ideas home. This new approach will make the concepts of differential equations and their uses more accessible to students.

We hope that our work, from the student's perspective, will help the students who use this text. Good luck!

Tiffany Arnal   '00
Claire Launay   '00
Joel Miller   '00

Contents

1 • First-Order Differential Equations and Models　　　**1**

1.1　A Modeling Adventure　1
1.2　Visualizing Solution Curves　10
1.3　The Search for Solution Formulas　15
1.4　Modeling with Linear ODEs　24
1.5　Introduction to Modeling and Systems　34
1.6　Separable Differential Equations　47
1.7　Planar Systems and First-Order ODEs　56
1.8　Cold Pills　67
1.9　Change of Variables and Pursuit Models　76
　　　Solution Formula Techniques Involving First-Order ODEs　85

2 • Initial Value Problems and Their Approximate Solutions　　　**87**

2.1　Existence and Uniqueness　87
2.2　Extension and Long-Term Behavior　96
2.3　Sensitivity　107
2.4　Introduction to Bifurcations　115
2.5　Approximate Solutions　122
2.6　Computer Implementation　131
2.7　Euler's Method, the Logistic ODE, and Chaos　137

3 • Second-Order Differential Equations　　　**147**

3.1　Springs: Linear and Nonlinear Models　147
3.2　Second-Order ODEs and Their Properties　157
3.3　Undriven Constant Coefficient Linear ODEs, I　167
3.4　Undriven Constant Coefficient Linear ODEs, II　177
3.5　Periodic Solutions and Simple Harmonic Motion　185
3.6　Driven Constant Coefficient Linear ODEs　190
3.7　The General Theory of Linear ODEs　202
　　　A Snapshot View of Polynomial Operators　212

4 • *Applications of Second-Order Differential Equations*   **213**

    4.1   Newton's Laws and the Pendulum   213
    4.2   Beats and Resonance   224
    4.3   Frequency Response Modeling   231
    4.4   Electrical Circuits   241

5 • *Systems of Differential Equations*   **251**

    5.1   First-Order Systems   251
    5.2   Properties of Systems   263
    5.3   Models of Interacting Species   276
    5.4   Predator-Prey Models   285
    5.5   The Possum Plague: A Model in the Making   294

6 • *The Laplace Transform*   **299**

    6.1   Introduction to the Laplace Transform   299
    6.2   Calculus of the Transform   307
    6.3   Applications of the Transform: Car Following   317
    6.4   Convolution   326
    6.5   Convolution and the Delta Function   331
         Tables of Laplace Transforms   336

7 • *Linear Systems of Differential Equations*   **339**

    7.1   Tracking Lead Through the Body   339
    7.2   Overview of Vectors and Matrices   345
    7.3   Systems of Linear Equations   352
    7.4   Eigenvalues and Eigenvectors of Matrices   362
    7.5   Undriven Linear Systems with Constant Coefficients   371
    7.6   Undriven Linear Systems: Complex Eigenvalues   381
    7.7   Orbital Portraits   389
    7.8   Driven Systems and the Matrix Exponential   400
    7.9   Steady States of Driven Linear Systems   409
   7.10   Lead Flow, Noise Filter: Steady States   418
    7.11   The Theory of General Linear Systems   425

8 • *Stability*   **433**

    8.1   Stability and Linear Systems   433
    8.2   Stability of a Nearly Linear System   442
         Stability of Perturbed Planar Systems   451
    8.3   Conservative Systems   452
    8.4   Lyapunov Functions   462

9 • *Cycles, Bifurcations, and Chaos* 471

    9.1   Cycles 471
    9.2   Long-Term Behavior 480
    9.3   Bifurcations 488
    9.4   Chaos 500

10 • *Fourier Series and Separation of Variables* 515

    10.1   Vibrations of a String 515
    10.2   Orthogonal Functions 525
    10.3   Fourier Series and Mean Approximation 531
    10.4   Fourier Trigonometric Series 539
    10.5   Half-Range and Exponential Fourier Series 547
    10.6   Sturm-Liouville Problems 553
    10.7   Separation of Variables 557
    10.8   The Heat Equation: Optimal Depth for a Wine Cellar 566
    10.9   Laplace's Equation 578

11 • *Series Solutions: Bessel Functions, Legendre Polynomials* 585

    11.1   Aging Springs and Steady Temperatures 585
    11.2   Series Solutions Near an Ordinary Point 592
    11.3   Legendre Polynomials 601
    11.4   Regular Singular Points 608
    11.5   Series Solutions Near Regular Singular Points, I 615
    11.6   Bessel Functions 622
    11.7   Series Solutions Near Regular Singular Points, II 634
    11.8   Steady Temperatures in Spheres and Cylinders 645

A • *Basic Theory of Initial Value Problems* 653

    A.1   Uniqueness 653
    A.2   The Picard Process for Solving an Initial Value Problem 655
    A.3   Extension of Solutions 663
    A.4   Sensitivity of Solutions to the Data 665

B • *Background Information* 671

    B.1   Engineering Functions 672
    B.2   Power Series 674
    B.3   Complex Numbers and Complex-Valued Functions 677
    B.4   Useful Algebra and Trigonometry Formulas 680
    B.5   Useful Results from Calculus 681
    B.6   Scaling and Units 685

• *Answers to Selected Problems* 691

• *Index* 699

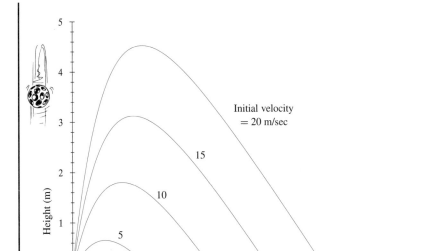

Initial velocity
= 20 m/sec

15

10

5

Throw a ball into the air. Does it take longer
to go up, or to come down? Check out
Example 1.5.5.

First-Order Differential Equations and Models

How many tons of fish can be harvested each year without killing off the population? When you double the dose of your cold medication, do you fall asleep in your math class? Does it take a ball longer to rise than to fall? In this chapter we model natural processes with differential equations in order to answer these and many other questions.

1.1 A Modeling Adventure

Differential equations provide powerful tools for explaining the behavior of dynamically changing processes. We will use them to answer questions about processes that are hard to answer in any other way.

Take a look, for example, at the fish population in one of the Great Lakes. What harvesting rates maintain both the population and the fishing industry at acceptable levels? We will use differential equations to find out how the population changes over time given birth, death, and harvesting rates.

The clue that a differential equation may describe what is going on lies in the words "birth, death, and harvesting rates." The key word here is "rates." Rates are derivatives with respect to time, but what quantity is to be differentiated in this case? Let's measure the population of living fish at time t by the total tonnage $y(t)$, and the time in years. Then the net rate of change of the fish population in tons of fish per year is $dy(t)/dt$, written as $y'(t)$ or simply y'. At any time t, we have

$$y'(t) = \text{ Birth rate } - \text{ Death rate } - \text{ Harvest rate} \tag{1}$$

where we measure all the rates in tons per year. We suppose that fish immigration and emigration rates from rivers that meet the lake cancel each other out, so they don't need to appear in (1). Close observation of many species over many years suggests that birth and death rates are each roughly proportional to the size of the population:

$$\text{Birth rate at time } t: \quad by(t)$$
$$\text{Death rate at time } t: \quad (m + cy(t))y(t)$$

where b, m, and c are nonnegative proportionality constants. The extra twist here is that the natural mortality coefficient m is augmented by the term $cy(t)$ which accounts for overcrowding. As a population increases in a fixed habitat, the death rate often increases much faster than can be accounted for by a single constant coefficient m. The overcrowding term is needed to model this accelerated mortality factor.

Now let's pull all of these bits and pieces together and create a model.

Making the Mathematical Model

Denoting the harvest rate by H and using the law given by (1), we have a differential equation for $y(t)$:

☞ If H is a positive constant then this model is *constant rate harvesting*.

$$y' = by - (m + c y)y - H$$

or

$$y' = ay - cy^2 - H \tag{2}$$

where $a = b - m$ is assumed to be positive. An equation like (2) that involves a to-be-determined function of a single variable and its derivatives is called an *ordinary differential equation* (ODE, for short).

Referring to our fishing model, we note that observation of an actual fish population gives us a fairly good idea of the birth and death rates (so we suppose that a and c are known), and the harvest rate H is under our control. That leaves the tonnage $y(t)$ to be determined from ODE (2). A function $y(t)$ for which

$$y'(t) = ay(t) - c(y(t))^2 - H$$

☞ We often say that $y(t)$ *satisfies* an ODE when we mean that $y(t)$ is a solution of that ODE.

for all t in an interval is called a *solution* of ODE (2). The value y_0 of $y(t)$ at some time t_0 can be estimated, and must surely be a critical factor in predicting later values of $y(t)$. The condition $y(t_0) = y_0$ is called an *initial condition*.

Measuring time forward from the time t_0, we have created a problem whose solution $y(t)$ is the predicted tonnage of fish at future times:

Mathematical Model for the Fish Population over Time

Given the constants a and c, the harvesting rate H, and the values t_0 and y_0, find a function $y(t)$ for which

$$y' = ay - cy^2 - H, \qquad y(t_0) = y_0 \tag{3}$$

on some t-interval containing t_0.

The ODE and the initial condition in (3) form an *initial value problem* (IVP) for $y(t)$. We will see in Chapter 2 that the general IVP (3) has a unique solution on some t-interval if the harvesting rate H is a constant, or if H is a continuous function of time. It is nice to know that we are dealing with a problem that has exactly one solution, even though we don't yet know how to construct that solution. It is like knowing in advance that the pieces of a jigsaw puzzle will indeed fit together.

So how do we describe the solution $y(t)$ of IVP (3)? Do we use words, graphs, or formulas? We will use all three.

A Solution Formula for IVP (3): No Overcrowding

We have put together a general model IVP for the fish tonnage. To describe the solution, it might be a good idea not to tackle the full-blown initial value problem, but to look at particular cases first.

Suppose that there is no overcrowding (so $c = 0$). Start the clock when the value y_0 is known. This gives us the following IVP: Find $y(t)$ so that

$$y' = ay - H, \qquad y(0) = y_0, \quad t \geq 0 \tag{4}$$

We assume that a, H, and y_0 are nonnegative constants. Here's a way to find a solution formula for IVP (4).

Suppose that $y(t)$ is a solution of IVP (4), that is,

$$y'(t) = ay(t) - H, \qquad y(0) = y_0 \tag{5}$$

Moving all the terms in the ODE of (5) to the left-hand side and multiplying through by e^{-at}, we have that

☞ We explain this approach in Section 1.3.

$$e^{-at}(y' - ay + H) = 0 \tag{6}$$

Since $(e^{-at})' = -ae^{-at}$ and $(e^{-at}y(t))' = e^{-at}y'(t) - ae^{-at}y(t)$, ODE (6) becomes

$$\left(e^{-at}y - \frac{H}{a}e^{-at}\right)' = 0$$

But from calculus we know that the only functions with zero derivatives everywhere are the constant functions. So for some constant C we have

$$e^{-at}y(t) - \frac{H}{a}e^{-at} = C \tag{7}$$

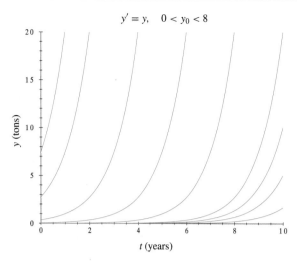

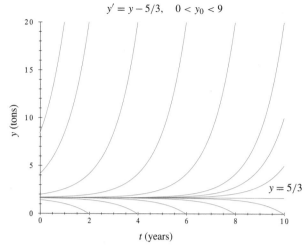

FIGURE 1.1.1 Exponential growth (no harvesting): IVP (4) with $a = 1$, $H = 0$.

FIGURE 1.1.2 Exponential growth and decline with harvesting: IVP (4) with $a = 1$, $H = 5/3$.

Setting $t = 0$ in formula (7), we can solve for C. We get

$$y_0 - \frac{H}{a} = C \qquad (8)$$

since $y(0) = y_0$. So, multiplying each side of formula (7) by e^{at}, using the value for C given in (8), and rearranging terms, we finally see that the solution of IVP (4) has the form

$$y(t) = \frac{H}{a} + \left(y_0 - \frac{H}{a}\right)e^{at}, \quad \text{for } t \geq 0 \qquad (9)$$

To complete the construction process you may want to verify that the function $y(t)$ given in (9) actually is a solution of IVP (4).

What does formula (9) tell us about the fish population? First, if initial tonnage y_0 is exactly H/a, then (9) yields $y(t) = H/a$ for all $t \geq 0$. This constant solution $y(t) = H/a$ is called an *equilibrium solution*. Second, note that if y_0 is the slightest bit greater than H/a, then exponential growth sets in. If y_0 is less than H/a, then the fish population becomes extinct since there is a time $t^* > 0$ such that $y(t^*) = 0$.

☞ Keep in mind: a *solution* is a function; a *solution curve* is the graph of a solution.

The graph in the ty-plane of a solution $y(t)$ of an ODE is called a *solution curve*. Figure 1.1.1 shows the exponential growth of the population if there isn't any fishing ($H = 0$). Figure 1.1.2 shows both exponential growth and decline away from equilibrium if there is fishing ($H = 5/3$ tons per year). These two figures can be generated directly by using formula (9) and graphing software.

If $y_0 < H/a$ we soon end up with extinction, but if $y_0 > H/a$, then the fish population grows without bound (which never happens in real life). So we need a better model. Maybe we need to put the overcrowding term back into play.

Overcrowding, No Harvesting

So let's temporarily drop the harvesting term from the ODE and put the overcrowding term back in to obtain the IVP

$$y' = ay - cy^2, \qquad y(0) = y_0, \quad t \geq 0 \tag{10}$$

where a, c, and y_0 are positive constants. Although there is a formula for the solution of IVP (10) (look ahead to Example 1.6.5), the formula isn't particularly easy to derive, so we need another way to describe the solution of IVP (10). There are computer programs called *numerical solvers* that compute very good approximations of the solution to an IVP like (10), even when there is no solution formula. Let's see what we can do with IVP (10) using a numerical solver.

Figure 1.1.3 shows approximate solution curves for IVP (10) with $a = 1, c = 1/12$:

$$y' = y - y^2/12, \qquad y(0) = y_0, \quad y_0 = \text{various positive values}, \quad t \geq 0 \tag{11}$$

We have set the computer solve-time interval at $0 \leq t \leq 10$ to predict future tonnage and the tonnage range at $0 \leq y \leq 20$; negative tonnage makes no sense here.

What does Figure 1.1.3 suggest about the evolving fish tonnage as time advances? First of all, there seem to be two equilibrium levels, $y(t) = 12$ for all $t \geq 0$ and $y(t) = 0$ for all $t \geq 0$. Are these actual solutions of the ODE in (11)? Yes, because the constant functions $y(t) = 12$ and $y(t) = 0$ satisfy the ODE, as can be verified by direct substitution. Intriguingly, the upper equilibrium seems to attract all other nonconstant solution curves in the *population quadrant* $y \geq 0$, $t \geq 0$. Left alone, the fish population tends toward this equilibrium level, no matter what the initial population might be.

☞ Warm up your numerical solver by doing Figure 1.1.3 yourself.

Since we will use numerical solvers often, let's see how they work.

Some Tips on Using a Numerical Solver

A numerical solver plots an approximate value of the solution $y(t)$ at hundreds of different instants of time and then connects these points on the computer screen with line segments. How well this graph approximates the true solution curve depends on the sophistication of the solver. Numerical analysts have done a remarkable job in coming up with reliable solvers; we have a great deal of confidence in ours.

For now we only need to concern ourselves with the basics of how to communicate with the solver. The first thing to do is to write the IVP in the form

$$y' = f(t, y), \qquad y(t_0) = y_0$$

because the numerical solver has to know the function $f(t, y)$ and the *initial point* (t_0, y_0). Since dy/dt is the time rate of change of the solution $y(t)$ of the IVP, the function $f(t, y)$ is often called a *rate function*. Next, the user needs to specify the *solve-time interval* as running from the initial point t_0 to the final point t_1. The IVP is said to be solved *forward* if $t_1 > t_0$, and *backward* if $t_1 < t_0$.

The solver must be told how to display solution curves. We like to select the screen size (i.e., the axis ranges) before telling our solver to find and plot solution curves. There are two reasons for this:

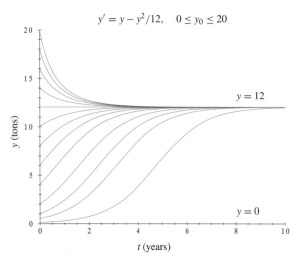

FIGURE 1.1.3 Overcrowding, no harvesting: equilibrium solutions $y = 0, 12$; IVP (11).

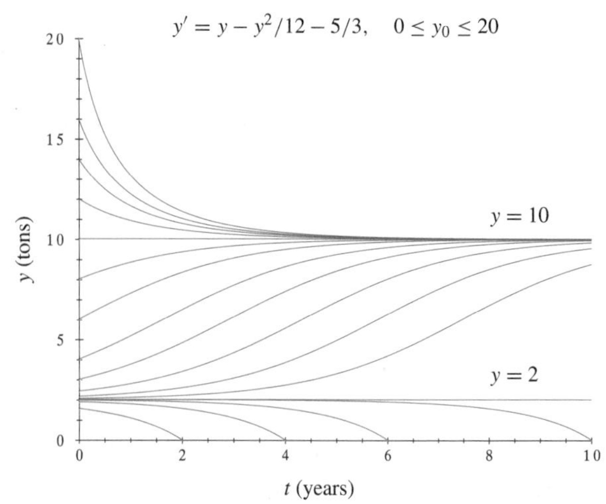

FIGURE 1.1.4 Overcrowding, harvesting: equilibrium solutions $y = 2, 10$; IVP (12).

☞ More tips on using solvers are in the Student Resource Manual.

- Well-designed solvers often shut down automatically when the solution curve gets too far beyond the specified screen area because of a poorly selected solve-time interval. This prevents the computer from working too hard (and perhaps crashing).

- Some solvers have a default setting that automatically scales the screen size to the solution curve over the solve-time interval. If you have a runaway solution curve, you won't see much on the screen.

Choosing the right screen size to bring out the features you wish to examine is often as much of an art as it is a science. Your skill at setting screen sizes will improve with experience.

Now we are ready to return to the fish population model. Let's put the fishing industry back in business and see what happens.

Overcrowding and Harvesting

Let's start out by including light harvesting, say $H = 5/3$ tons per year, so IVP (11) becomes

$$y' = y - \frac{y^2}{12} - \frac{5}{3} = -\frac{1}{12}(y - 2)(y - 10), \qquad y(0) = y_0 \geq 0 \qquad (12)$$

Let's use our solver to plot approximate solution curves to IVP (12) for positive values of y_0 (Figure 1.1.4). There are two equilibrium solutions: $y = 2$ and $y = 10$, all t. The upper equilibrium line still attracts solution curves, but now not all of them. Those starting out below the lower equilibrium line curve downward toward extinction. This model of a low harvesting rate flashes a yellow caution signal: light harvesting doesn't appear to be very harmful, at least if the initial tonnage y_0 is high enough, but even a light harvesting rate could drive a population to extinction if the population level is low

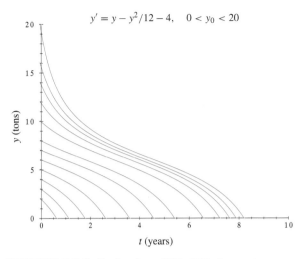

FIGURE 1.1.5 Extinction; IVP (13) for various y_0 values.

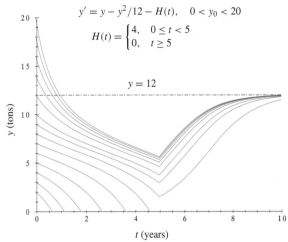

FIGURE 1.1.6 Ban on fishing over a five-year period restores fish population; IVP (14).

to begin with. Still, this is a scenario where both the fish population and the fishing industry do fairly well.

Now let's give the fishermen a free hand and suppose that the harvesting rate is much higher. Let's say the harvesting rate rises to 4 tons per year. We have the heavy harvesting IVP

$$y' = y - \frac{y^2}{12} - 4, \qquad y(0) = y_0, \quad t \geq 0 \tag{13}$$

This time if we search for equilibrium solutions by setting $y' = 0$ and using the quadratic formula to find the roots of $y - y^2/12 - 4$, we find that there are none. In fact, y' is always negative and Figure 1.1.5 shows the resulting catastrophe.

Ban on Fishing

We can't let the fish population die out. Let's see what happens in our model if, after five years of fishing at the rate of 4 tons per year, we ban fishing for five years. Now the harvest rate is given by the function

$$H(t) = \begin{cases} 4, & 0 \leq t < 5 \\ 0, & 5 \leq t \leq 10 \end{cases}$$

and the IVP is

$$y' = y - \frac{y^2}{12} - H(t), \qquad y(0) = y_0, \quad 0 \leq t \leq 10 \tag{14}$$

Fortunately, it is known that even if the harvesting rate is an on-off function like $H(t)$, an initial value problem such as (14) still has a unique solution $y(t)$ for each value of y_0. We don't have a formula for $y(t)$, but our numerical solver gives us a good idea of just how $y(t)$ behaves.

As you might expect, the fish population is rescued from extinction if y_0 is large. Figure 1.1.6 shows that after five years of heavy harvesting the surviving population heads toward the level of $y = 12$. We have saved the fish, but at the expense of the fishing industry.

Figure 1.1.6 shows a strange feature not seen in any of the other graphs: corners on the solution curves. These appear precisely at $t = 5$ when harvesting suddenly stops. So a discontinuity in the harvesting rate shows up in the graphs as a sudden change in the slope of a solution curve. That is not surprising because the slope of a solution $y(t)$ is the derivative $y'(t)$, and $y'(t)$ in ODE (14) involves the on-off harvesting rate.

Comments

We created a mathematical model using ODEs for changes in population size, a model that includes internal controls (the overcrowding factor) and external controls (the fishing rate). We found formulas for the solutions of the mathematical model in a simple case, used a numerical solver to graph solutions in more complex cases, and interpreted all of these solutions in terms of what happens to the fish population. The model introduced here has its flaws, as all models do. But the modeling process has allowed us to examine the consequences of various assumptions about the rate of change of the fish population.

There are many good solvers that require little or no programming skills. No specific solver is presumed in this text.

PROBLEMS

1. (*Exponential Growth*). Say that the model IVP for a fish population is given by $y'(t) = ay(t)$, $y(0) = y_0$, where a and y_0 are positive constants (no overcrowding and no harvesting).

   (a) Find a solution formula for $y(t)$.

   (b) What happens to the population as time advances? Is this a realistic model? Explain.

2. (*Control by Overcrowding and by Harvesting*). The IVP $y' = y - y^2/9 - 8/9$, $y(0) = y_0$, where y_0 is a positive constant, is a special case of IVP (3).

   (a) What is the overcrowding coefficient and its units? What is the harvesting rate?

   (b) Find the two positive equilibrium levels. [*Hint*: Find the roots of $y - y^2/9 - 8/9$.]

   (c) Graph solution curves of the IVP for various values of y_0. Use the axis ranges $0 \le t \le 10$, $0 \le y \le 15$. Interpret what you see in terms of the future of the fish population.

3. (*Restocking*). Restocking the fish population with R tons of fish per year leads to the model ODE $y' = ay - cy^2 + R$, where a and c are positive constants.

   (a) Explain each term in the model ODE.

   (b) Test the model on the IVP $y' = y - y^2/12 + 7/3$, $y(t_0) = y_0$, for various nonnegative values of t_0 and y_0. Carry solution curves forward and backward in time. Use the screen $0 \le t \le 10$, $0 \le y \le 25$. Interpret what you see.

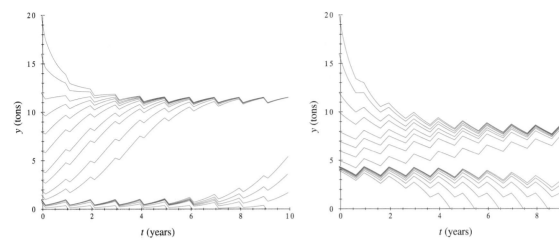

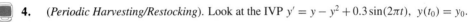

FIGURE 1.1.7 Short harvest season (Problem 7).   **FIGURE 1.1.8** Long harvest season (Problem 7).

4. (*Periodic Harvesting/Restocking*). Look at the IVP $y' = y - y^2 + 0.3\sin(2\pi t)$, $y(t_0) = y_0$.

    (a) Discuss the meaning of the ODE in terms of a fish population. Graph solution curves for $t_0 = 0$ and values of y_0 ranging from 0 to 2. Use the ranges $0 \le t \le 10$, $0 \le y \le 2$. Repeat for $t_0 = 1, 2, \ldots, 9$ and $y_0 = 0$. Interpret what you see in terms of the fish population.

    (b) Explain why the solution curves starting at (t_0, y_0) and $(t_0 + 1, y_0)$ look alike. In the rectangle $0 \le t \le 10$, $-1 \le y \le 2$, plot the solution curve through the point $t_0 = 0.5$, $y_0 = 0$. Why is this curve meaningless in terms of the fish population?

www 5. (*Constant Effort Harvesting*). The models in this section have a flaw. At low population levels a fixed high harvesting rate can't be sustained for long because the population dies out. A safer model (for the fish) is $y' = ay - cy^2 - H_0 y$, $y(0) = y_0$, where a, c, H_0, and y_0 are positive constants. In this model the lower the population, the lower the harvesting rate.

    (a) Interpret each term in the ODE. Why is this called constant effort harvesting?

    (b) For values of H_0 less than a, explain why this model produces solution curves similar to those in Figure 1.1.3, but possibly with a different stable equilibrium population.

6. (*Heavy Harvesting, Light Harvesting*). What happens when a five-year period of heavy harvesting is followed by five years of light harvesting? Combine IVPs (12) and (13) by supposing that $y' = y - y^2/12 - H(t)$, where

$$H(t) = \begin{cases} 4, & 0 \le t < 5 \\ 5/3, & 5 \le t \le 10 \end{cases}$$

Plot solution curves for $0 \le t \le 10$, $0 \le y \le 20$, and interpret what you see. Draw the lines $y = 10$ and $y = 2$ on your plot and explain their significance for the population for $t \ge 5$.

7. (*Seasonal Harvesting*). Say that harvesting is seasonal, "on" for the first few months of each year and "off" for the rest of the year. The ODE is $y' = ay - cy^2 - H(t)$, where $H(t)$ has value H_0 during the on-season, and value 0 during the off-season. The harvesting season is the first two months of each year in Figure 1.1.7 and the first eight months in Figure 1.1.8; in both figures $a = 1$, $c = 1/2$, $H_0 = 4$. Duplicate the graphs in Figures 1.1.7 and 1.1.8. Discuss what you see in terms of population behavior. [*Hint*: Try $H(t) = H_0 \operatorname{sqw}(t, d, 1)$, where $d = 100(2/12) = 50/3$ for a two-month season and $d = 100(8/12)$ for an eight-month season. See Appendix B.1 for more information about the on-off function sqw.]

☞ If you run into trouble, refer to "Tips" in the Student Resource Manual.

1.2 Visualizing Solution Curves

In the previous section we used a numerical solver to plot some solution curves for a few differential equations, and we accepted the results without question. This is a risky practice because every computer has its limitations, and that goes for software, too. It's always good to have more than one way to look at things so that results can be checked in as many ways as possible. In this section we show how to visualize the behavior of a solution curve based on the differential equation itself, rather than relying solely on computer output for the solution. First, though, we take a brief detour.

There are good reasons that most of the differential equations we have looked at so far have been written in the *normal form*

$$y' = f(t, y) \tag{1}$$

where $f(t, y)$ is a function defined on some portion (or all) of the ty-plane. For example, most numerical solvers only accept ODEs that have been written in normal form. In addition, the general theory of ODEs applies only to differential equations in this form.

Let's define what we mean by a solution of ODE (1). A function $y(t)$ defined on a t-interval I is a *solution* of ODE (1) if $f(t, y(t))$ is defined and $y'(t) = f(t, y(t))$ for all t in I. With this definition and the normal form in hand, we can begin a general discussion of the relation of the solutions $y(t)$ of ODE (1) to its solution curves.

Solution Curves

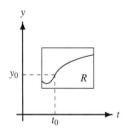

☞ Whoa—no solution formula?

Examples in the last section show that ODEs can have many solutions; we were fortunate in finding a formula for one of the simpler ODEs. We will see in Section 2.1 that under certain mild conditions, ODEs have solutions even though we can't always find explicit formulas for them. Armed with this knowledge, we can use a numerical solver to plot approximate solution curves. We will often use theory and geometric methods to examine properties of solutions without the benefit of solution formulas.

Look at the normal ODE $y' = f(t, y)$, where $f(t, y)$ is defined over a *closed rectangle* R, that is, a rectangle that contains its boundary lines. If (t_0, y_0) is any point in R and if f and $\partial f/\partial y$ are continuous on R, then (as we show in Section 2.1) there is exactly one solution $y(t)$ to the *initial value problem* (IVP)

$$y' = f(t, y), \qquad y(t_0) = y_0 \tag{2}$$

The point (t_0, y_0) in R is called the *initial point* for the IVP. The solution curve for $y(t)$ extends to the boundary of R both for $t > t_0$ and for $t < t_0$. In other words, the solution curve does not suddenly stop inside R. The unique solvability of IVP (2) for each point (t_0, y_0) in R implies that no two solution curves of the ODE $y' = f(t, y)$ can intersect inside R. We can think of R as being completely covered by solution curves, each point of R on exactly one solution curve.

Knowing that IVP (2) has a unique solution, we can use a numerical solver to compute and plot approximate solution curves. In Chapter 2 we will see that numerical solvers use a step-by-step process to generate approximate points on a solution curve, and then graph approximate solutions by connecting these points with line segments.

Let's turn now to some geometric interpretations.

Geometry of Solution Curves

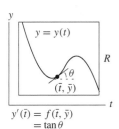

$$y'(\bar{t}) = f(\bar{t}, \bar{y})$$
$$= \tan \theta$$

☞ This is a handy way to recognize solution curves geometrically.

There is a way to view the solvability of IVP (2) that appeals to geometric intuition and lends itself to a graphical approach to finding solution curves. The ODE says that at each point $(\bar{t}, \bar{y})$ in R the number $f(\bar{t}, \bar{y})$ is the slope of the tangent line to the solution curve through that point (see margin figure).

On the other hand, suppose that the graph of a function $y(t)$ lies in R. Then $y(t)$ defines a solution curve of $y' = f(t, y)$ if at each point $(\bar{t}, \bar{y})$ on its graph the slope of the tangent line has the value $f(\bar{t}, \bar{y})$. This is the geometric way of saying that $y'(t) = f(t, y(t))$, that is, $y(t)$ is a solution for the ODE.

This change of viewpoint gives us an imaginative way to see solution curves for the ODE. By drawing short line segments with slopes $f(t, y)$ and centered at a grid of points (t, y) in R, we obtain a diagram, called a *direction field*. A direction field suggests curves in R with the property that at each point on each curve the tangent line to the curve at that point lies along the direction field line segment at the point. This process reveals solution curves in much the same way as iron filings sprinkled on paper held over the poles of a magnet reveal magnetic field lines.

Given the rate function $f(t, y)$, we can draw the line segments of a direction field by hand, but it's a lot easier to have a numerical solver do the work. Most solvers can do this, and many let the user choose the density of the grid points and the length of the segments.

Let's illustrate these ideas with an example.

EXAMPLE 1.2.1

A Direction Field and a Solution Curve

Figure 1.2.1 shows a direction field for the ODE

$$y' = y - t^2$$

You can almost see the solution curves. They rise wherever $y > t^2$ because y' is positive; the curves fall where $y < t^2$. The field line segments in the figure are all the same length, even if it doesn't appear so at first glance. The reason for this is that the computer screen length of a vertical unit is not the same as the screen length of a horizontal unit. The ratio of the former to the latter is the *aspect ratio* of the display.

☞ Aspect ratio makes plotted circles look like ellipses.

Now let's plot the solution curve through the initial point $t_0 = 0$, $y_0 = 1$. Figure 1.2.2 shows this curve (solid) extended forward and backward in time from $t_0 = 0$ until the curve leaves the rectangle defined by the computer screen. Notice how nicely the solution curve fits the direction field.

The *nullclines* (the curves of zero inclination) of $y' = f(t, y)$ are defined by $f(t, y) = 0$. For example, in the ODE $y' = y - t^2$ we see that $f(t, y) = y - t^2$, so the nullcline is the curve defined by $y = t^2$ (the dashed curve in Figure 1.2.2). The nullcline divides the ty-plane into the region above the parabola ($y > t^2$), where solution curves rise, and the region below ($y < t^2$), where they fall. Solution curves cross

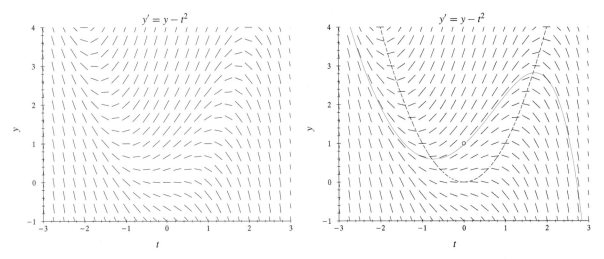

FIGURE 1.2.1 Direction field (Example 1.2.1).

FIGURE 1.2.2 Solution curve (solid) through the point $(0, 1)$ (Example 1.2.1); nullcline (dashed).

the nullcline with zero slope because the direction line segment centered at a point on the nullcline is horizontal.

Nullclines are usually *not* solution curves. But if y_0 is a root of a function $f(y)$, then the constant function $y = y_0$ is both a nullcline and a solution curve for the ODE

$$y' = f(y)$$

For example, the horizontal lines $y = 1$, $y = 2$, and $y = 3$ in the ty-plane are at the same time nullclines and solution curves for the ODE

$$y' = (y - 1)(y - 2)(y - 3)$$

☞ We take a longer look at equilibrium solutions in Section 2.2.

Such curves are *equilibrium solution curves* and the solutions that generate them are called *equilibrium solutions.*

Finding Equilibrium Solutions

To find the equilibrium solutions of the ODE $y' = f(y)$, first set $f(y) = 0$ and solve for y. If y_0 is a number such that $f(y_0) = 0$ then the constant function $y = y_0$, for all t, is an equilibrium solution.

EXAMPLE 1.2.2

Equilibrium Solutions
Let's look at the ODE $y' = 12y - y^2 - 20$ which models a fish population that is harvested at a constant rate. Now $f(y) = 12y - y^2 - 20$, and since the equation $12y - y^2 - 20 = 0$ has the two roots $y_0 = 2$ and $y_0 = 10$ we have two equilibrium solutions: $y = 2$, for all t and $y = 10$, for all t.

Here is another example where the direction field guides your eye along the solution curves.

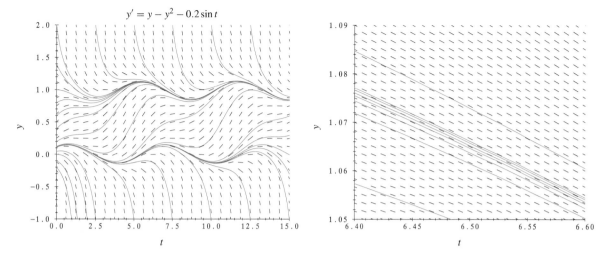

FIGURE 1.2.3 Direction field and some solution curves for ODE (3).

FIGURE 1.2.4 Zoom on a portion of Figure 1.2.3 centered at the point (6.50, 1.07).

EXAMPLE 1.2.3

☞ We looked at the harvested logistic population in Section 1.1.

Time Dependent Harvesting

Let's look at an ODE model of a fish population that is harvested and restocked sinusoidally:

$$y' = y - y^2 - 0.2 \sin t \qquad (3)$$

We used our solver to produce the direction field and solution curves in Figure 1.2.3 for this ODE. We can almost see the solution curves being traced out, guided along by the direction field. It appears that if the initial population is not too small, then the resulting population curve eventually looks like a sinusoid of period 2π, which is the period of the harvesting/restocking function $H(t)$. We haven't proven that this is so, but the visual evidence is fairly strong.

Compression, Expansion, and Zooming

The solution curves in Figure 1.2.3 were not as easy to graph as you might think. What initial point on the y-axis would you choose so that the solution curve passes through the "target point" $(7.5, -1)$? Choosing the exact initial point is very difficult since many solution curves intersect the y-axis very near our desired solution curve. If our choice of initial point is even slightly off we will end up with a solution curve which misses our target by a wide margin.

Why is is so hard to hit the target point? It appears from the graph that any two solution curves of ODE (3) either compress together or spread apart without bound as $t \to +\infty$. To plot the solution curve through $(7.5, -1)$, choose the initial point to be $(7.5, -1)$ and solve *backward* in time until the solution curve hits the y-axis.

The solution curves in the upper half of Figure 1.2.3 all seem to flow together and appear to touch inside the rectangle R defined by the screen. But we know this can't be

so for the following reason: compare ODE (3) with ODE (2) to find the rate function

$$f(t, y) = y - y^2 - 0.2 \sin t$$

The functions $f(t, y)$ and $\partial f/\partial y = 1 - 2y$ are both continuous on the entire ty-plane, and theory tells us that distinct solution curves of ODE (3) can never touch. So our comfortable world of theory collides with the practical problem of displaying data on a computer screen. Since the screen only contains a finite number of pixels, points that are less than a pixel apart are not distinguishable from one another, which explains the apparent contradiction. In Figure 1.2.4 we separate the apparently touching curves by zooming in on them. When we look at the direction field and the solution curves through a microscope this way, the field line segments seem to be parallel and the solution curves are very nearly parallel lines.

☞ Check this out by zooming with your own solver.

Comments

We have seen how to recognize solution curves of $y' = f(t, y)$ geometrically by look-ing at the direction field determined by the rate function f. We saw, too, that the null-clines [where $f(t, y)$ is zero] divide the ty-plane into regions where solution curves rise and regions where they fall. All of this shows the intimate relationship between the rate function and the way solution curves behave.

☞ Try repeating these conditions as a mantra.

We have also indicated conditions (the continuity of f and $\partial f/\partial y$ in a rectangle R) under which we can be assured that the IVP $y' = f(t, y)$, $y(t_0) = y_0$ has a unique solution curve in R, even when there is no known solution formula.

PROBLEMS

1. (*Ban on Fishing May Come Too Late*). Say that a fish population is modeled by the IVP $y' = y - y^2/12 - H(t)$, $y(0) = y_0$, where $H(t) = 4$ for $0 \le t \le 5$, $H(t) = 0$ for $5 \le t \le 10$. Estimate the smallest value of y_0 such that the fish population recovers after fishing is banned. Give reasons for your estimate. [*Hint*: See Figure 1.1.6.]

2. (*Equilibrium Solutions*). First, find all equilibrium solutions for each ODE. Then describe in words how the nonequilibrium solution curves behave in various regions in the ty-plane. Then plot a direction field and solution curves in the given rectangle. [*Hint*: Find the sign of y' above and below each equilibrium solution.]

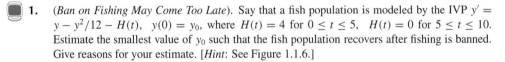

(**a**) $y' = y^2 - 11y + 10$, $|t| \le 1$, $-5 \le y \le 15$ (**b**) $y' = |y| - y^2$, $|t| \le 5$, $|y| \le 2$
(**c**) $y' = \sin(2\pi y/(1 + y^2))$, $|t| \le 5$, $|y| \le 2$

3. Use a numerical solver to plot solution curves of the ODEs. Use the four initial conditions $y(0) = -3, -1, 1, 3$ and the ranges $0 \le t \le 4$, $-4 \le y \le 4$. Shade (or describe in words) the regions where solution curves rise. [*Hint*: Plot the nullclines to find the boundaries of these regions. Before using your numerical solver, write the ODE in normal form.]

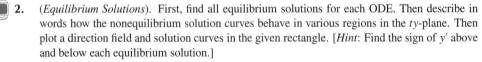

(**a**) $y' + y = 1$ (**b**) $y' + y = t$ (**c**) $y' + y = t + 1$ (**d**) $y' = \sin(3t) - y$

4. (*Direction Fields, Nullclines, Solution Curves*). Use a numerical solver with a direction field plotting option to plot a direction field for each ODE in the rectangle, $|t| \leq 2$, $|y| \leq 3$. Plot the nullclines on the same graph. Use the solver to plot the solution curves through the points $t = 0$, $y = 0, \pm 2$. Explain why you should expect that each solution curve can be extended from one edge of the rectangle to another. [*Hint*: Are the rate function $f(t, y)$ and its derivative $\partial f / \partial y$ continuous throughout the rectangle?]

    (a) $y' = -y + 3t$ **(b)** $y' = y + \cos t$ **(c)** $y' = \sin(ty)$

    (d) $y' = \sin t \sin y$ **(e)** $y' = \sin t + \sin y$ **(f)** $y' = \sin(t + y)$

www **5.** (*Zooming to Separate Solution Curves*). Use your numerical solver and plot solution curves of the IVP $y' = -2y + 3e^t$, $y(0) = y_0$, for $y_0 = -5, -4, \ldots, 4, 5$, over the *ty*-rectangle $0 \leq t \leq 3$, $-5 \leq y \leq 20$. To separate the apparently merging solution curves, zoom on the upper right corner of the *ty*-rectangle and see if this pulls the curves apart. Describe and interpret what you see in each graph.

The www icon means that the solution of this problem is on the web site at www.wiley. com/college/borrelli.

 6. (*Why Do Solution Curves Stop?*). Use your numerical solver to plot a direction field in the indicated rectangle and to plot solution curves through the given initial points as far forward into the rectangle as possible. If a solution curve suddenly stops or if your solver goes haywire, explain what's wrong. When are the nullclines not the only curves that separate regions where solution curves rise from regions where they fall? [*Hint*: Normalize each ODE to the form $y' = f(t, y)$, and look for places in the rectangle where f fails to be continuous.]

    (a) $2yy' = -1$, $y(0) = 1$; $0 \leq t \leq 1.5$, $-0.5 \leq y \leq 1.5$

    (b) $2yy' = -t$, $y(-2) = 1.4, 1.5$; $-2 \leq t \leq 1$, $-1 \leq y \leq 2$

 7. (*Do-It-Yourself ODEs*). Create your own first-order ODEs with direction fields and solution curves that are interesting, strange, or beautiful. Then make up some ODEs that crash your solver. Explain why your solver has difficulties.

The handshake icon indicates a team project.

1.3   The Search for Solution Formulas

Solutions of an ODE can be pursued on two levels: we can search for a formula that describes a solution, or we can search for solution curves with the use of a numerical solver. The process of finding solutions of differential equations at either level is often referred to as "solving a differential equation." In this section we will concentrate on finding solution formulas.

How can we find solution formulas for ODEs? Good guessing is one way. Guessing a solution has a long and honorable history as a way of finding solutions, but some differential equations defy anyone's ingenuity to guess a solution formula. We will see soon enough that our bag of tricks (including guessing) for finding solution formulas is rather small. That's where a numerical solver comes in handy. It finds and plots approximate numerical solutions even though there is no solution formula in sight.

But an important class of ODEs does have solution formulas, and we turn now to a description of that class.

First-Order ODEs: Linear or Nonlinear?

The *order* of an ODE is the order of the highest derivative of the to-be-determined function that appears in the equations. For example, $y' = y + \sin t$ is a first-order

ODE. All of the ODEs mentioned in the previous sections are first-order ODEs. Later in Sections 1.5 and 1.6 we will see some second-order ODEs like

$$y'' = -9.8 + 0.15y'$$

and in Section 1.7 the second-order ODE

$$y'' = \frac{-k}{(y+R)^2}$$

where k and R are positive constants.

Let's look more closely at a class of first-order ODEs which are encountered frequently in modeling natural processes.

A first-order ODE is *linear* if it can be written in the form

$$y' + p(t)y = q(t) \tag{1}$$

where $p(t)$ and $q(t)$ are functions that do not depend on y, but may depend on t. Linear ODEs written as in (1) are in *normal linear form*. The ODE $t^2 y' = e^t y + \sin 3t$ is a first-order ODE and it is linear because by dividing by t^2 it can be written in the normal linear form

$$y' - \frac{e^t}{t^2} y = \frac{\sin 3t}{t^2}$$

First-order ODEs which cannot be written in the form (1) are *nonlinear*

In Section 1.1 we looked at the first-order ODE

$$y' - ay = -H \tag{2}$$

where a and H are constants. ODE (2) is a first-order linear ODE which is written in normal linear form. We saw that all solutions of ODE (2) are described by the formula

$$y = \frac{H}{a} + \left(y_0 - \frac{H}{a} \right) e^{at}$$

where y_0 is any constant.

In Section 1.2 we used a numerical solver to come up with approximate solution curves of the first-order nonlinear ODE

$$y' = y - y^2 - 0.2 \sin t \tag{3}$$

Figure 1.2.3 displays the results. ODE (3) is nonlinear because the term y^2 prevents it from being written in the normal linear form (1). Whether a first-order ODE is linear or nonlinear, antidifferentiation is often used to construct a formula for the solutions.

Looking for Solution Formulas

☞ Theorem B.5.5 in Appendix B.5 has the traditional form of the Fundamental Theorem.

The Fundamental Theorem of Calculus turns out to be a key tool for finding solution formulas for ODEs. The basic concept in that theorem is antidifferentiation: an *antiderivative* of a function $f(t)$ is a function $F(t)$ such that $F'(t) = f(t)$.

Let's start by finding all solutions of the ODE $y' = f(t)$.

THEOREM 1.3.1

> Antiderivative Theorem. Suppose that $F(t)$ is an antiderivative of a continuous function $f(t)$ on a t-interval. Then all solutions of the ODE
>
> $$y' = f(t)$$
>
> are given on that t-interval by
>
> $$y = F(t) + C, \quad \text{where } C \text{ is any constant} \tag{4}$$

To see this, suppose that $y(t)$ is any solution of $y' = f(t)$, that $F(t)$ is any antiderivative of $f(t)$, and that t_0 is any point in the t-interval. Then integrating from t_0 to t and using the Fundamental Theorem of Calculus, we see that

$$y(t) - y(t_0) = \int_{t_0}^{t} y'(s)\, ds = \int_{t_0}^{t} f(s)\, ds = F(t) - F(t_0)$$

for all t in the interval, so $y(t)$ has the form $F(t) + $ a constant [the constant C in formula (4) is $y(t_0) - F(t_0)$ in this case]. On the other hand, $y = F(t) + C$ is a solution of the ODE $y' = f(t)$ for any value of the constant C, so we have captured all the solutions of the ODE.

The Antiderivative Theorem is the source of many methods for finding solution formulas for ODEs. You might say that it is the "mother of all methods." Let's use it to find all solutions of several ODEs.

EXAMPLE 1.3.1

Just Antidifferentiate!
Since $(\sin t)' = \cos t$, we see from Theorem 1.3.1 that the ODE

$$y' = \cos t$$

has all of its solutions given by the formula

$$y(t) = \sin t + C, \qquad C \text{ any constant}$$

Here is an example of how the Antiderivative Theorem is used to find a solution formula for a normal first-order linear ODE.

EXAMPLE 1.3.2

Preparing a Linear ODE for Antidifferentiation
Suppose that $y(t)$ is a solution of the linear ODE in normal linear form

$$y' - 2y = 2 \tag{5}$$

Multiply each side of the ODE by the function e^{-2t} to obtain the identity

$$e^{-2t} y'(t) - 2e^{-2t} y(t) = 2e^{-2t} \tag{6}$$

Since e^{-2t} is never zero, every solution of ODE (6) is also a solution of ODE (5), and the other way around. The product rule for derivatives shows that

$$[e^{-2t}y(t)]' = e^{-2t}y'(t) + (e^{-2t})'y(t) = e^{-2t}y'(t) - 2e^{-2t}y(t)$$

end so identity (6) can be written as

$$[e^{-2t}y(t)]' = 2e^{-2t} \qquad (7)$$

The trick of multiplying ODE (6) by e^{-2t} was not pulled out of a hat. Its purpose was to produce ODE (7). Now apply Theorem 1.3.1 to ODE (7) to obtain

$$e^{-2t}y(t) = -e^{-2t} + C$$

where C is any constant. Multiply through by e^{2t}:

$$y = -1 + e^{2t}C, \quad \text{all } t$$

We now have a formula for all solutions of ODE (5).

Multiplying both side of ODE (5) by e^{-2t} turned out to be an excellent strategy. You may wonder how we hit upon the magic factor e^{-2t}. Read on to see how to choose such "integrating factors" for linear ODEs such as (1).

The Integrating Factor Approach for Linear ODEs

Let's look at the first-order normal linear ODE

$$y' + p(t)y = q(t) \qquad (8)$$

where the *coefficient* $p(t)$ and *driving term* (or *input*) $q(t)$ are continuous on a t-interval I. If $q(t) = 0$ for all t in I, then ODE (8) is said to be *undriven*. Otherwise, ODE (8) is *driven* by the input $q(t)$. We now find a formula for all solutions of ODE (8).

Here's the approach that will give us a formula for all solutions. Suppose that $P(t)$ is any antiderivative of the coefficient $p(t)$ in ODE (8) on the interval I, that is, $P'(t) = p(t)$ on I. The exponential function $e^{P(t)}$ is called an *integrating factor* for ODE (8). Using the Chain Rule we have that $(e^{P(t)})' = e^{P(t)}P'(t) = e^{P(t)}p(t)$ and so we have the identity

☞ More on the Chain Rule in Theorem B.5.7, Appendix B.5.

$$
\begin{aligned}
[e^{P(t)}y(t)]' &= e^{P(t)}y'(t) + e^{P(t)}P'(t)y(t) \\
&= e^{P(t)}y'(t) + e^{P(t)}p(t)y(t) \\
&= e^{P(t)}[y'(t) + p(t)y(t)] \qquad (9)
\end{aligned}
$$

We will use identity (9) to help us solve ODE (8).

Suppose that $y(t)$ is any solution of ODE (8) on the interval I. Multiplying ODE (8) by the integrating factor $e^{P(t)}$, we have the identity

$$e^{P(t)}[y'(t) + p(t)y(t)] = e^{P(t)}q(t)$$

which, because of the identity (9), can be written as

$$[e^{P(t)}y(t)]' = e^{P(t)}q(t) \tag{10}$$

Suppose that $R(t)$ is any antiderivative of $e^{P(t)}q(t)$. Since $R'(t) = e^{P(t)}q(t)$, and applying Theorem 1.3.1 to ODE (10), we obtain

$$e^{P(t)}y(t) = R(t) + C$$

where C is some constant. After multiplying each side of this equality by $e^{-P(t)}$, we have a formula for the solution $y(t)$ of ODE (8) that we started with:

$$y = Ce^{-P(t)} + e^{-P(t)}R(t) \tag{11}$$

where C is any constant. Because $P(t)$ and $R(t)$ are defined on the t-interval I where $p(t)$ and $q(t)$ are continuous, we see that the solution is defined on I. So we have shown that every solution $y(t)$ of ODE (8) has the form (11) for some value of the constant C. But what values of C in formula (8) actually produce a solution of ODE (8)? The answer: *any* value of C! To see this, start with *any* constant C to define a function $y(t)$ via formula (11) and then reverse the above steps to show that this function $y(t)$ solves the ODE (8).

☞ Alternatively, just show directly that $y(t)$ in (11) solves ODE (8).

Summarizing, here are the steps we followed in finding the formula (11):

Solving a First-Order Linear ODE

☞ Forgetting this first step can be fatal!

1. **Write the ODE** in the normal linear form $y' + p(t)y = q(t)$ and identify the coefficient $p(t)$ and the driving term $q(t)$.

2. **Find an antiderivative** $P(t)$ for $p(t)$; any one will do.

3. **Multiply the ODE** by the integrating factor $e^{P(t)}$ and write the new ODE as

$$(e^{P(t)}y)' = e^{P(t)}q(t)$$

4. **Find an antiderivative** $R(t)$ for $e^{P(t)}q(t)$; any one will do.

5. **Apply the Antiderivative Theorem 1.3.1** to the new ODE, multiply, and rearrange terms to find the general solution formula (11).

Since formula (11) captures all solutions of ODE (8), it is called the *general solution* of the ODE. One implication of our constructive solution process is that after *any* choice of the antiderivatives $P(t)$ and $R(t)$ as described above, all solutions of ODE (8) always have the form (11). It is not obvious from formula (11) itself that this is so, but the construction process does not lie.

Since the constant C in the general solution formula (11) is arbitrary, we see that ODE (8) has infinitely many solutions, one for each value of the constant C. Now different values for the constant C in formula (11) give solution curves that never touch in I. To show this, insert distinct constants C_1 and C_2 into formula (11) to obtain two solutions, $y_1(t)$ and $y_2(t)$. Subtracting, we have

$$y_1(t) - y_2(t) = (C_1 - C_2)e^{-P(t)}$$

But exponential functions never have the value 0, and C_1 and C_2 are distinct. So $y_1(t)$ and $y_2(t)$ are never equal for any t in I, and the two solution curves don't touch.

When following the procedure above, we have to be fairly good at finding antiderivatives. In case you're a bit rusty at this, here is a short table of several of the antiderivatives you'll need in this text.

TABLE 1.3.1 Some Useful Antiderivatives

- For any constant $a \neq 0$:

$$\int e^{at}\,dt = \frac{1}{a}e^{at}, \qquad \int te^{at}\,dt = \frac{e^{at}}{a^2}(at-1), \qquad \int t^2 e^{at}\,dt = \frac{e^{at}}{a^3}(a^2 t^2 - 2at + 2)$$

$$\int t\sin at\,dt = -\frac{1}{a}t\cos at + \frac{1}{a^2}\sin at, \qquad \int t\cos at\,dt = \frac{1}{a}t\sin at + \frac{1}{a^2}\cos at$$

- For any constants a and b, not both zero:

$$\int e^{at}\cos bt\,dt = \frac{e^{at}}{a^2+b^2}(a\cos bt + b\sin bt), \qquad \int e^{at}\sin bt\,dt = \frac{e^{at}}{a^2+b^2}(a\sin bt - b\cos bt)$$

EXAMPLE 1.3.3

Use an Integrating Factor and Find the General Solution
Let's look at the ODE from Example 1.2.1:

$$y' - y = -t^2 \tag{12}$$

This ODE is already in normal linear form with $p(t) = -1$, and $q(t) = -t^2$, and these functions are continuous on the whole real line. Using the notation of the procedure just described we see that $P(t) = -t$, and the integrating factor is e^{-t}. Multiplying ODE (12) through by e^{-t}, we have

$$(e^{-t}y)' = -t^2 e^{-t} \tag{13}$$

So using Table 1.3.1 we see that $R(t) = e^{-t}(t^2 + 2t + 2)$ is an antiderivative of $-t^2 e^{-t}$. Applying Theorem 1.3.1 to ODE (13), we have

$$e^{-t}y = e^{-t}(t^2 + 2t + 2) + C \tag{14}$$

where C is any constant. Multiplying (14) through by e^t, we see that all solutions of ODE (12) are given by the general solution formula

$$y = Ce^t + t^2 + 2t + 2$$

where C is any constant.

Now let's use the general solution approach to solve an initial value problem.

Solving an Initial Value Problem

The constant C in formula (11) plays an important role in solving an IVP.

EXAMPLE 1.3.4

Find the General Solution, Solve an IVP
Here is a first-order ODE in normal linear form with $p(t) = 2$, $q(t) = 3e^t$:

$$y' + 2y = 3e^t \tag{15}$$

Since $2t$ is an antiderivative of 2, e^{2t} is an integrating factor. Multiply each side of ODE (15) by e^{2t} and apply Theorem 1.3.1 to obtain

$$e^{2t}y' + 2e^{2t}y = 3e^{3t}$$

Then, since $e^{2t}y' + 2e^{2t}y = (e^{2t}y)'$ by the formula for differentiating a product, we have

$$(e^{2t}y)' = 3e^{3t}$$
$$e^{2t}y = C + e^{3t}$$
$$y = Ce^{-2t} + e^t, \quad \text{all } t \tag{16}$$

where C is any constant. Formula (16) gives the general solution of ODE (15).

An initial condition will determine C. For example, to find the solution of the IVP

$$y' + 2y = 3e^t, \qquad y(0) = -3 \tag{17}$$

we set $y = -3$ and $t = 0$ in formula (16):

$$-3 = e^{-2 \cdot 0}C + e^0 = C + 1$$

and so $C = -4$. Replacing C in solution formula (16) by -4 gives the solution of IVP (17):

$$y = -4e^{-2t} + e^t, \qquad -\infty < t < \infty$$

If we had used the initial condition $y(1) = 0$ in IVP (17) instead of $y(0) = -3$, then we would have found that $C = -e^3$ and so the solution formula would be $y = -e^3 e^{-2t} + e^t$.

The formula $y = Ce^{-2t} + e^t$ for all solutions of $y' + 2y = 3e^t$ is very helpful if we want to describe the long-term behavior of solutions. Since $e^{-2t} \to 0$ as $t \to +\infty$, we see from (16) that all solutions look more and more like the solution $y = e^t$ as t increases.

The constructive process used in Example 1.3.4 suggests that an IVP involving a linear differential equation has exactly one solution. And that is exactly right!

THEOREM 1.3.2

Existence and Uniqueness. Suppose that $p(t)$ and $q(t)$ are continuous on a t-interval I, and that t_0 is any point in I. If y_0 is any number, then the IVP

$$y' + p(t)y = q(t), \qquad y(t_0) = y_0 \qquad (18)$$

has a solution $y(t)$, which is defined on all of I (existence), and there is no other solution (uniqueness).

Let's show this as follows: According to solution formula (11), the general solution of the ODE in (18) is

$$y(t) = Ce^{-P(t)} + e^{-P(t)}R(t), \qquad t \text{ in } I$$

where P and R are the respective antiderivatives of $p(t)$ and $e^{P(t)}q(t)$, and C is an arbitrary constant. To satisfy the condition $y(t_0) = y_0$, let's substitute t_0 and y_0 into the solution formula and obtain

$$y_0 = Ce^{-P(t_0)} + e^{-P(t_0)}R(t_0)$$

This algebraic equation for C has the unique solution $C = y_0 e^{P(t_0)} - R(t_0)$. This means that IVP (18) does indeed have a unique solution.

The Existence and Uniqueness Theorem provides the foundation for all of the applications of first-order linear IVPs because it guarantees that the IVP we are working with has exactly one solution.

Comments

Our search for solution formulas has paid off handsomely for first-order linear ODEs. The only tool we needed was the Fundamental Theorem of Calculus in the form of the Antiderivative Theorem (Theorem 1.3.1). So we see that first-order linear ODEs have explicit solution formulas. But there is a down side: We may not be able to find the required antiderivatives, so the solution formula (11) may not be especially helpful. You always have the option, though, of using a numerical solver to plot approximate solution curves of an ODE even if a solution formula is unrevealing.

PROBLEMS

1. Find the order of each ODE. Identify each ODE as linear or nonlinear in y, and write each linear ODE in normal linear form. [*Hint*: For part (d), find dy/dt.]

(a) $y' = \sin t - t^3 y$

(b) $e^t y' + 3y = t^2$

(c) $(t^2 + y^2)^{1/2} = y' + t$

(d) $dt/dy = 1/(t^2 - ty)$

(e) $y'' - t^2 y = \sin t + (y')^2$

(f) $y' = (1 + t^2)y'' - \cos t$

(g) $e^t y'' + (\sin t)y' + 3y = 5e^t$

(h) $(y''')^2 = y^5$

2. (*Finding Solutions*). Simple exponential functions like $y = e^{rt}$ are often solutions of ODEs. Find all values of the constant r so that $y = e^{rt}$ is a solution. [*Hint*: Insert e^{rt} into the ODE and find values of r that yield a solution. For example, $y = e^{rt}$ solves $y' - y = 0$ if $re^{rt} - e^{rt} = (r - 1)e^{rt} = 0$. Since $e^{rt} \neq 0$, we must have $r = 1$. So $y = e^{t}$ solves the given ODE.]

    (a) $y' + 3y = 0$ (b) $y'' + 5y' + 6y = 0$

☞ $y^{(n)}$ means the n-th derivative of $y(t)$

    (c) $y^{(5)} - 3y^{(3)} + 2y' = 0$ (d) $y'' + 2y' + 2y = e^{-t}$

3. (*Finding Solutions*). Sometimes multiples of a power of t solve an ODE. Find all values of the constant r so that $y = rt^3$ is a solution.

    (a) $t^2 y'' + 6ty' + 5y = 0$ (b) $t^2 y'' + 6ty' + 5y = 2t^3$ (c) $t^4 y' = y^2$

4. (*Finding Solutions*). Choosing the right value for r may give a solution $y = t^r$ of an ODE. Find all values of the constant r so that $y = t^r$ is a solution.

    (a) $t^2 y'' + 4ty' + y = 0$ (b) $t^4 y^{(4)} + 7t^3 y''' + 3t^2 y'' - 6ty' + 6y = 0$

5. (*Using the Antiderivative Theorem*). Use Theorem 1.3.1 to find all solutions in each case. [*Hint*: Write y'' as $(y')'$ and y''' as $(y'')'$.]

    (a) $y' = 5 + \cos t$ (b) $y' = t^2 + t + e^{-t}$ (c) $y' = e^{-t} \cos 2t$

    (d) $y'' = 0$ (e) $y'' = \sin t$, $y(0) = 0$, $y'(0) = 1$ (f) $y''' = 2$

    (g) $y'' + y' = e^{t}$ [*Hint*: Note that $(e^t y')' = e^t y'' + e^t y'$.]

6. (*Finding Solution Formulas*). Use Theorem 1.3.1 to find all the solutions. [*Hint*: In parts (a)–(c) multiply each side of the ODE by e^t.]

    (a) $y' + y = 1$ (b) $y' + y = t$ (c) $y' + y = t + 1$

    (d) $2yy' = 1$ [*Hint*: $(y^2)' = 2yy'$]   (e) $2yy' = t$

www **7.** (*Integrating Factors*). Find the general solution of each ODE by using an integrating factor.

    (a) $y' - 2ty = t$ (b) $y' - y = e^{2t} - 1$ (c) $y' = \sin t - y \sin t$

    (d) $2y' + 3y = e^{-t}$ (e) $t(2y - 1) + 2y' = 0$ (f) $y' + y = te^{-t} + 1$

8. (*Solving an IVP, Long-Term Behavior*). First find the general solution of the ODE. Then use the initial condition and find the solution of the IVP. Finally, discuss that solution's behavior as $t \to +\infty$.

    (a) $y' + y = e^{-t}$, $y(0) = 1$ (b) $y' + 2y = 3$, $y(0) = -1$

    (c) $y' + 2ty = 2t$, $y(0) = 1$ (d) $y' + (\cos t)y = \cos t$, $y(\pi) = 2$

9. Use a numerical solver to plot the solution curve of each IVP of Problem 8 on the given t-interval and compare with the graph of the "true" solution.
    (a) $[-2, 6]$ (b) $[-1, 5]$ (c) $[-5, 5]$ (d) $[-8, 8]$

10. (*Using an Integrating Factor*). Find the general solution of each ODE over the indicated t-interval by using an integrating factor. Discuss the behavior of the solution as $t \to 0^+$. Discuss the behavior of the ODEs in parts (a),(b), and (d) as $t \to +\infty$.

    (a) $ty' + 2y = t^2$, $t > 0$ (b) $(3t - y) + 2ty' = 0$, $t > 0$

    (c) $y' = (\tan t)y + t \sin 2t$, $|t| < \pi/2$ (d) $y' = y/t + t^n$, $t > 0$,   integer n

11. (*Solving an IVP*). Find the general solution of the ODE over the indicated t-interval. Then use the initial condition to find the solution of the IVP. Finally, discuss the behavior of the solution as t tends to the given value.

    (a) $ty' + 2y = \sin t$, $t > 0$; $y(\pi) = 1/\pi$; $t \to +\infty$

    (b) $(\sin t)y' + (\cos t)y = 0$, $0 < t < \pi$; $y(3\pi/4) = 2$; $t \to 0^+$

    (c) $y' + (\cot t)y = 2 \cos t$, $0 < t < \pi$; $y(\pi/2) = 3$; $t \to 0^+$

    (d) $y' + (2/t)y = (\cos t)/t^2$, $t > 0$; $y(\pi) = 0$; $t \to 0^+$, $t \to +\infty$

12. Use a numerical solver to plot the solution curve of each IVP of Problem 11 on the given interval and compare with the "true" solution.
  (a) $[0, 50]$ **(b)** $[0, 3]$ **(c)** $[0, 3]$ **(d)** $[0, 4]$

13. (*Do-It-Yourself Linear ODE*). Make up several linear ODEs for which a solution formula is not helpful in deciding how solutions behave. Use your numerical solver to find out what happens to solutions as t increases and as t decreases. For example, try the ODE $y' + 2ty = 1/(1 + t^2)$.

1.4 Modeling with Linear ODEs

First-order linear ODEs model natural processes that range from the changing concentration of a pollutant in a water tank to the rising temperature of an egg as it is being hard-boiled. We will model one of these processes and leave others to the problem set. Next we will take another look at the solution formula constructed in the last section. That formula reveals a simple structure for all solutions of a first-order linear ODE, and we will use this structure to understand what is going on in our models.

The Balance Law and a Compartment Model

First-order linear ODEs often arise in applications as the result of an underlying basic principle: If $y(t)$ denotes the size of a population or the amount of a substance in a *compartment* at time t, then the rate of change $y'(t)$ can be calculated as the "rate in" minus the "rate out" for the compartment. We formalize this principle as the *Balance Law*.

> Balance Law. Net rate of change = Rate in − Rate out.

Let's apply the Balance Law to a mixture process.

EXAMPLE 1.4.1

A Model IVP for Accumulation of a Pollutant

A tank contains 100 gallons of contaminated water in which y_0 pounds of pollutant are dissolved. Contaminated water starts to run into the tank at the rate of 10 gallons per minute. The concentration of pollutant in this incoming stream at time t is $c(t)$ pounds per gallon. The solution in the tank is thoroughly mixed, and contaminated water flows out at the rate of 10 gallons per minute. Find the IVP for the amount of pollutant $y(t)$ in the tank.

Think of the tank as the compartment that the pollutant occupies, and apply the Balance Law to find an expression for $y'(t)$. The initial time is 0, so y_0 is the amount of pollutant in the water at $t = 0$. Pollutant is added to the tank at the inflow rate

$$\left(10\frac{\text{gal}}{\text{min}}\right) \cdot \left(c(t)\frac{\text{lb}}{\text{gal}}\right) = 10c(t)\frac{\text{lb}}{\text{min}}$$

Pollutant leaves the tank at the rate

$$\left(10\frac{\text{gal}}{\text{min}}\right)\cdot\left(\frac{y(t)}{100}\frac{\text{lb}}{\text{gal}}\right)=0.1y(t)\frac{\text{lb}}{\text{min}}$$

since contaminated water drains at 10 gal/min, and the pollutant concentration in the tank and in the exit stream at time t is $(y(t)/100)$ lb/gal. The Balance Law says that

$$y'(t) = \text{Rate in} - \text{Rate out} = 10c(t) - 0.1y(t)$$

The corresponding IVP is

$$y'(t) + 0.1y(t) = 10c(t), \qquad y(0) = y_0, \quad t \geq 0 \qquad (1)$$

which is our model.

Mathematical formulas are a dime a dozen, but understanding their meaning and how they can be used takes thought and theory. One way to do this is to look at and interpret each part of a formula. Before solving the specific IVP (1) let's take some time out to examine the solution formula of a first-order linear ODE more closely.

Output Depends on Initial Data and Input

The output of a process depends on the initial data and the input. In the context of a first-order linear ODE we can describe the dependence in a quite precise way.

THEOREM 1.4.1

Response to Data and Input. **Suppose that** $p(t)$ and $q(t)$ are continuous on an interval I containing t_0 and that $P(t)$ is any antiderivative of $p(t)$. Then the IVP

$$y' + p(t)y = q(t), \qquad y(t_0) = y_0 \qquad (2)$$

has a unique solution, which can both be written as a formula and interpreted in terms of initial data y_0 and input $q(t)$:

$$y(t) = e^{P(t_0)}y_0 e^{-P(t)} + e^{-P(t)}\int_{t_0}^t e^{P(s)}q(s)\,ds, \qquad t \text{ in } I \quad (3)$$

Total Response $=$ Response to initial data $+$ Response to input

To show this, note that the general solution formula for $y' + p(t)y = q(t)$ is

$$y = Ce^{-P(t)} + e^{-P(t)}R(t) \qquad (4)$$

where C is any constant and $P(t)$ and $R(t)$ are any antiderivatives of $p(t)$ and $e^{P(t)}q(t)$, respectively. Let's take a specific antiderivative of $e^{P(t)}q(t)$:

$$R(t) = \int_{t_0}^t e^{P(s)}q(s)\,ds \qquad (5)$$

Now set $y = y_0$ and $t = t_0$:

$$y_0 = Ce^{-P(t_0)} + e^{-P(t_0)} \int_{t_0}^{t_0} e^{P(s)}q(s)\,ds = Ce^{-P(t_0)}$$

So $C = e^{P(t_0)}y_0$ and solution formula (3) follows.

We will use the response interpretation often, particularly in the applications, because we can see from formula (3) exactly how changes in the initial value y_0 and in the driving function $q(t)$ will affect the output $y(t)$.

Since the formula (3) defines a solution $y(t)$ for IVP (2) for *any* choice of y_0 and $q(t)$, we see that

$$\text{Response to initial data} \qquad y(t) = e^{P(t_0)}y_0e^{P(t)} \qquad (6)$$

solves the IVP

$$y' + p(t)y = 0, \qquad y(t_0) = y_0 \qquad (7)$$

[just put $q(t) = 0$ in formula (3)], and that

$$\text{Response to input} \qquad y(t) = e^{-P(t)} \int_{t_0}^{t} e^{P(s)}q(s)\,ds \qquad (8)$$

solves the IVP

$$y' + p(t)y = q(t), \qquad y(t_0) = 0 \qquad (9)$$

[just put $y_0 = 0$ in formula (3)]. Adding the right sides of (6) and (8) gives the total response.

Response Analysis for the Pollutant Model

Now that we know something about the structure of solutions for first-order linear ODEs, we will return to the pollutant accumulation model to analyze and to interpret its solutions.

EXAMPLE 1.4.2

How Much Pollutant Is in the Tank?

Now let's solve IVP (1). After multiplying through by the integrating factor $e^{0.1t}$, the ODE in (1) becomes

$$e^{0.1t}[y'(t) + 0.1y(t)] = [e^{0.1t}y(t)]' = e^{0.1t}10c(t)$$

So we see from Theorem 1.4.1 that

$$y(t) = y_0e^{-0.1t} + e^{-0.1t} \int_{0}^{t} e^{0.1s}[10c(s)]\,ds, \qquad t \geq 0 \qquad (10)$$

$$\text{Total Response} = \text{Response to initial amount} + \text{Response to inflow}$$

The first term on the right-hand side of formula (10) shows the declining amount of pollutant in the tank if pure water were to run in and contaminated water run out at the

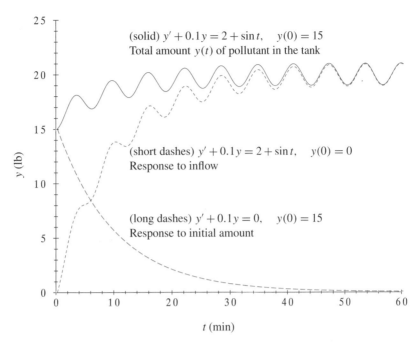

FIGURE 1.4.1 Accumulation of pollutant in tank (Example 1.4.2).

same rate. The second term shows the amount of pollutant in the tank at time t due solely to the incoming stream of contaminated water. In the long run we expect the second term to dominate.

It is the linearity of the ODE that allows the decoupling of the effect of the initial amount of pollutant from the effect of the incoming stream. We can illustrate all this by graphing the terms of formula (10) separately. To be specific, let's take $y_0 = 15$ lb and assume an input concentration $c(t)$ that varies sinusoidally about a mean of 0.2 lb of pollutant per gallon. For example, suppose that

$$c(t) = 0.2 + 0.1 \sin t \qquad (11)$$

☞ We could also have used a symbolic software package here.

We could have used an antiderivative in Table 1.3.1 to work out the integral in formula (10), but instead we used our numerical solver to solve IVP (1) directly. The results are displayed in Figure 1.4.1. We plotted the "Response to initial amount" and the "Response to inflow" curves by applying our solver to the respective IVPs (7) and (9). Over time, the amount of pollutant in the tank at the start becomes almost irrelevant, and it is pollutant in the inflow stream that determines the amount of pollutant in the outflow.

It looks as if the long-term response of the ODE to an oscillating input is an oscillatory output with the same frequency as the input. More (a lot more) on this curious behavior in Section 2.2.

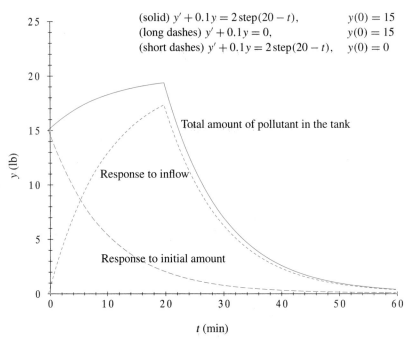

$$\text{(solid) } y' + 0.1y = 2\,\text{step}(20 - t), \qquad y(0) = 15$$
$$\text{(long dashes) } y' + 0.1y = 0, \qquad\qquad\qquad y(0) = 15$$
$$\text{(short dashes) } y' + 0.1y = 2\,\text{step}(20 - t), \quad y(0) = 0$$

FIGURE 1.4.2 Response to inflow stream clean-up (Example 1.4.3).

Cleaning Up the Water Supply: Step Functions

☞ Step functions are used to turn inputs and parameters "on" and "off."

Step functions can be used to switch midstream from one model ODE to another. The basic step function is defined like this:

$$\text{step}(t) = \begin{cases} 0, & t < 0 \\ 1, & t \geq 0 \end{cases} \tag{12}$$

We can work with this function as we would with any other. For example,

$$3\,\text{step}(t - 5) = \begin{cases} 0, & t < 5 \\ 3, & t \geq 5 \end{cases} \quad \text{and} \quad -4\,\text{step}(15 - t) = \begin{cases} -4, & t \leq 15 \\ 0, & t > 15 \end{cases}$$

☞ The engineering functions are described in Appendix B.1.

Step functions and other piecewise linear functions are commonly used to model on-off behavior in natural processes. We call them the *engineering functions*. Here's an example where a step function comes in handy.

EXAMPLE 1.4.3

Cleaning Up the Water Supply

In the tank model of Example 1.4.1 let's suppose that the inflow pollutant concentration $c(t)$ is 0.2 lb/gal, but that after 20 minutes all the pollutants are filtered out of the inflow stream. We can express $c(t)$ in terms of the step function in the following way:

$$c(t) = 0.2\,\text{step}(20 - t), \qquad t \geq 0 \tag{13}$$

So the model IVP

$$y' + 0.1y(t) = 10c(t), \qquad y(0) = y_0, \quad t \geq 0 \tag{14}$$

has, using formula (10), the solution

$$y = y_0 e^{-0.1t} + e^{-0.1t} \int_0^t e^{0.1s}[2\,\text{step}(20-s)]\,ds, \qquad t \geq 0 \qquad (15)$$

Notice that the integral in formula (15) changes its form at $t = 20$:

$$\int_0^t e^{0.1s}[10c(s)]\,ds = \begin{cases} \int_0^t e^{0.1s}[10 \cdot 0.2]\,ds = 20\left(e^{0.1t} - 1\right), & 0 \leq t \leq 20 \\[2mm] \int_0^{20} e^{0.1s}[10 \cdot 0.2]\,ds = 20(e^2 - 1), & t > 20 \end{cases}$$

We can put this integral into formula (15) and obtain the solution formula

$$y = y_0 e^{-0.1t} + \begin{cases} 20(1 - e^{-0.1t}), & 0 \leq t \leq 20 \\[2mm] 20(e^2 - 1)e^{-0.1t}, & t > 20 \end{cases}$$

We can graph solutions using this formula, or alternatively we can use a numerical solver to solve IVP (14) directly. We prefer the latter approach.

Figure 1.4.2 shows the changing pollutant concentration in the tank if $y_0 = 15$ lb. The corners on two of the solution curves are caused by the discontinuities at $t = 20$ in the step function used to model the inflow stream.

☞ Most numerical solvers accept step functions.

The set of solutions of a first-order linear ODE has a structure that is often used to find solutions without extensive use of integrating factors. Let's see how this approach works.

Another Approach to Finding a Solution Formula

There is a way to find a formula for all solutions of a first-order linear ODE other than the method of integrating factors presented in Section 1.3. In practice, engineers and scientists use this alternative approach in solving linear ODEs. Here is the result we need:

THEOREM 1.4.2

Structure of Solutions. **Suppose that** $p(t)$ and $q(t)$ are continuous on an interval I. Then the general solution $y(t)$ of the driven linear ODE

$$y' + p(t)y = q(t) \qquad (16)$$

is given by

$$y(t) = y_u(t) + y_d(t), \quad \text{all } t \text{ in } I \qquad (17)$$

where $y_u(t)$ is the general solution of the undriven linear ODE

$$y' + p(t)y = 0 \qquad (18)$$

and $y_d(t)$ is any particular solution of the driven ODE (16) itself (any one will do).

To see why this is true, take any particular solution $y_d(t)$ of the driven ODE (16); it can be any one. Now we show that $y(t)$ is a solution of ODE (16) if and only if the function $w = y - y_d$ is a solution of the undriven ODE (18). This fact results from the computation

$$w' + pw = (y - y_d)' + p(y - y_d)$$
$$= (y' + py) - (y_d' + py_d)$$
$$= q - q = 0$$

So this means that $y = y_u + y_d$, where y_u is the general solution of the undriven ODE (16). We saw earlier that $y_u = Ce^{-P(t)}$, where C is any constant and $P(t)$ is any single antiderivative of the coefficient $p(t)$.

Here's an example that uses Theorem 1.4.2 to find the general solution.

EXAMPLE 1.4.4

Good Guessing Works

The undriven form of the ODE

$$y' + y = 17 \sin 4t \tag{19}$$

is $y' + y = 0$, which has $y_u = Ce^{-t}$ as its general solution. So the general solution of the driven ODE (19) has the form

$$y = Ce^{-t} + y_d(t)$$

where C is any constant and y_d is any one particular solution.

We could use the procedure of the last section to find a particular solution, but there is another way. Because the input function is $17 \sin 4t$, a good guess for a particular solution would be

☞ We lucked out in having a sinusoidal input.

$$y_d = A \sin 4t + B \cos 4t \tag{20}$$

where A and B are constants to be determined. To see this, put this form of y_d into the left side of ODE (19) to obtain

$$y_d' + y_d = 4A \cos 4t - 4B \sin 4t + A \sin 4t + B \cos 4t$$
$$= (A - 4B) \sin 4t + (4A + B) \cos 4t$$

So for y_d in (20) to solve ODE (19) we must have

$$(A - 4B) \sin 4t + (4A + B) \cos 4t = 17 \sin 4t$$

The only way for this equality to hold for all t is for A and B to satisfy the equations

$$A - 4B = 17, \qquad 4A + B = 0$$

Solving for A by multiplying the second equation by 4 and then adding the two equations, we see that $A = 1$. So $B = -4$, and inserting these values into our trial solution (20), we have constructed a particular solution,

$$y_d = \sin 4t - 4 \cos 4t$$

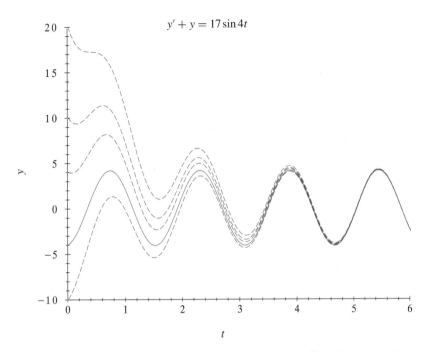

$$y' + y = 17\sin 4t$$

FIGURE 1.4.3 All solutions are attracted to the periodic solution $y_d(t) =$ $\sin 4t - 4\cos 4t$ (solid curve) (Example 1.4.5).

The general solution is then

$$y = Ce^{-t} + \sin 4t - 4\cos 4t, \qquad C \text{ any constant} \qquad (21)$$

so our guessing approach has paid off.

The structure formula (17) sometimes reveals interesting properties of solutions.

EXAMPLE 1.4.5

Attraction to a Unique Periodic Solution

In Example 1.4.4 we showed that all solutions of the ODE $y' + y = 17\sin 4t$ have the form $y = Ce^{-t} + y_d(t)$, where $y_d(t) = \sin 4t - 4\cos 4t$ and C is an arbitrary constant. Since $Ce^{-t} \to 0$ as $t \to +\infty$, we see that all solutions are attracted to the single solution $y_d(t)$, so the ODE has a unique periodic solution. This periodic solution $y_d(t)$ has period $\pi/2$, exactly the same period as the input $17\sin 4t$. Figure 1.4.3 shows the strong attraction of all solutions toward y_d.

Comments

In this section we have shown how to interpret the formulas for the general solution of a linear ODE and for the solution of a corresponding IVP. That is a crucial aspect when you use an IVP to model a natural process, and the application of the Balance Law to the pollutant concentration process bears this out.

PROBLEMS

www **1.** (*Pollution*). Contaminated waste water is pumped at the rate of 1 gal/min into a tank containing 1000 gal of clean water, and the well-stirred mixture leaves the tank at the same rate.

(a) Find the amount $y(t)$ of waste in the tank at time t. What happens as $t \to +\infty$?

(b) How long does it take the concentration of waste to reach 20% of its maximum level?

(c) After an hour, the waste inflow is stopped and clean water is pumped in at 1 gal/min. Use a step function and create a model IVP for this situation. Use a numerical solver to plot $y(t)$ over the time span $0 \le t \le 1500$. What happens as $t \to +\infty$? Explain what you see.

2. (*Pollution*). Water with pollutant concentration c_0 lb/gal starts to run at 1 gal/min into a vat holding 10 gal of water mixed with 10 lb of pollutant. The mixture runs out at 1 gal/min.

(a) Find the amount $y(t)$ of pollutant in the vat at time t. Find $\lim_{t \to +\infty} y(t)$.

(b) Plot $y(t)$ over the interval $0 \le t \le 40$ for values of c_0 in the range $0.1 \le c_0 \le 2$. Include $c_0 = 1.0$ and values of c_0 above and below 1.0. Explain why as $t \to +\infty$, $y(t) \to 10c_0$.

(c) Suppose that $c_0 = 1$ for the first 10 min, and then an efficient filter removes all of the pollutant from the inflow stream. Find the model IVP, using a step function to represent the input. As in Figure 1.4.2, use a numerical solver to plot the total response, the response to the initial data, and the response to the input. If possible, put all three curves on the same graph; otherwise use three graphs with the same t and y scales. What happens as $t \to +\infty$?

(d) Repeat part **(c)**, but with an inefficient filter that only removes 50% of the pollutant.

(e) Repeat part **(c)**, but with an efficient filter that is in place for just 10 minutes at the start of each hour. Plot for $0 \le t \le 150$, $-5 \le y \le 15$. What happens to each response as t increases? [*Hint*: Use the function $\text{sqw}(t, 100/6, 60)$.]

☞ Appendix B.1 has more on $\text{sqw}(t, d, T)$, or see "Tips" in the Student Resource Manual.

3. (*Salt Solution*). A solution containing 2 lb of salt per gallon starts to flow into a tank of 50 gal of pure water at a rate of 3 gal/min. After 3 minutes the mixture starts to flow out at 3 gal/min.

(a) How much salt is in the tank at $t = 2$ min? At $t = 25$ min? [*Hint*: Solve two IVPs: for $t_0 = 0$ and for $t_0 = 3$.]

(b) How much salt is in the tank as $t \to +\infty$? Can you guess without any calculation?

4. (*Good Guessing and Undetermined Coefficients*). Find a particular solution for the linear ODE by first guessing the form of the solution, and then determining the coefficients. Then find the general solution.

(a) $y' + y = t^2$ [*Hint*: Try $y_d = At^2 + Bt + C$.]   **(b)** $y' + ty = t^2 - t + 1$

(c) $y' + 2y = e^{-2t}$ [*Hint*: Try $y_d = Ate^{-2t}$.]   **(d)** $y' + 2y = 3e^{-t}$

(e) $y' + y = 5\cos 2t$ **(f)** $y' + y = e^{-t}\cos t$

5. (*Long-Term Behavior*). Describe the long-term behavior (i.e., as $t \to +\infty$) of all solutions of the corresponding ODE in parts **(a)**–**(f)** of Problem 4 by using the general solution formula. Justify your conclusions.

6. (*More Guessing and Undetermined Coefficients*). For all values of the constants a, b, c, find a particular solution of each ODE.

(a) $y' + ay = b\cos t + c\sin t$ **(b)** $y' + ay = bt + c$, $a \ne 0$

7. (*Long-Term Behavior*). Find all solutions of the undriven ODE $y' + 2ty = 0$, and describe what happens as $t \to +\infty$. Use your numerical solver to plot several solutions of the driven ODE $y' + 2ty = q(t)$ over a long enough time span that you can conjecture what happens to its general solution as $t \to +\infty$. Why is the general solution formula of little use here?

(a) $q(t) = 1$ **(b)** $q(t) = (1 + 2t^2)^{-1}$

(c) $q(t) = 1 + t^2$ **(d)** $q(t) = 1 + 2t^2$ [*Hint*: Guess a solution $y_d = A + Bt$.]

8. (*Long-Term Behavior*). The examples in this section may give the impression that as $t \to +\infty$ the general solution $y(t)$ of the undriven ODE $y' + p(t)y = 0$ tends to 0, so the general solution $y(t)$ of the driven ODE $y' + p(t)y = q(t)$ would tend to a particular solution $y_d(t)$ as $t \to +\infty$. Show that this is *not* always the case by creating your own linear ODE $y' + p(t)y = q(t)$ where there is some solution of the undriven ODE ($q = 0$) that tends to $+\infty$ as $t \to +\infty$, and there is some solution $y_d(t)$ of the driven ODE such that $y_d(t) \to 0$ as $t \to +\infty$.

9. Use a numerical solver in plotting solutions of the given IVPs.

    (a) Plot solutions of the IVPs $y' + y = 0$, $y(-5) = 0, \pm 2, \pm 4$, $-5 \le t \le 10$.

    (b) Plot the solution $y = y_d(t)$ of the IVP $y' + y = t\cos(t^2)$, $y(-5) = 0$, $-5 \le t \le 10$.

    (c) Plot the solutions of the IVPs $y' + y = t\cos(t^2)$, $y(-5) = 0, \pm 2, \pm 4$, where $-5 \le t \le 10$. Use (a) and (b) to explain the behavior of the solutions. What happens as $t \to +\infty$?

10. Plot solutions of the IVPs $y' = -(\cos t)y + \sin t$, $y(-10) = -6, -2, 0, 5$, $|t| < 10$, and explain why a general solution formula isn't any help here.

11. Use a numerical solver or a solution formula to plot solution curves.

    (a) $y' = (1 - y)(\sin t)$; $y(-\pi/2) = -1, 0, 1$; $-\pi/2 \le t \le 2\pi$

    (b) $y' + y = te^{-t} + 1$; $y(0) = -1, 0, 1$; $-1 \le t \le 3$

    (c) $ty' + 2y = t^2$; $y(2) = 0, 1, 2$; $0 < t < 4$. Repeat for $y(-2) = 0, 1, 2$; $-4 < t < 0$. What happens to each solution curve as $t \to 0$?

    (d) $y' = y\tan t + t\sin 2t$; $y(0) = -1, 0, 1$; $-\pi/2 < t < \pi/2$

12. (*Adjusting the Input*). The output $y(t)$ of the IVP $y'(t) + p(t)y(t) = q(t)$, $y(0) = y_0$, must be kept at the level y_0 for $t \ge 0$. Determine an input $q(t)$ that will accomplish the task.

13. (*Unbounded Solutions*). Suppose that $q(t)$ is continuous for all t, and that $y' + y = q(t)$ has a particular solution $y_d(t)$ with the property that $y_d(t) \to +\infty$ as $t \to +\infty$. Explain why all solutions have this property.

14. | (*Newton's Law of Cooling*). According to *Newton's Law of Cooling* (or *Warming*) the rate of change of the temperature of a body is proportional to the difference between the body's temperature and the surrounding medium's temperature.

☞ Shaded boxes contain modeling principles.

(a) Write a model ODE for the body's temperature, given the medium's temperature $m(t)$. Is the proportionality constant positive or negative?

(b) (*A Sick Horse*). A veterinarian wants to find the temperature of a sick horse. The readings on the thermometer follow Newton's Law. At the time of insertion the thermometer reads $82°$F. After 3 min the reading is $90°$F, and 3 min later $94°$F. A sudden convulsion destroys the thermometer before a final reading can be obtained. What is the horse's temperature?

(c) (*Cooling an Egg*). A hard-boiled egg is removed from a pot of hot water and set on the table to cool. Initially, the egg's temperature is $180°$F. After an hour its temperature is $140°$F. If the room's temperature is $65°$F, when will the egg's temperature be $120°$F, $90°$F, $65°$F?

(d) (*Cold Body Cools Hot Medium*). A cold egg is placed in a pot of hot water. As the egg warms up, the water cools down. Create a model ODE for the temperature of the egg and another for the temperature of the water. What happens to the two temperatures as $t \to +\infty$? Explain what you do.

15. (*Continuously Compounded Interest*). Some banks compound your savings continuously; for example, if the interest rate is 9%, then the ODE for the money in your account is $A' = 0.09A$.

    (a) What interest rate payable annually is equivalent to 9% interest compounded continuously?

    (b) What is the interest rate if continuously compounded funds double in eight years?

(c) How long will it take A dollars invested in a continuously compounded savings account to double if the interest rate is 5%; 9%; 12%?

16. (*Rule of 72*). Two common rules of thumb for bankers are the "Rule of 72" and the "Rule of 42." These rules say that the number of years for money invested at $r\%$ interest to increase by 100% or 50% is given by $72/r$ or $42/r$, respectively. Assuming continuous compounding, show that these rules overestimate the time required.

 17. (*Compound Interest*). At the end of every month Wilbert deposits a fixed amount in a savings account. He wants to buy a car that will cost $12 400. Currently, he has $5800 saved. If he earns 7% interest compounded continuously on his savings and has an income of $2600 per month, to what amount must Wilbert limit his monthly expenses if he is to have enough saved to buy the car in one year?

1.5 Introduction to Modeling and Systems

Modeling is a process of recasting a natural process from its natural environment into a form, called a model, that can be analyzed via techniques we understand and trust. A *model* is a device that helps the modeler to predict or explain the behavior of a phenomenon, experiment, or event. For example, say that a rocket is to be put into orbit. Physical intuition alone can't give us more than a rough idea of what guidance strategy to use once the rocket is launched. Since accuracy will be critical, a mathematical model of the problem is constructed using applicable natural laws. Then we can use the equations, constraints, and control elements in the model to give a reasonably precise description of the orbital elements of the rocket in its course.

Using models to explain outcomes of an observable situation is illustrated in Figure 1.5.1. Broad portions of the natural environment are given a mathematical form by a general model in which all possible outcomes are described by a few basic principles. A specific problem in the environment is translated into a specific mathematical problem in the general model. The mathematical problem is solved (often by a computer simulation) and the results are interpreted in the problem's natural environment.

Models often end up as a collection of ordinary differential equations in several to-be-determined functions, or for short, *differential systems*. When we use the word "system" by itself we will most always mean "differential system." The *order* of a system is the order of the highest derivative that occurs anywhere in the system. The first-order system

$$\begin{aligned} dx/dt &= f(t, x, y) \\ dy/dt &= g(t, x, y) \end{aligned} \qquad (1)$$

in the to-be-determined functions $x(t)$ and $y(t)$ is in *normal form* because the derivatives appear alone on the left-hand side of the ODEs. Numerical solvers prefer to have systems entered in normal form.

☞ Linear systems are used a lot in applications.

System (1) is a *linear differential system* if the given rate functions $f(t, x, y)$ and $g(t, x, y)$ have the form

$$\begin{aligned} f(t, x, y) &= a(t)x + b(t)y + h(t) \\ g(t, x, y) &= c(t)x + d(t)y + k(t) \end{aligned} \qquad (2)$$

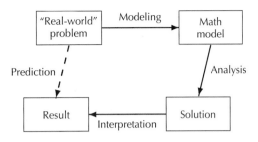

FIGURE 1.5.1 Problem solving via mathematical modeling.

where the *coefficients* $a(t)$, $b(t)$, $c(t)$, $d(t)$ and the *driving terms* $h(t)$, $k(t)$ are functions only of t. System (1) is *nonlinear* if the rate functions do not have the form (2).

Here's an example of a nonlinear system: If $\theta(t)$ denotes the angle of a moving pendulum bob subject to gravity and a damping force, and if $v(t) = \theta'$, then the pair $\theta(t)$, $v(t)$ solves the first-order system

$$\theta' = v, \qquad v' = -0.1v - 4\sin\theta \qquad (3)$$

System (3) is nonlinear because of the term $-4\sin\theta$.

In this section we'll look at two models:

- Radioactive decay and its use in dating old objects.
- Vertical motion: Does the ball take longer to rise or to fall?

The study of radioactivity and moving bodies revolutionized science. That is why we chose to model the basic principles of these two natural processes.

Modeling Radioactive Decay

Some elements are unstable, decaying into other elements by emission of alpha particles, beta particles, or photons. Such elements are said to be *radioactive*. For example, a radium atom might decay into a radon atom, giving up an alpha particle in the process. The decay of a single radioactive nucleus is a random event, and the exact time of decay can't be predicted with certainty. Nevertheless, something definite can be said about the decay process of a large number of radioactive nuclei.

For a collection of radioactive nuclei in a sample, we would like to know the number of radioactive nuclei present at any given time. There is plenty of experimental evidence to suggest that the following *decay law* is true.

> **Radioactive Decay Law.** In a sample containing a large number of radioactive nuclei, the rate of decrease in the number of radioactive nuclei at a given time is proportional to the number of nuclei present at the time.

Denoting the number of radioactive nuclei in the sample at time t by $N(t)$, the law translates to the mathematical equation

$$N'(t) = -kN(t) \tag{4}$$

where k is a positive coefficient of proportionality. Observation suggests that in most decay processes k is independent of t and N. A decay process of this type is said to be of *first order* with *rate constant k*.

☞ More on this difficulty in the Student Resource Manual.

This simple-looking law has all sorts of logical difficulties because we are using a "continuous" mathematical structure to describe discrete events. It's a remarkable fact that this law leads to a mathematical model that allows us to make accurate predictions.[1]

Let's apply the Radioactive Decay Law to create a mathematical model that predicts how many radioactive nuclei are present in a given sample at any time t.

EXAMPLE 1.5.1   **Exponential Decay**

Suppose that there are N_0 radioactive nuclei in a sample of an element at time t_0. How many radioactive nuclei $N(t)$ are present at a later time? The Radioactive Decay Law describing how $N(t)$ evolves is given by the forward IVP

$$N' = -kN, \qquad N(t_0) = N_0, \quad t \geq t_0 \tag{5}$$

Using the method of integrating factors (Section 1.3), we see that IVP (5) has precisely one solution for any given value for N_0:

$$N(t) = N_0 e^{-k(t-t_0)}, \qquad t \geq t_0 \tag{6}$$

Once k and N_0 are known, formula (6) can be used to predict the values of $N(t)$.

The rate constant k of a radioactive element is usually calculated from the element's *half-life*, which is the time τ required for half of the nuclei to decay. Curiously, τ is independent of both the time when the clock starts and the initial amount. For example, we see from formula (6) that at any times $t \geq t_0$ and $\tau > 0$,

$$\frac{N(t + \tau)}{N(t)} = \frac{N_0 e^{-k(t+\tau-t_0)}}{N_0 e^{-k(t-t_0)}} = e^{-k\tau}$$

So if we require that $N(t + \tau) = \frac{1}{2}N(t)$, then $e^{-k\tau} = \frac{1}{2}$, and (after taking logarithms)

$$\tau = \frac{\ln 2}{k} \quad \text{and} \quad k = \frac{\ln 2}{\tau} \tag{7}$$

and neither τ nor k depends on t_0 or on N_0. The half-lives of many radioactive elements have been determined experimentally. For example, the half-life τ of radium[2] is about 1600 years, so the rate constant k for radium is about 0.0004332 (years)$^{-1}$.

---

[1]See the article by E. Wigner, "The Unreasonable Effectiveness of Mathematics in the Physical Sciences," *Commun. Pure Appl. Math.* **13** (1960), pp. 1–14.

[2]The Polish scientist Marie Curie (1867–1934) received the Nobel Prize for Physics in 1903 and 1911 for her pioneering experiments with radium and other radioactive substances.

Formula (6) can be used to make predictions about the value of $N(t)$. These predictions can be checked against experimental determinations of $N(t)$. Given the logical gaps mentioned above, it might seem surprising that formula (6) provides a remarkably accurate description of radioactive decay processes. But it is a fact of contemporary experimental science that this is so, at least for time spans that are neither too long nor too short. The leap of faith made in ignoring the flaws of the law and the model is justified by the results.

Here is one way that radioactive decay is used to date events that occurred long ago, even before recorded history. Living cells absorb carbon from carbon dioxide in the air. The carbon in some of this carbon dioxide is radioactive carbon-14 (denoted by ^{14}C), rather than the common ^{12}C. Carbon-14 is produced by the collisions of cosmic rays with nitrogen in the atmosphere. The ^{14}C nuclei decay back to nitrogen atoms by emitting beta particles. All living things, or things that were once alive, contain some radioactive carbon nuclei. In the late 1940s, Willard Libby [3] showed how a careful measurement of the ^{14}C decay rate in a fragment of dead tissue can be used to determine the number of years since its death. This process is called *radiocarbon dating*. See Problem 10 for the way this technique was used to date the age of prehistoric wall paintings in a cave near Lascaux, France.

Now let's look at a completely different kind of natural process and model.

The Galilean Approach to Vertical Motion

Let's suppose that a ball is moving along a vertical line near the ground. Neglecting air resistance, Galileo[4] showed experimentally that the ball moves with constant acceleration g, where g is the earth's gravitational constant, and that this motion is independent of the size, shape, or mass of the ball. We will use this information to construct a model for the motion. It is known that the value of g is

$$g \approx 9.8 \text{ m/sec}^2 = 980 \text{ cm/sec}^2 \approx 32 \text{ ft/sec}^2$$

EXAMPLE 1.5.2

Differential Equation for a Moving Ball: No Air Resistance
Suppose that $y(t)$ measures the distance of a ball's center above the ground at time t, then the velocity of the ball at time t is $v(t) = y'(t)$. The acceleration of the ball is

---

[3] Willard Libby (1908–1980) was an American chemist who received the 1960 Nobel Prize for Chemistry for his work. His book *Radiocarbon Dating*, 2nd ed. (Chicago: University of Chicago Press, 1955) discusses the techniques used to do the dating. This process is most effective with material that is at least 200 years old, but not more than 70 000 years old.

[4] Galileo Galilei (1564–1642) was the first modeler of modern times. Although he is better known for his work in astronomy, his study of bodies dropped from a height and of motion on inclined planes led to a model for falling bodies relating the distance traveled to the time elapsed, and eventually to the law of acceleration. Later, Galileo turned his telescope toward the heavens and discovered four of the moons that orbit the planet Jupiter. His observations supported the Copernican model of the solar system. The connection between the physical and mathematical worlds is best said in Galileo's verse, "Philosophy is written in this grand book of the universe, which stands continually open to our gaze.... It is written in the language of mathematics."

Galileo Galilei

Ball

Ground

$a(t) = v'(t) = y''(t)$. Suppose the clock is started at $t = 0$, and at that instant $y(0) = y_0$ and $v(0) = v_0$. While the ball is in motion, Galileo's experimental result implies that $y''(t) = -g$ for all $0 \leq t \leq T$, where T is the time that the ball hits the ground. The minus sign arises because gravity acts downward in the direction of decreasing y. Summarizing, we see that $y(t)$ solves the IVP

$$y'' = -g, \qquad y(0) = y_0, \quad y'(0) = v_0 \tag{8}$$

over the interval $0 \leq t \leq T$. Let's call IVP (8) the *Galilean model* of vertical motion.

The ODE in (8) is not valid outside the interval $0 \leq t \leq T$ because we know nothing about the ball before $t = 0$ or after impact at $t = T$. Now let's solve IVP (8).

EXAMPLE 1.5.3

Solution to Galileo's Problem of the Moving Ball

Suppose that $y(t)$ solves IVP (8). Antidifferentiate each side of the ODE $y'' = -g$ to obtain $y' = -gt + C$ for some constant C. The condition $y'(0) = v_0$ implies that $C = v_0$, and so

$$y' = -gt + v_0$$

Antidifferentiate each side of $y' = -gt + v_0$ to obtain $y = -gt^2/2 + v_0 t + C$, where C is a constant. Since $y = y_0$ when $t = 0$, C must have the value y_0. The unique solution $y(t)$ of IVP (8) is

$$y(t) = y_0 + v_0 t - \frac{1}{2} g t^2, \qquad 0 \leq t \leq T \tag{9}$$

Figure 1.5.2 shows the parabolic solution curves, where $g = 9.8$ m/sec^2, $y_0 = 10$ m, and v_0 (in meters/second) takes on various values.

From formula (9) we see that if the earth had no atmosphere (and so no frictional forces to slow the body down), then all bodies moving vertically near the surface would move in the same way, regardless of their size, shape, or mass.

The Newtonian Approach to Vertical Motion

The Galilean model has flaws; for example, it ignores air resistance. When a vertically moving body is subjected to air resistance, Galileo's experimental results do not provide a model that accurately describes the body's motion. Actually, the motion of a body along the vertical is better governed by the following law:

☞ The general version is given in Section 4.1.

> Newton's Second Law (Restricted to Vertical Motion). The product of the constant mass and the acceleration of a body moving vertically is the sum of the external forces acting vertically on the body.

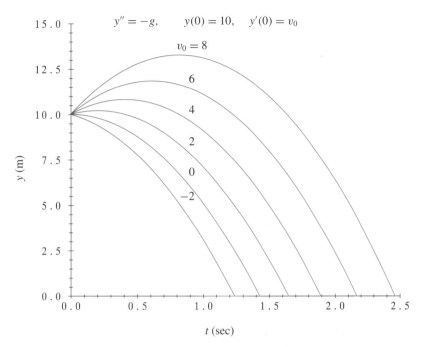

FIGURE 1.5.2 Solution curves of the moving ball problem (Example 1.5.3).

The celebrated general version of this law was formulated by Isaac Newton.[5]

The gravitational force on a body of mass m near the earth's surface is directed downward and has magnitude mg. If the gravitational force is the only force acting on the body, Newton's Second Law implies that the height $y(t)$ of the body above the earth's surface at time t satisfies the ODE $my'' = -mg$, since y'' is the body's acceleration. So $y'' = -g$, the same result we obtained in Example 1.5.2.

But common sense tells us that air resistance will dampen the motion. So let's build a better model that includes damping.

Sir Isaac Newton

[5]Isaac Newton (1642–1727) began his work in science and mathematics when he entered Trinity College in Cambridge, England in 1661. He graduated in 1665 without any special honors and returned home to avoid the plague, which was rapidly spreading through England that year. During the next two years, Newton discovered calculus, determined basic principles of gravity and planetary motion, and recognized that white light is composed of all colors—discoveries which he kept to himself. In 1667, Newton returned to Trinity College, obtaining a Master's degree and staying on as a professor. Newton continued work on his earlier discoveries, formulated the law of gravitation, basic theories of light, thermodynamics, and hydrodynamics, and invented the first good reflecting telescope. In 1687 he was finally persuaded to publish *Philosophae Naturalis Principia Mathematica*, which contains his basic laws of motion and is considered one of the most influential scientific books ever written. Despite his great accomplishments, Newton said of himself that, "I seem to have been only like a boy playing on the seashore and diverting myself in now and then finding a smoother pebble or prettier shell than ordinary, whilst the great ocean of truth lay all undiscovered before me."

Viscous Damping and the Motion of a Whiffle Ball

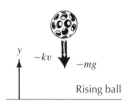

Rising ball

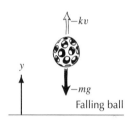

Falling ball

Experiments show that for a body of low density and extended rough surface (e.g., a feather, snowflake, or whiffle ball) the resistance of the air exerts a force on the body proportional to the magnitude of the velocity, but acting opposite to the direction of motion. This kind of resistive force is called *viscous damping*, and the constant of proportionality is the *viscous damping constant*. Suppose that y measures the distance along the local vertical, with up as the positive direction. Then $y' = v$ is the velocity of the body. The resistive force is modeled by $-kv$, where k is the viscous damping constant, and the minus sign reflects the fact that the force opposes motion.

Let's suppose that a whiffle ball of constant mass m is thrown straight up from ground level with initial velocity v_0. Then by Newton's Second Law for Vertical Motion, the location $y(t)$ of the object solves the IVP

$$my'' = -mg - ky', \qquad y(0) = 0, \quad y'(0) = v_0 \tag{10}$$

and we have the *viscous damping model* for the motion of the ball.

EXAMPLE 1.5.4

Tracking the Whiffle Ball

Since $y' = v$, we have $y'' = v'$; rearranging IVP (10) we obtain the following linear IVP in v:

$$v' + \left(\frac{k}{m}\right) v = -g, \qquad v(0) = v_0 \tag{11}$$

Now IVP (11) can be solved using the integrating factor $e^{kt/m}$ to obtain a formula for the whiffle ball's velocity:

$$v(t) = \left(v_0 + \frac{mg}{k}\right) e^{-kt/m} - \frac{mg}{k} \tag{12}$$

We can find the location $y(t)$ of the ball by replacing $v(t)$ by $y'(t)$ in formula (12) and then integrating. We obtain [using the fact that $y(0) = 0$]

$$
\begin{aligned}
y(t) &= \int_0^t \left[\left(v_0 + \frac{mg}{k}\right) e^{-ks/m} - \frac{mg}{k} \right] ds \\
&= \left[\frac{-m}{k} \left(v_0 + \frac{mg}{k}\right) e^{-ks/m} - \frac{mgs}{k} \right]_{s=0}^{s=t} \\
&= \frac{m}{k} \left(v_0 + \frac{mg}{k}\right) (1 - e^{-kt/m}) - \frac{mg}{k}t, \qquad 0 \le t \le T \tag{13}
\end{aligned}
$$

where T is the time the ball hits the ground.

From formula (12) we see that as $t \to +\infty$, $v(t)$ approaches the *limiting velocity* $v_\infty = -mg/k$. The whiffle ball's velocity quickly gets very close to v_∞ because the ball's mass is small and its damping constant is large. Just toss a whiffle ball into the air and watch its slow fall at an apparently constant speed.

Does the whiffle ball take longer to rise to its maximal height, or to fall back from that height? Formula (13) doesn't help at all to answer this question.

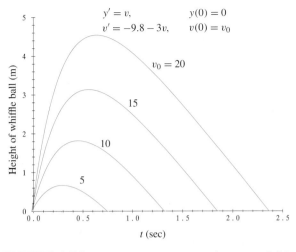

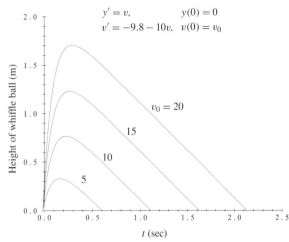

FIGURE 1.5.3 $k/m = 3$: Longer to rise or to fall? (Example 1.5.5).

FIGURE 1.5.4 $k/m = 10$: Longer to rise or to fall? (Example 1.5.5).

Longer to Rise or to Fall?

To decide whether the whiffle ball takes longer to rise or to fall, we can use a computer to graph $y = y(t)$ and answer the question by visual inspection. The graph can be produced by using graphing software and formula (13) for $y(t)$. A better way would be to use a numerical solver directly to solve the IVP

$$y'' = -g - \frac{k}{m}y', \qquad y(0) = 0, \quad y'(0) = v_0 \tag{14}$$

Although some solvers accept second-order ODEs, many do not. All solvers accept an IVP for a first-order system of ODEs equivalent to IVP (14). Here is how we convert IVP (14) into such a first-order system. Set $y' = v$ and use IVP (14) to obtain

☞ This is a first-order linear system.

$$\begin{aligned} y' &= v, & y(0) &= 0 \\ v' &= -g - \frac{k}{m}v, & v(0) &= v_0 \end{aligned} \tag{15}$$

If $y = y(t)$ and $v = v(t)$ solves IVP (15), then we see that

$$y'' = v' = -g - \frac{k}{m}y', \qquad y(0) = 0, \quad y'(0) = v_0$$

and so $y(t)$ solves IVP (14). In other words, IVPs (14) and (15) are equivalent. So let's specify values for g, k/m, and v_0 and apply a numerical solver to IVP (15).

EXAMPLE 1.5.5

It Takes Longer to Fall!

With $g = 9.8$ m/sec^2, $k/m = 3$ and then 10 sec^{-1}, and using various values of v_0, our solver produced the graphs shown in Figures 1.5.3 and 1.5.4. We see that the whiffle ball does indeed take longer to fall than it does to rise, and this is more apparent the higher the initial velocity. A whiffle ball with high initial velocity and high k/m ratio quickly reaches its maximal height, and then very slowly falls back. Use a straightedge and determine the slope of the flat side of the top graph in Figure 1.5.4. The measured slope is close to the limiting velocity $-mg/k$, which has the value -0.98 m/sec.

☞ Graphs aren't proofs, but they are the next best thing!

This doesn't prove that the fall time is greater than the rise time, but the graphical evidence strongly suggests it. Toss a whiffle ball into the air, and check it out.

Dynamical Systems

These examples bring to light some essential components of ODE models.

Elements of a Model

Natural Variables: A natural process is described by a collection of variables called *natural variables* that depend on a single independent variable. For the viscous damping problem, the independent variable is time t, and the natural variables are the position, velocity, and acceleration of the ball at time t.

Natural Laws: A natural process evolves in time according to *natural laws* or *principles* involving the natural variables. Sometimes these laws arise empirically (e.g., the viscous damping law), and sometimes they have an intrinsic significance, such as Newton's Second Law. Sometimes, a natural law is expressed in commonsense terms, such as the Balance Law in Section 1.4. Using appropriate notation, a mathematical structure can be given to the natural variables and the natural laws.

Natural Parameters: The natural laws often contain *parameters* that must be experimentally determined; for example, the viscous damping constant k is a parameter.

In many situations the natural laws describe a process that evolves over time, so we almost always think of time as the independent variable.

☞ Sometimes we use this format to highlight important definitions.

❖ **State Variables, Dynamical Systems**. *State variables* are natural variables whose given values at an instant, together with the natural laws of the process, uniquely determine values of these variables for all times when the laws apply. A natural process (or its mathematical representation) described by state variables is a *dynamical system*. Values of state variables at any instant are said to describe the *state* of the dynamical system at that instant.

For example, position and velocity are state variables for the dynamical system of a vertically moving body. The dynamical systems approach to modeling, involving as it does the evolution of the state of the system through time, is a concept that is modern in its scope but has been around in some form for centuries. It is our basic approach.

Some conventional terminology is used in speaking about a dynamical system. The effects of the external environment are referred to as *input data* to the system. Input data are often called *driving terms* or *source terms*, and such systems are said to be *driven*. Values of the state variables at a given initial time are *initial data*. The behavior of the state variables due to the input and initial data is the *response* (or

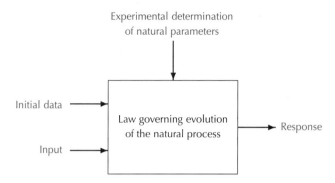

FIGURE 1.5.5 Schematic of a dynamical system.

output) of the dynamical system. Figure 1.5.5 summarizes the basic components of a dynamical system. We need to say something more about experimental determination of natural parameters, an often overlooked component of dynamical systems. The proportionality coefficients in our models play a very important role in the modeling process and someone has to determine their values. This important task usually falls upon scientists and engineers.

The Modeling Process

We need tools, variables, and natural laws to build a model, but one important step in the modeling process is often overlooked. This last step in the modeling process is usually called *validation* of the model. The modeler needs to make definitions and sim-plifying assumptions and discover some laws or principles that govern or explain the behavior of the phenomenon at hand. The goal of the modeler is to generate a model that is general enough to explain the phenomenon at hand, but not so complicated that analysis is impossible. There are trade-offs in this process. To have confidence in the model, the modeler often solves special problems and checks the results against experimental evidence. In this way the modeler learns something about the limits of applicability of the model. For example, if the differential equation $y'' = -g$ were used to model the motion of a whiffle ball when dropped from a height of 20 meters, the inadequacy of the model would quickly be discovered. Modelers speak of *ranges of validity* to describe these limitations. Schematics such as Figure 1.5.6 describe the modeling process, but modelers rarely follow them rigidly.

As experienced practitioners know, models enjoy only a transitory existence. A model based on the empirical data of the day may become inadequate when better data are available. In any field, models are constantly being examined for accuracy in predicting the phenomena modeled. This involves a careful reconsideration of the basic assumptions that produced the model as well as an analysis of the mathematical approximations used in the course of computation. When models are found to be deficient they are modified or supplanted by other models. The progress of science depends on this process.

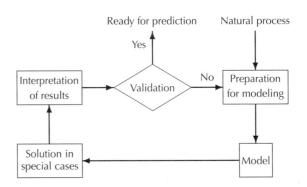

FIGURE 1.5.6 Modeling process schematic.

Initial Value Problems

State variables evolve over time, and rates of change are modeled by derivatives. After selecting state variables for a dynamical system, a mathematical model arises that consists of an ODE and conditions giving values of the state variables at a specific time t_0. As we have noted, these models are *initial value problems* (IVPs), and conditions involving the initial data are *initial conditions*. For example,

$$y'' = -g - \frac{k}{m}y', \qquad y(0) = 0, \quad y'(0) = v_0$$

is an IVP. The number of initial conditions (two) in this IVP agrees with the order of the ODE, and this agreement is typical of IVPs.

The model for the motion of the whiffle ball tells you what happens after the ball is thrown upward. This is a *forward initial value problem*. The initial conditions in such problems are expressed in terms of the limiting value of the state variables as $t \to t_0^+$. Other phenomena lead to *two-sided initial value problems* where t_0 is in the interior of a time interval. To complete our characterization of initial value problems, we note that in a *backward initial value problem*, a description of the state variables is sought for $t < t_0$. Problem 10(**a**) has an example of a backward IVP.

Art is the lie that helps us to see the truth.

Picasso

Comments

A mathematical model of a natural process is a portrait in the language of mathematics. Like all portraits, the model will emphasize some features of the original and distort others. So modeling is as much an art as it is a logical procedure. A skillfully constructed model often can provide more insight than observation of the natural process itself. Accurate direct analysis of many natural processes is impossible, and our only perception of their realities is by mathematical or other models based on partial data, scientific common sense, experience, and intuition.

PROBLEMS

1. (*Radioactive Decay*). A substance decays at a rate proportional to the amount present, and in 25 years 1.1% of the initial amount N_0 has decomposed. What is the half-life of the substance?

2. (*Radioactive Decay*). What percentage of a substance remains after 100 years if its half-life is 1000 years?

3. (*Radioactive Decay*). Radioactive phosphorus with a half-life of 14.2 days is used as a tracer in biochemical studies. After an experiment with 8 grams of phosphorus, researchers must safely store the material until only 10^{-5} grams remain. How long must the contents be stored?

4. (*Population Growth*). A population grows exponentially for T months with growth constant 0.03 per month. Then the growth constant suddenly increases to 0.05 per month. After a total of 20 months, the population has doubled. At what time T did the growth constant change? [*Hint*: Solve $y' = 0.03y$, $y(0) = y_0$, over the interval $0 \leq t \leq T$. Then use $y(T)$ as the initial value for a similar problem, replacing 0.03 by 0.05 and solving over the interval $T \leq t \leq 20$.]

5. (*No Damping*). A body whose mass is 600 grams is thrown vertically upward from the ground with an initial velocity of 2000 cm/sec. Use $g = 980$ cm/sec^2 and ignore air resistance.

(a) Find the highest point and the time to reach that point. [*Hint*: See Example 1.5.3.]

(b) Find the height of the body and the velocity after 3 sec. When does the body hit the ground?

6. (*Vertical Motion of a Ball*). Review Examples 1.5.2 and 1.5.3 and answer the questions.

(a) Why do the solution curves in Figure 1.5.2 bend downward? [*Hint*: Recall the connection between the sign of the second derivative of a function and the concavity of its graph.]

(b) Graph solution curves of the ODE $y'' = -9.8$ in the ty-plane for some values of y_0, v_0.

(c) Thrown vertically from an initial height of 75 m, what is the initial velocity of the ball so that it stays in the air 5 sec before it hits the ground? Verify with a graph.

(d) If $y_0 = 0$ and the ball reaches 30 m, find v_0. Verify with a graph.

7. (*No Damping*). A person drops a stone from the top of a building, waits 1.5 sec, then hurls a baseball downward with an initial speed of 20 m/sec. Ignore air resistance.

(a) If the ball and stone hit the ground at the same time, how high is the building?

(b) Show that if you wait too long before throwing the ball downward, the ball can't catch up with the stone. Show that the maximum waiting time for a catch-up is independent of the building's height.

8. (*No Damping: Longer to Rise or to Fall?*). Suppose that a ball is thrown vertically upward with initial velocity v_0 from ground level. Ignore air resistance. What is the time required for the ball to reach its maximum height? Does the ball spend as much time going up as it does coming down? What is the velocity of the ball on impact with the ground? Give reasons.

www **9.** (*Viscous Damping: Longer to Rise or to Fall?*). Suppose that a whiffle ball of mass m is thrown straight up with velocity v_0 from height h and is subject to viscous air resistance.

(a) Show that the IVP $y' = v$, $y(0) = h$, $v' = -g - (k/m)v$, $v(0) = v_0$, governs the whiffle ball's height y and velocity v.

(b) Suppose that $h = 2$ m, $g = 9.8$ m/sec^2, and $k/m = 5$ sec^{-1} for the system in part **(a)**. Use a numerical solver to estimate the rise time and the fall time (back to the initial height) for $v_0 = 10, 30, 50, 70$ m/sec. Repeat with $k/m = 1, 10$. Explain what you see.

(c) (*Proof That It Takes Longer to Fall than to Rise*). Suppose that $k/m = 1$ in the IVP in part **(a)**. Find the time T required for the whiffle ball to reach its highest point. Let the time $\tau > 0$ be such that $y(\tau) = h$. Show that $\tau > 2T$. Why do you think the fall-time is longer than the rise-time? Compare with the results of Problem 8. [*Hint*: Let $f(v_0) = y(2T) - h$, express $f(v_0)$ explicitly in terms of g and v_0, and show that $f(0) = 0$, $df(v_0)/dv_0 > 0$ for $v_0 > 0$.]

10. (*Age of the Lascaux Cave Paintings*). In 1950 a Geiger counter was used to measure the decay rate of ^{14}C in charcoal fragments found in a cave near Lascaux, France, where there are prehistoric wall paintings of various animals. The counter recorded about 1.69 disintegrations per minute per gram of carbon, while for living tissue such as the wood in a tree the number of disintegrations was 13.5 per minute per gram of carbon. Follow the outline below to find out when the wood was burned to make the charcoal (and so determine the age of the cave paintings).

In any living organism the ratio of the amount of ^{14}C to the total amount of carbon in the cells is the same as that in the air. After the organism is dead, ingestion of CO_2 ceases, and only the radioactive decay continues. The half-life τ of ^{14}C is known to be about 5568 years. Suppose that $q(t)$ is the amount of ^{14}C per gram of carbon at time t in the charcoal sample; $q(t)$ is dimensionless because it is the ratio of masses. Suppose that $t = 0$ is now, and that $T < 0$ is the time that the wood was burned. Then $q(t) = q(T)$ for $t \leq T$.

(a) Suppose that q_0 is the amount of ^{14}C per gram of carbon in the sample at $t = 0$. Verify that on the interval $T \leq t \leq 0$, $q(t)$ is the unique solution of the backward IVP

$$q' = -kq, \qquad q(0) = q_0, \qquad T \leq t \leq 0$$

(b) Solve the IVP in part **(a)** and show that

$$T = -\frac{1}{k} \ln \frac{q_T}{q_0} = -\frac{\tau}{\ln 2} \ln \frac{q'(T)}{q'(0)}$$

where k is the rate constant and τ is the half-life for ^{14}C.

(c) The reading of a Geiger counter at time t is proportional to $q'(t)$, the rate of decay of radioactive nuclei in a sample. Using the given data in the problem statement, find T.

11. (*Radiocarbon Dating: Scaling*). Follow the outline below to obtain a graphical solution for the radiocarbon dating problem of the Lascaux paintings. [*Hint*: Read Problem 10.]

(a) Define the new state variable $Q(t) = q(t)/q_0$. Show that this change of variables converts the IVP of that example into the form $Q' = -kQ$, $Q(0) = 1$. Show that the dating problem may be reformulated as follows: Find a value $T < 0$ such that $Q(T) = 13.5/1.69$.

(b) Plot the solution of the IVP of part **(a)** backward in time and find the value of T.

12. (*Dating Stonehenge*). In 1977, the rate of ^{14}C radioactivity of a piece of charcoal found at Stonehenge in southern England was 8.2 disintegrations per minute per gram of carbon. Given that in 1977 the rate of ^{14}C radioactivity of a living tree was 13.5 disintegrations per minute per gram, estimate the date of construction. [*Hint*: Read Problem 10.]

13. (*Dating a Sea Shell*). An archeologist finds a sea shell that contains 60% of the ^{14}C of a living shell. How old is the shell? [*Hint*: Read Problem 10.]

14. (*The Bones of Olduvai: Potassium-Argon Dating*). Olduvai Gorge, in Kenya, cuts through volcanic flows, volcanic ash, and sedimentary deposits. It is the site of bones and artifacts of early hominids, considered by some to be precursors of man. In 1959, Mary and Louis Leakey uncovered a fossil hominid skull and primitive stone tools of obviously great age. Carbon-14 dating methods being inappropriate for a specimen of that age and nature, dating had to be based on the ages of the underlying and overlying volcanic strata. The method used was that of potassium-argon decay. The potassium-argon clock is an accumulation clock, in contrast to the ^{14}C dating method. Follow the outline below to model this accumulation clock.

The potassium-argon method depends on measuring the accumulation of "daughter" argon atoms, which are decay products of radioactive potassium atoms. Specifically, potassium-40 (^{40}K) decays to argon-40 (^{40}Ar) and to calcium-40 (^{40}Ca) at rates proportional to the amount of potassium but with respective constants of proportionality k_1 and k_2.

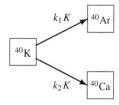

The model for this decay process may be written in terms of the amounts $K(t)$, $A(t)$, and $C(t)$ of potassium, argon, and calcium in a sample of rock. Using the Balance Law, we have

$$K' = -(k_1 + k_2)K, \qquad A' = k_1 K, \qquad C' = k_2 K \tag{16}$$

where time t is measured *forward* from the time the volcanic ash was deposited around the skull.

(a) Solve the system to find $K(t)$, $A(t)$, and $C(t)$ in terms of k_1, k_2, and $k = k_1 + k_2$. Set $K(0) = K_0$, $A(0) = C(0) = 0$. Why is $K(t) + A(t) + C(t) = K_0$ for all $t \geq 0$? Show that $K(t) \to 0$, $A(t) \to k_1 K_0 / k$, and $C(t) \to k_2 K_0 / k$ as $t \to \infty$.

(b) The age T of the volcanic strata is the current value of the time variable t because the potassium-argon clock started when the volcanic material was laid down. This age is estimated by measuring the ratio of argon to potassium in a sample. Show that this ratio is $A/K = (k_1/k)(e^{kT} - 1)$. Show that the age of the sample (in years) is $(1/k) \ln[(k/k_1)(A/K) + 1]$.

(c) When the actual measurements were made at the University of California at Berkeley, T (the age of the bones) was estimated to be 1.75 million years. The values of the constants of proportionality are known to be $k_1 = 5.76 \times 10^{-11}$/yr and $k_2 = 4.85 \times 10^{-10}$/yr. What was the value of the measured ratio A/K?

 15. (*Shoveling Snow*). During a steady snowfall, a man starts clearing a sidewalk at noon, shoveling the snow at a constant rate and clearing a path of constant width. He shovels two blocks by 2 P.M., one block more by 4 P.M. When did the snow begin to fall? Explain your modeling process. [*Hint*: Additional assumptions are required to solve the problem.]

1.6 Separable Differential Equations

We have just about wrapped up the topic of first-order linear ODEs in that we know how to find and interpret a formula for all solutions and how to interpret solution curves produced with a numerical solver. Now it's time to look at first-order nonlinear ODEs. In this section we will look at ODEs of the form, $N(y)y'(t) + M(t) = 0$, which are called *separable* ODEs because the variables y and t are separated as indicated.

As we will see later on, there are advantages to using x as a variable name instead of t. So let's find a solution formula for the separable ODE

$$N(y)y'(x) + M(x) = 0 \tag{1}$$

This ODE is often written as $N(y)y' = -M(x)$ with the equal sign "separating" the y-terms and the x-terms. Let's suppose that $N(y)$ and $M(x)$ are continuous on the respective y and x intervals J and I.

Here is a procedure for solving ODE (1).

Solving a Separable ODE

☞ Follow this procedure to solve a separable ODE.

1. **Separate variables** to write the ODE in the form $N(y)y' + M(x) = 0$, and identify the coefficients $N(y)$ and $M(x)$.

2. **Find any antiderivative** $G(y)$ for $N(y)$ and any antiderivative $F(x)$ for $M(x)$.

3. **All solutions** $y = y(x)$ satisfy the equation $G(y) + F(x) = C$, where C is constant.

After you use this procedure a few times, it will become second nature and you won't have to refer to it again.

Here is an explanation of why the procedure works. Suppose that $G(y)$ is any antiderivative of $N(y)$ on J and that $F(x)$ is any antiderivative of $M(x)$ on I. In other words,

$$dG/dy = N(y), \quad \text{for all } y \text{ in } J, \qquad dF/dx = M(x), \quad \text{for all } x \text{ in } I$$

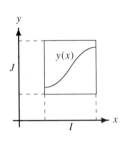

If $y(x)$ is any solution of ODE (1) whose solution curve remains in the xy-rectangle defined by I and J, then ODE (1) becomes

$$N(y)y'(x) + M(x) = [G(y(x)) + F(x)]' = 0 \tag{2}$$

because the Chain Rule says that

$$\frac{dG(y(x))}{dx} = \frac{\partial G}{\partial y}\frac{dy}{dx} = N(y)y'(x)$$

Antidifferentiation of (2) then implies that

$$G(y(x)) + F(x) = C \tag{3}$$

for some constant C and for all x on the interval where $y(x)$ is defined.

On the other hand, suppose that $y(x)$ is a continuously differentiable function whose graph lies in the rectangle determined by I and J and that $y(x)$ satisfies equation (3) for some constant C. Then differentiation of (3) via the Chain Rule shows that $y(x)$ is a solution of ODE (1). So equation (3) can be regarded as implicitly defining the *general solution* of ODE (1) for a range of values of the constant C. Formula (3) is implicit because it has not been solved to express y explicitly as a function of x.

Here's an example that uses the separable variables method.

EXAMPLE 1.6.1   **Separating the Variables and Solving**
The ODE

$$y' = -x/y$$

in separated form is

$$yy' + x = 0$$

and so $N(y) = y$ and $M(x) = x$. Then $G(y) = y^2/2$ and $F(x) = x^2/2$ are antiderivatives of N and M. Using formula (3), the general solution of the ODE is given by

$$\frac{[y(x)]^2}{2} + \frac{x^2}{2} = C \tag{4}$$

where C is a constant. For any value $C > 0$, we can solve equation (4) for y in terms of x. There are two solutions

$$y = \pm\sqrt{2C - x^2} \tag{5}$$

each defined on the interval $|x| < \sqrt{2C}$. Direct substitution in ODE (4) shows that formula (5) does define the solutions.

It's worth noting that if we think of x as a function of y, instead of the other way around, then we can write ODE (1) as

$$N(y) + M(x)\frac{dx}{dy} = 0$$

If we don't want to choose between x and y we can write ODE (1) in the *differential form*

$$N(y)\,dy + M(x)\,dx = 0$$

Now let's look at the geometric structure of the solution curves of a separable ODE.

Integrals and Integral Curves

Our approach to solving separable ODEs suggests the need for some new terms.

> ❖ **Integrals and Integral Curves.** A nonconstant function $H(x, y)$ on a rectangle R is an *integral* of the ODE $Ny' + M = 0$ if for any solution $y(x)$ whose graph lies in R, $H(x, y(x)) = C$ for some constant C. The set of points in R satisfying $H(x, y) = C$ is an *integral curve* for the ODE. If C is an unspecified constant, then $H(x, y) = C$ is the *general implicit solution* of the ODE.

For the antiderivatives $G(y)$ and $F(x)$ defined above, the function $H(x, y) = G(y) + F(x)$ is an integral for ODE (1), so the equation $H(x, y) = C$ defines an integral curve of ODE (1). These integral curves are *level sets* or *contours* of H. Every solution curve of the ODE is an arc of some integral curve for that ODE.

EXAMPLE 1.6.2

Integral Curves of $yy' + x = 0$
As we saw in Example 1.6.1, $H(x, y) = y^2/2 + x^2/2$ is an integral for the separable ODE $yy' + x = 0$. The integral curves are circles defined by $y^2/2 + x^2/2 = \text{constant}$.

☞ An integral curve may have several distinct branches.

The ODE of Examples 1.6.1 and 1.6.2 shows that solving the equation $H(x, y) = C$ for y in terms of x may lead to more than one solution of the ODE for each value of C. The next example shows that integral curves may have several branches.

EXAMPLE 1.6.3

Four Solution Curves on the Two Branches of an Integral Curve
Let's look at the IVP

$$yy' - x = 0, \qquad y(2) = -1$$

Following the procedure above with $N(y) = y$ and $M(x) = -x$, we have (after carrying out the integrations)

$$y^2/2 - x^2/2 = C \qquad\qquad (6)$$

☞ When solving an IVP be careful in choosing the arc or branch of an integral curve.

The function $H = y^2/2 - x^2/2$ is an integral of the ODE. The constant C is determined from the initial condition $x = 2$ when $y = -1$, and we see that the implicit solution is

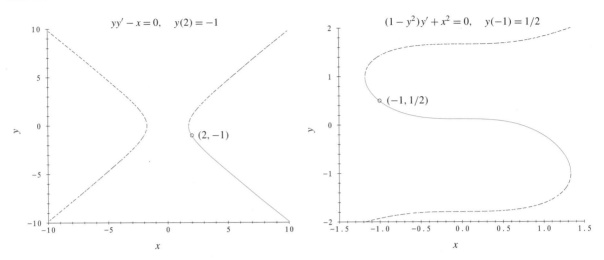

FIGURE 1.6.1 Solution curve (solid) on an integral curve $y^2/2 - x^2/2 = -3/2$ (dashed) (Example 1.6.3).   **FIGURE 1.6.2** Solution curve (solid) on an integral curve (dashed) (Example 1.6.4).

$y^2/2 - x^2/2 = -3/2$, which defines a hyperbola through the initial point $(2, -1)$. The explicit solution $y(x)$ is given by

$$y = -(x^2 - 3)^{1/2}, \qquad x > 3^{1/2}$$

which defines the solid arc passing through $(2, -1)$ and shown in Figure 1.6.1. The hyperbolic integral curve defined by (6) contains three other solution curves that are also shown in Figure 1.6.1: $y = (x^2 - 3)^{1/2}$, $x > 3^{1/2}$ (upper right), and $y = \pm(x^2 - 3)^{1/2}$, $x < -3^{1/2}$ (on the left).

Sometimes it is not practical to find y as an explicit function of x.

EXAMPLE 1.6.4

A Solution Curve on an Integral Curve
The separable ODE

$$(1 - y^2)y' + x^2 = 0$$

has the integral $H(x, y) = y - y^3/3 + x^3/3$ and the general solution

$$y - y^3/3 + x^3/3 = C \tag{7}$$

where C is a constant. There is no simple way to solve (7) for y in terms of x and C.

Suppose now that we want to find the integral curve through the point $(-1, 1/2)$. Inserting $x = -1$ and $y = 1/2$ into formula (7), we see that $C = 1/8$. The solution $y(x)$ of the IVP

$$(1 - y^2)y' + x^2 = 0, \qquad y(-1) = 1/2 \tag{8}$$

is defined implicitly by the formula

$$y - y^3/3 + x^3/3 = 1/8 \tag{9}$$

See Figure 1.6.2 for the S-shaped integral curve defined by formula (9). The solid arc on this integral curve is the graph of the solution $y(x)$ of IVP (8) extended forward and backward from $x = -1$ as far as it will go.

To find the largest x-interval on which the graph of $y(x)$ in Example 1.6.4 is defined, note that y' is infinite when $y = \pm 1$ [because $y' = x^2/(y^2 - 1)$]. Putting $y = \pm 1$ into the equation

$$y - y^3/3 + x^3/3 = 1/8$$

we obtain $x^3 = -13/8, 19/8$. So, $-13^{1/3}/2$ and $19^{1/3}/2$ are the endpoints of the largest x-interval on which this solution curve is defined.

☞ Impatient? Look ahead to Example 1.7.1.

One small mystery still remains: How was the integral curve in Figure 1.6.2 plotted? We did not use a contour plotter! In Section 1.7 all will be revealed.

Examples 1.6.3 and 1.6.4 illustrate the power of the integral curves to help us to visualize the solution curves of an ODE. In fact, if we plot all of the integral curves that lie in a rectangle, then we can immediately see all of the solution curves in the rectangle. In this sense integral curves give a *global* description of the solution curves.

We end the section with two applications that involve separable ODEs.

The Logistic ODE

In Section 1.1 we looked at a model ODE

☞ More on the logistic equation in Section 1.1 of the Student Resource Manual.

$$dy/dt = ay - cy^2$$

where a and c are positive constants. This ODE was used to describe a fish population, taking overcrowding into account. The more common form of this ODE arises by putting $r = a$ and $K = a/c$ to obtain the *logistic equation*

$$dy/dt = ry(1 - y/K) \tag{10}$$

The constant r is called the *intrinsic growth constant* and measures the difference between the birth and death rates per population unit if there is no overcrowding. For example, $r = 0.05$ corresponds to a net growth rate of 5% per unit time. The positive constant K is the *saturation constant* or *carrying capacity*. As we will soon see, every solution $y(t)$ approaches K as $t \to +\infty$ if $y(0)$ is positive. Let's derive a solution formula for the logistic ODE (10).

EXAMPLE 1.6.5

Solving the Logistic ODE
Let's write the logistic equation in the differential form

$$dy - ry(1 - y/K)\, dt = 0 \tag{11}$$

This ODE isn't linear, but it is separable. Assume that y is not 0 or K to avoid division by 0, and separate the variables:

$$\frac{K}{(K - y)y}\, dy - r\, dt = 0 \tag{12}$$

In the notation of formula (3), we need an antiderivative $G(y)$ of $K/[y(K - y)]$ and the antiderivative $F(t) = -rt$ of the function $-r$. We use partial fractions to find $G(y)$:

$$\frac{K}{(K - y)y} = \frac{1}{K - y} + \frac{1}{y}$$

Integrate to obtain

$$G(y) = -\ln|K - y| + \ln|y|$$

We have the general solution of ODE (11):

$$-\ln|K - y| + \ln|y| - rt = c, \quad \text{or} \quad \ln\left|\frac{y}{K - y}\right| = rt + c$$

where c is any constant. Exponentiate to obtain

$$\left|\frac{y}{K - y}\right| = Ce^{rt}, \quad \text{where } C = e^c \tag{13}$$

If the absolute value sign in (13) is dropped, then C can be either positive or negative. Now let's solve $y/(K - y) = Ce^{rt}$ for y:

$$y = (K - y)Ce^{rt}$$

$$y(1 + Ce^{rt}) = KCe^{rt}$$

$$y = \frac{KCe^{rt}}{1 + Ce^{rt}} \tag{14}$$

Evaluating C from the initial condition $y(0) = y_0 \geq 0$, we have from (13) that $y_0 = KC/(1 + C)$. Solving for C in terms of y_0 we have $C = y_0/(K - y_0)$. Inserting this value for C in (14) we have (after a lot of algebra)

$$y(t) = \frac{y_0 K}{y_0 + (K - y_0)e^{-rt}} \tag{15}$$

which is the formula for the solutions of the logistic ODE (10). Note that if $y_0 > 0$, then $y(t) \to K$ as $t \to +\infty$ since $e^{-rt} \to 0$

Let's put in some numbers for r and K.

EXAMPLE 1.6.6

Logistic Population Change

Suppose that the rate coefficient r is 1 (corresponding to a 100% growth rate per unit time) and the carrying capacity is 12. From formula (15), the solution formula for the IVP $y' = (1 - y/12)y$, $y(0) = y_0$, is

$$y(t) = \frac{12y_0}{y_0 + (12 - y_0)e^{-t}}$$

Since $e^{-t} \to 0$ as $t \to +\infty$, we see that $y(t) \to 12y_0/(y_0 + 0) = 12$, so the population tends to the carrying capacity $K = 12$. Turn back to Figure 1.1.3 for a picture of the solution curves.

The S-shaped population curves of the logistic equation are called *logistic curves*. Figure 1.1.3 illustrates the appropriateness of the terms "carrying capacity" and "saturation population" for $y = K$. The resources of the community can support a population of size K, which is the asymptotic limit of the population curves as $t \to +\infty$.

Newtonian Damping and a Sky Diver

Careful measurements reveal that when a dense body rises or falls through the air the magnitude of the damping force is proportional to the square of the magnitude of the velocity. This is known as *Newtonian damping*; compare this with viscous damping defined in Section 1.5. Suppose that y measures the distance along the vertical with "up" as the positive direction. Taking into account the forces acting on the sky diver, we have the initial value problem

$$my'' = -mg - kv|v|, \qquad y(0) = h, \quad v(0) = v_0 \qquad (16)$$

where $v = y'$ and the minus sign in the term $-kv|v|$ indicates that the drag acts opposite to the motion. The positive constant k in (16) is the *Newtonian damping constant*. Note that for a falling body the velocity $v = y'$ is negative so $-kv|v| = kv^2$.

EXAMPLE 1.6.7

Newtonian Damping: Limiting Velocity

A free-falling sky diver jumps from a plane h feet above the ground, and is acted upon by Newtonian air resistance. Because the sky diver is falling, the Newtonian damping force acts upward (in the positive direction) and so is given by kv^2. Replacing y'' by v' in IVP (16) and dividing by m, we obtain the first-order IVP for the velocity

$$v' = -g + \left(\frac{k}{m}\right) v^2, \qquad v(0) = 0 \qquad (17)$$

The ODE in (17) is separable. After separating the variables and integrating, it can be shown (Problem 7) that the solution of IVP (17) is

$$v(t) = \left(\frac{mg}{k}\right)^{1/2} \frac{e^{-At} - 1}{e^{-At} + 1} \qquad (18)$$

☞ Compare with the limiting velocity under viscous damping in Example 1.5.4.

where $A = 2(gk/m)^{1/2}$. As $t \to +\infty$, we see that $v(t)$ tends to the *limiting velocity* $v_\infty = -(mg/k)^{1/2}$. This velocity is not actually reached because the sky diver hits the ground at some finite time, and the model IVP (17) loses its validity.

The motion of a sky diver has been studied in detail, and Newtonian damping provides a good model. If the mass of the equipped sky diver is 120 kg, then the Newtonian damping constant k is about 0.1838 kg/m, and the limiting velocity has magnitude about 80 m/sec. This value compares favorably with the observed limiting velocities of free-falling sky divers.

Comments

Historically, the ODE $N(dy/dx) + M = 0$ was usually written in the differential form $Ndy + Mdx = 0$, and that is why the subject is named "differential equations" rather than "derivative equations." Because of the convenience of the differential form, we sometimes use it, especially in the problems.

The case where N and M are functions of both x and y is taken up in Problem 10 for the special case where $\partial N/\partial x = \partial M/\partial y$.

PROBLEMS

1. (*Losing a Solution*). Sometimes when you rewrite an ODE to separate the variables you may inadvertently lose a solution. Find all solutions of the ODE $y' = 2xy^2$, but watch out that you don't lose a solution.

2. Find all solutions $y = y(x)$ of each ODE. [*Hint*: Watch out for solutions that are lost when you separate the variables. See Problem 1.]

    (a) $dy/dx = -4xy$ **(b)** $2y\,dx + 3x\,dy = 0$ **(c)** $y' = -xe^{-x+y}$

    (d) $(1-x)y' = y^2$ **(e)** $y' = -y/(x^2 - 4)$ **(f)** $y' = xe^{y-x^2}$

3.   For each IVP find a solution formula and the largest x-interval on which it is defined. For **(b)** and **(e)**, find solution formulas and also solve the IVPs with a numerical solver.

    (a) $y' = (y+1)/(x+1)$, $y(1) = 1$ **(b)** $y' = y^2/x$, $y(1) = 1$

    (c) $y' = ye^{-x}$, $y(0) = e$ **(d)** $y' = 3x^2/(1+x^3)$, $y(0) = 1$

    (e) $y' = -x/y$, $y(1) = 2$ **(f)** $2xyy' = 1 + y^2$, $y(2) = 3$

4. For each ODE find the general solution formula, but don't try to solve that formula for y as an explicit function of x.

    (a) $y' = (x^2 + 2)(y+1)/xy$ **(b)** $(1 + \sin x)\,dx + (1 + \cos y)\,dy = 0$

    (c) $(\tan^2 y)\,dy = (\sin^3 x)\,dx$ **(d)** $(3y^2 + 2y + 1)y' = x\sin(x^2)$

5. (*Logistic Change*). The problems below involve logistic growth and decay.

    (a) Solve the IVPs $y' = (1 - y/20)y$, $y(0) = 5, 10, 20, 30$. [*Hint*: See Formula (15).]

     (b) Plot the solution curves in **(a)** on the interval $0 \le t \le 10$ and highlight the carrying capacity.

    (c) (*Harvesting*). The ODE $y' = 3(1 - y/12)y - 8$ models population changes for a harvested, logistically changing species. Find the two equilibrium levels, and discuss the fate of the species if $y(0) = 2, 4, 6, 8$, or 10. [*Hint*: Follow the analysis of IVP (12) in Section 1.1]

    (d) Plot the solution curves in **(c)** on the interval $0 \le t \le 5$ and highlight the equilibrium levels.

    (e) A colony of bacteria grows according to the logistic law, with a carrying capacity of 5×10^8 individuals and natural growth coefficient $r = 0.01$ day^{-1}. What will the population be after 2 days if it is initially 1×10^8 individuals?

6. (*Projectile Motion: Newtonian Damping*). A spherical projectile weighing 100 lb is observed to have a limiting velocity of -400 ft/sec.

    (a) Show that the velocity v of the projectile undergoing either upward or downward vertical motion and acted upon by Newtonian air resistance is given by $v' = -g - (gk/w)v|v|$, where the magnitude of k is 1/1600 and $w = mg$ is the weight of the projectile.

    (b) Model the projectile motion as a differential system in state variables y and v, where $v = y'$.

(c) If the projectile is shot straight upward from the ground with an initial velocity of 500 ft/sec, what is its velocity when it hits the ground? [*Hint*: Use a numerical solver to solve the system of part **(b)** with suitable initial conditions. Use the solver to plot the solution $v = v(t)$ and $y = y(t)$ as a parametric curve in the vy-plane. Estimate the impact velocity from the graph.]

 (d) (*Longer to Fall?*). Does the projectile in **(c)** take longer to rise or to fall? Explain. Repeat for $v_0 = 100, 200, \ldots, 1000$ ft/sec. [*Hint*: Use a numerical solver and graph $y(t)$.]

www 7. (*The Newtonian Sky Diver's Velocity*). In Example 1.6.7 it was shown that the sky diver's velocity $v(t)$ satisfies the IVP $v' = -g + kv^2/m$, $v(0) = 0$. Separate variables, use partial fractions and algebraic manipulation to derive the solution formula

☞ If you get stuck, look at the Student Resource Manual.

☞ If you need help with partial fractions, see Table 6.2.1.

$$v(t) = \left(\frac{mg}{k}\right)^{1/2} \frac{e^{-At} - 1}{e^{-At} + 1}, \quad \text{where } A = 2\left(\frac{gk}{m}\right)^{1/2}$$

8. (*Sky Diver: Newtonian Damping*). A sky diver and equipment weigh 240 lb. [Note: weight $=$ mg.] In free fall the sky diver reaches a limiting velocity of 250 ft/sec. Some time after the parachute opens the sky diver reaches the limiting velocity of 17 ft/sec. Suppose that later the sky diver jumps out of an airplane at 10 000 feet. Use $g = 32.2$ ft/sec^2, and answer the following questions about the second jump.

(a) How much time must elapse before the free-falling sky diver falls at a speed of 100 ft/sec? [*Hint*: Use the data to determine the coefficient k in IVP (16); then use formula (18).]

(b) When falling at 100 ft/sec the sky diver pulls the rip cord and the chute opens instantaneously. How much longer does it take for the speed of the descent to drop to 25 ft/sec?

(c) How long does the jump last? What is the velocity on landing?

9. (*A Harvested Logistic Population*). An initial value problem for a logistic population harvested at a constant rate is given by

$$y' = y(1 - y/10) - 9/10, \qquad y(0) = y_0$$

(a) Find the solution formula for this IVP. [*Hint*: Allow enough time for a lot of algebraic manipulation.]

(b) Plot some solution curves for the IVP.

10. (*Solution Formula for Exact ODEs*). The first-order ODE

$$N(x, y)y' + M(x, y) = 0$$

is *exact* in a rectangle R of the xy-plane if M and N are continuously differentiable in R and if there is a function $H(x, y)$ such that $\partial H/\partial x = M$, $\partial H/\partial y = N$ for all (x, y) in R. Such a function H is called an *integral* of the ODE. If the ODE is exact, then its general solution is $H(x, y) = C$, where the constant C is chosen appropriately. The level curves of an integral are called *integral curves*. It can be shown that the *exactness condition*

☞ This generalizes the definition of an integral for a separable ODE.

$$\partial N/\partial x = \partial M/\partial y$$

for all (x, y) in R guarantees that a function H with the desired properties exists. So if the ODE satisfies the exactness condition, all we need to do to construct the general solution $H(x, y) = C$ is to calculate H. This can be accomplished by completing the following steps: **(1)** Check that the exactness condition holds. **(2)** Think of y as a constant and evaluate the x-antiderivative $\int M(x, y)\, dx$; write $H = \int M(x, y)\, dx + g(y)$ in terms of an unknown function g. **(3)** Then

$$g'(y) = \frac{\partial H}{\partial y} - \frac{\partial}{\partial y}\int M(x, y)\, dx = N - \frac{\partial}{\partial y}\int M(x, y)\, dx$$

which follows from the fact that $\partial H/\partial y = N$ (definition of exactness). **(4)** The exactness condition implies that $N - (\partial/\partial y)\int M(x, y)\, dx$ is independent of x (despite appearances) since

$$\frac{\partial}{\partial x}\left[N - \frac{\partial}{\partial y}\int M(x, y)\, dx\right] = \frac{\partial N}{\partial x} - \frac{\partial}{\partial y}\left[\frac{\partial}{\partial x}\int M(x, y)\, dx\right]$$

$$= \frac{\partial N}{\partial x} - \frac{\partial}{\partial y}M(x, y) = 0$$

So $g(y)$ is given by any antiderivative of $g'(y)$. **(5)** So $H = \int M(x, y)\, dx + g(y)$ is an integral of the ODE, and the general solution is $H(x, y) = C$.

(a) Show that if the ODE is exact, then it may be written as $d(H(x, y(x)))/dx = 0$. Conclude that $y(x)$ is a solution of the ODE if and only if its solution curve lies on a level set of H.

☞ Shows that exact ODEs generalize separable ODEs.

(b) Show that the separable ODE $N(y)y' + M(x) = 0$ is exact if N and M are continuous.

(c) Show that $(2y - x)y' = y - 2x$ is exact. Find an integral H and the general solution. Plot integral curves in the rectangle $|x| \leq 5$, $|y| \leq 4$. Highlight some solution curves $y = y(x)$.

☞ Check out Figure 1.7.1.

(d) (*Teddy Bears*). Show that $[\sin y - 2\sin(x^2)\sin(2y)]y' + \cos x + 2x\cos(x^2)\cos(2y) = 0$ is exact. Find an integral H, and plot several integral curves for $|x| \leq 6$, $|y| \leq 10$. What is the general solution?

(e) Show that $xe^y(y^2 + 2y)y' + y^2e^y + 2x = 0$ is exact on the xy-plane. Find an integral H and the general solution; plot several integral curves in the rectangle $|x| \leq 5$, $|y| \leq 4$.

(f) Show that $y'\sin x\sin y - \cos x\cos y = 0$ is exact on the xy-plane. Find an integral H and the general solution; plot several integral curves in the rectangle $|x| \leq 5$, $|y| \leq 4$.

(g) An inexact ODE can be made exact by multiplying the ODE by an integrating factor. Show that $\cos x$ is an integrating factor for the inexact ODE $\cos x\, dy - (2y\sin x - 3)\, dx = 0$. Then find the general solution and plot some integral curves in the rectangle $|x| \leq 5$, $|y| \leq 5$.

 11. (*Newtonian versus Viscous Damping*). Design an experiment to determine whether Newtonian or viscous damping better describes the motion of the sky diver of Problem 8. Carry out a computer simulation and discuss the results. [*Hint:* See Examples 1.5.4, 1.5.5, and 1.6.7. The damping constant can be determined from the terminal velocity.]

1.7 Planar Systems and First-Order ODEs

Most of the dynamical systems considered so far have been modeled by single, first-order ODEs in one state variable. Some dynamical systems require two state variables, say x and y, and the governing laws lead to two rate equations:

☞ This is a *normal system* since it has been solved for x' and y'.

$$\frac{dx}{dt} = x' = f(t, x, y), \qquad \frac{dy}{dt} = y' = g(t, x, y) \tag{1}$$

The pair of differential equations is called a *first-order planar differential system*; the xy-plane is the *state space*. When the rate functions f and g do not depend on t, system (1) is said to be *autonomous*.

Systems first appeared in (15) of Section 1.5, where we modeled a whiffle ball of mass m moving along the local vertical, using position y and velocity v as state variables, to obtain the planar autonomous system

$$y' = v, \qquad v' = -g - \frac{k}{m}v \tag{2}$$

where g is the earth's gravitational constant and k is the viscous damping constant. We find solutions of system (2) by observing that the second ODE (the v' equation) decouples from the first. We can solve the decoupled first-order ODE for $v(t)$ and insert it into the other first-order ODE, which can then be solved for $y(t)$.

It is not always the case that one ODE of a first-order planar system decouples from the other ODE, so we need to search for other solution techniques. But first, here is some important terminology for planar systems.

❖ **Solutions, Orbits, Component Curves**. The functions $x = x(t)$, $y = y(t)$ define a *solution* of system (1) if $x'(t) = f(t, x(t), y(t))$, $y'(t) = g(t, x(t), y(t))$ for all t in an interval I. The parametric plot of $x(t)$ versus $y(t)$ in the xy-state plane is the *orbit* of the solution. The graphs of $x(t)$ versus t in the tx-plane and $y(t)$ versus t in the ty-plane are the *component curves* of the solution.

Examples of orbits and component curves appear in the figures that follow.

Now we come to one reason why x is often used instead of t in a first-order ODE.

The System Way to Plot Solution Curves

Solution curves of the first-order ODE

☞ Solution formulas can sometimes be found by using the technique given in Problem 10 in Section 1.6.

$$N(x, y)\frac{dy}{dx} + M(x, y) = 0 \tag{3}$$

can be obtained by first plotting orbits of the planar autonomous system

$$\frac{dx}{dt} = N(x, y), \qquad \frac{dy}{dt} = -M(x, y) \tag{4}$$

We can explain this process by reasoning as follows: The parametric plot in xy-space of a solution $x = x(t)$, $y = y(t)$ of system (4) is an orbit of the system. The slope dy/dx of the orbit at a point (x, y) is $-M(x, y)/N(x, y)$ since $dy/dx = (dy/dt)/(dx/dt) = -M/N$. Solution curves of ODE (3), and so of $dy/dx = -M/N$, are arcs of orbits of the system (4), which contain no vertical tangents.

The ODE $dy/dx = -M/N$ may be hard to solve numerically in regions where the denominator $N(x, y)$ vanishes. The technique described above gets around this problem of the vanishing denominator, but at the cost of introducing another state variable. The examples below illustrate this approach.

EXAMPLE 1.7.1

Solution Curves of a First-Order ODE as Arcs of Orbits of a System

Let's find the solution curve through the point $(-1, 1/2)$ of the ODE

$$(1 - y^2)\frac{dy}{dx} + x^2 = 0 \tag{5}$$

☞ This approach beats most contour plotting software. Problem 12 has more.

where $N = 1 - y^2$ and $M = x^2$, by first plotting the orbit of the system IVP

$$\frac{dx}{dt} = N = 1 - y^2, \qquad x(0) = -1$$

$$\frac{dy}{dt} = -M = -x^2, \qquad y(0) = 1/2 \tag{6}$$

using the initial point $(-1, 1/2)$. This is done by using a numerical solver to solve IVP (6) and letting t run forward and then backward from $t = 0$ until the orbit exits the rectangle $|x| \leq 2$, $|y| \leq 3$.

The orbit is the S-shaped curve shown in Figure 1.6.2. The solution curve of ODE (5) that passes through $(-1, 1/2)$ is the longest arc of this orbit that contains the point $(-1, 1/2)$ and has no vertical tangents. The solution curve is the solid arc shown in Figure 1.6.2.

Here's a second example that uses the same approach, but with many initial points and orbits.

EXAMPLE 1.7.2　**Teddy Bears**

Consider the first-order ODE

$$(\sin y - 2\sin x^2 \sin 2y)\frac{dy}{dx} + \cos x + 2x\cos x^2 \cos 2y = 0 \qquad (7)$$

The first-order system (4) that corresponds to ODE (7) is

$$\begin{aligned}
\frac{dx}{dt} &= \sin y - 2\sin x^2 \sin 2y \\
\frac{dy}{dt} &= -\cos x - 2x\cos x^2 \cos 2y
\end{aligned} \qquad (8)$$

The orbits of system (8) shown in Figure 1.7.1 were produced by a numerical solver using several initial points (x_0, y_0) in the rectangle $|x| \leq 6$, $|y| \leq 10$. For example, the curve outlining the lower torso and legs of the teddy bears is the orbit of system (8) that passes through the point $(0, \pi/2)$ at $t = 0$. In each case, the orbit is run forward and backward in time from the initial point selected until the orbit exits the rectangle or returns to its starting point (as happens with orbits inside a bear).

In this example we have not singled out the arcs on each orbit that correspond to solutions $y = y(x)$ of the original ODE (7). The fact that the rate functions for $x(t)$ and $y(t)$ are periodic functions of y with period 2π suggests that orbital shapes might repeat with every increase of 2π in y, so that's why we see the teddy bear triplets.

The teddy bear example shows how the system approach helps us to visualize solution curves of a first-order ODE over an extended region, that is, globally. Since solution curves are arcs of orbits with no vertical tangents, we clearly see from Figure 1.7.1 that some solution curves live only over very short x-intervals (see, for example, the many "eyes" in a teddy bear).

Here's a way to use the systems approach to construct formulas for some second-order ODEs that occur quite often in the applications.

Reduction Methods and Planar Systems

Some second-order differential equations can be solved by reducing them to first-order systems by a suitable choice of new variables. For example, the second-order ODE

$$y'' = F(t, y') \qquad (9)$$

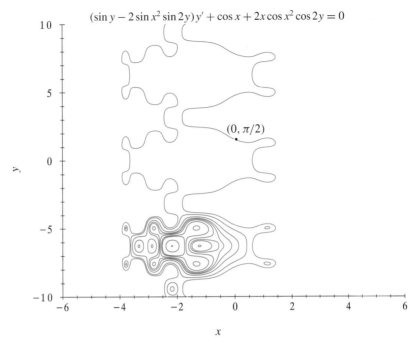

$$(\sin y - 2\sin x^2 \sin 2y)\, y' + \cos x + 2x\cos x^2 \cos 2y = 0$$

FIGURE 1.7.1 Orbits of the first-order system (8) (Example 1.7.2).

where the derivatives are with respect to t, is transformed to the system

$$\begin{aligned} y' &= v \\ v' &= F(t, v) \end{aligned} \tag{10}$$

by introducing a new state variable $v = y'$. The second ODE in system (10) decouples from the first ODE, so it can be solved first. The falling bodies problems in Sections 1.5 and 1.6 illustrate this reduction-of-order technique. Here is another example.

☞ We lucked out here. Solving this ODE involves just solving two first-order ODEs in succession.

EXAMPLE 1.7.3

Solving $y'' = F(t, y')$

The ODE

$$y'' = y' - t$$

can be solved by setting $v = y'$ and rewriting the ODE as

$$v' - v = -t$$

which is a linear first-order ODE in v with integrating factor e^{-t}. We have $(ve^{-t})' = -te^{-t}$, which can be solved to obtain $v = C_1 e^t + 1 + t$, where C_1 is an arbitrary constant. Recalling that $y' = v$, an antidifferentiation yields the solutions

$$y = C_1 e^t + t + t^2/2 + C_2$$

where C_1 and C_2 are arbitrary constants.

Another method applies when the second-order ODE is autonomous:

$$y'' = F(y, y') \tag{11}$$

☞ Subtle use of the Chain Rule here:

$$v(t) \to v(y(t))$$

$$\frac{dv}{dt} = \frac{dv}{dy} \cdot \frac{dy}{dt}$$

Since t does not appear explicitly in the differential equation, an independent variable other than t might be suitable. We introduce y itself as the new independent variable and $v = y'$ as a new dependent variable! Using the Chain Rule, we have

$$y'' = \frac{d^2 y}{dt^2} = \frac{dv}{dt} = \frac{dv}{dy}\frac{dy}{dt} = \frac{dv}{dy}v$$

☞ Tricky. Everyone has trouble with this.

and ODE (11) is transformed to a pair of first-order ODEs quite different from (10):

$$
\begin{aligned}
v\frac{dv}{dy} &= F(y, v) \\
\frac{dy}{dt} &= v(y)
\end{aligned}
\tag{12}
$$

The first ODE in (12) decouples from the second, so it can be solved for $v(y)$ separately as a first-order ODE. Once the solution $v(y)$ is known, the second ODE in (12) can be solved for $y(t)$ by separating variables and integrating. Here is an example.

EXAMPLE 1.7.4

Solving $y'' = F(y, y')$

The IVP

$$y'' = \frac{1}{y}(y')^2 - \frac{y'}{y}, \qquad y = 1 \quad \text{and} \quad y' = 2 \quad \text{when } t = 0$$

can be solved for $y(t)$ by the method given above. Comparing this ODE with (11) we see that $F(y, y') = [(y')^2 - y']/y$. We assume that $y > 0$. Introducing y and $v = y'$ as variables, we have from (12) and the initial data the two first-order IVPs:

$$
\begin{aligned}
v\frac{dv}{dy} &= \frac{1}{y}v^2 - \frac{v}{y}, & v = 2 \quad \text{when } y = 1 & \tag{13a} \\
\frac{dy}{dt} &= v(y), & y = 1 \quad \text{when } t = 0 & \tag{13b}
\end{aligned}
$$

The solution $v = 0$ of the ODE in (13a) is not of interest, so we divide out the factor v and solve the resulting linear IVP $dv/dy = (v - 1)/y$ by the integrating factor technique to find

$$v = 1 + y$$

So IVP (13b) becomes

$$\frac{dy}{dt} = v(y) = 1 + y, \qquad y = 1 \quad \text{when } t = 0$$

This linear ODE can be solved by using an integrating factor to obtain $y = Ce^t - 1$. Using the initial data $y = 1$ when $t = 0$, we have the desired solution:

$$y(t) = 2e^t - 1$$

We must have that $t > -\ln 2$ because we assumed at the start that y is positive.

Next, we present a surprising application of this reduction technique.

Inverse Square Law of Gravitation: Escape Velocities

Using the extensive astronomical work and empirical laws of Tycho Brahe (1546–1601) and Johannes Kepler (1571–1630) concerning the orbits of the moon and the planets, Newton focused attention on just one force: gravity. His law of universal gravitation deals with the gravitational effect of one body upon another.

> Newton's Law of Universal Gravitation.   The force **F** between two particles having masses m_1 and m_2 and separated by a distance r is attractive, acts along the line joining the particles, and has magnitude
>
> $$|\mathbf{F}| = \frac{m_1 m_2 G}{r^2} \tag{14}$$
>
> where G is a constant. This is the *Inverse Square Law of Gravitation.*

G is a universal constant independent of the masses m_1 and m_2; in SI units,

$$G = 6.67 \times 10^{-11} \text{ Nm}^2/\text{kg}^2$$

Newton showed that bodies affect one another as if the mass of each were concentrated at its center of mass, provided that the mass of each body is distributed in a spherically symmetric way. In this case r is the distance between the centers of mass.

Suppose that a spherical projectile of mass m is hurled straight up from the surface of the earth. Can it escape into outer space? As it will turn out, we don't need a solution formula to answer this question. Suppose that the projectile moves along a y-axis perpendicular to the earth's surface with the positive direction upward ($y = 0$ at the surface). Suppose that air resistance is negligible, so the only significant force acting on the body is the gravitational attraction of the earth. This force acts downward and, according to Newton's Law of Universal Gravitation, is given by

$$F = \frac{-mMG}{(y + R)^2}$$

where M is the mass and R the radius of the earth. For values of y close to 0, the gravitational force F is close to the constant $-mMG/R^2 = -mg$, where $g = MG/R^2$. This approximation to the gravitational force is quite good near the ground, but it is not appropriate here because we want to see what happens when the projectile is far above the earth's surface.

Using Newton's Second Law (see Section 1.5), and the information given above, the projectile's upward motion is modeled by the IVP

$$my'' = \frac{-mMG}{(y + R)^2}, \qquad y(0) = 0, \quad y'(0) = v_0 > 0 \tag{15}$$

If the initial velocity v_0 is small enough, we know from experience that the body will rise to a high point and then fall back to earth. Is there a smallest value for v_0 such that the body does *not* fall back?

Since the differential equation in IVP (15) doesn't involve t explicitly, we may use the second method of reduction of order. If we set $y'' = v\, dv/dy$, we have that

$$v\frac{dv}{dy} = -\frac{MG}{(y+R)^2}, \qquad v = v_0 \quad \text{when } y = 0 \tag{16}$$

When the variables are separated and IVP (16) is solved, we have

$$v^2 = \left(v_0^2 - \frac{2MG}{R}\right) + \frac{2MG}{y+R} \tag{17}$$

From formula (17) we see that v^2 remains positive for all $y \geq 0$ as long as $v_0^2 \geq 2MG/R$.

So for any value $v_0 > 0$ such that $v_0 \geq (2MG/R)^{1/2}$, it follows that $v(y) > 0$ for all $y \geq 0$. Reason: if $v(\bar{y}) \leq 0$ for some value $\bar{y} > 0$, then there would be a height h between 0 and $\bar{y}$ such that $v(h) = 0$. But formula (17) tells us that v^2 is the sum of a nonnegative term $[v_0^2 - (2MG/R)]$ and a positive term $[2MG/(y+R)]$ so v can't be zero. On the other hand, if $0 < v_0 < (2MG/R)^{1/2}$, then from (17) there is a value of y that makes $v^2 = 0$ and from that point on the projectile falls back toward the earth. The velocity

$$v_0 = (2MG/R)^{1/2} \tag{18}$$

☞ Numerical solvers can't do everything!

is called the *escape velocity* of the body because it is the smallest value of v_0 for which the body never falls back. Try to show this fact with a numerical solver! Note that we have answered our original question about the existence of an escape velocity without completing all the steps in the reduction process.

The escape velocity $(2MG/R)^{1/2}$ from the surface of a body depends only on the body's mass and radius. The escape velocity from the earth's surface turns out to be roughly 11.179 km/sec. Escape velocities for all the larger satellites in the solar system have been calculated and are listed in *Handbook of Chemistry and Physics*, 75th ed. (D. R. Lide, ed., Boca Raton, Fla.: CRC Press, 1994).

Sometimes we can work this process backwards and solve a system by looking at a first-order ODE. We illustrate this process by a very different (and controversial) model.

Destructive Competition

In 1916, Lanchester[6] described some mathematical models for air warfare. These have been extended to a general combat situation, not just air warfare. We look at a particular case of the models.

---

[6]Frederick William Lanchester (1868–1946) was an English engineer, mathematician, inventor, poet, and musical theorist who designed and produced some of the earliest automobiles and wrote the first theoretical treatise of any substance on flight, the book *Aerial Flight*, (Constable, London, 1907–08). His book *Aircraft in Warfare* (Constable, London, 1916), contains the square law of conventional combat discussed in this section. Twenty-five years later (during the Second World War) Lanchester's scientific approach to military questions became the basis of a new science, Operations Research, which is now applied to problems of industrial management, production, and government procedure, as well as to military problems.

An x-force and a y-force are engaged in combat. Suppose that $x(t)$ and $y(t)$ denote the respective strengths of the forces at time t, where t is measured in days from the start of the combat. It is not easy to quantify strength, including, as it does, the numbers of combatants, their battle readiness, the nature and number of the weapons, the quality of leadership, and a host of psychological and other intangible factors difficult even to describe, much less to turn into state variables. Nevertheless, we will suppose that the strengths can be quantified, that $x(t) > y(t)$ means the x-force is stronger than the y-force, and that $x(t)$ and $y(t)$ are differentiable functions. The pair of values $x(t)$, $y(t)$ defines the state of this system at time t.

Suppose that we can estimate the *noncombat loss rate* of the x-force (i.e., the loss rate due to diseases, desertions, and other noncombat mishaps), the *combat loss rate* due to encounters with the y-force, and the *reinforcement rate*. Then the net rate of change in $x(t)$ is given by the Balance Law:

$$x'(t) = \text{Reinforcement rate} - \text{Noncombat loss rate} - \text{Combat loss rate}$$

A similar ODE applies to the y-force. The problem is to analyze the solutions $x(t)$ and $y(t)$ of the resulting system to determine who wins the combat.

According to Lanchester, a model system of ODEs for a pair of *conventional combat forces* operating in the open with negligible noncombat losses is

$$\begin{aligned} x'(t) &= R_1(t) - by(t) \\ y'(t) &= R_2(t) - ax(t) \end{aligned} \tag{19}$$

where a and b are positive constants and R_1 and R_2 are the rates of reinforcement. The reinforcement rates are assumed to depend only on time and not on the strength of either force (a dubious assumption). The combat loss rates $by(t)$ and $ax(t)$ introduce actual combat into the model. Lanchester argues for the specific form of the combat loss rate terms in system (19) as follows. Every member of the conventional force is assumed to be within range of the enemy. It is also assumed that as soon as a conventional force suffers a loss, fire is concentrated on the remaining combatants. This implies that the combat loss rate of the x-force is proportional to the number of the enemy, and so is given by $by(t)$. The coefficient b is a measure of the average effectiveness in combat of each member of the y-force. A similar argument applies to the term $ax(t)$.

☞ The interplay between first-order ODEs and planar autonomous systems goes both ways.

Earlier in this section we showed how the solution curves of the ODE $N(y)y' + M(x) = 0$ can be characterized in terms of the orbits of the differential system

$$dx/dt = N(y), \qquad dy/dt = -M(x)$$

The next example shows the usefulness of the reverse process.

EXAMPLE 1.7.5

Conventional Combat: From a System to a First-Order ODE
Combat between two forces with no reinforcements may be modeled by the system

$$x' = -by, \qquad y' = -ax \tag{20}$$

Let $x = x(t)$ and $y = y(t)$ solve system (20), and suppose that the orbit lies inside the first quadrant of the xy-plane (i.e., $x > 0$, $y > 0$). We have from (20) that

$$\frac{dy}{dx} = \frac{dy/dt}{dx/dt} = \frac{-ax}{-by}$$

or

$$by\,dy - ax\,dx = 0 \tag{21}$$

which is a separable ODE.

So the integral curves of the "reduced" ODE (21) are just the orbits of system (20). The only assumptions made are that $x > 0$ and $y > 0$. Now let's solve ODE (21) and see what happens.

EXAMPLE 1.7.6

Who Wins?

Let's suppose that $x_0 > 0$ and $y_0 > 0$ are the strengths of the two forces at the start of combat. Integrating (21), we have

$$by^2(t) - ax^2(t) = by_0^2 - ax_0^2 \tag{22}$$

Equation (22) is known as the *square law of conventional combat*. Although we can't tell just how $x(t)$ and $y(t)$ change in time, (22) implies that the point $(x(t), y(t))$ moves along an arc of a hyperbola as time advances. These hyperbolas are orbits of system (20) and are plotted in Figure 1.7.2 for $a = 0.064$, $b = 0.1$, $x_0 = 10$, and various values of y_0. The arrowheads on the curves show the direction of changing strengths as time passes. Since $x'(t) = -by(t)$, $x(t)$ decreases as time advances if $y(t)$ is positive. Similarly, $y(t)$ decreases with the advance of time if $x(t)$ is positive.

☞ The x-force wins if $y_0 < (a/b)^{1/2}x_0$. So very small changes in initial strengths (x_0, y_0) might affect the outcome of a conflict.

One force wins if the other force vanishes first. For example, y wins if $by_0^2 > ax_0^2$, since from (22), $y(t)$ would never vanish, while the x-force is gone by the time $y(t)$ has decreased to $\left((by_0^2 - ax_0^2)/b\right)^{1/2}$. So the y-force wants a combat setting in which $by_0^2 > ax_0^2$; that is, the y-force needs a large enough initial strength y_0 so that

$$y_0 > (a/b)^{1/2}x_0 \tag{23}$$

Figure 1.7.3 shows components of the top orbit in Figure 1.7.2 ($a = 0.064$, $b = 0.1$). The combatants start out even in Figure 1.7.3 at $x_0 = y_0 = 10$, but the y-force wins because $b = 0.1$, which is larger than $a = 0.064$.

The simplified model solved above is unrealistic. If noncombat loss rates and reinforcement rates are included, a certain element of realism enters and one might actually compare the model with historical battles. Studies along these lines for the Battle of the Ardennes and the Battle of Iwo Jima in the Second World War have been carried out.[7] These studies give results reasonably close to the actual combat statistics once the coefficients a and b are determined. Whether these coefficients could ever be estimated with any accuracy *before* combat is an open question.

---

[7]See, for example, C. S. Coleman, "Combat Models," *Modules of Applied Mathematics*, vol. 1, Chap. 8, W. Lucas, M. Braun, C. Coleman, and D. Drew, eds. (New York: Springer-Verlag, 1983).

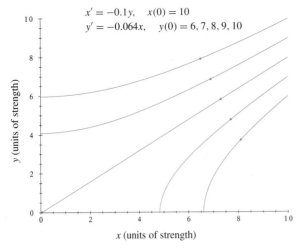

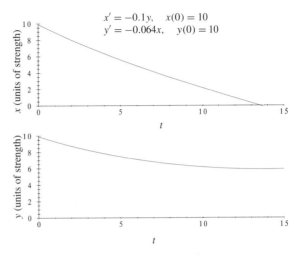

FIGURE 1.7.2 Arcs of hyperbolas of the square law of conventional combat (Example 1.7.6).

FIGURE 1.7.3 Component curves of the top orbit in Figure 1.7.2.

The Dark Side of Modeling

Combat models raise ethical questions about the uses of mathematics. G. H. Hardy, a noted mathematician of the early twentieth century, wrote:

> *So a real mathematician has his conscience clear; there is nothing to be set against any value his work may have; mathematics is... a harmless and innocent occupation.*[8]

One wonders if it is that simple. Mathematics is a part of the cultures and societies of the mathematicians who create or discover it. Since war and the preparation for war continue to be a preoccupation of humankind, it is not surprising that mathematics is applied to its study and analysis.

PROBLEMS

1. (*The System Way to Plot Integral Curves*). Find an integral for each ODE and plot representative integral curves (which in this case are also orbits) in the given rectangle. [*Hint*: Follow the method used in Example 1.7.1 to plot curves, choose initial points in the rectangle, and solve forward and backward in time.]

   (a) $(1 - y^2)y' + x^2 - 1 = 0$; $|x| \le 3$, $|y| \le 3$

   (b) $(1 - y^2)y' + x^2 = 0$; $|x| \le 1.5$, $|y| \le 2$

---

[8]See G. H. Hardy, *A Mathematician's Apology*, 2nd ed. (Cambridge: Cambridge University Press, 1967), pp. 140–141.

2. (*The System Way to Plot Solution Curves*). Plot the solution curve of each IVP on the largest possible interval.

(a) $x^2 - 1 + (1 - y^2)y' = 0$; $y(-1) = -2$ (b) $(1 - y^2)y' + x^2 = 0$; $y(-1) = 0.5$

(c) $(2x^2 + 2xy)y' = 2xy + y^2$; $y(-0.5) = 1$

 3. (*Combat Model: Who Wins the Battle?*).

(a) Reproduce Figure 1.7.2 as closely as possible. Plot the orbit with initial condition $x_0 = 10$, $y_0 = 7$. Does y win or lose?

(b) How much time does it take for the conflict to be resolved?

4. (*Reduction of Order*). Solve the following ODEs by reduction of order. Plot solution curves $y = y(t)$ in the *ty*-plane using various sets of initial data. [*Hint*: If y does not appear explicitly, set $y' = v$, $y'' = v'$. If t does not appear explicitly, set $y' = v$ and $y'' = v\,dv/dy$.]

(a) $ty'' - y' = 3t^2$ (b) $y'' - y = 0$ (c) $yy'' + (y')^2 = 1$ (d) $y'' + 2ty' = 2t$

www 5. (*Reduction of Order*). Solve the given initial value problems.

(a) $2yy'' + (y')^2 = 0$, $y(0) = 1$, $y'(0) = -1$

(b) $y'' = y'(1 + 4/y^2)$, $y(0) = 4$, $y'(0) = 3$

(c) $y'' = -g - y'$, $y(0) = h$, $y'(0) = 0$, g, h are positive constants

6. (*Reduction of Order by Varied Parameters*). Suppose that $z(t)$ is a solution of the linear second-order *undriven* ODE $z'' + a(t)z' + b(t)z = 0$, where $a(t)$ and $b(t)$ are continuous on an interval I. The function $z(t)$ can be used to find solutions of the *driven* ODE $y'' + a(t)y' + b(t)y = f(t)$, where $f(t)$ is also continuous on I.

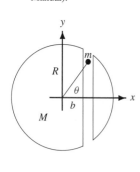

Albert Einstein (1879–1955)

☞ We have just developed a truly marvelous problem involving Einstein's field equations which this margin is too narrow to contain. (It's in the Student Resource Manual).

(a) Show that $y(t) = u(t)z(t)$ is a solution on I of the ODE $y'' + a(t)y' + b(t)y = f(t)$ if $u'(t)$ solves the first-order linear ODE $zu'' + (2z' + az)u' = f$.

(b) If $z(t) \neq 0$ on I, show that the linear ODE for u' in part (a) can be put in normal linear form by dividing by z. Find $u'(t)$ by using the integrating factor $z^2 e^{A(t)}$, where $A(t)$ is an antiderivative of $a(t)$.

(c) Find another solution of $tz'' - (t + 2)z' + 2z = 0$ for $t > 0$, given that $z = e^t$ is one solution.

7. Using the method described in Problem 6, find a solution of the ODE $t^2 y'' + 4ty' + 2y = \sin t$, $t > 0$, where it is known that $z(t) = t^{-2}$ solves the ODE $t^2 y'' + 4ty' + 2y = 0$. [*Hint*: First put the ODE in normal linear form by dividing by t^2.]

8. A tunnel is bored through the earth and an object of mass m is dropped into the tunnel.

(a) Neglecting friction and air resistance, write an ODE which describes the motion of the object given that the component of the force of gravity in the direction of motion is given by $f = GmM \sin\theta/(b^2 + y^2)$.

(b) Find the general solution of the equation of motion. [*Hint*: Use a reduction-of-order technique. Your answer should be left in implicit form with an integral.]

9. (*Escape Velocity in an Inverse Cube Universe*). Suppose that on a planet in another universe the magnitude of the force of gravity obeys the Inverse Cube Law:

$$|\mathbf{F}| = \frac{\tilde{G}mM}{(y + R)^3}$$

where m is the mass of the object, y its location above the surface of the planet, M the mass of the planet, $\tilde{G}$ a new universal constant, and R the radius of the planet.

(a) What is the escape velocity $\tilde{v}_0$ from this planet?

(b) What is the ratio of $\tilde{v}_0$ to the Inverse Square Law escape velocity v_0 [equation (18)]?

10. (*Conventional Combat with Reinforcements*). Consider the conventional combat model (19) where R_1 and R_2 are positive constants.

- Follow the procedure of Example 1.7.5 to obtain a separable first-order ODE in the variables x and y. Solve the ODE and show that $x(t)$ and $y(t)$ satisfy the condition

$$b\left(y(t) - \frac{R_1}{b}\right)^2 - a\left(x(t) - \frac{R_2}{a}\right)^2 = C$$

where C is a constant determined by the initial data.

- Take $a = b = 1$, $R_1 = 2$, and $R_2 = 3$ and plot the orbits when (i) $x(t_0) = 3$, $y(t_0) = 1$; (ii) $x(t_0) = 2$, $y(t_0) = 2$; (iii) $x(t_0) = 4$, $y(t_0) = 2$.

- Assign arrowheads to the orbits indicating the direction of increasing time.

- Who wins in each of the cases described above?

- What happens when $x(t_0) = R_2/a$ and $y(t_0) = R_1/b$? Explain in terms of the development of the combat.

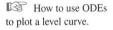

11. (*Rescaling and Escape Velocity*). This problem examines the notion of an escape velocity from a geometrical viewpoint. Rescale IVP (15) by defining $z = y/R$ and $t = s(R/MG)^{1/2}$, and derive a new IVP using the new variables z and s. Let $z'(0) = a$, and use a as a scaled "velocity" parameter. Use a numerical solver to find as good a lower bound for the initial escape velocity a as you can.

12. (*The System Way to Plot a Level Curve*). Suppose that $f(x, y)$ is continuously differentiable in a region R where $\partial f/\partial x$ and $\partial f/\partial y$ have no common zeros.

☞ How to use ODEs to plot a level curve.

- For any point (x_0, y_0) in R show that the orbit of the autonomous planar IVP

$$x' = \partial f/\partial y, \qquad x(0) = x_0$$
$$y' = -\partial f/\partial x, \qquad y(0) = y_0$$

is a level curve, $f(x, y) = f(x_0, y_0)$, of the function $f(x, y)$ through the point. [*Hint:* Show that $H(x, y) = f(x, y)$ is an integral of the system]

- Consider the function $f(x, y) = -x + 2xy + x^2 + y^2$. Use the above method to plot several level curves for f in the rectangle $|x| \leq 6$, $|y| \leq 6$.

1.8 Cold Pills

What do you do when you catch a cold? If you are like many of us, you take cold pills. The pills contain a decongestant to relieve stuffiness and an antihistamine to stop the sneezing and to dry up a runny nose. The pill dissolves and releases the medications into the gastrointestinal tract. The medications diffuse from there into the blood, and the bloodstream takes each medication to the site where it has therapeutic effect. Both medications are gradually cleared from the blood by the kidneys and the liver.

Pharmaceutical companies do a lot of testing to determine the flow of a medication through the body. This flow is modeled by treating the parts of the body as compartments, and then tracking the medication as it enters and leaves each compartment. A typical cold medication leaves one compartment (e.g., the GI tract) and moves into another (such as the bloodstream) at a rate proportional to the amount present in the first compartment. The constant of proportionality depends upon the medication, the compartment, and the age and general health of the individual.

Now let's build a system of ODEs that models the passage of one of the medications, say antihistamine, through the body compartments.[9]

One Dose of Antihistamine

Let's see what happens to a dose of antihistamine once it lands in the GI tract.

EXAMPLE 1.8.1

Modeling the Rates

Suppose that there are A units of antihistamine in the GI tract at time 0 and that $x(t)$ is the number of units remaining at any later time t. The Balance Law applies:

$$\text{Net rate} = \text{Rate in} - \text{Rate out}$$

Since we start with A units and the medication moves out of the GI tract and into the blood at a rate proportional to the amount in the GI tract, we have the IVP

$$\frac{dx(t)}{dt} = -k_1 x(t), \qquad x(0) = A \tag{1}$$

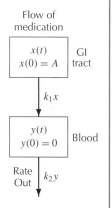

Flow of medication

$x(t)$ $x(0) = A$	GI tract

$k_1 x$

$y(t)$ $y(0) = 0$	Blood

Rate Out $k_2 y$

where k_1 is a positive constant. The top part of the margin sketch pictures the IVP. Time is measured in hours and k_1 in hours^{-1}.

We assume that the compartment labeled "blood" includes the tissues where the medication does its work. The level $y(t)$ of antihistamine in the blood builds up from zero but then falls as the kidneys and liver do their job of clearing foreign substances from the blood. The Balance Law applied to the antihistamine in the blood compartment leads to the IVP

$$\frac{dy(t)}{dt} = k_1 x(t) - k_2 y(t), \qquad y(0) = 0 \tag{2}$$

The first term on the right side of the rate equation in (2) models the fact that the exit rate of antihistamine from the GI tract equals the entrance rate into the blood. The second rate term models the clearance of antihistamine from the blood. The clearance constant k_2 is measured in hours^{-1}. The lower part of the margin sketch illustrates IVP (2).

Putting (1) and (2) together, we have a system of two first-order ODEs with initial data:

$$\begin{aligned} \frac{dx}{dt} &= -k_1 x, & x(0) &= A \\ \frac{dy}{dt} &= k_1 x - k_2 y, & y(0) &= 0 \end{aligned} \tag{3}$$

Edward Spitznagel

[9]The cold pill model is based on the work of the contemporary applied mathematician, Edward Spitznagel, Professor of Mathematics at Washington University, St. Louis, Missouri. His current research areas are pharmacokinetics (study of the flow of medications through the body) and bioequivalence (study of the efficacy of medications), fields that use compartment models extensively. Professor Spitznagel started off his career in mathematics because he could not limit his interest to just one area of science. He does consulting for a wide variety of clients, principally the pharmaceutical industry and medical schools. His advice to aspiring applied mathematicians: Learn as much mathematics as you can because it will make you versatile and able to pick up new ideas easily. Seek out opportunities to learn how to apply mathematics in the real world by on-the-job training.

IVP (3) is our mathematical model for the flow of a single dose of A units of medication through the GI tract and blood compartments.

Now let's solve IVP (3) and find formulas for the amounts of medication in each compartment.

EXAMPLE 1.8.2

Growth and Decay of Antihistamine Levels

Integrating factors are used to solve the linear differential equations of IVP (3) one at a time, starting with the first. Rearrange the first rate equation, multiply by the integrating factor $e^{k_1 t}$, integrate, and then use the initial data to obtain

$$x(t) = Ae^{-k_1 t}$$

Insert this formula for $x(t)$ into the second rate equation, which becomes

$$\frac{dy}{dt} + k_2 y = k_1 A e^{-k_1 t}, \qquad y(0) = 0$$

This linear IVP is solved using the integrating factor $e^{k_2 t}$. After some calculation, we obtain the formula

$$y(t) = \frac{k_1 A}{k_1 - k_2} \left(e^{-k_2 t} - e^{-k_1 t} \right)$$

Summarizing, we see that the antihistamine levels in the GI tract and blood are given by the formulas

$$x(t) = Ae^{-k_1 t}, \qquad y(t) = \frac{k_1 A}{k_1 - k_2} \left(e^{-k_2 t} - e^{-k_1 t} \right) \tag{4}$$

☞ See if you can find the solution when $k_1 = k_2$.

We have assumed that $k_1 \neq k_2$, an assumption that is justified by the pharmaceutical data, as we will soon see.

From the formulas in (4), we see that the antihistamine levels in the GI tract and the blood tend to zero as time increases. The level in the blood reaches a maximum value at some positive time (see Problem 4) and then drops back.

The symbolic solution formulas (4) are abstract, so let's introduce some numbers.

EXAMPLE 1.8.3

The Rate Constants

One pharmaceutical company estimates that the values of the rate constants for the antihistamine in the cold pills it makes are

$$k_1 = 0.6931 \, (\text{hour})^{-1}, \qquad k_2 = 0.0231 \, (\text{hour})^{-1} \tag{5}$$

Because k_2 is so much smaller than k_1, antihistamine stays at a high level a lot longer in the blood than in the GI tract. Figure 1.8.1 shows the levels over a six-hour period as predicted by system (3) if $A = 1$ and k_1 and k_2 are given by (5).

The values of the rate constants given in (5) are for the "average" person, but what about someone who isn't average?

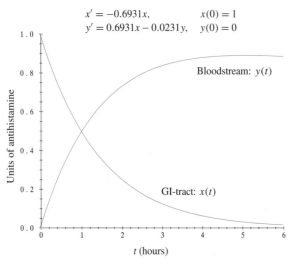

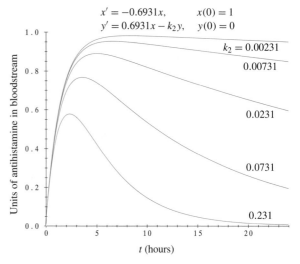

FIGURE 1.8.1 Effect of one dose of antihistamine (Example 1.8.3).

FIGURE 1.8.2 Sensitivity to the clearance coefficient (Example 1.8.4).

EXAMPLE 1.8.4

Sensitivity to the Clearance Coefficient

The *clearance coefficient* k_2 of medication from the blood is often much lower for the old and sick than it is for the young and healthy. This means that for some people medication levels in the blood may become and then remain excessively high, even with a standard dosage. Figure 1.8.2 displays the results of a parameter study in which antihistamine levels in the blood are plotted over a 24-hour period. This is done for $k_1 = 0.6931$ and five values of k_2, corresponding to five people of different ages and states of health. The variations in the levels are a measure of the *sensitivity* of the medication levels to changes in the value of k_2.

Now let's determine the effect of repeating the dose.

Repeated Doses

Few of us stop with taking just one cold pill. We take several doses until we feel better. Most cold pills dissolve quickly and release their medications in the GI tract at a constant rate over half an hour. The dose is then repeated every six hours to maintain the medication levels in the blood. Typically, the release rate of antihistamine from the pill into the GI tract is constant for a short time, but then cuts off entirely until the next dose. We will use an on-off function to model this kind of repeated dosage.

EXAMPLE 1.8.5

One Dose Every Six Hours: Square Wave Delivery Rate

Let's suppose that 6 units of antihistamine are delivered to the GI tract at a constant

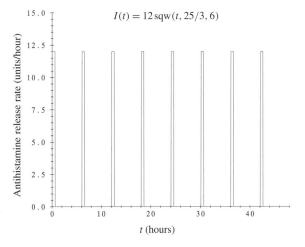

FIGURE 1.8.3 Square wave antihistamine delivery rate into the GI tract (Example 1.8.5).

FIGURE 1.8.4 Antihistamine levels: one dose every six hours (Example 1.8.6).

rate and over a half-hour time span, then repeated every six hours. The model IVP is

$$
\begin{aligned}
\frac{dx}{dt} &= I(t) - k_1 x, & x(0) &= 0 \\
\frac{dy}{dt} &= k_1 x - k_2 y, & y(0) &= 0 \\
k_1 &= 0.6931\,(\text{hour})^{-1}, & k_2 &= 0.0231\,(\text{hour})^{-1} \\
I(t) &= 12\,\text{sqw}(t, 25/3, 6)
\end{aligned}
$$

(6)

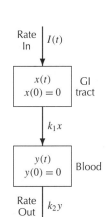

The periodic on-off function $12\,\text{sqw}(t, 25/3, 6)$ denotes the square wave of amplitude 12 and period 6 hours, which is "on" for half an hour at the start of each period (and so delivers 6 units of medication in that time span), and is "off" otherwise. Note that one half hour is $100/12 = (25/3)\%$ of the six-hour period. See Figure 1.8.3 for the graph of $I(t)$.

Each ODE in (6) is linear, but the strange form of the input function $I(t)$ is somewhat daunting. Is there a way to study solutions of IVP (3) without having to find solution formulas?

EXAMPLE 1.8.6

☞ Appendix B.1 has more on square waves and other on-off functions.

Rising Levels of Antihistamine: Using a Numerical Solver

Although the integrating factor technique used before also applies to system (6), there is an awkward integration to carry out because $I(t)$ repeatedly turns on and off. So we used a numerical solver that handles on-off functions to produce Figure 1.8.4. The figure shows that the amount of antihistamine in the GI tract rises quite rapidly as the cold pill dissolves, but then drops back almost to the zero-level before the next dose is taken. It is a different story in the blood, where the low value of the clearance coefficient k_2 keeps antihistamine levels high. Even after 48 hours the levels in the blood show little sign of approaching any kind of equilibrium.

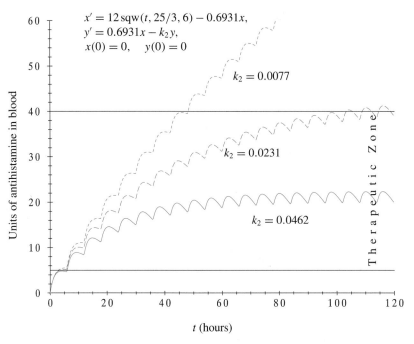

$$x' = 12\,\mathrm{sqw}(t, 25/3, 6) - 0.6931x,$$
$$y' = 0.6931x - k_2 y,$$
$$x(0) = 0, \quad y(0) = 0$$

FIGURE 1.8.5 Therapeutic zone for antihistamine (Example 1.8.7).

When you take cold pills, you want the medication in the bloodstream to reach therapeutic levels quickly and to stay in a safe range. The rising levels in the blood shown in Figure 1.8.4 are alarming. What happens if they get too high?

| EXAMPLE 1.8.7 | **Asleep in Math Class** |

High levels of antihistamine cause drowsiness, but low levels are ineffective. Suppose that the therapeutic but safe range for antihistamine in the blood is from 5 to 40 units. Figure 1.8.5 shows the amounts of antihistamine in the blood of three math students with quite different clearance coefficients k_2. From the graphs we see that each of the three gets relief from cold symptoms within six hours. But with repeated doses the levels are markedly different. Which of the three students fall asleep in math class?

The cold pill model is just one of many examples of a compartment model. Let's take a brief look at some general compartment models.

Compartment Models

A *compartment model* consists of a finite number of compartments (or boxes) connected by arrows. Each arrow means that the substance being tracked leaves the box at the foot of the arrow and enters the box at the arrowhead.

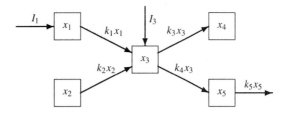

FIGURE 1.8.6 A joining–branching linear cascade (Example 1.8.8).

☞ There are also nonlinear cascades.

The simplest compartment models are the linear cascades. A compartment model is a *linear cascade* if

(a) no directed chain of arrows and boxes begins and ends at the same box, and

(b) the substance exits a box at a rate proportional to the amount in the box and, if it enters another box, does so at that same rate.

An arrow pointing toward one box, but not out of another, indicates an external source of the substance (i.e., an *input*). These arrows are labeled with an "*I*," where *I* is the input rate of the substance. An arrow that points away from a box, but not toward another box, means that the substance exits the system from that box. The state variable $x_i(t)$ denotes the amount of the substance in box i at time t. The symbols kx_i by an arrow leaving box i and entering box j mean that the substance exits i and enters j at the rate kx_i.

The models for the flow of cold medication are examples of linear cascades. Here is a more complex example.

EXAMPLE 1.8.8

From Boxes and Arrows to a System of ODEs

Figure 1.8.6 shows a linear cascade with two inputs. The corresponding ODEs can be constructed directly from the boxes and arrows. The first-order linear system of ODEs is based on the Balance Law applied to each box:

$$\begin{aligned}
x_1' &= I_1 - k_1 x_1 \\
x_2' &= -k_2 x_2 \\
x_3' &= I_3 + k_1 x_1 + k_2 x_2 - k_3 x_3 - k_4 x_3 \\
x_4' &= k_3 x_3 \\
x_5' &= k_4 x_3 - k_5 x_5
\end{aligned} \tag{7}$$

System (7) can be solved from the top down, one ODE at a time.

Turning this around, we can construct a linear cascade model of boxes and arrows as in Figure 1.8.6 from a system of linear ODEs like (7).

The ODEs that model a linear cascade are called a *linear cascade of ODEs*; the system can always be solved from the top down, as in Example 1.8.8.

Linear cascades may be used to model other physical phenomena besides flow through compartments. For example, in a radioactive decay process one element decays into another, which in turn decays into a third, and so on, until the process terminates in a stable, nonradioactive element. In this case each compartment corresponds to a distinct element.

Comments

☞ The "Tips" in Section 1.1 of the Student Resource Manual tell more about the internal step-size.

A numerical solver is useful in solving a cascade of IVPs, particularly if some intake rates are on-off functions such as square waves. If you do use a solver in this situation, you may have to set the maximal internal step-size quite low so that the solver detects each of the on-off times.

Although on-off intake rate functions are discontinuous, the solution formulas of Sections 1.3 and 1.4 for first-order linear IVPs still apply. Imagine a sequence of IVPs, one for each interval between the break points of the rate function. The final value for one interval becomes the initial value for the next. The resulting solution curves may have corners, but the curves are continuous. See Figure 1.8.4 for an example.

PROBLEMS

1. (*From Boxes and Arrows to ODEs*). Write the system of first-order linear IVPs that models each of the following linear cascades, solve, and describe what happens in each compartment as $t \to \infty$. [*Hint*: See Example 1.8.8.]

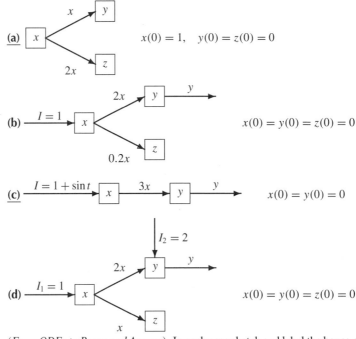

(a) $x(0) = 1, \quad y(0) = z(0) = 0$

(b) $I = 1$ $x(0) = y(0) = z(0) = 0$

(c) $I = 1 + \sin t$ $x(0) = y(0) = 0$

(d) $I_2 = 2$, $I_1 = 1$ $x(0) = y(0) = z(0) = 0$

2. (*From ODEs to Boxes and Arrows*). In each case sketch and label the boxes and arrows diagram.

    (a) $x' = 5 - x, \quad y' = x - 5y$

    (b) $x' = -x/2, \quad y' = 1 - y/3, \quad z' = x/2 + y/3$

    (c) $x' = -x, \quad y' = x/2 - 3y, \quad z' = x/2 + 3y - 2z$

3. (*One Dose of Antihistamine*). Suppose that A units of antihistamine are present in the GI tract and B units in the blood at time 0.

(a) Solve IVP (3) with the condition $y(0) = 0$ replaced by $y(0) = B$.

 (b) Use a numerical solver to solve IVP (3) but with $x(0) = 1$, $y(0) = 1$, $k_1 = 0.6931$, $k_2 = 0.0231$, $0 \le t \le 6$. Graph the levels of antihistamine in the GI tract and in the blood. Estimate the highest level of the antihistamine in the blood and the time it reaches that level.

4. (*Time of Maximum Dosage*). Show that the medication in the bloodstream reaches its maximum after a single dose when $t = (\ln k_1 - \ln k_2)/(k_1 - k_2)$, $k_1 \ne k_2$. [*Hint*: Use the model developed in Examples 1.8.1 and 1.8.2.]

www **5.** (*One Dose: Sensitivity to k_1*). Let $A = 1$, keep $k_2 = 0.0231$ as in Example 1.8.3, but let k_1 vary.

(a) Using a numerical solver for system (3), display the effects on the antihistamine levels in the bloodstream if $k_1 = 0.06931, 0.11, 0.3, 0.6931, 1.0$, and 1.5. Plot the graphs over a 24-hour period. Why do the graphs for larger values of k_1 cross the graphs for smaller values?

(b) You need to keep medication levels within a fixed range in order to be both therapeutic and safe. Suppose that the desired range for antihistamine levels in the blood is from 0.2 to 0.8 for a unit dose taken once. With $k_2 = 0.0231$, find upper and lower bounds on k_1 so that the antihistamine levels in the blood reach 0.2 within 2 hours and stay below 0.8 for 24 hours.

6. (*Cold Medication: Decongestant*). Most cold pills contain a decongestant as well as an antihistamine. The form of the rate equations for the flow of decongestant is the same as that for antihistamine, but the rate constants are different. The values of these rate constants for one brand of cold pills now on the market have been determined by the manufacturer to be $k_1 = 1.386$ (hour)$^{-1}$ for the passage from the GI tract into the blood, and $k_2 = 0.1386$ (hour)$^{-1}$ for clearance from the blood. The respective amounts of decongestant in the GI tract and the blood are denoted by $x(t)$, $y(t)$. Use Examples 1.8.1–1.8.6 as guides as you solve the following problems.

(a) (*A Single Dose of Decongestant*). Suppose that A units of decongestant are in the GI tract at time 0, while the blood is free of decongestant. Construct a labeled boxes and arrows diagram and the corresponding system of IVPs for the flow of decongestant.

 (b) Solve the IVPs of part (a). Then set $A = 1$, plot $x(t)$ and $y(t)$ over a six-hour time span, and describe what happens to the decongestant levels. Compare with Figure 1.8.1.

 (c) (*Sensitivity to Clearance Coefficient*). Plot decongestant levels in the blood if $A = 1$, keeping the value of k_1 fixed at 1.386 but setting $k_2 = 0.01386, 0.06386, 0.1386, 0.6386, 1.386$. Describe what you see.

 (d) (*One Dose Every Six Hours*). Suppose decongestant is released in the GI tract at the constant rate of 12 units/hour for 1/2 hour, and then repeated every 6 hours. Construct the model system of IVPs and plot the decongestant levels in the GI tract and blood over 48 hours (use $k_1 = 1.386$, $k_2 = 0.1386$). Compare and contrast your plots with Figure 1.8.4.

 (e) (*Take the Pills for Five Days*). Using the data of part (d), plot decongestant levels in the GI tract and in the blood over a 5-day period. Repeat with $k_2 = 0.06386, 0.01386$. Will there be excessive decongestant accumulation in the blood of the old and the sick?

7. (*Continuous Doses of Medication*). If a medication is embedded in beads of resins that dissolve at varying rates, a constant flow rate I of medication can be assured. See the margin sketch.

(a) Use the information in the compartment diagram to construct IVPs for $x(t)$ and $y(t)$.

(b) Solve the system found in part (a) from the top down (assume that $k_1 \ne k_2$).

(c) (*Approach to Equilibrium*). What happens to the levels of medication in the GI tract and in the blood as $t \to \infty$?

 (d) (*Antihistamines and Decongestants*). Use a numerical solver to plot $x(t)$ and $y(t)$ for

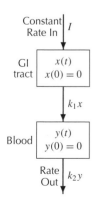

200 hours. Use $I = 1$ unit/hour, $k_1 = 0.6931$ (hour)$^{-1}$, and $k_2 = 0.0231$ (hour)$^{-1}$ for anti-histamine. Are the antihistamine levels in the blood close to equilibrium as $t \to 200$? Repeat with the decongestant: $k_1 = 1.386$ (hour)$^{-1}$, $k_2 = 0.1386$ (hour)$^{-1}$, $0 \le t \le 100$.

(e) (*The Old and the Sick*). The coefficients k_1 and k_2 for the old and sick may be much less than those for the young and healthy. Plot decongestant and antihistamine levels in the GI tract and the blood if the values of k_1 and k_2 are one-third of those in (d). Interpret your graphs.

(f) (*Safe and Effective Zone*). Assume that the clearance coefficients k_1 and k_2 have one-third of the values given in part (d). Suppose that the levels of antihistamine in the blood are designed to reach and remain between 25 and 50 units for a dose taken continuously at the rate of 1 unit per hour. Is this possible during the second through the fifth days?

1.9 Change of Variables and Pursuit Models

A change of variable may convert an apparently intractable ODE into another that can be solved by one of the techniques of this chapter. We will show how to convert an ODE with a certain kind of rate function (defined below) to a separable ODE, which can be solved by the techniques of Section 1.6. Problems 9 and 10 show how to reduce some other nonlinear first-order ODEs to linear form. We will also show how scaling the variables in an ODE has all sorts of computational and modeling pay-offs.

Homogeneous Rate Functions of Order Zero

Some types of rate functions appear so often in applications that it is worthwhile singling them out for special consideration. One of these types is defined here. A continuous function $f(x, y)$ is said to be *homogeneous of order zero* if

$$f(kx, ky) = f(x, y) \tag{1}$$

for all $k > 0$, x, y for which $f(x, y)$ and $f(kx, ky)$ are defined. Here's an example.

EXAMPLE 1.9.1   A Homogeneous Function of Order Zero
The function $f = (x^2 + y^2)^{1/2}/(x + y)$ is homogeneous of order zero because we have $f(kx, ky) = ((kx)^2 + (ky)^2)^{1/2}/(kx + ky) = f(x, y)$. But $g = (x + 1)/(x + y)$ is *not* homogeneous of order zero because $g(kx, ky) = (kx + 1)/(kx + ky) \ne g(x, y)$.

Now consider the ODE $y' = f(x, y)$, where $f(x, y)$ is homogeneous of order zero. The change of variable $y = xz$ converts y' to $(xz)' = xz' + z$ and f to

$$f(x, y) = f(x, xz) = f(x \cdot 1, x \cdot z) = f(1, z)$$

since f is homogeneous of order zero (x plays the role of k in this setting). We have a separable ODE in x and z:

$$xz' + z = f(1, z), \quad \text{or} \quad xz' = f(1, z) - z, \quad \text{or} \quad \frac{1}{f(1, z) - z} z' = \frac{1}{x} \tag{2}$$

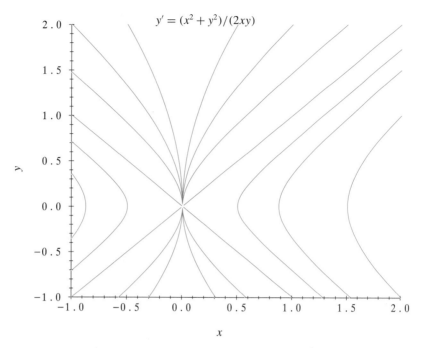

FIGURE 1.9.1 Integral curves of an ODE with a homogeneous rate function of order zero (Example 1.9.2).

The last ODE in (2) can be solved by integrating each side: $F(z) = \ln|x| + C$, where $F(z)$ is an antiderivative of $[f(1, z) - z]^{-1}$ and C is any constant. Since $z = y/x$,

$$F\left(\frac{y}{x}\right) = \ln|x| + C \tag{3}$$

So, a solution $y(x)$ of the first-order ODE $y' = f(x, y)$ satisfies (3) on some x-interval and for some choice of C. Let's see how this works in practice.

EXAMPLE 1.9.2

An ODE with a Homogeneous Rate Function of Order Zero
The function $f(x, y) = (x^2 + y^2)/2xy$ is homogeneous of order zero. To solve

$$y' = \frac{x^2 + y^2}{2xy}, \qquad x \neq 0, \quad y \neq 0 \tag{4}$$

we introduce the new variable z by setting $y = xz$ to obtain the equation

$$xz' + z = \frac{x^2 + x^2 z^2}{2zx^2} = \frac{1 + z^2}{2z}$$

Separating the variables, we have

$$\frac{2z}{1 - z^2} z' = \frac{1}{x}, \qquad z \neq \pm 1$$

Antidifferentiate both sides to obtain

$$-\ln|1 - z^2| = \ln|x| + C, \qquad z \neq \pm 1$$

where C is the constant of integration. Rearranging and returning to the original variables y and x by setting $z = y/x$, we have

$$\ln|x| + \ln\left|1 - \frac{y^2}{x^2}\right| = \ln\left|x\left(1 - \frac{y^2}{x^2}\right)\right| = -C, \qquad y \neq \pm x \tag{5}$$

Exponentiate each side of the last equality in (5) to get

$$\left|x\left(1 - \frac{y^2}{x^2}\right)\right| = e^{-C}, \qquad y \neq \pm x \tag{6}$$

We remove the absolute value signs from (6), replace e^{-C} by a constant K (which can have any real value), multiply by x, and obtain the following equation:

$$x^2 - y^2 = Kx, \qquad x \neq 0, \quad y \neq 0 \tag{7}$$

Figure 1.9.1 shows some of the hyperbolic curves defined by (7) for various values of K. For $K \neq 0$, solution curves $y = y(x)$ of ODE (4) are the hyperbolic arcs above or below the x-axis. Finally, note that $y = \pm x$ are also solutions of ODE (4).

Ordinary differential equations with homogeneous rate functions of order zero come up in pursuit problems. Let's see how this happens.

Curves of Pursuit

In models of pursuit, the pursuer chases a target (whose motion is known) by using a predetermined strategy, for example, by deliberately aiming toward it. Let's say a ferryboat is set to sail across the river to a dock, but a current complicates the captain's decision making. The captain decides to aim the ferry toward the dock at all times. Will the ferry make it? Although not at all obvious, the problem of finding the path of pursuit where the target is at rest eventually comes down to solving the first-order ODE $dy/dx = f(x, y)$, where $f(x, y)$ is a homogeneous function of order zero. Let's see why this is so by looking at a specific example.

EXAMPLE 1.9.3

A Goose Flies to Its Nest: The Mathematical Model

A goose attempts to fly back to its nest, which is directly west of its position, but a steady wind is blowing from the south. The goose keeps heading to its nest, and the wind blows it off course. What is the goose's flight path? Can the goose get home?

Suppose that the path of the goose is given by the parametric equations $x = x(t)$, $y = y(t)$, where t is time, $(x(t), y(t))$ is the location of the goose at time t, $(x(0), y(0)) = (a, 0)$, and $(x(T), y(T)) = (0, 0)$. Time T is the unknown time of arrival at the nest. Suppose that the bird can fly at a rate of b miles per hour, and that the south wind is blowing at a rate of w miles per hour. The heading angle θ (see Figure 1.9.2) will change as the bird's position changes.

The rate dx/dt is the component of the goose's velocity in the x direction:

$$\frac{dx(t)}{dt} = -b\cos\theta = \frac{-bx}{(x^2 + y^2)^{1/2}} \tag{8}$$

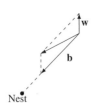

☞ The goose's resultant velocity is the vector sum indicated above.

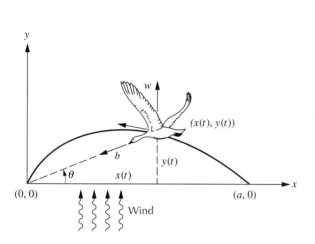

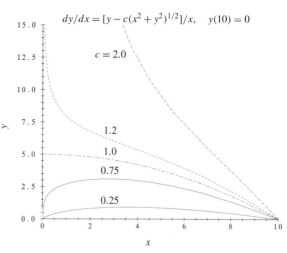

$$dy/dx = [y - c(x^2 + y^2)^{1/2}]/x, \quad y(10) = 0$$

$c = 2.0$

1.2

1.0

0.75

0.25

FIGURE 1.9.2 Flight path of the goose with wind from the south (Example 1.9.3).

FIGURE 1.9.3 Flight path of the goose for various c values (Examples 1.9.3, 1.9.4).

The rate dy/dt is obtained similarly, taking the wind into account:

$$\frac{dy(t)}{dt} = -b\sin\theta + w = \frac{-by}{(x^2 + y^2)^{1/2}} + w \tag{9}$$

The differential system (8), (9) can be treated as we did the combat model system in Section 1.7 because the rate functions don't contain time explicitly.

Dividing ODE (9) by ODE (8), we have a first-order ODE in x and y, where x is now the independent variable. Setting the constant $c = w/b$, we have the IVP

$$\frac{dy}{dx} = \frac{by - w(x^2 + y^2)^{1/2}}{bx} = \frac{y - c(x^2 + y^2)^{1/2}}{x}, \qquad y(a) = 0 \tag{10}$$

The ODE in (10) has a rate function $f(x, y)$ that is homogeneous of order zero, since

$$f(kx, ky) = \frac{ky - c(k^2x^2 + k^2y^2)^{1/2}}{kx} = \frac{y - c(x^2 + y^2)^{1/2}}{x} = f(x, y)$$

So now that we have the model, let's solve ODE (10) and see what happens to the goose.

EXAMPLE 1.9.4

Going Home, or Gone with the Wind?
We set $y = xz$ and from ODE (10) we get a separable ODE

$$\frac{dy}{dx} = x\frac{dz}{dx} + z = \frac{zx - c(x^2 + x^2z^2)^{1/2}}{x} = z - c(1 + z^2)^{1/2}, \quad \text{or}$$

$$x\frac{dz}{dx} = -c(1 + z^2)^{1/2}$$

$$(1 + z^2)^{-1/2}\frac{dz}{dx} = -\frac{c}{x}$$

with solutions defined by

$$\ln[z + (1 + z^2)^{1/2}] = -c \ln x + C$$

where C is a constant of integration. We do not need absolute value signs inside the logarithms since $x > 0$ and $z > 0$. Note that $C = c \ln a$ because $z = y = 0$ if $x = a$. So, $\ln[z + (1 + z^2)^{1/2}] = -c \ln x + c \ln a = \ln(x/a)^{-c}$; exponentiating, we obtain

$$z + (1 + z^2)^{1/2} = \left(\frac{x}{a}\right)^{-c} \tag{11}$$

Write (11) as $(1 + z^2)^{1/2} = (x/a)^{-c} - z$, square, and solve for z:

$$z = \frac{1}{2}\left[\left(\frac{x}{a}\right)^{-c} - \left(\frac{x}{a}\right)^{c}\right]$$

Since $z = y/x$, the equation of the path followed by the goose is

$$y = \frac{a}{2}\left[\left(\frac{x}{a}\right)^{1-c} - \left(\frac{x}{a}\right)^{1+c}\right] \tag{12}$$

☞ Why can't the bird reach its nest if $c > 1$, that is, if $w > b$?

In Figure 1.9.3 this path is plotted for $a = 10$ and several values of $c = w/b$. If the wind's speed is less than the bird's (i.e., if $c < 1$), the bird will reach its nest (solid curves) because the terms in formula (12) have positive exponents and so tend to 0 as $x \to 0^+$. But if $c > 1$, then the exponent of the first term inside the brackets is negative, and so that term blows up as $x \to 0^+$: the goose is gone with the wind.

Preparing for Computation: Scaling the Variables

Sometimes we change variables, not to reduce an ODE to a form where there is a known solution formula, but to reduce the number of symbolic coefficients appearing in the ODE. This is usually done by scaling the variables, that is, replacing y and t in the ODE $y' = f(t, y)$ by $s = t/t_1$, $w = y/y_1$, for suitably chosen constants y_1 and t_1. The example below shows how this is done.

EXAMPLE 1.9.5

☞ This ODE was derived in Section 1.6.

☞ More on the chain rule in Theorem B.5.7 in Appendix B.5.

Scaling the Velocity and Time Variables
The velocity $v(t)$ of a dense body of mass m falling along a vertical line against a Newtonian damping force is the solution of the IVP

$$\frac{dv}{dt} = -g - \frac{k}{m}v|v|, \qquad v(0) = v_0 \tag{13}$$

where g and k are the gravitational and damping constants. Let's show that by scaling time and velocity we can change the number of parameters in IVP (13) from four (g, k, m, and v_0) to just one.

Say that $t = t_1 s$ and $v = v_1 w$ (the positive constants t_1 and v_1 to be determined). Then the chain rule converts IVP (13) to

$$\frac{dv}{dt} = \frac{dv}{dw}\frac{dw}{ds}\frac{ds}{dt} = \frac{v_1}{t_1}\frac{dw}{ds} = -g - \frac{k}{m}v_1 w|v_1 w|, \qquad w(0) = \frac{v(0)}{v_1}$$

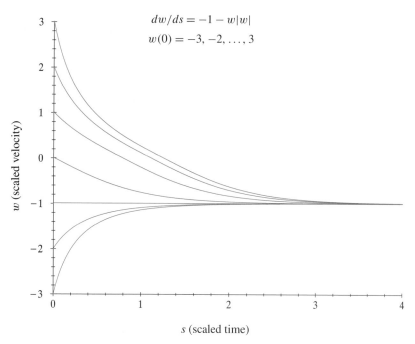

$$\frac{dw}{ds} = -1 - w|w|$$
$$w(0) = -3, -2, \ldots, 3$$

FIGURE 1.9.4 Scaled velocity profiles (Example 1.9.5).

Multiply by t_1/v_1 and replace $|v_1 w|$ by $v_1|w|$ (valid since $v_1 > 0$):

$$\frac{dw}{ds} = -\frac{t_1 g}{v_1} - \frac{t_1 k v_1}{m} w|w|, \qquad w(0) = \frac{v(0)}{v_1} \qquad (14)$$

The IVP (14) looks more complicated than IVP (13), but we haven't yet chosen values for t_1 and v_1. In particular, let's choose the scaling constants t_1 and v_1 so that the coefficients in the ODE have value 1. To find the values of t_1 and v_1 that will do the job, we work backwards:

$$t_1 g / v_1 = 1, \qquad t_1 k v_1 / m = 1, \quad \text{and so}$$
$$t_1 = v_1/g, \qquad k v_1^2/mg = 1, \quad \text{and so}$$
$$v_1 = (mg/k)^{1/2}, \qquad t_1 = (m/kg)^{1/2}$$

With these values of t_1 and v_1, the only parameter left in the scaled IVP

$$\frac{dw}{ds} = -1 - w|w|, \qquad w(0) = w_0 = (k/mg)^{1/2} v_0 \qquad (15)$$

is w_0. Scaling has reduced IVP (13) with four parameters to IVP (15) with only one!

Some solution curves for ODE (15) are plotted in Figure 1.9.4 for various values of w_0. All the solution curves tend to the equilibrium solution $w = -1$ as s increases. This means that for any values of v_0, g, k, and m, the solution of IVP (13) tends toward the limiting value $v = -(mg/k)^{1/2}$.

☞ Appendix B.6 has
more examples of scaling.

There are various reasons for scaling an ODE before solving. Sometimes only combinations of parameters can be measured, not the individual parameters. In Example 1.9.5, rescaling has reduced the set of four original parameters, g, k, m, and v_0 to the single parameter $(k/mg)^{1/2}v_0$. There is another advantage of working with scaled variables: you don't have to worry about units. For example, in Example 1.9.5 the scaled variable s has no units (i.e., it is *dimensionless*). This follows from the fact that $(m/kg)^{1/2}$ has units of time since k has units of mass/distance and g has units of distance/(time)^2. Similarly, w has no units, which means that whether the various quantities in IVP (13) are measured in feet or meters, hours or seconds, grams or slugs, it makes no difference at all if we use IVP (15) for our analysis and our computing.

Comments

Changing variables to simplify an ODE is an art form and has been around for a long time. To see several hundred examples, check out the handbook *Exact Solutions for Ordinary Differential Equations* by Andrei D. Polyanin and Valentin Zaitsev (Boca Raton, Fla.: CRC Press, 1995).

PROBLEMS

1. (*Homogeneous Rate Functions of Order Zero*). For each ODE find a formula that defines solution curves implicitly. [*Hint*: Change variables from y to z by $y = xz$, and solve as in Example 1.9.2.]

    (a) $y' = (y+x)/x$ (b) $(x-y)\,dx + (x-4y)\,dy = 0$

    (c) $(x^2 - xy - y^2)\,dx - xy\,dy = 0$ (d) $(x^2 - 2y^2)\,dx + xy\,dy = 0$

    (e) $x^2y' = 4x^2 + 7xy + 2y^2$

 2. Use the approach in Example 1.7.2 to plot solution curves for parts (a)–(e) of Problem 1. Highlight arcs of orbits that are solution curves.

3. (*From Nonseparable to Separable*). Nonseparable ODEs may become separable by changing a variable. For each of the following cases, demonstrate this process and solve the new ODE. Then find the solution of the original ODE. [*Hint*: Let $z = x + y$ in parts (a), (d); let $z = 2x + y$ in part (b).]

    (a) $dy/dx = \cos(x+y)$ (b) $(2x+y+1)\,dx + (4x+2y+3)\,dy = 0$

    (c) $(x+2y-1)\,dx + 3(x+2y)\,dy = 0$ (d) $e^{-y}(y'+1) = xe^x$

4. Use the substitution $y = z^{1/2}$ to solve the IVP $yy'' + (y')^2 = 1$, $y(0) = 1$, $y'(0) = 0$.

5. (*Alternative Derivation of Solution Formula (15) in Section 1.6*). The logistic ODE given by $y' = r(1 - y/K)y$ can be solved by making the change of dependent variable $z = 1/y$, which transforms the logistic ODE into the linear ODE $z' = -rz + r/K$.

    (a) Show that the ODE for z is as claimed.

    (b) Solve the ODE for z, and then show that $y(t)$ becomes solution formula (15) in Section 1.6 if $y(0) = y_0$ and $y = 1/z$.

6. (*Reduction to Linear Form*). Look at the ODE

$$y'(t) = (a + by)(c(t) + d(t)y)$$

where a and b are constants, $b \neq 0$, and $c(t)$ and $d(t)$ are continuous on some t-interval I.

(a) Show that the variable change $y = (1/z - a)/b$ converts the given ODE into the linear ODE

$$dz/dt = [ad(t) - bc(t)]z - d(t)$$

(b) Find all solutions of the ODE

$$y' = (3 - y)(2t + ty)$$

7. (*Rescaling the Logistic ODE*). Show that the logistic ODE $P'(t) = r(1 - P(t)/K)P(t)$ can be rescaled to $x'(s) = (1 - x(s))x(s)$ if we set $x = P/K$ and $t = s/r$. Why would you want to do this rescaling before using a computer to examine the long term behavior of some logistically changing population?

8. (*Scaling the Whiffle Ball IVP*). The velocity of a vertically moving whiffle ball subject to viscous damping is given by the ODE $v' = -g - (k/m)v$, where g, k, and m are positive constants. Rescale the state and time variables so that the scaled ODE is free of these constants. Solve the scaled ODE and draw a conclusion about the limiting velocity for the original ODE. [*Hint*: Let $t = aT$, $v = bV$; see Example 1.9.5.]

www 9. (*Bernoulli's ODE*). The ODE $dy/dt + p(t)y = q(t)y^b$ is *Bernoulli's ODE*.[10]

(a) Show that the change of variable $z = y^{1-b}$ changes the Bernoulli ODE, where b is a constant, $b \neq 0, 1$, to the linear ODE $dz/dt + (1-b)p(t)z = (1-b)q(t)$.

(b) Show that the logistic ODE $y' = r(1 - y/K)y$ is a Bernoulli ODE with $b = 2$.

(c) Find all solutions of $dy/dt + t^{-1}y = y^{-4}$, $t > 0$. Plot solutions for $t > 0$, $|y| \leq 5$.

(d) Find all solutions of $dy/dt - t^{-1}y = -y^{-1}/2$, $t > 0$. Plot solutions for $t > 0$, $1 \leq y \leq 2$.

10. (*Riccati's ODE*). The Riccati[11] ODE is $dy/dt = a(t)y + b(t)y^2 + F(t)$. If $F(t) = 0$, the ODE is a special case of Bernoulli's ODE (Problem 9). Riccati's ODE may be reduced to a first-order linear ODE if one solution is known [see part **(a)** below]. Parts **(b)**–**(e)** contain examples.

(a) Let $g(t)$ be one solution of Riccati's ODE. Let $z = [y - g]^{-1}$. Show that $dz/dt + (a + 2bg)z = -b$, which is a first-order linear ODE in z. If $z(t)$ is the general solution of the linear ODE, show that the general solution $y(t)$ of the Riccati ODE is $y = g + 1/z$.

(b) Show that the ODE $dy/dt = (1 - 2t)y + ty^2 + t - 1$ has a solution $y = 1$. Let $z = (y - 1)^{-1}$ and show that $dz/dt = -z - t$. Find the general solution $y(t)$ of the original ODE.

(c) Find all solutions of $dy/dt = e^{-t}y^2 + y - e^t$. [*Hint*: First show that $y = e^t$ is a solution.]

(d) Show that $y = t$ is a solution of $dy/dt = t^3(y - t)^2 + yt^{-1}$, $t > 0$, and then find all solutions.

(e) Show that the harvesting model of Problem 9, Section 1.6, is a Riccati ODE and solve it.

---

[10]Jacques Bernoulli (1654–1705) introduced the ODE and his brother Jean (1667–1748) solved it, but Gottfried Leibniz (1646–1717) solved it the same way we do. The two Bernoullis were members of a remarkable Swiss family which produced eight famous mathematicians over a span of four generations. Leibniz was a German mathematician and philosopher who discovered the calculus independently of Newton, but at roughly the same time. The two later quarreled over who should get the credit for the discovery.

[11]The Italian mathematician Jacopo Riccati (1676–1754) discussed particular cases of the ODE now named in his honor, but it was the Bernoulli brothers who actually worked out the solutions.

11.   (*Flight Path of a Goose, Wind from the Southeast*).

(**a**) Set up the flight path problem for the goose, flying at speed b, if the wind is blowing from the southeast at a speed of $w = b/\sqrt{2}$ and the goose starts at $x = a > 0$, $y = 0$. Solve the IVP in implicit form (don't attempt to solve for y in terms of x). [*Hint*: See Example 1.9.3. Note that $x' = -b\cos\theta - b/2$, $y' = -b\sin\theta + b/2$. Use a table of integrals.]

 (**b**) Set $b = 1$, $a = 1, 2, \ldots, 9$ and plot the paths. [*Hint*: Use a numerical solver on the system $x' = \ldots$, $y' = \ldots$ that models the flight path.] Does the goose reach the nest? Does it overshoot?

 12.   (*The Goose and a Moving Nest*). On a windless day a goose sees its gosling aboard a raft in the middle of a river that is moving at 8 yd/sec. When the raft is directly opposite the goose, the raft is 30 yd distant, and the goose instantly takes flight to save her gosling from going over a waterfall 60 yd downstream. If the goose flies directly toward the raft at the constant speed of 10 yd/sec, does she rescue her gosling before it tumbles over the falls? Follow the outline below for this problem.

At $t = 0$ place the raft and the goose in the xy-plane at the origin and at $(30, 0)$, respectively. Let the river flow in the positive y-direction. The parametric path $(x(t), y(t))$ followed by the goose in the xy-plane has the following properties: the goose's velocity vector at time t, $(x'(t), y'(t))$, always points toward the raft, and $((x')^2 + (y')^2)^{1/2} = 10$ yd/sec at all times. So, if the goose is at (x, y) at time t, then the raft is at $(0, 8t)$, and there is a factor $k > 0$ (which may depend on x, y, and t) such that $x' = k(-x)$, $y' = k(8t - y)$. Since $(x')^2 + (y')^2 = 100$, we find that $k = 10/(x^2 + (8t - y)^2)^{1/2}$, and we have the IVP

$$x' = \frac{-10x}{(x^2 + (8t - y)^2)^{1/2}}, \qquad x(0) = 30$$

$$y' = \frac{10(8t - y)}{(x^2 + (8t - y)^2)^{1/2}}, \qquad y(0) = 0$$

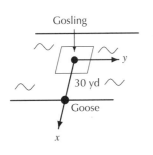

Gosling

30 yd

Goose

x

y

☞ A nonautonomous planar system.

Since the rate functions in this system depend on t, we can't directly apply the technique used in Example 1.9.3. But the system is written in normal form, so a numerical solver can be used to solve and plot an orbit of this system to determine if the goose rescues the gosling. If the goose does reach the gosling in time, how long does it take?

Solution Formula Techniques Involving First-Order ODEs

Explicit Techniques

1. **Linear ODE** $y' + p(t)y = q(t)$	Multiply the ODE through by an integrating factor $e^{P(t)}$, where $P(t) = \int p(t)\,dt$, and then use the Antiderivative Theorem. (See Example 1.3.3 and the procedure on page 19.)
2. **Linear Cascade** $x' = k_1 x + f(t)$ $y' = k_2 x + k_3 y + g(t)$	The coefficients k_1, k_2, and k_3 may depend on t. Solve the first linear ODE for $x(t)$, insert $x(t)$ into second linear ODE, which then can be solved as a linear ODE for $y(t)$. (See Examples 1.8.1, 1.8.2.)

Implicit Techniques

1. **Variables Separate** $N(y)y' + M(x) = 0$	Find antiderivatives $F(x) = \int M(x)\,dx$, $G(y) = \int N(y)\,dy$. The level curves defined by $F(x) + G(y) = C$, for a constant C, are integral curves of the ODE in the xy-plane. An arc of an integral curve with no vertical tangents is a solution curve of the ODE. (See Examples 1.6.1 and 1.6.3. Also see the procedure on page 47.) For another way to plot integral curves see Example 1.7.1.
2. **Exact ODE** $N(x, y)y' + M(x, y) = 0$ $\dfrac{\partial M}{\partial y} = \dfrac{\partial N}{\partial x}$ in a rectangle R	Find a function $F(x, y)$ with $\partial F/\partial x = M$, $\partial F/\partial y = N$ in R; then put $F(x, y) = C$, a constant, and solve for y. Such a function $F(x, y)$ is an integral of the ODE and solution curves can be visualized using integral curves. (See Problem 10, Section 1.6.)
3. **Differential Form of ODE** $M(x, y)\,dx + N(x, y)\,dy = 0$	Convenient way to write as an ODE for $y(x)$, or as an ODE for $x(y)$: $$N\frac{dy}{dx} + M = 0, \quad \text{or} \quad M\frac{dx}{dy} + N = 0$$

System Techniques

1. **Planar Autonomous System** $\dfrac{dx}{dt} = N(x, y), \quad \dfrac{dy}{dt} = -M(x, y)$	Arcs of orbits of the system in the xy-plane with no vertical tangent lines are solution curves of the first-order ODE $dy/dx = -M/N$. (See Example 1.7.5.)
2. **General ODE** $N(x, y)y' + M(x, y) = 0$	Convert the first-order ODE to a planar autonomous system $$\frac{dx}{dt} = N(x, y)$$ $$\frac{dy}{dt} = -M(x, y)$$ whose orbits are composed of solution curves of the ODE. (See Example 1.7.2.)

Change of Variables

1. **Bernoulli Equation** $y' + p(t)y = q(t)y^b$ $b \neq 0, 1$	Change the state variable to $z = y^{1-b}$ to obtain the new ODE $z' + (1-b)pz = (1-b)q$, which is linear in z. (See Problem 9, Section 1.9.)
2. **Riccati Equation** $y' = a(t)y + b(t)y^2 + F(t)$	If one solution $g(t)$ of the Riccati equation is known, then every solution $y(t)$ has the form $y(t) = g(t) + 1/z(t)$, where $z(t)$ solves the linear ODE $z' + (a + 2bg)z = -b$. (See Problem 10, Section 1.9.)
3. The ODE $y' = (a + by)(\alpha(t) + \beta(t)y)$ where a, b are constants, $b \neq 0$	Change the state variable to $z = (a + by)^{-1}$ to obtain the new ODE $z' = [a\beta(t) - b\alpha(t)]z - \beta(t)$, which is linear in z. (See Problem 6, Section 1.9.)
4. **Homogeneous Rate Function of Order Zero** $y' = f(x, y)$, where $f(kx, ky) = f(x, y)$	Change the state variable to $z = y/x$, to get a new ODE, $xz' = f(1, z) - z$, whose variables separate. (See Example 1.9.2.)
5. **ODEs in Polar Coordinates** Write $y' = f(x, y)$ in polar form; look for solutions, $r = r(\theta)$, of the new ODE	Change the variables x and y to polar form using $x = r\cos\theta$, $y = r\sin\theta$. If $r = r(\theta)$ is a solution curve of the ODE, then $r(\theta)$ satisfies the ODE (see The Student Resource Manual, Section 1.9): $$\frac{dr}{d\theta}\sin\theta + r\cos\theta = f(r\cos\theta, r\sin\theta)\left(\frac{dr}{d\theta}\cos\theta - r\sin\theta\right)$$

Reduction to First-Order ODEs

1. **Method of Varied Parameters** $y'' + a(t)y' + b(t)y = f(t)$	Let $z(t)$ be a known solution of $z'' + a(t)z' + b(t)z = 0$. Then every solution $y(t)$ of $y'' + a(t)y' + b(t)y = f(t)$ is given by $y = uz$, where $u(t)$ solves the ODE $zu'' + (2z' + az)u' = f$, which is a linear first-order ODE for $v = u'$. (See Problem 6, Section 1.7.)
2. $y'' = F(t, y')$ F is independent of y	The state variable $v = y'$ solves the first-order ODE $v' = F(t, v)$. The state variable $y = \int v(t)\,dt$. (See Example 1.7.3.)
3. $y'' = F(y, y')$ F is independent of t	Introduce y as a new independent variable and consider the state variable $v = y'$ as a function of y: $v = v(y)$. Since $y'' = (dv/dy)v$, it follows that v solves the first-order ODE $v(dv/dy) = F(y, v)$. Note that $y(t)$ solves the ODE $dy/dt = v(y)$. (See Example 1.7.4.)

Forced Oscillation

$y' + p_0 y = q(t)$ $0 \neq p_0 = $ constant, $q(t)$ is piecewise continuous, periodic with period T	Has a unique periodic solution with period T generated by the initial condition $y(0) = y_0$, given by (see Theorem 2.2.4.) $$y_0 = [e^{p_0 T} - 1]^{-1} \int_0^T e^{p_0 s} q(s)\,ds$$

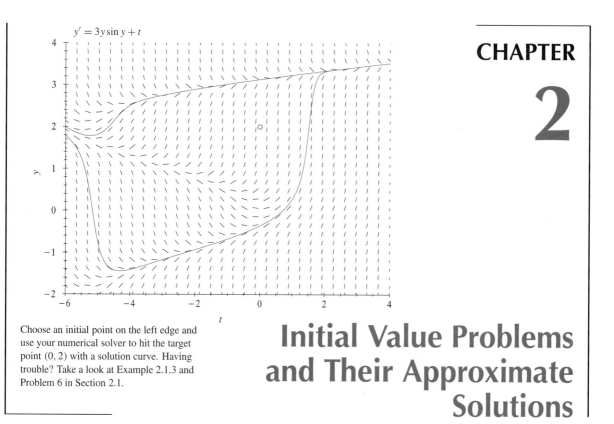

$y' = 3y \sin y + t$

Choose an initial point on the left edge and use your numerical solver to hit the target point $(0, 2)$ with a solution curve. Having trouble? Take a look at Example 2.1.3 and Problem 6 in Section 2.1.

CHAPTER

2

Initial Value Problems and Their Approximate Solutions

The Existence and Uniqueness Theorem in Section 2.1 tells us why natural dynamics match the mathematical description given by an initial value problem so well. Basic questions about solutions of initial value problems, their long-term behavior and sensitivity, and their numerical approximations are answered in this chapter. Chaos tumbles out in the last section and shows some intriguing properties of dynamical systems.

2.1 Existence and Uniqueness

The intimate connection between evolving natural processes and initial value problems is the reason for the central importance of initial value problems. Modelers exploit this connection by using their mathematical knowledge of initial value problems to predict the behavior of natural processes. That is the reason why we need to learn as much as possible about IVPs and the mathematical foundations of this connection. Stated another way, the analysis of a mathematical model amounts to an examination of the associated IVP.

☞ Approximate
numerical solutions and
their graphs to the rescue!

Much of Chapter 1 was devoted to finding solution formulas for the first-order IVP $y' = f(t, y)$, $y(t_0) = y_0$, for different types of rate functions f. But how can we describe a solution and its behavior when no solution formula can be found? How do we even know that the IVP has a solution if we can't construct a formula for it? How do we know that there is only one solution? In this section we start to give answers that do not require solution formulas.

Basic Questions for Initial Value Problems

Let's start by posing the *basic questions* that this chapter is all about. We will refer to the rate function $f(t, y)$ and the initial value y_0 together as *data* for an IVP. In practice, data arise as the result of measurements and observations.

Basic Questions for the IVP $y' = f(t, y)$, $y(t_0) = y_0$

Existence: Under what conditions will the IVP have at least one solution?

Uniqueness: Under what conditions will the IVP have at most one solution?

Extension and Long-Term Behavior: How far ahead into the future and back into the past can a solution be extended? How does a solution behave as t gets large?

Sensitivity: How much does a solution change if the data f, y_0 are changed?

Description: How can a solution be described?

Calculo ergo sum

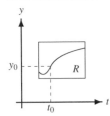

Simple conditions on the rate function f produce satisfactory answers to all the basic questions. Existence and uniqueness are treated in this section, extension and long-term behavior in Section 2.2, and sensitivity in Section 2.3. The answers to the first four questions are summarized by the Fundamental Theorem at the end of Section 2.3. Sections 2.5–2.7 outline one way to answer the fifth question by showing how to describe solutions of IVPs by numerical approximations.

Existence and Uniqueness

The starting point of any general discussion of initial value problems is the question of whether there is anything to discuss, that is, whether an IVP has a solution at all. And if it has a solution, does it have just one? Let's answer both questions.

THEOREM 2.1.1

Existence and Uniqueness Theorem. Suppose that the functions $f(t, y)$ and $\partial f(t, y)/\partial y$ are continuous on a closed rectangle R of the ty-plane and that (t_0, y_0) is a point inside R. Then the IVP

$$y' = f(t, y), \qquad y(t_0) = y_0 \qquad (1)$$

has a solution $y(t)$ on some t-interval I containing t_0 in its interior (*Existence*), but no more than one solution in R on any t-interval containing t_0 (*Uniqueness*).

☞ A full proof of the theorem is in Appendix A.1–A.2.

Here's a sketch of the main idea behind the proof that a solution exists: Integrating each side of the ODE in (1) from t_0 to t and using the initial condition $y(t_0) = y_0$, we find that a function $y(t)$ is a solution of IVP (1) on some interval I containing t_0 if and only if $y(t)$ solves the *integral equation*

$$y(t) - y(t_0) = \int_{t_0}^{t} y'(s)\, ds = \int_{t_0}^{t} f(s, y(s))\, ds$$

which can be written as

$$y(t) = y_0 + \int_{t_0}^{t} f(s, y(s))\, ds, \qquad t \text{ in } I \qquad (2)$$

In the integrals above we denote the variable of integration by s in order not to confuse it with the independent variable t in the upper limit of the integrals.

There is a neat algorithm that homes in on a solution $y(t)$ of equation (2). The first step of the algorithm is to set $y_0(t) = y_0$. Now suppose that the algorithm has generated $y_1(t), \ldots, y_n(t)$. To find the iterate $y_{n+1}(t)$, just insert $y_n(s)$ for $y(s)$ into the right-hand side of (2), integrate, and denote the result by $y_{n+1}(t)$. So we have

$$y_0(t) = y_0$$
$$y_1(t) = y_0 + \int_{t_0}^{t} f(s, y_0)\, ds$$
$$y_2(t) = y_0 + \int_{t_0}^{t} f(s, y_1(s))\, ds$$
$$\vdots$$
$$y_{n+1}(t) = y_0 + \int_{t_0}^{t} f(s, y_n(s))\, ds$$
$$\vdots$$

$$(3)$$

The functions $y_0(t)$, $y_1(t)$, $y_2(t)$, ... are called *Picard iterates*[1] and the process of generating them is the *Picard Iteration Scheme*. The iterates $y_n(t)$ converge to a solution $y(t)$ of integral equation (2) and so also of IVP (1), but the proof of that is tricky and is deferred to Appendix A.2. The proof of uniqueness is given in Appendix A.1.

When you apply the theorem be sure to check that the functions $f(t, y)$ and $\partial f(t, y)/\partial y$ are both continuous on a common rectangle R containing (t_0, y_0). In particularly simple cases, the Picard iterates may be calculated by hand and the exact solution may be identified from the first few iterates, as below. In more complicated cases the integrals in the Picard Scheme (3) can be evaluated numerically and plotted using a general purpose mathematics package like Maple.[2]

Charles Emile Picard

[1]Charles Emile Picard (1856–1941), not to be confused with Jean-Luc Picard, was an eminent French mathematician and the permanent secretary of the Paris Academy of Science. His mathematical work includes deep results in complex analysis, PDEs, and ODEs.

[2]Registered trademark of Maple Waterloo, Inc., Waterloo, Canada.

EXAMPLE 2.1.1

Picard Iterates for a Simple IVP
The solution of the IVP

$$y' = -y, \qquad y(0) = 1$$

is $y(t) = e^{-t}$. Let's apply Picard to the IVP. The first few Picard iterates are

$$y_0(t) = 1$$

$$y_1(t) = 1 - \int_0^t y_0 \, ds = 1 - \int_0^t 1 \, ds = 1 - t$$

$$y_2(t) = 1 - \int_0^t y_1(s) \, ds = 1 - \int_0^t (1 - s) \, ds = 1 - t + t^2/2$$

$$y_3(t) = 1 - \int_0^t y_2(s) \, ds = 1 - \int_0^t (1 - s + s^2/2) \, ds = 1 - t + t^2/2 - t^3/6$$

Soon it dawns on us that each Picard iterate is just a partial sum of the Taylor series

$$1 - t + t^2/2 - t^3/6 + \cdots$$

of the function e^{-t}. Figure 2.1.1 shows the graphs of several iterates over the interval $0 \le t \le 5$, as well as the graph of the actual solution $y = e^{-t}$. Note the alternating signs in the series and the way the graphs of the Picard iterates alternately turn upward and downward away from the graph of the solution. Is there a connection?

Picard iterates cannot always be produced by a symbolic integration because the integrands may be much too complicated. Instead, an interval I is selected and the iterates of the Picard Iteration Scheme (3) are generated by numerical integration over I. Here's an example where we can't find a simple formula even for $y_1(t)$.

EXAMPLE 2.1.2

☞ By "elementary functions" we mean functions like
$t^2 + \sin t$,
$t^3 \cos(3t + 5)$, $e^{t^2} - 5$
and so on.

Picard Iterates for a Complicated IVP
There is no formula in terms of elementary functions for the solution of the IVP

$$y' = (t^2 + y^2 + 1)^{1/2} \sin(ty), \qquad y(0) = 1, \quad 0 \le t \le 2.5$$

So let's set up the Picard Iteration Scheme.

$$y_0(t) = 1$$

$$y_1(t) = 1 + \int_0^t (s^2 + y_0^2(s) + 1)^{1/2} \sin(s y_0(s)) \, ds = 1 + \int_0^t (s^2 + 2)^{1/2} \sin s \, ds$$

$$y_2(t) = 1 + \int_0^t (s^2 + y_1^2(s) + 1)^{1/2} \sin(s y_1(s)) \, ds$$

$$\vdots$$

Figure 2.1.2 shows the graphs of some of the approximate iterates. The iterate $y_{20}(t)$ is so close to the solution $y(t)$ generated by our numerical solver over the interval $0 \le t \le 2.5$ that it isn't possible to separate their graphs.

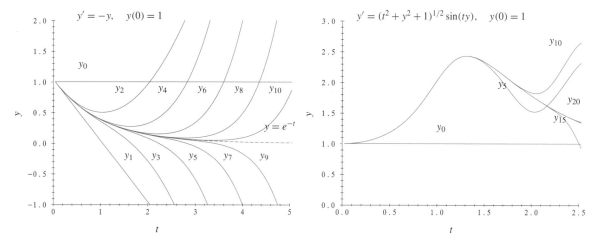

FIGURE 2.1.1 Picard iterates (solid) home in on the true solution (dashed). See Example 2.1.1.

FIGURE 2.1.2 Convergence of some Picard iterates (Example 2.1.2).

The Picard Iteration Scheme is not an efficient way to generate approximate numerical solutions of an IVP because of the need to approximate integrals at every step. In Section 2.5 we introduce practical approximation methods that do not require an integration to be carried out but still converge to the solution (usually).

The Existence and Uniqueness Theorem has an important consequence.

THEOREM 2.1.2

> Solution Curves Don't Meet. If f and $\partial f/\partial y$ are continuous on a rectangle R, then two different solution curves of $y' = f(t, y)$ can't meet in R.

Here's how we verify this result. Say that two distinct solution curves do meet at some point (t_0, y_0) in R. Then the IVP $y' = f(t, y)$, $y(t_0) = y_0$, would have two distinct solutions on some interval I containing t_0, contradicting the uniqueness part of Theorem 2.1.1. So the solution curves can't meet after all.

If the rate function f satisfies the conditions of the Existence and Uniqueness Theorem in the rectangle defined by your computer screen, then any apparent meeting of solution curves on the screen is only due to the finite resolution capability of the device. Let's illustrate Theorems 2.1.1 and 2.1.2.

EXAMPLE 2.1.3

Solution Curves for the Chapter Cover Figure

The function $f = 3y \sin y + t$ is continuous for all values of t and y, as is $\partial f/\partial y = 3 \sin y + 3y \cos y$. So Theorems 2.1.1 and 2.1.2 tell us that through every point (t_0, y_0) of the ty-plane there passes exactly one solution curve of the ODE

$$y' = 3y \sin y + t \tag{4}$$

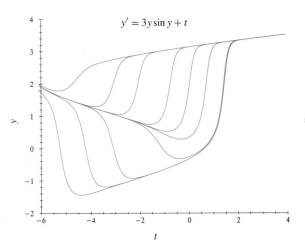

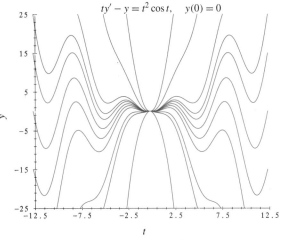

FIGURE 2.1.3 Solution curves of ODE (4) with various initial values (Example 2.1.3). Do the curves really merge?

FIGURE 2.1.4 An IVP with infinitely many solutions (Example 2.1.4). Graph the curves by solving forward or backward from edge points.

Figure 2.1.3 shows several solution curves of ODE (4). The two solution curves starting at $t_0 = -6$, $y_0 = 1.8$, 2.0, appear to enclose a large region on the screen. Notice that solution curves that start close together on the y-axis at $t = -6$ seem to spread apart as t increases. This makes it nearly impossible to use a numerical solver to generate the curves shown inside the region by starting at $t_0 = -6$. To plot these curves we started with initial points inside the region and solved forward and backward in time until the solution curves left the srceen. The solution curves appear to merge, but that is an illusion caused by finite pixel size.

The Exceptional Cases

If the existence and uniqueness conditions aren't satisfied (i.e., $f(t, y)$ or $\partial f(t, y)/\partial y$ fail to be continuous), all bets are off. An initial value problem may have no solution, just one solution, or even infinitely many.

EXAMPLE 2.1.4

An IVP with an Infinite Number of Solutions, Another with None
The linear first-order ODE

$$ty' - y = t^2 \cos t \tag{5}$$

has the normal linear form on the interval $t > 0$ (or $t < 0$)

$$y' - y/t = t \cos t \tag{6}$$

Here, the function $f(t, y) = t \cos t + y/t$ is discontinuous at $t = 0$, so Theorem 2.1.1 can't be applied in any rectangle that intersects the y-axis. Multiply ODE (6) by the integrating factor $e^{-\int (1/t)\,dt} = e^{-\ln t} = 1/t$ to obtain

$$(y' - y/t)/t = (y/t)' = \cos t$$

Antidifferentiation gives $y/t = \sin t + C$, where C is any constant, or

$$y = t \sin t + Ct \tag{7}$$

Formula (7) defines the general solution of ODE (5) for all t (including $t = 0$), as we verify by substitution in the ODE:

$$ty' - y = t(t \sin t + Ct)' - (t \sin t + Ct) = t^2 \cos t$$

All solutions (7) are defined for all t and satisfy $y(0) = 0$. So the IVP

$$ty' - y = t^2 \cos t, \qquad y(0) = 0 \tag{8}$$

has infinitely many solutions (Figure 2.1.4). But if we replace the initial condition, $y(0) = 0$, by $y(0) = 1$, there is no solution at all!

  Sometimes a discontinuity in the rate function doesn't cause any difficulty.

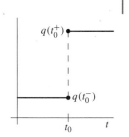

  ❖ **Piecewise Continuity.** A function $q(t)$ is *piecewise continuous* on the interval I, $a \le t \le b$, if the one-sided limits[3] $q(t_0^+)$ and $q(t_0^-)$ exist for all t_0 in I and coincide for all but a finite number of points in I. The function $q(t)$ is piecewise continuous on the real line if it is piecewise continuous on every interval. The discontinuities of these functions are called *jump discontinuities*

A continuous function $q(t)$ is also piecewise continuous, but the one-sided limits $q(t^+)$ and $q(t^-)$ have the same value for every t. The step function is an example of a piecewise continuous function that is not continuous.

☞ Appendix B.1 has more on these and other on-off functions.

  With a slight generalization of the definition of a solution, the linear IVP

$$y' + p(t)y = q(t), \qquad y(t_0) = y_0$$

☞ The response graph in Figure 2.1.5 qualifies as a solution of IVP (9) in this generalized sense.

is still uniquely solvable if $p(t)$ is continuous, but $q(t)$ is only piecewise continuous. We allow as a solution a continuous function $y(t)$ whose derivative $y'(t)$ is piecewise continuous, and $y(t)$ satisfies the ODE at all points of continuity of $q(t)$. The solution formula in Theorem 1.4.1 is still valid in this case. Here's an example.

EXAMPLE 2.1.5

Response to Step-Function and Pulse Inputs
The upper graph in Figure 2.1.5 displays the step input for the IVP

$$y' + y = \text{step}(t - 1) = \begin{cases} 0, & t < 1 \\ 1, & t \ge 1 \end{cases}, \qquad y(0) = 1/2, \quad t \ge 0 \tag{9}$$

To solve IVP (9), use the integrating factor e^t and Theorem 1.4.1 to obtain

$$y(t) = e^{-t}/2 + e^{-t} \int_0^t e^s \, \text{step}(s - 1) \, ds$$

---

[3] The symbols $q(t_0^+)$ [respectively, $q(t_0^-)$] denote the limit of $q(t)$ as t approaches t_0 through values of t greater than [less than] t_0.

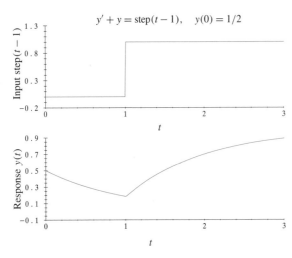

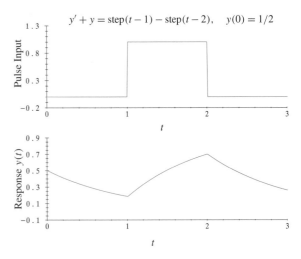

FIGURE 2.1.5 Response $y(t)$ (bottom) to step input (top) (Example 2.1.5).

FIGURE 2.1.6 Response $y(t)$ (bottom) to pulse input (top) (Example 2.1.5).

$$= \begin{cases} e^{-t}/2, & 0 \le t \le 1 \\ e^{-t}/2 + e^{-t} \displaystyle\int_1^t e^s \, ds = e^{-t}/2 + 1 - e^{1-t}, & t \ge 1 \end{cases}$$

where we have used the fact that $\text{step}(t-1)$ only turns on at $t = 1$. Figure 2.1.5 shows how the continuous solution curve suddenly changes its slope and concavity at $t = 1$. Figure 2.1.6 shows the response of the IVP

$$y' + y = \text{step}(t-1) - \text{step}(t-2), \qquad y(0) = 1/2, \quad t \ge 0$$

where the input is the pulse function shown in the upper graph.

So, as this example suggests, IVPs that involve linear ODEs can accept an on-off function as input without much difficulty, but the output curve may have corners.

Comments on the Existence and Uniqueness Theorem

Observation leads us to believe that natural processes are repeatable in the sense that if the process is rerun with exactly the same dynamics and initial conditions, then the outcome will be the same. We want the model IVP for the process to have the same property, and Theorem 2.1.1 gives us conditions under which that is the case. Also, before using a numerical solver to approximate a solution of an IVP, we want to be sure that there is exactly one solution. Otherwise, the solver's output would have little meaning. The solver's approximation may be good or bad, but Theorem 2.1.1 guarantees that there is a unique solution to be approximated.

PROBLEMS

1. (*The Existence and Uniqueness Theorem Applies*). Verify that there exists exactly one solution to each of the following IVPs by checking the hypotheses of the Existence and Uniqueness Theorem. [*Hint*: Don't bother finding a solution formula.]

(a) $y' = e^t y - y^3$, $y(0) = 0$ (b) $y' = |t|y^2 - 1/(3y + t)$, $y(0) = 1$

(c) $y' = |t||y|$, $y(0) = 1$

2. The Existence and Uniqueness Theorem 2.1.1 does not apply to the IVPs below.

(a) Why does Theorem 2.1.1 not apply to $2tyy' = t^2 + y^2$, $y(0) = 1$?

(b) Show that the $y_1 = t^2$ and $y_2 = t^2 \operatorname{step}(t)$ are solutions of IVP $ty' = 2y$, $y(0) = 0$, $-\infty < t < \infty$. Why doesn't this contradict the Theorem 2.1.1?

<u>3.</u> Find all the solutions of the IVP $ty' - y = t^{n+1}$, $y(0) = 0$, for any positive integer n.

4. Show that the IVP $y' = |y|$, $y(0) = 0$, has a unique solution even though the hypotheses of the Existence and Uniqueness Theorem are not satisfied. [*Hint*: First find the general solution for $y \le 0$, and then for $y \ge 0$.]

5. (*Solvability of an IVP*). The IVPs below do not satisfy the conditions of the Existence and Uniqueness Theorem. If the IVP has solutions, find them all. If there are none, why not?

(a) $y' = y \operatorname{step}(t)$, $y(0) = 1$ (b) $2yy' = -1$, $y(1) = 0$ (c) $y' = (1 - y^2)^{1/2}$, $y(0) = 1$

 6. (*Chapter 2 Cover Figure: No Solution Formula*). For the nonlinear ODE $y' = 3y \sin y + t$ of the chapter cover figure there is no known explicit solution formula so we must turn to a direction field or to a numerical ODE solver to see how solution curves behave.

(a) Plot a direction field over the rectangle, $-6 \le t \le 3$, $-2 \le y \le 4$. Sketch some solution curves suggested by the direction field. Redraw your direction field with a finer grid, and see if your curves need any adjustment. Use a numerical solver to check the accuracy of your sketches. Record your observations about the use of direction fields to sketch solution curves.

(b) Take initial points $t_0 = -6$, $1.8 \le y_0 \le 2$, solve forward 10 units of time, and try to hit the target point $(0, 2)$. What's the difficulty? Is there another way to win?

7. (*Discontinuity in the Normalized ODE*). The linear ODE $ty' + y = 2t$ has normal form $y' + y/t = 2$, and $p(t) = 1/t$ is discontinuous at $t = 0$.

(a) Find all the solutions of $ty' + y = 2t$.

(b) Show that the IVP $ty' + y = 2t$, $y(0) = 0$, has exactly one solution, but if $y(0) = y_0 \ne 0$, there is no solution at all. Why doesn't this contradict the Existence and Uniqueness Theorem?

 (c) Plot several solutions of the ODE over the interval $|t| \le 5$.

 8. (*Jump Discontinuity in the Rate Function*). Plot the solution of each IVP on the interval $0 \le t \le 2$. Explain any odd features of the graph in terms of a feature of the IVP.

(a) $y' + 2y = q(t)$, $y(0) = 0$,   where $q(t) = \operatorname{step}(1 - t)$

(b) $y' + p(t)y = 0$, $y(0) = 2$,   where $p(t) = 1 + \operatorname{step}(1 - t)$

www 9. (*Picard Iterates*). Find the Picard iterates $y_0(t)$, $y_1(t), \ldots, y_4(t)$ for the IVP $y' = y$, $y(0) = 1$. Show that $y_4(t)$ consists of the first five terms of the Taylor series for the solution $y = e^t$.

 10. (*Trouble with the Rate Function*). Here are two IVPs with difficulties of their own. What is going on?

(a) Examine the IVP $ty' - 2y = t^3 \operatorname{step}(t)$, $y(1) = 0$. Show that there is exactly one solution defined on the interval $t \ge 0$ but infinitely many on the entire t-axis. Why doesn't this contradict the Existence and Uniqueness Theorem?

(b) Consider the IVP $y' = 3y^{2/3}$, $y(0) = 0$. Find formulas for infinitely many solutions on $-\infty < t < \infty$. Why doesn't the Existence and Uniqueness Theorem apply?

☞ Trouble? Look ahead to problem 4 in section 2.2

2.2   Extension and Long-Term Behavior

The Existence and Uniqueness Theorem is a little vague about the length of the time intervals on which the solutions are defined, but our numerical solver is not bothered by this. We just tell the solver to plot the solution of an IVP over larger and larger time intervals and the resulting solution curve appears to extend itself to these larger and larger time spans until it eventually exits the screen (or the solver crashes). By rescaling our screen we can see what happens to solution curves far into the future. We are particularly interested in seeing if solution curves remain bounded for large t, or if they blow up. A solution is *bounded* on $t \geq 0$ if there is a constant $M \geq 0$ such that $|y(t)| \leq M$, for $t \geq 0$. We can look for cases where graphs approach equilibrium values as t gets large, and other cases where solution curves are attracted to a periodic solution curve. Sometimes, these questions are answered by using a formula for solutions, but it is a remarkable fact that we can often answer questions like these without a solution formula. In this section we lay the foundation for this approach.

Extension of a Solution

Redundancy is built into the definition of a solution of an ODE. If $y(t)$, defined on an interval I, is a solution of an ODE, then $y(t)$ restricted to any subinterval of I is also a solution. It is not useful to distinguish between these solutions since they are all "pieces" of the original solution $y(t)$ on I. But what if there is a solution $z(t)$ on a larger interval that contains I with $y(t) = z(t)$ for t in I? In this case the solution $z(t)$ is said to be an *extension* of the solution $y(t)$.

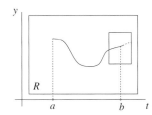

    Can a solution of the ODE $y' = f(t, y)$ be extended? With no solution formula, here's what we can do: Suppose that the rate function $f(t, y)$ and its derivative $\partial f/\partial y$ are continuous on a closed rectangle R in the ty-plane (think of R as the portion of the ty-plane displayed on your computer screen). Now suppose that $y(t)$, $a \leq t \leq b$, defines a solution curve of $y' = f(t, y)$ that lies entirely in R. If the endpoint $(b, y(b))$ is inside R, then the Existence and Uniqueness Theorem (Theorem 2.1.1) implies that the unique solution $z(t)$ to the IVP

$$z' = f(t, z), \qquad z(b) = y(b)$$

extends the solution $y(t)$ to the right of $t = b$. A similar construction extends $y(t)$ to the left of $t = a$ if the point $(a, y(a))$ doesn't lie on the boundary of R. This procedure is *not* recommended for finding extensions by using numerical solvers.

☞ The tips on using solvers in Section 2.6 of the Student Resource Manual have more on this.

    Now we can begin to answer the question of how far a solution can be extended.

THEOREM 2.2.1

☞ Appendix A.3 has a full proof.

> Extension Principle. Suppose that f and $\partial f/\partial y$ are continuous on a closed and bounded rectangle R. If (t_0, y_0) is inside R, then the solution curve of the IVP
>
> $$y' = f(t, y), \qquad y(t_0) = y_0 \tag{1}$$
>
> can be extended backward and forward in time until it meets the boundary of R.

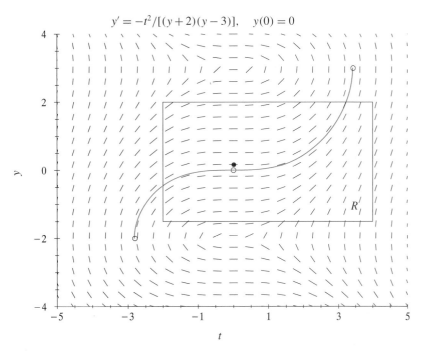

$$y' = -t^2/[(y+2)(y-3)], \quad y(0) = 0$$

FIGURE 2.2.1 Maximally extended solution curve (Example 2.2.1).

In other words, the solution curve of IVP (1) can be extended so far that if it "dies" at all it does so outside of R. We give a name to solutions that have been extended forward and backward in time as far as possible.

❖ **Maximally Extended Solution.** A solution $y(t)$ of an ODE on an interval I is *maximally extended* if it can't be extended to an interval larger than I.

From now on, when we refer to a solution we mean a maximally extended solution.

Direction fields (see Section 1.2) are an important tool in understanding the behavior of solution curves, and, in particular, give us clues about how far a solution curve can be extended. The first example illustrates these ideas.

EXAMPLE 2.2.1

☞ Computer tip: when entering the rate function, don't forget the extra parentheses in the denominator.

☞ Problem 2 has more on this example.

Death of a Solution

We used a numerical solver to plot the solution curve of the IVP

$$y' = -t^2/[(y+2)(y-3)], \qquad y(0) = 0 \qquad (2)$$

by solving forward and backward from $t_0 = 0$ as far as we could go (Figure 2.2.1). Our solver stopped when the solution curve reached the lines $y = -2, 3$. Why did it stop? The denominator of the rate function in IVP (2) is zero if $y = -2$ or 3, so the slope is infinite and the conditions of the Extension Principle fail. The conditions *are* satisfied in the rectangle R shown in Figure 2.2.1, and the solution curve of the IVP extends from edge to edge of R. But the extended solution curve dies outside R when it hits the lines $y = -2$ and $y = 3$. The nearly vertical direction field lines in Figure 2.2.1 near the ends of the curve are visual clues that the solution curve is about to die.

When the rate function $f(t, y)$ in ODE (1) satisfies the conditions of the Extension Principle on the entire ty-plane, you might think that every solution can be extended to the whole real line (i.e., $-\infty < t < \infty$), but you would be wrong. Some innocent-looking ODEs have solutions that blow up (i.e., become large without bound) as t approaches a finite value. When this happens we say that the solution has a *finite escape time*

EXAMPLE 2.2.2

From Here to Infinity in the Twinkling of an Eye
We will find the maximally extended solution of the IVP

$$y' = y^2, \qquad y(0) = 1 \tag{3}$$

Suppose that $y(t)$ is the solution of IVP (3), where $y(t) \neq 0$ on some interval I containing $t = 0$. Separating the variables in the ODE, we have

$$y^{-2}(t)y'(t) = (-y^{-1})' = 1, \quad \text{all } t \text{ in } I \tag{4}$$

Applying the Antiderivative Theorem, we have the general solution of $y' = y^2$:

$$-y^{-1} = t + C, \quad \text{all } t \text{ in } I \tag{5}$$

where C is any constant. Using the initial condition $y(0) = 1$, we see that $C = -1$. Inserting $C = -1$ in (5) and solving for y, we obtain the solution

☞ We say that this solution "escapes to infinity in finite time."

$$y(t) = \frac{1}{1-t}, \quad \text{all } t \text{ in } I \tag{6}$$

From formula (6) we see that the largest interval I on which the solution makes sense is $-\infty < t < 1$. So from its initial point $t_0 = 0$, $y_0 = 1$, the solution escapes to infinity in just 1 unit of time since $y(t) \to +\infty$ as $t \to 1^-$. See Figure 2.2.2.

There is an important class of first-order ODEs where we can say a lot about the extension and long-term behavior of solutions without having a solution formula to fall back on. Let's look at these ODEs now.

Sign Analysis, Equilibrium Solutions, and State Lines

☞ Autonomous ODEs first appeared in Section 1.7.

An ODE is *autonomous* if the rate function does not depend on the independent variable t. Consider the autonomous ODE $y' = f(y)$, and suppose that $f(y_0) = 0$. Then a direction field line segment centered at any point of the horizontal line $y = y_0$ has zero slope. But the line $y = y_0$ also has zero slope everywhere, and it fits a direction field. So the constant function $y = y_0$, for all t, is a solution of the ODE $y' = f(y)$, a so-called *equilibrium solution*. Notice that equilibrium solutions are defined on $-\infty < t < \infty$, so they are maximally extended.

☞ Figures 2.2.2 and 2.2.3 show these bands.

Let's assume now that the rate function $f(y)$ and its derivative $\partial f/\partial y$ are continuous for all y. Then from the Existence and Uniqueness Theorem (Theorem 2.1.1) we see that every point (t_0, y_0) in the ty-plane has a unique solution curve passing through it. The question is, What do the maximally extended solutions look like? The equilibrium solution lines divide the ty-plane into horizontal bandlike regions. We see that each nonequilibrium solution curve is confined to a single band, because otherwise some solution curve must cross an equilibrium line, violating the uniqueness property of Theorem 2.1.1. So every solution curve is trapped and must always remain within a single band, no matter how far the curve is extended. If a band is bounded above and below by equilibrium solution lines, then no solution curve in that band can escape to infinity in finite time. In fact, every solution curve in such a band can be extended to all of $-\infty < t < \infty$ (which follows directly from the Extension Principle). However, the solution curves above the highest equilibrium line (or below the lowest) may escape to infinity in finite time, as we saw in Example 2.2.2.

For an autonomous ODE $y' = f(y)$, the direction field lines centered at points on a horizontal line ($y = $ constant) are all parallel. This means that if a solution curve is shifted to the right or the left along the t-axis, then this shifted curve is also a solution curve because it still fits the direction field. In other words, the shape of a solution curve of an autonomous ODE does not depend on when the clock is started. This is the *translation property* for solution curves of an autonomous ODE.

☞ The *translation property* works only for autonomous ODEs.

EXAMPLE 2.2.3

Translating Solution Curves in Time: Two Examples
In Figure 2.2.2 notice that for the ODE

$$y' = y^2$$

all direction field segments centered on any horizontal line are parallel. This means that the time translation of any solution curve is another solution curve. Lay two rulers horizontally above the equilibrium line $y = 0$ in Figure 2.2.2. The solution curves visible between the rulers are just time translations of one another. As we saw in Example 2.2.2, one of these curves escapes to infinity in finite time, so by the translation property, all solution curves above $y = 0$ escape to infinity in finite time.

The ODE

☞ The factor 0.1 was put in the rate function to produce an attractive graph.

$$y' = 0.1(y-3)(y-1)(y+1)$$

has three equilibrium lines: $y = 3$, $y = 1$, and $y = -1$. So each solution curve of the ODE starting within one of the horizontal bands in the ty-plane created by the equilibrium lines must remain in that band. Notice again that the solution curves in a band all look alike (Figure 2.2.3).

The vertical lines with dots and arrows to the left of Figure 2.2.2 and to the right of Figure 2.2.3 will be explained in the next example.

It is a remarkable fact that by knowing just the algebraic sign of a continuous function $f(y)$ on the y-axis, we can sketch qualitatively accurate solution curves of the ODE $y' = f(y)$ in the ty-plane. This technique, called *sign analysis*, is described below.

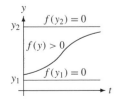

Suppose that $y_1 < y_2$ are two consecutive zeros of $f(y)$, so $f(y_1) = f(y_2) = 0$, but $f(y) \neq 0$ between y_1 and y_2. Since $f(y)$ is continuous, $f(y)$ cannot change sign in the band $y_1 < y < y_2$. The sign can be discovered by evaluating $f(y)$ at any convenient point between y_1 and y_2. Suppose that f is positive in the band, so the solution $y(t)$ that passes through a point (t_0, y_0), $y_1 < y_0 < y_2$, must be increasing as t increases.

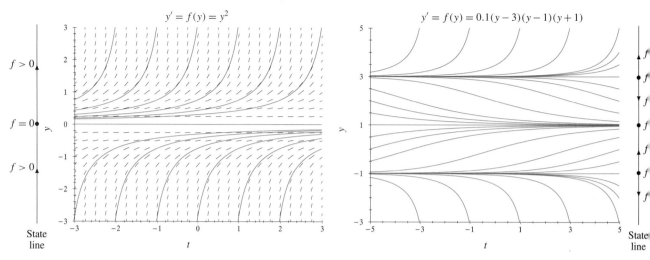

FIGURE 2.2.2 Direction field and solution curves for an autonomous ODE (Examples 2.2.2 and 2.2.3).

FIGURE 2.2.3 Four bands, one stable and two unstable equilibrium solutions (Examples 2.2.3, 2.2.4).

Now we know that the maximally extended solution curve is defined on the entire t-axis and remains in the band. It can be shown that the solution $y(t)$ has the limiting values:

☞ The Student Resource Manual shows why this is true.

$$\lim_{t \to -\infty} y(t) = y_1 \quad \text{and} \quad \lim_{t \to +\infty} y(t) = y_2$$

☞ Sign analysis is used a lot, so learn it well.

A similar result holds if $f(y) < 0$ for $y_1 < y < y_2$. The long-term behavior of all solution curves in the band is determined by the sign of $f(y)$, and that's where the term *sign analysis* comes from.

| **EXAMPLE 2.2.4** | **Sign Analysis and State Lines** |

Let's apply sign analysis to the first ODE of Example 2.2.3:

$$y' = y^2$$

The autonomous rate function $f(y) = y^2$ is zero only at $y = 0$. Since $f(y)$ is positive whenever $y \neq 0$, sign analysis tells us that solution curves in the upper half plane rise upwards and away from $y = 0$ as t increases, while those in the lower half plane rise and approach $y = 0$. We can summarize this behavior in a *state line*, which is a vertical y-line with the equilibrium points identified by dots and with $f(y)$ identified as positive or negative at nonequilibrium points. Take a look at the state line to the left of Figure 2.2.2. The arrowheads on the state line indicate that the corresponding point $(t, y(t))$ on a solution curve moves upward as time advances.

Now let's apply sign analysis and draw the state line for the second ODE

$$y' = 0.1(y - 3)(y - 1)(y + 1)$$

in Example 2.2.3. The rate function $f(y) = 0.1(y - 3)(y - 1)(y + 1)$ is zero at the y-values 3, 1, and -1. The state line for this ODE can be seen to the right of Figure 2.2.3. Sign analysis tells us that as t increases, nonequilibrium solutions move as directed by the corresponding arrow on the state line for each band. The motion is either toward an equilibrium solution (but never reaching it), or toward $+\infty$ or $-\infty$. As a nonequilibrium solution point $y(t)$ moves along its interval on the state line, the point $(t, y(t))$ tracks that motion along the solution curve in the band corresponding to that interval.

As you might guess from the above examples, sign analysis can give us a good idea of the long-term behavior of a bounded solution of an autonomous ODE. Sign analysis is the heart of the proof of the following fundamental result:

THEOREM 2.2.2

☞ The Student Resource Manual has a proof of this fact.

Long-Term Behavior. Suppose that $f(y)$ and df/dy are continuous for all y and that $y(t)$ is a solution of the autonomous ODE

$$y' = f(y)$$

which is bounded for $t \geq 0$ (respectively, $t \leq 0$). Then as $t \to +\infty$ ($t \to -\infty$), $y(t)$ approaches an equilibrium solution of the ODE.

Take another look at Figure 2.2.3 for an illustration of this theorem. Solution curves in the band $-1 < y < 1$ rise toward the equilibrium line $y = 1$ as $t \to +\infty$, and they fall toward the equilibrium line $y = -1$ as $t \to -\infty$.

So it appears that the equilibrium solutions of an autonomous first-order ODE determine limiting values of the past and the future of each bounded solution. We will see in Chapter 9 that equilibrium solutions also play a very important role in analyzing the long-term behavior of bounded solutions of a planar autonomous system of ODEs.

Next, we show that linear first-order ODEs still have a few surprises left.

Long-Term Behavior for Solutions of Linear ODEs: Steady States

What happens to the solutions of a first-order linear ODE as $t \to +\infty$? In the setting we are about to describe, we can say exactly what the long-term behavior is. Let's start with a definition: A constant or periodic solution of a first-order linear ODE is a *steady state* if the solution attracts all other solutions as $t \to +\infty$.

Here is an example of a constant steady-state solution.

EXAMPLE 2.2.5

A Constant Steady State

Using integrating factors, we see that the general solution of the linear ODE

$$y' + y = 1$$

is given by the formula

$$y = Ce^{-t} + 1$$

where C is an arbitrary constant. The equilibrium solution $y = 1$ is a steady-state solution since it attracts all other solutions as $t \to +\infty$ (Figure 2.2.4).

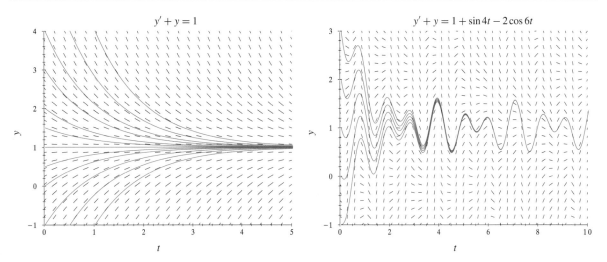

FIGURE 2.2.4 The steady state is the equilibrium solution $y = 1$ (Example 2.2.5).

FIGURE 2.2.5 The steady state is a nonconstant periodic solution (Example 2.2.6).

The above example illustrates the following result about constant steady states.

THEOREM 2.2.3

> Constant Steady State. Suppose that $p_0 > 0$ and q_0 are constants. Then the ODE
>
> $$y' + p_0 y = q_0 \qquad (7)$$
>
> has the unique steady-state solution $y = q_0/p_0$.

☞ We showed this back in Section 1.4.

This result follows from the fact that $y_u = Ce^{-p_0 t}$ is the general solution of the undriven ODE $y' + p_0 y = 0$, while $y_d = q_0/p_0$ is a particular solution of the driven ODE. So the general solution of ODE (7) is $y = y_u + y_d$, which is

$$y = Ce^{-p_0 t} + q_0/p_0$$

from which we see that $y(t) \to q_0/p_0$ as $t \to +\infty$.

Note that if $p_0 < 0$, then q_0/p_0 is still an equilibrium solution, but it is *not* a steady state because it doesn't attract solutions as $t \to +\infty$ (it repels them).

Let's look at an ODE with a periodic input.

EXAMPLE 2.2.6

A Periodic Steady State

According to the solution formulas of Sections 1.3 and 1.4, the general solution of the linear ODE with periodic input $q(t) = 1 + \sin 4t - 2\cos 6t$

$$y' + y = 1 + \sin 4t - 2\cos 6t$$

is given by

$$y = Ce^{-t} + e^{-t} \int_0^t e^s [1 + \sin 4s - 2\cos 6s]\, ds \qquad (8)$$

Although we could have used a table of integrals to express the integral in formula (8) as a sum of sines and cosines, we chose to use a numerical solver to plot approximate solutions. Figure 2.2.5 shows that after a while all solutions look periodic. In fact, the ODE has a unique attracting periodic steady state of period $T = \pi$ (the period of the input).

Example 2.2.6 illustrates a special case of the following general result:

THEOREM 2.2.4

☞ A proof of this result is outlined in Problem 12.

The Periodic Forced Oscillation.   If p_0 is a positive constant and $q(t)$ is a piecewise-continuous periodic function of period T, then the driven linear ODE

$$y' + p_0 y = q(t) \qquad (9)$$

has a unique periodic solution $y_d(t)$ with period T whose initial value is

$$y_d(0) = (e^{p_0 T} - 1)^{-1} \int_0^T q(s) e^{p_0 s} ds \qquad (10)$$

This solution $y_d(t)$ attracts all solutions as $t \to +\infty$, so it is the steady state.

A solution of ODE (9) that has the same period as $q(t)$ is called a *periodic forced oscillation*

Formula (10) for the initial value of the steady-state periodic forced oscillation is often too complicated to be practical, so a numerical solver is usually the best way to home in on the steady state. We did this in Figure 2.2.5. If p_0 is a negative constant, ODE (9) still has a periodic solution whose initial point is given by formula (10), but this solution repels all other solutions as $t \to +\infty$, so it is not a steady state.

The final example of a steady-state periodic forced oscillation illustrates the way a first-order linear ODE responds to a periodic, on-off input function. The example is adapted by permission from an article by A. Felzer, in CODEE, Winter 1993, pp. 7–9.

EXAMPLE 2.2.7

☞ You can follow this example without knowing electrical circuits, but Section 4.4 has an explanation of the terms.

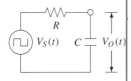

Will the Message Get Through?

A bread-and-butter problem in electrical engineering is to find the frequencies that can be transmitted by a communications channel (such as a coaxial cable) and still be recognizable at the other end. One of the simplest channels is the electrical circuit diagrammed in the margin. The source voltage $V_S(t)$ is the message sent and the output voltage $V_O(t)$ is the message received.

The model linear ODE for the determination of the message received, given the message sent, is the *RC circuit* equation

$$V_O' + \frac{1}{RC} V_O = \frac{1}{RC} V_S$$

The variables R and C are the resistance and the capacitance of the circuit. Suppose that the value of RC is 1 and that V_O initially has value 0. We have the IVP

$$V_O' + V_O = V_S(t), \qquad V_O(0) = 0 \qquad (11)$$

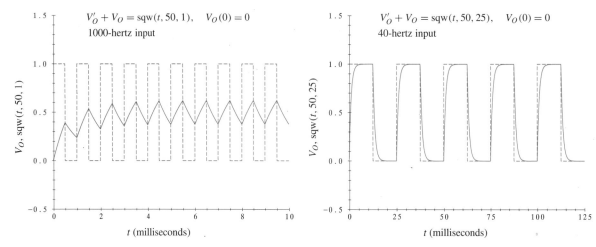

FIGURE 2.2.6 A high-frequency input (dashed) and the distorted response (solid) (Example 2.2.7).

FIGURE 2.2.7 A low-frequency input (dashed) and the response (solid) (Example 2.2.7).

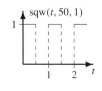

☞ The circuit acts as a *low-pass filter*

Let's send a message $V_S(t)$ of square waves of various frequencies. The challenge is to choose a range of frequencies so that the message received $V_O(t)$ is as close as possible to the message sent.

We set $V_S(t) = \text{sqw}(t, 50, T)$, so the period is T and the wave is "on" for the first 50% of each period. What values of the period T will produce an output message $V_O(t)$ that looks like the message sent? Figures 2.2.6 and 2.2.7 show the message sent (dashed) and the message received (solid) if T is, respectively, 1.0 and 25 milliseconds, corresponding to the respective frequencies of 1000 and 40 *hertz* (i.e., cycles per second). Although the solution formulas of Section 1.4 could be applied, they would involve an awkward integration because of the square wave, so we used a numerical solver. It is the long period (low-frequency) square waves that emerge from the channel almost unchanged. So if we want the message to get through, we should translate it into a combination of low-frequency square waves.

This example shows the Periodic Forced Oscillation Theorem in action. The attraction of the solutions to the periodic steady state solution is clearly visible in each of Figures 2.2.6 and 2.2.7. Note the large change in the steady state response due to the change in the input frequency. This sensitivity of an output to a parameter change is the subject of the next section.

Comments

When an engineer designs a part for a machine, it is of critical importance that the long-term dynamic behavior of the part be well understood. Since behavior is often modeled by a differential equation, what happens to solutions of ODEs over long time spans has become a fundamental research area both in engineering and in mathematics.

PROBLEMS

1. (*Maximally Extended Solutions*). Find a solution formula for the maximally extended solution of each IVP below. Describe the t-interval on which the maximally extended solution is defined. Sketch the graph of this solution. Verify your sketch using a numerical solver.

   (a) $yy' = -t$, $y(0) = 1$ (b) $2y' + y^3 = 0$, $y(0) = -1$

2. (*Extension Principle*). Maximally extended solutions are discussed below.

   (a) (*Solution Formula for Example 2.2.1*). By separating the variables, show that the solution $y = y(t)$ of the IVP

   $$y' = -t^2/[(y+2)(y-3)],   \qquad y(0) = 0$$

   satisfies the equation $2y^3 - 3y^2 - 36y = -2t^3$. Show that $t = -(22)^{1/3}$ if $y = -2$ and that $t = (81/2)^{1/3}$ if $y = 3$. Explain why the solution $y(t)$ can't be extended beyond the interval $-(22)^{1/3} < t < (81/2)^{1/3}$.

   (b) Use a computer solver to plot the maximally extended solution of the IVP

   $$(3y^2 - 4)y' = 3t^2,   \qquad y(0) = 0$$

   Find the t-interval on which this maximally extended solution is defined. [*Hint*: Find the exact t-interval by first integrating each side of the ODE and using the initial data.]

3. (*Finite Escape*). Show that the solution of the IVP $y' = 1 - y^2$, $y(0) = y_0$, where $y_0 < -1$, escapes to infinity in finite time. Find the escape time as a function of y_0.

4. (*Using the Translation Property*). Consider the IVP $y' = 3y^{2/3}$, $y(t_0) = 0$.

   (a) Does the IVP satisfy the conditions of the Existence and Uniqueness Theorem for any value of t_0? Give reasons.

   (b) Show that $y_1 = 0$ and $y_2 = t^3$ are solutions of the IVP $y' = 3y^{2/3}$, $y(0) = 0$.

   (c) Find as many other solutions of the IVP as you can. [*Hint*: Translate $y_2(t)$.]

5. (*Sign Analysis*). Find all equilibrium solutions of the ODEs below and sketch their graphs. Sketch representative solution curves above, below, and between the equilibrium curves. For every bounded solution $y(t)$, find $\lim_{t \to +\infty} y(t)$ and $\lim_{t \to -\infty} y(t)$.

   (a) $y' = (1-y)(y+1)^2$ (b) $y' = \sin(y/2)$ (c) $y' = y(y-1)(y-2)$ (d) $y' = 3y - ye^{y^2}$

6. (*Rise, Fall*). Sketch solution curves of the ODE below.

   (a) Let R be the rectangle $|t| \le 6$, $|y| \le 8$. Imagine R to be filled with the solution curves of the ODE $y' = -y\cos t$. Where in R are the curves rising? Falling?

   (b) Make a rough sketch in R of the solution curves of the ODE. Sketch the nullclines.

www

7. Use direction fields and sign analysis to sketch nullclines and some solution curves for the following ODEs. Verify your sketch by using a numerical solver to plot solution curves.

   (a) $y' = (y+3)(y-2)$ (b) $y' = 2t - y$ (c) $y' = ty - 1$

   (d) $y' = (1-t)y$ (e) $y' + (\sin t)y = t\cos t$ (f) $y' = y - t^2$

☞ Quadratic rate functions can lead to solutions that escape to infinity in finite time.

8. (*Finite Escape*). Suppose that $y(t)$ is a nonzero solution of $y' = f(t, y)$ and that, for some constant $c > 0$, f satisfies the inequality $f(t, y) \ge cy^2$, for all $t \ge 0$ and $y \ge 0$. Show that $y(t)$ escapes to infinity in finite time. [*Hint*: Suppose $y(0) > 0$. Write $y' - cy^2 \ge 0$ as $y^{-2}y' - c \ge 0$ for some interval $0 \le t \le T$. Then $(-y^{-1} - ct)' \ge 0$; what does this say about $-y^{-1} - ct$?]

9. (*Finite Escape Time and Picard Iterates*). Consider the IVP $y' = y^2$, $y(0) = 5$.

   (a) Find the maximally extended solution for the IVP. Describe the behavior of this solution as t approaches the endpoints of the interval where this solution is defined.

   (b) Graph the Picard Iterates $y_0, y_1, \ldots, y_{12}$ for the IVP. Describe these iterates in the light of your answer to part (a). [*Hint*: Use numerical integration software, not symbolic calculation.]

10. (*Will the Message Get Through?*). Plot the solutions of $V_O' + V_O = \text{sqw}(t, 50, T)$, $V_O(0) = 0$, $T = 0.1, 10, 50$ milliseconds. How closely does the message received match the message sent? [*Hint*: See Example 2.2.7.]

11. (*Long-Term Behavior*). Plot several solutions for each of the ODEs. Describe the long-term behavior of the solutions, and give reasons why the behavior is expected.

(a) $y' + y = 3\,\text{sqw}(t, 25, 3)$, $0 \le t \le 20$ (b) $y' + 0.75y = 3\,\text{sqw}(t, 50, 1)$, $0 \le t \le 20$

(c) $y' = 3\,\text{sqw}(t, 50, 2)$, $0 \le t \le 5$ (d) $y' + 0.2y = \text{sqw}(t, 50, 1)$, $0 \le t \le 25$

(e) $y' + y = \sin 3t$, $0 \le t \le 10$ (f) $y' + 0.5y = 2\,\text{trw}(t, 50, 2)$, $0 \le t \le 20$

12. (*Periodic Forced Oscillations*). Complete the first bulleted item below to verify Theorem 2.2.4. Then do the other three items to see what happens in a variety of periodic situations.

More about on-off functions like $\text{sww}(t, d, T)$ in Appendix B.1.

- Show that there is exactly one value y_0 for which the IVP $y' + p_0 y = q(t)$, $y(0) = y_0$, has a periodic solution of period T for any constant $p_0 \ne 0$ and any periodic continuous function $q(t)$ with period T. [*Hint*: Use a solution formula from Section 1.4, and the fact that a continuous function $y(t)$ has period T if and only if $y(t + T) = y(t)$ for all t.]

- Plot the solution of the IVP $y' + 0.5y = \text{sww}(t, 50, 4)$, $y(0) = 2$, over the interval $0 \le t \le 24$. Even though $y_0 = 2$ in this IVP is not the correct value to generate the periodic forced oscillation, one can see the oscillation in the graph of the solution for large t. Why is this? Would this still be true for any nonzero value of the constant p_0? Why?

- Replace p_0 in the IVP with a continuous periodic function $p(t)$ with the same period T as $q(t)$. If $\int_0^T p(s)\,ds \ne 0$, then show that there is a unique periodic forced oscillation of period T. [*Hint*: Proceed as above but using these facts: if $P_0(t) = \int_0^t p(s)\,ds$, then $P_0(t + T) = P_0(t) + P_0(T)$, and $\int_T^{t+T} e^{P_0(s)} q(s)\,ds = e^{P_0(T)} \int_0^t e^{P_0(s)} q(s)\,ds$ for any t-value.]

- Discuss the existence of periodic forced oscillations for the driven ODE $y' + p(t)y = q(t)$, where p and q have a common period T, except that $\int_0^T p(s)\,ds = 0$. Give some examples to illustrate your conclusions.

13. (*S.O.S.*). The abbreviation S.O.S. is the universal code for "Send help!" In International Morse Code, S.O.S. is represented by three short pulses, three long pulses, and then another three short pulses. Use combinations of square waves to code a message of repeated S.O.S. appeals.

More about on-off functions in Appendix B.1.

- Simulate sending your message down the coaxial cable modeled by $V_O' + V_O = V_S(t)$, where $V_S(t)$ is your message.

- Adjust periods (frequencies) so that your message is understood at the receiving end.

- Now use the coaxial cable circuit modeled by $RCV_O' + V_O = V_S(t)$ and choose various values of the positive constant RC. Does your appeal for help get through? What values of RC work best?

When you write your report, be sure to describe what you did and why, and attach graphs of the messages sent and received.

14. (*Sensitivity to Input Frequency*). In Example 2.2.7 we saw that the solution $V_O(t)$ of the IVP $V_O' + V_O = \text{sqw}(t, 50, T)$ is quite sensitive to changes in the value of T. Explore the sensitivity using a numerical solver and other input functions that are periodic of period T. In particular, replace $\text{sqw}(t, 50, T)$ by $\sin(Tt/2\pi)$, $\cos(Tt/2\pi)$, $\text{trw}(t, 10, T)$, and $\sin(Tt/2\pi) + \sin(2Tt/\pi)$.

2.3   Sensitivity

A dynamical system such as an electrical circuit, a chemical reactor, or a space probe must be designed and engineered so that it can absorb small shocks and disturbances and still remain within specified operational tolerances. This design problem can be phrased in the following way:

Controlling a System

A persistent disturbance or a change in the expected initial conditions, even though small in magnitude, may have a cumulative effect that over time will force the output to unacceptably high values. Can this response be predicted and then avoided?

The generic term for the changes in the output of an initial value problem due to changes in the data is *sensitivity*. How to measure sensitivity is the fourth of the basic questions asked at the begining of the chapter.

Most of what we do in this section relates to the sensitivity of first-order linear IVPs, but at the end of the section we extend the results to the general first-order IVP and summarize everything by a fundamental theorem.

Let's begin with a sensitivity study that a numerical solver can handle.

What Happens When a Parameter Changes?

Systems often have parameters that can be adjusted to "tune" the behavior to some desired response. For example, in the last section the frequency of an input signal was adjusted to achieve optimal response in a communications channel (Example 2.2.7). A numerical solver can be put to good use in showing just how the response of a first-order IVP changes as parameters in the rate function are changed.

Most solvers can handle systems, and the first example illustrates how systems can be used to visualize sensitivity. If your solver is not set up to do sensitivity analysis, there is a way to coax it into being more cooperative. The first step is to convert all parameters into state variables and then enlarge the single ODE into a system.

EXAMPLE 2.3.1

Coaxing Your Solver into Doing a Parameter Study
How does the solution of the linear IVP

$$y' = -y \sin t + ct \cos t, \qquad y(-4) = -6 \tag{1}$$

change when the parameter c changes?

ODE solvers like state variables, but some can't handle system parameters because they don't know what to do with them. Let's trick the solver by listing the parameter c as a state variable whose ODE is $c' = 0$. If we choose $c(t_0) = c_0$, then $c(t) = c_0$ for all t. Now let's set the solver to work on the system:

$$
\begin{aligned}
y' &= -y \sin t + ct \cos t, & y(-4) &= -6 \\
c' &= 0, & c(-4) &= c_0
\end{aligned}
\tag{2}
$$

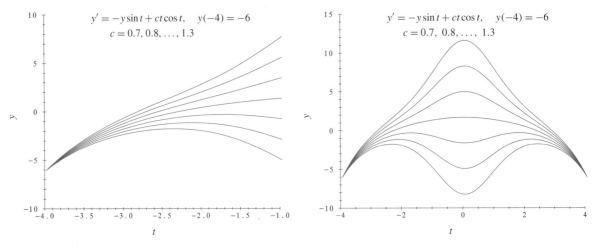

FIGURE 2.3.1 Sensitivity of output to changes in a parameter (Example 2.3.1).

FIGURE 2.3.2 Things look different over a longer time span (Example 2.3.1).

for various values of c_0. The component $y = y(t)$ of the solution pair $y(t)$, $c(t)$ of (2) solves IVP (1) for the value $c = c_0$. If we overlay graphs of $y(t)$ versus t for several values of c_0, we will see how sensitive the solution of IVP (1) is to changes in c_0.

Figure 2.3.1 shows graphs of $y(t)$ for the values $c_0 = 0.7, 0.8, \ldots, 1.3$ (reading the graphs from the bottom up). For $-4 \le t \le -1$, and for c-values in the interval $0.7 \le c \le 1.3$, it appears that the solutions of IVP (1) move away from each other as t increases through the interval $-4 \le t \le -1$. But sensitivity depends a lot on the time interval. Look at what happens over the longer time span, $-4 \le t \le -4$ (Figure 2.3.2); and for a surprise take a look at the solution behavior over the time span $-4 \le t \le 20$ (Problem 1). So sensitivity depends very much on the interval under consideration.

Examples are a good way to introduce sensitivity, but theory is needed to deepen our understanding, so let's develop that theory. The starting point is to estimate the magnitude of the response of a linear IVP.

How Big Is the Response of a Linear IVP?

We will answer this question for the IVP

$$y' + p(t)y = q(t), \qquad y(0) = y_0, \quad t \text{ in } I \tag{3}$$

where throughout this section we assume that $p(t)$ and $q(t)$ are continuous on the interval I, which is either the interval $0 \le t \le T$ for some positive number T, or else the infinite interval $0 \le t < +\infty$. The formula for the solution of IVP (3) was derived in Sections 1.3 and 1.4:

$$y(t) = e^{-P(t)}y_0 + e^{-P(t)} \int_0^t e^{P(s)}q(s)\,ds, \qquad P(t) = \int_0^t p(s)\,ds, \quad t \text{ in } I \tag{4}$$

We can find an upper bound for $|y(t)|$ with only minimal information about the coefficient $p(t)$ and the input $q(t)$.

THEOREM 2.3.1

Bounded Input–Bounded Output (BIBO). Suppose there are positive constants p_0 and M such that the following inequalities hold:

$$p(t) \geq p_0, \qquad |q(t)| \leq M, \qquad \text{all } t \geq 0$$

☞ The proof of this upper bound estimate appears in the Student Resource Manual.

Then the solution $y(t)$ of IVP (3) satisfies the inequality

$$|y(t)| \leq e^{-p_0 t}|y_0| + \frac{M}{|p_0|}|1 - e^{-p_0 t}|, \quad \text{all } t \geq 0 \qquad (5)$$

In particular, $|y(t)|$ is bounded by a constant:

$$|y(t)| \leq |y_0| + \frac{M}{p_0}, \quad \text{all } t \geq 0 \qquad (6)$$

In fact, inequality (5) holds even if p_0 is negative, and there is a similar (but simpler) bound if $p_0 = 0$. Here is how to derive inequality (6) from inequality (5). Since the lower bound p_0 on the value of $p(t)$ is assumed to be positive, we see that the values of $e^{-p_0 t}$ and $1 - e^{-p_0 t}$ are both between 0 and 1 for all $t \geq 0$. This means that

$$e^{-p_0 t}|y_0| \leq |y_0|, \qquad \frac{M}{p_0}(1 - e^{-p_0 t}) \leq \frac{M}{p_0}, \qquad \text{all } t \geq 0 \qquad (7)$$

From (7) and inequality (5), the desired inequality (6) follows.

BIBO says that, under the stated conditions, if the input $q(t)$ is bounded in magnitude for all $t \geq 0$, then $|y(t)|$ can't "blow-up," which is certainly reassuring and is of vital importance in engineering design. Bounds on $|y(t)|$ given in (6) can be used as a design tool, as shown in the next example.

EXAMPLE 2.3.2

How to Keep the Reaction Under Control

A chemical A is converted into a chemical B in a reactor. The reactor is a tank containing a volume V_0 of a solution of A and B. How much of A remains in the tank at time t after the reaction begins?

Let's suppose that $y(t)$ is the *concentration* in grams of A per liter of solution, and that a *first-order rate law* models the reaction $A \rightarrow B$:

$$\frac{dy}{dt} = -ky \qquad (8)$$

where k is a positive rate constant.[4] With the passage of time, the concentration $y(t)$ will decay exponentially from its initial value: $y(t) = y(0)e^{-kt}$.

---

[4]To a chemist, (8) is a *first-order* rate law, while rate laws such as $dy/dt = -ky^2$ are *second-order*

In real life we have to contend with leaky valves, open stopcocks, and so on, that permit small amounts of A to drain into the mixture even while the reaction is taking place. Suppose that $r(t)$ is the unknown rate at which A drips into the reactor and that $V(t)$ is the volume of solution in the tank. Then from the Balance Law,

$$\text{Rate of accumulation} = \text{Rate in} - \text{Rate out}$$

we conclude that

$$\frac{dy}{dt} = \frac{r(t)}{V(t)} - ky$$

where r/V is the rate of increase of the concentration of the chemical A in the tank due to the inflow. We assume that $V(t) > V_0$ for some positive number V_0 (i.e., the tank doesn't run dry). If we estimate that the inflow rate $r(t)$ never exceeds a positive constant r_0, then the BIBO estimate (6) implies that

$$|y(t)| \leq y(0) + \frac{r_0}{kV_0} \tag{9}$$

since $p_0 = k$, $y(0) \geq 0$, and $0 < r(t)/V(t) \leq r_0/V_0$.

Let's suppose that the reactor's design criteria specify that the concentration $y(t)$ in the tank must never exceed a fixed value K. Using inequality (9), we see that operational and design restrictions such as

$$y(0) \leq \frac{1}{2}K, \qquad r_0 \leq \frac{kV_0}{2}K$$

will guarantee that the design criteria are met.

The results show that we can find upper bounds for the magnitude of the response of a linear IVP, given some information (but not much) about $p(t)$ and the input $q(t)$.

Next, we find criteria for keeping the response within an operational region about an "ideal" response, even if the input and initial value are not precisely known.

How to Keep the Response Within Specified Limits

The design specifications for a system require that the response deviate from an ideal response by no more than a specified tolerance. We must design our model so that the uncertainties in the initial value and the input are kept small enough to ensure that the output meets the specifications. But how small is "small enough"?

Let's suppose that the ideal design results in the response $y(t)$, which is the solution of the IVP

$$y' + p(t)y = q(t), \qquad y(0) = a, \quad t \text{ in } I \tag{10}$$

where I is the interval $0 \leq t \leq T$. Operational uncertainties in $q(t)$ and a, however, produce the actual response $Z(t)$, which is the solution of a different IVP:

$$Z' + p(t)Z = m(t), \qquad Z(0) = b, \quad t \text{ in } I \tag{11}$$

where b is the actual initial value and $m(t)$ is the actual input.

The theorem below gives upper bounds on the deviation $|y(t) - Z(t)|$ from the ideal in terms of upper bounds on the data differences $|a - b|$ and $|q(t) - m(t)|$.

THEOREM 2.3.2

Estimating the Deviation. Suppose there are positive constants p_0 and M such that the following inequalities hold:

$$p(t) \geq p_0 \quad \text{and} \quad |q(t) - m(t)| \leq M, \qquad t \text{ in } I$$

If $y(t)$ and $Z(t)$ are the respective solutions of IVPs (10) and (11), then we have the following estimate for $|y(t) - Z(t)|$:

$$|y(t) - Z(t)| \leq e^{-p_0 t}|a - b| + \frac{M}{|p_0|}|1 - e^{-p_0 t}|, \qquad t \text{ in } I \qquad (12)$$

As we can see by subtracting the terms in IVP (11) from the corresponding terms in IVP (10), the function $y(t) - Z(t)$ solves the IVP

$$(y - Z)' + p(t)(y - Z) = q(t) - m(t), \quad (y - Z)(0) = a - b, \quad t \text{ in } I$$

Then inequality (12) follows from inequality (5) with y replaced in (5) by $y - Z$ and y_0 by $a - b$.

The next example shows how to use the estimates on the deviation $|y(t) - Z(t)|$ to meet prescribed tolerances.

EXAMPLE 2.3.3

Design Specifications for a Mixture Model

A salt solution runs into a tank, mixes with the solution in the tank, and the mixture runs out. Under ideal operating conditions, suppose that we have

> *Inflow rate*: 10 gal/min   *Volume of solution in tank*: 100 gal
> *Outflow rate*: 10 gal/min   *Initial amount of salt in tank*: 15 lb
> *Concentration of salt in inflow stream*: 0.2 lb/gal

If $y(t)$ is the amount of salt in the tank at time t, then $y(t)$ solves the "ideal" IVP

☞ Use the Balance Law to get this ODE.

$$y' = 0.2 \cdot 10 - \frac{y}{100} \cdot 10 = 2 - 0.1y, \qquad y(0) = 15 \qquad (13)$$

But suppose that in reality the inflow salt concentration isn't exactly 0.2 lb/gal and that we can't be certain about the initial amount of salt in the tank. So the actual amount $Z(t)$ of salt in the tank satisfies another IVP:

$$Z' = m(t) - 0.1Z, \qquad Z(0) = b \qquad (14)$$

where $m(t)$ and b may be different from the values 2 and 15, respectively. The design specifications require that for a prescribed constant K,

$$|y(t) - Z(t)| \leq K, \quad \text{all } t \geq 0 \qquad (15)$$

Can we meet the specifications?

In order to meet the design criteria we must put operational bounds on the deviations of the inflow concentration and of the initial value from the ideal. The next example shows one way to do this.

Meeting the Design Specifications

Continuing with the problem presented in Example 2.3.3, we see from Theorem 2.3.2 that the deviation $y(t) - Z(t)$ in the output satisfies the inequality

$$|y(t) - Z(t)| \leq e^{-0.1t}|15 - b| + \frac{|2 - m(t)|}{0.1}|1 - e^{-0.1t}|$$

$$\leq |15 - b| + 10|2 - m(t)|, \quad \text{all } t \geq 0$$

because $0 < e^{-0.1t} \leq 1$ for all $t \geq 0$. We can guarantee that $|y(t) - Z(t)|$ is no larger than the allowable deviation K by requiring that

$$|15 - b| + 10|2 - m(t)| \leq K, \quad \text{all } t \geq 0$$

and this can be accomplished by requiring that for some number r, $0 \leq r \leq 1$,

$$|15 - b| \leq rK \quad \text{and} \quad 10|2 - m(t)| \leq (1 - r)K \tag{16}$$

In Figure 2.3.3 we have taken $K = 2$ and $r = 1/2$, so half of the allowable deviation of $K = 2$ is allocated to uncertainty in the initial amount and half to uncertainty in the inflow concentration. The solution curve of the ideal solution $y(t)$ of IVP (13) is the solid curve and the dashed curves $y(t) \pm 2$ bound the "tolerance band." The short dashed solution curve corresponds to the deviating solution $Z(t)$ of IVP (14) where $m(t)$ is taken to be $2 - 0.09\sin(t/3)$ and $b = 14.1$. Note how the solution curve stays in the tolerance band.

Figure 2.3.4 shows the same kind of behavior but with a more stringent bound $K = 1$ and the tighter constraint on the initial data corresponding to $r = 1/10$ in (16) and $1 - r = 9/10$. The short dashed curve is the solution of IVP (14) where $b = 15.09$ and $m(t) = 2 + 0.08\,\text{sqw}(t, 50, 20)$.

These theorems and examples give us a measure of the sensitivity of a linear first-order IVP to data changes. But what about the sensitivity of a general first-order IVP?

Sensitivity and the Fundamental Theorem for First-Order ODEs

Examples 2.3.3 and 2.3.4 illustrate the following property of linear IVPs: Given any tolerance bound for the output, we can impose bounds on the deviation in the data (input and initial value in this case) so that output deviations meet the required tolerances over a given time interval. So small changes in the data can't produce sudden jumps in the output. Sensitivity to data changes of this nature is often called *continuity in the data*.

A similar sensitivity property holds under appropriate conditions for the solutions of the general IVP

$$y' = f(t, y), \quad y(t_0) = y_0 \tag{17}$$

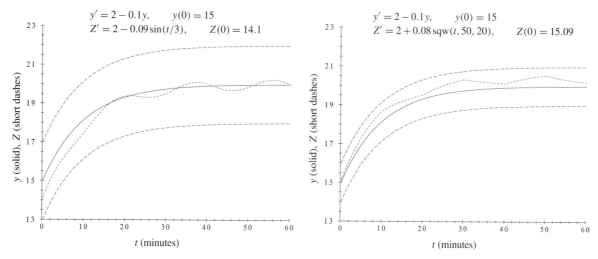

FIGURE 2.3.3 Staying within a wide tolerance band of width 2 (Examples 2.3.3, 2.3.4).

FIGURE 2.3.4 Staying within a tight tolerance band of width 1 (Examples 2.3.3, 2.3.4).

☞ Appendix A.4 takes up the sensitivity/continuity question for IVP (17).

The sensitivity property states that the solution of IVP (17) remains within prescribed bounds over a small enough t-interval containing t_0 if the values of the rate function $f(t, y)$ stay within imposed bounds for all (t, y) in a rectangle R containing (t_0, y_0), and the initial data stay within an imposed y-interval which contains y_0. This kind of sensitivity is also called *continuity in the data*. If the solution of IVP (17) varies continuously in the data, then we have answered the fourth Basic Question about IVPs stated in Section 2.1.

The final theorem summarizes the answers to the four questions described here and in Sections 2.1 and 2.2.

THEOREM 2.3.3

> Fundamental Theorem for First-Order ODEs. Let's look at the IVP
>
> $$y' = f(t, y), \qquad y(t_0) = y_0$$
>
> Suppose that the functions $f(t, y)$ and $\partial f(t, y)/\partial y$ are continuous in a closed rectangle R in the ty-plane, and that the point (t_0, x_0) is inside R.
>
> **Existence, Uniqueness, Extension:** The IVP has exactly one solution $y(t)$ on a t-interval I containing t_0, and I can be extended so that the endpoints of the solution curve lie on the boundary of R.
>
> **Sensitivity/Continuity:** The solution $y(t)$ is continuous in the data f and y_0 for a sufficiently small time interval that contains t_0.

This theorem will be reformulated in later chapters to apply to higher-order ODEs and to systems of first-order ODEs.

Comments

Sensitivity/continuity in the data is a feature of most natural processes, and that is one reason it is included in Theorem 2.3.3. We say that an initial value problem is *well-posed* if it possesses the existence, uniqueness, and sensitivity/continuity properties. Our intuition and experience tell us that many IVPs that arise as models of natural processes should be well-posed so that they can be used to give accurate information about the evolution of the process over time. This is also important in simulating a dynamical process by using a numerical solver since the solver introduces "errors" of its own, that is the solver approximates the rate functions and (often) the initial data in order to construct an approximate solution. The sensitivity/continuity property is essential if we are to have any confidence in the solver's output.

PROBLEMS

 1. (*Example 2.3.1 Continued*). Consider the IVPs $y' = -y \sin t + ct \cos t$, $y(-4) = -6$, $c = 0.7$, $0.8, \ldots, 1.3$. Graph the seven solution curves for $|t| \le 4$, and then for $-4 \le t \le 20$. Make conjectures about the long-term behavior of the solutions if $|c - 1| \le 0.3$.

2. (*Restocking*). A population is replenished from time to time with new stock. Suppose that the rate law is $P'(t) = -r(t)P(t) + R(t)$, where all we know about r and R is that for $t \ge 0$, $0 < r_0 \le r(t)$ and $|R(t)| \le R_0$ for some positive constants r_0 and R_0. If $P(0) = P_0 > 0$, find a reasonable upper bound for $P(t)$. [*Hint*: Use BIBO.]

3. (*Salt Solution*). A vat contains 100 gallons of brine in which initially 5 pounds of salt is dissolved. More brine runs into the vat at a rate of r gallons per minute with a concentration of $c(t)$ pounds of salt per gallon. The solution is thoroughly mixed and runs out at a rate of r gallons per minute. For safety, the concentration in the tank must never exceed 0.1 pound of salt per gallon. Suppose for $t \ge 0$, and some positive constants r_0, r_1, and c_0, that $0 < r_0 \le r \le r_1$ and $0 \le c(t) \le c_0$. What conditions must r_0, r_1, and c_0 satisfy to ensure safe operation?

 4. (*Sensitivity Depends on Parameter Range, Time Span*). Plot the solution curves of the IVPs $y' = -cy/t$, $y(10) = 3$, $0 < t \le 30$, where $c = -1.5, -1.0, -0.5, \ldots, 1.5$. Discuss the sensitivity to changes in c for each of the given ranges for c and for t: $|c + 1| \le 0.5, 0 < t \le 10$; $|c + 1| \le 0.5$, but $10 \le t \le 30$; $|c - 1| \le 0.5$, $0 < t \le 10$; $|c - 1| \le 0.5$, $10 \le t \le 30$.

☞ Don't forget to set bounds for the display of your solution curves.

5. (*Sensitive or Insensitive?*). Solve each IVP below. Is the solution sensitive or insensitive on the given time interval to changes in the initial value a or parameter c? Explain your answers.

(**a**) The solution of $y' = -y + e^{-t}$, $y(0) = a$, is denoted by $y(t, a)$. Find $|y(t, a) - y(t, b)|$ for $t \ge 0$, and answer the sensitivity questions. [*Hint*: First show that $y(t, a) = ae^{-t} + te^{-t}$.]

(**b**) The solution of $y' = -y + c$, $y(0) = 1$, is denoted by $y_c(t)$. It is required that $|c - 5| \le 0.1$. Find $|y_c(t) - y_5(t)|$ for $t \ge 0$, and answer the above questions about sensitivity.

(**c**) The solution of $y' = -y + ct$, $y(0) = a$, is denoted by $y_c(t, a)$. Find $|y_c(t, a) - y_1(t, b)|$ for $0 \le t \le 10$ and answer the above questions about sensitivity.

 6. (*Sensitivity to Parameter Changes*). Plot the respective solutions of $V_O' + V_O = \text{trw}(t, 50, T)$, $V_O(0) = 0$, $T = 0.1, 10, 50$ milliseconds over the respective time intervals $0 \le t \le S$, $S = 1$, $100, 500$. How closely does the message received, $V_0(t)$, match the message sent, $\text{trw}(t, 50, T)$? [*Hint*: See Appendix B.1 for the function trw.]

7. (*Estimate (5) Can't Be Improved*). Show that the inequality

$$|y(t)| \le e^{-p_0 t}|y_0| + \frac{M}{|p_0|}|1 - e^{-p_0 t}|$$

of estimate (5) can't be improved. [*Hint*: Solve the IVP $y' + p_0 y = M$, $y(0) = y_0$, where p_0, M, and y_0 are positive constants.]

www **8.** (*Limited Validity of BIBO*). Show that the hypothesis $p(t) \ge p_0 > 0$ in BIBO can't be extended to the condition $p(t) > 0$. [*Hint*: Show that the problem $y' + (t+1)^{-2}y = e^{1/(t+1)}$, $t \ge 0$, has unbounded solutions even though $p(t) > 0$ and $|q(t)| \le e$ for $t \ge 0$.]

9. (*Limited Validity of BIBO*). Show that the equation $y' + y = t$ has unbounded solutions on the interval $0 \le t$, so showing that the hypothesis $|q(t)| \le M$ can't be dropped from BIBO.

10. (*Failure of BIBO for a Nonlinear ODE*). The following shows how badly things may go wrong in the presence of a nonlinearity even though all solutions of the undriven ODE tend to 0 as $t \to +\infty$ and the driving term is bounded.

    (**a**) Show that $y(t) \to 0$ as $t \to +\infty$ if $y(t)$ solves the ODE $y' = -y/(1+y^2)$.

    (**b**) Show that every solution $y(t)$ of the driven ODE $y' = -y/(1+y^2) + 1$ satisfies the inequality $y'(t) \ge 1/2$, for all $t \ge 0$.[*Hint*: Use calculus to find the minimum value of $-y/(1+y^2) + 1$.]

    (**c**) Integrating the inequality in part (**b**), we have $y(t) \ge t/2 + C$, where C is a constant. Show that if $y(t)$ solves the ODE in part (**b**), then $y(t)$ becomes unbounded as $t \to +\infty$.

    (**d**) Plot some solution curves for the ODEs in parts (**a**) and (**b**) for $0 \le t \le 10$, $|y| \le 5$.

2.4 Introduction to Bifurcations

Equilibrium solutions play a critical role in determining the behavior of all solutions of a first order autonomous ODE, as we saw in the last section. Now let's see what happens when the autonomous rate function depends on a parameter c:

$$y' = f(y, c) \tag{1}$$

☞ For fixed c, the equilibrium solution curves are horizontal lines.

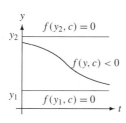

For each value of c the equilibrium solutions of ODE (1) are the zeros of the rate function $f(y, c)$, and the equilibrium solution curves are straight lines that divide the ty-plane into horizontal bands. Inside each band f has a fixed sign, and all solution curves rise or fall with time's advance away from one of the bounding lines and toward the other. If c changes a little, then the bands will widen or narrow a little, but one expects the general appearance of the solution curves to be much as before. This expectation is reasonable, but often wrong.

As we change the value of c, the equilibrium solutions of ODE (1) change. At critical c-values, an equilibrium solution may split (i.e., bifurcate) into several equilibrium solutions, or merge, or even disappear entirely. After a bifurcation occurs, the long-term behavior of nonequilibrium solutions may be drastically altered. In a natural process modeled by ODE (1), the parameter c may be externally controlled by the environment, so natural conditions may trigger a bifurcation and cause the natural process to undergo a change in character. Tracking these changes as the parameter c changes is called *bifurcation analysis*. Here are the steps of the analysis.

Bifurcation Analysis of the ODE $y' = f(y, c)$

The steps of a *bifurcation analysis* of the ODE $y' = f(y, c)$ are:

Part (i): Track the equilibrium solutions as they move, merge, split up, or disappear with changes in c.

Part (ii): Describe the effects of these changes on the long-term behavior of non-equilibrium solutions.

Part (iii): Summarize behavior in a bifurcation diagram.

Many patterns of bifurcation can occur when the parameter c in ODE (1) moves along the c-axis. We present two types in this section: the saddle-node bifurcation and the pitchfork bifurcation.[5]

Harvesting/Restocking a Population: A Saddle-Node Bifurcation

Ocean fishing is under intense scrutiny because it is believed that overfishing has brought stocks of several species of food fish such as cod (in the Atlantic) and salmon (in the Pacific) to dangerously low levels. How can the situation be remedied? Three strategies are currently being tested: lower the allowable limit of fish caught (fishermen don't like this at all), restrict fishing to a fixed season each year (acceptable to most fishermen, but with considerable grumbling), or develop ways to restock the fish population (fishermen like this approach). The decline of the fish populations is not just because of overfishing; pollution of streams, rivers, and the ocean itself is also a major factor. The bifurcation model outlined below is one starting point for thinking about the long-term effects of various harvesting and restocking policies.

A simple model for a logistically changing population undergoing harvesting or restocking is

$$P' = r\left(1 - \frac{P}{K}\right)P + Q, \qquad P(0) = P_0 \tag{2}$$

☞ More on harvesting and logistic populations in Sections 1.1 and 1.6.

where $P(t)$ is the population at time t, and $r > 0$, $K > 0$, $P_0 \geq 0$, and Q are constants. The population is being harvested if Q is negative, restocked if Q is positive. What happens to the population levels if the *harvesting/restocking rate* Q is changed? If this rate goes through a critical value, the nature of the solution curves changes dramatically. At the critical value a *bifurcation* is said to occur. To bifurcate means "to split into two branches," and we will soon see just how closely the term describes what happens to the population.

Keeping track of several parameters is confusing, so the first step in the study of the behavior of solutions of IVP (2) as Q is changed is to get rid of as many of the other parameters as possible. This can be done by scaling the time and the population variables. Let's do this now.

---

[5]The strange name "saddle-node" is explained in Section 9.3, while the choice of the term "pitchfork" becomes clear at the end of this section.

| EXAMPLE 2.4.1 | **Scaling the Variables** |

Suppose that $P = ay$ and $t = bs$, where a and b are positive scaling constants to be determined, and y and s are the new scaled population and time variables. Inserting these changes into IVP (2) and using the Chain Rule, we have the transformed IVP for $y(s)$:

☞ More on scaling in Appendix B.6.

$$\frac{dP}{dt} = \frac{dP}{dy}\frac{dy}{ds}\frac{ds}{dt} = \frac{a}{b}\frac{dy}{ds} = r\left(1 - \frac{a}{K}y\right)ay + Q$$
$$P(0) = P_0 = ay(0) \tag{3}$$

where we have used the fact that $dP/dy = a$ and $ds/dt = d(t/b)/dt = 1/b$. After multiplying the ODE in (3) by b/a, and the initial condition by $1/a$, we have the IVP

$$\frac{dy}{ds} = br\left(1 - \frac{a}{K}y\right)y + \frac{b}{a}Q, \qquad y(0) = \frac{P_0}{a} \tag{4}$$

We can simplify IVP (4) for the scaled population y by setting

$$a = K, \qquad b = \frac{1}{r}, \qquad c = \frac{bQ}{a} = \frac{Q}{rK}, \qquad y_0 = \frac{P_0}{K}$$

to obtain the transformed IVP

$$\frac{dy}{ds} = (1 - y)y + c, \qquad y(0) = y_0 \tag{5}$$

with just the two parameters c and y_0, and not the four parameters r, K, Q, and P_0 of IVP (2). Note that c plays the role of the harvesting/restocking term in the rescaled system.

From here on we carry out a bifurcation analysis for the scaled population problem IVP (5), and our conclusions will carry over to the original ODE. First, let's see how the equilibrium populations depend on the harvesting parameter c.

| EXAMPLE 2.4.2 | **Bifurcation Analysis of IVP (5); Part (i)** |

The zeros of the rate function in IVP (5) are the equilibrium populations. Using the quadratic formula we find that the zeros y_1, y_2 of the rate function $(1 - y)y + c$ are

$$y_1 = \frac{1}{2} - \frac{1}{2}(1 + 4c)^{1/2}, \qquad y_2 = \frac{1}{2} + \frac{1}{2}(1 + 4c)^{1/2} \tag{6}$$

We see from the formulas (6) that

- There are no equilibria if $c < -0.25$.
- A single equilibrium appears at $c = -0.25$.
- There are two equilibria if $c > -0.25$.

So the bifurcation event occurs at $c = -0.25$.

This completes the first part of our bifurcation analysis. Now let's see how the population curves behave as we slide the parameter c upward to $c = 0$.

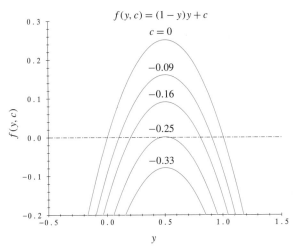

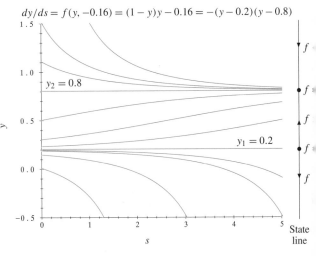

FIGURE 2.4.1 Plots of the rate function $f(y, c)$ for five values of c: saddle-node bifurcation at $c = -0.25$ (Example 2.4.3).

FIGURE 2.4.2 Light fishing ($c = -0.16$), so solution curves above $y_1 = 0.2$ tend to saturation level $y_2 = 0.8$ (Example 2.4.3).

EXAMPLE 2.4.3

Bifurcation Analysis of IVP (5): Part (ii)

Figure 2.4.1 gives a plot of the rate function $f(y) = (1 - y)y + c$ against y for various values of c. From this plot we can read off the properties of the solution curves of $y' = (1 - y)y + c$ as c changes. Extinction always occurs if the harvesting rate $c < -0.25$ since the rate function in this case is always negative. At $c = -0.25$ there is a single equilibrium line $y = 0.5$. Solution curves above the line fall toward it and those below fall away and cross the extinction line $y = 0$. This is a risky scenario for the fish population because a disturbance could force the population below the equilibrium level and then extinction is inevitable.

Let's set $c = -0.16$ (a value above the bifurcation level of $c = -0.25$) and do some harvesting. The equilibrium lines now are $y_1 = 0.2$ and $y_2 = 0.8$ since the rate function $(1 - y)y - 0.16$ factors to $-(y - 0.2)(y - 0.8)$. Population curves above $y_1 = 0.2$ rise toward the saturation equilibrium level $y_2 = 0.8$ but those below $y_1 = 0.2$ fall away and eventually the population becomes extinct (Figure 2.4.2). This behavior is encoded in the state line to the right of Figure 2.4.2.

There isn't any harvesting or restocking if $c = 0$, and the equilibrium lines are $y_1 = 0$ and $y_2 = 1$. Figure 2.4.3 shows these equilibrium lines and other solution curves. The upper line attracts nearby solution curves, but the lower line repels. The solution curves in Figure 2.4.3 below the line $y = 0$ have no physical meaning (negative fish?). The state line to the left of the figure summarizes this behavior.

The sudden appearance of an equilibrium point and its splitting into two as the parameter c crosses a critical value is an example of a *saddle-node bifurcation*. It is one of a class of *tangent bifurcations* called by that name because at the value of c where bifurcation occurs the graph of f in the yf-plane is tangent to the y-axis (see the parabola corresponding to $c = -0.25$ in Figure 2.4.1). Now we will look at a

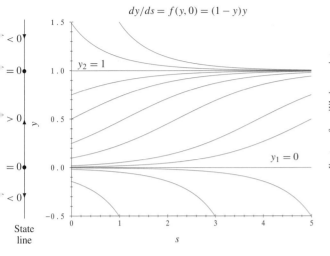

$dy/ds = f(y, 0) = (1 - y)y$

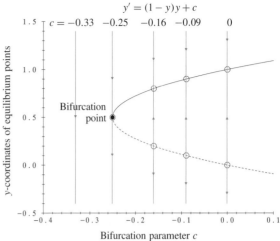

$y' = (1 - y)y + c$

FIGURE 2.4.3 No fishing ($c = 0$) means solution curves inside population quadrant approach saturation level $y = 1$ (Example 2.4.3).

FIGURE 2.4.4 State lines and the saddle-node bifurcation diagram for $dy/ds = (1 - y)y + c$ (Example 2.4.4).

different kind of diagram: a saddle-node bifurcation diagram.

EXAMPLE 2.4.4

Saddle-Node Bifurcation Analysis: Part (iii)

Let's summarize what we've learned so far in a *bifurcation diagram* (Figure 2.4.4). The graphs of the curves of equilibrium points $y_1 = 1/2 - (1/2)(1 + 4c)^{1/2}$, $y_2 = 1/2 + (1/2)(1 + 4c)^{1/2}$ in this diagram follow the convention that solid arcs denote attractors and dashed arcs denote repellers. Each equilibrium point $y_2(c)$ on the solid arc "attracts" all points on the vertical line through y_2 and above the dashed arc. The greater the vertical separation between the solid and the dashed arcs, the larger the "region" of attraction of y_2. The five vertical state lines correspond to five values of c.

This bifurcation diagram tells the whole story of bifurcation, harvesting and logistic change. At the left we see disaster for the fish and in the long run for the fishermen, but in the middle of the diagram we see good times for fish and fishermen. What's the fish story at the far right where c is positive?

Finally, let's go back to the original IVP (2) involving $P(t)$ and the parameters r, K, Q, and P_0. The bifurcation point $c = -0.25$ corresponds to the value $Q = -rK/4$, which defines the *critical harvesting rate* of $rK/4$ units of population per unit of time. The analysis up to this point has been mathematical, but it has profound implications for the harvested species. It implies that if the harvesting rate exceeds the critical value of $rK/4$, then the species is doomed. But if the harvesting rate is subcritical, then the species has a chance of survival.

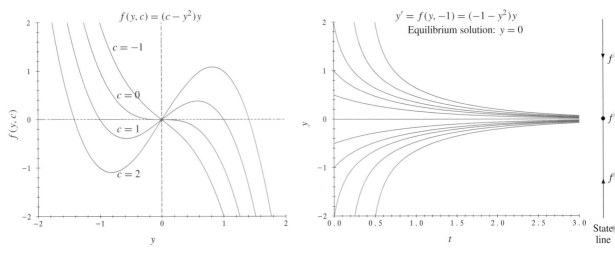

FIGURE 2.4.5 Graphs of $f(y, c)$ for four c-values.   **FIGURE 2.4.6** Before the bifurcation: $c = -1$.

The Pitchfork Bifurcation

Equilibrium solutions can split apart, merge, or disappear in many ways as a parameter changes. One way is the saddle-node bifurcation just discussed. Another way is a *pitchfork bifurcation*, which is another kind of tangent bifurcation. Here's an example.

The nonlinear autonomous ODE

$$y' = (c - y^2)y \tag{7}$$

contains a parameter c. The equilibrium solutions of ODE (7) are given by

$$y_1 = 0, \quad y_2 = c^{1/2}, \quad \text{and} \quad y_3 = -c^{1/2} \tag{8}$$

For negative values of c the only equilibrium solution is y_1. As the value of c increases through 0, the equilibrium $y = 0$ bifurcates into three equilibria: y_1, y_2, y_3.

Figure 2.4.5 shows graphs of the rate function $f(y, c) = (c - y^2)y$ for $c = -1$, 0, 1, 2. For each positive c the graph of f cuts the y-axis at the three points $y_1 = 0$, $y_2 = c^{1/2}$, and $y_3 = -c^{1/2}$. At the bifurcation value $c = 0$ the graph of f is tangent to the y-axis at $y_1 = 0$.

Figures 2.4.6–2.4.7 show equilibrium lines and other solution curves for $c = -1$ (before the bifurcation), and $c = 2$ (beyond the bifurcation). The equilibrium $y = 0$ is an attractor before the bifurcation, but it turns into a repeller at the bifurcation and transfers its attracting character to the two new outlying equilibria $y = y_2$, y_3.

Figure 2.4.8 is the *pitchfork bifurcation diagram* for ODE (7). It shows the equilibrium values as functions of the bifurcation parameter c. The solid arcs correspond to attracting equilibrium solutions, the dashed line to repelling equilibrium solutions. Note, for example, that a point P on the lower parabolic arc $y_3 = -c^{1/2}$ attracts all points on a vertical line through P and below the dashed line $y_1 = 0$. We can see in Figure 2.4.8 the reason for the name "pitchfork bifurcation." We leave it to the reader to draw the vertical state lines in Figure 2.4.8.

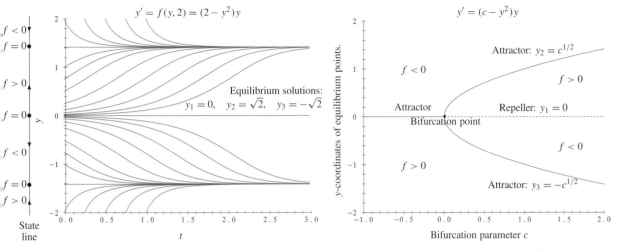

FIGURE 2.4.7 After the bifurcation: $c = 2$. **FIGURE 2.4.8** The pitchfork bifurcation diagram.

Comments

☞ Section 9.3 has examples of bifurcations for a system of autonomous ODEs.

It is no coincidence that at the c-value where bifurcation occurs the graph of $f(y, c)$ in the yf-plane is tangent to the y-axis. This is true for both the saddle-node and the pitchfork bifurcations. In fact, that kind of tangent behavior is usually the clue that a bifurcation of some kind has occurred.

The radical changes in the behavior of solution curves as a parameter goes through a bifurcation value are clearly observable, but only after a rather long span of time. This is because over a short time span solutions change continuously as functions of the data, so small changes in the data mean only small changes in the solutions.

PROBLEMS

1. (*Saddle-Node Bifurcations*). Explain why there is a saddle-node bifurcation at some value of the parameter c. In each case sketch the saddle-node bifurcation diagram, using solid arcs for attracting equilibria and dashed arcs for repelling equilibria. [*Hint*: See the discussion in Examples 2.4.2–2.4.3, and the paragraphs following those examples.]

  (a) $y' = c - y^2$ **(b)** $y' = c - 2y + y^2$ **(c)** $y' = c + 2y + y^2$

 2. (*Saddle-Node Bifurcations*). Plot several solution curves of the ODEs in Problem 1**(a)**–**(c)** for values of c above, at, and below the saddle-node bifurcation value, and describe the long-term behavior of the curves as t increases.

3. (*Pitchfork Bifurcations*). For each ODE below, explain why there is a pitchfork bifurcation at some value of the parameter c. Sketch the pitchfork bifurcation diagram using dashed arcs for repelling equilibria, solid arcs for attracting equilibria.

  (a) $y' = (c - 2y^2)y$ **(b)** $y' = -(c + y^2)y$ **(c)** $y' = (c - y^4)y$

 4. **(a)**–**(c)** (*Pitchfork Bifurcations*). Plot several solution curves of the ODEs in Problem 3 **(a)**–**(c)** for values of c above, at, and below the pitchfork bifurcation value, and describe the long-term behavior of the curves as t increases.

www

☞ Yet another kind of bifurcation.

5. (*Transcritical Bifurcations*). In a *transcritical bifurcation*, as the parameter c in the rate function for the ODE $y' = f(y, c)$ is changed, a pair of equilibrium solutions, one an attractor and the other a repeller, merge, and then separate, exchanging their attracting or repelling properties in the process. That is, one equilibrium solution passes through the other but changes its attracting/repelling character as it does. Explain why each of the following ODEs has a transcritical bifurcation. Draw a bifurcation diagram (dashed arcs for repelling equilibria, solid arcs for attracting equilibria). Then graph several solution curves for the ODEs for values of c below, at, and above the bifurcation values.

   (a) $y' = cy - y^2$ (b) $y' = cy + 10y^2$

6. (*Too Late to Save a Population from Extinction?*). Referring to the text and the discussion of the harvested population model ODE $P' = r(1 - P/K)P - H$, $H > rK/4$, explain why if the value of P is near 0, then restricting the harvest rate to slightly below the bifurcation value of $rK/4$ will not save the population from extinction. What if you ban harvesting altogether in this case? Can you save the species?

7. (*How Many Hunting Licenses Should Be Issued?*). The duck population around a hunting lodge is modeled by the ODE $P' = (1 - P/1000)P - H$, where H is the harvesting rate.

   (a) How many licenses can be issued per year so that the duck population has a chance of survival? Each hunter is allowed to shoot up to 20 ducks per year.

   (b) Suppose N licenses are issued, where N is less than the maximal number found in part (a). What values of the initial duck population lead to total extinction of the species? Explain.

2.5 Approximate Solutions

From the Existence and Uniqueness Theorem 2.1.1 we know that the IVP

$$y' = f(t, y), \qquad y(t_0) = y_0 \tag{1}$$

has a unique solution on an interval containing t_0 if the rate function $f(t, y)$ is well enough behaved. How do we go about describing this solution? The collection of rate functions $f(t, y)$ for which a solution formula for IVP (1) can be found is remarkably small, so the solution formula approach is usually not a realistic option. As we have seen, even when a solution formula can be found, it is not always very informative. Lacking a solution formula, how do we describe the solution?

☞ At last we'll see how numerical solvers do their work.

We present some basic numerical procedures for finding approximate values for the solution $y(t)$ of IVP (1) at a discrete set of times near t_0.

Euler's Method

The direction field approach used in Section 1.2 to characterize solution curves of a first-order ODE suggests techniques for finding approximate numerical solutions for IVP (1). Euler's Method is the simplest of these approximation methods.

Say we wish to approximate the value of the solution of IVP (1) at some future time T. First, partition the interval $t_0 \leq t \leq T$ with N equal *steps* of step size h:

$$h = (T - t_0)/N$$

$$t_n = t_0 + nh, \qquad n = 0, 1, 2, \ldots, N$$

We know that (t_0, y_0) is on the solution curve. To find an approximation to $y(t_1)$, just follow the tangent line to the solution curve through (t_0, y_0) out to t_1. Since the slope of the tangent line to the solution curve at (t_0, y_0) is $f(t_0, y_0)$, we see that

$$y_1 = y_0 + hf(t_0, y_0)$$

is a reasonable approximation to $y(t_1)$ if h is small. Using (t_1, y_1) as a base point and pretending that (t_1, y_1) is on the desired solution curve, we may construct an approximation y_2 to $y(t_2)$ in the same way:

$$y_2 = y_1 + hf(t_1, y_1)$$

Since (t_1, y_1) is most likely not precisely on the desired solution curve, the calculated value y_2 also acquires an error from this source. This calculation can be repeated N times to produce an approximation y_N to the value $y(T)$ of the true solution of IVP (1) at $t = T$. We call this process *Euler's Method*.[6]

❖ **Euler's Method.** For the IVP $y' = f(t, y)$, $y(t_0) = y_0$, the recursive scheme

$$y_n = y_{n-1} + hf(t_{n-1}, y_{n-1}), \qquad t_n = t_{n-1} + h, \quad 1 \le n \le N \qquad (2)$$

is called *Euler's Method* with step size h.

Connecting the points (t_0, y_0), (t_1, y_1), ... , (t_N, y_N) by a broken line produces the *Euler Solution* approximation to the true solution curve of IVP (1). Here are two examples of an Euler Solution.

EXAMPLE 2.5.1

A Simple IVP and an Euler Solution
Figure 2.5.1 illustrates the geometry of Euler's Method with $h = 1$ for the IVP

$$y' = y, \qquad y(0) = 1 \qquad (3)$$

In this case $f(t, y) = y$ so we have

$$y_n = y_{n-1} + hf(t, y_{n-1}) = y_{n-1} + hy_{n-1} = 2y_{n-1}, \qquad n = 0, \dots, N \qquad (4)$$

The numbers y_n produced by Euler's Method consistently underestimate the values of the true solution, $y = e^t$. In fact, we see that $y_1 = 2y_0 = 2$, $y_2 = 2y_1 = 4$, ..., $y_n = 2y_{n-1} = 2^n$, but the true value is $y(n) = e^n \approx (2.71828)^n$, so the error $e^n - 2^n$ grows dramatically as n increases.

Leonhard Euler

[6]Born in Switzerland, Leonhard Euler (1707–1783) was one of the greatest mathematicians of all time. He was also one of the most prolific writers in any field, producing a flood of papers in every area of pure and applied mathematics. His mathematical abilities were immense, leading one physicist to remark that "he calculated without apparent effort, as men breathe, or as eagles sustain themselves in the wind." He was blind the last 17 years of his life, but dictated his seemingly endless flow of new mathematical results until the day he died. The Swiss government is nearing the end of a monumental project to publish all of Euler's work—100 massive volumes so far. Incidentally, his name is pronounced "oiler," not "youler."

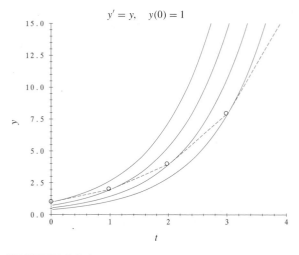

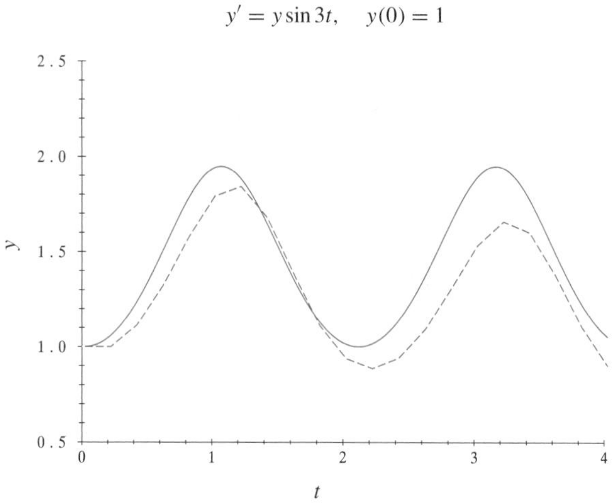

FIGURE 2.5.1 True solution curves (solid) through the Euler points (circles); broken line Euler Solution of IVP (3) with $h = 1$ (Example 2.5.1).

FIGURE 2.5.2 The true solution curve (solid) and an Euler Solution (dashed) with $h = 0.2$ (Example 2.5.2).

The solid curves in Figure 2.5.1 are the true solution curves of the corresponding IVPs,

$$y' = y, \qquad y(t_n) = y_n, \quad n = 0, 1, 2, 3$$

where y_n is the Euler estimate for e^{t_n}. The inaccuracy of the Euler Solution in this case is largely due to the fact that the step size $h = 1$ is much too big.

EXAMPLE 2.5.2

Another IVP and Euler Solution

We seek an approximate solution curve for the IVP

$$y' = y \sin 3t, \qquad y(0) = 1, \quad 0 \le t \le 4 \tag{5}$$

Let's take $h = 0.2$, $N = 20$, so $t_n = (0.2)n$, $n = 0, 1, \ldots, 20$. Then Euler's Method becomes

$$y_n = y_{n-1} + 0.2 y_{n-1} \sin 3t_{n-1}, \qquad n = 1, 2, \ldots, 20, \quad \text{with } y_0 = 1 \tag{6}$$

The linear IVP (5) has the unique solution $y = \exp[(1 - \cos 3t)/3]$. Figure 2.5.2 displays the true solution (solid) and the Euler Solution (dashed). We see again that Euler's Method may provide only a rough approximation to the true solution of an IVP if h is not small.

In spite of the roughness of the Euler approximation, it can be shown that if f and $\partial f / \partial y$ are continuous (as in these examples), then the error can be made as small as desired by taking the step size small enough. Let's see what happens when we take smaller step sizes in Examples 2.5.1 and 2.5.2.

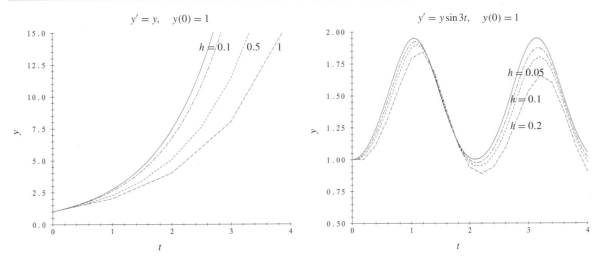

FIGURE 2.5.3 True solution (solid) and Euler solutions (dashed) (Example 2.5.3).

FIGURE 2.5.4 True solution (solid) and Euler solutions (dashed) (Example 2.5.3).

EXAMPLE 2.5.3

Take Smaller Steps

Let's use Euler's Method to approximate solutions of the IVPs in Example 2.5.1 and 2.5.2. Figures 2.5.3 and 2.5.4 show that for these IVPs the smaller the step size the better the Euler approximation.

One-Step Methods

Euler's Method for producing approximate solutions of IVP (1) is an example of a *one-step method*. Such methods produce an approximate value for the solution of IVP (1) at a selected point T in the following way. Suppose that $T > t_0$. Select an increasing sequence $t_1, t_2, \ldots, t_N$ with $t_N = T$ and $t_1 > t_0$, and define the *step size* $h_n = t_n - t_{n-1}$ at step n for $n = 1, 2, \ldots, N$. From a given y_0 and a given function $A(t, y, h)$, a one-step method computes an approximation y_n to $y(t_n)$ using the *discretization scheme*

$$y_n = y_{n-1} + h_n A(t_{n-1}, y_{n-1}, h_n), \qquad n = 1, 2, \ldots, N \qquad (7)$$

To compute y_n, only the value of y_{n-1} is required (so the name "one-step method"). Method (7) uses the given value y_0 to generate y_1, y_1 to generate y_2, and so on, until the process terminates with the calculation of y_N, which is an approximation of $y(T)$.

The function A is called an *approximate slope function* for $y(t)$ at t_{n-1} since

$$\frac{y_n - y_{n-1}}{t_n - t_{n-1}} = \frac{y_n - y_{n-1}}{h_n} = A(t_{n-1}, y_{n-1}, h_n)$$

where the last equality comes from (7). As we will see, the slope function $f(t_{n-1}, y_{n-1})$ used in Euler's Method is not always the best choice for $A(t_{n-1}, y_{n-1}, h_n)$.

The approximation to the solution of IVP (1) by the discrete one-step scheme (7) has a simple interpretation. For each $j = 1, \ldots, N$, join the point (t_{j-1}, y_{j-1}) to (t_j, y_j)

by a line segment, to form a broken-line path from (t_0, y_0) to (t_N, y_N). This path approximates the graph of the solution $y(t)$ (see Figure 2.5.5 for a schematic).

Errors

Let's estimate how much the approximations $y_1, y_2, \ldots, y_N$ generated by a one-step method deviate from the exact values $y(t_1), y(t_2), \ldots, y(t_N)$. If precise arithmetic (i.e., no rounding off or chopping of decimal strings) is used, the deviation

$$E_n = |y(t_n) - y_n|, \qquad n = 1, \ldots, N$$

is the *global discretization error* at the n-th step.

There is a local version of the error due to discretization. By the time we reach the point (t_{n-1}, y_{n-1}) on the broken line path of approximation, the scheme (7) has "forgotten" previously computed results. In the next step all we can hope for is to estimate the difference between y_n as given by (7) and $\tilde{y}(t_n)$, which is the value at t_n of the "true" solution of the IVP

$$\tilde{y}' = f(t, \tilde{y}), \qquad \tilde{y}(t_{n-1}) = y_{n-1}$$

The magnitude

$$e_n = |\tilde{y}(t_n) - y_n|, \qquad n = 1, \ldots, N$$

☞ Software designers use this local version of errors to control global errors because it's easier to estimate local errors than global ones.

is called the *local discretization error* at step n. The global discretization error E_n is due to the local errors $e_1, \ldots, e_{n-1}$ that produce the inexact values $y_1, \ldots, y_{n-1}$. See Figure 2.5.5.

One-step methods may be classified according to the order of magnitude of the global discretization errors incurred when the method is applied.

❖ **Order of a One-Step Method**. Suppose that the solution of the IVP

$$y' = f(t, y), \qquad y(t_0) = y_0, \qquad t_0 \le t \le T$$

is approximated by scheme (7) with fixed step $h = (T - t_0)/N$. If there are positive constants M and p such that for every N,

$$E_N \le Mh^p$$

then method (7) is of *order p*

It is known that Euler's Method is a first-order method. Note that whatever the order of a method, the smaller the step size h the smaller the bound on E_N. It is often easy to adjust the step size when implementing a one-step method on a computer. For a fourth-order method (i.e., $p = 4$), cutting the step size in half results in a 16-fold drop in the upper bound on E_N since $(h/2)^4 = h^4/16$. But halving the step size of a first-order method ($p = 1$) gives only a twofold decrease in the upper bound. This suggests that the higher the order of a method, the more accurately it will approximate the solution of IVP (1). In specific instances this may not hold true since the constant

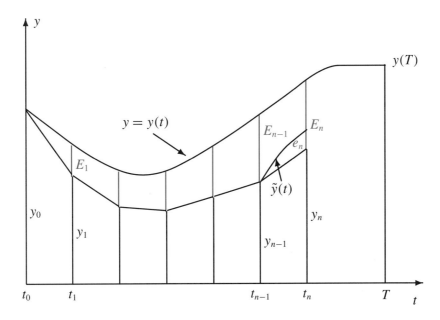

FIGURE 2.5.5 Broken-line approximation to true solution: discretization errors.

M may be larger for a higher-order method than for a lower-order algorithm. Also, higher-order methods usually involve more calculations and function evaluations, and the accompanying round-off errors may undo the advantages of the higher order. Still, the order of a method is a good indication of its accuracy.

Let's describe some other approximation methods.

Heun's Method

Euler's Method may be converted to a second-order method if we compute the approximate slope function by averaging the slopes at t_{n-1} and at t_n. Euler's Method is used to find a first approximation to y_n so that a slope at (t_n, y_n) can be calculated. We have

☞ Also called
Improved Euler's
Method.

❖ **Heun's Method.** *Heun's Method*[7] is the second-order one-step method with constant step size h for the IVP $y' = f(t, y)$, $y(t_0) = y_0$, given by

$$y_n = y_{n-1} + \frac{h}{2}[f(t_{n-1}, y_{n-1}) + f(t_n, y_{n-1} + hf(t_{n-1}, y_{n-1}))] \qquad (8)$$

See Example 2.5.4 and Figure 2.5.6 for an illustration of the method.

[7]Karl Heun (1859–1929) worked in classical mechanics and applied mathematics.

Runge–Kutta Methods

One-step algorithms that use averages of the slope function $f(t, y)$ at two or more points over the interval $[t_{n-1}, t_n]$ to calculate y_n are *Runge–Kutta Methods*.[8] Heun's method is a second-order Runge–Kutta Method. The fourth-order method given below is the most widely used of any of the one-step algorithms. It involves a weighted average of slopes at the midpoint $t_{n-1} + h/2$ and at the endpoints t_{n-1} and t_n.

❖ **Fourth-Order Runge–Kutta Method (RK4).** For the IVP $y' = f(t, y)$, $y(t_0) = y_0$, the *Fourth-Order Runge-Kutta Method* is the one-step method

$$y_n = y_{n-1} + \frac{h}{6}(k_1 + 2k_2 + 2k_3 + k_4) \tag{9}$$

where h is fixed, $t_n = t_{n-1} + h$, and

$$k_1 = f(t_{n-1}, y_{n-1}), \qquad k_2 = f\left(t_{n-1} + \frac{h}{2}, y_{n-1} + \frac{h}{2}k_1\right)$$

$$k_3 = f\left(t_{n-1} + \frac{h}{2}, y_{n-1} + \frac{h}{2}k_2\right), \qquad k_4 = f(t_n, y_{n-1} + hk_3)$$

It may be shown that RK4 generalizes Simpson's Formula for approximating an integral (see Problem 5).

The next example compares the three approximation methods discussed above.

EXAMPLE 2.5.4

Comparison of Numerical Methods

The initial value problem $y' = y$, $y(0) = 1$, has the unique solution $y = e^t$. Approximations that use the Euler, Heun, and RK4 methods are plotted in Figure 2.5.6 on the interval $0 \le t \le 4$. As expected, the higher-order methods give more accuracy than those of lower order. Accuracy improves as the step size is reduced from 0.5 to 0.25.

So far we have only talked about approximating the solution of a single first-order ODE. How can we approximate solutions of a system of several ODEs?

Euler's Method and RK4 for Systems

C. D. T. Runge

Numerical methods for approximating solutions of a first-order ODE can be extended to approximate solutions of a system of first-order ODEs. We extend Euler's Method and RK4 to the IVP

$$\begin{aligned} x' &= f(t, x, y), & x(t_0) &= x_0 \\ y' &= g(t, x, y), & y(t_0) &= y_0 \end{aligned} \tag{10}$$

[8]The German applied mathematican C. D. T. Runge (1856–1927) did notable work in numerical analysis and diophantine equations. M. W. Kutta (1867–1944) was a German applied mathematician and contributed to the early theory of airfoils.

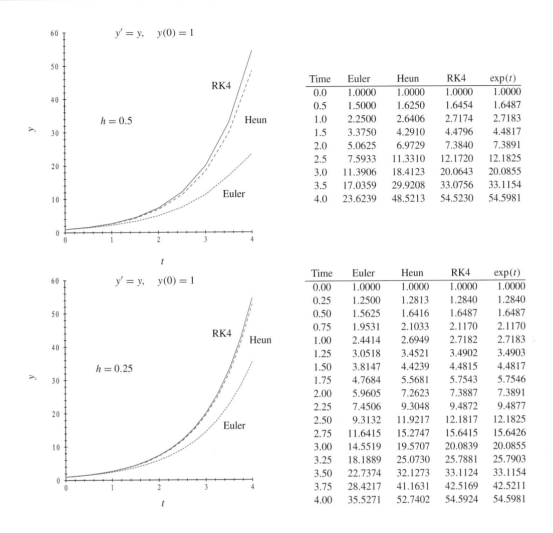

Time	Euler	Heun	RK4	$\exp(t)$
0.0	1.0000	1.0000	1.0000	1.0000
0.5	1.5000	1.6250	1.6454	1.6487
1.0	2.2500	2.6406	2.7174	2.7183
1.5	3.3750	4.2910	4.4796	4.4817
2.0	5.0625	6.9729	7.3840	7.3891
2.5	7.5933	11.3310	12.1720	12.1825
3.0	11.3906	18.4123	20.0643	20.0855
3.5	17.0359	29.9208	33.0756	33.1154
4.0	23.6239	48.5213	54.5230	54.5981

Time	Euler	Heun	RK4	$\exp(t)$
0.00	1.0000	1.0000	1.0000	1.0000
0.25	1.2500	1.2813	1.2840	1.2840
0.50	1.5625	1.6416	1.6487	1.6487
0.75	1.9531	2.1033	2.1170	2.1170
1.00	2.4414	2.6949	2.7182	2.7183
1.25	3.0518	3.4521	3.4902	3.4903
1.50	3.8147	4.4239	4.4815	4.4817
1.75	4.7684	5.5681	5.7543	5.7546
2.00	5.9605	7.2623	7.3887	7.3891
2.25	7.4506	9.3048	9.4872	9.4877
2.50	9.3132	11.9217	12.1817	12.1825
2.75	11.6415	15.2747	15.6415	15.6426
3.00	14.5519	19.5707	20.0839	20.0855
3.25	18.1889	25.0730	25.7881	25.7903
3.50	22.7374	32.1273	33.1124	33.1154
3.75	28.4217	41.1631	42.5169	42.5211
4.00	35.5271	52.7402	54.5924	54.5981

FIGURE 2.5.6 Approximate solutions (Example 2.5.4).

Euler's Method for IVP (10) has the same interpretation as for the case of the single ODE. Calculate $f(t_0, x_0, y_0)$ and $g(t_0, x_0, y_0)$ and move away from the point (x_0, y_0) along the field vector $(f(t_0, x_0, y_0), g(t_0, x_0, y_0))$ of the system in IVP (10) to obtain an Euler approximation (x_1, y_1) to $(x(t_0 + h), y(t_0 + h))$, where h is the fixed step size.

The *Euler algorithm* for IVP (10) is given by

$$t_n = t_{n-1} + h$$
$$x_n = x_{n-1} + hf(t_{n-1}, x_{n-1}, y_{n-1})$$
$$y_n = y_{n-1} + hg(t_{n-1}, x_{n-1}, y_{n-1})$$

As before, this is a first-order algorithm.

The *RK4* algorithm for IVP (10) is given by

$$t_n = t_{n-1} + h$$

$$x_n = x_{n-1} + \frac{h}{6}(k_1 + 2k_2 + 2k_3 + k_4)$$

$$y_n = y_{n-1} + \frac{h}{6}(p_1 + 2p_2 + 2p_3 + p_4)$$

where h is the fixed step size and

$$k_1 = f(t_{n-1}, x_{n-1}, y_{n-1}) \qquad\qquad p_1 = g(t_{n-1}, x_{n-1}, y_{n-1})$$

$$k_2 = f\left(t_{n-1} + \frac{h}{2}, x_{n-1} + \frac{h}{2}k_1, y_{n-1} + \frac{h}{2}p_1\right) \qquad p_2 = g\left(t_{n-1} + \frac{h}{2}, x_{n-1} + \frac{h}{2}k_1, y_{n-1} + \frac{h}{2}p_1\right)$$

$$k_3 = f\left(t_{n-1} + \frac{h}{2}, x_{n-1} + \frac{h}{2}k_2, y_{n-1} + \frac{h}{2}p_2\right) \qquad p_3 = g\left(t_{n-1} + \frac{h}{2}, x_{n-1} + \frac{h}{2}k_2, y_{n-1} + \frac{h}{2}p_2\right)$$

$$k_4 = f(t_{n-1} + h, x_{n-1} + hk_3, y_{n-1} + hp_3) \qquad p_4 = g(t_{n-1} + h, x_{n-1} + hk_3, y_{n-1} + hp_3)$$

As the abbreviation suggests, RK4 is a fourth-order method. Euler's Method and RK4 can be extended to systems with any number of first-order ODEs.

Comments

There are many methods for finding approximate numerical solutions of IVPs.[9] Euler's Method has the virtue of being simple to visualize and implement, but it is not the most practical choice for numerical solvers. RK4 provides a combination of accuracy and efficiency that makes it an excellent choice for a one-step method.

Many commercial solvers use multistep methods, the approximate slope function at each step being computed from the slopes at several of the previously computed solution points. Some solvers are adaptive, adjusting the step size and the numerical method automatically to meet the immediate computational needs. Many solvers will allow the user to select a particular method from a list that includes Euler's Method, RK4, and an adaptive method. The numerical solver we use is based on LSODA, an adaptive multistep method whose origins go back to C. W. Gear's DIFFSUB and ODE-PACK developed by Alan Hindmarsh at Lawrence Livermore National Laboratories.

PROBLEMS

www   **1.** Use the indicated method to estimate $y(1)$ if $y' = -y$, $y(0) = 1$, $h = 0.1$. In **(a)** and **(c)** plot the approximating polygons.

     (a) Euler's Method   **(b)** Heun's Method   **(c)** RK4

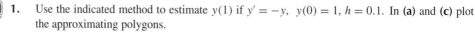

---

[9]For a highly readable source on approximation methods for IVPs, see the book by the noted American numerical analyst, L. F. Shampine, *Numerical Solution of Ordinary Differential Equations* (New York: Chapman & Hall, 1994).

2. *(Comparison of Euler and RK4).* Use the indicated method to estimate $y(1)$ if $y' = -y$, $y(0) = 1$, $h = 0.01, 0.001$, and 0.0001, and plot the approximating polygons.

   (a) Euler's Method   **(b)** RK4

  3. *(Comparison of Euler and RK4).* Use the indicated method to estimate $y(1)$ if $y' = -y^3 + t^2$, $y(0) = 0$, $h = 0.1, 0.01, 0.001$, and plot the approximating polygons.

   (a) Euler's Method.   **(b)** RK4.

4. Show that for each of the following IVPs, $y_N(T) \to y(T)$ as $N \to \infty$, where $T > 0$ is fixed, $h = T/N$, and $\{y_N(T) : N = 1, 2, \ldots\}$ is the sequence of Euler approximations to $y(T)$.

   (a) $y' = 2y$, $y(0) = 1$ **(b)** $y' = -y$, $y(0) = 1$

5. Simpson's Formula for approximating the integral $\int_a^b f(t) \, dt$ is

$$\frac{b-a}{6}\left[f(a) + 4f\left(\frac{a+b}{2}\right) + f(b)\right]$$

Show that RK4 for the IVP $y' = f(t)$, $y(t_0) = y_0$, gives Simpson's Formula at each step.

2.6 Computer Implementation

Although the calculations needed to implement a numerical algorithm for solving ODEs can be carried out by hand, that is hardly an efficient way to do it. Numerical solver packages are preferable. These packages often contain sophisticated ODE solvers, based on the Runge–Kutta algorithms of Section 2.5 and multistep methods that use the computed approximations from several steps before the current step. These solvers have many automatic features for error control. The solution graphs in this text can be reproduced reasonably well by most of these solvers.

Using Solvers

In using a numerical solver for the IVP

$$y' = f(t, y), \qquad y(t_0) = y_0$$

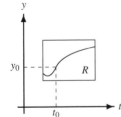

we need to input the function $f(t, y)$, the initial data t_0 and y_0, and a solution time interval. Our solver also asks us for the dimensions of a rectangle R in the ty-plane containing the point (t_0, y_0). Many solvers do not require a step size to be specified, nor the dimensions of the rectangle R. Instead, the user specifies an error tolerance and the solver calculates a step size adaptively at each step that will achieve this tolerance. The rectangle R might be constructed automatically to fit the graph after the solution process has finished. To prevent the use of too much processing time, the user may be able to specify a minimum allowable step size or a maximum number of function evaluations. If accuracy requirements are too high, these constraints may be exceeded and an error message returned.

   In addition to the actual numerical solution, there may be output available from the solver that provides important information that could suggest changing some solver parameter. This information may include the actual step size used, the number of function evaluations, the number of interval steps used in one call to the solver, or an estimate of the local error in the solution accumulated during a call. Looking at these outputs may suggest that error tolerances be raised or lowered if the actual errors differ significantly from worst-case estimates, or that a different technique be implemented that uses a different number of function evaluations.

Should You Believe Your Solver?

Most of the commercial numerical solvers produce quite accurate approximations for most IVPs. But even a sophisticated numerical solver delivers approximate, not exact, values. The user needs to be aware of how things can sometimes go quite wrong.

Nonexact Arithmetic

A major source of errors in any step-by-step process arises because precise arithmetic and precise evaluation of functions is not possible in general. Every computation made by a machine (or a human) chops long decimal strings, for example the fraction 1/3 might be carried as 0.33333, π as 3.14159, and e might be evaluated as 2.71828. These are examples of *round-off errors*, individually small but often devastating in their cumulative effect. The *total error* of a computation includes the errors due to the approximation formula being used and the errors due to round-off. For example, the values y_1, y_2, ... generated in a one-step method can't be computed with infinite precision, so the total error must be viewed as having a component due to round-off error as well as to the choice of the particular approximate slope function used in the discretization process (discussed in the last section). Round-off has the nasty feature that the more operations that are to be performed, the more likely it is for the error to grow as it propagates through the scheme.

So if we want to compute an approximate value y^* for the solution $y(t)$ of an IVP at $t = t^*$, we are faced with the following trade-off. By taking the step size very small, we can make the discretization error at t^* as small as desired. But since we have many more operations to perform in order to approximate $y(t^*)$, this will frequently be at the expense of a buildup of round-off errors. So, it is not easy to say just what step size will minimize the total error.

Problems in Choice of Step Size

Too small a step size not only makes the computer work too hard, but it might also lead to large round-off errors. But as the next example shows, too large a step size may lead to sizable global errors that obliterate features of the solution.

EXAMPLE 2.6.1 **A Misleading Euler Solution for the Logistic ODE**
Let's approximate the solution of the logistic IVP

$$y' = 9(1 - y)y, \qquad y(0) = 1.25 \qquad (1)$$

Euler's Method for IVP (1) generates the sequence of values y_1, y_2, ... given by

$$y_n = y_{n-1} + 9h(1 - y_{n-1})y_{n-1}, \qquad n = 1, 2, \ldots \qquad (2)$$

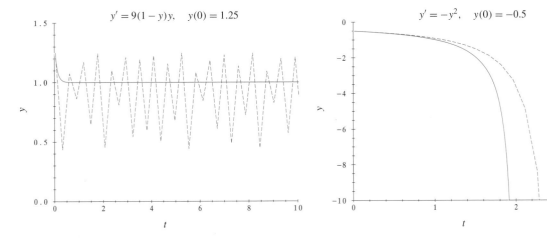

FIGURE 2.6.1 Euler Solution (dashed), true solution (solid): $h = 0.29$ (Example 2.6.1).

FIGURE 2.6.2 Euler Solution (dashed), true solution (solid): $h = 0.15$ (Example 2.6.2).

where h is the step size. Figure 2.6.1 shows the Euler Solution (dashed) for IVP (1) when $h = 0.29$, and the (solid) true solution (or rather the very accurate numerical solution produced by our solver LSODA). If we didn't already know from Examples 1.6.5 and 1.6.6 that the solution of IVP (1) tends to $y = 1$ as $t \to +\infty$, we would certainly never predict that behavior from the Euler Solution.

In the next section we will have more to say about the apparently chaotic behavior of some Euler approximations to the solution of a logistic IVP.

For an initial value problem whose solution is not well behaved, the discretization process may introduce some errors that should make the modeler wary. For example, when the solution escapes to infinity in finite time, the basic methods considered so far may not be able to detect this fact.

EXAMPLE 2.6.2

Tracking Solutions That Escape to Infinity in Finite Time
Let's use Euler's Method with step size $h = 0.15$ to solve the IVP

$$y' = -y^2, \qquad y(0) = -0.5 \tag{3}$$

The result is the dashed curve in Figure 2.6.2. Since the ODE seems to be well behaved, we are inclined to accept this as a reasonably accurate approximate solution of IVP (3). But the ODE $y' = -y^2$ is separable and can be solved exactly. The maximally extended solution of IVP (3) is

$$y(t) = 1/(t-2), \qquad t < 2$$

☞ Example 2.2.2 has more on finite escape time.

which escapes to $-\infty$ as $t \to 2^-$ (the solid curve in Figure 2.6.2). Our numerical solution is not aware of that fact.

What happened is that the Euler Solution stepped across $t = 2$ and connected up with neighboring solution curves. Inspection of the ODE reveals nothing about a finite

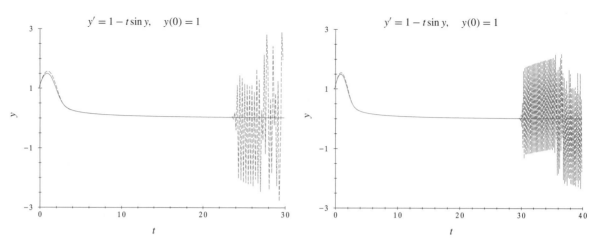

FIGURE 2.6.3 True solution (solid), Euler Solution (dashed): $h = 0.15$ (Example 2.6.3).

FIGURE 2.6.4 True solution (solid), Euler Solution (dashed): $h = 0.1$ (Example 2.6.3).

escape time, and if we did not know the true solution, we would have been inclined to accept the approximate solution of Figure 2.6.2. The implications of this observation are sobering, because the behavior of the computed solution may give no clue about the long-term behavior of the true solution.

Long-Term Degradation

Step-by-step solvers may not track solution curves very well toward the end of long t-intervals. Sometimes the accumulated error grows so fast that it obscures the long-term features of the solution curve. The next example illustrates this fact.

EXAMPLE 2.6.3

A Good Euler Solution Goes Wild

If Euler's Method is used with $h = 0.15$, $0 \le t \le 30$, to solve the IVP

$$y' = 1 - t \sin y, \qquad y(0) = 1$$

and the solution graphed, a strange phenomenon is observed (Figure 2.6.3). Comparing the Euler Solution (dashed) to the actual solution (solid), we see that the agreement is reasonably good, if not perfect, for $0 \le t \le 3$, but then it is dramatically good for a long time interval. Finally, at about $t = 24$ the accuracy of the Euler Solution seems to degenerate badly. Choosing a smaller step size seems only to delay the onset of the bad behavior of the Euler Solution (Figure 2.6.4). The fact that Euler's Method is a first-order method tells us nothing about its *long-term behavior*.

☞ It would be revealing to use RK4 in Figures 2.6.1–2.6.4 instead of Euler's Method.

Comments

Although we have illustrated computational "lies" only with Euler's Method, these and other inaccuracies may occur whichever numerical solver and algorithm is used. To the previous observations we can add the following:

- Computers can only recognize finitely many numbers. So a computer is confused by a number whose magnitude is smaller (or greater) than the numbers the computer recognizes—a condition called *underflow* (or *overflow*).

- The methods described in Section 2.5 produce approximations to IVPs that converge to the true solution when exact arithmetic is used and the step size tends to zero. But because exact arithmetic is impossible, practical techniques for finding such approximations require considerable ingenuity.

- Although, in theory, moving forward along the solution of an IVP and then backtracking returns the exact initial value, that may be far from true when a numerical solver is used (see Problems 2 and 3).

The reader should not be discouraged by the bad behavior of the elementary solvers described in the examples above. It *is* often possible to achieve satisfactory results using the approximation algorithms described in this chapter, but a good general-purpose solver uses much more than a single one-step method. As noted earlier, the plots of solutions of differential equations in this text were produced by a very powerful solver and are quite accurate (we think!).

PROBLEMS

www **1.** (*High Rates*). Consider the IVP $y' = y^3$, $y(0) = 1$.

(a) Use separation of variables to find a formula for the maximally extended solution of the IVP. On what t-interval is the solution defined?

(b) Plot an Euler Solution in the rectangle $0 \leq t \leq 1$, $0 \leq y \leq 20$. Use $h = 0.05$.

(c) Plot the exact solution found in part (a) and the Euler Solution found in part (b) in the same rectangle. Explain what you see near $t = 1/2$.

2. (*Reversibility of an Approximation Method*). If a numerical solver is used to solve an IVP forward in time, and then backward, it may return a number that is nowhere near the initial value. This problem looks at this unhappy fact of computational life.

☞ Dynamical systems are reversible, but solutions from a numerical solver may not be!

(a) Assume that $f(t, y)$ satisfies the conditions of the Existence and Uniqueness Theorem (Theorem 2.1.1) and that $y = y(t)$, $t_0 \leq t \leq t_1$, solves the forward IVP $y' = f(t, y)$, $y(t_0) = y_0$. Show that this same function $y(t)$ defines a solution $z = y(t)$ of the backward IVP $z' = f(t, z)$, $z(t_1) = y(t_1)$, $t_0 \leq t \leq t_1$, and that therefore $z(t_0)$ must be y_0.

(b) Use RK4 with step size $h = 0.1$ to find and plot an approximate solution of the IVP $y' = 3y \sin y - t$, $y(0) = 0.4$, over the interval $0 \leq t \leq 8$.

(c) (*A Good Solver Does Bad Things*). Using the value for $y(8)$ found in part (b), use RK4 to solve the backward IVP $z' = 3z \sin z - t$, $z(8) = y(8)$, from t-initial = 8 to t-final = 0. Plot the solution. How close is $z(0)$ to 0.4? Explain any significant difference.

3. (*More on Reversibility*). As noted in Problem 2, if $f(t, y)$ satisfies the conditions of the Existence and Uniqueness Theorem, then the IVP $y' = f(t, y)$, $y(t_0) = y_0$, $t_0 \leq t \leq t_1$, is reversible in theory, but experience with numerical solvers shows that actual practice is different. In this problem, you will experiment with a linear IVP and show that the practical difficulties of running an IVP backwards can be resolved (but only in part) by shortening the time step.

(a) Find all solutions of the linear ODE $y' + 2y = \cos t$. Show that as $t \to +\infty$, all solutions approach the particular solution $y_p = 0.4 \cos t + 0.2 \sin t$.

(b) Plot solution curves in the rectangle $0 \le t \le 20$, $|y| \le 1.5$ using as initial points $(0, 0.4)$, $(0, \pm 1.5)$, and several other points on the top and bottom sides of the rectangle. Observe that solutions converge to the solution $y_p(t)$ as t increases.

(c) Use RK4 with step size $h = 0.1$ to solve the IVP with $y(0) = 0.4$ and plot an approximation to $y_p(t)$ for $0 \le t \le 20$.

(d) (*Good Solver, Bad Behavior*). Using the value for $y_p(20)$ found in part **(c)**, solve the IVP $z' + 2z = \cos t$, $z(20) = y_p(20)$, backward from t-initial $= 20$ to t-final $= 0$, again using RK4 with $h = 0.1$. What happens? How would you explain the difficulty? [*Hint*: Exact solutions converge on $y_p(t)$ in forward time but diverge from $y_p(t)$ in backward time. The approximate nature of computed solutions suggests that the computed point $(20, y_p(20))$ is not quite on the solution curve of $y = y_p(t)$.]

(e) Repeat part **(d)**, but with $h = 0.01, 0.001$. Any improvement?

4. (*Long-Term Behavior*). The problems below show how shortening the step size affects the long-term behavior of approximate solutions.

    (a) Use Euler's Method with step size $h = 0.1$ to plot an approximate solution of the IVP $y' = 1 + ty \cos y$, $y(0) = 1$, $0 \le t \le 28$. Repeat with $h = 0.01$. Describe what you see.

    (b) Repeat part **(a)** using RK4.

5. (*Qualitative Behavior of Numerical Approximations*). The problems show that approximate solutions do not always faithfully describe the qualitative properties of exact solutions.

☞ Shows that you should be careful not to read too much from a numerical solution.

    (a) (*Logistic ODE*). Use separation of variables to solve the IVP $y' = (1 - y)y$, $y(0) = y_0$. What happens to the solution as $t \to +\infty$ if $y_0 > 0$? Use a grapher to plot this solution over the interval $0 \le t \le 8$ when $y_0 = 2$. Then apply your best numerical solver to the same IVP. How does this approximate solution compare to the true solution?

    (b) Solve the IVP in part **(a)** using Euler's Method with $h = 0.75$ and $y_0 = 2$, and plot the Euler Solution on $0 \le t \le 8$. Does the Euler Solution behave qualitatively the same as the solutions in part **(a)**? Explain differences.

    (c) Repeat part **(b)**, but with $h = 1.5$ and $y_0 = 1.4$.

    (d) Repeat part **(b)**, but with $h = 2.5$ and $y_0 = 1.3$.

6. (*Convergent Euler Sequences*). Consider the logistic IVP $y' = r(1 - y)y$, $y(0) = y_0 > 0$, where r is a positive constant. The Euler iterates are generated by $y_n = y_{n-1} + rh(1 - y_{n-1})$, $n = 1, 2, \ldots$, and $h > 0$. The Euler Solution depends on the product parameter $a = rh$, rather than on r or h separately. For each of the parameter intervals $0 < a \le 1$ and $1 < a \le 2$, the Euler Solutions exhibit qualitatively different properties in the long term. Perform the following simulations.

    (a) Let $r = 10$, $h = 0.05$ (so $a = rh = 0.5$). Construct and plot Euler Solutions for each of the initial values $y_0 = 0.3, 2.0$. Compare the Euler Solutions to the exact solution curve over $0 \le t \le 1$ (for $y_0 = 0.3$) and over $0 \le t \le 0.5$ (for $y_0 = 2.0$). Explain qualitative differences.

    (b) Let $r = 100$, $h = 0.015$ (so $a = rh = 1.5$). Construct and plot Euler Solutions for each of the initial values $y_0 = 0.5, 1.5$. Compare the Euler Solutions to the exact solution curve over $0 \le t \le 0.1$ (for $y_0 = 0.5$) and over $0 \le t \le 0.15$ (for $y_0 = 1.5$). Explain qualitative differences.

7. (*Sign Analysis and Approximate Solutions*). Consider the IVP $y' = -y(1 - y)^2$, $y(0) = 1.5$.

    (a) Show by sign analysis (Section 2.2) that solution curves above the solution line $y = 1$ fall toward that line, while solution curves below the line fall away with increasing time.

    (b) Graph the approximate solutions given by Euler's method with step size $h = 0.1, 0.5, 1, 1.5$, and 2, and comment on any noteworthy properties.

8. (*Sampling Rates and Aliasing*). All the solutions of the autonomous linear system of ODEs, $x' = y$, $y' = -4x$, are given by the solution formulas $x = A \sin(2t + \phi)$, $y = 2A \cos(2t + \phi)$,

☞ If you get stuck, peek at Example 3.5.2.

where A and ϕ are arbitrary constants. In this problem we show that if we "sample" a computed solution only at relatively long time intervals, the resulting graphs may be quite misleading.

(a) Plot the tx- and ty-component graphs and the xy-orbits of the solutions with initial points $x_0 = 0$, $y_0 = 1$, and then with $x_0 = 0$, $y_0 = 2$. Describe these graphs.

(b) Using RK4 for systems (Section 2.5), solve the IVP $x' = y$, $x(0) = 0$, $y' = -4x$, $y(0) = 1$, over the time interval $0 \le t \le 100$ using step size $h = 0.1$. Plot $x(t)$ in the tx-plane.

(c) Using the step size $h = 0.1$ of part **(b)**, plot only every fifth computed point of $x(t)$. Repeat, but plot only every tenth point, and then every twentieth point. Explain the differences in these plots from the plot in part **(b)**.

2.7 Euler's Method, the Logistic ODE, and Chaos

We discussed Euler's Method and the logistic ODE in several sections of Chapters 1 and 2, but what do the Method and the ODE have to do with chaos? A chaotic state is a condition of confusion and disorder, and it is hard to find anything chaotic in the precise formulas of Euler's Method or in the logistic ODE and its solutions. The only hint that there might be a connection appears in Figure 2.6.1, where an Euler Solution of a particular logistic ODE seems to wander around a lot. In this section we will examine the long-term behavior and the chaotic wandering of Euler Solutions of a logistic ODE for a range of step sizes.

Euler's Method for the Scaled Logistic ODE

☞ The general logistic ODE $P' = (1 - P/K)P$ was scaled to $y' = (1 - y)y$ in Section 2.4.

Euler's Method with step h applied to the scaled logistic IVP

$$y' = (1 - y)y, \qquad y(0) = y_0 > 0 \tag{1}$$

where y' denotes dy/ds, gives us the specific Euler algorithm

$$y_n = y_{n-1} + h(1 - y_{n-1})y_{n-1}, \qquad n = 1, 2, \ldots \tag{2}$$

Just as described in Section 2.5, the *Euler values* y_n give us the *Euler points* (nh, y_n) at the vertices of the broken-line Euler Solution that approximates the exact solution of IVP (1). If h is small, then the Euler Solution gives a fairly good approximation to the true solution. At least that seems to be so if y_0 is not too far away from the equilibrium value of 1 and if we take a large number of steps so that any initial wandering of the Euler Solution damps out.

Period Doubling

The following sequence of examples shows a strange pattern of events as we apply Euler's Method for increasing values of h. The solid curve in Figures 2.7.1–2.7.4 is the graph of the true solution of the ODE $y' = (1 - y)y$, with $y(0)$ as given.

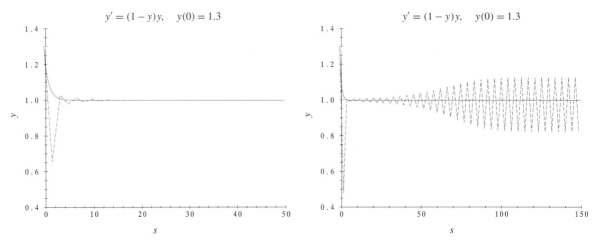

FIGURE 2.7.1 Euler Solution (dashed) tends to $y = 1$ for $h = 1.65$ (Example 2.7.1).

FIGURE 2.7.2 Euler Solution (dashed) tends to a 2-cycle Euler Solution: $h = 2.1$ (Example 2.7.2).

EXAMPLE 2.7.1

Euler Solution Approaches True Solution

Figure 2.7.1 shows the Euler Solution (dashed) of IVP (1) if $h = 1.65$, $y_0 = 1.3$. After some initial uncertainty, the Euler values y_n settle down to repeating the number 1. The values $y_n = 1$, for all n, define a periodic solution

$$1, \quad 1, \quad 1, \quad \dots$$

of the Euler algorithm (2). This is an example of a *1-cycle*

It is only for values of h beyond 2 that interesting behavior occurs, so let's try the values $h = 2.1, 2.5, 2.56, 2.567$, then chaos at 2.65.

EXAMPLE 2.7.2

A 2-Cycle Appears: $h = 2.1$

Figure 2.7.2 shows what happens to the Euler values y_n when $h = 2.1$ and $y_0 = 1.3$. The coordinates of the high and low points of the Euler Solution (dashed) shown in Figure 2.7.2 are $t = nh$, $y = y_n$, $n = 1, 2, \dots$. These Euler values seem to settle down to a periodic sequence, after some initial confusion, eventually swinging back and forth between the values $y_n \approx 1.13$ and $y_{n+1} \approx 0.83$ for sufficiently large n. The periodic sequence

$$1.13, \quad 0.83, \quad 1.13, \quad 0.83, \dots$$

is said to be a *2-cycle* since it repeats two values over and over again, forever. There is a second 2-cycle,

$$0.83, \quad 1.13, \quad 0.83, \quad 1.13, \dots$$

which is just the first cycle shifted by a half-period.

After some experimentation, it appears that for any y_0 near 1 (but not equal to 1)

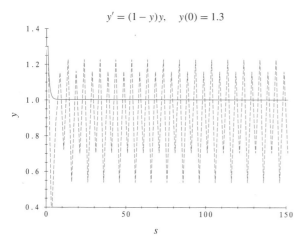

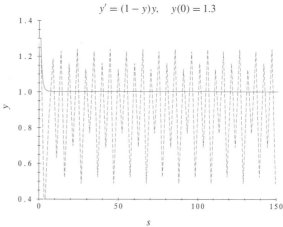

FIGURE 2.7.3 Approach to a 4-cycle Euler Solution (dashed): $h = 2.5$ (Example 2.7.3).

FIGURE 2.7.4 Approach to an 8-cycle Euler Solution (dashed): $h = 2.56$ (Example 2.7.4).

the resulting Euler Solution for $h = 2.1$ is attracted by a 2-cycle Euler Solution. This attraction property is the reason that we can see these solutions.

We have defined a 2-cycle, now let's define a k-cycle. An Euler Solution for the logistic IVP (1) generates a *k-cycle* if it repeats the same k Euler values $y_0, y_1, \ldots, y_{k-1}$ in that order, forever:

$$y_0, \quad y_1, \ldots, y_{k-1}, \quad y_0, \quad y_1, \ldots, y_{k-1}, \ldots$$

If k is the smallest integer for which this repetition occurs, then k is the *period* of the cycle. For example, for small h the Euler algorithm (2) has a unique positive 1-cycle, $y_n = 1$, for all n, and all Euler Solutions with y_0 near 1 are attracted to the 1-cycle Euler Solution. The Euler Solution of Figure 2.7.2 is attracted to a 2-cycle. Now let's increase the step size and watch the Euler Solution approach a 4-cycle Euler Solution.

EXAMPLE 2.7.3

Appearance of a 4-Cycle: $h = 2.5$

Figure 2.7.3 shows that the Euler Solution (dashed) with $h = 2.5$ and $y_0 = 1.3$ is attracted to a 4-cycle Euler Solution. Although not shown in the figure, the Euler Solution starting with any value of y_0 close enough to 1.3 is attracted to this 4-cycle.

We have chosen various time intervals (corresponding to various numbers of Euler steps) over which to plot the Euler Solutions. In each case we go out far enough so that we can see what the Euler Solution settles down to, but not so far that the graphics blur. Let's try to double the period again (from 4 to 8) by increasing h just a little.

EXAMPLE 2.7.4

Now an 8-Cycle Appears: $h = 2.56$

Set $h = 2.56$ and $y_0 = 1.3$ (again) and the Euler Solution is attracted to an 8-cycle Euler Solution (see Figure 2.7.4).

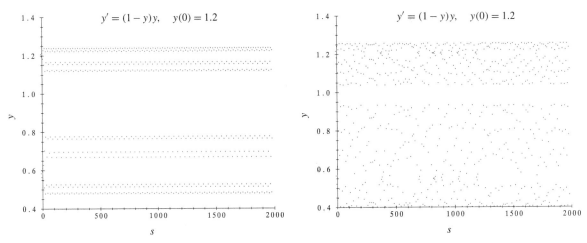

FIGURE 2.7.5 Approach to the Euler points of a 16-cycle Euler Solution: $h = 2.567$ (Example 2.7.5).   **FIGURE 2.7.6** Chaotic wandering of Euler points: $h = 2.65$ (Example 2.7.6).

The cycles are getting hard to detect because of all of the up-and-down segments. So from this point on we plot only the Euler points (t_n, y_n), where y_n are the Euler values, and leave out the connecting line segments. Figure 2.7.5 shows these points after another period doubling as h increases from 2.56 to 2.567.

EXAMPLE 2.7.5

A 16-Cycle: $h = 2.567$

Figure 2.7.5 shows that if $h = 2.567$ and $y_0 = 1.2$, then the Euler Solution tends to a 16-cycle Euler Solution. You have to look hard to see all 16 levels of the Euler points. Note the empty band around the line $y = 1$. Apparently the Euler points stay well away from the true solution of the IVP

$$y' = (1 - y)y, \qquad y(0) = 1.2$$

Where does the period doubling end? What happens after it ends? It is known that there is a convergent sequence of step sizes

$$2 = h_1 < h_2 < \cdots < h_k < \cdots$$

with $h_k \to h^*$ as $k \to \infty$, where h^* is a specific number a little less than 2.65, such that as h is increased through each h_k, an attracting 2^k-cycle bifurcates into an attracting 2^{k+1}-cycle. This is an example of a sequence of *period-doubling bifurcations*. The values of h used in the last four examples were selected from the intervals (h_1, h_2), $\dots$, (h_4, h_5). So let's see what happens for a value of h greater than h^*.

☞ Chaos begins for $h > h^*$.

EXAMPLE 2.7.6

Chaotic Behavior: $h = 2.65$, $y_0 = 1.2$.

Figure 2.7.6 shows Euler points that clearly wander around somewhat chaotically. The only distinctive feature of the figure is that (as in Example 2.7.5) the Euler points stay out of a narrow band centered at the line $y = 1$.

The Bifurcation Diagrams for Chaos

In recent years mathematicians, scientists, engineers, and economists have tried to develop a theory of chaotic dynamics that would offer a way to model apparently random, yet precisely defined, behavior. In trying to make sense out of all this, the example of the Euler algorithm for the logistic map has played a central role.

As we saw in Section 2.4, bifurcation diagrams can be used to show pictorially what happens when a parameter changes. We can also do this for the Euler algorithm with parameter h. What we want to see in the diagram are the Euler values (for a given h) that represent the long-term (i.e., large values of n) behavior, whether it's 2-cyclic, 4-cyclic, chaotic, or whatever. Here's how to carry out the construction.

Construction of a Bifurcation Diagram

The behavior of the Euler values y_n for each of a sequence of h values can be investigated by constructing a graph in the hy-plane as follows:

- **Select a value for** y_0 near 1.2.

- **Select a value of** h in the range $0 < h \leq 3$.

- **Select** N **large enough** so that the Euler values $y_1, \ldots, y_N$ reveal any attracting cycle.

- **Select a positive integer** J, and compute the Euler values $y_1, \ldots, y_{N+J}$.

- **Plot the points** (h, y_i) in the hy-plane for $N < i \leq N + J$.

- **Choose another value of** h and repeat the process.

The values $y_1, \ldots, y_N$ are discarded in the fifth step of the construction so that any initial wandering around is suppressed. This means that if we do see wandering in the Euler values, it really does represent chaotic long-term behavior.

☞ Figure 2.7.7 changes little if other values of y_0 near 1.2 are chosen.

Figure 2.7.7 shows what happens if this process is carried out over the interval $1.6 \leq h \leq 3$ for 1400 equally spaced values of h with $N = 500$ and $J = 200$. The attracting 1-cycles, 2-cycles, 4-cycles, and 8-cycles are clearly visible at the left of the diagram, but then we move into a blurred region where it is not clear just what is happening to the Euler values for a particular h. There seems to be an attracting 6-cycle for $h \approx 2.628$, and an attracting 5-cycle for $h \approx 2.740$. These cycles are more clearly visible in Figure 2.7.8, which zooms on the region $2.6 \leq h \leq 2.9$.

But the most startling feature is the broad vertical white band beginning at $h = \sqrt{8}$ (this value has been shown to be precise), where there is an attracting 3-cycle. It has been proven that wherever there is a 3-cycle (attracting or not) there are also cycles of every period. This result is true not just for the Euler algorithm, but for any algorithm $y_n = F(y_{n-1})$, where F is continuous on the real line. Since some people define *chaos* to be the presence of cycles of every period, the existence and exact location of 3-cycles

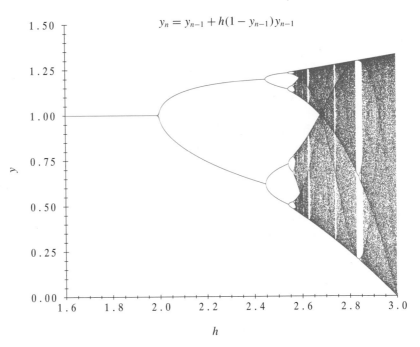

$$y_n = y_{n-1} + h(1 - y_{n-1})y_{n-1}$$

FIGURE 2.7.7 Bifurcation diagram for the Euler-logistic algorithm.

is of considerable importance. For a given value of h, it is known that there is at most one attracting cycle. Although there may be many cycles present, computationally we see only the attracting cycles.

It is widely believed that what we see on each vertical slice of the blurry region of the graphs in Figures 2.7.7 and 2.7.8 is the result of chaotic, nonperiodic wandering. By the way, the "curved spokes" in Figure 2.7.8 that emanate from the point $h \approx 2.68$, $y = 1$ are *not* computational artifacts. We suggest you look in the references listed in the Student Resource Manual for more information on the behavior of Euler values and the values generated by other discrete algorithms.

Period Three Implies Chaos

Let's explore the importance of the cycles of period 1, 2, 4, 8, ...introduced in this section and just what role a period-3 cycle plays in the story of chaos. We begin with an unusual ordering of the positive integers:

☞ This is the
Šarkovskii Sequence of
the positive integers.

$$3 \succ 5 \succ 7 \succ \cdots$$
$$2 \cdot 3 \succ 2 \cdot 5 \succ 2 \cdot 7 \succ \cdots$$
$$2^2 \cdot 3 \succ 2^2 \cdot 5 \succ 2^2 \cdot 7 \succ \cdots$$
$$\vdots$$
$$\cdots \cdots \succ 2^4 \succ 2^3 \succ 2^2 \succ 2 \succ 1$$

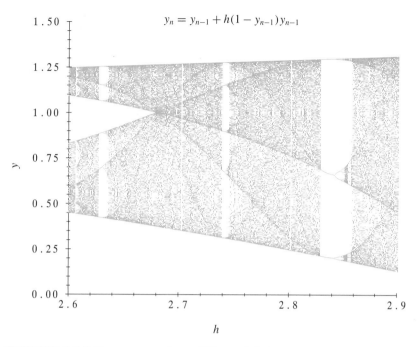

$$y_n = y_{n-1} + h(1 - y_{n-1})y_{n-1}$$

FIGURE 2.7.8 Zoom on portion of Figure 2.7.7.

The odd integers larger than 1 are listed in ascending order at the head of the sequence. Then each of these odd numbers is multiplied by 2 and listed, then the odd integers are multiplied by 2^2, and so on until all positive numbers *except* the powers of 2 are listed. We then list the powers of 2 in descending order, $\cdots \succ 2^4 \succ 2^3 \succ 2^2 \succ 2^1 \succ 2^0 = 1$. What does this sequence have to do with chaos? In 1964 the Russian mathematician A. N. Šarkovskii proved a remarkable theorem about continuous functions and this sequence.

THEOREM 2.7.1

> Šarkovskii's Theorem. **Suppose that** $f(y)$ **is any real-valued continuous function,** $-\infty < y < \infty$. **If the algorithm** $y_n = f(y_{n-1})$, $y_0 = a$, $n = 1, 2, \ldots$ **has a cycle of least period** k **and if** p **is any integer that** *follows* k **in the Šarkovskii Sequence, the algorithm also has a cycle of period** p.

The startling aspect of all this is that with such a weak hypothesis (f is continuous) we get such a strong conclusion. In our work here, we used the Euler algorithm with

$$f(y) = y + h(1 - y)y$$

which, for each value of h, is continuous for all values of y. Figure 2.7.7 starts out on the left with values of h that give us cycles with powers-of-2 periods (see the tail of the Šarkovskii Sequence). But we see evidence of an attracting cycle of period 6 if $h \approx 2.628$. Locating the number 6 in the sequence, we conclude that for this value of h, the function $f(y)$ has cycles of period 10, 14, 18, and indeed cycles of all periods (except perhaps 3, 5, 7, . . . which precede 6 in the sequence). Why don't these cycles show up in Figure 2.7.7? It is known that for each value of h the Euler algorithm has at most one attracting cycle. In order to see one of these infinitely many other cycles with other periods, we would have to choose exactly the right initial value since they aren't attractors. We can make a similar analysis for the value $h \approx 2.740$ where f apparently has an attracting 5-cycle. In this case, f has cycles of *all* other periods (except perhaps 3).

That leaves us with the three-cycle, which is where James Yorke[10] enters the story of chaos. In 1975 Yorke and his collegue T. Y. Li published a paper with the title "Period Three Implies Chaos." Yorke was the first to use the word "chaos" in this setting. Since that seminal paper appeared, mathematicians, scientists, engineers, economists, and many others have tried to interpret, understand and modify the meaning of chaos. Alternate definitions appear regularly in the literature and the whole subject remains in a state of chaos.

Comments

In recent years mathematicians and scientists around the world have studied the appearance of chaos in deterministic dynamical systems that model everything from the long-term behavior of planetary motion to the turbulent dynamics beneath a thunderhead. Scientists and mathematicians are interested in dynamical systems that show signs of chaos and have sought explanations for this strange behavior. The remarkable aspect of all this is that dynamical systems are deterministic in the sense that any future state of the system is completely determined by its initial state. So how can chaos occur? We have given some insight into this question in the context of the discrete logistic equation. Here is a short list of references that provide more details about chaotic dynamics.

- R. L. Devaney, *An Introduction to Chaotic Dynamical Systems*, 2nd ed. (New York: Addison-Wesley, 1989).

- T. Y. Li and J. A. Yorke, "Period Three Implies Chaos," *American Mathematical Monthly 82* (1975), pp. 985–992.

- M. Martelli, *Discrete Dynamical Systems and Chaos* (New York: Longman/Wiley, 1992).

James Yorke

[10]James Yorke is a professor of mathematics at the University of Maryland's Institute for Physical Science and Technology. In 1972, a colleague gave Yorke a copy of Lorenz's 1963 paper "Deterministic Nonperiodic Flow," which deeply impressed him and, led to Yorke's highly influential work in chaotic dynamics. Over the years, Yorke has had the chance to work with scientists and investigate fascinating problems from a wide variety of disciplines. Among other projects, he convinced the federal government to change the ways it controls the spread of disease, argued correctly that the even-odd gasoline rationing system of the 1970's would backfire, and proved that a photograph of an antiwar demonstration had actually been taken half an hour later than claimed by the government. Asserting the importance of studying the deterministic disorder he calls chaos, Yorke said, "People say, what use is disorder. But people have to know about disorder if they are going to deal with it."

- P. Saha and S. H. Strogatz, "The Birth of Period 3," *Mathematics Magazine 68* (1995), pp. 42–47.

- P. D. Straffin, "Periodic Points of Continuous Functions," *Mathematics Magazine 51* (1978), pp. 99–105.

PROBLEMS

1. (*2-Cycles*). Show that for any $h > 2$ the recursion relation

$$y_{n+1} = y_n + hy_n(1 - y_n), \qquad y_0 \text{ given}$$

has precisely two values of y_0, $(h + 2 \pm \sqrt{h^2 - 4})/2h$, in the interval $0 < y_0 < (1 + h)/h$ that give rise to 2-cycles. [*Hint*: Let u be one of these values of y_0 and put $v = u + hu(1 - u)$. Then for u to give rise to a 2-cycle, we must have $u = v + hv(1 - v)$.]

2. (*Period Doubling*). Let $h = 0.023$. Construct the Euler Solution for $y' = 100y(1 - y)$, $y(0) = 1.3$, $0 \le t \le 1$. Compare this Euler Solution to the exact solution curve. Repeat this simulation with $h = 0.025$. Do the values of these Euler Solutions eventually settle down to any pattern? If so, how would you describe that pattern?

www **3.** (*Chaotic Wandering*). Investigate the Euler Solutions for $y' = ry(1 - y)$, $y(0) = 1.2$, with step size h, where $rh = 2.65$.

(**a**) Let $r = 1$ and $h = 2.65$. Plot the Euler polygon starting at $y(0) = 1.2$ over four intervals: $[0, 100]$, $[100, 200]$, $[200, 300]$, $[300, 400]$. Do you see any patterns? Interpret what you see.

(**b**) Let $r = 2.65$ and $h = 1$. Plot the Euler polygon starting at $y(0) = 1.2$ over the four intervals $[0, 100/2.65]$, $[100/2.65, 200/2.65]$, $[200/2.65, 300/2.65]$, $[300/2.65, 400/2.65]$. Why do these graphs look exactly like the corresponding graphs in part (**a**)? What would the Euler polygons corresponding to $r = r_0 > 0$, $h = h_0 > 0$, $y(0) = 1.2$ look like if $r_0 h_0 = 2.65$ and the graphs are plotted over $[0, 100/r_0]$, $[100/r_0, 200/r_0]$, and so on? Explain.

4. (*Chaotic Wandering*). Construct the Euler iterates with $h = 2.75$ for the logistic IVP $y' = y(1 - y)$, $y(0) = y_0$, over the interval $0 \le t \le 2000$, where $y_0 = 0.25, 0.5, 0.75, 1.2$. Plot the graphs for each value of y_0. Do you see any pattern in the iterates? [Warning: Do this problem only if your solver will plot iterates without the connecting lines.]

5. (*Cobweb Diagrams*). In this section, we applied Euler's Method with step size h to the logistic IVP $dy/ds = (1 - y)y$, $y(0) = y_0$, giving the algorithm $y_n = y_{n-1} + h(1 - y_{n-1})y_{n-1}$, $n = 1, 2, \ldots, N$. We plotted the sequence $y_1, \ldots, y_N$ to obtain the graphs illustrated in the section. Cobweb diagrams provide a different, simple geometric way of "seeing" what the Euler solution will look like. Follow the outline below to learn about how cobweb diagrams work.

- Rewrite the right-hand side of $y_n = y_{n-1} + h(1 - y_{n-1})y_{n-1}$ by completing the square in the quadratic expression for y_{n-1} to obtain

$$y_n = -h\left(y_{n-1} - \frac{1 + h}{2h}\right)^2 + \frac{(1 + h)^2}{4h}$$

- Plot y_n versus y_{n-1}. For what positive value y_{n-1}^* of y_{n-1} is $y_n = 0$? Show that the value $y_{n-1} = (1 + h)/h$ produces the maximum value of y_n. What is this maximum value? What values $y_{n-1}^{(1)}$ and $y_{n-1}^{(2)}$ make $y_n = 1$?

- Show that for $0 \le h \le 3$, $y_n^* \le$ the maximum value of y_n. Conclude that for $0 \le h \le 3$, if $0 \le y_0 \le (1 + h)/h$, then $y_1, y_2, \ldots$ all lie in the interval $[0, (1 + h)/h]$.

- Superimpose the line $y_n = y_{n-1}$ over the parabola obtained by plotting y_n versus y_{n-1}, and show that the Euler sequence $y_1, y_2, \ldots$ can be constructed as follows:

  1. Draw a vertical dashed line upward from y_0 to the parabola to find y_1.

2. Then draw a horizontal dashed line over to the line $y_n = y_{n-1}$.
3. Draw a vertical dashed line to the parabola again to find y_2.
4. Repeat these steps.

This process is illustrated in the following graphs for the cases $0 < h < 1$, $1 < h < 2$, and $2 < h < 3$, respectively. A great deal about the behavior of the Euler sequences y_1, $y_2, \ldots$ can be inferred from these graphs, often called *cobweb diagrams*. Describe the behavior of the Euler solution predicted by each graph. Why does the third graph indicate approach to a 4-cycle?

(1) $h = 0.75$, $y_0 = 2$

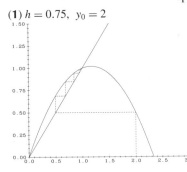

(2) $h = 1.5$, $y_0 = 1.4$

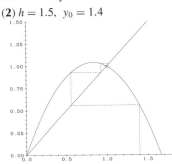

(3) $h = 2.5$, $y_0 = 1.3$

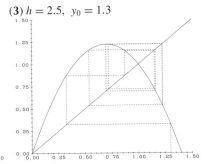

- Choose a few values for h. Draw the cobweb diagrams by hand and predict the solution behavior. Plot solution curves and compare with your predictions.

 6. (*Behavior of Euler Solutions*). Give a convincing justification for the following general description of the behavior of Euler Solutions for the IVP $y' = y(1 - y)$, $y(0) = y_0$, in the indicated parameter regimes. What does this behavior imply about the choice of step size h? [*Hint:* You may wish to use cobweb diagrams (see Problem 5).]

- For $0 < h \le 1$ the Euler sequence $\{y_n\}$ generated by any choice of initial point y_0 in the interval $0 < y_0 < (1+h)/h$ behaves as follows:

  (a) If $0 < y_0 < 1$, then $\{y_n\}$ rises steadily toward 1 as $n \to \infty$.
  (b) If $y_0 = 1$ or $1/h$, then $y_n = 1$, for $n \ge 1$.
  (c) If $1 < y_0 < 1/h$, then $\{y_n\}$ decreases steadily toward 1 as $n \to \infty$.
  (d) If $1/h < y_0 < (1+h)/h$, then $0 < y_1 < 1$ and $\{y_n\}$ rises toward 1 as $n \to \infty$.

- For $1 < h \le 2$ the Euler sequence $\{y_n\}$ generated by the initial choice of y_0 in the interval $0 < y_0 < (1+h)/h$ behaves as follows:

  (a) If $0 < y_0 < 1/h$, the Euler sequence $\{y_n\}$ rises steadily until $y_N > 1$ for some N; for $n > N$, $\{y_n\}$ alternates about 1 while steadily approaching 1 as $n \to \infty$.
  (b) If $y_0 = 1/h$ or 1, then $y_n = 1$ for all $n \ge 1$.
  (c) If $1/h < y_0 < 1$, then $\{y_n\}$ alternates about 1 while approaching 1 as $n \to \infty$.
  (d) If $1 < y_0 < (1+h)/h$, then $0 < y_1 < 1$. Thereafter $\{y_n\}$ behaves as described in parts (a), (b), or (c), depending on the value of y_1.

- (*Chaotic Wandering and Period Doubling*). For $h > 2$ the Euler sequence $\{y_n\}$ generated by any initial value y_0 in the interval $0 < y_0 < (1+h)/h$ does not approach 1 as $n \to \infty$. How would you describe the behavior of $\{y_n\}$?

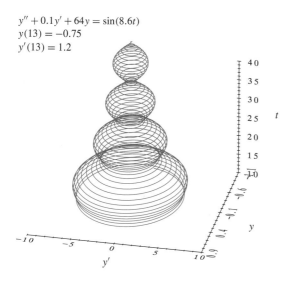

$$y'' + 0.1y' + 64y = \sin(8.6t)$$
$$y(13) = -0.75$$
$$y'(13) = 1.2$$

The graph tracks the motion of a weight on a spring in the three-dimensional space of position y, velocity y', and time t. Describe and interpret the projections onto the ty-plane and the yy'-plane. Check out Example 3.2.2.

Second-Order Differential Equations

The oscillations of a weighted spring, the swinging motion of a pendulum, the alternating current in an electrical circuit—all can be modeled by second-order ODEs. In doing the modeling here and in the next chapter we will see that nonlinear ODEs come into play, but approximating linear ODEs with explicit solution formulas are useful for many applications.

3.1 Springs: Linear and Nonlinear Models

We bounce on springs, we sleep on springs, we play with springs; car travel is smooth and comfortable because of springs. We will use ODEs to model the oscillations of several kinds of springs and relate the solutions of our model ODEs to our experiences with physical springs.

To focus our attention, let's suppose that a weight is attached to one end of a massless spring, and the other end is fastened to the ceiling. If the weight is pulled downward a little ways and then released, the spring-mass system oscillates up and down. The nature of the motion depends on the strength of the spring force, the frictional forces, the gravitational force, and any other external forces acting on the body.

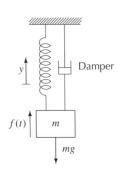

Let's measure the vertical position of the weight by a y-axis pointing upward with $y = 0$ at the neutral position where the spring is unstretched and uncompressed. Then $y < 0$ indicates a stretched spring and $y > 0$ a compressed spring. We use Newton's Second Law (Section 1.5) to describe the position $y(t)$ of the body, so we need to know what forces are acting on the body at any time. The four forces that we consider are: the spring force, the damping force, gravity, and an outside driving force. The margin figure shows a stretched spring-body system ($y < 0$), a driving force $f(t)$ acting parallel to the y-axis, and a frictional force represented by a damper (shown as a cylinder of fluid through which a piston moves). The gravitational force mg on the body of mass m acts vertically downward.

The following models for the spring and damping forces are based on observations of real springs.

Spring and Damping Forces

Spring Force: The force $S(y)$ exerted by a spring on a body acts parallel to the spring's axis (the y-axis) in a direction to return the spring to the neutral state $y = 0$.

☞ Viscous damping is introduced in Section 1.5.

Viscous Damping Force: The viscous damping force on a body has magnitude proportional to the velocity and acts in the direction opposite to the direction of motion.

☞ You may have seen "free body diagrams" used for this purpose.

First, we express the forces acting on the body in terms of the state variables y and y'. Next, we take into account the orientation of the y-axis ("up" is the positive direction) when combining the forces. Keep in mind that $S(y)$ acts in a direction opposite to the displacement y and that the damping force acts in a direction opposite to the velocity y'. Doing all this, we see that by Newton's Second Law the position $y(t)$ of the body at time t must satisfy the second-order ODE

$$my'' = \text{sum of forces} = S(y) - cy' - mg + f(t) \tag{1}$$

where c is the positive *viscous damping constant* ($c = 0$ if there is no damping).

Careful measurements suggest that if the displacement from the neutral state is small, then the spring force $S(y)$ is an odd function, that is, $S(-y) = -S(y)$. The commonly used models for $S(y)$ are

Hooke's Law spring	$S(y) = -ky$	(2)
Hard spring	$S(y) = -ky - jy^3$	(3)
Soft spring	$S(y) = -ky + jy^3$	(4)
Aging spring	$S(y, t) = -k(t)y$	(5)

where k and j are positive constants and $k(t)$ is a positive function. These forces act in the opposite direction to displacement (for small displacements from the neutral state). Compared to the Hooke's Law force, the hard spring force strengthens (and the soft spring force weakens) with extension or compression, and the aging spring coefficient $k(t)$ weakens with time (see margin figures). The constant k is often called the *spring constant*. Real springs can be compressed or stretched only so much, and so the range of validity of the model ODE (1) is limited.

Let's begin by taking a look at some model ODEs for the motion of a spring.

Linear Models

A *linear second-order* ODE written in *normal linear form* is given by

$$y'' + a(t)y' + b(t)y = c(t) \tag{6}$$

where the functions $a(t)$, $b(t)$, and $c(t)$ are defined on a common *t*-interval. Choosing $S(y) = -ky$ (Hooke's Law) in ODE (1), dividing through by the mass m, and rearranging some terms, we can write ODE (1) in normal linear form:

$$y'' + \frac{c}{m}y' + \frac{k}{m}y = -g + \frac{1}{m}f(t) \tag{7}$$

Now let's show how we can "get rid of gravity" in ODE (7).

Suppose that $h > 0$ is the *static deflection* when we attach the body of mass m to the spring (margin figure). To find h, note that for the body to hang at rest at the end of the spring, the force kh exerted upward by the spring must be equal to the downward gravitational force, that is, $kh = mg$. Replacing y in ODE (7) by $y = z - h$, we obtain an equivalent ODE in the new state variable z:

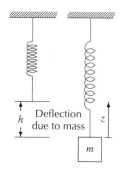

h Deflection due to mass z

$$z'' + \frac{c}{m}z' + \frac{k}{m}(z-h) = -g + \frac{1}{m}f(t)$$

$$z'' + \frac{c}{m}z' + \frac{k}{m}z = \frac{1}{m}f(t) \tag{8}$$

Note that the gravitational constant has disappeared from our model ODE. We conclude that our spring-body system would behave the same way (relative to the static deflection position) on the moon or on Mars as it would on the earth. However, since $h = mg/k$, the static deflection position itself *does* depend upon the value of g.

The example below shows how to do the modeling and how to handle the units.

EXAMPLE 3.1.1

☞ More on units in Appendix B.6.

A Damped and Driven Hooke's Law Spring: From the Data to the ODE

A Hooke's Law spring is suspended from a support that is attached to a motor. With the motor off, a 1-lb weight is hung from the spring's free end. The resulting static deflection is 15.36 in. A damper provides a damping force of $-cv$, where v is the velocity of the weight relative to the support, and the value of the damping constant c is 1.30×10^{-4} lb·sec/in. The motor is then turned on and exerts an oscillating vertical driving force of frequency 5.6 radians per second and amplitude 0.26

$$f(t) = 0.26\sin(5.6t) \text{ lb}$$

The weight's displacement $z(t)$ is measured from the static deflection position.

Let's use the given data to find the terms in ODE (8) for $z(t)$. Since the weight mg of the body is given, we first multiply ODE (8) by mg:

$$mgz'' + cgz' + kgz = gf(t)$$

We have $mg = 1$ lb, $c = 1.30 \times 10^{-4}$ lb·sec/in, $h = 15.36$ in, $g = 32$ ft/sec^2 = 384 in/sec^2, and $k = mg/h$, so

$$cg = 1.30 \times 10^{-4} \times 384 = 0.05 \text{ lb/sec}$$

$$kg = (mg/h)g = (1/15.36)384 = 25 \text{ lb/sec}^2$$

$$gf(t) = 384(0.26 \sin 5.6t) = 100 \sin(5.6t) \text{ in·lb/sec}^2$$

So the IVP that describes the motion of the weight from an initial rest state is

$$z'' + 0.05z' + 25z = 100\sin(5.6t), \qquad z(0) = 0, \quad z'(0) = 0$$

where z is measured in inches. We show how to find an exact solution formula for any IVP of this kind in Section 3.6. For now, we use a numerical solver to find and graph the solution $z = z(t)$ (Figure 3.1.1). The figure shows that the weight suspended by this spring is in for a wild ride!

We know from experience that a weighted spring stays at rest in an equilibrium position if no external force acts on it. This rest state is determined from an equilibrium solution of the model ODE. So now let's look at equilibrium states for various spring models.

Autonomous ODEs: Equilibrium Solutions

☞ We need autonomous ODEs to talk about equilibrium solutions. Peek back to Section 2.2 to refresh your memory.

A normalized *autonomous* second-order ODE has the form

$$y'' = F(y, y') \tag{9}$$

where the acceleration function F does not depend upon t. Suppose that y_0 is such that $F(y_0, 0) = 0$. Then the constant $y = y_0$, for all t, is called an *equilibrium solution*. In other words, if the system starts out at rest [i.e., $y'(0) = 0$] when $y(0) = y_0$, then the system stays at rest at this point forever. Here's why. Suppose that $y(t) = y_0$ for all t. Then $y'(t) = 0$ and $y''(t) = 0$. But $F(y_0, 0) = 0$ by assumption, so $y(t) = y_0$, for all t, is indeed a solution of IVP (9). The point $(y_0, 0)$ in the yy'-plane is called an *equilibrium point*

EXAMPLE 3.1.2

Equilibrium Solution for a Hooke's Law Linear Spring Model
The Hooke's Law linear spring ODE (7) with $f(t) = 0$, all t, is an autonomous ODE. Writing this ODE as

$$y'' = -\frac{k}{m}y - \frac{c}{m}y' - g$$

we see from comparison with ODE (9) that

$$F(y, y') = -(k/m)y - (c/m)y' - g$$

There is only one equilibrium solution because $F(y_0, 0) = 0$ implies that $y_0 = -mg/k$, which agrees with the static deflection value h computed earlier.

☞ Appendix B.6 takes up the messy issue of competing systems of units.

Here's a nonlinear autonomous ODE with three equilibria. Note that we use metric units here, rather than the British Engineering units of Example 3.1.1.

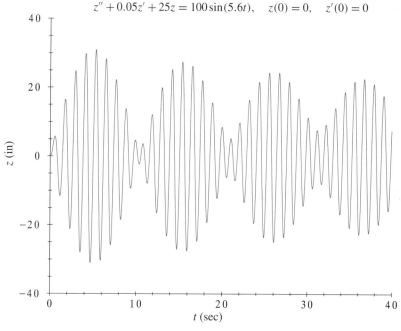

$$z'' + 0.05z' + 25z = 100\sin(5.6t), \quad z(0) = 0, \quad z'(0) = 0$$

☞ The odd shape
of this solution curve
results from combining
two oscillatory
functions. More on this
in Section 4.2.

FIGURE 3.1.1 Damped, driven Hooke's Law spring (Example 3.1.1).

EXAMPLE 3.1.3

Equilibrium Solutions for a Soft Spring Model
The damped and undriven soft spring ODE with $g = 9.8$ m/sec², $c/m = 0.2$ sec⁻¹,
$k/m = 10$ sec⁻², $j/m = 0.2$ m⁻² sec⁻² is

$$y'' = -10y + 0.2y^3 - 0.2y' - 9.8$$

Using the notation of ODE (9) we see that

$$F(y, y') = -10y + 0.2y^3 - 0.2y' - 9.8$$

So the equilibrium solutions y satisfy the condition

$$F(y, 0) = 0.2y^3 - 10y - 9.8 = 0$$

After a bit of experimentation, we discover that $y = -1$ is a root, and this cubic equation can be factored as

$$(y + 1)(0.2y^2 - 0.2y - 9.8) = 0$$

It appears that $F(y, 0)$ has three roots: $y = -1$ and (using the quadratic formula) $y = (1 \pm \sqrt{197})/2 \approx 7.52, -6.52$.
 So the corresponding equilibrium solutions are

$$y_1 = -1, \quad y_2 \approx 7.52, \quad y_3 \approx -6.52, \qquad \text{for all } t$$

while the equilibrium points in the yy'-plane are $(y_1, 0)$, $(y_2, 0)$, $(y_3, 0)$. The first point corresponds to the position of static equilibrium for the soft spring, but the two outlying equilibrium points are strange. More about this later in the section.

Some numerical solvers will not accept second-order ODEs, but the simple trick below (first seen in Section 1.5) will make such ODEs acceptable to most solvers.

From a Second-Order ODE to a First-Order System

Suppose $y(t)$ solves the IVP

$$y'' = F(t, y, y'), \qquad y(t_0) = y_0, \quad y'(t_0) = v_0 \tag{10}$$

If we let $v = y'$, then we see that $v(t)$ satisfies the ODE $v' = F(t, y, v)$ and the initial condition $v(t_0) = v_0$. So the pair of functions $y(t)$, $v(t)$ solves the system IVP

$$\begin{aligned} y' &= v, & y(t_0) &= y_0 \\ v' &= F(t, y, v), & v(t_0) &= v_0 \end{aligned} \tag{11}$$

☞ This interplay between a single higher-order ODE and first-order systems comes in handy.

The other way around, if the pair of functions $y(t)$, $v(t)$ solves IVP (11), then we see that $y(t)$ also solves IVP (10).

The *state space* for the differential system in IVP (11) is the yv-plane. The solution $y = y(t)$, $v = v(t)$ of IVP (11) for specific initial data y_0, v_0 generates a curve in the yv-plane parametrically, a curve that is called an *orbit* of the differential system in IVP (11). Orbits show graphically the evolution of the state of a physical system modeled by the ODE, and we will use them a lot for this purpose. A collection of orbits is known as a *state portrait*. Here's an example.

EXAMPLE 3.1.4

State Portrait for a Soft Spring ODE
Let's take still another look at the IVP for a soft spring:

$$y'' = -10y + 0.2y^3 - 0.2y' - 9.8, \qquad y(0) = y_0, \quad y'(0) = v_0$$

Converting this IVP to an equivalent IVP for a first-order system, we obtain

$$\begin{aligned} y' &= v, & y(0) &= y_0 \\ v' &= -10y + 0.2y^3 - 0.2v - 9.8, & v(0) &= v_0 \end{aligned} \tag{12}$$

See Figure 3.1.2 for a state portrait of IVP (12) for various values of y_0 and v_0. The three equilibrium points obtained in Example 3.1.3 are indicated by the dots on the line $v = 0$.

The state portrait shows that if the initial data are chosen near the static equilibrium values $y = -1$ and $v = 0$, then the solution $y = y(t)$, $v = v(t)$ tends to oscillate about the equilibrium point $(-1, 0)$ with amplitude that steadily decreases toward zero. But orbits originating near the other two equilibrium points are very sensitive to the actual choice of initial data. The orbits that escape as time advances demonstrate the limits of the validity of the soft spring model. For example, the two orbits at the right of Figure 3.1.2 that escape upward as time increases would correspond to the soft spring undergoing unbounded compression. But the coils of a compressed spring soon bunch up against one another, so infinite compression is nonsense. In spite of difficulties near

☞ Any guesses about soft spring behavior near the left equilibrium point in Figure 3.1.2?

$$y' = v, \quad v' = -10y + 0.2y^3 - 0.2v - 9.8$$

☞ The arrowheads
show the advance of time
on the orbits.

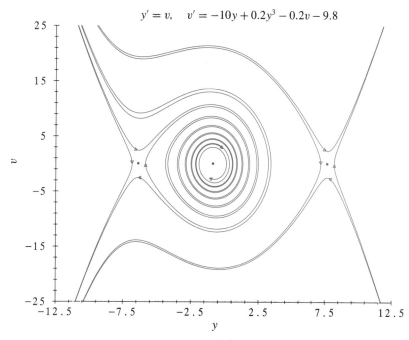

FIGURE 3.1.2 State portrait of orbits for a soft spring ODE (Example 3.1.4).

the outer two equilibria, the model predicts the behavior of the vibrating soft spring quite accurately near the inner equilibrium point at $y = -1$, $v = 0$.

After modeling a physical situation and plotting orbits and solution curves, it is always important to think about the physical interpretation of the curves. As the examples showed, it may be necessary to restrict the validity of a model or to modify the model equations. Experience with model building indicates that no matter what natural process is considered, the best models overall have a nonlinear mathematical formulation. So how do we get by with using so many linear models?

Nonlinear Model ODEs with Good Linear Approximations

Practitioners sometimes describe the nonlinear behavior of natural processes by saying that "nature is inherently nonlinear." Interestingly enough, however, we will see that linear ODEs do arise in the modeling of natural processes, and the results that they produce often agree rather well with physical reality. So, linear ODEs occupy a very special place in the construction of mathematical models, and that is the reason why the bulk of this chapter and the next is dedicated to them. Linear ODEs are also important because a knowledge of the behavior of their solutions is useful in the investigation of the behavior of solutions of more general ODEs.

For our next example we will use the definition of the linear approximation of a function $F(u, v)$ at a base point (u_0, v_0).

❖ **Linear Approximation**. Suppose that $F(u, v)$ is continuously differentiable in a rectangle R in the uv-plane and (u_0, v_0) is a point inside R. Then the function $G(u, v)$ given by

☞ The value of the partial derivative $\partial F(u, v)/\partial u$ at the point (u_0, v_0) is often written as $F_u(u_0, v_0)$.

$$G(u, v) = F(u_0, v_0) + F_u(u_0, v_0)(u - u_0) + F_v(u_0, v_0)(v - v_0) \qquad (13)$$

is called the *linear approximation* to $F(u, v)$ at the base point (u_0, v_0).

☞ Theorem B.5.15 is Taylor's Theorem.

$G(u, v)$ is obtained by using Taylor's Theorem for $F(u, v)$ at the base point (u_0, v_0) and keeping only the constant and linear terms in $u - u_0$ and $v - v_0$. We can use these ideas to approximate the nonlinear soft spring ODE by a linear ODE, where we use an equilibrium point as the base point.

EXAMPLE 3.1.5

Linearization of the Nonlinear Soft Spring ODE

Let's look again at the IVP for a soft spring given in Example 3.1.3:

$$y'' = -10y + 0.2y^3 - 0.2y' - 9.8, \qquad y(0) = y_0, \quad y'(0) = v_0 \qquad (14)$$

The ODE has the form $y'' = F(y, y')$, where

$$F(y, y') = -10y + 0.2y^3 - 0.2y' - 9.8$$

Let's linearize IVP (14) near the equilibrium point $y = -1$, $y' = 0$ in the yy'-plane. To calculate the linear approximation $G(y, y')$ to $F(y, y')$ at $(-1, 0)$, observe that $F(-1, 0) = 0$ and that

☞ The notation $(\cdots)|_{(-1,0)}$ means put $y = -1$ and $y' = 0$ for the y's and y''s inside the parentheses.

$$F_y(-1, 0) = (-10 + 0.6y^2)\big|_{(-1,0)} = -9.4, \qquad F_{y'}(-1, 0) = -0.2$$

So we see that the linear approximation to F at $(-1, 0)$ is

$$G(y, y') = -9.4(y + 1) - 0.2y'$$

The IVP

$$y'' = -9.4(y + 1) - 0.2y', \qquad y(0) = y_0, \quad y'(0) = v_0 \qquad (15)$$

☞ Computer tip: Problem 7 shows how to coax your solver to plot solution curves of different ODEs on the same graph.

is called the *linearization* (or *linear approximation*) *of IVP* (14) at the static equilibium point $(-1, 0)$. Figure 3.1.3 shows how well the solution curve (dashed) of the linearized IVP (15) with $y_0 = -1$ and $v_0 = 5$ tracks the solution curve (solid) of the nonlinear IVP (14). The slight phase difference (i.e., time lag) between the two curves is a consequence of linearization.

Both curves oscillate with decaying amplitude about the position $y = -1$ of static equilibrium. This is the kind of motion we would expect for a damped spring near the static equilibrium since friction gradually dissipates energy and (in this case) there is no driving force.

Since Taylor's Theorem was used to linearize IVP (14), we should expect the approximation to provide good results only if the initial data y_0, v_0 is close enough to the static equilibrium values $y_0 = -1$, $v_0 = 0$. Figure 3.1.4 shows that if the initial point is far away from the equilibrium point $(-1, 0)$, the solution (dashed) of the

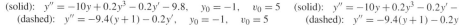

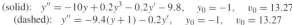

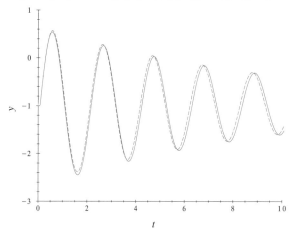

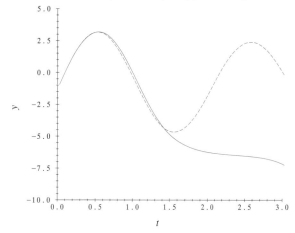

FIGURE 3.1.3 Solution curve (solid) of nonlinear IVP (14) with $y_0 = -1$, $v_0 = 5$; solution curve (dashed) of its linear approximation IVP (15). See Example 3.1.5.

FIGURE 3.1.4 Solution curve (solid) of nonlinear IVP (14) with $y_0 = -1$, $v_0 = 13.27$; solution curve (dashed) of its linear approximation IVP (15). Note how bad the approximation is for $t \geq 1.5$.

linearized IVP may eventually deviate quite a bit from the solution curve (solid) of the nonlinear IVP. In this case, we should be cautious about using the linearized ODE to predict the behavior of the spring.

Comments

The second-order ODEs that model linear and nonlinear springs are widely used in science and engineering as guides to understanding the motions of real springs. The ODEs themselves may be used to model a host of other physical phenomena, and so there is good reason to study the ODEs independent of any particular interpretation.

PROBLEMS

1. (*Hooke's Law Spring*). Let's consider a 1-lb weight suspended from an undamped, undriven Hooke's Law spring. Suppose that z denotes the deflection of the weight from its equilibrium position.

   (a) If the weight's static deflection is 24 in, show that $z(t)$ satisfies the ODE $z'' + 16z = 0$, where z is measured in inches. [*Hint*: Weight is mg and g is 384 in/sec^2.]

   (b) Suppose that the weight is pushed up z_0 inches and released from rest, where z_0 takes in turn each of the values 15, 10, 5, and 2 inches. Plot the resulting orbits in zz'-state space, inserting arrowheads to indicate increasing time.

   (c) Plot the tz- and tz'-component graphs for each of the orbits in part **(b)**. What can you say about the motion? If it's periodic, what is the period?

2. (*Damped Hooke's Law Spring*). Consider a 1-lb weight suspended from a Hooke's Law spring. Suppose that z denotes the deflection (in inches) of the weight from its equilibrium position. Assume a viscous damping force but no driving force acting on the weight.

(a) If the static deflection is 24 in and the damping coefficient $c = 2.60 \times 10^{-4}$ lb·sec /in, show that $z(t)$ satisfies the ODE $z'' + 0.1z' + 16z = 0$. [*Hint:* Weight $= mg$, and $g = 384$ in/sec^2.]

(b) Suppose that the weight is pushed up z_0 inches and released from rest, where z_0 takes, in turn, the values 15, 10, and 2 inches. Plot the resulting orbits in zz'-state space, inserting arrowheads to show the direction of increasing time.

(c) Plot the tz- and tz'-component curves for each of the orbits in part (b). What can you say about the body's motion and the sequence of transit times through the t-axis?

www **3.** (*Nonlinear Springs*). A body of mass m is attached to a wall by means of a spring and moves back and forth along its axis on a horizontal frictionless surface. Suppose that y measures the displacement of the body along the spring's axis from the equilibrium position; $y < 0$ corresponds to a stretched spring, $y > 0$ to a compressed spring. The margin figure shows the body with viscous damping forces represented by a damper. The downward gravitational force on the body is exactly canceled by an upward force exerted by the supporting surface, so the mass moves only horizontally, unaffected by gravity. Under these assumptions, Newton's Second Law (interpreted for horizontal motion) implies that

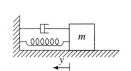

$$my''(t) = S(y(t)) - cy'(t)$$

where $c > 0$ is the damping constant. Some possible formulas for the *spring restoring force* $S(y)$ are given by equations (2)–(5). In parts (a)–(c) below first write the equivalent system as in IVP (11). Plot the orbits through the given initial points in the yv-plane, and plot the corresponding solution curves in the ty-plane. Interpret your graphs in terms of the motion of a spring, and discuss the "reality" of each orbit. Estimate the periods of periodic solutions; does the period increase, decrease, or stay the same as the magnitude of the initial velocity increases?

(a) (*Undamped Hard Spring*). $y'' = -0.2y - 0.02y^3$, $0 \le t \le 25$; $y_0 = 0$, $v_0 = 0, 1, 3, 9$.

(b) (*Undamped Soft Spring*). $y'' = -0.2y + 0.02y^3$, $0 \le t \le 30$; $y_0 = 0$, $v_0 = 0, 0.4, 0.9$; $y_0 = -5$, $v_0 = 1.49, 1.51$; $y_0 = 5$, $v_0 = -1.49, -1.51$.

(c) (*Damped Soft Spring*). $y'' = -y + 0.1y^3 - 0.1y'$, $0 \le t \le 40$; $y_0 = 0$, $v_0 = 0, 2.44, 2.46$.

(d) (*Linearization*). Linearize the ODE in part (c) at the static equilibrium $y = 0$, $y' = 0$. Plot orbits of the original and of the linearized system near this equilibrium state, using the same initial data sets for the two systems. Explain the difference between the two state portraits. Using the same data sets, plot ty-solution curves of the two systems, and explain what you see.

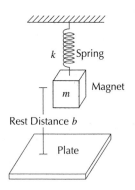

4. (*Modeling a Magnetically Driven Spring*). As in the margin figure, a magnet is suspended by a spring above a fixed iron plate. Suppose that z is the vertical displacement of the magnet from its rest position at distance b from the plate. Assume that the force of magnetic attraction is inversely proportional to the square of the distance between the magnet and the plate, and that the spring force is directly proportional to the displacement z. Find the ODE for the magnet's motion. Is it linear?

5. (*Effect of Gravity on Spring Motion*). Does gravity affect the vertical motion of a weight on a hard spring? On a soft spring? Contrast your conclusions with the Hooke's Law spring, and give reasons.

6. (*Driven Hooke's Law Spring*). In a Hooke's Law spring model system, suppose that the spring constant is 1.01 newtons/meter and the mass of the body is 1 kg. The body is acted on by viscous damping with a coefficient equal to 0.2 newton · second/meter, and driven by the periodic force (measured in newtons) described by ODE (8) with $f(t) = $ sqw$(t, 50, 2\pi)$. Graph the response $z(t)$ of the system if it is at rest in its equilibrium state at $t = 0$, and interpret what you see. [*Hint:* Use a numerical solver to solve the IVP that you construct to model the motion.]

☞ Appendix B.1 defines the square wave function sqw(t, d, T).

7. (*More Solver Tricks, Another Way to Generate Figures 3.1.3 and 3.1.4*). Some solvers may not be able to overlay plots from two different ODEs on the same set of axes. Follow the outline below to trick your solver into reproducing Figures 3.1.3 and 3.1.4.

- Let's consider the IVPs

$$y'' = (1 - c)(-10y + 0.2y^3 - 0.2y' - 9.8) + c[-9.4(y + 1) - 0.2y']$$

$$y(0) = -1, \quad y'(0) = 5, 13.27$$

where c is a constant. Explain why plotting solutions of these IVPs for $c = 0$ and $c = 1$ generates the graphs in Figures 3.1.3 and 3.1.4.

- Show that by denoting the state variable y' by v, the IVPs above can be converted to the equivalent IVPs for a first-order system:

$$y' = v, \qquad\qquad\qquad\qquad\qquad\qquad\qquad y(0) = -1$$
$$v' = (1 - c)(-10y + 0.2y^3 - 0.2v - 9.8) + c[-9.4(y + 1) - 0.2y'], \qquad v(0) = 5, 13.27$$

For $c = 0$ and $c = 1$ graph orbits of these IVPs.

- Suppose that $c = c(t)$ is another state variable and consider the following IVPs for a first-order system with three state variables:

$$y' = v, \qquad\qquad\qquad\qquad\qquad\qquad\qquad y(0) = -1$$
$$v' = (1 - c)(-10y + 0.2y^3 - 0.2v - 9.8) + c[-9.4(y + 1) - 0.2y'], \qquad v(0) = 5, 13.27$$
$$c' = 0, \qquad\qquad\qquad\qquad\qquad\qquad\qquad c(0) = c_0$$

Solver tip: This is how our solver produced Figure 3.1.3.

Show that if these IVPs are solved for $c_0 = 0$ and $c_0 = 1$ and y plotted against t in both cases, then Figures 3.1.3 and 3.1.4 result. Graph the orbits.

- Plot the solution curves of $y'' = -y + cy^3$, $y(0) = 1$, $y'(0) = 0$, for $c = 0, -0.1, 0.1$, $0 \leq t \leq 50$ on the same screen. Interpret in terms of particular types of springs.

3.2 Second-Order ODEs and Their Properties

We will state a general existence/uniqueness/sensitivity/continuity theorem that applies to almost all of the second-order initial value problems one is ever likely to meet in modeling natural processes. In this theorem we take up the same issues for second-order IVPs that we encountered in Sections 2.1–2.3 for first-order IVPs.

Second-order ODEs need two state variables (such as position and velocity in mechanical models), which means we need three dimensions in order to graph solution behavior, one dimension for time and two for the state variables. Curves can do things in three dimensions that aren't possible in two, so we expect to see some strange and unusual pictures. Toward the end of this section we introduce a descriptive vocabulary for solution graphs and list some of the properties that solution graphs must have.

The Fundamental Theorem for a Second-Order IVP

The *normal form* for a second-order IVP is

$$y'' = F(t, y, y'), \qquad y(t_0) = y_0, \quad y'(t_0) = v_0 \qquad\qquad (1)$$

A function $y(t)$ defined on a t-interval I containing t_0 is a *solution* of IVP (1) if y is twice continuously differentiable in t and if $y(t)$ satisfies (1) for all t in I. As t runs through I, the point $(t, y(t), y'(t))$ traces out a curve in tyy'-space called a *time-state curve*. The *data* of IVP (1) are the function F and the initial values y_0, v_0.

Now we can state the following theorem.

THEOREM 3.2.1

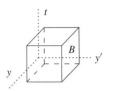

☞ A box is *closed* if it contains its bounding faces.

Fundamental Theorem for Second-Order ODEs.   Suppose that F, $\partial F/\partial y$, and $\partial F/\partial y'$ are continuous in a closed box B in tyy'-space, and the point (t_0, y_0, v_0) lies inside B. Then

Existence/Uniqueness: The IVP

$$y'' = F(t, y, y'), \qquad y(t_0) = y_0, \quad y'(t_0) = v_0$$

has a unique solution $y(t)$ on a t-interval I containing t_0.

Extension: The interval I on which the solution of the IVP is defined can be extended until the endpoints of the time-state curve hit the boundary of B.

Sensitivity/Continuity: The solution of the IVP is a continuous function of the data over a sufficiently small t-interval that contains t_0.

This important theorem is a direct extension of Theorem 2.3.3 for first-order IVPs. The existence/uniqueness property assures us that a second-order IVP that we construct to model a natural process has the same feature of repeatability as that process. So if we solve the IVP twice with exactly the same data, then we obtain exactly the same solution. The sensitivity/continuity property assures us that even with changes in the data, the solution of the changed IVP stays within a specified tolerance of the solution of the original IVP if the data changes are small and if we confine our attention to a small enough t-interval about t_0.

☞ This is called *continuity in the data*

We assume from this point on that all solutions are extended forward and backward in time as far as possible, that is, they are *maximally extended*. With this point in mind, the uniqueness conclusion of Theorem 3.2.1 implies that no two distinct time-state curves of the ODE $y'' = F(t, y, y')$ can touch in box B. Moreover, no time-state curve can die inside the box, so each curve reaches from one rectangular face of the box to another. Since the initial point (t_0, y_0, v_0) can be placed anywhere inside the box, we see that the time-state curves fill up B, each point being on one curve.

We can say more if the ODE is in the normal linear form

$$y'' + a(t)y' + b(t)y = f(t)$$

but let's postpone that discussion until Sections 3.3–3.7.

Now let's look at some graphs associated with second-order ODEs.

Time-State Curves, Solution Curves, Component Curves, Orbits

Suppose that $y(t)$ is a solution of the differential equation

$$y'' = F(t, y, y')$$

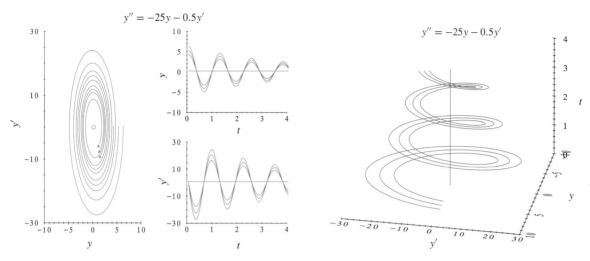

FIGURE 3.2.1 Orbits and component curves for the damped spring model (Example 3.2.1).

FIGURE 3.2.2 Time-state curves for the damped spring model (Example 3.2.1).

for t in an interval I. We have already defined the time-state curve associated with the solution $y(t)$. Here are some other graphs. The graph of $y(t)$ in the ty-plane is called a *solution curve*. The solution curve in the ty-plane and the graph of $y'(t)$ in the ty'-plane are the *component curves*. The *orbit* (or *trajectory*) is the parametric graph of $y = y(t)$, $y' = y'(t)$ in the yy'-plane (the *state plane*), where t is the parameter. The first example shows some of these curves.

<table>
<tr><td>**EXAMPLE 3.2.1**</td></tr>
</table>

Curves for a Damped, Hooke's Law Spring Model

Let's look at the curves corresponding to the solutions of the four IVPs

$$y'' = -25y - 0.5y', \qquad \begin{cases} y(0) = 0, & y'(0) = 0 \\ y(0) = 0, 4, 5, 6, & y'(0) = 0 \end{cases} \qquad (2)$$

Here y is the position relative to the static equilibrium of a weight suspended by a damped Hooke's Law spring, where we assume that t, y, and the coefficients in the ODE are measured in some consistent set of units. The acceleration function

$$F(t, y, y') = -25y - 0.5y'$$

is continuous for all t, y, and y', as are $\partial F/\partial y = -25$ and $\partial F/\partial y' = -0.5$. So the conditions of Theorem 3.2.1 are satisfied in any box B that contains the initial points.

The left graph in Figure 3.2.1 shows the *point-orbit* at the origin corresponding to the static equilibrium, as well as the other three orbits that spiral inward toward the origin as time increases. The corresponding solution curves $y = y(t)$ are shown in the ty-plane at the top right of Figure 3.2.1, while the other component curves, $y' = y'(t)$ in the ty'-plane, appear at the lower right. The horizontal lines $y = 0$, $y' = 0$ are the component curves for the equilibrium solution. The four time-state curves appear in Figure 3.2.2 (the vertical line corresponds to equilibrium).

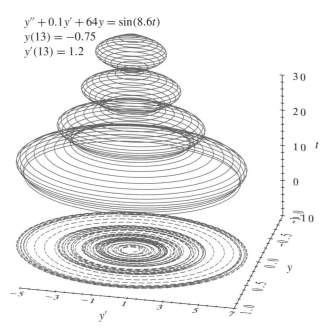

FIGURE 3.2.3 Time-state curve (solid) of the chapter cover IVP and its orbit (dashed) (Example 3.2.2).

Each of the numerically computed graphs in Figures 3.2.1 and 3.2.2 tells us something about the behavior of the solutions of the IVPs in (2), even though we don't have any solution formulas. For example, the solution curves in Figure 3.2.1 show that successive transits through the equilibrium position $y = 0$ are separated by about 0.63 of a unit of time, a number independent of the initial conditions. All of the graphs suggest that the amplitudes of the vibrations decay exponentially to zero as $t \to +\infty$.

Properties of the Curves and Orbits

Theorem 3.2.1 implies that if F, $\partial F/\partial y$, and $\partial F/\partial y'$ are continuous, then the curves associated with the second-order ODE

$$y'' = F(t, y, y') \tag{3}$$

have several important properties. We have already mentioned some of these, but we list them all here for ease of reference.

- **Time-state curves of the ODE never intersect.** This follows from uniqueness.
- **Solution curves may intersect.** A solution curve is the projection of a time-state curve onto the ty-plane. So solution curves might intersect, even though time-state curves never do. If two solution curves intersect, they do so with different slopes; otherwise uniqueness would be violated. For example, the five solution curves shown at the upper right of Figure 3.2.1 intersect one another many times, but always with different slopes.

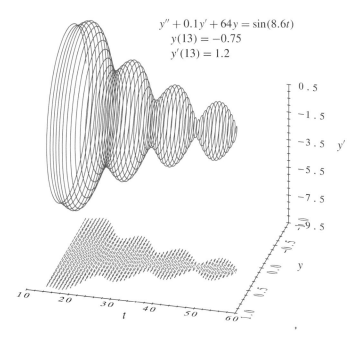

$$y'' + 0.1y' + 64y = \sin(8.6t)$$
$$y(13) = -0.75$$
$$y'(13) = 1.2$$

FIGURE 3.2.4 Another view of the time-state curve (solid); the solution curve (dashed) is the projection onto the ty-plane (Example 3.2.2).

- **Orbits may intersect.** An orbit is the projection of a time-state curve onto the yy'-state plane, so two orbits may intersect. But the two orbits must pass through an intersection point at different times; otherwise the uniqueness property would be violated. An orbit may even intersect itself.

The next example shows a self-intersecting orbit.

| EXAMPLE 3.2.2 | **A Damped and Driven Hooke's Law Spring Model** |

The time-state curve of the IVP

$$y'' + 0.1y' + 64y = \sin(8.6t), \qquad y(13) = -0.75, \quad y'(13) = 1.2 \qquad (4)$$

☞ This answers the questions posed in the chapter cover figure.

is shown in the cover figure for this chapter. If we project this curve down into a yy'-plane, we get the orbit. With different scales on the axes, Figure 3.2.3 shows the same time-state curve (solid) and also the orbit (dashed) over the time span $13 \leq t \leq 54$. The orbit is plotted in the yy'-state plane at $t = -10$ so that the orbit and the time-state curve are visually separated. Note that the orbit intersects itself.

The time and velocity axes are interchanged and rescaled again in Figure 3.2.4. The time-state curve is solid, and the solution curve (dashed) is the projection down onto a ty-plane. This plane is taken to be the $y' = -9.5$ plane to avoid visually mixing the curves.

We very carefully chose the axis scales, the starting time $t_0 = 13$, position $y_0 = -0.75$, and velocity $v_0 = 1.2$ to get the unusual graphs shown here. Other choices lead to curves that look quite different.

The following properties apply only to the curves of an *autonomous* ODE

$$y'' = F(y, y') \tag{5}$$

where F does *not* depend on time. We assume that F, $\partial F/\partial y$, and $\partial F/\partial y'$ are continuous and that every solution of ODE (5) is maximally extended.

☞ First-order
autonomous ODEs have
this time-shifting
property, too. See
Example 2.2.5.

- **Time-shifting a solution curve of the autonomous ODE (5).** If $y(t)$ solves ODE (5) over the interval $t_0 \leq t \leq t_1$, then for any constant T the function $y(t+T)$ solves the same ODE over $t_0 - T \leq t \leq t_1 - T$. The solution curve for $y(t+T)$ is just $y(t)$ shifted T units backward along the t-axis if $T > 0$, and forward along the t-axis if $T < 0$. So the solution curve of an autonomous ODE over a t-interval of specified length would look the same no matter when the clock is started, except that it would be shifted one direction or the other along the t-axis to accommodate the reset clock. The same property holds for the time-state curves in tyy'-space. For example, the time-state curves in Figure 3.2.2 would be shifted along the t-axis if a new starting time other than $t_0 = 0$ is actually used.

- **Initial time for an orbit of an autonomous ODE.** An orbit of an autonomous ODE does not depend on when the clock is started (i.e., on t_0), but only on the duration of time for which the orbit is defined. The reason is that time-state curves merely shift along the t-axis when the initial time is changed, and an orbit is just a projection of a time-state curve onto the yy'-plane.

- **Distinct orbits of an autonomous ODE do not intersect.** Suppose that the *distinct* orbits $(z(t), z'(t))$ and $(w(t), w'(t))$ *do* intersect at a point (y_0, v_0). Then $z(t_1) = w(t_2) = y_0$ and $z'(t_1) = w'(t_2) = v_0$ for t_1 and t_2. By the item above, the orbit $w(t)$ could be generated using *any* starting time (so long as the solution time interval has the same length), so let's select the initial conditions for w to be $w(t_1) = y_0$ and $w'(t_1) = v_0$. Then we find that $z(t)$ and $w(t)$ solve the same IVP:

$$y'' = F(y, y'), \qquad y(t_1) = y_0, \quad y'(t_1) = v_0$$

  and so by uniqueness the two orbits must be identical, a contradiction. This shows that distinct orbits of an autonomous system can't intersect.

- **A self-intersecting orbit of an autonomous system is periodic.** If a nonconstant orbit of an autonomous system intersects itself after T units of time (but not before), then by restarting the clock at that point we see that the orbit repeats itself after every T units of time, so the orbit represents a periodic solution of period T.

The following examples illustrate the last property.

EXAMPLE 3.2.3

☞ The ODE
$y'' + \omega^2 y = 0$ is called
the *harmonic
oscillator*, and will be
looked at in detail in
section 3.5.

Harmonic Oscillator

The autonomous ODE

$$y'' + 4y = 0 \tag{6}$$

models the displacement from static equilibrium of an undamped, undriven Hooke's Law spring, where all of the quantities are measured in some consistent set of units. Figure 3.2.5 shows the orbits (at the left) and the component curves (at the right) that

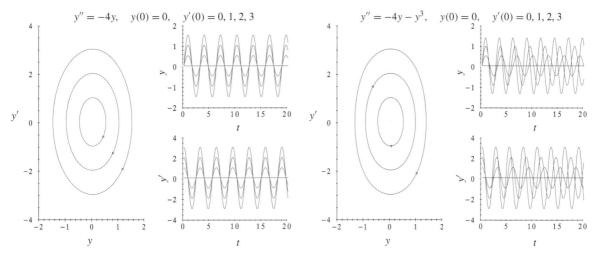

$y'' = -4y, \quad y(0) = 0, \quad y'(0) = 0, 1, 2, 3$

$y'' = -4y - y^3, \quad y(0) = 0, \quad y'(0) = 0, 1, 2, 3$

FIGURE 3.2.5 Orbits and component curves of a harmonic oscillator (Example 3.2.3).

FIGURE 3.2.6 Orbits and component curves of a hard spring model (Example 3.2.4).

correspond to the initial points $y(0) = 0$, $y'(0) = 0, 1, 2, 3$. The nonconstant solutions seem to be periodic with a constant period T of about 3.1 (in fact, $T = \pi$, as we see in Section 3.5). Would you expect this behavior for undamped, undriven spring motion?

Now let's take the linear Hooke's Law spring of the harmonic oscillator and turn it into a nonlinear hard spring.

EXAMPLE 3.2.4

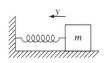

☞ The vertical forces on m balance out and so gravity doesn't appear in the ODE.

Motion of a Hard Spring

Suppose that a block of mass m is attached to a wall by means of a hard spring and moves back and forth without air resistance along its axis on a frictionless horizontal surface. Then, assuming no damping, the displacement $y(t)$ satisfies the autonomous nonlinear ODE

$$y'' = -ky - jy^3$$

where k and j are positive constants. The only equilibrium solution is $y(t) = 0$ for all t, so if we were to stretch or compress the spring away from equilibrium and then release it, we expect to see the spring-mass system move back toward equilibrium. Figure 3.2.6 shows the orbits and the component curves if $k = 4$ and $j = 1$. We use the initial points of the last example, $y(0) = 0$, $y'(0) = 0, 1, 2, 3$.

In Figure 3.2.6 the nonconstant solutions again seem to be periodic, but now the periods of the solutions are *not* constant as was the case in Figure 3.2.5. In fact, it appears that the larger the velocity y', the shorter the period (and so the higher the frequency). For example, the solution curve corresponding to $y(0) = 0$, $y'(0) = 3$ has eight "peaks" in the time span $0 \leq t \leq 20$, but the curve corresponding to $y_0 = 0$, $y'(0) = 1$ has only seven peaks.

Any guesses about the period of a small amplitude oscillation? How about the period of a large amplitude oscillation? Would you expect the oscillations of a hard spring with the extra snap-back force modeled by $-jy^3$ to have these properties?

The figures suggest many properties, and in the rest of this chapter, the next chapter, and the problem sets we will discuss the validity of these properties.

Comments

The graphs of this section were created by using a numerical solver for the system of first-order differential equations

$$y' = v, \qquad\qquad y(t_0) = y_0$$
$$v' = F(t, y, v), \qquad v(t_0) = v_0$$

which is equivalent to the second-order IVP

$$y'' = F(t, y, y'), \qquad y(t_0) = y_0, \quad y'(t_0) = v_0$$

This approach was described in Section 3.1, and will be used throughout the chapter.

The Fundamental Theorem for Second-Order ODEs (Theorem 3.2.1) supports all of the theory, and most of the models and examples in the remainder of this chapter and Chapter 4, although we don't often refer to the theorem directly. When we talk about the solution of an IVP for a second-order ODE, we have the properties given in this theorem in mind. If we change a parameter or initial value a little or let $t \to +\infty$, we are implicitly using the continuity and extension properties of a solution.

So far, we have had to rely on computer simulations in our study of second-order ODEs, but in the remaining sections we shift our focus to the construction of some actual solution formulas for linear second-order ODEs.

Figure 3.2.4 is a wonderful example of how graphical displays sometimes produce misleading optical illusions. The "pawn" display in Figure 3.2.4 is just another view of the pawn in Figure 3.2.3, but it looks like its base slants the wrong way. This appears to contradict the projection of this pawn into the ty-plane. Try rotating Figure 3.2.4 so that its pawn is standing on its base. Do you notice anything different?

PROBLEMS

 1. Verify that each IVP satisfies the conditions of Theorem 3.2.1 throughout tyy'-space. Plot component curves, orbits, and time-state curves. If the ODE is autonomous, find all equilibrium solutions. Make conjectures about the long-term behavior of solutions.

(a) (*Undamped Hooke's Law Spring*). $y'' + y = 0$, $y(0) = 0, 0.5, 1.0, 1.5$, $y'(0) = 0$; $0 \le t \le 20$, $|y| \le 2$, $|y'| \le 3$.

(b) (*Damped Hooke's Law Spring*). $y'' + 0.1y' + 4y = 0$, $y(0) = 0, 0.5, 1.0, 1.5$, $y'(0) = 0$; $0 \le t \le 40$, $|y| \le 2$, $|y'| \le 4$.

(c) (*Going Backward in Time*). $y'' - 0.1y' + 4y = 0$, $y(0) = 0, 1, 2$, $y'(0) = 0$; $-100 \le t \le 0$, $|y| \le 2, |y'| \le 4$.

(d) (*Vertical Oscillations of a Damped Hard Spring*). $y'' + 0.2y' + 10y + 0.2y^3 = -9.8$, $y(0) = -4, 0$, $y'(0) = 0$; $0 \le t \le 40$, $-5 \le y \le 3, |y'| \le 9$.

(e) (*Nonlinear ODE*). $y'' + (0.1y^2 - 0.08)y' + y^3 = 0$, $y(0) = 0$, $y'(0) = 0.5$; $0 \le t \le 100$, $-1.9 \le y \le 2.1, |y'| \le 2.5$.

(f) (*Driven and Damped Hooke's Law Spring*). $y'' + 0.05y' + 25y = \sin(5.5t)$, $y(0) = 0$, $y'(0) = 0$; $0 \le t \le 40$, $|y| \le 0.4, |y'| \le 2.1$.

2. (*Concavity*). The sign of the second derivative of $y(t)$ determines whether the graph of $y = y(t)$ is concave up ($y'' > 0$) or down ($y'' < 0$). For each of the ODEs determine the regions in the ty-plane where solution curves are concave down. Then plot solutions that meet given initial conditions and verify that the solution curves are concave down in the proper regions.

(a) $y'' = -1$; $y(0) = 1$; $y'(0) = 2, 0, -2$; use the window $0 \le t \le 10$, $-27 \le y \le 3$.

(b) $y'' = -y$; $y(0) = 1, 0, -1$; $y'(0) = 0$; use the window $0 \le t \le 20$, $|y| \le 1$.

(c) (*Painlevé Transcendent*).[1] The ODE $y'' = y^2 - t$ is one of the *Painlevé transcendents* (see Figure 3.2.7), a family of nonlinear ODEs studied by Paul Painlevé (1863–1933). Use the initial values $y(0) = 0$, $y'(0) = 0.75, 0.925$; use the window $0 \le t \le 10$, $|y| \le 5$. [*Hint*: Plot the parabola $t = y^2$; the concavity changes when a solution curve crosses the parabola. See also Problem 10.]

 3. Consider the nonlinear IVP

$$t[y''y + (y')^2] + y'y = 1, \qquad y(1) = 1, \quad y'(1) = v_0$$

using the screen size $0 \le t \le 3$, $|y| \le 5$. Plot several solution curves using several values of v_0 with $|v_0| \le 10$.

4. (*Oscillating Springs*).

(a) Reproduce the graphs in Figure 3.2.6. Estimate the periods of the solutions. What do you think will happen to the periods if the initial data is $y(0) = 0$, $y'(0) = v_0$ and v_0 is taken very large and positive?

(b) Plot the orbit, component curves, and the time-state curve for the IVP $y'' + y = \cos(1.1t)$, $y(0) = 5$, $y'(0) = 0$. Solve over the interval $0 \le t \le 75$. Describe what you see. Is the solution periodic? If so, what is the period?

www **5.** Find the solution formula for the IVP $y'' = 2y'y$, $y(0) = 1$, $y'(0) = 1$. Plot the solution curve and orbit of this IVP. What is the largest t-interval on which the solution of this IVP is defined? [*Hint*: Write $2y'y$ as $(y^2)'$.]

6. Suppose that $F(y)$ and $F'(y)$ are continuous. Describe a method for finding all orbits of the ODE $y'' = F(y)$. [*Hint*: Multiply the ODE by y' and write $y'y''$ as $(y'^2)'/2$.] Repeat for $y'' = y'F(y)$.

7. (*An Euler ODE*). Show that the IVP

$$t^2y'' - 2ty' + 2y = 0, \qquad y(0) = 0, \quad y'(0) = 0$$

has infinitely many solutions. Why does this fact not contradict the assertion of uniqueness in Theorem 3.2.1? [*Hint*: Look for solutions in the form Ct^α, where C and α are constants.]

8. (*Closed Curve Orbits of Harmonic Oscillator*). Show that the nonconstant orbits of the general harmonic oscillator ODE $y'' + \omega^2 y = 0$, where ω is a positive constant, are ellipses. [*Hint*: Use the technique described in Problem 6.]

9. (*Closed Curve Orbits of Undamped Hard Spring*). Show that the nonequilibrium orbits of the ODE $y'' = -ky - jy^3$, where j and k are positive constants, are simple closed curves. [*Hint*: Use the technique of Problem 6 or a reduction of order method from Section 1.7 and show that the orbits are given by $y' = v = \pm(C - ky^2 - jy^4/2)^{1/2}$, where C is any positive constant.]

---

[1]Material adapted from V. Anne Noonburg, "The Painlevé Transcendent," CODEE (Spring 1993).

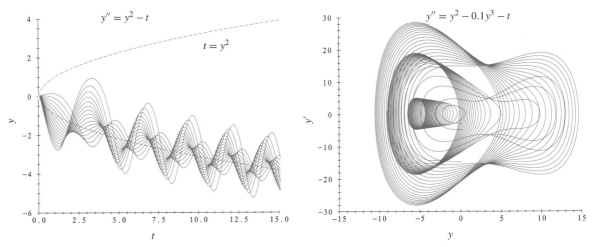

FIGURE 3.2.7 Solution curves of a Painlevé transcendent and graph of $t = y^2$ [Problems 2(c), 10(a)].

FIGURE 3.2.8 Three orbits of the perturbed Painlevé transcendent [Problem 10 (b)].

10. Here are examples of second-order nonlinear ODEs with striking solution curves.

(a) (*Painlevé Transcendent Again*). The ODE $y'' = y^2 - t$, is the first Painlevé transcendent [see also Problem 2(c)]. Duplicate the solid solution curves in Figure 3.2.7, where $y(0) = 0$, $-3.5 \leq y'(0) \leq 0.5$. What do solution curves do when they cross the parabola $t = y^2$ (dashed curve)? Why do the curves turn back to cross and recross the parabola? Now use initial data, $y(0) = 0$, $y'(0) = -4, 1$ and guess what happens to these two curves as t increases.

(b) (*Stereo Speakers? Bells?*). The ODE $y'' = y^2 - 0.1y^3 - t$ is a perturbation by the term $-0.1y^3$ of the Painlevé transcendent of part (a). Duplicate the orbital graphs shown in Figure 3.2.8 using $y(0) = 0$, $y'(0) = -3, -5, -15$, $0 \leq t \leq 50$. What do the time-state and component curves look like?

(c) Plot the solution curves of the transcendent and of the perturbed transcendent for $y(0) = 0$, $y'(0) = -3$, $0 \leq t \leq 15$. Explain why the two curves are close together. [*Hint*: See Problem 7 in Section 3.1 for a way to overlay solution curves of different ODEs.]

11. Suppose that the functions $F(t, y, y')$, $\partial F/\partial y$, and $\partial F/\partial y'$ are continuous for all t, y, y'.

(a) (*Incompatible Functions*). Explain why $y_1 = \sin t$ and $y_2 = t$, $-\infty < t < \infty$, can't both be solutions of the same ODE $y'' = F(t, y, y')$, regardless of the choice of F.

(b) (*More Incompatibility*). Repeat part (a), but with $y_1 = e^t$ and $y_2 = 1 + t + t^3/3$.

12. (*One Way to Generate Figure 3.2.3*). Explain why the procedure below generates the graphs in Figure 3.2.3. Consider the first-order differential system with three state variables:

$$y' = v$$

$$v' = -64y - 0.1v + \sin(8.6t)$$

$$z' = f(z), \quad \text{where} \quad f(z) = \begin{cases} 0, & z = 0 \\ 1, & z \neq 0 \end{cases}$$

Computer tip: We used this approach to produce Figure 3.2.3.

- Solve the system over $13 \leq t \leq 54$ under the conditions $y(13) = -0.75$, $v(13) = 1.2$, $z(13) = 13$, and plot the solution curve in xyz-space. [*Hint*: Note that $z(t) = t$.]

- Repeat with $z(13) = -10$, and plot the orbit in xyz-space. [*Hint*: $z(t) = -10$, all t.]

- Reproduce the graphs in Figure 3.2.3 by following the above procedure.

- Find a procedure that will reproduce the graphs in Figure 3.2.4.

3.3 Undriven Constant Coefficient Linear ODEs, I

We have seen how second-order ODEs are used to model natural systems, for example, the oscillating spring of Section 3.1, and we outlined the theory of second-order ODEs in Section 3.2. Now it's time to search for some solution formulas, and we will be successful in a very important special case.

In this and the next section we will derive formulas that capture all solutions for the undriven linear ODE

$$y'' + ay' + by = 0 \tag{1}$$

where a and b are real numbers. According to Theorem 3.2.1, the IVP

$$y'' + ay' + by = 0, \qquad y(t_0) = y_0, \quad y'(t_0) = v_0 \tag{2}$$

has a unique solution $y(t)$ for each set of values of t_0, y_0, and v_0. We will find a formula for that solution.

Finding Solution Formulas

One way to find a solution formula for ODE (1) is to make a lucky guess. What should we guess? Undriven first-order linear ODEs like $y' + ay = 0$, where a is a constant, have solutions of the form Ce^{rt} for a certain constant r, and any constant C. So let's try to find a solution of ODE (1) in the same form.

Substituting $y = Ce^{rt}$ into ODE (1), we obtain

$$(Ce^{rt})'' + a(Ce^{rt})' + b(Ce^{rt}) = r^2 Ce^{rt} + arCe^{rt} + bCe^{rt}$$
$$= Ce^{rt}(r^2 + ar + b) = 0 \tag{3}$$

The last equality in (3) is true if and only if $C = 0$ (which gives the *trivial solution* $y = 0$) or r is a root of $r^2 + ar + b$. The polynomial $r^2 + ar + b$ is the *characteristic polynomial* of ODE (1) and its roots r_1 and r_2 are the *characteristic roots* of the ODE. In this section we assume that r_1 and r_2 are real; we will consider the complex case in the next section.

It follows from (3) that $C_1 e^{r_1 t}$ and $C_2 e^{r_2 t}$ are solutions of ODE (1) for any constants C_1 and C_2. As a matter of fact, for any values of C_1 and C_2 the sum

$$y = C_1 e^{r_1 t} + C_2 e^{r_2 t} \tag{4}$$

is also a solution of ODE (1) because if we put $y_1 = e^{r_1 t}$ and $y_2 = e^{r_2 t}$, then

$$(C_1 y_1 + C_2 y_2)'' + a(C_1 y_1 + C_2 y_2)' + b(C_1 y_1 + C_2 y_2)$$
$$= [(C_1 y_1)'' + a(C_1 y_1)' + b(C_1 y_1)] + [(C_2 y_2)'' + a(C_2 y_2)' + b(C_2 y_2)]$$
$$= 0 + 0 = 0$$

where we have used the fact that $C_1 y_1$ and $C_2 y_2$ both satisfy ODE (1). Let's use this approach to find a solution formula for the following example.

EXAMPLE 3.3.1

One Positive, One Negative Characteristic Root
The characteristic polynomial for the ODE

$$y'' + y' - 2y = 0 \qquad (5)$$

is $r^2 + r - 2 = (r - 1)(r + 2)$ with roots $r_1 = 1$, $r_2 = -2$. So we see that for any constants C_1 and C_2,

$$y = C_1 e^t + C_2 e^{-2t}$$

solves ODE (5). Figure 3.3.1 shows some solution curves with a common value of $y(0)$ but distinct values of $y'(0)$. As t increases from 0, the positive exponential dominates and the negative exponential decays; as t decreases from 0, it is reversed.

We know that the general solution of a first-order undriven linear ODE has one arbitrary constant. It appears that we can find solutions for a second-order undriven linear ODE with two arbitrary constants. If $r_1 = r_2$, however, we seem to lose a constant because $y = C_1 e^{r_1 t} + C_2 e^{r_1 t}$ becomes $(C_1 + C_2)e^{r_1 t}$. But $C_1 + C_2$ is just a single arbitrary constant. Let's find a replacement for the solution that was "lost".

If $r_1 = r_2$, we can show by direct substitution that $C_2 t e^{r_1 t}$ is a solution of ODE (1) for any value of the constant C_2. In fact, in the case of a repeated root

☞ The "lost" term is replaced by $C_2 t e^{r_1 t}$

$$y = C_1 e^{r_1 t} + C_2 t e^{r_1 t} \qquad (6)$$

solves ODE (1) for any constants C_1 and C_2. Let's look at an example.

EXAMPLE 3.3.2

Repeated Characteristic Roots
The characteristic polynomial of the ODE

$$y'' + y'/2 + y/16 = 0 \qquad (7)$$

is $r^2 + r/2 + 1/16$ with roots $r_1 = r_2 = -1/4$. We claim that

$$y = C_1 e^{-t/4} + C_2 t e^{-t/4} \qquad (8)$$

where C_1 and C_2 are arbitrary real numbers, is a solution of ODE (7).
To verify this, substitute $y = C_1 e^{-t/4} + C_2 t e^{-t/4}$ into ODE (7) to obtain

$$\left((C_1 + C_2 t)e^{-t/4}\right)'' + \left((C_1 + C_2 t)e^{-t/4}\right)' + (C_1 + C_2 t)e^{-t/4}/16$$

$$= C_1 e^{-t/4}\left(\frac{1}{16} - \frac{1}{8} + \frac{1}{16}\right) + C_2 e^{-t/4}\left(-\frac{1}{2} + \frac{1}{2} + \frac{1}{16}t - \frac{1}{8}t + \frac{1}{16}t\right) = 0$$

So formula (8) defines a solution of ODE (7). Figure 3.3.2 shows solution curves with a common value of $y(0)$ and distinct values of $y'(0)$. The negative exponent $-t/4$ in formula (8) eventually makes all solutions tend to 0 as $t \to +\infty$.

We leave it to the reader to check out the general repeated root case. If you decide to do the calculations, notice that in this case $r^2 + ar + b = (r - r_1)^2 = r^2 - 2r_1 r + r_1^2$, and so $a = -2r_1$ and $b = r_1^2$.

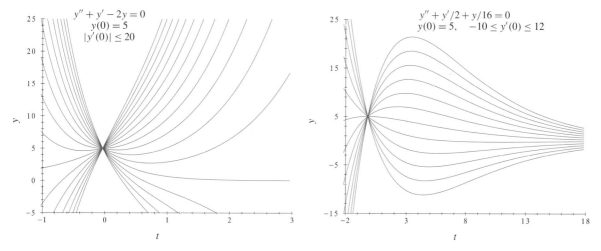

FIGURE 3.3.1 Solution curves; characteristic polynomial has roots 1 and -2 (Example 3.3.1).

FIGURE 3.3.2 Solution curves; characteristic polynomial has repeated root $-1/4$ (Example 3.3.2).

It is a remarkable fact that when the characteristic roots r_1 and r_2 are real, formula (4) defines *all* solutions of ODE (1) if $r_1 \neq r_2$, and formula (6) defines *all* solutions of ODE (1) if $r_1 = r_2$. We'll put off justifying these facts until the end of this section. Let's record these results for ease of reference.

THEOREM 3.3.1

Solution Formulas. Suppose that a and b are constants and that the two roots r_1 and r_2 of the characteristic polynomial $r^2 + ar + b$ are real numbers. Then formulas (9) give solutions of the ODE $y'' + ay' + by = 0$:

$$
\begin{aligned}
y &= C_1 e^{r_1 t} + C_2 e^{r_2 t}, && \text{if } r_1 \neq r_2 \\
y &= C_1 e^{r_1 t} + C_2 t e^{r_1 t}, && \text{if } r_1 = r_2
\end{aligned}
\tag{9}
$$

where C_1 and C_2 are arbitrary real numbers. There are no other solutions.

Theorem 3.3.1 has some important consequences for the ODE $y'' + ay' + by = 0$:

- **All solutions live on the entire t-axis.** This follows by inspection of the solution formulas in (9).

- **The trivial solution $y = 0$, all t, is the only solution** of the IVP

$$
y'' + ay' + by = 0, \qquad y(t_0) = 0, \quad y'(t_0) = 0
\tag{10}
$$

To see this, we again use the formulas in (9). Say that $r_1 \neq r_2$. Then the solution of the ODE $y'' + ay' + by$ has the form $y = C_1 e^{r_1 t} + C_2 e^{r_2 t}$. From the initial conditions $y(t_0) = 0$, $y'(t_0) = 0$ we have

$$
C_1 e^{r_1 t_0} + C_2 e^{r_2 t_0} = 0
$$

$$
r_1 C_1 e^{r_1 t_0} + r_2 C_2 e^{r_2 t_0} = 0
$$

If $r_2 = 0$, then the second equation implies that $C_1 = 0$, and from the top equation it follows that $C_2 = 0$ as well. If $r_2 \neq 0$, then multiplying the first equation by $-r_2$ and adding it to the second, we obtain $C_1(-r_2 + r_1)e^{r_1 t_0} = 0$ Since $r_1 \neq r_2$ and $e^{r_1 t_0} \neq 0$, it follows that $C_1 = 0$, so $C_2 = 0$ as well. This shows that the only solution of IVP (10) is $y = 0$, for all t. The case $r_1 = r_2$ proceeds analogously.

Using Theorem 3.3.1, we can find all solutions of IVP (2) very quickly once the characteristic roots r_1 and r_2 are known. With the formulas in (9) in hand, we can solve any IVP. Here is an example.

EXAMPLE 3.3.3

Finding All Solutions, Solving an IVP
To find all solutions of $y'' + y' - 2y = 0$ note that the polynomial $r^2 + r - 2$, has roots 1 and -2. So Theorem 3.3.1 says that a formula for all solutions is

$$y = C_1 e^t + C_2 e^{-2t}$$

where C_1 and C_2 are arbitrary constants.

Now let's find values for C_1 and C_2 such that $y = C_1 e^t + C_2 e^{-2t}$ satisfies the initial conditions $y(0) = 0$, $y'(0) = 3$. We have that

$$0 = y(0) = C_1 e^0 + C_1 e^0 = C_1 + C_2$$
$$3 = y'(0) = C_1 e^0 - 2C_2 e^0 = C_1 - 2C_2$$

So the constants C_1 and C_2 must be $C_1 = 1$ and $C_2 = -1$, and the solution of the IVP

$$y'' + y' - 2y = 0, \qquad y(0) = 0, \quad y'(0) = 3$$

is given by the formula

$$y = e^t - e^{-2t}$$

☞ OK, so we don't always follow our own advice!

It's always a good idea to use direct substitution to check if you really have found a solution of the IVP.

The rest of this section is taken up with introducing operator methods, which we will use freely in later sections. A byproduct of all this will be the verification that the formulas in Theorem 3.3.1 do indeed capture *all* solutions of the ODE $y'' + ay' + by = 0$ if the roots of the characteristic polynomial are real. For this reason the solutions in (9) are called *general solutions*

The Differentiation Operator D

Let's look at a new approach for treating ODE (1) which also simplifies solving *driven* constant coefficient linear ODEs.

First let's write the expression $y'' + ay' + by$ as

$$\frac{d^2}{dt^2} y + a \frac{d}{dt} y + by \tag{11}$$

Now by "factoring" y off to the right, let's rewrite this expression as

$$\left(\frac{d^2}{dt^2} + a\frac{d}{dt} + b \right)[y] \tag{12}$$

We can always go back to the expression (11) by "multiplying out" expression (12) again. The square brackets remind us to operate on y by taking the appropriate number of derivatives.

This notation becomes awkward to write after a while, so let's use the symbol D instead of d/dt and the symbol D^2 instead of d^2/dt^2. When D is applied to y, it acts like d/dt, and when D^2 is applied to y it acts like d^2/dt^2. Finally, we can write expression (11) as

☞ Read this as "the operator D applied to $y(t)$ produces $y'(t)$."

$$(D^2 + aD + b)[y]$$

We call D the *differentiation operator*. Newton used the symbol D, and we do too. We'll see that D makes things easier later on.[2] Notice that for any constant c we have

$$D[cy] = (cy)' = cy' = cD[y], \qquad D^2[cy] = cy'' = cD^2[y]$$

☞ When $r = 0$, $D - r$ becomes just D.

Now let's define the operators $D - r$ and $D^2 + aD + b$ for any constants r, a, and b by the operational formulas

$$(D - r)[y] = D[y] - ry = y' - ry \tag{13}$$

$$(D^2 + aD + b)[y] = D^2[y] + aD[y] + by = y'' + ay' + by \tag{14}$$

Let's see what the D operator does to functions.

EXAMPLE 3.3.4

Using the D Operator

Following formulas (13) and (14), we see that if $y = \sin 3t$ then

$$(D - 2)[\sin 3t] = D[\sin 3t] - 2\sin 3t$$
$$= 3\cos 3t - 2\sin 3t$$
$$(D^2 - D - 2)[\sin 3t] = D^2[\sin 3t] - D[\sin 3t] - 2\sin 3t$$
$$= -9\sin 3t - 3\cos 3t - 2\sin 3t$$
$$= -11\sin 3t - 3\cos 3t$$

Notice that applying $D - 2$ and $D + 1$ in succession to $\sin 3t$ gives us

$$(D + 1)\big[(D - 2)[\sin 3t]\big] = (D + 1)[3\cos 3t - 2\sin 3t] = -11\sin 3t - 3\cos 3t$$

---

[2]The notation dy/dt to denote the derivative of y was introduced by Leibniz and was used by the German school in the early days of calculus. The English school on the other hand followed Newton and used D applied to y to denote the derivative of y. Leibniz's notation can be misleading, but the Germans were much more successful in developing new results in calculus, so that's why we still use that notation today. It is ironic that we return to Newton's D at this point, but it has a purpose. Newton would be pleased.

Operator Identities

Here are three operator identities that we will find useful:

$$(D - r)[e^{st}] = (s - r)e^{st} \tag{15}$$

$$(D - r)[h(t)e^{st}] = e^{st}(D + s - r)[h] \tag{16}$$

$$(D^2 + aD + b)[e^{st}] = e^{st}(s^2 + as + b) \tag{17}$$

for any constants a, b, r and s and any differentiable function $h(t)$. Direct computation using (13) and (14) shows that these identities are valid (try doing it yourself to sharpen your skill in working with operators). To remember formulas (13)–(17), just "multiply" each of the polynomials $D - r$ and $D^2 + aD + b$ into the quantity in square brackets in the usual way, and then recall that $D[y] = y'$ and $D^2[y] = y''$.

We will often use the notation

$$P(D) = D^2 + aD + b$$

to write ODE (1) in the operator form $P(D)[y] = 0$. We denote these operators simply by $P(D)$ and call them *polynomial operators* because they look like polynomials in the basic operator D. The *characteristic polynomial* of $P(D)$ is $P(r) = r^2 + ar + b$.

Polynomial operators have two useful properties that can be useful in solving driven linear constant coefficient ODEs.

☞ This turns out to be the same characteristic polynomial that we introduced at the beginning of the section.

THEOREM 3.3.2

> **Polynomial Operator Theorem.** The operator $P(D) = D^2 + aD + b$, where a and b are constants, has the following properties:
>
> *Linearity Property.* For any constants C_1 and C_2, and any twice differentiable functions $y_1(t)$ and $y_2(t)$, we have
>
> $$P(D)[C_1y_1 + C_2y_2] = C_1P(D)[y_1] + C_2P(D)[y_2] \tag{18}$$
>
> *Closure Property.* If $y_1(t)$ and $y_2(t)$ solve the ODE $P(D)[y] = 0$, then so does $C_1y_1 + C_2y_2$ for any constants C_1 and C_2.

☞ This property is also called the *Principle of Superposition*

To show the Linearity Property, notice that

$$D[C_1y_1 + C_2y_2] = C_1D[y_1] + C_2D[y_2]$$

$$D^2[C_1y_1 + C_2y_2] = C_1D^2[y_1] + C_2D^2[y_2]$$

These properties follow directly from the properties of the derivative, so

$$P(D)[C_1y_1 + C_2y_2] = D^2[C_1y_1 + C_2y_2] + aD[C_1y_1 + C_2y_2] + b(C_1y_1 + C_2y_2)$$

$$= C_1(D^2 + aD + b)[y_1] + C_2(D^2 + aD + b)[y_2]$$

$$= C_1P(D)[y_1] + C_2P(D)[y_2]$$

The Closure Property follows from the linearity of $P(D)$ because

$$P(D)[C_1y_1 + C_2y_2] = C_1P(D)[y_1] + C_2P(D)[y_2] = 0$$

So $y = C_1y_1 + C_2y_2$ is a solution of $P(D)[y] = 0$ if y_1 and y_2 are solutions.

Operators which satisfy the Linearity Property are called *linear operators*. Writing ODE (1) in operator form as $P(D)[y] = 0$ we have just seen that $P(D)$ is a linear operator. This is the actual reason that we use the term *linear* ODE.

The expression $C_1 w + C_2 z$, where w and z are functions and C_1 and C_2 are constants, arises so often that it deserves a name: a *linear combination* of w and z. The Closure Property can then be restated as follows: Any linear combination of solutions of $P(D)[y] = 0$ is again a solution of $P(D)[y] = 0$.

☞ We will use this fact over and over.

Factoring a Polynomial Operator

For any constants r and s we can apply a polynomial operator, say $D - s$, to a twice differentiable function $y(t)$ and then follow by applying another one, say $D - r$. We will define the *operator product* $(D - r)(D - s)$ by the operational formula

$$(D - r)(D - s)[y] = (D - r)\big[(D - s)[y]\big] \tag{19}$$

If $r = s$, we write $(D - r)(D - r)$ as $(D - r)^2$. Now for any constants r and s,

$$(D - r)(D - s) = (D - s)(D - r) \tag{20}$$

☞ It doesn't matter what order we apply $D - r$ and $D - s$

so the order of the factors in this product is immaterial. We can show that identity (20) is valid by just evaluating each side using the definition (19). First notice that

$$(D - r)(D - s)[y] = (D - r)[y' - sy] = D[y' - sy] - r(y' - sy)$$
$$= (y' - sy)' - r(y' - sy)$$
$$= y'' - (r + s)y' + rs$$

So we see that

$$(D - r)(D - s) = D^2 - (r + s)D + rs \tag{21}$$

Interchanging r and s in (21) gives the same result for $(D - s)(D - r)$.

Suppose that the characteristic polynomial $P(r) = r^2 + ar + b$ of the operator $P(D) = D^2 + aD + b$ has the two *real* roots r_1 and r_2 (which might be equal). Then $P(r)$ can be written as

$$P(r) = (r - r_1)(r - r_2) = r^2 - (r_1 + r_2)r + r_1 r_2$$

and we see that $a = -(r_1 + r_2)$ and $b = r_1 r_2$. Using identity (21) we see that $P(D)$ can be factored as

$$P(D) = (D - r_1)(D - r_2)$$

As noted before, the order of the factors is immaterial.

Now let's use the properties of operators to complete the verification of Theorem 3.3.1.

Why Theorem 3.3.1 Gives All Solutions

We use a constructive approach to show that the formulas in Theorem 3.3.1 describe *all* solutions of

$$P(D)[y] = (D^2 + aD + b)[y] = 0 \qquad (22)$$

for the indicated cases. Suppose that r_1 and r_2 are the (not necessarily distinct) characteristic roots of $P(r)$.

Then ODE (22) takes the operator form

$$P(D)[y] = (D - r_1)\big[(D - r_2)[y]\big] = 0 \qquad (23)$$

Suppose that $y(t)$ is a solution of ODE (23). We want to show that $y(t)$ must have the form of one of the solutions listed in Theorem 3.3.1. Let's apply $D - r_2$ to y and give the name v to the result:

$$v = (D - r_2)[y] \qquad (24)$$

Then we know by substitution into (23) that v satisfies the *first-order linear* ODE

$$(D - r_1)[v] = v' - r_1 v = 0 \qquad (25)$$

since $(D - r_1)[v] = (D - r_1)\big[(D - r_2)[y]\big] = 0$.

But ODE (25) is a first-order linear ODE and we can use the method of integrating factors to find all functions $v(t)$ which solve it. The solutions are given by

$$v = C_1 e^{r_1 t}, \qquad C_1 \text{ any real number}$$

So the given solution $y(t)$ of ODE (23) must solve the *first-order linear* ODE

$$(D - r_2)[y] = y' - r_2 y = C_1 e^{r_1 t} \qquad (26)$$

which we obtain by inserting $C_1 e^{r_1 t}$ for v in (24). Now assume that $r_1 \neq r_2$. The method of integrating factors applied to (26) shows that

$$(ye^{-r_2 t})' = C_1 e^{(r_1 - r_2)t}$$

Integrating, we have that

$$ye^{-r_2 t} = \frac{C_1}{r_1 - r_2} e^{(r_1 - r_2)t} + C_2$$

where C_2 is an arbitrary real number. Since C_1 is arbitrary we can replace $C_1/(r_1 - r_2)$ by simply C_1. Solving for $y(t)$ gives the result stated in Theorem 3.3.1 if $r_1 \neq r_2$.

Now if $r_1 = r_2$, then multiplying ODE (26) through by the integrating factor $e^{-r_1 t}$ we obtain

$$(ye^{-r_1 t})' = C_1$$

Integrating we have

$$ye^{-r_1 t} = C_1 t + C_2$$

where C_2 is an arbitrary real constant. Solving this equation for y gives the result stated in Theorem 3.3.1 for this case of a repeated root. And so we have shown that the formulas of Theorem 3.3.1 really capture all solutions of $P(D)[y] = 0$ if the roots of the characteristic polynomial are real.

Using the Operator Approach to Solve a Driven ODE

The operator approach can also be used to solve driven ODEs as the next example shows.

EXAMPLE 3.3.5

Solving a Driven ODE
Let's look at the driven ODE

$$y'' + y' - 2y = \sin t \tag{27}$$

Since $D^2 + D - 2 = (D+2)(D-1)$, we write the ODE as

$$(D+2)\big[(D-1)[y]\big] = \sin t$$

So if we first solve

$$(D+2)[v] = v' + 2v = \sin t \tag{28}$$

for v, and then solve

$$(D-1)[y] = y' - y = v$$

for y, we will have our general solution formula for ODE (27).

Let's first tackle the first-order linear ODE (28). Multiplying through by the integrating factor e^{2t}, this ODE becomes

$$(ve^{2t})' = e^{2t}\sin t$$

Using Table 1.3.1 to find an antiderivative of the right-hand side and integrating both sides of this ODE, we obtain

$$ve^{2t} = (2/5)e^{2t}\sin t - (1/5)e^{2t}\cos t + C_1 \tag{29}$$

where C_1 is an arbitrary constant.

Solving equation (29) for v and inserting it into the ODE $(D-1)[y] = v$, we obtain the first-order ODE

$$y' - y = (2/5)\sin t - (1/5)\cos t + C_1 e^{-2t}$$

Multiplying through by the integrating factor e^{-t}, this ODE becomes

$$(ye^{-t})' = (2/5)e^{-t}\sin t - (1/5)e^{-t}\cos t + C_1 e^{-3t}$$

Using Table 1.3.1 again to find an antiderivative of the right-hand side and integrating both sides of this ODE, we obtain

$$ye^{-t} = -(3/10)e^{-t}\sin t - (1/10)e^{-t}\cos t - (C_1/3)e^{-3t} + C_2$$

☞ In Section 3.6 we'll see a better way to find this formula.

where C_2 is an arbitrary constant. Replacing $-C_1/3$ by just C_1 (OK, since C_1 is arbitrary) and solving for $y(t)$, we have

$$y(t) = C_1 e^{-2t} + C_2 e^t - (3/10)\sin t - (1/10)\cos t$$

which is the general solution formula for ODE (27).

Note that we have replaced the second-order linear ODE (27) by an equivalent system of two first-order linear ODEs

$$(D+2)[v] = v' + 2v = \sin t$$
$$(D-1)[y] = y' - y = v$$

which we then solved one at a time.

Comments

☞ Operator identities are summarized for handy reference at the end of the chapter.

Linear ODEs with constant coefficients often occur in applications (see Chapter 4, for example). We have seen in this section that finding a general solution formula for *undriven* constant coefficient linear ODEs just comes down to finding the roots of a polynomial, the characteristic polynomial associated with the ODE.

PROBLEMS

1. Find the general solution of the ODEs below.

(a) $y'' = 0$ (b) $y'' + y' - 2y = 0$ (c) $y'' - 4y' + 4y = 0$

(d) $y'' - 4y = 0$ (e) $5y'' - 10y' = 0$ (f) $2y'' + 12y' + 18y = 0$

(g) $y'' - 6y' + 9y = 0$ (h) $y'' + 4y' - y = 0$ (i) $y'' + 2y' + y = 0$

(j) $4y'' - 4y' + y = 0$ (k) $y'' - 2y' + y = 0$ (l) $y'' - 10y' + 25y = 0$

 2. *(IVPs)*. Solve the IVPs below. Plot the solution curves.

(a) $y'' + y' = 0$, $y(0) = 1$, $y'(0) = 2$ (b) $y'' + 3y' + 2y = 0$, $y(0) = 0$, $y'(0) = 1$

(c) $y'' - 9y = 0$, $y(0) = 2$, $y'(0) = -1$ (d) $y'' - 4y' + 4y = 0$, $y(0) = 1$, $y'(0) = 1$

(e) $y'' - 25y = 0$, $y(1) = 0$, $y'(1) = 1$ (f) $y'' + y' - 6y = 0$, $y(0) = 1$, $y'(0) = -1$

www **3.** *(Given a Solution, What's the ODE?)*. Find an ODE of the form $y'' + ay' + by = 0$, where $a^2 > 4b$ for which the given function is a solution, or else explain why no such ODE exists.

(a) $e^t - e^{-t}$ (b) $e^t - te^t$ (c) $e^{-t} + e^{-2t}$ (d) $1 + e^{-3t}$

(e) $e^{2t} + 10000e^{3t}$ (f) $e^{\sqrt{2}t} + e^{-\sqrt{2}t}$ (g) $e^{\pi t} - 3$ (h) e^{t^2}

(i) $t + 2$ (j) $t^2 e^{-t}$

4. Find the solution for which $y(-1) = 2$, $y'(-1) = 0$. Graph the solution curve.

(a) $y'' = 0$ (b) $y'' + y' - 2y = 0$ (c) $y'' - 4y' + 4y = 0$

5. Apply each operator to the function $y(t) = e^{-t}$.

(a) $D - 2$ (b) $D + 3$ (c) $D^2 + D - 6$

(d) $(D-2)(D+3)$ (e) $(D+3)(D-2)$ (f) $(D+1)^2$

6. Find a polynomial operator $P(D) = D^2 + aD + b$ that gives 0 when applied to the indicated function $y(t)$.

   (a) $y(t) = 2e^{5t}$ **(b)** $y(t) = e^{2t} + e^{-t}$ **(c)** $y(t) = te^t$ **(d)** $y(t) = \sin t$

7. Find a function $y(t)$ such that $(D-2)[y(t)] = f(t)$, where $f(t)$ is the indicated function.

   (a) $f(t) = e^{-t}$ **(b)** $f(t) = t - 1$ **(c)** $f(t) = \sin t + 4$

8. Find a general solution formula for each ODE below: [*Hint*: Follow Example 3.3.5.]

   (a) $y'' - y = \sin 2t$ **(b)** $y'' - y' - 2y = \cos t$ **(c)** $y'' + 2y' + y = e^{-t}$

9. Find a general solution formula for the ODE $(D-1)(D+2)(D-3)[y] = 0$.

10. (*Uniqueness Theorem*). Show the following result: For any constants a, b, t_0, y_0, v_0, the IVP $y'' + ay' + by = 0$, $y(t_0) = y_0$, $y'(t_0) = v_0$, can't have more than one solution on any t-interval containing the point t_0. [*Hint*: If $u(t)$ and $v(t)$ are two solutions of the IVP, then show that $w(t) = u(t) - v(t)$ solves the IVP $y'' + ay' + by = 0$, $y(t_0) = 0$, $y'(t_0) = 0$.]

3.4 Undriven Constant Coefficient Linear ODEs, II

In this section we show how to find a formula for all the solutions of the undriven constant coefficient ODE

$$P(D)[y] = y'' + ay' + by = 0 \tag{1}$$

☞ This approach is not as strange as it might seem at first.

when the characteristic roots are complex numbers rather than real numbers. To do this efficiently it turns out to be helpful to let ODE (1) have complex-valued solutions. This might seem like a complicated way to solve a simple problem, but bear with us and you will see a big payoff.

IVPs based on ODE (1) come up often in applications. Let's look at an example and illustrate the various graphical displays generated by a solution.

EXAMPLE 3.4.1

The Curves of a Damped, Hooke's Law Spring Model
As we saw in Section 3.1, the second-order IVP for the position $y(t)$ (relative to the static equilibrium position) of a hanging weight supported by a Hooke's Law spring is

$$y'' + \frac{k}{m}y + \frac{c}{m}y' = 0, \qquad y(t_0) = y_0, \quad y'(t_0) = v_0 \tag{2}$$

where k, m, and c are the respective spring constant, mass of the weight, and damping constant, respectively. Let's use a numerical solver to solve IVP (2) with the specific data (measured in some consistent set of units)

$$k/m = 65, \quad c/m = 0.4, \quad t_0 = 20, \quad y_0 = 9, \quad v_0 = 0 \tag{3}$$

Figure 3.4.1 shows the inward spiraling time-directed orbit in the yy'-plane on the left, the solution curve on the upper right, and the y'-component curve on the lower right. Figure 3.4.2 shows the time-state curve (solid) in tyy'-space and its orbit (dashed) lying below in the yy'-plane.

☞ Problem 12 in Section 3.2 describes how we plotted Figure 3.4.2 with our solver.

The graphs in Figures 3.4.1 and 3.4.2 suggest that the spring oscillates about its equilibrium state with exponentially decaying amplitudes. This has a lot to do with the fact that the characteristic roots of the ODE in (2), with the coefficients given in (3), are *not* real numbers.

Complex-Valued Functions and Solutions

☞ Appendix B.3 has a lot more on complex-valued functions.

If the ODE $y'' + ay' + by = 0$ arises in a mathematical model of a natural process, then the coefficients a and b are real numbers, and the solutions we ultimately seek are real-valued. But what if we could find complex-valued solutions of this ODE? Would that help us find real-valued solutions? We answer both these questions affirmatively in the paragraphs below. We will see that the concept of complex-valued solutions of linear ODEs can be very useful as a computational device in much the same way that complex numbers make it easier to discuss the roots and factorability of polynomials.

What does it mean to say that a complex-valued function $y(t)$ of a real variable t is a solution of the linear ODE (1), where the coefficients are real or complex numbers? To answer this, we show that D acts on complex functions just as on real functions.

A complex-valued function $y(t)$ of the real variable t can be uniquely written as a sum $y(t) = u(t) + iv(t)$, where the functions $u(t)$ and $v(t)$ are real-valued. The functions $u(t)$ and $v(t)$ are called the *real part* and the *imaginary part*, respectively, of $y(t)$, and are denoted by Re[y] and Im[y]. Using an overbar to denote *complex conjugation*, notice that if $y(t) = u(t) + iv(t)$ then $\overline{y(t)} = u(t) - iv(t)$.

Next, we need to define the derivative of a complex-valued function of a real variable. It can be shown that if $y(t) = u(t) + iv(t)$, then

$$\lim_{h \to 0} \frac{y(t+h) - y(t)}{h} = \lim_{h \to 0} \frac{u(t+h) - u(t)}{h} + i \lim_{h \to 0} \frac{v(t+h) - v(t)}{h}$$

if and only if the limits all exist. So it is reasonable to define the differentiation operator D for complex-valued functions as follows:

❖ **Derivative of a Complex-Valued Function.** Let $y(t) = u(t) + iv(t)$ be a complex-valued function with real and imaginary parts $u(t)$ and $v(t)$. Then

$$y' = D[y] = D[u + iv] = D[u] + iD[v]$$

All of the usual differentiation rules are valid for the operators D and $P(D)$. It can be shown that the Linearity Property and the Closure Property of Theorem 3.3.2 also hold for complex quantities. If $y(t) = u(t) + iv(t)$ is a solution of

$$P(D)[y] = (D^2 + aD + b)[y] = 0 \tag{4}$$

where a and b are real numbers, then $u(t)$ and $v(t)$ are both *real*-valued solutions of ODE (4). We can show this by using the Linearity Property

☞ If you can find all complex-valued solutions, then real-valued solutions fall right out.

$$P(D)[u + iv] = P(D)[u] + iP(D)[v] = 0 + 0i \tag{5}$$

But two complex quantities are equal if and only if the real parts match up and the imaginary parts match up. So $P(D)[u] = 0$ and $P(D)[v] = 0$.

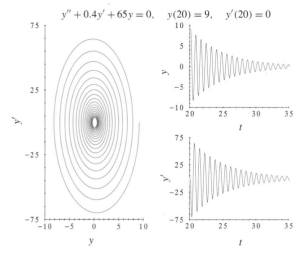

FIGURE 3.4.1 Orbit and component curves for the damped spring model (Example 3.4.1).

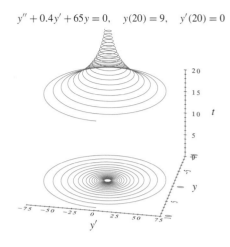

FIGURE 3.4.2 Time-state curve (solid) and its orbit (dashed) (Example 3.4.1).

All the polynomial operator identities in Section 3.3 hold when the coefficients and roots are complex numbers. In particular, even if the characteristic roots r_1 and r_2 for the polynomial operator $D^2 + aD + b$ are complex numbers, then we have

$$D^2 + aD + b = (D - r_1)(D - r_2)$$

where the order of the factors on the right is immaterial.

The Exponential Function $e^{(\alpha + i\beta)t}$

The exponential function e^{rt}, where r is a complex number and t is a real number, is a very useful complex-valued function. As the notation suggests, we would like this function to have the usual properties enjoyed by exponential functions. So if r is the complex number $\alpha + i\beta$, then we would expect the exponent law to hold:

$$e^{(\alpha + i\beta)t} = e^{\alpha t + i\beta t} = e^{\alpha t} e^{i\beta t} \tag{6}$$

Now we need to see how $e^{i\gamma}$ is to be defined, where γ is any real number. A reasonable way to do this would be to replace x by $i\gamma$ in the Taylor series

$$e^x = 1 + x + \frac{1}{2!}x^2 + \frac{1}{3!}x^3 + \cdots + \frac{1}{n!}x^n + \cdots$$

and separate out the real and imaginary terms to obtain

$$e^{i\gamma} = \left(1 - \frac{\gamma^2}{2!} + \cdots + (-1)^k \frac{\gamma^{2k}}{(2k)!} + \cdots\right) + i\left(\gamma - \frac{\gamma^3}{3!} + \cdots + (-1)^k \frac{\gamma^{2k+1}}{(2k+1)!} + \cdots\right)$$

The series in parentheses are Taylor series for $\cos \gamma$ and $\sin \gamma$. It is reasonable to define $e^{i\gamma} = \cos \gamma + i \sin \gamma$, and setting $\gamma = \beta t$ we have *Euler's formula*

$$e^{i\beta t} = \cos \beta t + i \sin \beta t, \quad \text{for any real numbers } \beta, t \tag{7}$$

Using (6) and (7) we arrive at the formula

$$e^{(\alpha+i\beta)t} = e^{\alpha t}\cos\beta t + ie^{\alpha t}\sin\beta t \tag{8}$$

It follows that

$$\operatorname{Re}[e^{(\alpha+i\beta)t}] = e^{\alpha t}\cos\beta t, \qquad \operatorname{Im}[e^{(\alpha+i\beta)t}] = e^{\alpha t}\sin\beta t$$

Since $\bar{r} = \alpha - i\beta$ for any complex number $r = \alpha + i\beta$, we see from formula (8) that

$$\overline{e^{rt}} = \overline{e^{(\alpha+i\beta)t}} = e^{\alpha t}\cos\beta t - ie^{\alpha t}\sin\beta t = e^{\bar{r}t} \tag{9}$$

Let's apply the operator D to the exponential function defined in formula (8).

EXAMPLE 3.4.2

The Derivative Formula for $e^{(\alpha+i\beta)t}$

As we saw in formula (7), for any real number β the complex-valued function $e^{i\beta t}$ is defined for real t-values as $e^{i\beta t} = \cos\beta t + i\sin\beta t$. So we have

$$\begin{aligned} D[e^{i\beta t}] &= D[\cos\beta t] + iD[\sin\beta t] = -\beta\sin\beta t + i\beta\cos\beta t \\ &= i\beta(\cos\beta t + i\sin\beta t) = i\beta e^{i\beta t} \end{aligned} \tag{10}$$

Now suppose that $r = \alpha + i\beta$ is any complex number. From (8), $e^{rt} = e^{(\alpha+i\beta)t} = e^{\alpha t}e^{i\beta t}$, and using (10) and the product rule for differentiation, we have that

$$\begin{aligned} D[e^{rt}] &= D[e^{\alpha t}e^{i\beta t}] = D[e^{\alpha t}]e^{i\beta t} + e^{\alpha t}D[e^{i\beta t}] \\ &= \alpha e^{\alpha t}e^{i\beta t} + e^{\alpha t}i\beta e^{i\beta t} = re^{rt} \end{aligned} \tag{11}$$

which is the same differentiation formula as for real values of r.

For any polynomial operator $P(D) = D^2 + aD + b$, and any complex number r, we see from formula (11) that, just as in the real case,

$$\begin{aligned} P(D)[e^{rt}] &= D^2[e^{rt}] + aD[e^{rt}] + be^{rt} \\ &= r^2 e^{rt} + are^{rt} + be^{rt} \\ &= e^{rt}(r^2 + ar + b) = e^{rt}P(r) \end{aligned} \tag{12}$$

Let's bring together some of these properties of the complex exponential function.

Properties of the Exponential Function, e^{rt}

The complex-valued function e^{rt} defined in (8) for any complex number $r = \alpha + i\beta$ has the following properties:

- $\operatorname{Re}[e^{rt}] = e^{\alpha t}\cos\beta t, \quad \operatorname{Im}[e^{rt}] = e^{\alpha t}\sin\beta t$
- $\overline{e^{rt}} = e^{\bar{r}t}$, where the overbar denotes complex conjugation.
- $P(D)[e^{rt}] = e^{rt}P(r)$, for any polynomial operator $P(D)$.
- If $r_1 \neq r_2$, and if the constants K_1 and K_2 are such that $K_1 e^{r_1 t} + K_2 e^{r_2 t} = 0$, for all t, then $K_1 = K_2 = 0$.

To show the last property above, apply the operator D to the identity $K_1e^{r_1t} + K_2e^{r_2t} = 0$ to produce the new identity $K_1r_1e^{r_1t} + K_2r_2e^{r_2t} = 0$, for all t. Evaluate these identities at $t = 0$ to get the two equations

$$K_1 + K_2 = 0 \quad \text{and} \quad K_1r_1 + K_2r_2 = 0$$

Solving the first equation for K_2 and substituting into the second equation, we get $K_1(r_1 - r_2) = 0$. Since $r_1 \neq r_2$, it follows that $K_1 = 0$, and so also $K_2 = 0$.

Now let's use the properties of e^{rt} to find more solution formulas.

Finding Solution Formulas

Suppose that the real polynomial operator $P(D) = D^2 + aD + b$ has the pair of conjugate roots $\alpha \pm i\beta$. Here's how we solve $P(D)[y] = 0$. Identity (12) says that

$$P(D)\left[e^{(\alpha+i\beta)t}\right] = e^{(\alpha+i\beta)t}P(\alpha + i\beta) = 0$$

so $e^{(\alpha+i\beta)t}$ is a complex-valued solution of $P(D)[y] = 0$. Now using (8) let's put

$$u(t) = \text{Re}[e^{(\alpha+i\beta)t}] = e^{\alpha t}\cos\beta t, \qquad v(t) = \text{Im}[e^{(\alpha+i\beta)t}] = e^{\alpha t}\sin\beta t$$

Then from (5) we see that

$$P(D)[u + iv] = P(D)[u] + iP(D)[v] = 0, \quad \text{for all } t$$

and since $P(D)$ has real coefficients, it follows that $e^{\alpha t}\cos\beta t$ and $e^{\alpha t}\sin\beta t$ are both real-valued solutions of $P(D)[y] = 0$. Because of the Closure Property for $P(D)$ we see that

$$y = C_1e^{\alpha t}\cos\beta t + C_2e^{\alpha t}\sin\beta t \tag{13}$$

is a solution of $P(D)[y] = 0$ for arbitrary real numbers C_1 and C_2. It is a remarkable fact that (13) describes *all* the real-valued solutions of $P(D)[y] = 0$, but we put off justifying this fact till the end of this section.

For reference purposes let's give a general solution formula for all real-valued solutions of $P(D)[y] = 0$ regardless of what the characteristic roots are. We formulate this as a theorem, but on a practical level it is actually a procedure for constructing the general solution.

THEOREM 3.4.1

☞ These formulas summarize everything we've done in Sections 3.3 and 3.4.

Real-Valued General Solution Theorem. Suppose the coefficients of $P(D) = D^2 + aD + b$ are real numbers and that the characteristic polynomial $P(r)$ has the roots r_1 and r_2. Then the general real-valued solution of the ODE $P(D)[y] = 0$ is given by

$$
\begin{aligned}
y &= C_1e^{r_1t} + C_2e^{r_2t}, & &\text{if } r_1, r_2 \text{ real}, \quad r_1 \neq r_2 \\
y &= C_1e^{\alpha t}\cos\beta t + C_2e^{\alpha t}\sin\beta t, & &\text{if } r_1 = \bar{r}_2 = \alpha + i\beta, \quad \beta \neq 0 \\
y &= C_1e^{r_1t} + C_2te^{r_1t}, & &\text{if } r_1 = r_2
\end{aligned} \tag{14}
$$

where C_1 and C_2 are arbitrary real numbers.

Here's an example of how to find all real-valued solutions when the characteristic roots are complex numbers.

EXAMPLE 3.4.3

Real-Valued Solutions from Complex-Valued Solutions
The characteristic polynomial of the linear ODE

$$y'' + y' + 100.25y = 0 \qquad (15)$$

is $r^2 + r + 100.25$, which has the roots $r_1 = -1/2 + 10i$ and $r_2 = -1/2 - 10i$. By Theorem 3.4.1, the general solution is

$$y = C_1 e^{-t/2} \cos 10t + C_2 e^{-t/2} \sin 10t$$

where C_1 and C_2 are arbitrary reals. Figure 3.4.3 shows some solution curves of ODE (15). Note the declining amplitudes as t increases from 0 (because of the negative exponentials) and the oscillations (because of the sinusoids of period $\pi/5$).

Justification of Solution Formula (14)

When the coefficients a and b are real-valued and the characteristic roots r_1 and r_2 are complex-valued, it is useful first to find all the complex-valued solutions of the un-driven ODE $P(D)[y] = 0$. The methods used in Section 3.3 to establish Theorem 3.3.1 also hold when the characteristic roots are complex (that's why we defined the operator D as we did for complex-valued functions). Here's the result:

THEOREM 3.4.2

> **Complex-Valued General Solution Theorem.** Suppose that a and b are real (or complex) constants and that r_1 and r_2 are the roots of the polynomial $P(r) = r^2 + ar + b$. Then the general complex-valued solution of the ODE $P(D)[y] = 0$ is given by
>
> $$y = K_1 e^{r_1 t} + K_2 e^{r_2 t}, \quad \text{if } r_1 \neq r_2 \qquad (16a)$$
>
> $$y = K_1 e^{r_1 t} + K_2 t e^{r_1 t}, \quad \text{if } r_1 = r_2 \qquad (16b)$$
>
> where K_1 and K_2 are arbitrary complex constants.

Care should be taken to distinguish the *general complex-valued solution* from the *general real-valued solution*. Of course, the real-valued solutions must appear among the collection of all complex-valued solutions. Here's how we can recognize them: $y(t)$ is a real-valued solution if and only if $\overline{y(t)} = y(t)$, for all t, or alternatively, if and only if $\text{Im}[y(t)] = 0$ for all t. Now let's justify Theorem 3.4.1.

If $P(D) = D^2 + aD + b$ has real coefficients and complex characteristic roots r_1 and r_2, then $r_1 = \alpha + i\beta$ and $r_2 = \alpha - i\beta$, where $\beta \neq 0$. Since $r_1 \neq r_2$, Theorem 3.4.2 gives the general complex-valued solution

☞ If a, b are real, then complex roots of $r^2 + ar + b$ occur in conjugate pairs.

$$y = K_1 e^{(\alpha + i\beta)t} + K_2 e^{(\alpha - i\beta)t} \qquad (17)$$

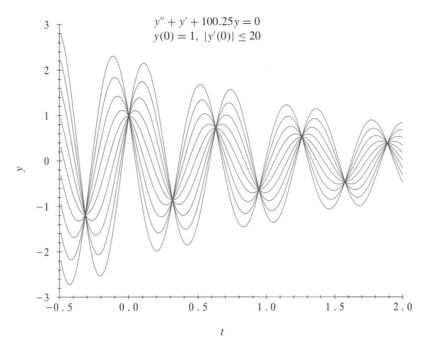

$$y'' + y' + 100.25y = 0$$
$$y(0) = 1, \ |y'(0)| \le 20$$

FIGURE 3.4.3 Oscillating solution curves: roots of $P(r)$ are $-1/2 \pm 10i$ (Example 3.4.3).

where K_1 and K_2 are arbitrary complex numbers. To recognize real-valued solutions in formula (17) we set $y(t) = \overline{y(t)}$ for all t, and obtain (where for convenience we set $\alpha + i\beta = r$)

$$K_1 e^{rt} + K_2 e^{\bar{r}t} = \overline{K_1} e^{\bar{r}t} + \overline{K_2} e^{rt}$$

or, after rearranging terms,

$$(K_1 - \overline{K_2})e^{rt} + (K_2 - \overline{K_1})e^{\bar{r}t} = 0 \tag{18}$$

But, as noted earlier, equation (18) implies that $K_1 - \overline{K_2} = 0$ and $K_2 - \overline{K_1} = 0$. So all real solutions look like

$$y = K_1 e^{rt} + \overline{K_1} e^{\bar{r}t} = 2 \, \text{Re}\left[K_1 e^{(\alpha + i\beta)t} \right] \tag{19}$$

If we set $K_1 = k_1 + ik_2$ where k_1 and k_2 are arbitrary reals, then from formula (19) we have

$$y = 2k_1 e^{\alpha t} \cos \beta t - 2k_2 e^{\alpha t} \sin \beta t \tag{20}$$

☞ Just replace $2k_1$ by C_1 and $-2k_2$ by C_2.

which, recalling that k_1 and k_2 are arbitrary, justifies our claim that all real-valued solutions have the form stated in (14).

Comments

☞ Operator identities are summarized for handy reference at the end of the chapter.

Now we know how to find all real-valued solutions of the undriven ODE $y'' + ay' + by = 0$, where a and b are real constants. In Section 3.6 we show how to find a particular solution of the driven ODE $y'' + ay' + by = f(t)$, where $f(t)$ has the special form seen in most applications.

PROBLEMS

1. Find all real-valued solutions of $P(D)[y] = 0$ where

(a) $P(D) = D^2 - 2D + 2$ (b) $P(D) = D^2 - 4D + 5$ (c) $P(D) = D^2 - 4D + 4$

2. (*Using Solutions to Construct the ODE*). For each case below write the normalized linear ODE of lowest order with constant real coefficients for which the given expression is a solution.

(a) $e^{-5t} - e^{-2t}$ [*Hint*: Look at $(D+5)(D+2)$.]   (b) $3e^{4t} + 2te^{4t}$ [*Hint*: Look at $(D-4)^2$.]

(c) $8e^{-4t} \sin t$ (d) $e^{(3-5i)t}$

(e) $\cos t + e^{(2-i)t}$ [*Hint*: First find $P(D)$ so that $P(D)[\cos t] = 0$. Then find $Q(D)$ so that $Q(D)[e^{(2-i)t}] = 0$. Then look at $P(D)Q(D)$.]

www **3.** Find the general real-valued solution of each ODE. Then plot solution curves in the ty-plane for $-1 \leq t \leq 5$, where $y(0) = 1$, $y'(0) = -6, -3, 0, 3, 6$. Plot the corresponding orbits in a rectangle in the yy'-plane that shows the main features of the orbits. Discuss your results.

(a) $y'' + y' = 0$ (b) $y'' + 2y' + 65y = 0$ (c) $y'' + 3y' + 2y = 0$

(d) $y'' + 10y = 0$ (e) $y'' - y/9 = 0$ (f) $y'' - 3y'/4 + y/8 = 0$

 4. (*IVPs*). Find the real solution and plot the solution curve. What happens as $t \to +\infty$?

(a) $y'' + 2y' + 2y = 0$, $y(0) = 1$, $y'(0) = 0$

(b) $y'' + 4y' + 4y = 0$, $y(1) = 2$, $y'(1) = 0$

(c) $y'' + y' - 6y = 0$, $y(-1) = 1$, $y'(-1) = -1$

5. (*Decaying Solutions*). Consider the ODE $y'' + ay' + by = 0$, where a and b are real constants.

(a) Show that if $a > 0$, $b > 0$, then every solution tends to zero as $t \to +\infty$.

(b) Show that if every solution tends to zero as $t \to \infty$, then $a > 0$ and $b > 0$.

6. (*Positively Bounded Solutions*). A solution $y(t)$ of $y'' + ay' + by = 0$, a and b real numbers, is said to be *positively bounded* if there is a positive constant M such that $|y(t)| \leq M$ for all $t \geq 0$.

(a) Show that all solutions are positively bounded if $a \geq 0$, $b \geq 0$, but not both a and b are zero.

(b) Show that if all solutions of the ODE are positively bounded, then $a \geq 0$, $b \geq 0$, but not both a and b are zero.

7. (*Complex Coefficients: What to Do When Theorem 3.4.1 Does Not Apply*).

(a) Find the general solution of the ODE $y'' + iy' + 2y = 0$.

(b) Does the ODE in part (a) have nontrivial real-valued solutions?

☞ De Moivre's Formula is in Appendix B.3.

(c) Find the general solution of $y'' + iy = 0$. [*Hint*: Use De Moivre's Formula to find the roots of the characteristic polynomial.]

8. (*More ODEs with Complex Coefficients*). Give all the real-valued solutions (if any) of the following ODEs. [*Hint*: Factor the characteristic polynomial by inspection.]

(a) $y'' + (1+i)y' + iy = 0$ (b) $y'' + (-1+2i)y' - (1+i)y = 0$

3.5 Periodic Solutions and Simple Harmonic Motion

Periodic solutions come up quite often in ODEs used in the design and analysis of systems such as machinery and electronic devices. Let's first review some properties of periodic functions and how they arise as solutions of second-order linear ODEs. Then we'll see how the strange phenomena of aliasing arises when tracking a periodic function numerically.

Periodic Functions

We have talked about periodic functions informally before, but now it's time to get all the definitions and properties down in print.

A nonconstant function $f(t)$ defined on the real line is a *periodic function* if there exists a value $T > 0$ (called a *period*) such that $f(t + T) = f(t)$ for all t. Any integer multiple of a period is also a period. The smallest positive period of a periodic function is said to be the *fundamental period*. From this point on the term "period" will always mean "fundamental period." A periodic function with period T is said to complete one *cycle* over any time span T. In other words, the period T of a periodic function is the amount of time required for the function to go through one cycle. For example, the function $\sin t$ completes one cycle over any time span of length 2π. If $f(t)$ has period T, then $f(\omega t)$ has period T/ω because that's the minimal time needed for ωt to run through a time span of length T. For example, $\sin \omega t$ and $\cos \omega t$ have period $2\pi/\omega$. The *amplitude* A of a periodic function is given by half the difference between the function's maximum and minimum values. For example, the periodic function $1 + 2\cos 5t$ has amplitude 2.

If $f(t)$ and $g(t)$ are periodic with the same period T, then $f(t)g(t)$ and $af(t) + bg(t)$, for arbitrary constants a and b (not both zero), are both periodic with the same period T. Although it is less obvious, if $f(t)$ has period T and $g(t)$ has period S, then $f(t) + g(t)$ is periodic if and only if T/S is a rational number. In fact, if two positive integers m and n without common factors (other than 1) produce the common value $mT = nS$, then this common value is the period of $f + g$. For example, $\cos 2t$, $\cos 3t$, and $\cos \pi t$ have the respective periods $T_1 = \pi$, $T_2 = 2\pi/3$, and $T_3 = 2$. Since $2T_1 = 3T_2 = 2\pi$, the function $\cos 2t + \cos 3t$ has period 2π. But $\cos 3t + \cos \pi t$ is *not* periodic because $T_2/T_3 = \pi/3$, which is an irrational number. There are many functions other than sinusoids that are periodic. One that often occurs in applications is the square wave sqw(t, d, T).

A term closely related to the period of a periodic function is "frequency." The *frequency* f of a periodic function with period T is defined to be the number of cycles per unit time, so $f = 1/T$. For example, the function $\sin t$ has period $T = 2\pi$ and frequency $f = 1/2\pi$, because that is the fraction of a cycle executed in one unit of time. If time is measured in seconds, the unit *hertz* is often used to denote one cycle per second. For sinusoids such as $\sin \omega t$ and $\cos \omega t$ another widely used measure of frequency is *circular frequency*, which is the number of radians covered per unit time. For example, the *circular* frequency of $\sin \omega t$ is ω, while the frequency is $\omega/2\pi$.

☞ From now on, when we say "period," we mean "fundamental period"!

☞ We also need to stipulate that fg and $af + bg$ are not constant functions.

☞ We first used hertz in Example 2.2.7.

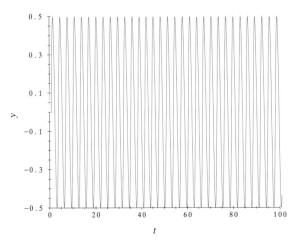

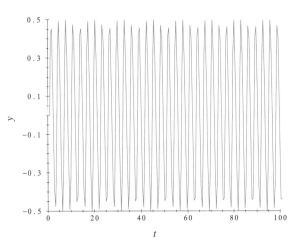

FIGURE 3.5.1 Sampling simple harmonic motion at 10 points per unit time (Example 3.5.2).

FIGURE 3.5.2 Sampling simple harmonic motion at 5 points per unit time (Example 3.5.2).

Circular frequencies are used only with trigonometric functions. From this point on we use the terms "period" and "frequency" with roughly equal frequency when we are talking about periodic functions.

Now let's examine an ODE, all of whose nonconstant solutions are periodic.

EXAMPLE 3.5.1

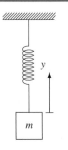

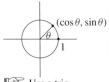

☞ Use a trig identity from Appendix B.4.

Simple Harmonic Motion

For any positive number ω, the undamped, undriven Hooke's Law spring ODE

$$P(D)[y] = y'' + \omega^2 y = 0 \tag{1}$$

has the general solution

$$y = C_1 \sin \omega t + C_2 \cos \omega t \tag{2}$$

where C_1 and C_2 are arbitrary reals. This is because $\pm i\omega$ are the roots of $P(r) = r^2 + w^2$. If C_1 and C_2 are not both zero, then $A = (C_1^2 + C_2^2)^{1/2} > 0$, and the general solution has the equivalent form

$$y = A \left(\frac{C_1}{A} \sin \omega t + \frac{C_2}{A} \cos \omega t \right) \tag{3}$$

We define θ to be any angle such that $\cos \theta = C_1/A$, $\sin \theta = C_2/A$. Then we see that the general solution of $y'' + \omega^2 y = 0$ is

$$y = A(\cos \theta \sin \omega t + \sin \theta \sin \omega t) = A \sin(\omega t + \theta)$$

where the *amplitude A* is an arbitrary nonnegative constant and the *phase angle θ* is an arbitrary constant (positive or negative) since C_1 and C_2 are arbitrary. Equivalently, $y = A \cos(\omega t + \phi)$ with phase angle ϕ, where $\phi = \theta - \pi/2$. So every solution of the ODE is a sinusoid of period $2\pi/\omega$. The ODE $y'' + \omega^2 y = 0$ is called the *simple harmonic oscillator* and its solution graphs depict *simple harmonic motion*

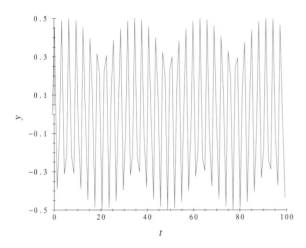
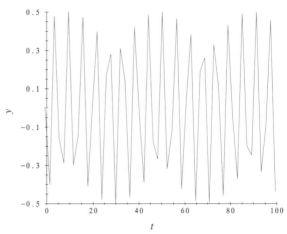

FIGURE 3.5.3 Sampling simple harmonic motion at 1 point per unit time (Example 3.5.2).

FIGURE 3.5.4 Sampling simple harmonic motion at 0.5 point per unit time: aliasing (Example 3.5.2).

Tracking Periodic Solutions with a Numerical Solver

Detecting periodic solutions of an ODE using a numerical solver can take a strange twist. For simplicity, we will look only at simple harmonic motion.

EXAMPLE 3.5.2

Sampling Rates: Aliasing

From Example 3.5.1 we see that the general solution of the ODE

$$y'' + 4y = 0$$

is the sinusoid

$$y = A\sin(2t + \theta)$$

with arbitrary amplitude $A \geq 0$ and arbitrary phase angle θ. All solutions have period π. Figures 3.5.1–3.5.4 show four plots of the solution of the IVP

$$y'' + 4y = 0, \qquad y(0) = 0, \quad y'(0) = 1 \tag{4}$$

over the interval $0 \leq t \leq 100$. In Figure 3.5.1, fourth-order Runge–Kutta (RK4) is used to approximate the y-component curve of the system equivalent to IVP (4), where

$$y' = z, \qquad z' = -4y, \qquad y(0) = 0, \qquad z(0) = 1$$

at 1000 equally spaced points in the interval $0 \leq t \leq 100$. The result appears to be a sinusoid of period π (as it should be).

The plots in Figures 3.5.2–3.5.4 are made using the exact same data points as in Figure 3.5.1, but plotting, respectively, every 5th, 10th, and 20th point. The graphs degrade as fewer and fewer solution points are plotted. In Figure 3.5.4, the period of the approximate solution curve appears to have doubled to 2π.

Since these graphs are all produced by sampling from the same list of 1000 fairly good approximate values, the accuracy of the solver is not to blame for this strange behavior. In fact, this phenomenon always occurs when an oscillatory curve is under-sampled and the sample points are connected by line segments. The sinusoidal amplitudes appear to worsen as the number of sample points per unit time decreases. Finally, as Figure 3.5.4 shows, when there are fewer than two sample points per period the graphical representation appears to have a larger period (i.e., lower frequency). Engineers call this "aliasing," one frequency taking the name of the other.

Practically speaking, what can we do to correct the misrepresentation inherent in representing a continuous periodic function by a discrete point set? If the experimenter knows in advance the period of the oscillation, then the sampling rate can be set high enough to mitigate the problem. But in practice these periods are not known, so some experimentation may be called for before accepting the validity of a graphical result.

Now let's take a look at sinusoidal solutions of a *driven* harmonic oscillator.

Finding Sinusoidal Solutions of a Driven ODE

☞ Example 3.3.5 used operators to find the solutions of a particular driven ODE.

Although we don't "officially" find solution techniques for driven second-order linear ODEs until the next section, we have already sneaked in some examples. Here is another.

EXAMPLE 3.5.3

The Driven Harmonic Oscillator: A Valentine

Let's drive the harmonic oscillator of Example 3.5.1 with the sinusoidal driving term $3 \sin kt$, where $k \neq \omega$, and find a formula for all solutions of the ODE

$$y'' + \omega^2 y = 3 \sin kt \tag{5}$$

We'll adapt an approach we used for first-order ODEs in Example 1.4.4.

Say that $y_d(t)$ is a solution of ODE (5). If $y(t)$ is any other solution, then the difference $z = y - y_d$ solves ODE (1), and so z has the form $C_1 \sin \omega t + C_2 \cos \omega t$ [see formula (2)]. It follows that all solutions of ODE (5) are given by the formula

$$y = y_d + z = y_d(t) + C_1 \sin \omega t + C_2 \cos \omega t$$

So now we just have to find $y_d(t)$; any solution of ODE (5) could be taken for $y_d(t)$.

☞ Another instance of good guessing as a solution technique.

Why not try to guess such a solution? Since we want to end up with $3 \sin kt$, we could start out with $y_d(t) = A \sin kt + B \cos kt$ for unknown constants A and B, and after substituting y_d into ODE (5), just match up the coefficients of $\sin kt$ and $\cos kt$:

$$-Ak^2 \sin kt - Bk^2 \cos kt + \omega^2(A \sin kt + B \cos kt) = 3 \sin kt$$

So the constants A and B must satisfy the equations

$$-Ak^2 + \omega^2 A = 3 \quad \text{and} \quad -Bk^2 + \omega^2 B = 0$$

Solving, we have $A = 3/(\omega^2 - k^2)$ and $B = 0$. So we put it all together and see that

$$y = \frac{3}{\omega^2 - k^2} \sin kt + C_1 \sin \omega t + C_2 \cos \omega t \tag{6}$$

where C_1 and C_2 are arbitrary constants, is the general solution of ODE (5).

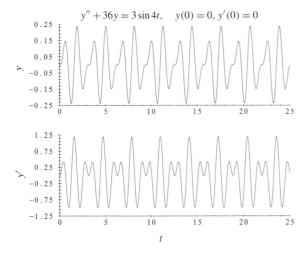

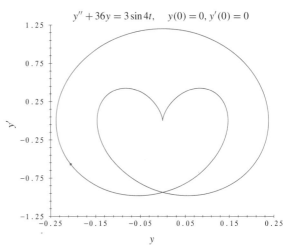

FIGURE 3.5.5 The graphs of $y(t)$ and $y'(t)$: period π (Example 3.5.3).

FIGURE 3.5.6 The periodic orbit in the yy'-plane: period π (Example 3.5.3).

From formula (6) we see that when $C_1 = C_2 = 0$ ODE (5) has a solution of period $2\pi/k$. If k/ω is an irrational number, then ODE (5) has no other periodic solutions. If $k/\omega = m/n$ for some positive integers m and n with no common factors, then every solution of ODE (5) other than $y_d(t) = (3/(\omega^2 - k^2)) \sin kt$ is periodic with period

$$m \cdot \frac{2\pi}{k} = n \cdot \frac{2\pi}{\omega}$$

For example, if $\omega = 6$ and $k = 4$, then we can take $m = 2$ and $n = 3$, so $2\pi m/k = \pi = 2\pi n/\omega$, and the solutions of ODE (5) have period π. Figures 3.5.5 and 3.5.6 show component curves and the orbit if $y(0) = 0$ and $y'(0) = 0$.

PROBLEMS

www 1. (*Periodic Function Facts*).

(a) Show that if $f(t)$ has period T, then $f(\omega t)$ has period T/ω.

(b) Suppose $f(t)$ and $g(t)$ have periods T and S, respectively, with T/S rational. Show that $h(t) = f(t) + g(t)$ is periodic with period equal to the smallest value k such that $k = mT = nS$ for positive integers m and n.

(c) Verify the claim in part **(b)** by graphing $f(t) = \sin(2t)$, $g(t) = \sin(5t)$, and $f(t) + g(t)$ on the same plot. What is the period of each function?

(d) Explain why $h(t) = f(t) + g(t)$ is *not* periodic if the ratio T/S of the periods of f and g is irrational.

2. Show that the ODE $(D^2 + aD + b)[y] = 0$, for real constants a and b, has a periodic solution if and only if $a = 0$ and $b > 0$.

3. Find a formula for all solutions of the ODE $(D^2 + 9)[y] = -3\cos 2t$. Does the ODE have any periodic solutions? Explain.

4. Does the ODE $(D^2 + 4)[y] = \sin 2t$ have any periodic solutions? [*Hint*: Use the approach in Example 3.5.3 with a guess of $y_d(t) = At \sin 2t + Bt \cos 2t$, for constants A and B.]

5. (*Amplitude Modulation*). Explain the modulation phenomenon in Figures 3.5.2 and 3.5.3. Formulate a prediction about how the modulation varies with the sampling rate, and verify your
prediction experimentally.

3.6 Driven Constant Coefficient Linear ODEs

A body with unit mass is suspended by a Hooke's Law spring and oscillates with
viscous damping about its equilibrium position. The clock reads $t = t_0$ as the spring
support suddenly begins to move up and down in a prescribed way. We see from
Section 3.1 that the response of this spring-mass system to the driving term $f(t)$ solves
an IVP of the form

$$y'' + ay' + by = f(t), \qquad y(t_0) = y_0, \quad y'(t_0) = y_0' \tag{1}$$

☞ Section 3.7 gives
the general theory of
linear ODEs

for constants a and b. Theorem 3.2.1 can be strengthened to say that IVP (1) has a
unique solution on any t-interval I containing t_0 on which $f(t)$ is continuous. It is
good to know that IVP (1) has a unique solution, but this doesn't help us to find a
formula for that solution. In this section we describe a process that leads to a solution
formula for IVP (1), at least for driving terms $f(t)$ in a class that includes sinusoids,
polynomials, and some exponentials.

Our approach uses polynomial operators as a computational device, so first we
remind ourselves of the properties of these operators that we will find useful.

Properties of Polynomial Operators

In Section 3.3 we introduced the differentiation operator D and used it to define the
first-order and second-order linear polynomial operators $D - r$ and $D^2 + aD + b$,
where a, b, and r are given constants:

$$(D - r)[y] = y' - ry, \qquad (D^2 + aD + b)[y] = y'' + ay' + by$$

where $y(t)$ has the necessary number of continuous derivatives. In Section 3.4 we saw
that these operators also make sense when the coefficients r, a, and b are complex
numbers and the function $y(t)$ is complex-valued.

Let's review some identities for constant-coefficient polynomial operators

$$P(D) = D^2 + aD + b$$

☞ These identities are
summarized on page 212
at the end of the chapter.

- If c_1, c_2 are any constants and y_1, y_2 are any real or complex-valued twice differentiable functions then we have the Linearity Property

$$P(D)[c_1 y_1(t) + c_2 y_2(t)] = c_1 P(D)[y_1] + c_2 P(D)[y_2] \tag{2}$$

- If the roots of the characteristic polynomial $P(r) = r^2 + ar + b$ are r_1 and r_2, then
  $P(D)$ can be factored as $(D - r_1)(D - r_2)$, and in addition

$$(D - r_1)(D - r_2) = (D - r_2)(D - r_1) \tag{3}$$

- For any complex number s

$$P(D)[e^{st}] = e^{st}P(s) \qquad (4)$$

- Identity (4) can be extended to

☞ Here's a new and very useful identity.

$$P(D)[h(t)e^{st}] = e^{st}P(D+s)[h] \qquad (5)$$

for any complex number s and any twice differentiable function $h(t)$.

To show that the operator identity in (5) is valid, notice first that by the product rule for differentiation

$$D[h(t)e^{st}] = e^{st}D[h] + se^{st}h = e^{st}(D+s)[h] \qquad (6)$$

So applying D to both sides of this identity we have

$$D^2[he^{st}] = D\big[D[he^{st}]\big] = D\big[D[h]e^{st} + hse^{st}\big] = D\big[e^{st}(D+s)[h]\big]$$
$$= e^{st}(D+s)\big[(D+s)[h]\big] = e^{st}(D+s)^2[h] \qquad (7)$$

☞ Use of these operator identities to solve $P(D)[y] = f$ is summarized for handy reference at the chapter's end.

Since $P(D)[he^{st}] = D^2[he^{st}] + aD[he^{st}] + bhe^{st}$, identities (6) and (7) imply (5).

Now let's find some ways to put these polynomial operators and identities to work in finding the general solution for a driven linear ODE such as $P(D)[y] = f$.

The General Solution for a Driven Linear ODE

The Linearity Property (2) of polynomial operators $P(D)$ turns out to be very useful in characterizing solutions of the driven linear ODE $P(D)[y] = f(t)$.

THEOREM 3.6.1

General Solution Theorem. Suppose that $P(D) = D^2 + aD + b$, where a and b are constants, and suppose that $f(t)$ is continuous for all t. The general solution of the ODE

$$P(D)[y] = f \qquad (8)$$

can be written as

$$y = y_d + y_u$$

☞ Note that y_d is a particular solution of the driven ODE, which is the reason for the subscript d.

where y_d is a particular solution of the driven ODE (8) (any solution will do), and y_u is the general solution of the undriven ODE $P(D)[y] = 0$.

To see this, say that by hook or by crook we have found a particular solution $y_d(t)$ of this driven ODE so that $P(D)[y_d] = f(t)$. Now if $y(t)$ is any other solution of the ODE, then from the Linearity Property of $P(D)$ we see that

$$P(D)[y - y_d] = P(D)[y] - P(D)[y_d] = f(t) - f(t) = 0$$

So $y - y_d = w$ is a solution of the undriven ODE $P(D)[y] = 0$. On the other hand, if $w(t)$ is any solution of $P(D)[y] = 0$, then

$$P(D)[y_d + w] = P(D)[y_d] + P(D)[w] = f(t) + 0 = f(t)$$

So we see that the general solution of the *driven* ODE $P(D)[y] = f(t)$ is just $y_d(t)$ plus the general solution y_u of the *undriven* ODE $P(D)[y] = 0$.

Doesn't this remind you of a similar result for driven *first-order* linear ODEs given in Section 1.4?

Let's summarize what we have just learned about solutions of $P(D)[y] = f$.

General Solution of the ODE $(D^2 + aD + b)[y] = f$

Put $P(D) = D^2 + aD + b$, where a and b are real numbers. Two things are needed to find a formula for all solutions of the ODE $P(D)[y] = f$:

- Any one solution y_d of $P(D)[y] = f$.
- A formula for all solutions y_u of $P(D)[y] = 0$; do this by finding the roots of $P(r)$ and applying Theorem 3.4.1.

Then the general solution of $P(D)[y] = f$ is given by $y = y_d + y_u$.

EXAMPLE 3.6.1

Finding a Particular Solution

To find a particular solution y_d of the ODE

$$(D^2 - 2D + 1)[y] = 3e^{-t} \tag{9}$$

we could use identity (4) with $s = -1$ and $P(D) = D^2 - 2D + 1$ to guess that y_d has the form Ce^{-t}. Because $P(D)[Ce^{-t}] = CP(-1)e^{-t} = 4Ce^{-t}$, taking $4C = 3$ gives us the result that $y_d = (3/4)e^{-t}$ solves ODE (9).

Now let's change the driving term slightly and look at the ODE

$$(D^2 - 2D + 1)[y] = 3e^t \tag{10}$$

The approach above now fails because $P(1) = 0$. So instead, let's use identity (5) and guess that $y_d = h(t)e^t$. Since $P(D)[h(t)e^t] = e^t P(D + 1)[h]$ we see that if $h(t)$ solves the ODE $P(D + 1)[h] = 3$, then $y_d = h(t)e^t$ solves ODE (10). Since $P(D + 1) = (D + 1)^2 - 2(D + 1) + 1 = D^2$ we have that $D^2[h] = 3$, which has as a solution $h(t) = (3/2)t^2$. So $y_d = (3/2)t^2 e^t$ solves ODE (10).

EXAMPLE 3.6.2

Finding the General Solution of a Driven ODE

Let's find the general solution of the driven ODE

$$y'' - y' - 2y = 4t \tag{11}$$

Writing this ODE in operator form $P(D)[y] = 4t$ with $P(D) = D^2 - D - 2$, we see that the polynomial $P(r) = r^2 - r - 2$ has the real roots $r_1 = 2$ and $r_2 = -1$. So from Theorem 3.4.1, the general solution of the undriven ODE $P(D)[y] = 0$ is $y_u = C_1 e^{2t} + C_2 e^{-t}$, where C_1 and C_2 are arbitrary constants.

Let's try to guess a solution y_d. For constants A and B put $y_d = At + B$. Note that

$$P(D)[At + B] = -2At - A - 2B$$

so $y_d = At + B$ solves ODE (11) if $-2A = 4$ and $-A - 2B = 0$. Solving, we find $A = -2$ and $B = 1$, so we have $y_d = -2t + 1$. So for arbitrary constants C_1 and C_2

$$y = y_d + y_u = -2t + 1 + C_1 e^{2t} + C_2 e^{-t}$$

is the general solution of ODE (11).

Often the driving function is the sum of several terms, but in the search for a particular solution we can treat each summand separately as the next result shows.

THEOREM 3.6.2

Summing Particular Solutions. Given the polynomial operator $P(D)$, suppose that $z(t)$ is a particular solution of $P(D)[y] = f(t)$ and that $w(t)$ is a particular solution of $P(D)[y] = g(t)$. Then it follows that $z(t) + w(t)$ is a particular solution of $P(D)[y] = f(t) + g(t)$.

The Linearity Property (2) for $P(D)$ says that

$$P(D)[z(t) + w(t)] = P(D)[z(t)] + P(D)[w(t)] = f(t) + g(t)$$

so our claim is valid.

Because of Theorem 3.6.2 we need only look at driving functions with a one summand. The following example illustrates the summing property.

EXAMPLE 3.6.3

Summing Solutions

Let's find a particular solution y_d of the ODE

$$y'' - y' - 2y = 4t + e^t \tag{12}$$

by first finding particular solutions of the two ODEs

$$u'' - u' - 2u = 4t \tag{13}$$

$$v'' - v' - 2v = e^t \tag{14}$$

We saw in Example 3.6.2 that $u_d = -2t + 1$ is a particular solution of ODE (13).

To find a solution v_d of (14) we use the identity (4) with $P(D) = D^2 - D + 2$ and $s = 1$. Guessing that $v_d = Ce^t$ for some constant C, and using (4) we have

☞ This shows how to solve $P(D)[y] = Ae^{st}$ when $P(s) \neq 0$. Just take $y_d = (A/P(s))e^{st}$.

$$P(D)[Ce^t] = Ce^t P(1) = Ce^t(1 - 1 + 2) = 2Ce^t$$

So taking $C = 1/2$, we have that $v_d = e^t/2$. Finally

$$y_d = u_d + v_d = -2t + 1 + e^t/2$$

is a particular solution of ODE (12).

As we will soon see, the guessing approach used in Example 3.6.2 to find a particular solution y_d of the driven linear ODE $P(D)[y] = f(t)$ only works when $f(t)$ is the sum and product of polynomials, sine and cosine functions, and exponential functions. The approach we describe below is called the *Method of Undetermined Coefficients*; it is a practical, if often-times a labor intensive, approach.

Polynomial-Exponential Functions

Let's find a particular solution y_d for the driven ODE

$$(D^2 + aD + b)[y] = f(t) \tag{15}$$

where a and b are *any* constants and $f(t)$ is *any* polynomial-exponential function.

A *polynomial-exponential* function is any finite sum of terms that look like $At^n e^{st}$, where A and s are real or complex numbers, and n is a nonnegative integer. For example, $f(t) = -3t^2 e^{-t} + 2te^{2t} - 5$ is a polynomial-exponential function but why is $t^2 e^{-t} \cos 2t$ a polynomial-exponential function? Since $e^{i\theta} = \cos\theta + i\sin\theta$, for any real number θ, it follows that $e^{-i\theta} = \cos\theta - i\sin\theta$, and solving for $\sin\theta$, $\cos\theta$ we have the formulas

☞ Some people call these functions "exponomials."

$$\cos\theta = \frac{1}{2}(e^{i\theta} + e^{-i\theta}), \qquad \sin\theta = -\frac{i}{2}(e^{i\theta} - e^{-i\theta})$$

for all real θ. So we can write

$$t^2 e^{-t} \cos 2t = t^2 e^{-t}\left(\frac{1}{2}e^{2it} + \frac{1}{2}e^{-2it}\right)$$

which is evidently a polynomial-exponential function.

Because of Theorem 3.6.2, to find a particular solution of ODE (15) for *any* polynomial-exponential we only need to find a solution of

$$P(D)[y] = (D^2 + aD + b)[y] = At^n e^{st} \tag{16}$$

where A and s are real or complex constants and n is a nonnegative integer.

Method of Undetermined Coefficients

By identity (5) we can guess that ODE (16) has a particular solution of the form

$$y_d = h(t)e^{st}$$

In fact, from identity (5) we see that

$$P(D)[h(t)e^{st}] = At^n e^{st}$$

which implies that $y_d = h(t)e^{st}$ solves ODE (16) if

$$P(D + s)[h] = At^n$$

☞ This is where the guessing comes in.

It is reasonable to guess that $h(t)$ is also a polynomial. Let's look at an example.

EXAMPLE 3.6.4

Using Operator Methods to Find a Particular Solution

Let's find a particular solution y_d for the ODE

$$(D-1)(D-2)[y] = te^t$$

Substituting $h(t)e^t$ for y and using the operator identity (5) we see that

$$(D-1)(D-2)[h(t)e^t] = e^t D(D-1)[h] = te^t$$

So $y_d = h(t)e^t$ solves the given ODE if

$$D(D-1)[h] = t$$

☞ Why no constant term in h? Because the factor D would kill it.

Guessing that $h(t) = At^2 + Bt$, we have

$$(D-1)D[At^2 + Bt] = (D-1)[2At + B] = 2A - (2At + B)$$

so by matching coefficients we must have $-2A = 1$ and $2A - B = 0$. Solving, we see that $A = -1/2$ and $B = -1$, so

$$y_d = \left(-(1/2)t^2 - t\right)e^t$$

Our approach uses the operator identity (5) and the Method of Undetermined Coefficients to find a polynomial solution to a constant coefficient linear ODE with a polynomial driving term. Here's the procedure for accomplishing this last step.

Guessing a Polynomial Solution to $(D^2 + aD + b)[y] = t^n$

For constants a and b, and any nonnegative integer n, here are some successful guesses for a solution $y_d = h(t)$ of the ODE $(D^2 + aD + b)[y] = t^n$.

- If $b \neq 0$, guess that $h(t) = \sum_{k=0}^n A_k t^k$.
- If $b = 0$, but $a \neq 0$, guess that $h(t) = \sum_{k=1}^{n+1} A_k t^k$.
- If $a = b = 0$, then $h(t) = t^{n+2}/[(n+2)(n+1)]$ does the job.

So we can find a solution y_d for $(D^2 + aD + b)[y] = Ct^n e^{rt}$ for *any* constants a, b, C, r, and *any* nonnegative integer n. Just use identity (5) and follow up with an entry in the list above.

EXAMPLE 3.6.5

Finding a Particular Solution

Let's find a particular solution of the ODE

$$(D^2 + 2D + 4)[y] = -12t^2 e^{(-1+2i)t} \tag{17}$$

The characteristic roots of $P(D) = D^2 + 2D + 4$ are $r_1 = -1 + 2i$ and $r_2 = -1 - 2i$, so we can write $P(D) = (D-r_1)(D-r_2)$. Since by identity (5)

$$(D-r_1)(D-r_2)[h(t)e^{r_1 t}] = e^{r_1 t} D(D + r_1 - r_2)[h]$$

we see that $y_d(t) = h(t)e^{r_1 t}$ solves the second ODE in (17) if

$$D(D + r_1 - r_2)[h] = (D + 4i)D[h] = -12t^2 \tag{18}$$

Writing the operator in ODE (18) in the form $D^2 + aD + b$ we see that $b = 0$, so by the second item in the Guessing Procedure above we guess that $h(t)$ has the form

$$h(t) = At^3 + Bt^2 + Ct$$

Inserting this expression for h in ODE (18), we obtain

$$(D + 4i)D[At^3 + Bt^2 + Ct] = (D + 4i)[3At^2 + 2Bt + c]$$
$$= D[3At^2 + 2Bt + C] + 4i(3At^2 + 2Bt + C)$$
$$= 6At + 2B + 12iAt^2 + 8iBt + 4iC$$

So for h to solve ODE (18) we have, by matching coefficients,

$$12iA = -12, \qquad 6A + 8iB = 0, \qquad 2B + 4iC = 0$$

Solving, we find that $A = i$, $B = -3/4$, and $C = -(3/8)i$, so the function

$$y_d(t) = \left(it^3 - \frac{3}{4}t^2 - \frac{3}{8}it\right)e^{(-1+2i)t} \tag{19}$$

is a particular complex-valued solution of ODE (17).

Now let's find a solution y_d of the ODE $P(D)[y] = f(t)$, where $f(t)$ has the form $t^n e^{\alpha t}\sin\beta t$ or $t^n e^{\alpha t}\cos\beta t$, α and β are real numbers, and n is a nonnegative integer.

Real-Valued Solutions from Complex-Valued Solutions

Let's show how to get real-valued solutions of $P(D)[y] = f$, where f is real-valued, from complex-valued solutions of a related ODE. The next result holds the key.

THEOREM 3.6.3

Real- and Complex-Valued Solutions. Suppose that the polynomial operator $P(D) = D^2 + aD + b$ has real coefficients, and suppose that $g(t)$ and $h(t)$ are given real-valued continuous functions. Then

$$y(t) = u(t) + iv(t)$$

for real-valued functions $u(t)$ and $v(t)$ is a solution of

$$P(D)[y] = g(t) + ih(t)$$

if and only if

$$P(D)[u] = g \quad\text{and}\quad P(D)[v] = h \tag{20}$$

Here is why this is true. The Linearity Property for the operator $P(D)$ says that $P(D)[u + iv] = P(D)[u] + iP(D)[v]$. Since the coefficients in $P(D)$ are real-valued, it follows that $P(D)[u]$ and $P(D)[v]$ are real-valued functions. Matching the real and imaginary parts of $P(D)[u + iv]$ and $f + ig$ yields (20).

So if $P(D)$ has real coefficients, any solution of the undriven ODE $P(D)[y] = 0$ has the property that its real and imaginary parts are real-valued solutions of the ODE. The next example shows how to use Theorem 3.6.3 in practice.

| EXAMPLE 3.6.6 | **Extracting Real-Valued Solutions from Complex-Valued Ones** |

Let's find the general real-valued solution of the ODE

$$y'' + 25y = \sin 4t \tag{21}$$

which models the vertical displacement y of an undamped Hooke's Law spring with a sinusoidal driving term. By Theorem 3.6.3 if we had a solution z_d of the ODE

$$z'' + 25z = e^{4it} \tag{22}$$

then $y_d = \text{Im}[z_d]$ solves ODE (21) because $\text{Im}[e^{4it}] = \sin 4t$. Writing ODE (22) as $P(D)[z] = e^{4it}$, with $P(D) = D^2 + 25$ we can use identity (4) to guess that z_d is of the form $z_d = Ce^{4it}$. Now $P(D)[Ce^{4it}] = CP(4i)e^{4it}$, and since $P(4i) \neq 0$ we can take $C = 1/P(4i)$, so

$$z_d = \frac{e^{4it}}{(4i)^2 + 25} = \frac{1}{9}e^{4it}$$

Since the imaginary part of $e^{4it}/9$ is $(\sin 4t)/9$, we see by Theorem 3.6.3 that a particular solution of ODE (21) is $y_d = (\sin 4t)/9$. Since the characteristic roots of $D^2 + 25$ are $r_1 = 5i$, $r_2 = -5i$, we see from Theorem 3.4.1 that the general solution of $(D^2 + 25)[y] = 0$ is $C_1 \cos 5t + C_2 \sin 5t$ for arbitrary reals C_1 and C_2. Finally, putting everything together, we see that the general solution of ODE (21) is

☞ All
solutions (23) of
ODE (21) are periodic
with period 2π. The
discussion of periodic
solutions in Section 3.5
shows why.

$$y = \frac{1}{9}\sin 4t + C_1 \sin 5t + C_2 \cos 5t \tag{23}$$

where C_1 and C_2 are any real constants.

Now let's impose an initial condition and see what happens.

| EXAMPLE 3.6.7 | **Another Valentine IVP** |

The solution of the IVP $y'' + 25y = \sin 4t$, $y(0) = 0$, $y'(0) = 0$, is found from formula (23) by using the initial data to find the constants C_1 and C_2:

☞ How can the
orbit in Figure 3.6.2
have a cusp? Is this a
mistake?

$$y = -\frac{4}{45}\sin 5t + \frac{1}{9}\sin 4t \tag{24}$$

Figure 3.6.1 shows the component curves and Figure 3.6.2 shows the surprising orbit in the yy'-plane. Why is solution (24) periodic with period 2π? It's because $\sin 5t$ has period $2\pi/5$, $\sin 4t$ has period $2\pi/4$, and $5(2\pi/5) = 4(2\pi/4) = 2\pi$.

Here's an example that has a cosine function in the driving term.

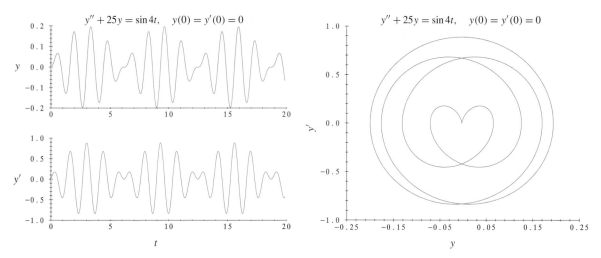

FIGURE 3.6.1 Component curves (Example 3.6.7).   **FIGURE 3.6.2** An orbit (Example 3.6.7).

EXAMPLE 3.6.8

Hooke's Law Spring Reprise

Let's look at the IVP with a cosine driving term

$$y'' + y = \cos(1.1t), \qquad y(50) = 5, \quad y'(50) = 0 \tag{25}$$

Using $P(D) = D^2 + 1$, if z_d solves the related ODE $P(D)[z] = e^{1.1it}$ then $y_d = \mathbb{R}[z_d]$ solves the ODE in (25). Guessing the solution $z_d = Ce^{1.1it}$ we see from the identity $P(D)[Ce^{1.1it}] = CP(1.1i)e^{1.1it}$ that

$$z_d = \frac{1}{P(1.1i)}e^{1.1it} = \frac{1}{(1.1i)^2 + 1}e^{1.1it} = \frac{-1}{0.21}e^{1.1it}$$

solves the ODE $P(D)[z] = e^{1.1it}$. So $y_d = \text{Re}[z_d] = \frac{1}{0.21}\cos(1.1t)$ solves the ODE in (25). The general solution of the ODE in (25) is

$$y = C_1 \cos t + C_2 \sin t - \frac{1}{0.21}\cos 1.1t$$

The initial conditions were chosen at the initial time $t_0 = 50$ to produce the nice time-state curve and associated orbit plot that appears in Figure 3.6.3. We won't bother to find C_1 and C_2. The time-state curve is plotted in tyy'-space over the interval $50 \leq t \leq 125$. Notice that the time-state curve does not intersect itself. When this curve is projected along the t-axis into the yy'-plane, the resulting orbit *does* intersect itself, but this is no surprise since the ODE is nonautonomous.

Now let's find a real-valued solution for the driven ODE $P(D)[y] = At^n e^{\alpha t} \cos \beta t$, where n is a nonnegative integer, A, α, and β are real constants, and $P(D)$ has *real* coefficients: First note that $At^n e^{\alpha t} \cos \beta t$ is the real part of $At^n e^{st}$, where $s = \alpha + i\beta$. Then from the relationship between real and complex solutions (see Theorem 3.6.3), we see that if z_d is any solution of the ODE $P(D)[z] = At^n e^{st}$, then $y_d = \text{Re}[z_d]$ is a real-valued solution of the ODE $P(D)[y] = At^n e^{\alpha t} \cos \beta t$.

$$y'' + y = \cos(1.1t), \quad y(50) = 5, \ y'(50) = 0$$

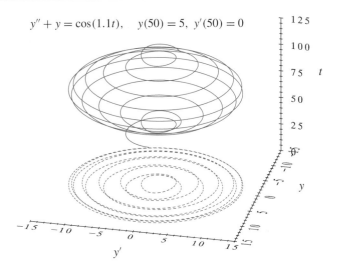

☞ Computer tip: Problem 12 in Section 3.2 describes how we produced this graph.

FIGURE 3.6.3 A time-state curve (solid) of a nonautonomous ODE and its orbit (dashed) (Example 3.6.8).

EXAMPLE 3.6.9

Response of Hooke's Law Spring to Oscillatory Driving Force

Consider the ODE that models the response of a damped Hooke's Law spring to an oscillatory driving force:

$$(D^2 + 2D + 4)[y] = -12t^2 e^{-t} \cos 2t \tag{26}$$

Note that $-12t^2 e^{-t} \cos 2t = \text{Re}[-12t^2 e^{(-1+2i)t}]$, so let's look at the ODE

$$(D^2 + 2D + 4)[z] = -12t^2 e^{(-1+2i)t}$$

But this is precisely the ODE considered in Example 3.6.5, and we found the particular solution

$$z_d = \left(it^3 - \frac{3}{4}t^2 - \frac{3}{8}it\right) e^{(-1+2i)t}$$

So all we need to do is extract the real part of z_d to find a particular real-valued solution y_d of ODE (26). Doing this we have

$$y_d = \text{Re}[z_d] = \text{Re}\left[(it^3 - \frac{3}{4}t^2 - \frac{3}{8}it)e^{-t}(\cos 2t + i\sin 2t)\right]$$

$$= e^{-t}(-t^3 \sin 2t - \frac{3}{4}t^2 \cos 2t + \frac{3}{8}t \sin 2t)$$

which is a particular solution of ODE (26).

Comments

If $f(t)$ is not a polynomial-exponential, or if $P(D)$ has nonconstants coefficients, then the method of this section for solving the ODE $P(D)[y] = f$ won't work. Instead, the Method of Variation of Parameters in Problem 8 of Section 3.7 must be used.

We have tacitly assumed in our approach that the roots of the characteristic polynomial $P(r)$ are known precisely, together with their multiplicities. Such precise information is, in general, not possible if $\deg P \geq 5$. So what is the value of our solution formulas and procedures if we can't explicitly find the precise information about the characteristic polynomial required by our approach? If we have close bounds on the characteristic roots, the solution formulas are often useful in obtaining approximate solutions or in obtaining information on the qualitative properties of solutions of $P(D)[y] = f$. Many commercial software packages contain efficient solvers for finding close approximations of all roots of a polynomial. So we consider the case where $P(D)$ has constant coefficients and f is a polynomial-exponential as solved.

PROBLEMS

1. (*Polynomial-Exponentials*). Show that the functions below are polynomial-exponentials.

    (a) $\cos 2t - \sin t$ (b) $t \sin^2 t$ (c) $t^2 \sin 2t - (1+t)\cos^2 t$

    (d) $\sin^3 t$ (e) $(1-t)e^{it} \cos 3t$ (f) $(i + t - t^2)e^{(3+i)t} \sin^2 3t$

2. Express each function in Problem 1(a)–(c) as the real part of some polynomial-exponential.

3. (*Annihilators*). For each function below find a polynomial operator with constant real coefficients $P(D)$ such that the function solves the ODE $P(D)[y] = 0$. Such a polynomial operator is said to be an *annihilator* for the function.

    (a) $t^2 e^{-it}$ (b) $te^{-t} \cos 2t$ (c) $t + \sin t$ (d) $\sin t + \cos 2t + t^2$

4. Find an undriven ODE with constant coefficients that includes e^t and te^{-3t} among its solutions. [*Hint*: What must the roots and multiplicities of the characteristic polynomial be? Look at the hint in Problem 2 **(e)** in Section 3.4.]

5. Solve each of the following initial value problems.

    (a) $y'' - 4y = 2 - 8t$, $y(0) = 0$, $y'(0) = 5$

    (b) $y'' + 9y = 81t^2 + 14\cos 4t$, $y(0) = 0$, $y'(0) = 3$

    (c) $y'' + y = 10e^{2t}$, $y(0) = 0$, $y'(0) = 0$

    (d) $y'' - y = e^{-t}(2\sin t + 4\cos t)$, $y(0) = 1$, $y'(0) = 1$

    (e) $y'' - 3y' + 2y = 8t^2 + 12e^{-t}$, $y(0) = 0$, $y'(0) = 2$

6. (*Finding General Solution Formulas*). Find all solutions of the following differential equations. Plot solution graphs of the corresponding IVPs with $y(0) = 0$, $y'(0) = -1, 0, 1$ for **(a)–(h)**.

    (a) $y'' - y' - 2y = 2\sin 2t$ (b) $y'' - y' - 2y = t^2 + 4t$ (c) $y'' - 2y' + y = -te^t$

    (d) $y'' - 2y' + y = 2e^t$ (e) $y'' + 2y' + y = e^t \cos t$ (f) $y'' + y' + y = \sin^2 t$

    (g) $y'' + 4y' + 5y = e^{-t} + 15t$ (h) $y'' + 4y = e^{2it}$

7. (*IVPs*). Solve the IVPs below.

    (a) $y'' - y' - 2y = 2e^t$, $y(0) = 0$, $y'(0) = 1$

    (b) $y'' + 2y' + 2y = 2t$, $y(1) = 1$, $y'(1) = 0$

8. (*Operator Identities*). The problems below give some useful identities.

    (a) Show that if n is any positive integer, r_0 is any constant, and h is any n-times differentiable function, then $(D - r_0)^n[he^{r_0 t}] = e^{r_0 t} D^n[h]$. [*Hint*: Do for $n = 1$ and iterate.]

    (b) Show for any polynomial $p(t)$ of degree $n - 1$ (or less) that $y = p(t)e^{r_0 t}$ is a solution for the nth-order ODE $(D - r_0)^n[y] = 0$ for any constant r_0.

(c) For the polynomial operator $P(D) = D^2 + aD + b$, show that $P(D)[h(t)e^{st}] = e^{st}P(D + s)[h]$, where s is any real or complex number and $h(t)$ is any real- or complex-valued function that is sufficiently differentiable.

www **9.** (*Hearts and Eyes*). Find a solution formula for $y'' + 25y = \sin \omega t$, where $\omega \neq 5$. Plot the solution curve of the IVP with $y(0) = y'(0) = 0$, where $\omega = 4$. Then plot the orbit for $0 \leq t \leq 20$ in the rectangle $|y| \leq 0.1$, $-0.5 \leq |y'| \leq 0.3$. Repeat with $\omega = 1$. Try to overlay the two graphs. [*Hint*: See Example 3.6.7.]

 10. (*Undetermined Coefficients*). Consider the constant coefficient ODE $P(D)[y] = f(t)$, where $P(D) = D^2 + aD + b$ and a and b are constants. Suppose that r_1 and r_2 are the roots of $P(r)$.

(a) Suppose $f(t) = Ae^{st}$, where A and s are real or complex constants. Show that a particular solution y_d of $P(D)[y] = Ae^{st}$ has the indicated form:

- If $s \neq r_1, r_2$, then $y_d = [A/P(s)]e^{st}$.
- If $s = r_1$, $r_2 \neq r_1$, then $y_d = [A/(r_1 - r_2)]te^{r_1 t}$.
- If $s = r_1$, $r_2 = r_1$, then $y_d = (A/2)t^2 e^{r_1 t}$.

[*Hint*: When s is a root of the characteristic polynomial, then look for y_d in the form $y_d = h(t)e^{st}$ and apply the identity $P(D)[h(t)e^{st}] = e^{st}P(D+s)[h]$ to conclude that $P(D+s)[h] = A$. Solve for $h(t)$ using undetermined coefficients.]

(b) Show that the ODE $P(D)[y] = p(t)e^{st}$, where p is a polynomial and s a constant, has a particular solution $y_d = q(t)e^{st}$, with polynomial q as below.

If $\deg(p) = n$, and:	Then take
$s \neq r_1, s \neq r_2$	$q(t) = \sum_{k=0}^{n} a_k t^k$ and $q'' + (a + 2s)q' + P(s)q = p$
$s = r_1 \neq r_2$	$q(t) = t\sum_{k=0}^{n} a_k t^k$ and $q'' + (r_1 - r_2)q' = p$
$s = r_1 = r_2$	$q(t) = t^2 \sum_{k=0}^{n} a_k t^k$ and $q'' = p$

(c) Find a particular solution of $(D^2 + 2D + 10)[y] = te^{-t}\cos 3t$.

 11. (*Periodic Solutions of a First-Order ODE*). Follow the outline below to show that the ODE $y' + ay = A\cos \omega t$, where a, A, and ω are nonzero constants, has a unique periodic solution.

(a) First, show that if $z_d(t)$ is a particular solution of the ODE $z' + az = Ae^{i\omega t}$, then $y_d(t) = \text{Re}[z_d(t)]$ is a particular solution of the given ODE $y' + ay = A\cos \omega t$.

(b) Use the operator formula $(D + a)[Ce^{i\omega t}] = C(i\omega + a)e^{i\omega t}$ to get the particular solution

$$z_d = \frac{A(a - i\omega)}{a^2 + \omega^2}e^{i\omega t}$$

of the ODE in part (a).

(c) Let ϕ be an angle such that $\cos \phi = a/\sqrt{a^2 + \omega^2}$, $\sin \phi = \omega/\sqrt{a^2 + \omega^2}$ and show that

$$y_d = \frac{A}{\sqrt{a^2 + \omega^2}}\cos(\omega t - \phi)$$

is a particular solution of the ODE $y' + ay = A\cos \omega t$.

(d) Write out the general solution of the ODE $y' + ay = A\cos \omega t$. How many periodic solutions are in its solution set?

(e) Now adapt the process described above to the ODE $y' + ay = A\sin \omega t$. Repeat parts (a)–(d), suitably modified to handle the driving term $A\sin \omega t$.

3.7 The General Theory of Linear ODEs

We now look closely at the solution set of the general *linear second-order ODE* in *normal linear form*

$$y'' + a(t)y' + b(t)y = f(t) \tag{1}$$

where the *coefficients* $a(t)$ and $b(t)$, and the *driving term* $f(t)$ are all continuous on a common interval I on the t-axis. Although the general linear ODE (1) arises quite often in applications, we will rarely be able to find an explicit solution formula when the coefficients $a(t)$ and $b(t)$ are not constants. The case where a and b are constants is an important exception, as we saw in the previous sections. Euler's equation is another important exception (see Problem 4). Nevertheless, we can say a lot about the structure of the solution set of ODE (1), even in the general case. Of course, numerical solvers can always be used to produce solution curves.

To define a solution of an IVP without using awkward wording, we will define classes of differentiable functions that are known as *continuity sets*

❖ **Continuity Sets.**   Suppose that I is an interval on the t-axis. The symbol $\mathbf{C}^0(I)$ denotes the set of all continuous functions on I. The symbol $\mathbf{C}^1(I)$ denotes the set of functions in $\mathbf{C}^0(I)$ with a continuous derivative on I. The symbol $\mathbf{C}^2(I)$ denotes the set of all functions in $\mathbf{C}^1(I)$ whose second derivative is continuous on I. For each integer $n > 2$, $\mathbf{C}^n(I)$ is defined in a similar way.

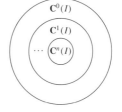

Note that $\mathbf{C}^n(I)$ is contained in $\mathbf{C}^m(I)$ if $m \leq n$. Also, if $y_1(t)$ and $y_2(t)$ are any two functions in $\mathbf{C}^n(I)$, and if c_1 and c_2 are any two constants, then from calculus, we see that the linear combination $c_1 y_1 + c_2 y_2$ is a function in $\mathbf{C}^n(I)$ as well. In other words, the result of applying the operations of addition and multiplication by a constant to functions in $\mathbf{C}^n(I)$ returns a function in $\mathbf{C}^n(I)$.

Before going on, let's look at what the theory says about solutions of ODE (1).

The Fundamental Theorem for a Linear ODE

We can sharpen Theorem 3.2.1 in the case of a linear ODE.

THEOREM 3.7.1

☞ Note that *every* solution of the linear ODE can be extended to *all* of I.

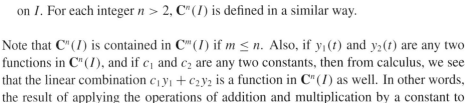

Fundamental Theorem for a Linear ODE. For any functions $a(t)$, $b(t)$, and $f(t)$ that are continuous on a common interval I, for any choice of t_0 in I, and for any choice of the values y_0, v_0, the IVP

$$y'' + a(t)y' + b(t)y = f(t), \qquad y(t_0) = y_0, \quad y'(t_0) = v_0 \tag{2}$$

has a unique solution $y(t)$ in $\mathbf{C}^2(I)$, which is a continuous function of t_0, y_0, v_0.

Theorem 3.7.1 has several important features:

- The unique solution of IVP (2) is defined on the entire interval I. Solutions of the ODE in IVP (2) do not *escape to infinity in finite time* within a closed interval I where $a(t)$, $b(t)$, and $f(t)$ are continuous.

- The values of t_0 in I and of y_0 and v_0 can be chosen arbitrarily.

- Theorem 3.7.1 requires the ODE in IVP (2) to be normalized.

The next theorem is a consequence of Theorem 3.7.1 and is often used to simplify calculations.

THEOREM 3.7.2

☞ The solution $y(t) = 0$ for all t is called the *trivial solution*

Vanishing Data Theorem. The IVP $y'' + a(t)y' + b(t)y = 0$, $y(t_0) = 0$, $y'(t_0) = 0$, where $a(t)$ and $b(t)$ are continuous on an interval I containing t_0, has the unique solution $y(t) = 0$, all t in I.

To see this, note that $y(t) = 0$ for all t in I solves IVP (2) in this case. The uniqueness part of Theorem 3.7.1 says that it is the only solution.

The Vanishing Data Theorem implies that no nontrivial solution curve for the ODE $y'' + a(t)y' + b(t)y = 0$ is tangent to the t-axis. For if there were such a solution $y(t)$, then at the point t_0 of tangency, we would have $y(t_0) = 0$, $y'(t_0) = 0$, and then Theorem 3.7.2 would imply that $y(t)$ is the trivial solution, a contradiction. The following example shows what may happen if $a(t)$ or $b(t)$ *isn't* continuous.

EXAMPLE 3.7.1

Loss of Uniqueness

The linear ODE $t^2 y'' - 2ty' + 2y = 0$ has infinitely many solutions, $y = Ct^2$, where C is an arbitrary constant. These solutions also solve the IVP

$$t^2 y'' - 2ty' + 2y = 0, \qquad y(0) = 0, \quad y'(0) = 0$$

☞ This example shows why Theorem 3.7.1 is stated for normal ODEs.

because the function Ct^2 satisfies the initial conditions. This appears to contradict Theorem 3.7.1, but it doesn't. When written in normal linear form, the ODE is

$$y'' - 2t^{-1}y' + 2t^{-2}y = 0$$

Since the coefficients $a(t) = -2t^{-1}$ and $b(t) = 2t^{-2}$ are not continuous for any interval I containing the origin, Theorem 3.7.1 doesn't apply, so there is no contradiction.

Next we introduce the concept of operators acting on continuity sets as a convenient device to describe the structure of solution sets of linear ODEs.

Operator Formulation of ODEs

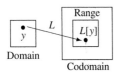

An operator L *acts* on any element y in its *domain* to produce an element $L[y]$ in another set called its *codomain*. An operator is fully defined when its domain and codomain are given and its action is known. The *range* of an operator is the set of all elements in the codomain that are actually produced by the operator acting on its domain. For example, suppose that the domain and codomain of an operator L are the continuity sets $\mathbf{C}^2(I)$ and $\mathbf{C}^0(I)$, respectively, and that the action of L is given by

☞ An operator is fully defined when its domain, codomain, and action are known.

$$y \to L[y] = y'' + a(t)y' + b(t)y, \quad \text{for } y \text{ in } \mathbf{C}^2(I)$$

where $a(t)$, $b(t)$ are continuous on the interval I. Notice that square brackets enclose the function that L acts upon. $L[y]$ is a function in $\mathbf{C}^0(I)$ for any y in $\mathbf{C}^2(I)$; this is often indicated by the schematic $L: \mathbf{C}^2(I) \to \mathbf{C}^0(I)$. Theorem 3.7.1 shows that ODE (1) has a solution $y(t)$ in $\mathbf{C}^2(I)$ for every function $f(t)$ in $\mathbf{C}^0(I)$, so this means that $\mathbf{C}^0(I)$ is actually the range of the operator L. As expected, we use the symbol D for the operator that acts on an element $y(t)$ in $\mathbf{C}^1(I)$ and produces the element $y'(t)$ in $\mathbf{C}^0(I)$. So, $D: \mathbf{C}^1(t) \to \mathbf{C}^0(I)$. By the Fundamental Theorem of Calculus, every continuous function $f(t)$ has an antiderivative $F(t)$ [i.e., $D[F(t)] = f(t)$]. This means that the range of the operator D is all of $\mathbf{C}^0(I)$. If $y(t)$ is in $\mathbf{C}^2(I)$, then D can be applied twice in succession to produce the action $y \to D[D[y]] = y''$. This operator is denoted by D^2, and $D^2: \mathbf{C}^2(I) \to \mathbf{C}^0(I)$.

☞ The Fundamental Theorem of Calculus is Theorem B.5.5.

For any functions $a(t)$ and $b(t)$ in $\mathbf{C}^0(I)$ we use the notation

$$P(D) = D^2 + a(t)D + b(t)$$

to denote the *polynomial operator* $P(D): \mathbf{C}^2(I) \to \mathbf{C}^0(I)$ for which

$$P(D)[y] = D^2[y] + a(t)D[y] + b(t)y = y'' + a(t)y' + b(t)y$$

Using this operator $P(D)$, ODE (1) takes the operator form $P(D)[y] = f(t)$.

It's a remarkable fact that the conclusions of the Polynomial Operators Theorem (Theorem 3.3.2) still hold even when the polynomial operator $P(D)$ has nonconstant coefficients. To see this, go back and look how we justified that theorem in Section 3.3. Nowhere did we assume that $P(D)$ had constant coefficients. So we can freely use the linearity property and the Superposition Principle for the operator $P(D)$ whenever we need it to solve the ODE $P(D)[y] = f$.

As we saw in Sections 3.3–3.6, if the coefficients of $P(D)$ are constants and the function $f(t)$ is a polynomial-exponential, then we can write out an explicit formula for all solutions of $P(D)[y] = f(t)$. In doing this we saw that the set of all solutions of the undriven ODE $P(D)[y] = 0$ plays a key role. We will see that this solution set is also useful in characterizing all solutions of $P(D)[y] = f(t)$, even when the coefficients of $P(D)$ are not constants and $f(t)$ is not a polynomial-exponential.

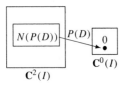

The set of all solutions of the undriven ODE $P(D)[y] = 0$ is called the *null space* and is denoted by $N(P(D))$. So we see that for $P(D) = D^2 + a(t)D + b(t)$, the set of all functions in $\mathbf{C}^2(I)$ carried into the zero function by the action of $P(D)$ is the null space of $P(D)$.

Basic Solution Sets and the Null Space of $P(D)$

First we define a quantity with a strange name.

❖ **Wronskian.** For any two functions f and g in $\mathbf{C}^1(I)$, the function

$$W[f, g](t) = f(t)g'(t) - g(t)f'(t) \qquad (3)$$

is called the *Wronskian* of f and g.[3]

Using Wronskians we can identify pairs of functions in $N(P(D))$ that will help us describe all of $N(P(D))$.

☞ The trivial solution $y = 0$, for all t, can't belong to any basic solution set.

❖ **Basic Solution Set.** A pair of solutions y_1 and y_2 of the ODE $P(D)[y] = y'' + a(t)y' + b(t)y = 0$, where a and b are in $\mathbf{C}^0(I)$, is called a *basic solution set* if $W[y_1, y_2](t) \neq 0$ for t in I.

Actually, we only need to check the Wronskian at a single point in I to see if a pair of solutions of $P(D)[y] = 0$ is a basic solution set, as the following theorem shows.

THEOREM 3.7.3

Abel's[4] Theorem. Suppose that $y_1(t)$, $y_2(t)$ are any two solutions of the ODE

$$P(D)[y] = 0$$

where $a(t)$, $b(t)$ are continuous on a common interval I. Then the Wronskian $W(t) = W[y_1, y_2](t) = y_1 y_2' - y_2 y_1'$ solves the first-order linear ODE

$$W' + a(t)W = 0$$

So, either $W = 0$ for all t in I, or else $W(t)$ is never zero for t in I. In the latter case $\{y_1, y_2\}$ is a basic solution set.

To see this, differentiate W as given in formula (3) and use the facts $P(D)[y_1] = 0$ and $P(D)[y_2] = 0$, which imply that $y_i'' = -ay_i' - by_i$, for $i = 1, 2$:

$$W' = y_1' y_2' + y_1 y_2'' - y_2' y_1' - y_2 y_1'' = y_1 y_2'' - y_2 y_1''$$

$$= y_1(-ay_2' - by_2) - y_2(-ay_1' - by_1) = -aW$$

---

[3] Höene Wronski (1778–1853) started out as a soldier and went on to become, successively, a mathematician, a philosopher, and insane.

[4] The Norwegian Niels Henrik Abel (1802–1829) was one of the best mathematicians of the nineteenth century. Living in poverty and dying of tuberculosis at the age of 27, Abel made all of his mathematical contributions in the short span of six years. Although his work was not widely appreciated during his life, Abel is honored now by the appearance of his name alongside the mathematical concepts he developed: abelian groups, abelian integrals, Abel's series, Abel summability, ...

Solving the ODE $W' + aW = 0$ by integrating factors, we have Abel's Formula $W(t) = Ce^{-A(t)}$, where C is any constant and $A(t)$ is any antiderivative of $a(t)$. The only way W can be zero is for $C = 0$, and then W is zero on all of I.

EXAMPLE 3.7.2

Using Abel's Theorem

The functions $y_1 = \sin 2t$ and $y_2 = \cos 2t$ are solutions of the linear ODE $y'' + 4y = 0$, for all t on the real line, and $W = y_1 y_2' - y_2 y_1' = -2\sin^2 t - 2\cos^2 t = -2$, for all t. From Abel's Theorem we see that $\{\sin 2t, \cos 2t\}$ is a basic solution set for the ODE.

Here's the reason why basic solution sets are so important.

THEOREM 3.7.4

Basic Solution Sets and the Null Space. Consider the undriven ODE

$$P(D)[y] = y'' + a(t)y' + b(t)y = 0 \qquad (4)$$

where a and b are in $\mathbf{C}^0(I)$. If $\{y_1, y_2\}$ is a basic solution set, then any solution y of the ODE may be written as a linear combination of y_1 and y_2,

$$y = c_1 y_1 + c_2 y_2 \qquad (5)$$

for some constants c_1 and c_2, so every element of the null space $N(P(D))$ has the form given in (5).

This result was established at the end of Section 3.3 for $P(D)$ with constant co-efficients. A reexamination of that result shows that the coefficients a and b are not required to be constants. If c_1 and c_2 are arbitrary constants, then $y = c_1 y_1 + c_2 y_2$ is the *general solution* of ODE (4).

Does a basic solution set always exist? The answer to this question is not only "yes," but "yes" with a vengeance. For any $a(t)$, $b(t)$ in $\mathbf{C}^0(I)$ the ODE $P(D)[y] = 0$ has infinitely many basic solution sets. Here's one way to find a basic solution set (not a very practical way, though). Suppose that t_0 is in I, y_1 is the solution of the IVP

$$P(D)[y] = 0, \qquad y(t_0) = 1, \quad y'(t_0) = 0$$

and y_2 is the solution of the IVP

$$P(D)[y] = 0, \qquad y(t_0) = 0, \quad y'(t_0) = 1$$

Because $W[y_1, y_2](t_0) = y_1(t_0)y_2'(t_0) - y_1'(t_0)y_2(t_0) = 1$, it follows from Abel's Theorem that $W[y_1, y_2](t) \neq 0$, for all t in I; so $\{y_1, y_2\}$ is a basic solution set.

EXAMPLE 3.7.3

Finding a Basic Solution Set

Theorem 3.7.1 says that solutions of

$$t^2 y'' - 2y = 0 \qquad (6)$$

are defined at least over the interval $-\infty < t < 0$, or the interval $0 < t < +\infty$, be-cause, after writing ODE (6) in normal linear form $y'' - (2/t^2)y = 0$, the coefficient

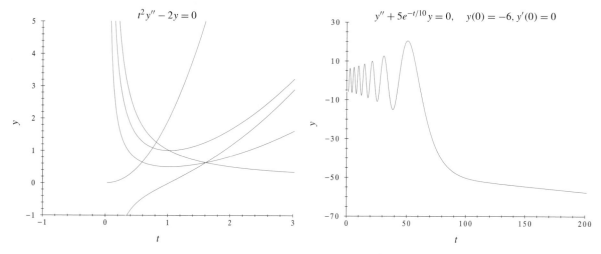

FIGURE 3.7.1 Some solution graphs of ODE (6) in Example 3.7.3.

FIGURE 3.7.2 Long-term aging spring motion (Example 3.7.4).

$b(t) = -2/t^2$ is continuous on any interval not containing the origin. We see by direct substitution of $y = t^\alpha$, α a constant, into ODE (6) that

$$t^2 y'' - 2y = t^2 \alpha(\alpha - 1)t^{\alpha-2} - 2t^\alpha$$

$$= t^\alpha(\alpha^2 - \alpha - 2) = t^\alpha(\alpha - 2)(\alpha + 1)$$

So if $\alpha = 2$ or $\alpha = -1$ then $y = t^\alpha$ is a solution of ODE (6). Put $y_1(t) = t^2$ and $y_2(t) = 1/t$. Notice that $y_2(t)$ is a solution on either $-\infty < t < 0$ or on $0 < t < +\infty$, but that $y_1(t)$ is a solution of ODE (6) on the entire t-axis. Since $W[y_1, y_2](t) = t^2(-t^{-2}) - 2tt^{-1} = -3$, for $t \neq 0$, the pair of solutions is a basic solution set, and the general solution of ODE (6) on the interval $0 < t < +\infty$ (to be specific) is

$$y = c_1 t^2 + c_2/t$$

where c_1 and c_2 are arbitrary constants.

    The constants can be determined from initial data at some point, say $t_0 = 1$. So if $y(1) = y_0$ and $y'(1) = v_0$, then $c_1 + c_2 = y_0$ and $2c_1 - c_2 = v_0$. Solving, we get $c_1 = (y_0 + v_0)/3$ and $c_2 = (2y_0 - v_0)/3$. If initial data are chosen randomly at $t = 1$, then chances are that the solution will escape to $\pm\infty$ as $t \to 0$ (because $c_2 \neq 0$ unless v_0 is precisely $2y_0$). Figure 3.7.1 shows some solution graphs of the ODE for $t > 0$, including the graph of the solution $y = 2t^2$, which arises from the initial conditions $y(1) = 2$, $y'(1) = 4$, and does *not* escape to infinity as $t \to 0$. Your solver may have difficulty in generating the graph of this solution because a small numerical error may cause your solver to track a solution that *does* escape to infinity as $t \to 0$.

Let's look at an ODE which models the motion of an aging spring.

EXAMPLE 3.7.4

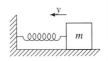

The Aging Spring Model

Suppose that an aging spring is fastened to a wall and to a 1-kg mass which slides without friction or damping on a horizontal platform. Further suppose that the mass moves in a straight line and that the position of the mass is located by its displacement from the equilibrium position. The coordinate y is used for the displacement with $y > 0$ for the compressed spring. Let's measure displacement in meters and time in seconds. The aging spring force [see (5) in Section 3.1] is given by $S(y) = -5e^{-t/10}y$ newtons, and there are no other forces acting on the weight. The displacement of this spring is modeled by the linear ODE

$$1 \cdot y'' + 5e^{-t/10}y = 0$$

which does *not* have constant coefficients. Figure 3.7.2 shows what happens if the weight is pulled out 6 meters and released from rest. Eventually, the spring becomes so weak that the weight appears to depart for regions unknown.

Finally, we turn to the problem of characterizing the solutions of a driven ODE.

Describing All Solutions of $P(D) = f(t)$

We can use the linearity of the operator $P(D)$ to describe all the solutions of the driven ODE $P(D)[y] = f(t)$ in exactly the same way as was done in Section 3.6 for the constant-coefficients case. Here it is again, but stated for nonconstant coefficients.

THEOREM 3.7.5

> General Solution of $P(D)[y] = f(t)$.   Suppose that $\{y_1, y_2\}$ is a basic solution set for the undriven linear ODE
>
> $$y'' + a(t)y' + b(t)y = 0$$
>
> where $a(t)$ and $b(t)$ are continuous on an interval I. For a given function $f(t)$ in $\mathbf{C}^0(I)$, suppose that $y_d(t)$ is a particular solution of the driven ODE
>
> $$y'' + a(t)y' + b(t)y = f(t) \qquad (7)$$
>
> Then all solutions of the driven ODE (7) are given by the solution formula
>
> $$y = y_d + c_1y_1 + c_2y_2, \qquad c_1, c_2 \text{ arbitrary constants}$$

☞ Real-analytic functions are defined in Appendix B.2, item 8.

In Chapter 11 we show how to find a basic solution pair $\{y_1, y_2\}$ for $P(D)[y] = 0$ when the coefficients $a(t)$ and $b(t)$ are real-analytic at $t = t_0$. Until then, we are pretty much on our own when searching for solution formulas. There are a few tricks shown in the problem set that work in special cases.

It's a remarkable fact, though, that we can write out a formula for a particular solution of a driven ODE $P(D)[y] = f(t)$ in terms of a basic solution set. This approach is described in Problem 8 and is called the *Method of Variation of Parameters*

Comments

Linear ODEs occupy a special place in the theory of differential equations. Ordinary differential equations that arise from many different modeling environments turn out to be linear or can be closely approximated by a linear ODE (as we saw in Section 3.1). In addition, the theory of linear ODEs is wonderfully straightforward and without the strange behavior associated with theories of nonlinear ODEs.

PROBLEMS

1. (*Null Space, Uniqueness*). Consider the polynomial operator $P(D) = tD^2 + D$.

   (a) Find the null space of $P(D)$ over $0 < t < \infty$. [*Hint*: Write $P(D)[y]$ as $(ty')'$.]

   (b) Show that the IVP $P(D)[y] = 2t$, $y(0) = a$, $y'(0) = b$, has a unique solution on any interval I containing the origin, but only provided that a and b are chosen suitably. [*Hint*: $t^2/2$ is a particular solution.]

   (c) Why doesn't the result in part (b) contradict Theorem 3.7.1?

2. Find the null space of the operator $P(D) = (1 - t^2)D^2 - 2tD$ and the general solution of $P(D)[y] = 0$ on the interval $|t| < 1$. [*Hint*: $(1 - t^2)D^2 - 2tD = D[(1 - t^2)D]$.]

3. (*IVPs*). First verify that the given functions y_1 and y_2 are solutions of the ODE. Then use y_1 and y_2 to find all solutions of the IVP. Plot the solution curves.

   (a) $t^2 y'' + ty' - y = 0$, $\quad y(1) = 0$, $\quad y'(1) = -1$; $\quad y_1 = t^{-1}$, $\quad y_2 = t$

   (b) $t^2 y'' - ty' + y = 0$, $\quad y(1) = 1$, $\quad y'(1) = 0$; $\quad y_1 = t$, $\quad y_2 = t\ln t$

   (c) $t^2 y'' + ty' - 4y = 0$, $\quad y(0) = 0$, $\quad y'(0) = 0$; $\quad y_1 = t^2$, $\quad y_2 = t^{-2}$

4. (*Euler ODE*). If p and q are constants, then $t^2 y'' + pty' + qy = 0$, $t > 0$, is an *Euler ODE*.

   ☞ This is one of the few linear ODEs with nonconstant coefficients for which we can find a simple solution formula.

   (a) Show that Euler's ODE has the solution $y = t^r$ if r solves the quadratic equation $Q(r) = r^2 + (p - 1)r + q = 0$. Show that $y_1 = t^{r_1}$, $y_2 = t^{r_2}$ form a basic solution set if r_1 and r_2 are distinct real roots of $Q(r)$. If $r = \alpha + i\beta$ is a complex root, show that $y_1 = t^\alpha \cos(\beta \ln t)$, $y_2 = t^\alpha \sin(\beta \ln t)$ form a basic solution set for the Euler ODE. [*Hint*: Write $t^{\alpha + i\beta}$ as $e^{(\alpha + i\beta)\ln t}$.]

   (b) Searching for solutions on $t > 0$, show that the change of independent variable $t = e^s$ converts the Euler equation into the constant coefficient ODE

   $$\frac{d^2 y}{ds^2} + (p - 1)\frac{dy}{ds} + qy = 0$$

   (c) Use the techniques presented in part (a) to solve the IVP $t^2 y'' - 2ty' + 2y = 0$, $y(1) = 0$, $y'(1) = -1$. What is the largest interval on which this solution is defined?

   (d) Use the technique in part (b) to solve the IVP $t^2 y'' + 2ty' + 2y = 0$, $y(1) = 0$, $y'(1) = 1$.

5. (*Reduction of Order Technique*). Suppose that $y = u(t)$ is a solution of $y'' + a(t)y' + b(t)y = 0$, where $a(t)$ and $b(t)$ are in $\mathbf{C}^0(I)$.

   (a) Show that $y = u(t)z(t)$ solves that ODE if $z(t)$ solves the ODE $uz'' + (2u' + a(t)u)z' = 0$.

   (b) Show that

   $$z(t) = \int_{t_0}^t \left\{ \frac{1}{(u(s))^2} \exp\left[-\int^s a(r)\,dr \right] \right\} ds$$

   is a solution of the z-equation in part (a) if $u(t) \neq 0$ for any t in I.

(c) Show that the pair of solutions $\{u,\ uz\}$ constructed in parts **(a)** and **(b)** is a basic solution set for the given ODE.

(d) Find all solutions of $ty'' - (t+2)y' + 2y = 0$, for $t > 0$, given that e^t is one solution.

6. *(Wronskian Reduction of Order)*. Suppose that $P(D) = D^2 + a(t)D + b(t)$, where $a(t)$, $b(t)$ are in $\mathbf{C}^0(I)$. Suppose that $W(t) = W[y_1, y_2](t)$ is the Wronskian of a pair of solutions $\{y_1, y_2\}$ of $P(D)[y] = 0$.

(a) Show that $W(t)$ satisfies the first-order linear equation $W' = -a(t)W$, so W is given by

Abel's Formula

$$W(t) = W(t_0)\exp\left[-\int_{t_0}^{t} a(s)\,ds\right] \quad \text{for } t_0,\ t \text{ in } I$$

[*Hint*: Differentiate W directly, and look at the proof of Theorem 3.7.3.]

(b) Use Abel's Formula to show that if $u(t) \neq 0$ is a solution of $P(D)[y] = 0$, then a second solution $v(t)$ of the ODE can be found by solving the following first-order linear ODE for v,

$$u(t)v' - u'(t)v = \exp\left[-\int a(s)\,ds\right]$$

Show that the pair $\{u(t),\ v(t)\}$ is a fundamental set for $P(D)[y] = 0$.

(c) Given that e^t is a solution of $ty'' - (t+2)y' + 2y = 0$, find a second solution v such that $\{e^t, v\}$ is a fundamental set. [*Hint*: Remember to normalize the ODE first.] Notice that by an appropriate choice of the limits of integration, the solution v can be made identical to the solution uz in Problem 5**(b)**.

www 7. *(Interlacing of Zeros of a Basic Solution Pair)*. Suppose that $a(t)$ and $b(t)$ are continuous on $-\infty < t < \infty$, and suppose that $\{u(t),\ v(t)\}$ is a basic solution set for the ODE $y'' + a(t)y' + b(t)y = 0$. Show that between any two consecutive zeros of one solution there is precisely one zero of the other solution. Why are the margin graphs *not* of a basic solution pair?

8. *(The Method of Variation of Parameters)*. Suppose that $P(D) = D^2 + a(t)D + b(t)$ with $a(t)$, $b(t)$ in $\mathbf{C}^0(I)$, that $\{y_1, y_2\}$ is a given basic solution pair for $P(D)[y] = 0$, that t_0 is any point in I, and that f is any function in $\mathbf{C}^0(I)$. Follow the outline below to derive a formula for the unique solution of the driven IVP

$$P(D)[y] = f, \quad y(t_0) = 0, \quad y'(t_0) = 0$$

(a) Assume that the solution has the form

$$y_d = c_1(t)y_1(t) + c_2(t)y_2(t)$$

and substitute y_d into the driven ODE $P(D)[y] = f$. Assuming that $c_1(t)$ and $c_2(t)$ are chosen so that $c_1'y_1 + c_2'y_2 = 0$, show that if y_d solves $P(D)[y] = f$, then $c_1'y_1' + c_2'y_2' = f$.

(b) If $W(t) = W[y_1, y_2](t)$ is the Wronskian of the basic solution set $\{y_1, y_2\}$, then show that the *varied parameter* functions

☞ The value of this solution formula is more theoretical than practical.

$$c_1(t) = \int_{t_0}^{t} \frac{-y_2(s)f(s)}{W(s)}\,ds, \quad \text{and} \quad c_2(t) = \int_{t_0}^{t} \frac{y_1(s)f(s)}{W(s)}\,ds$$

satisfy the conditions imposed on $c_1(t)$ and $c_2(t)$ in part **(a)**, and that

$$y_d = c_1(t)y_1(t) + c_2(t)y_2(t)$$

solves the IVP $P(D)[y] = f,\ y(t_0) = 0,\ y'(t_0) = 0$.

(c) Show that $y_d(t)$ given in part **(b)**, may be written in the form

$$y_d(t) = \int_{t_0}^{t} K(t, s)f(s)\,ds$$

Green's kernel
$K(t, s)$ looks like it
depends on the basic
solution set $\{y_1, y_2\}$, but
it is exactly the same for
every basic solution set.

where $K(t, s)$ is the *kernel function* (also called *Green's kernel*).

$$K(t, s) = \frac{y_1(s)y_2(t) - y_1(t)y_2(s)}{y_1(s)y_2'(s) - y_2(s)y_1'(s)} = \frac{y_1(s)y_2(t) - y_1(t)y_2(s)}{W[y_1, y_2](s)}$$

(d) Use the *Method of Variation of Parameters* derived in parts **(a)**–**(c)** to solve the IVP

$$y'' + y = \tan t, \quad y(0) = 0, \quad y'(0) = 0$$

(e) Use the Method of Variation of Parameters to solve

$$y'' + 2y' + 10y = 3\cos 3t, \quad y(0) = y_0, \quad y'(0) = y_0'$$

(f) It is known that the ODE

$$y'' + 3y' + 2y = 4\sin 2t$$

has a unique periodic solution (a justification of this claim is outlined in Problem 9 of Section 4.3). Use the Method of Variation of Parameters to find it. Then use the Method of Undetermined Coefficients to find it. Which method is easier?

 (g) The Method of Variation of Parameters will accept on-off driving functions without any difficulty since piecewise-continuous functions are integrable. Since integrals that involve an on-off function such as a square wave are awkward to evaluate, it is better to use a numerical solver if you want to see how solutions behave. Apply your numerical solver to the ODE

$$y'' + 0.4y' + y = 20\,\mathrm{sqw}(t, 50, 2\pi) - 10$$

with various initial conditions and then describe the long-term behavior of solutions. [The figures below show some ty- and ty'-curves (solid) and the square wave input (dashed).]

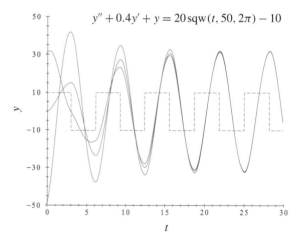

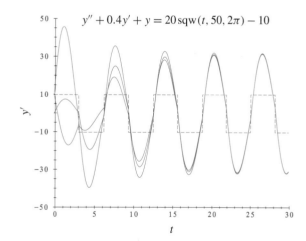

(h) Repeat part **(g)** with the ODE

$$y'' + y = 10\,\mathrm{trw}(t, 25, 2\pi)$$

Do you see something unusual in the long-term behavior of the solutions? Any explanations?

A Snapshot Look at Constant-Coefficient Polynomial Operators

Definitions

- The operator D acts on a function $y(t)$ to produce dy/dt. Notationally: $D[y] = y'$. Applying D to $D[y]$ produces y''. Notationally: $D\big[D[y]\big] = D^2[y] = y''$.

- The polynomial operators $D + p$ and $D^2 + aD + b$, for constants a, b, and p, are

$$(D + p)[y] = y' + py, \qquad (D^2 + aD + b)[y] = y'' + ay' + by$$

- For constants p and q, the operator applied to a function $y(t)$ which produces $(D + p)\big[(D + q)[y]\big]$ is denoted by the product $(D + p)(D + q)$.

Properties

Polynomial operators $P(D) = D^2 + aD + b$ satisfy the following identities.

- **Factorization:** If r_1 and r_2 are roots of $r^2 + ar + b = 0$, then

$$P(D) = (D - r_1)(D - r_2) = (D - r_2)(D - r_1)$$

- **Identities:** For the polynomial operator $P(D)$ we have that

$$P(D)[e^{st}] = e^{st} P(s), \quad \text{for any constant } s$$

$$P(D)[h(t)e^{st}] = e^{st} P(D + s)[h], \quad \text{for any constant } s, \text{ any differentiable } h(t)$$

- **Linearity:** For any constants C_1, C_2, and any differentiable y_1 and y_2,

$$P(D)[C_1 y_1 + C_2 y_2] = C_1 P(D)[y_1] + C_2 P(D)[y_2]$$

- **Closure:** For any constants C_1, C_2, and differentiable functions y_1, y_2,

$$P(D)[y_1] = 0, \ P(D)[y_2] = 0 \quad \text{imply that} \quad P(D)[C_1 y_1 + C_2 y_2] = 0$$

A Solution Formula for $y'' + ay' + by = Ct^n e^{\alpha t} \cos \beta t$

☞ These steps also work for $Ct^n e^{\alpha t} \sin \beta t$, but taking $y_d = \text{Im}[z_d]$.

Follow these steps to find a formula for a particular solution $y_d(t)$ of the driven ODE

$$P(D)[y] = y'' + ay' + by = Ct^n e^{\alpha t} \cos \beta t$$

where a, b, α, β, C are real constants and n is a nonnegative integer.

1. Notice that $\text{Re}[Ct^n e^{(\alpha + i\beta)t}] = Ct^n e^{\alpha t} \cos \beta t$.

2. Look for a solution $z_d(t)$ of the ODE $P(D)[z] = Ct^n e^{(\alpha + i\beta)t}$.

3. If $n = 0$ and $P(\alpha + i\beta) \neq 0$, then take $z_d = (C/P(\alpha + i\beta))e^{(\alpha + i\beta)t}$, and $y_d = \text{Re}[z_d]$ (because of Step 1).

4. If $P(\alpha + i\beta) = 0$, then write $z_d(t) = h(t)e^{(\alpha + i\beta)t}$ and note that $P(D)[h(t)e^{(\alpha + i\beta)t}] = P(D + \alpha + i\beta)[h]$. So to find $z_d = he^{(\alpha + i\beta)t}$ as a solution of $P(D)[z] = Ct^n e^{(\alpha + i\beta)t}$ we must have that $P(D + \alpha + i\beta)[h] = Ct^n$.

5. Guess $h(t)$ to be a polynomial with undetermined coefficients whose values are determined by substitution into the ODE for h at the end of Step 4. After finding h, put $z_d = he^{(\alpha + i\beta)t}$ and then $y_d = \text{Re}[z_d]$.

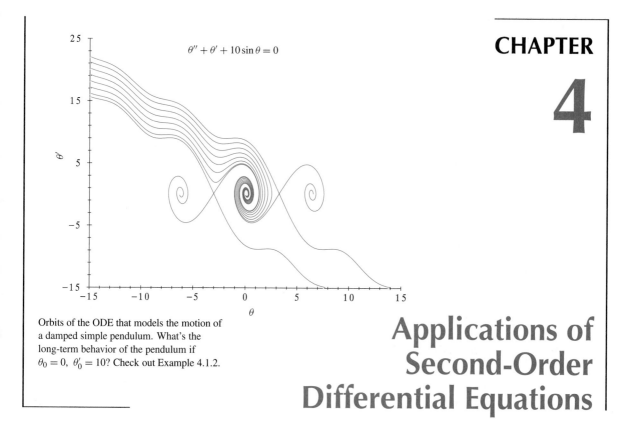

$$\theta'' + \theta' + 10\sin\theta = 0$$

CHAPTER

4

Orbits of the ODE that models the motion of a damped simple pendulum. What's the long-term behavior of the pendulum if $\theta_0 = 0$, $\theta_0' = 10$? Check out Example 4.1.2.

Applications of Second-Order Differential Equations

The models treated in this chapter lead to second-order ODEs, both linear and nonlinear, and involve motion in a force field and the flow of electrical energy in a circuit. Along the way we give a brief introduction to the vector concept in mechanics and to Kirchhoff's laws for circuits. Some important properties of driven linear ODEs are presented: beats, resonance, and frequency modeling. The principal application in mechanics is to the motion of a pendulum.

4.1 Newton's Laws and the Pendulum

The motion of a swinging pendulum has a certain hypnotic fascination to it. Galileo, Newton, the Bernoulli brothers and the Dutch scientist Christian Huyghens (1629–1695) all observed pendulum motion, thought deeply about what they saw, and created mathematical models for the motion. We use the tools of differential equations, geometric vectors, and numerical solvers to carry out our own investigations of a swinging pendulum. Let's begin by describing geometric vectors and how they are used to model the location, velocity, and acceleration of a moving body.

Geometric Vectors

Basic physical concepts such as velocity and acceleration can be pictured as directed line segments, or arrows, each with a head and a tail. We will use boldface letters in this section to denote these *geometric vectors*. Two vectors **v** and **w** are considered equivalent if and only if they can be made to coincide by translations (translations preserve length and direction of vectors). Equivalent vectors with different initial points can share the same name. The *length* (or *magnitude*) of a vector **v** is denoted by $\|\mathbf{v}\|$. We define the *zero vector* (denoted by **0**) to be the vector of zero length, but no direction.

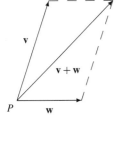

The *resultant* or *sum* **v** + **w** of **v** and **w** is defined by the parallelogram law as follows: Find a vector equivalent to **v** whose tail coincides with **w**'s head; then **v** + **w** is the vector whose tail is **w**'s and whose head is **v**'s. The vector **v** + **w** is a diagonal of the parallelogram formed by **v** and **w** (margin figure). If r is any real number, then the *product* $r\mathbf{v}$ is the vector of length $|r|\,\|\mathbf{v}\|$ that points in the direction of **v** if $r > 0$, and in the opposite direction to **v** if $r < 0$.

Dot Product: The *dot product* (or *scalar product*) of two vectors **u** and **v**, denoted by $\mathbf{u} \cdot \mathbf{v}$, is the real number $\|\mathbf{u}\|\,\|\mathbf{v}\|\cos\theta$ if neither **u** nor **v** is the zero vector and θ is the angle between them. When either **u** or **v** is the zero vector, the dot product is zero. Two nonzero vectors are *perpendicular* (or *orthogonal*) if and only if their dot product is zero. See Problem 1 for additional properties.

Derivative: If the vector $\mathbf{u} = \mathbf{u}(t)$ depends on t in an interval of the real axis, then the *derivative* $d\mathbf{u}/dt$ [or $\mathbf{u}'(t)$] is defined as the limit of a difference quotient:

$$\mathbf{u}'(t) = \frac{d\mathbf{u}}{dt} = \lim_{h \to 0} \frac{\mathbf{u}(t+h) - \mathbf{u}(t)}{h} \tag{1}$$

if the limit exists.[1] If **u** is a constant vector, then $\mathbf{u}' = \mathbf{0}$. If $\mathbf{u}(t)$ and $\mathbf{v}(t)$ are two differentiable vector functions and $r(t)$ is a differentiable function, then we have the identities

$$(\mathbf{u} \cdot \mathbf{v})' = \mathbf{u}' \cdot \mathbf{v} + \mathbf{u} \cdot \mathbf{v}', \qquad [r\mathbf{u}]' = r'\mathbf{u} + r\mathbf{u}'$$

which resemble the usual product differentiation rule.

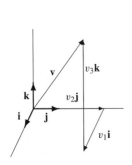

A *coordinate frame* in three dimensions is a triple of vectors, denoted by $\{\mathbf{i}, \mathbf{j}, \mathbf{k}\}$, which are mutually orthogonal and all of unit length. A little thought (and some trigonometry) reveals that every vector can be uniquely written as the sum of vectors parallel to **i**, **j**, and **k**. So for each vector **v** there is a unique set of real numbers v_1, v_2, and v_3 such that $\mathbf{v} = v_1\mathbf{i} + v_2\mathbf{j} + v_3\mathbf{k}$. The elements of the ordered triple (v_1, v_2, v_3) are called the *coordinates* (or *components*) of **v** (margin figure) in the frame $\{\mathbf{i}, \mathbf{j}, \mathbf{k}\}$. There are many frames of reference in the three-dimensional space of our experience.

---

[1] In the physics and engineering literature one often sees the dot notation, $\dot{\mathbf{u}}$, instead of the prime notation, $\mathbf{u}'$, to denote the derivative of the vector function of time, $\mathbf{u}(t)$.

We can imagine frames that move in space, or frames that are fixed. Suppose that $\{\mathbf{i}, \mathbf{j}, \mathbf{k}\}$ is a fixed frame and that the location of a moving particle is described by the *position vector*

$$\mathbf{R} = \mathbf{R}(t) = x(t)\mathbf{i} + y(t)\mathbf{j} + z(t)\mathbf{k}$$

If $\mathbf{R}$ is differentiable, then

$$\mathbf{R}'(t) = x'(t)\mathbf{i} + y'(t)\mathbf{j} + z'(t)\mathbf{k} \tag{2}$$

$\mathbf{R}'(t)$ is the *velocity vector* $\mathbf{v}(t)$ of the particle with mass m at time t, and $\mathbf{v}(t)$ is tangent to the path of the particle's motion at the point $\mathbf{R}(t)$. If $\mathbf{R}'(t)$ is differentiable, then

$$\mathbf{R}''(t) = x''(t)\mathbf{i} + y''(t)\mathbf{j} + z''(t)\mathbf{k} \tag{3}$$

is the *acceleration vector* $\mathbf{a}(t)$ (or $\mathbf{v}'(t)$) for the particle (margin figure).

Now let's express Newton's Laws in the language of vectors.

Forces and Newton's Laws

Newton developed Galileo's idea that the environment creates forces that act on bodies causing them to accelerate. The Galileo/Newton principle is now known as Newton's First Law.

> Newton's First Law. A body remains in a state of rest, or of uniform motion in a straight line (i.e., with constant velocity), if there is no net external force acting on it.

Scientists of Newton's time were able to show experimentally that forces behave like geometric vectors, that is, they satisfy the parallelogram law. The effect of Newton's First Law is to identify coordinate frames that are either fixed in space or undergoing a translation at constant velocity with respect to a fixed frame. Such frames are called *inertial*. According to Newton's First Law, a frame is inertial if and only if a body is unaccelerated with respect to that frame whenever the sum of all the forces acting on the body is the zero vector $\mathbf{0}$.

Next, we have the following basic principle in dynamics.

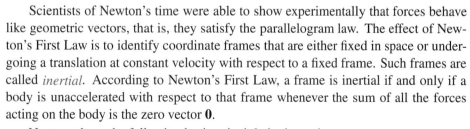

> Newton's Second Law. For a body with acceleration $\mathbf{a}$ and constant mass m,
>
> $$\mathbf{F} = m\mathbf{a} \tag{4}$$
>
> where $\mathbf{F}$ is the sum of all external forces acting on the body.

Since $\mathbf{a} = \mathbf{v}'$, Newton's Second Law[2] can be written in the form

$$\mathbf{F} = (m\mathbf{v})' \qquad (5)$$

where $\mathbf{v}$ is the velocity of the body. The product $m\mathbf{v}$ is the *momentum* of the body. Newton himself expressed the Second Law in terms of the rate of change of the "quantity of motion," his term for momentum. Indeed, $\mathbf{F} = (m\mathbf{v})'$ holds whether or not the mass is constant. So in this form Newton's Second Law states that "the sum of the external forces acting on a body equals the rate of change of the momentum of the body."

The terms in Newton's Second Law must be quantified in some system of units. In the *MKS system* (meter, kilogram, second), also called the *SI system*, the unit of mass is the kilogram. The relation in (4) between mass, acceleration, and force is then used to define one unit of force as that amount of force which will accelerate a 1-kg mass at 1 m/sec^2. This unit of force is called a *newton*. In the *CGS system* (centimeter, gram, second) the unit of force is a *dyne* (1 dyne $= 10^{-5}$ newton) and is that force which accelerates a 1-g mass at 1 cm/sec^2. In these metric units, mass, length, and time are considered to be fundamental quantities, while force is secondary. In the *British Engineering system* mass is measured in slugs, and the pound (the weight, not the monetary unit) is the unit of force. We often avoid the use of any particular system of units. When units must be used, they will be explicitly labeled.

☞ Appendix B.6 has more on unit systems.

Newton's Third Law gives some insight on how to treat multiple body systems, such as the earth, moon, sun system and the mutual gravitational forces acting upon them. For example, the earth and the moon exert equal but opposite gravitational forces on each other.

> Newton's Third Law. If body A exerts a force $\mathbf{F}$ on body B, then body B exerts a force $-\mathbf{F}$ on body A.

Let's now apply Newton's Laws to model the motion of a pendulum.

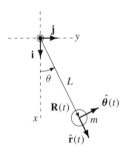

☞ Note that the positive x-direction is down.

The Simple Pendulum

The *simple pendulum* consists of a bob of mass m hanging on a rigid rod of fixed length L (and negligible mass) attached to a support. The pendulum is in equilibrium when the bob and rod are hanging downward at rest. If the pivot on the top of the rod is positioned at the origin of the standard xy-coordinate system and the initial position and velocity of the center of mass of the bob are in the xy-plane (margin figure), the bob will oscillate back and forth in the xy-plane or whirl up and over the pivot. The angle θ tracks the motion of the bob.

---

[2]The form of Newton's Second Law in (4) is for so-called point masses. For some distributed masses it is permissible to apply (4) as if all the mass of the body were concentrated at its center of mass.

☞ Viscous damping
was introduced in
Section 1.5.
Suppose that the moving pendulum experiences a viscous damping force **f** proportional to its velocity vector and opposing the motion. Suppose that the other forces acting on the bob are the gravitational force $mg\mathbf{i}$ and the tension **T** in the rod. We will derive an expression for the acceleration of the bob relative to the **ij**-frame, compute the sum of the external forces acting on the bob, and then substitute these quantities into Newton's Second Law. Let's suppose that $\mathbf{R}(t)$ is the position vector from the pivot to the center of mass of the bob. Introducing polar coordinates in the plane of oscillation with θ measured positively in the counterclockwise direction from the downward position, we can write

$$\mathbf{R}(t) = r(t)\hat{\mathbf{r}}(t)$$

where $r(t)$ is the length L of $\mathbf{R}(t)$ and $\hat{\mathbf{r}}(t)$ is a unit vector directed along $\mathbf{R}(t)$. Denote by $\hat{\boldsymbol{\theta}}(t)$ the unit vector orthogonal to $\mathbf{R}(t)$ that points in the direction of increasing θ. Note that the unit vectors $\hat{\mathbf{r}}$, $\hat{\boldsymbol{\theta}}$, **i**, and **j** are related by the equations

$$\hat{\mathbf{r}} = \cos\theta\mathbf{i} + \sin\theta\mathbf{j}, \qquad \hat{\boldsymbol{\theta}} = -\sin\theta\mathbf{i} + \cos\theta\mathbf{j}$$
$$\mathbf{i} = \cos\theta\hat{\mathbf{r}} - \sin\theta\hat{\boldsymbol{\theta}}, \qquad \mathbf{j} = \sin\theta\hat{\mathbf{r}} + \cos\theta\hat{\boldsymbol{\theta}} \tag{6}$$

Using the formulas in (6) and the differentiation rule in (2), we see that

$$\hat{\mathbf{r}}' = (-\sin\theta\mathbf{i} + \cos\theta\mathbf{j})\theta' = \theta'\hat{\boldsymbol{\theta}}$$
$$\hat{\boldsymbol{\theta}}' = (-\cos\theta\mathbf{i} - \sin\theta\mathbf{j})\theta' = -\theta'\hat{\mathbf{r}} \tag{7}$$

Using (7) and the fact that $\mathbf{R} = L\hat{\mathbf{r}}(t)$, where L is a constant, we see that

$$\mathbf{v} = \mathbf{R}' = (L\hat{\mathbf{r}})' = L\theta'\hat{\boldsymbol{\theta}}$$
$$\mathbf{a} = \mathbf{R}'' = (L\theta'\hat{\boldsymbol{\theta}})' = L\theta''\hat{\boldsymbol{\theta}} + L\theta'\hat{\boldsymbol{\theta}}' = L\theta''\hat{\boldsymbol{\theta}} - L(\theta')^2\hat{\mathbf{r}} \tag{8}$$

For some positive constants T and c, the external forces on the bob are

$$(\text{Tension}) \qquad \mathbf{T} = -T\hat{\mathbf{r}}$$
$$(\text{Viscous damping}) \quad \mathbf{f} = -c\mathbf{v} = -c\mathbf{R}' = -cL\theta'\hat{\boldsymbol{\theta}} \tag{9}$$
$$(\text{Gravity}) \qquad mg\mathbf{i} = mg(\cos\theta\hat{\mathbf{r}} - \sin\theta\hat{\boldsymbol{\theta}})$$

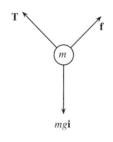

where we have used (6) and (8) to obtain expressions for **v** and **i** in terms of θ and r. See the margin figure for a geometric visualization of these forces. Now substitute (8) and (9) into Newton's Second Law [formula (4)]:

$$m\mathbf{R}'' = \text{sum of forces} = \mathbf{T} + \mathbf{f} + mg\mathbf{i}$$
$$mL\theta''\hat{\boldsymbol{\theta}} - mL(\theta')^2\hat{\mathbf{r}} = -T\hat{\mathbf{r}} - cL\theta'\hat{\boldsymbol{\theta}} + mg(\cos\theta\hat{\mathbf{r}} - \sin\theta\hat{\boldsymbol{\theta}})$$

Equating coefficients of $\hat{\mathbf{r}}$ and $\hat{\boldsymbol{\theta}}$ on both sides of the equation, we obtain

$$-mL(\theta')^2 = -T + mg\cos\theta \qquad (\text{from } \hat{\mathbf{r}})$$
$$mL\theta'' = -cL\theta' - mg\sin\theta \qquad (\text{from } \hat{\boldsymbol{\theta}}) \tag{10}$$

Once a solution $\theta(t)$ of the second ODE in (10) has been found, it can be put into the first ODE and the corresponding tension $T(t)$ can be determined. We focus entirely on the second ODE.

If the pendulum is driven by an external force **h** of the form $h(t)\hat{\boldsymbol{\theta}}$, the second ODE in (10) is

$$mL\theta'' + cL\theta' + mg\sin\theta = h(t) \qquad (11)^{\bullet}$$

called the *ODE of the driven and damped simple pendulum*. Although the polar angle θ is often restricted to the range $0 \le \theta \le 2\pi$, or to $|\theta| \le \pi$, we make no such restriction here. Indeed, we want to consider situations in which the pendulum is given so much energy that it rotates end over end about its pivot, so we must allow the angular variable θ to increase or decrease without bound.

The ODE in (11) is nonlinear in θ due to the term $\sin\theta$ and can't be solved in terms of elementary functions. But our numerical solver can tell a lot about a pendulum's behavior, as the first example for an undamped and undriven simple pendulum shows.

EXAMPLE 4.1.1

Orbits, Solution Curves of an Undamped and Undriven Simple Pendulum

The ODE in (11) with $c = 0$ and $h = 0$ describes the motion of an undamped, undriven simple pendulum. Figure 4.1.1 shows some orbits in the state space (i.e., $\theta\theta'$-space) of the corresponding ODE

$$\theta'' + \frac{g}{L}\sin\theta = 0$$

where we suppose that the value of g/L is 10. The equilibrium solutions $\theta = 2n\pi$, $\theta' = 0$, for all t, where n is any integer, correspond to the pendulum hanging at rest straight down, and give us equilibrium points in $\theta\theta'$-state space. Surrounding each of these equilibrium points is a region of closed oval orbits that correspond to periodic solutions of the ODE. These are the familiar back-and-forth motions of the pendulum about a downward equilibrium position.

If the pendulum starts out with high enough angular velocity θ'_0, the pendulum will forever tumble end over end about the pivot. The corresponding orbits are the wavy lines above and below the regions of periodic solutions. For every integer n the equilibrium solution $\theta = (2n + 1)\pi$, $\theta' = 0$, for all t, corresponds to the pendulum at rest and perfectly balanced over the pivot.

The orbits that are hardest to find are those that separate the periodic orbits from the tumbling orbits. These orbits in Figure 4.1.1 (called *separatrices*) tend from above the θ-axis to the equilibrium points $\theta = (2n + 1)\pi$, $\theta' = 0$ as $t \to +\infty$ and $\theta = (2n - 1)\pi$, $\theta' = 0$ as $t \to -\infty$, and the other way around from below the θ-axis. Six complete separatrices are visible in Figure 4.1.1, three above the θ-axis and three below. Parts of another four can be seen at the left and right edges. Separatrices represent the pendulum moving away from one upended position and toward another, but never actually arriving. Since separatrices are actually orbits, they cannot touch the equilibrium points which are also orbits; to do so would violate uniqueness.

☞ Doesn't Figure 4.1.1 remind you of the graph on the cover of the book?

☞ Can you match the solution curves in Figure 4.1.2 with their orbits in Figure 4.1.1?

Figure 4.1.2 shows the solution curves of the IVPs $\theta'' + 10\sin\theta = 0$, $\theta(0) = 0$, $\theta'(0) = 3, 5$ (periodic back-and -forth motions), $7, 8$ (tumbling motions over the pivot), and $\sqrt{40}$ (separatrix motion). The number $\sqrt{40}$ is exact but computer arithmetic is not, so we had to start a little below $\sqrt{40}$ to even come close to the separatrix solution shown. Note that the two periodic solutions in Figure 4.1.2 have *different* periods.

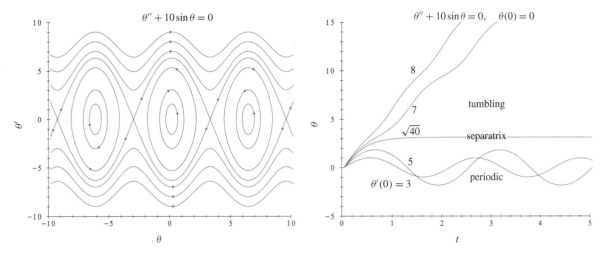

FIGURE 4.1.1 Orbits of an undamped simple pendulum (Example 4.1.1).

FIGURE 4.1.2 Periodic, separatrix, and tumbling solution curves (Example 4.1.1).

The discussion of orbits and solution curves is based on an examination of the graphs in Figures 4.1.1 and 4.1.2, graphs generated by a numerical solver. Computer graphics suggest, but can't prove. So how do we know that the properties described above are real? Fortunately, these properties can be shown mathematically (Problems 6, 8).

What happens to the orbits if the pendulum is subject to viscous damping? The chapter cover figure shows orbits for the model ODE of a viscously damped pendulum:

$$\theta'' + \theta' + 10 \sin \theta = 0 \qquad (12)$$

Let's take a closer look at that state portrait.

EXAMPLE 4.1.2

Orbits of the Damped Pendulum: Chapter 4 Cover Figure
The figure shows that damping slows the tumbling motion of the pendulum and, with the exceptions noted below, leads eventually to decaying oscillations about the equilibrium points $(2n\pi, 0)$ as $t \to +\infty$. These equilibrium points correspond to the pendulum hanging straight down at rest. The exceptional orbits head toward the equilibrium points $((2n+1)\pi, 0)$ that correspond to the pendulum at rest standing straight up from its support. These exceptional orbits are *separatrices* because each one separates orbits that eventually decay toward a specific downward point $(2n\pi, 0)$ from nearby orbits that do not. The separatrices in the chapter cover figure appear to dead-end at the equilibrium points $(\pm\pi, 0)$ as $t \to +\infty$. The "tendril" orbits correspond to pendulum motions that leave an upended position as t increases from $-\infty$ and move toward a downward position as $t \to +\infty$

The *basin of attraction* of the equilibrium point $(0, 0)$ is the collection of all orbits that tend to $(0, 0)$ as $t \to +\infty$. It is the wavy band in the chapter cover figure that extends from the upper left to the lower right. This pattern of orbital behavior is repeated periodically as one moves along the θ-axis.

The Linearized Pendulum

Writing the undriven damped simple pendulum ODE in normalized form, we have

☞ This ODE is just a form of ODE (11) with $h(t)$ replaced by zero.

$$\theta'' = -\frac{g}{L}\sin\theta - \frac{c}{m}\theta' \qquad (13)$$

The ODE in (13) is autonomous and has the equilibrium solutions $\theta = n\pi$, $\theta' = 0$, where n is any integer. So if the pendulum starts out from rest at $\theta = n\pi$, then the bob will remain in that position forever. The orbital portrait in the chapter cover figure indicates that motions starting near the equilibrium point $\theta = 0$, $\theta' = 0$ stay near that point (this point corresponds to the bob hanging straight down forever). As we did for the soft spring in Example 3.1.5, we will look at small amplitude motions near the equilibrium point $(0,0)$ by linearizing ODE (13) near that point.

Denoting the right hand side of ODE (13) by $H(\theta, \theta')$, let's expand $H(\theta, \theta')$ in a Taylor series about the base point $(0,0)$, keeping only the linear terms:

☞ More on Taylor Series in Theorem B.5.15.

$$H(0,0) + H_\theta(0,0)\theta + H_{\theta'}(0,0)\theta' = 0 - \frac{g}{L}\theta - \frac{c}{m}\theta'$$

So we arrive at the *linearization of the pendulum ODE with damping*:

$$\theta'' = -\frac{g}{L}\theta - \frac{c}{m}\theta' \qquad (14)$$

Note the similarity of ODE (14) with ODE (8) in Section 3.1 for the Hooke's Law spring. We see again the power of the modeling approach: Even though the linearized pendulum is a different physical system from the Hooke's Law spring-body system considered earlier, both have ODE models of the same form. So the methods of Chapter 3 allow us to solve the linearized ODEs and explain the motion of both physical systems.

How closely do the solutions of the linear and nonlinear models agree?

| **EXAMPLE 4.1.3** | **Solution Curves, Orbits of Nonlinear and Linearized Undamped Pendulum** |

Figures 4.1.3 and 4.1.4 show the solution curves and orbits of the IVPs for the undamped simple pendulum (solid), and for its linearization about $\theta = 0$, $\theta' = 0$ (dashed):

☞ Problem 7 of Section 3.1 explains one way to plot solution curves of different ODEs on the same graph.

$$\theta'' + 10\sin\theta = 0, \qquad \theta(0) = 1, \quad \theta'(0) = 0$$
$$\theta'' + 10\theta = 0, \qquad \theta(0) = 1, \quad \theta'(0) = 0$$

The solution $\theta = \cos 2t$ of the linearized problem is quite close to the solution of the nonlinear problem (Figure 4.1.3), but there is a noticeable time lag. The solution curves of the linearized IVP cover more ground in a fixed time than the curves of the nonlinear IVP.

☞ We only consider linearized ODEs in Sections 4.2–4.4.

This example suggests that the linear and the nonlinear pendulum models give similar results. This is true if $|\theta(0)|$ and $|\theta'(0)|$ are small, but false if $|\theta(0)|$ or $|\theta'(0)|$ are large (see Problem 5).

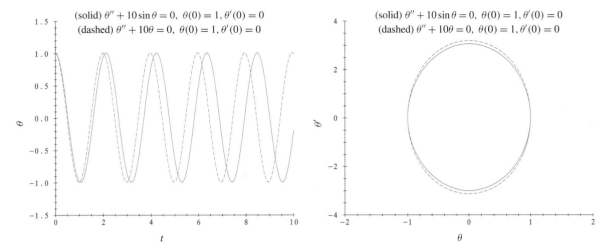

FIGURE 4.1.3 Solution curves of the undamped simple pendulum IVP (solid) and of the linearized IVP (dashed) (Example 4.1.3).

FIGURE 4.1.4 Orbits of the undamped simple pendulum IVP (solid) and of the linearized IVP (dashed) (Example 4.1.3).

Comments

Just as in Section 3.1 where the second-order ODE modeling the motion of a spring was also written as an equivalent system of two first-order ODEs, we can do the same thing for the ODE of the driven and damped simple pendulum. In terms of state variables θ and $\theta' = v$, we have the system equivalent to ODE (11):

$$
\begin{aligned}
\theta' &= v \\
v' &= -\frac{g}{L}\sin\theta - \frac{c}{m}v + \frac{1}{mL}h(t)
\end{aligned}
\tag{15}
$$

It is often convenient to use the equivalent system rather than the ODE to study the motion of the pendulum. For example, the chapter cover figure and Figures 4.1.1 and 4.1.4 are most easily interpreted in terms of systems.

PROBLEMS

1. (*Coordinate Version of the Dot Product*). Let $\mathbf{u} = u_1\mathbf{i} + u_2\mathbf{j} + u_3\mathbf{k}$ and $\mathbf{v} = v_1\mathbf{i} + v_2\mathbf{j} + v_3\mathbf{k}$.

    (a) Show that $\|\mathbf{u}\|^2 = u_1^2 + u_2^2 + u_3^2$ for any vector $\mathbf{u}$.

    (b) Show that $\mathbf{u} \cdot \mathbf{v} = u_1v_1 + u_2v_2 + u_3v_3$ for any vectors $\mathbf{u}$ and $\mathbf{v}$. [*Hint*: Imagine $\mathbf{u}$ and $\mathbf{v}$ to have their tails at the origin, then use the Law of Cosines, the definition of $\mathbf{u} \cdot \mathbf{v}$, and part **(a)**.]

    (c) (*Symmetry*). Show that $\mathbf{u} \cdot \mathbf{v} = \mathbf{v} \cdot \mathbf{u}$ for all $\mathbf{u}$, $\mathbf{v}$.

    (d) (*Bilinearity*). Show that $(\alpha\mathbf{u} + \beta\mathbf{w}) \cdot \mathbf{v} = \alpha\mathbf{u} \cdot \mathbf{v} + \beta\mathbf{w} \cdot \mathbf{v}$ for all scalars α, β and vectors $\mathbf{u}$, $\mathbf{v}$, and $\mathbf{w}$.

    (e) (*Positive Definiteness*). Show that $\mathbf{u} \cdot \mathbf{u} \geq 0$ for all $\mathbf{u}$ and that $\mathbf{u} \cdot \mathbf{u} = 0$ if and only if $\mathbf{u} = \mathbf{0}$.

    (f) (*Cauchy-Schwarz Inequality*). Show that $|\mathbf{u} \cdot \mathbf{v}| \leq \|\mathbf{u}\| \|\mathbf{v}\|$ for all vectors $\mathbf{u}$ and $\mathbf{v}$.

2. (*Vector Addition*). An airplane flies from point P in space. It flies 20 mi due south, turns left $90°$ and goes into a climb 8 mi long at an angle of $10°$ with the horizontal, turns left again, and flies horizontally 42 mi due north. Let $\mathbf{i}, \mathbf{j}, \mathbf{k}$ point north, west, and upward from P. What is the final position of the airplane relative to P?

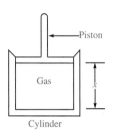

Piston

Gas

Cylinder

3. (*Ideal Gas Law*). According to the *Ideal Gas Law*, the pressure P, volume V, temperature T, and the number of moles (1 mole $= 6.02 \times 10^{23}$ molecules) n of a gas in a closed container satisfy the equation $PV = nRT$, where R is a universal constant. Suppose that a cylinder contains an ideal gas with a piston of mass m on top (see the margin figure). Assume that the temperature is constant and that the only forces acting on the piston are gravity and gas pressure. Find the ODE for the position of the piston measured from the bottom of the cylinder. Note that P is defined to be the magnitude of the gas pressure per unit area of the piston.

4. (*Linearized Pendulum*). Consider a simple, undamped, linearized and undriven pendulum described by $\theta'' + g\theta/L = 0$.

(a) Find the period T of the pendulum in terms of L and g.

(b) Find the length of a pendulum whose period is exactly 1 sec.

(c) If the pendulum is 1 meter long and swings with an amplitude of 1 radian, compute the angular velocity of the pendulum at its lowest point. Find the accelerations at $\theta = \pm 1$.

 5. (*Damped Simple Pendulum and Its Linearization*). The ODE for the simple undriven pendulum with viscous damping is $mL\theta'' + cL\theta' + mg\sin\theta = 0$.

(a) Set $m = 1$, $c = 0$, $g/L = 10$ and plot a portrait of the orbits of the system in θ, θ' variables, $\theta' = v$, $v' = -10\sin\theta$. Use the screen size $|\theta| \le 15$, $|\theta'| \le 19$. [*Hint*: See Figure 4.1.1.]

(b) (*Chapter Cover Figure*). Set $c = 1$ so that the ODE becomes $\theta'' + \theta' + 10\sin\theta = 0$. Plot some orbits for the equivalent system $\theta' = v$, $v' = -10\sin\theta - v$ with screen size $|\theta| \le 15$, $|v| \le 19$. Compare this portrait with those in part **(a)** and the cover figure. Explain the differences.

(c) Now repeat parts **(a)** and **(b)** but with the linearized ODE $mL\theta'' + cL\theta' + mg\theta = 0$. Use $m = 1$, $g/L = 10$, $c = 0$ (first), and $c = 1$ (second). Compare your graphs with those obtained earlier for the nonlinear ODE and explain any differences.

6. (*Variable-Length Pendulum*). Show that if the length $L(t)$ of a pendulum is a function of time, then the equation of motion of the pendulum is $mL\theta'' + (2mL' + cL)\theta' + mg\sin\theta = F$. [*Hint*: See ODE (11) and use $\mathbf{R}' = (L\hat{\mathbf{r}})'$, but don't assume L constant.]

www 7. (*Orbits of Undriven, Undamped Simple Pendulum*). A formula for the orbits of the undriven, undamped simple pendulum may be derived directly from the ODE $mL\theta'' + mg\sin\theta = 0$.

 (a) Multiply each side of the ODE by θ' to get $mL\theta'\theta'' + mg\theta'\sin\theta = \left[\frac{1}{2}mL(\theta')^2 - mg\cos\theta\right]'$. Show that $(1/2)mL(\theta')^2 - mg\cos\theta = c$, where c is a constant, are the equations of the orbits. Let $g/L = 4$, and reproduce Figure 4.1.1 by plotting orbits for several values of c.

(b) (*Conservation of Energy*). The *kinetic energy* (KE) of the moving pendulum is $(1/2)m(L\theta')^2$ and the *potential energy* (PE) is $mgL(1 - \cos\theta)$. Show that, given θ and θ' at time 0, the total energy $E(t) = KE + PE$ at time t is the same as $E(0)$. Explain why $E(t) = E(0)$ is the equation of an orbit.

 8. (*Closed Orbits of an Undriven, Undamped Simple Pendulum*). The ODEs

$$mL\theta'' + mg\sin\theta = 0 \quad \text{and} \quad mL\theta'' + mg\theta = 0$$

model the motion of an undriven, undamped simple pendulum and of a linearized pendulum, respectively. State portraits of each ODE show a region of closed orbits encircling the origin (see Figure 4.1.1, for example). These closed orbits (or *cycles*) correspond to periodic solutions because the ODEs are autonomous. Find and compare the periods of the cycles for the simple and for the linearized pendulum. Follow the outline below in proving the existence of cycles and in studying the periods. [*Hint*: See also Problem 6.]

☞ This is more than you may want to know about the simple pendulum ODE.

- (*Linearized Pendulum*). The equation of the linearized pendulum is $mL\theta'' + mg\theta = 0$. Show that all nonconstant solutions are periodic of period $2\pi\sqrt{L/g}$. Show that the corresponding orbits in the $\theta\theta'$-state space are elliptical cycles of the same period. Choose a value for L, plot a portrait of cycles in the state space, plot component graphs, and verify graphically the formula for the period.

- (*Closed Orbits of Simple (Nonlinear) Pendulum*). Suppose that the simple pendulum modeled by $mL\theta'' + mg\sin\theta = 0$ is released from rest when $\theta = \theta_0$, where $0 < \theta_0 < \pi$. Show that the subsequent motion is periodic. [*Hint*: Use Problem 7 to show that orbits are described by the relation $(\theta')^2 = (2g/L)(\cos\theta - \cos\theta_0)$, and use symmetries in this relation to show that the orbits are closed, and so represent periodic solutions.]

- (*Periods of Simple Pendulum*). Let T be the period of the orbit with $\theta(0) = \theta_0$, $\theta'(0) = 0$. Show that T is given by

$$T = 4\sqrt{\frac{L}{2g}} \int_0^{\theta_0} \frac{d\theta}{\sqrt{\cos\theta - \cos\theta_0}}$$

  [*Hint*: Since $\theta(t)$ initially decreases as t increases, $\theta' = -(2g/L)^{1/2}(\cos\theta - \cos\theta_0)$. Note that θ continues to decrease until the time $t = t_1$ for which $\theta(t_1) = -\theta_0$.]

- (*Elliptic Integrals and the Periods of the Simple Pendulum*). Show that the change of variables $k = \sin(\theta_0/2)$, $\sin\phi = (1/k)\sin(\theta/2)$ give

$$T = 4\sqrt{\frac{L}{g}} \int_0^{\pi/2} \frac{d\phi}{\sqrt{1 - k^2\sin^2\phi}}$$

  The integral is called an *elliptic integral of the first kind*. Its approximate values have been tabulated[3] for various values of k. For example, if $\theta_0 = 2\pi/3$, then $k = \sqrt{3}/2$, and the value of the integral is ≈ 2.157. The corresponding period is $\approx 8.628\sqrt{L/g}$, quite different from the period of the linearized pendulum, which is $2\pi\sqrt{L/g} \approx 6.282\sqrt{L/g}$. Why would you expect the period of the nonlinear pendulum to be greater than the period of the linearized pendulum? Choose various values for L and use an ODE solver to verify the above estimate for the periods.

- (*Asymptotic Values of the Periods of the Simple Pendulum*). Argue that $T \to 2\pi\sqrt{L/g}$ as $\theta_0 \to 0$. Why would one expect this to be true? It is known that $T \to \infty$ as $\theta_0 \to \pi$, although a complete mathematical proof of this fact is not given here. Why is this result expected on physical grounds?

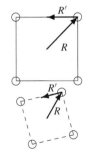

 9. (*Creeping Bugs*). Four bugs are at the corners of a square table whose sides are of length a. The bugs begin to move at the same instant, each crawling at the same constant speed directly toward the bug on its right. Find the path of each bug. Do the bugs ever meet? If so, when? [*Hint*: Let $\mathbf{R}(t) = r(t)\hat{\mathbf{r}}$ point from the center of the table to one of the bugs, where $r(0) = a/\sqrt{2}$ and $\theta(0) = 0$. Explain why the velocity vector $\mathbf{R}'(t)$ makes a 135° angle with $\mathbf{R}(t)$ for $t \geq 0$.]

---

[3]M. Abramowitz and I. A. Stegun, eds., *Handbook of Mathematical Functions* (Washington, D.C.: National Bureau of Standards, 1964); Dover (reprint), New York, 1965.

4.2 Beats and Resonance

☞ Spring motion is
explained in Section 3.1. The damped vertical motion of a weight attached to a driven Hooke's Law spring is
modeled by the linear ODE

$$a_2 y'' + a_1 y' + a_0 y = f(t) \tag{1}$$

where a_0, a_1, and a_2 are positive constants, and $f(t)$ is the driving force. We found
in the previous section that the motion of the damped linearized pendulum also satisfies a differential equation that has the same form as ODE (1). We will show in the
final section of this chapter that the voltage and the current in an electrical circuit are
each modeled by ODEs that look like ODE (1). So, once again, one ODE has many
interpretations.

The techniques of Chapter 3 show that all possible solutions of ODE (1) are determined when all solutions of the undriven ODE

$$a_2 y'' + a_1 y' + a_0 y = 0$$

that is, the *free solutions*, and a single particular solution (a *forced solution*) of ODE (1)
are known. In this section we explore the consequences of this observation. Let's start
by looking at the free solutions.

Free Oscillations: Periodic or Damped

☞ See Example 3.5.1. Free solutions behave differently when damping is present than when it is not. In the
undamped case free solutions satisfy the *harmonic oscillator* ODE

$$y'' + k^2 y = 0 \tag{2}$$

which arises from ODE (1) with $f(t) = 0$ by putting $a_1 = 0$, dividing through by a_2,
and replacing a_0/a_2 by k^2. All solutions of ODE (2) can be written as sinusoids:

$$y(t) = A\cos(kt + \delta)$$

where A and δ are any real numbers; $\omega_0 = k$ is called the *natural circular frequency*
of the harmonic oscillator. Motions described by ODE (2) are periodic with period
$T = 2\pi/\omega_0$, and they persist forever. Such motion is called *simple harmonic motion*,
or a *free periodic oscillation*

No physical spring or linearized pendulum can sustain simple harmonic motion
since some damping is always present because of frictional forces. When damping is
present $a_1 \neq 0$, and ODE (1) can be written as

$$y'' + 2cy' + k^2 y = F(t) \tag{3}$$

where we have put $a_1/a_2 = 2c$, $a_0/a_2 = k^2$, and $F(t) = f(t)/a_2$. The free solutions
satisfy the undriven ODE

$$y'' + 2cy' + k^2 y = 0 \tag{4}$$

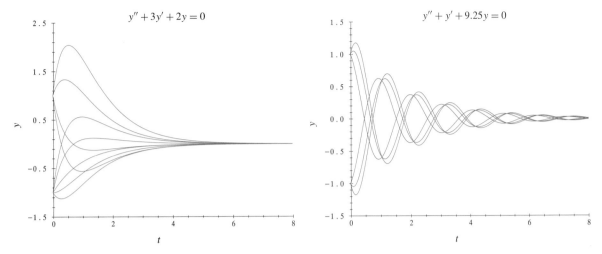

FIGURE 4.2.1 Solution curves for an overdamped and undriven ODE.

FIGURE 4.2.2 Some free damped oscillations for an underdamped and undriven ODE.

and can be found by computing the roots r_1 and r_2 of the characteristic polynomial

$$P(r) = r^2 + 2cr + k^2$$

☞ The relative values of c and k give rise to the cases shown in Figures 4.2.1–4.2.2.

Using the quadratic formula, we have

$$r_1, r_2 = -c \pm \sqrt{c^2 - k^2} \tag{5}$$

Since c and k are positive real numbers, the real parts of r_1 and r_2 are always negative, whatever the actual values of c and k. To see this, note that if $c < k$, then r_1 and r_2 are complex conjugates with real part $-c < 0$. If $c \geq k$, then $0 \leq c^2 - k^2 \leq c^2$, and r_1 and r_2 are real and negative. So all damped free solutions are *transients* (i.e., they die out with increasing time), but their exact nature depends on the relative sizes of the constants c and k.

There are three cases for the free solutions, that is, the solutions of ODE (4).

☞ Springs with heavy damping don't oscillate.

- *Overdamped.* If $c > k$, then r_1 and r_2 are real, negative, and distinct. Every free solution is a linear combination of the decaying real exponentials $e^{r_1 t}$ and $e^{r_2 t}$. See Figure 4.2.1 for an example of overdamped solution curves.

- *Critically damped.* If $c = k$, then $r_1 = r_2 = -c < 0$ and every free solution is the product of the decaying exponential $e^{r_1 t}$ and a polynomial $k_1 + k_2 t$, where k_1 and k_2 are constants.

☞ Springs with light damping have decaying oscillations about the equilibrium point.

- *Underdamped.* If $c < k$, then $r_1 = \alpha + i\beta = \bar{r}_2$, where $\alpha = -c < 0$ and $\beta = \sqrt{k^2 - c^2}$. Every solution is the product of a decaying exponential $Ae^{\alpha t}$ and a sinusoid $\cos(\beta t + \gamma)$, where A and γ are constants. In this case the solutions are *free damped oscillations* with circular frequency β. See Figure 4.2.2 for some underdamped solution curves.

Let's move on now and look at the forced solutions.

Forced Oscillations: Beats and Resonance in Undamped Systems

☞ You may want to review the discussion on periodic functions in Section 3.5.

When the free solutions of the linear ODE are oscillatory in character, sinusoidal driving forces may produce peculiar responses that scientists call *resonance*. The excitation of a tuning fork in response to the vibrations of another tuning fork is an example of resonance. Tuning the circuits in radio receivers is also a familiar example. The phenomenon of resonance is central to the operation of a great many devices and instruments. For simplicity, we take up the undamped case first.

Let's consider the ODE

$$y'' + \omega_0^2 y = A \cos \omega t \tag{6}$$

where ω_0, A, and ω are positive constants. The ODE in (6) models an undamped spring-mass system driven by a sinusoidal force of circular frequency ω. The ODE can arise in many other contexts. Since the analysis will use only the ODE and no other feature of the physical system, the conclusions apply to any system modeled by the ODE.

Every free solution $y = C_1 \cos \omega_0 t + C_2 \sin \omega_0 t$ is periodic with circular frequency ω_0. The driving term of ODE (6) has circular frequency ω, and the character of the solutions of the ODE depends dramatically on whether or not $\omega = \omega_0$.

Case I: Pure Resonance: $\omega = \omega_0$

We look for a real-valued particular solution of ODE (6) with $\omega = \omega_0$ by first finding a complex-valued particular solution of the ODE

$$z'' + \omega_0^2 z = A e^{i\omega_0 t} \tag{7}$$

Here's the reason why. As we saw in Section 3.6, $\text{Re}[A e^{i\omega_0 t}] = A \cos \omega_0 t$, so if $z(t)$ is any solution of ODE (7), then $y(t) = \text{Re}[z(t)]$ is a real-valued solution of ODE (6) with $\omega = \omega_0$.

As in Example 3.6.7, we see that the driven ODE (7) has a solution z_d of the form

$$z_d = Cte^{i\omega_0 t} \tag{8}$$

for some constant C. Setting $P(D) = D^2 + \omega_0^2$ and using operator formula (5) in Section 3.6, we see that

$$\begin{aligned}
P(D)[z_d] &= P(D)[Cte^{i\omega_0 t}] \\
&= e^{i\omega_0 t} P(D + i\omega_0)[Ct] = e^{i\omega_0 t}((D + i\omega_0)^2 + \omega_0^2)[Ct] \\
&= e^{i\omega_0 t} D(D + 2i\omega_0)[Ct] = e^{i\omega_0 t}(D + 2i\omega_0)D[Ct] \\
&= e^{i\omega_0 t}(D + 2i\omega_0)[C] = 2iC\omega_0 e^{i\omega_0 t}
\end{aligned}$$

Since we want $P(D)[z_d] = A e^{i\omega_0 t}$, we must take $2Ci\omega_0 = A$; so $C = A/2i\omega_0$. From (8) we see that

$$z_d = \frac{A}{2i\omega_0} te^{i\omega_0 t} = -\frac{Ai}{2\omega_0} t (\cos \omega_0 t + i \sin \omega_0 t)$$

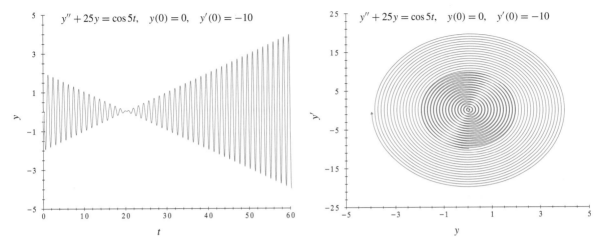

FIGURE 4.2.3 A solution curve of ODE (6) (with $\omega = \omega_0 = 5$) showing pure resonance.

FIGURE 4.2.4 The orbit of the solution curve of Figure 4.2.3.

Now since we seek a real-valued solution of ODE (6), we see that

$$y_d = \text{Re}[z_d] = \frac{A}{2\omega_0}t \sin \omega_0 t$$

is a particular solution of ODE (6). So the general solution of ODE (6) is

$$y = C_1 \cos \omega_0 t + C_2 \sin \omega_0 t + \frac{A}{2\omega_0}t \sin \omega_0 t$$

where C_1 and C_2 are arbitrary real constants. Every solution of ODE (6) is the super-position of a free periodic oscillation, $C_1 \cos \omega_0 t + C_2 \sin \omega_0 t$, with a natural circular frequency ω_0 (and period $T = 2\pi/\omega_0$), and a forced oscillation (which is *not* periodic), $y_d = (At)/(2\omega_0) \sin \omega_0 t$ which has the form of a sinusoid of circular frequency ω_0 and amplitude $At/(2\omega_0)$ that grows over time. Observe that the forced oscillation is the response of the system to a periodic external force whose frequency exactly matches the natural frequency of the system.

When an undamped system is driven with a sinusoidal function having the system's natural frequency, the unbounded oscillation that results is said to be due to *pure resonance*. In this case the system responds to a bounded input with an unbounded output. Figure 4.2.3 shows what the pure resonant response looks like. Figure 4.2.4 shows the corresponding orbit, which starts out spiraling inward but then spirals outward and becomes unbounded as $t \to +\infty$. Can you think of any explanation for this initial decaying oscillation even though there is no damping?

Case II: Beats: $\omega \neq \omega_0$

Just as in Case I, we look for a real-valued particular solution of

$$y'' + \omega_0^2 y = A \cos \omega t \tag{9}$$

by first finding a complex-valued particular solution of the ODE $z'' + \omega_0^2 z = A e^{i\omega t}$. Again putting $P(D) = D^2 + \omega_0^2$, we note that $P(i\omega) = -\omega^2 + \omega_0^2 \neq 0$, so we can look for a particular solution of ODE (9) in z by using the operational formula $P(D)[e^{rt}] = e^{rt} P(r)$ for any complex number r. Taking $z_d = C e^{i\omega t}$ for some constant C to be determined, we see that

☞ This operator identity appears in Section 3.6.

$$P(D)[C e^{i\omega t}] = C e^{i\omega t} P(i\omega)$$

Since we want $P(D)[C e^{i\omega t}]$ to be $A e^{i\omega t}$, we take $C P(i\omega) = A$, or $C = A/P(i\omega) = A/(\omega_0^2 - \omega^2)$ and

$$z_d = \frac{A}{\omega_0^2 - \omega^2} e^{i\omega t} \tag{10}$$

So the particular real solution y_d of ODE (9) that we are looking for is

$$y_d = \text{Re}[z_d] = \frac{A}{\omega_0^2 - \omega^2} \cos \omega t$$

The general solution of ODE (9) is given by

$$y(t) = C_1 \cos \omega_0 t + C_2 \sin \omega_0 t + \frac{A}{\omega_0^2 - \omega^2} \cos \omega t \tag{11}$$

where C_1 and C_2 are arbitrary real constants.

The solution $y(t)$ in formula (11) is the superposition of the free periodic oscillation $C_1 \cos \omega_0 t + C_2 \sin \omega_0 t$ with natural circular frequency ω_0 and a periodic forced oscillation $(A/(\omega_0^2 - \omega^2)) \cos \omega t$ representing the system's response to the driving term, $A \cos \omega t$. Observe that the circular frequency ω of the periodic forced oscillation exactly matches that of the driving term. Also note that if the ratio ω/ω_0 is a rational number, then every solution of ODE (6) is periodic. Here's why. If m and n are positive integers such that $\omega/\omega_0 = m/n$, then $m/\omega = n/\omega_0$. In this case each term in the general solution (11) has the period $(2\pi/\omega)m = (2\pi/\omega_0)n$.

☞ This argument appeared earlier in Example 3.5.3

Let's find the unique solution that starts out from rest at $y = 0$. Using the general solution formula (11), we see that the initial conditions $y(0) = 0$, $y'(0) = 0$ imply that $C_1 = A(\omega^2 - \omega_0^2)^{-1}$, $C_2 = 0$ and we obtain the solution

$$y(t) = \frac{A}{\omega_0^2 - \omega^2} (\cos \omega t - \cos \omega_0 t)$$

Using the trigonometric identity

☞ Appendix B.4 has more trig identities.

$$\cos \alpha - \cos \beta = 2 \sin \frac{\beta - \alpha}{2} \sin \frac{\beta + \alpha}{2}$$

to rewrite this solution, we have

$$y(t) = \left(\frac{2A}{\omega_0^2 - \omega^2} \sin \frac{\omega_0 - \omega}{2} t \right) \sin \frac{\omega_0 + \omega}{2} t \tag{12}$$

which looks like a sinusoid of circular frequency $(\omega_0 + \omega)/2$ and amplitude that varies periodically with circular *beat frequency* $|(\omega_0 - \omega)/2|$.

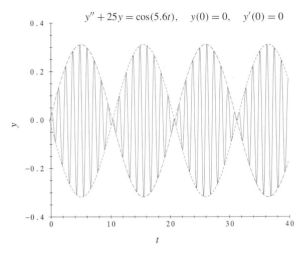

$$y'' + 25y = \cos(5.6t), \quad y(0) = 0, \quad y'(0) = 0$$

$$y'' + 25y = \cos(5.6t), \quad y(0) = 1, \quad y'(0) = 0$$

FIGURE 4.2.5 Beats with slow circular beat frequency 0.3 and long period $2\pi/0.3 \approx 20.94$ (Example 4.2.1).

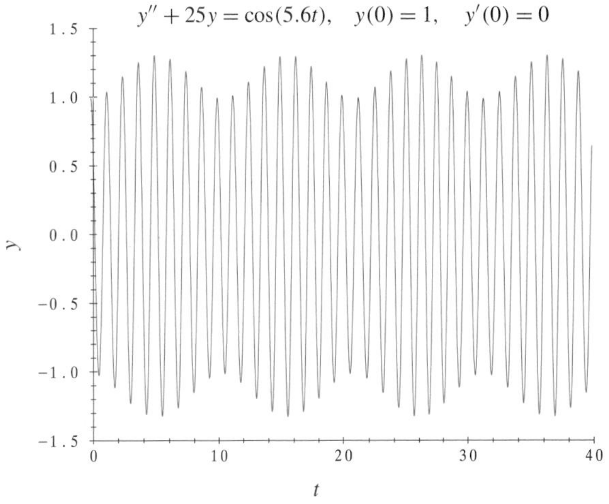

FIGURE 4.2.6 Another beat pattern with slow circular beat frequency 0.3 and long period $2\pi/0.3$ (Example 4.2.2).

There is an interesting interpretation when $|\omega_0 - \omega|$ is small compared to $\omega_0 + \omega$, the phenomenon of beats. In this case, the system's response is a sinusoid of circular frequency $(\omega_0 + \omega)/2$, but with a slowly varying amplitude, called a *beat*, which is a sinusoid with the slow circular beat frequency $|\omega_0 - \omega|/2$ (and long period $T = 4\pi/|\omega_0 - \omega|$). One can hear this in the alternate swelling and fading sounds generated by a pair of tuning forks: measuring time in seconds, say that one fork is tuned to middle C (258 hertz) and the other is slightly off pitch (260 hertz). So the beat frequency is $(260 - 258)/2 = 1$ hertz, while the pitch you hear is $(260 + 258)/2 = 259$ hertz.

Let's take a graphical look at the beat phenomenon.

EXAMPLE 4.2.1

Beats from Initial Rest State

Following the approach described above for producing beats, consider the IVP

$$y'' + 25y = \cos(5.6t), \qquad y(0) = 0, \quad y'(0) = 0 \qquad (13)$$

where $\omega_0 = 5.0$, $\omega = 5.6$, and $A = 1$. This sinusoidally driven IVP has a response that shows the beat phenomenon. By formula (12), the solution of IVP (13) is

$$y(t) = \left(\frac{1}{3.18} \sin(0.3t) \right) \sin(5.3t)$$

The solution curve is plotted in Figure 4.2.5 (solid curve). The beat is given by

$$\frac{1}{3.18} \sin(0.3t)$$

with circular frequency 0.3, period $2\pi/0.3 \approx 20.94$, and amplitude $1/3.18$.

Graphs of $\pm(\sin 0.3t)/3.18$ (dashed curves in Figure 4.2.5) envelop the solution curve of IVP (13). Each hump in the solution curve has time width $\pi/0.3 \approx 10.47$.

The shape of the beats depends on the initial data, as the next example shows.

EXAMPLE 4.2.2

Beats from a Nonequilibrium Initial State

When a harmonic oscillator, not initially at equilibrium, is driven by a sinusoid, then beats have a different appearance than those in Figure 4.2.5. We can see this by looking at the solution curve (Figure 4.2.6) of the IVP

$$y'' + 25y = \cos(5.6t), \qquad y(0) = 1, \quad y'(0) = 0 \tag{14}$$

The basic structure of the beats is the same as in Figure 4.2.5. The humps are approximately 10.47 apart, but the oscillations are not pinched off between the humps.

A little experimentation shows that the response appears to translate along the t-axis, and the difference between the height of the peaks and the depth of the valleys changes as the initial data in IVP (14) change. The modulation of the response, however, always retains its most important feature, the circular beat frequency $|\omega_0 - \omega|/2$.

As noted earlier, beats can be experienced as the alternate fading and swelling in amplitude heard when two tuning forks vibrate at not quite the same frequency. This simple device permits the human ear to detect frequency differences as low as 0.06%. The beat phenomenon is used in radio signal detection, where an incoming radio signal is mixed with the signal of an oscillator in the receiver to give a new signal with a frequency down in the audio range. That signal is then applied to a detector whose output will drive a loudspeaker. Finally, observe that the maximum amplitude $|2A/(\omega_0^2 - \omega^2)|$ of the sinusoid in formula (12) tends to ∞ as $\omega \to \omega_0$; this is another indication of the phenomenon of resonance.

Comments

Although we have illustrated the occurrences of resonance and beats only with the driving force $A\cos\omega t$, the phenomena appear with any driving force of the form $A\cos\omega t + B\sin\omega t$. If the driving force has the form $F(t) + A\cos\omega t + B\sin\omega t$, you will certainly see some kind of beats or resonance if ω is near the natural frequency, although the effects may be somewhat hidden by the response to the term $F(t)$.

PROBLEMS

1. *(Beats and Resonance)*. Find the solutions of the IVPs. Any resonance or beats?

  (a) $y'' + 9y = 5\cos 2t, \ y(0) = 0, \ y'(0) = 0$

  (b) $y'' + 4y = \cos 3t, \ y(0) = 1, \ y'(0) = -1$

  (c) $y'' + y = \cos t, \ y(0) = 1, \ y'(0) = 0$

2. *(Beats and Resonance)*. Use a numerical solver to study solutions of the ODE $y'' + 25y = \sin\omega t$ for initial points $(y(0), y'(0))$ of your choice and the given values of ω. Plot the results on separate graphs. Compare these graphs and record your observations. Choose initial points that give quite different graphs.

  (a) $\omega = 5.6$ (b) $\omega = 5.2$ (c) $\omega = 5.0$ (d) $\omega = \pi$

3. (*Beats*). For $\omega \neq \omega_0$ show that the solution of the IVP $y'' + \omega_0^2 y = A \sin \omega t$, $y(0) = 0$, $y'(0) = -A/(\omega_0 + \omega)$, exhibits the phenomenon of beats similiar to the graph in Figure 4.2.5. [*Hint*: Use the trigonometric identity $\sin \alpha - \sin \beta = 2 \sin((\alpha - \beta)/2) \cos((\alpha + \beta)/2)$.]

4. (*Archimedes' Buoyancy Principle*). According to Archimedes' Principle, the buoyant force acting on a body wholly or partially immersed in a fluid equals the weight of the fluid displaced. At equilibrium, the mass of the body equals the mass of the displaced water. So a wooden ship with the identical dimensions as an iron ship would float higher on the water.

(a) Ignoring friction, show that a flat wooden block of face L^2 square feet and thickness h feet floating half submerged in water will act as a harmonic oscillator of period $2\pi\sqrt{h/2g}$ if it is depressed slightly. [*Hint*: Show that if x is a vertical coordinate from the midpoint of the block, then $(L^2 h\rho/2)x'' + L^2 \rho g x = 0$, where ρ is the density of water.]

(b) Assuming no friction, replace the block in part **(a)** by a buoyant sphere of radius R floating half submerged, and show that if the initial displacement is small, the sphere will act as a harmonic oscillator of period $2\pi\sqrt{2R/3g}$. Show that if the displacement is large, the nonlinearities become important in the ODE modeling the motion, and so can't be ignored. In this case, the ODE of a soft spring is obtained (see Section 3.1).

www   **5.** (*Detecting a Periodic Forced Oscillation*). Solutions of the ODE $y'' + \omega_0^2 y = A \cos \omega t$, where A, ω_0, ω ($\omega_0 \neq \omega$) are positive constants, are superpositions of sinusoids of period $2\pi/\omega_0$ and period $2\pi/\omega$. Given the graph of a solution, is it possible to detect the periodic forced oscillation (i.e., the unique solution of period $2\pi/\omega$)? Does the periodic forced oscillation have a phase shift? Are the nonconstant solutions periodic, and if they are, what is their period? In addressing these questions, start out by solving graphically the two IVPs $y'' + 4y = \cos \omega t$, $y(0) = 0$, $y'(0) = 0$, $\omega = 3, \pi$. You may also want to find solution formulas.

 6. Release a hollow ball under water and see what happens. Model the dynamics and explain. [*Hint*: Try using an approach similar to that for Problem 4]

 7. Explore the phenomenon of beats for a lightly damped and driven oscillator. Begin your study with the ODE for a damped and driven Hooke's Law spring in Example 3.1.1.

4.3 Frequency Response Modeling

Structures such as bridges or buildings may have resonant frequencies even though they are damped systems. An external oscillating force at this resonant frequency could conceivably induce an oscillation of destructive amplitude. One theory of the collapse of Tacoma Narrows Bridge in 1940 while a gentle wind was blowing is that the support work along the roadway created sinusoidal eddies whose frequency matched the natural frequency of the bridge. Tachometers, seismometers, and vibrometers all depend on the resonance phenomenon for their operation. For example, the *Frahm tachometer* is a box containing rows of cantilever-mounted steel reeds of varying frequencies. When the box is placed on a vibrating machine, the reeds with frequencies close to the machine's frequency start to vibrate.

In the last section we looked at forced oscillations and resonance in undamped systems. In this section we present a method for analyzing oscillations in damped systems, a method that engineers call *frequency response modeling*.

Periodic Forced Oscillations in Damped Systems: Examples

Let's consider the ODE

$$y'' + 2cy' + k^2 y = F_0 \sin \omega t \tag{1}$$

where the constants c and k are such that $0 < c < k$ (the underdamped case). Setting $P(D) = D^2 + 2cD + k^2$, we see that the roots of $P(r) = r^2 + 2cr + k^2$ are

$$r_1 = -c + i\sqrt{k^2 - c^2}, \qquad r_2 = -c - i\sqrt{k^2 - c^2}$$

So the general solution of the undriven ODE $y'' + 2cy' + ky^2 = 0$ has the form

$$y_u = C_1 e^{-ct} \cos\left((k^2 - c^2)^{1/2} t\right) + C_2 e^{-ct} \sin\left((k^2 - c^2)^{1/2} t\right) \tag{2}$$

where C_1, C_2 are arbitrary constants. The solutions defined by formula (2) are *free* solutions because they are solutions of the undriven ODE. These free solutions are also *transients* because they tend to zero as time advances.

Suppose that we can show that ODE (1) has a *periodic forced oscillation*, that is, a particular solution $y_d(t)$ that is periodic with the same circular frequency ω as the driving term. By Theorem 3.6.1 the general solution of ODE (1) looks like

$$y = y_u + y_d \tag{3}$$

where y_u is given by formula (2). Any solution of ODE (1) which is periodic must appear in the general solution formula (3) for some value of the constants C_1 and C_2. If C_1 and/or C_2 is nonzero, then the solution $y = y_u + y_d$ cannot be periodic because y_u is a transient and does not repeat. Therefore, the only periodic solution is given when $C_1 = C_2 = 0$, so ODE (1) can have only one periodic solution y_d.

Before we construct the periodic solution of ODE (1), let's look at two examples.

| EXAMPLE 4.3.1 | **A Periodic Forced Oscillation** |

Using the methods of Section 3.6, we see that the driven linear ODE with damping,

$$y'' + y' + y = 13 \sin 2t \tag{4}$$

has the periodic forced oscillation

$$y_d = -3 \sin 2t - 2 \cos 2t$$

which has the same circular frequency as the driving function. From formula (2) with $c = 1/2$ and $k = 1$ we see that the free solutions have the form $e^{-t/2}[C_1 \cos(\sqrt{3}t/2) + C_2 \sin(\sqrt{3}t/2)]$, and they are transients. The general solution of ODE (4) is the sum of the free solutions and the periodic forced oscillation:

$$y = e^{-t/2}\left[C_1 \cos(\sqrt{3}t/2) + C_2 \sin(\sqrt{3}t/2)\right] - 3 \sin 2t - 2 \cos 2t$$

The transients oscillate with circular frequency $\sqrt{3}/2$, and their amplitudes decay exponentially. So the transients are *not* periodic. On the other hand, the forced oscillation *is* periodic and attracts all solutions as $t \to +\infty$. See Figure 4.3.1 for the graphs of

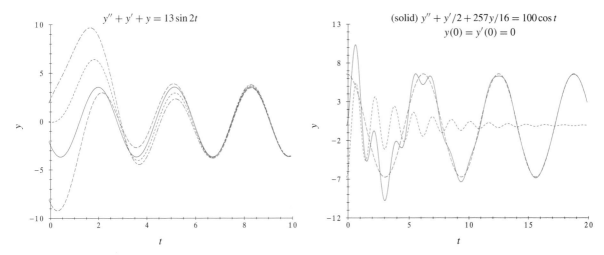

FIGURE 4.3.1 Solution curves (dashed) approach the unique periodic forced oscillation (solid curve) (Example 4.3.1).

FIGURE 4.3.2 Solution curve for an IVP (solid), steady state (long dashes), and transient (short dashes) (Example 4.3.2).

the periodic forced oscillation [initial data, $y(0) = -2$, $y'(0) = -6$] and three other solution curves [initial points are $(y(0), y'(0)) = (-8, -6)$, $(0, 0)$, $(2, 6)$]; all of these curves tend to the periodic forced oscillation y_d as t increases.

EXAMPLE 4.3.2

Response of a Damped System to a Sinusoidal Input
It can be verified that $y_d = (60\cos t + 2\sin t)/9.01$ is a periodic solution of the ODE

$$y'' + y'/2 + 257y/16 = 100\cos t$$

and as noted earlier there can be no other periodic solution. Note from formula (2) with $c = 1/4$ and $k^2 = 257/16$ that all free solutions are given by the oscillatory transients

$$y_u = C_1 e^{-t/4}\cos 4t + C_2 e^{-t/4}\sin 4t$$

In Figure 4.3.2 we plot y_d (long-dashed curve) and the transient y_u (short-dashed curve) which has been selected so that $y = y_d + y_u$ solves the IVP

$$y'' + y'/2 + 257y/16 = 100\cos t, \quad y(0) = y'(0) = 0$$

The solid curve in Figure 4.3.2 is the corresponding solution curve.

Constructing the Periodic Forced Oscillation

Now that we have studied periodic forced oscillations for two particular ODEs driven by sinusoids we are ready to tackle the general case. Let's construct the periodic forced oscillation y_d for the general linear driven ODE

$$y'' + 2cy' + k^2 y = F_0 \sin \omega t \tag{5}$$

We will use the methods of Section 3.6. The first step is to replace the driving term $F_0 \sin \omega t$ in ODE (1) by the complex exponential function $F_0 e^{i\omega t}$ and find a particular solution z_d of the ODE

$$P(D)[z] = F_0 e^{i\omega t} \tag{6}$$

where $P(D) = D^2 + 2cD + k^2$. Since $\text{Im}[F_0 e^{i\omega t}] = F_0 \sin \omega t$, it follows that $y_d = \text{Im}[z_d]$ is a solution of ODE (5). Since

$$P(D)[e^{rt}] = P(r)e^{rt}, \quad \text{for any constant } r$$

and since $P(i\omega) \neq 0$, we are led to look for a particular solution of ODE (6) in the form

$$z_d(t) = A e^{i\omega t}$$

where A is a constant to be determined. Observe that $P(D)[Ae^{i\omega t}] = AP(i\omega)e^{i\omega t}$. By comparison with (6), we see that A should be chosen such that

$$AP(i\omega) = F_0$$

so $A = F_0/P(i\omega)$. Note that

$$P(i\omega) = (i\omega)^2 + 2c(i\omega) + k^2$$
$$= k^2 - \omega^2 + 2ic\omega$$

which is nonzero unless $k = \omega$ and $c = 0$. Except for that special case, we see that if we define $H(r)$ by

$$H(r) = 1/P(r)$$

then

$$z_d(t) = A e^{i\omega t} = F_0 e^{i\omega t}/P(i\omega) = H(i\omega) F_0 e^{i\omega t} \tag{7}$$

solves ODE (6). The function $H(r)$ is called the *transfer function* for ODE (1) and makes the transfer from the input $F_0 e^{i\omega t}$ to the output $z_d(t)$.

To get the imaginary part of $z_d(t)$, we rationalize the denominator to write $H(i\omega)$ in the form $a + ib$:

$$H(i\omega) = \frac{1}{k^2 - \omega^2 + 2ic\omega}$$
$$= \frac{1}{k^2 - \omega^2 + 2ic\omega} \cdot \frac{k^2 - \omega^2 - 2ic\omega}{k^2 - \omega^2 - 2ic\omega}$$
$$= \frac{k^2 - \omega^2}{(k^2 - \omega^2)^2 + 4c^2\omega^2} + i \frac{-2c\omega}{(k^2 - \omega^2)^2 + 4c^2\omega^2} = a + ib \tag{8}$$

We can also write $H(i\omega)$ in polar form as follows:

$$H(i\omega) = M(\omega)e^{i\varphi(\omega)} \tag{9}$$

where the amplitude M and the polar angle φ are given by

$$M(\omega) = \sqrt{a^2 + b^2} = \frac{1}{\sqrt{(k^2 - \omega^2)^2 + 4c^2\omega^2}}$$

$$\varphi(\omega) = \cot^{-1}\left(\frac{a}{b}\right) = \cot^{-1}\left(\frac{\omega^2 - k^2}{2c\omega}\right)$$

(10)

Since by formula (8) the number $\text{Im}[H(i\omega)]$ is negative, the complex number $H(i\omega)$ always points downward as a vector in the complex plane, so the polar angle $\varphi(\omega)$ is in the range $-\pi \leq \varphi(\omega) \leq 0$.

Now using (9) in formula (7), we see that

$$z_d = H(i\omega)F_0 e^{i\omega t} = F_0 M(\omega)e^{i(\omega t + \varphi(\omega))}$$

$$= F_0 M(\omega)[\cos(\omega t + \varphi(\omega)) + i\sin(\omega t + \varphi(\omega))]$$

Extracting the imaginary part of z_d, we find that the sinusoid

$$y_d = F_0 M(\omega)\sin[\omega t + \varphi(\omega)]$$

(11)

☞ The forced oscillation represents the long-term behavior of all solutions

is a real-valued solution of ODE (5). This is the *periodic forced oscillation* with circular frequency ω that we wanted to find. All other solutions of ODE (5) approach this solution as $t \to +\infty$, so y_d is known as the *steady-state solution*

Now that we have found a formula for the periodic forced oscillation, let's take another approach and plot the amplitude and polar angle of that solution as functions of the driving frequency.

Frequency Response Modeling and Bodé Plots

The amplitude and phase shift functions $M(\omega)$ and $\varphi(\omega)$ as defined in (10) don't depend on the initial data. Comparing the formula for y_d in (11) with the input $F_0 \sin \omega t$, we see that the steady-state solution has the same sinusoidal form as the input, but with amplitude $F_0 M$ and phase φ. The ratio of the steady-state amplitude $F_0 M(\omega)$ to the input amplitude F_0 is $M(\omega)$ and is called the *gain*. If the gain M is larger than 1 then the response has a magnified amplitude, but if M is less than 1 the response has a diminished amplitude. The steady-state solution is shifted in time by $|\varphi(\omega)|/\omega$ radians to the right, so it is common to refer to $\varphi(\omega)$ as the *phase shift*

The graphs of gain and phase shift against ω (using a $\log_{10}$ scale on the ω-axis) are called *Bodé plots* and give valuable information about the system. Engineers call the quantity $20\log_{10} M(\omega)$ the *gain in decibels* (dB) and use decibel units on the gain axis. Degrees are used on the φ-axis instead of radians. These two graphs constitute the *frequency response curves* for the system.

Here's an example showing the characteristic shapes of Bodé plots.

EXAMPLE 4.3.3

Bodé Plots

Let's plot the frequency response curves for the ODE

$$y'' + 0.2y' + y = F_0 \sin \omega t$$

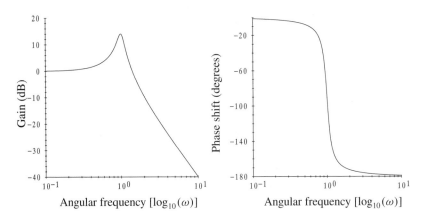

FIGURE 4.3.3 Bodé plots: gain and phase shift (Example 4.3.3).

Comparing this ODE to ODE (1), we see that $c = 0.1$ and $k = 1.0$. Figure 4.3.3 shows Bodé plots for this ODE.

Bodé plots are often used in engineering and science because they give instant visual information about the way the gain and the phase shift change as the input frequency changes. In the words of the Fundamental Theorem for Second-Order ODEs (Theorem 3.2.1), these graphs give information about the sensitivity of the response to changes in the parameter ω.

Parameter Identification

Up to this point, we have used the form of the ODE, solution formulas, and numerical solvers to give us information about the behavior of solutions. Now let's turn the process upside down and suppose that we already know something about the solution behavior of some unknown seond-order ODE with known sinusoidal driving terms. Can we use that information to tell us about the ODE itself? It's like solving a mystery. We have some clues, so can we find the ODE that "did it"?

Sometimes it is known that a system can be modeled by an ODE

$$y'' + 2cy' + k^2 y = 0 \tag{12}$$

but the system parameters c and k can't be accurately measured. If $\sin \omega t$ for a known value of ω is used to drive the ODE in (12), and if the gain $M(\omega)$ and phase shift $\varphi(\omega)$ in the steady-state output formula (11) can be experimentally determined, then the formulas in (10) form a basis for determining the system constants c and k. The next example shows how this is done, but first we need some trigonometric facts.

The graph of the phase-shifted sinusoid $\sin(\omega t + \varphi)$ can be obtained from the graph of $\sin \omega t$ as follows: If φ is positive (respectively, negative), then shifting the graph of $\sin \omega t$ to the left (respectively, to the right) by φ/ω radians results in the graph of $\sin(\omega t + \varphi)$. This follows from writing $\sin(\omega t + \varphi) = \sin(\omega(t + \varphi/\omega))$.

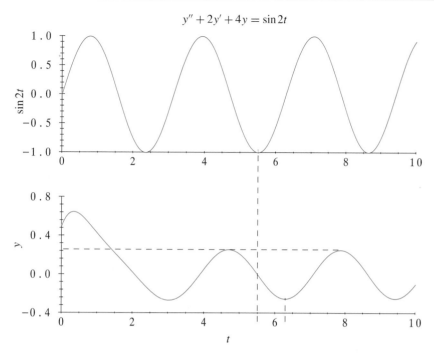

$$y'' + 2y' + 4y = \sin 2t$$

FIGURE 4.3.4 Finding c and k from the steady-state response of a system (Example 4.3.4).

EXAMPLE 4.3.4

An Example of Parameter Identification

Consider the IVP with unknown parameters c and k:

$$y'' + 2cy' + k^2 y = \sin 2t, \qquad y(0) = 0.5, \quad y'(0) = 1 \qquad (13)$$

Suppose that the graphs of the input $\sin 2t$ and the output $y(t)$ are as shown in Figure 4.3.4. Let's use these graphs to compute $M(2)$ and $\varphi(2)$ and compare these values to the theoretical values given by (10). Observe that the output quickly settles down to a periodic steady state. Using the vertical dashed line, we estimate that the output is shifted to the right by about 0.77 units. Since $\omega = 2$ and $\varphi/\omega \approx 0.77$, $\varphi(2)$ is about -1.54 radians. Using the horizontal dashed line, we estimate that the gain $M(2)$ is approximately 0.26.

Using $\omega = 2$ and these estimates for $M(2)$ and $\varphi(2)$, we conclude from the formulas in (10) (after a fair amount of algebraic manipulation) that $k \approx 2.029$ and $c \approx 0.961$. These values compare quite well with the actual values we used ($k = 2$, $c = 1$) when we numerically solved IVP (13) and plotted the solution curve shown in Figure 4.3.4.

Resonance in Damped Systems

Damped systems, like undamped systems, may show resonance. So let's use the Bodé plots to see what happens. The shapes of the Bodé plots in Figure 4.3.3 change as the parameters c and k of the system change. Notice from the formulas in (10) (under the general assumption that $0 < c < k$) that $M(\omega)$ is a bounded continuous function on $\omega \geq 0$. It can be shown (see Problem 7) that if $k^2 > 2c^2$, then the plot of $M(\omega)$ versus ω looks like the one on the left in Figure 4.3.3 with a unique value ω_r where $M(\omega)$ achieves a maximum value:

$$M(\omega_r) = \frac{1}{[2c\sqrt{k^2 - c^2}]}, \qquad \omega_r = \sqrt{k^2 - 2c^2} \qquad (14)$$

The value ω_r is called the *resonant frequency* of the damped system.

If no damping is present (i.e., $c = 0$), then formula (10) implies that $M(\omega) \to +\infty$ as $\omega \to k$, where k is the natural frequency of the undamped system. If damping is present in the system, then $M(\omega_r)$ can be made arbitrarily large if c is made small enough.

Comments

We have used only sinusoidal driving terms in this section for ODE (1), but a unique periodic forced oscillation for ODE (1) actually exists for any periodic driving term. In Problem 9 we show that if a and b are constants and the driving term $f(t)$ is periodic with period T, then the ODE $y'' + ay' + by = f(t)$ has a unique periodic response with period T, if there are no periodic solutions of the ODE $y'' + ay' + by = 0$ of period T. This surprising result gives valuable information about the long-term behavior of solutions of linear ODEs that are driven by periodic functions. Compare this with a similar result for a first-order ODE in Section 2.2.

PROBLEMS

1. (*Periodic Forced Oscillation*). Find the unique periodic solution of the ODE $y'' + 2cy' + k^2 y = F_0 \cos \omega t$, where c, k, F_0, and ω are positive constants.

2. (*Gain and Phase Shift*). Use the solution of the IVP

$$y'' + 2y' + 4y = 3 \sin 2t, \qquad y(0) = 0.5, \quad y'(0) = 0.5$$

to estimate the gain $M(2)$ and phase shift $\varphi(2)$ for the steady-state forced response. [*Hint*: See Figure 4.3.4 and Example 4.3.2.] Compare these estimates with the exact values calculated from (10). Find the system transfer function.

3. (*Gain and Phase Shift*). Consider the ODE $y'' + y' + 2y = F_0 \sin \omega t$.

(**a**) Use an ODE solver to find enough values of the gain $M(\omega)$ and the phase shift $\varphi(\omega)$ to make rough sketches of M and φ versus ω.

(**b**) Use the formulas in (10) to plot the graphs of $M(\omega)$ and $\varphi(\omega)$. Compare these plots with the sketches you drew in part (**a**).

4. (*Parameter Identification*). Find the parameters c and k of the system whose Bodé plots are shown in Figure 4.3.3. [*Hint*: Pick a frequency ω and find M and φ from the Bodé plots. The equations in (10) are then to be solved for c and k. The exact values are $c = 0.1$ and $k = 1$. How close did you get to these values using the data you read off the graphs in Figure 4.3.3?]

www **5.** (*Parameter Identification*). In a damped Hooke's Law spring system modeled by

$$y'' + 2cy' + ky = 0$$

the static deflection is 0.0127 m, while damped vibrations decay from amplitude 0.01016 m to amplitude 0.00254 m in 20 cycles. Assume a mass of 1 kg.

(a) Find the damping constant $2c$. [*Hint*: First find k, then find c such that amplitudes decay by a factor of 4 over a $20T$ time span, where $T = 2\pi/(k - c^2)^{1/2}$.]

(b) Find the resonant frequency ω_r and calculate the maximal amplitude of the steady-state response of the system to a driving force $A \cos \omega_r t$.

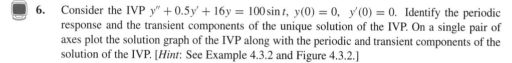 **6.** Consider the IVP $y'' + 0.5y' + 16y = 100 \sin t$, $y(0) = 0$, $y'(0) = 0$. Identify the periodic response and the transient components of the unique solution of the IVP. On a single pair of axes plot the solution graph of the IVP along with the periodic and transient components of the solution of the IVP. [*Hint*: See Example 4.3.2 and Figure 4.3.2.]

7. (*Gain and Phase Shift*). The steady-state response of the ODE $y'' + 2cy' + k^2 y = F_0 \sin \omega t$, with $0 < c \le k$, can be characterized in terms of an amplitude magnification and a phase shift applied to the sinusoidal driving term $f(t)$. Explore this connection by performing the tasks below:

- Verify [using the formulas in (10)] that the general features of the graphs of $M(\omega)$ and $\varphi(\omega)$, for fixed positive constants k and c with $k^2 > 2c^2$, are similiar to the plots in Figure 4.3.3. What is the value ω_r (the resonant frequency) where $M(\omega)$ achieves a maximum? What are $M(0)$, $M(\omega_r)$? What is the value ω_0 where $\varphi(\omega)$ changes inflection?

- What happens to the frequency ω_r and the maximum value $M(\omega_r)$ determined in part **(a)** for fixed $k > 0$ as $c \to 0$? Interpret this observation for a vibrating mechanical system generated by ODE (1) with very small damping constant c and a sinusoidal driving term with frequency close to a natural (i.e., undamped) frequency of the system.

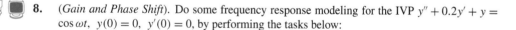 **8.** (*Gain and Phase Shift*). Do some frequency response modeling for the IVP $y'' + 0.2y' + y = \cos \omega t$, $y(0) = 0$, $y'(0) = 0$, by performing the tasks below:

- Let $\omega = 0.5$. Plot the solution of the IVP over a time sufficiently long that the transient part of the solution has become negligible and the steady-state response is clearly visible. From the graph, determine the amplitude of the steady-state response, compare it to the amplitude of the driving function, and then determine the gain M.

- Now determine the phase shift φ between the driving function and the steady-state response. [*Hint*: See Figure 4.3.4.]

- Explain why the amplitude M and the phase shift φ do not depend on initial conditions. Confirm this fact by repeating the above steps for several sets of initial conditions.

- Determine the amplitude M and the phase shift φ for various other values of ω for the IVP. For instance, choose $\omega = 0.25$, 0.75, 1, 1.5, 2, and 3. Try some values of ω very close to 1. Sketch the graphs of M versus ω and φ versus ω.

- Solve the ODE analytically and determine the steady-state response. Find M and φ as functions of ω. Compare these results with those found above.

9. (*Periodic Forced Oscillations*).Look at the ODE $y'' + ay' + by = f(t)$, where a and b are constants and $f(t)$ is a periodic continuous function on the real line with period T. Show that if the ODE $y'' + ay' + by = 0$ does not have a periodic solution with period T, then the driven ODE $y'' + ay' + by = f(t)$ has a unique periodic solution of period T. [*Hint*: Follow the outline below.]

☞ This problem proves the existence of a periodic forced oscillation; it uses the Method of Variation of Parameters (see Problem 8 in Section 3.7). Compare with Theorem 2.2.4.

- Show that a solution $y(t)$ of the ODE is a periodic solution with period T if and only if $y(0) = y(T)$ and $y'(0) = y'(T)$. [*Hint*: To show that $y(t)$ repeats in the time interval $T \leq t \leq 2T$ when the conditions hold, replace t by $s + T$ in the ODE and start the clock at $s = 0$.]

- Suppose that y_1, y_2 are solutions of the undriven ODE $y'' + ay' + by = 0$ with $y_1(0) = 1$, $y_1'(0) = 0$; $y_2(0) = 0$, $y_2'(0) = 1$. Suppose that $y_d(t)$ is the particular solution of the driven ODE $y'' + ay' + by = f(t)$ given by the formula $y_d = c_1(t)y_1(t) + c_2(t)y_2(t)$ with

$$c_1(t) = \int_0^t \frac{-y_2(s)f(s)}{W(s)}\, ds, \qquad c_2(t) = \int_0^t \frac{y_1(s)f(s)}{W(s)}\, ds$$

where $W(s)$ is the Wronskian of y_1 and y_2, $W = y_1 y_2' - y_1' y_2$. These formulas are given in Problem 8 of Section 3.7. Show that $y = y_0 y_1 + y_0' y_2 + y_d$ is the solution of the IVP

$$y'' + ay' + by = f(t), \qquad y(0) = y_0, \qquad y'(0) = y_0'$$

- From the above, we know that the conditions $y(0) = y(T)$ and $y'(0) = y'(T)$ applied to the solution $y(t)$ above guarantee that $y(t)$ is periodic, and vice versa. Show that these conditions on y are equivalent to the matrix equation

$$(I - M(T))\begin{bmatrix} y_0 \\ y_0' \end{bmatrix} = \begin{bmatrix} y_d(T) \\ y_d'(T) \end{bmatrix}$$

where I is the matrix $\begin{bmatrix} 1 & 0 \\ 0 & 1 \end{bmatrix}$ and the matrix $M(T)$ is $\begin{bmatrix} y_1(T) & y_2(T) \\ y_1'(T) & y_2'(T) \end{bmatrix}$.

- Suppose that the matrix $I - M(T)$ is *not* invertible. Then there exists a nonzero vector v with components α and β such that $(I - M(t))v = 0$, so

$$\begin{bmatrix} \alpha \\ \beta \end{bmatrix} = M(T)\begin{bmatrix} \alpha \\ \beta \end{bmatrix}$$

Show that $z(t) = \alpha y_1(t) + \beta y_2(t)$ is a nontrivial periodic solution of the undriven ODE $y'' + ay' + by = 0$, contradicting the hypothesis of the theorem.

- The contradiction above to our hypothesis establishes that $I - M(T)$ *is* invertible. Show that the ODE $y'' + ay' + by = f(t)$ has a unique periodic solution with period T.

- Show that the above result holds even when $f(t)$ is a piecewise continuous periodic function, provided that we allow the second derivative of a solution to be only piecewise continuous.

- Show that the ODE $y'' + 4y' + 3y = \text{sqw}(t, 50, 4)$ has a unique periodic solution of period $T = 4$. Without actually calculating the values of y_0 and y_0' necessary to generate the periodic solution, try to produce a graph approximating this periodic solution. How do you know your graph stays close to the periodic solution for large t?

4.4 Electrical Circuits

One of the most important physical processes that can be modeled by second-order linear ODEs is the flow of electrical energy in a circuit. The two basic measures of energy flow in a circuit are current and voltage.

Current and Voltage

The *current I* in a circuit is proportional to the number of positive charge carriers that move per second through any given point in the conductor. By analogy, think of a river and the amount of water moving past a given point per second. Current is measured in *amperes*,[4] in terms of which all other units are defined. One ampere corresponds to 6.2420×10^{18} charge carriers moving past a given point in one second. The unit of charge is the *coulomb*, defined to be the amount of charge that flows through a cross section of wire in one second when a current of one ampere is flowing, so one ampere is one coulomb per second. If $I(t)$ is the current at time t, then the amount of charge flowing past a point over the time interval $t_1 \leq t \leq t_2$ is given by $\int_{t_1}^{t_2} I(t)\, dt$.

As current moves through a circuit, the charge carriers exchange energy with the circuit elements. This process is described by defining a function V (called a *potential function*) throughout a circuit. The energy per coulomb of charge that has been exchanged by the carriers as they flow from point a to point b is computed as

$$V_{ab} = V_a - V_b$$

where V_a and V_b are the values of the potential function V at points a and b of the circuit. The difference V_{ab} is called the *voltage drop* or *potential difference* from a to b; it is measured in joules per coulomb, or *volts*. Using the water flow analogy, the voltage drop corresponds to the difference in water pressure at points a and b.

Batteries and electric generators have the property that they can maintain a voltage drop, denoted by E, between two terminals. For batteries, the terminal with the higher potential is labeled with a plus sign and the lower potential terminal with a minus sign. The internal chemical energy supplied by the battery imparts a constant amount of energy per coulomb as charge carriers move through the battery, and this raises the potential function V by the voltage rating of the device.

Now let's introduce the elements of a circuit.

---

[4] André Marie Ampère (1775–1836), a French mathematician and natural philosopher, is known for his contributions to electrodynamics. The names of many early researchers in electricity are used for electrical units. We have ohms [Georg Simon Ohm (1787–1854), German physicist], henries [Joseph Henry (1797–1878), American physicist], farads [Michael Faraday (1791–1867), English scientist], volts [Alessandro Volta (1745–1827), Italian physicist], and coulombs [Charles Auguste de Coulomb (1736–1806), French physicist].

Circuit Elements

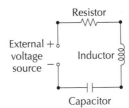

Three basic elements can be used to build a simple circuit (margin sketch): the resistor, the capacitor, and the inductor, which we describe below.

Resistor

As current flows through a segment of a circuit, electrical energy is lost, so the potential at one end of the segment is lower than the potential at the other end. A circuit segment between points *a* and *b* where a lot of energy is lost is called a resistor. Heating elements and lightbulb filaments, for example, are good resistors, converting electrical energy into heat and light. The voltage drop across a resistor and the current flowing through it are usually modeled by Ohm's Law.

> Ohm's Law. The voltage drop V_{ab} between the endpoints *a* and *b* of a resistor is proportional to the current *I* flowing through the resistor:
>
> $$V_{ab} = RI$$
>
> The constant *R* is called the *resistance* of the resistor.

If the current is directed from *a* to *b*, then the voltage at *b* is lower than the voltage at *a*, and $V_{ab} = RI$ is positive. Resistance is measured in *ohms* (denoted by the capital Greek letter omega, Ω) if current is measured in amperes and voltage in volts.

Inductor

A changing electrical current $I(t)$ through a circuit segment creates a changing magnetic field that induces a voltage drop between the ends of the segment. This effect can be quite large in circuit segments arranged in certain geometries (such as coils). These devices are called *inductors*.

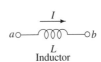

> Faraday's Law. The voltage drop V_{ab} across an inductor is proportional to the rate of change of the current:
>
> $$V_{ab} = L\frac{dI}{dt}$$
>
> The constant *L* is called the *inductance* of the inductor.

Faraday's Law says that the potential drop does not directly depend on the current *I*, but on its rate of change, dI/dt. Inductance is measured in *henries* (denoted by H) if the voltage is measured in volts and dI/dt in amperes per second.

Capacitor

A *capacitor* consists of two plates separated by an insulator such as air. If the terminals *a* and *b* of a capacitor are hooked up to a voltage source, charges of opposite sign will build up on the two plates. We speak of the total charge $q(t)$ on the capacitor and observe that if $q(t_0)$ is the initial charge, then

$$q(t) = q(t_0) + \int_{t_0}^{t} I(s)\,ds, \quad \text{for } t \geq t_0 \tag{1}$$

A capacitor is like a water tank used to store water and provide a source of pressure.

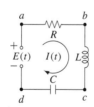

$a \circ \!\!-\!\!| \!\!-\!\! \circ b$
C
Capacitor

> **Coulomb's Law.** The voltage drop V_{ab} as the current flows from a to b across a capacitor is proportional to the charge on the capacitor:
>
> $$V_{ab}(t) = \frac{1}{C}q(t) = \frac{1}{C}\left\{ q(t_0) + \int_{t_0}^{t} I(s)\,ds \right\}$$
>
> The constant C is called the *capacitance* of the capacitor.

Capacitance is measured in farads (denoted by F) if charge is in coulombs, voltage is in volts, and current is in amperes. Because the coulomb is a rather large amount of charge, a typical capacitor will store only a small fraction of a coulomb at typical voltages, so C is usually very small (on the order of 10^{-5} or 10^{-6} F).

The Laws of Ohm, Faraday, and Coulomb are empirical and are based on many observations of simple circuits.

Kirchhoff's Laws

Electrical circuits consist of one or more closed loops, each containing resistors, inductors, capacitors, or voltage sources. In order to model the current flowing in a circuit we need a formula relating the voltage drops across the various components of a circuit. The following conservation law has been observed to hold for each closed loop in a circuit.

> **Kirchhoff's Voltage Law.** Select points $a_1, a_2, \ldots, a_n$ in the circuit and assume that the current flows from a_i to a_{i+1}, $i = 1, 2, \ldots, n$ $(a_{n+1} = a_1)$.
> Then
>
> $$V_{a_1 a_2} + V_{a_2 a_3} + \cdots + V_{a_n a_1} = 0$$
>
> where $V_{a_i a_{i+1}} = V_{a_i} - V_{a_{i+1}}$ is the voltage drop between points a_i and a_{i+1}.

For example, look at the schematic in the margin for an RLC loop. Let's suppose that the positive direction of current flow is given by the sequence of nodes $abcda$. So we have $V_{ab} + V_{bc} + V_{cd} + V_{da} = 0$, or $E(t) = V_{ab} + V_{bc} + V_{cd}$ since $E(t) = -V_{da}$.

Because of the way voltage is defined, Kirchhoff's Voltage Law is a conservation of energy law. Kirchhoff[5] proposed a conservation law for current as well.

---

[5] Gustav Robert Kirchhoff (1824–1887) was a German physicist who studied the properties of the first electrical circuits.

Gustav Robert Kirchhoff

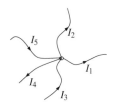

For example, in the sketch to the left we have $I_3 + I_5 = I_1 + I_2 + I_4$. We are now ready to start modeling circuits with ODEs.

Model ODEs for Charge and Current in an *RLC* Loop

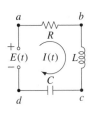

Let's apply Kirchhoff's Voltage Law to a simple *RLC* circuit to derive ODEs that model how the charge on the capacitor and the current in the circuit evolve with time. First we introduce reference points *a*, *b*, *c*, and *d* as shown in the margin and label the polarity of the external source with plus and minus signs in order to be precise about the way the external source is connected to the circuit. Applying Kirchhoff's Current Law, the current entering each point equals the current leaving that point, which is why we can use a single symbol *I* for the current at every point in this circuit. So current in the single loop circuit is independent of location, but may change over time. Since the current flows in the same direction and has the same value through every circuit element, we can assume that the positively directed current $I(t)$ moves clockwise around the circuit (as shown). Now let's look at the voltage drops.

As we follow the current through the external voltage source, the voltage increases, but the voltage decreases on each of the other circuit elements. Ohm's Law, Faraday's Law, and Coulomb's Law yield the respective voltage drops

$$V_{ab} = RI(t), \qquad V_{bc} = L\frac{dI}{dt}, \qquad V_{cd} = \frac{1}{C}\left[q(t_0) + \int_{t_0}^{t} I(s)\,ds\right]$$

Kirchhoff's Voltage Law yields an equation for $I(t)$:

$$RI(t) + L\frac{dI}{dt} + \frac{1}{C}\left[q(t_0) + \int_{t_0}^{t} I(s)\,ds\right] = E(t) \qquad (2)$$

We can convert equation (2) to an ODE in two ways.

Since $q = q(t_0) + \int_{t_0}^{t} I(s)\,ds$, so $q' = I$, and $q'' = I'$. Now we can write ODE (2) in terms of the charge q:

$$Lq'' + Rq' + \frac{1}{C}q = E(t) \qquad (3)$$

Alternatively, if $E(t)$ is differentiable, we can differentiate ODE (3) with respect to t and obtain (since $q' = I$)

$$LI'' + RI' + \frac{1}{C}I = E' \qquad (4)$$

The relevant IVP with $t_0 = 0$ for the current ODE (4) is derived as follows: If the initial charge q_0 and the initial current I_0 are known, then from (2) we have

$$RI_0 + LI'(0) + (1/C)q_0 = E(0)$$

which immediately determines $I'(0)$ in terms of I_0, q_0, and $E(0)$. We are thus led to solve the IVP

$$LI'' + RI' + \frac{1}{C}I = E'(t)$$

$$I(0) = I_0, \quad I'(0) = \frac{E(0) - RI_0 - q_0/C}{L} \tag{5}$$

The ODE in (5) has the same form as the driven and damped Hooke's Law spring ODE and also as the driven and damped linearized pendulum ODE.

The Current in the Circuit: An Example

Let's find all solutions of ODE (4) when the external voltage source is periodic.

EXAMPLE 4.4.1

Finding the Current of a Driven RLC Circuit

We will determine all solutions $I(t)$ of the simple RLC series circuit ODE in (4) when $L = 20$ H, $R = 80\ \Omega$, $C = 10^{-2}$ F, and the external voltage is $E(t) = 50 \sin 2t$. After division by $L = 20$, ODE (4) becomes

$$I'' + 4I' + 5I = 5 \cos 2t \tag{6}$$

The characteristic polynomial is $P(r) = r^2 + 4r + 5$, which has the roots $r_1 = -2 + i$ and $r_2 = -2 - i$. The solutions of $I'' + 4I' + 5I = 0$ have the form

$$k_1 e^{-2t} \cos t + k_2 e^{-2t} \sin t \tag{7}$$

where k_1 and k_2 are arbitrary real constants.

Next we find a particular solution for ODE (6). Observe that $5\cos 2t = \text{Re}[5e^{2it}]$, so if z_d is any solution of the ODE

$$z'' + 4z' + 5z = 5e^{2it} \tag{8}$$

then $\text{Re}[z_d] = I_d$ is a real-valued solution of ODE (6). A solution z_d of ODE (8) may be found by the methods of Section 3.6:

$$z_d = \frac{5e^{2it}}{P(2i)} = \frac{5e^{2it}}{(2i)^2 + 4(2i) + 5} = \frac{5e^{2it}}{1 + 8i}$$

$$= \frac{5}{1 + 8i} \cdot \frac{1 - 8i}{1 - 8i} e^{2it} = \frac{1}{13}(1 - 8i)e^{2it}$$

So we see that

$$I_d(t) = \text{Re}[z_d] = \frac{1}{13}(\cos 2t + 8\sin 2t) \tag{9}$$

is a particular solution of ODE (6). Note that I_d is a periodic forced oscillation.

At any time t the current $I(t)$ has the form

$$I(t) = k_1 e^{-2t} \cos t + k_2 e^{-2t} \sin t + \frac{1}{13}(\cos 2t + 8\sin 2t) \tag{10}$$

where k_1 and k_2 are constants. The terms $k_1 e^{-2t} \cos t + k_2 e^{-2t} \cos t$ are transients and tend to 0 as $t \to +\infty$. The other two terms in formula (10) have the same frequency as the input $5 \cos t$; they represent the steady-state periodic current in the circuit.

Constructing the Steady-State Current in an RLC Circuit

Using Example 4.4.1 as a guide, we can find all solutions $I(t)$ of ODE (4), where $R \neq 0$ and the driving term $E'(t)$ is given by $A \cos \omega t$. Since the roots of the characteristic polynomial $P(r) = Lr^2 + Rr + 1/C$ are complex conjugates with negative real parts, the solutions of the undriven ODE are transients. Now we need only a single driven solution to complete our analysis.

Using the approach given in Example 4.4.1, we find a particular solution z_d of

$$Lz'' + Rz' + \frac{1}{C}z = Ae^{i\omega t} \tag{11}$$

and then observe that $I_d = \text{Re}[z_d]$ is a particular real-valued solution of ODE (4) with $E'(t) = A \cos \omega t$. Since the roots of the characteristic polynomial have negative real parts, $P(i\omega) \neq 0$, so

$$z_d = Ae^{i\omega t}/P(i\omega) \tag{12}$$

is a particular solution of ODE (11). Since $P(i\omega) = -L\omega^2 + 1/C + iR\omega$, we can rationalize the denominator [see (8) in Section 4.3] and write $1/P(i\omega)$ in the form $a + ib$ and then use (12) to find I_d:

$$1/P(i\omega) = [(-L\omega^2 + 1/C) - iR\omega]/[(-L\omega^2 + 1/C)^2 + R^2\omega^2]$$

$$I_d = \text{Re}[z_d] = \frac{A}{(1/C - L\omega^2)^2 + \omega^2 R^2}[(1/C - L\omega^2)\cos \omega t + \omega R \sin \omega t] \tag{13}$$

$$= \frac{A}{[(1/C - L\omega^2)^2 + \omega^2 R^2]^{1/2}} \cos(\omega t + \varphi) \tag{14}$$

where we have also used the trigonometric identity

$$a \cos \theta + b \sin \theta = (a^2 + b^2)^{1/2} \cos(\theta + \varphi)$$

$$\sin \varphi = \frac{a}{(a^2 + b^2)^{1/2}}, \quad \cos \varphi = \frac{b}{(a^2 + b^2)^{1/2}}, \quad -\pi < \varphi \leq \pi$$

The steady-state current I_d is a periodic forced oscillation with the same circular frequency ω as the driving term but shifted by φ in phase. The denominator $[(1/C - L\omega^2)^2 + \omega^2 R^2]^{1/2}$ in formula (14) is called the *impedance* of the circuit; it plays a role similar to the one played by the resistance R in Ohm's Law.

For a fixed resistance R, the amplitude

$$\frac{A}{[(1/C - L\omega^2)^2 + \omega^2 R^2]^{1/2}} \tag{15}$$

of the steady-state response current in (14) to a given driving frequency ω depends on the values of ω, L, and C. For a fixed positive R and ω, the current has its maximum amplitude if the term $1/C - L\omega^2$ in the denominator in (15) is zero, that is, when

$$\frac{1}{LC} = \omega^2 \tag{16}$$

This observation makes it possible to tune the *RLC* circuit to receive a signal of a given frequency ω. The next example illustrates this property.

EXAMPLE 4.4.2

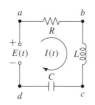

Tune Mozart In

Suppose that a classical music station broadcasts at one frequency and a hard rock station at another. When you tune your radio to your favorite station, you are actually selecting a carrier wave which transmits audible sound (this is called amplitude modulation). Let's see how we can tune in a desired carrier frequency by severely reducing the amplitudes of all other frequencies. In the circuit shown in the margin take $L = 1$ H, $R = 0.1\ \Omega$, $E(t) = -\cos t - (4/5)\cos 5t$, $I(0) = 0$, $I'(0) = 0$. Using the two carrier frequencies, $\sin t$ and $\sin 5t$ IVP for the current becomes

$$I'' + 0.1I' + \frac{1}{C}I = E'(t) = \sin t + 4\sin 5t, \qquad I(0) = 0, \quad I'(0) = 0$$

where the capacitance C is the tuning parameter. The input $E'(t)$ is the sum of the two circular frequencies 1 and 5. The IVP is solved over $0 \leq t \leq 75$ but solutions are plotted for $50 \leq t \leq 75$ after any transients have died out. The component of frequency 5 is evident in the upper graph of Figure 4.4.1; the frequency 1 component is there as well, but it's not so obvious. If we "tune" the circuit badly by setting $C = 1/81$ (so $\sqrt{1/C} = 9$, which is far away from the frequencies of 1 and 5), we would not expect to hear either input, and indeed we do not [see the nearly flat line (solid) in the lower graph of Figure 4.4.1].

If we tune the circuit correctly by choosing C so that $\sqrt{1/C} = \omega$, we can easily pick up a component of frequency ω in the external signal. The long-dashed curve of frequency 1 in the lower graph of Figure 4.4.1 corresponds to setting $C = 1$ farad and represents the circuit's response to the input $\sin t$. In the same way, if C is set at $1/25$ farad so that $\sqrt{1/C} = 5$, then the circuit picks up the frequency 5 component of the input and amplifies it (see the short-dashed curve in Figure 4.4.1).

The frequency 1 and frequency 5 signals are far apart on the frequency spectrum. Will changing the resistance or the inductance allow us to tune in on Mozart even though a hard rock station is close by in frequency (see Problems 7, 9)? Now let's look at more complicated circuits.

Multiloop Circuits

Electrical circuits with several loops can be modeled with ODEs for the currents flowing in the individual loops, as in the following example.

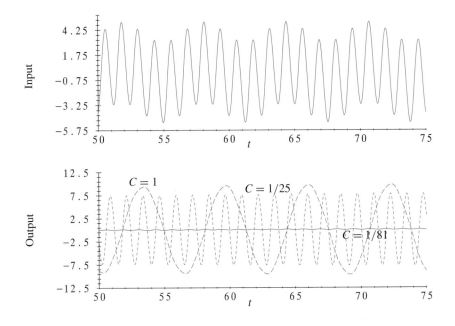

FIGURE 4.4.1 Vary capacitance C to tune a circuit (Example 4.4.2).

EXAMPLE 4.4.3

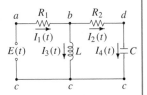

A Two-Loop Circuit

Consider the two-loop circuit pictured in the margin where I_1, I_2, I_3, and I_4 are assumed to have the directions shown (if some of the actual currents are oppositely directed, then minus signs will appear). From Kirchhoff's Current Law at node points b and d, the current I_3 through the inductor must equal $I_1 - I_2$, while the currents I_2 and I_4 must be equal. Applying Kirchhoff's Voltage Law to the two inner loops, we have (assuming $q_0 = 0$ on the capacitor)

$$R_1 I_1 + L(I_1' - I_2') = E(t)$$

$$R_2 I_2 + \frac{1}{C} \int_0^t I_2(s)\, ds + L(I_2' - I_1') = 0$$

Rearrange both equations and differentiate the second to get rid of the integral:

$$L(I_1' - I_2') + R_1 I_1 = E(t)$$

$$L(I_2'' - I_1'') + R_2 I_2' + \frac{1}{C} I_2 = 0$$

We leave the mathematical treatment of such coupled linear systems to later chapters.

Comments

Second-order linear constant coefficient ODEs model *RLC* circuits quite well. They can be extended successfully to multiloop circuits as suggested by Example 4.4.3 and Problems 3 and 8.

However, there are some difficulties with the ODE approach. For example, what if the circuit elements are not passive but change in time? What if the elements are distributed through the circuit and can't be lumped into distinct elements such as resistors as we have assumed? The latter case is exemplified by the current I and voltage drop V in a long transmission line; a model that involves partial differential equations is required in this case because I and V depend on location as well as time.

PROBLEMS

1. (*Damped Oscillation*). Let's look at IVP (5) for current in an RLC-circuit.

   (a) Solve IVP (5) if $R = 20\ \Omega$, $L = 10$ H, $C = 0.05$ F, $E(t) = 12$ volts, $q_0 = 0.6$ coulomb, and $I_0 = 1$ ampere.

   (b) Show that a damped oscillation occurs. Plot the graph of $I = I(t)$.

   (c) To what value must the resistance be increased in order to reach the overdamped case?

2. Suppose that a simple RLC circuit is charged up by a battery so that the charge on the capacitor is $q_0 = 10^{-3}$ coulomb, while $I_0 = 0$. The battery is then removed. In each case find the charge $q(t)$ on the capacitor and the current $I(t)$ in the circuit for $t \ge 0$. [*Hint*: Use IVP (5) with $I_0 = 0$, $E(t) = 0$, $t_0 = 0$, $q_0 = 10^{-3}$, and find the current. Then use Formula (1) for the charge.]

   (a) $L = 0.3$ H, $R = 15\ \Omega$, $C = 3 \times 10^{-2}$ F

   (b) $L = 1$ H, $R = 1000\ \Omega$, $C = 4 \times 10^{-4}$ F

   (c) $L = 2.5$ H, $R = 500\ \Omega$, $C = 10^{-6}$ F

www **3.** Two capacitors ($C_1 = 10^{-6}$ F and $C_2 = 2 \times 10^{-6}$ F) and a resistor ($R = 3 \times 10^6\ \Omega$) are arranged in a circuit as shown. The capacitor C_1 is initially charged to E_0 volts with polarity as shown. The switch S is closed at time $t = 0$. Determine the current that flows through the capacitor C_1 as a function of time. [*Hint*: Apply Kirchhoff's Voltage Law to the left loop, then to the right loop. Note that $q_1(0) = C_1 E_0$, $q_2(0) = -C_2 E_0$, where q_1 and q_2 are the respective charges on the capacitors C_1 and C_2.]

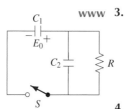

4. Find the charge $q(t)$ on the capacitor of the simple RLC circuit in the margin. Assume zero initial charge and current, $R = 20\ \Omega$, $L = 10$ H, $C = 0.01$ F, and $E(t) = 30\cos 2t$ volts.

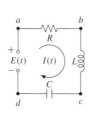

5. A simple RLC circuit has a capacitor with $C = 0.25$ F, a resistor with $R = 7 \times 10^4\ \Omega$, and an inductor with $L = 2.0$ H. The initial charge on the capacitor is zero, as is the initial current. If an impressed voltage of 60 volts is connected to the circuit, and the circuit is closed at $t = 0$, determine the charge on the capacitor for $t > 0$. Estimate the charge when $t = 0.1$ sec. [*Hint*: First solve the current ODE for $I(t)$, then integrate to get $q(t)$. Use IVP (5).]

6. (*Resonance*). Let's look at ODE (3) for the charge in an RLC circuit.

   (a) Show that if $R = 0$ in the simple RLC circuit and the impressed voltage is of the form $E_0 \cos \omega t$, the charge on the capacitor will become unbounded as $t \to \infty$ if $\omega = 1/\sqrt{LC}$. This is the phenomenon of resonance. [*Hint*: Look at the form of particular solutions.]

   (b) Show that the charge will always be bounded, no matter what the choice of ω, provided that there is some resistance in the circuit.

7. (*Tuning a Circuit: Sensitivity to Resistance*). Using the ODE for $I(t)$ from Example 4.4.2 with $L = 1$ H, $C = 1/25$ F, $R = 1.0, 0.1, 0.01\ \Omega$, graphically study the effect of the resistance on the output current when tuning the circuit to the input frequency 5. What do you conclude? Explain. [*Hint*: Look at formulas (10) and (14) in Section 4.3.]

8.   Set up, but do not solve, three ODEs for I_1, I_2, I_3 for the circuit below.

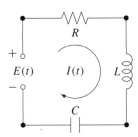

  9.   Referring to Example 4.4.2, tune in the Grateful Dead by varying the resistance, inductance, or capacitance of a circuit. Look at the case where the two stations broadcast at almost the same frequency.

10. (*On-Off Voltages*). In this problem you will start with IVP (17) for the charge $q(t)$ on the capacitor of the simple circuit shown below:

$$Lq'' + Rq' + \frac{1}{C}q = E(t), \qquad q(0) = 10^{-2}, \quad q'(0) = 0 \qquad (17)$$

where

$$L = 20\,\text{H}, \quad R = 80\,\Omega, \quad C = 10^{-2}\,\text{F}, \quad E = 50\sin 2t$$

and time is measured in seconds, charge in coulombs, and voltage E in volts.

Use a numerical solver to carry out each of the following procedures.

- Graph the solution $q(t)$ of IVP (17).
- Replace E by each of the functions (refer to Appendix B.1) given below and graph the solution $q = q(t)$ of IVP (17). Compare and explain what you see.
    - $E = 50\,\text{sqw}(t, 50, \pi)$
    - $E = 50\,\text{trw}(t, 50, \pi)$
    - $E = 50\,\text{sww}(t, 50, \pi)$
    - $E = 50\,\text{step}(t - 10) + 20\,\text{step}(30 - t)$
    - $E = 50\,\text{sqp}(t, 10)$
    - $E = 50\,\text{trp}(t, 10)$
    - $E = 50\,\text{swp}(t, 10)$
- Carry out a parameter study of the effects on $q(t)$ of varying the resistance R over the range $0 \le R \le 200$. Explain what you see.

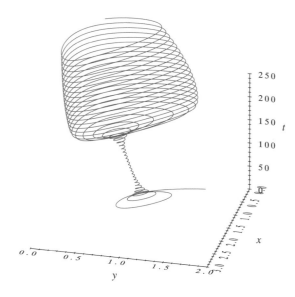

Chemical reaction in a wineglass? Take a
look at Example 5.1.5.

Systems of
Differential Equations

*Systems of ODEs that involve the first, but no higher, derivatives of the state
variables are called first-order systems. In this chapter we will focus entirely
on these systems, their theory and the natural processes that are modeled by
these systems. We will use numerical solvers to see how solutions change
as the initial data and the coefficients of the ODEs change.*

5.1 First-Order Systems

Many natural processes are modeled by systems of first-order ODEs. In this section
we will use differential systems to model a chemical reaction and the vibrations of
coupled springs and masses. Let's start by reviewing how we can convert a single n-th
order ODE into an equivalent system of n first-order ODEs.

From a Single ODE to a System

There is a way to find a first-order system equivalent to the n-th order ODE

$$y^{(n)} = g(t, y, y', \ldots, y^{(n-1)}) \tag{1}$$

Introduce new state variables $x_1, x_2, \ldots, x_n$ by setting

$$x_1 = y, \quad x_2 = y', \quad x_3 = y'', \quad \ldots, \quad x_{n-1} = y^{(n-2)}, \quad x_n = y^{(n-1)} \tag{2}$$

Differentiate each side of each equation in (2) and use the equations (2) again to find the first $n - 1$ ODEs below. Then use ODE (1) to obtain the last ODE:

$$
\begin{aligned}
x_1' &= y' = x_2 \\
x_2' &= y'' = x_3 \\
&\;\;\vdots \\
x_{n-1}' &= y^{(n-1)} = x_n \\
x_n' &= y^{(n)} = g(t, x_1, x_2, \ldots, x_n)
\end{aligned}
\tag{3}
$$

The state variables of this system are $x_1, x_2, \ldots, x_n$. So we see that if $y(t)$ solves ODE (1), then the functions $x_1(t), \ldots, x_n(t)$ defined via (2) solve system (3). The other way around, if $x_1(t), \ldots, x_n(t)$ solve system (3), then the function $y(t) = x_1(t)$ solves ODE (1), because system (3) shows that the relations (2) are valid and this implies that $y(t)$ solves ODE (1). System (3) and ODE (1) are equivalent because each solution of one generates a unique solution of the other through the variable naming process (2).

Since most solvers only accept first-order ODEs or systems of first-order ODEs, this procedure for replacing ODE (1) by system (3) is crucial in preparing a higher-order ODE for computing. Here's an example of how the process works.

EXAMPLE 5.1.1

From a Second-Order ODE to a System

Let's convert the second-order ODE

$$y'' + 2y' + 3y - y^3 = \sin t \tag{4}$$

into an equivalent system of two first-order ODEs following these steps:

1. Rewrite ODE (4) with the highest derivative isolated on the left side.

$$y'' = -2y' - 3y + y^3 + \sin t \tag{5}$$

2. Introduce two new state variables.

$$x_1 = y, \qquad x_2 = y' \tag{6}$$

3. Differentiate each of the new variables.

$$x_1' = y', \qquad x_2' = y'' \tag{7}$$

4. In (7) replace y'' by the expression in (5):

$$x_1' = y', \qquad x_2' = -2y' - 3y + y^3 + \sin t$$

5. Use (6) to replace y by x_1 and y' by x_2.

$$
\begin{aligned}
x_1' &= x_2 \\
x_2' &= -2x_2 - 3x_1 + x_1^3 + \sin t
\end{aligned}
\tag{8}
$$

So system (8) is equivalent to ODE (4). This process can be extended to an ODE of any order, but it requires isolating the highest derivative as in step (5).

Besides making ODEs acceptable to solvers, systems are important because many natural processes are modeled quite naturally by systems of ODEs. Let's look at a physical process that can be modeled by a first-order system of ODEs in four state variables.

EXAMPLE 5.1.2

Modeling the Motion of Coupled Springs and Masses

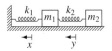

Attach a coupled system of masses and springs to the wall and let the masses slide back and forth on a smooth table. At equilibrium the springs are neither stretched nor compressed. Let's measure the respective displacements of the masses from their equilibrium positions by x and y, with the positive direction indicated by the arrows.

Suppose that air resistance and sliding friction are negligible, so the only forces acting on each mass are gravity, the upward force of the table, and the spring forces. Gravity and the table force are equal and opposite, so we can ignore them. Let's suppose that each spring force is proportional to the spring's displacement from its unstretched and uncompressed length (Hooke's Law) and acts in a direction that would restore the spring to its equilibrium length.

In particular, if the displacements of the masses are x and y, then the spring force acting on the body of mass m_2 is $-k_2(y - x)$, where k_2 is the positive Hooke's Law constant for the spring. To check that the algebraic sign of the spring force is correct, note that if $y > x$ then the second spring is compressed and the spring force acts to uncompress the spring. So in this case the spring force on m_2 should be directed to the right and should be negative (which it is). On the other hand, we see that the spring connecting the two masses exerts a force $k_2(y - x)$ on m_1 that is equal and opposite in sign to the force it exerts on m_2 [by Newton's Third Law (Section 4.1)]. In addition, Hooke's Law implies that the mass m_1 is also subjected to the spring force $-k_1 x$.

Applying Newton's Second Law to each mass, we obtain a pair of coupled second-order ODEs.

$$\begin{aligned}
m_1 x'' &= -k_1 x + k_2(y - x) = -(k_1 + k_2)x + k_2 y \\
m_2 y'' &= -k_2(y - x) = k_2 x - k_2 y
\end{aligned} \tag{9}$$

But we need a system of first-order ODEs, so let's divide out the masses and introduce the four state variables

$$x_1 = x, \qquad x_2 = x', \qquad x_3 = y, \qquad x_4 = y'$$

Then we can replace system (9) by the equivalent first-order system

$$\begin{aligned}
x_1' &= x' = x_2 \\
x_2' &= x'' = -\left(\frac{k_1 + k_2}{m_1}\right)x_1 + \frac{k_2}{m_1}x_3 \\
x_3' &= y' = x_4 \\
x_4' &= y'' = \frac{k_2}{m_2}x_1 - \frac{k_2}{m_2}x_3
\end{aligned} \tag{10}$$

and we are done.

Let's introduce some terms and notation for systems.

States, Systems, and Solutions

The mathematical model of Example 5.1.2 is a system containing four ODEs in four state variables. More complex natural processes may lead to more state variables and more complicated ODEs. An astounding variety of processes can be modeled by *first-order systems* of ODEs that have the *normal form*

$$
\begin{aligned}
x_1' &= f_1(t, x_1, x_2, \ldots, x_n) \\
x_2' &= f_2(t, x_1, x_2, \ldots, x_n) \\
&\vdots \\
x_n' &= f_n(t, x_1, x_2, \ldots, x_n)
\end{aligned}
\tag{11}
$$

in the *state variables* $x_1, \ldots, x_n$ and the independent variable t, representing time. All derivatives are with respect to time, and the *rate functions* f_i are assumed to be known functions of time and state. Initial conditions can be given for each of the state variables, so we have the *initial value problem* (IVP)

$$
\begin{aligned}
x_1' &= f_1(t, x_1, x_2, \ldots, x_n), & x_1(t_0) &= a_1 \\
x_2' &= f_2(t, x_1, x_2, \ldots, x_n), & x_2(t_0) &= a_2 \\
&\vdots \\
x_n' &= f_n(t, x_1, x_2, \ldots, x_n), & x_n(t_0) &= a_n
\end{aligned}
\tag{12}
$$

where $a_1, \ldots, a_n$ are given constants. A collection of functions $x_1(t), \ldots, x_n(t)$ defined on a t-interval I containing t_0 that satisfy the equations of IVP (12) for all t in I is called a *solution* of IVP (12). We assume that I is the largest interval on which the solution $x(t)$ is defined, so the solution is *maximally extended*. The space $\mathbb{R}^n$ of the state variables $(x_1, \ldots, x_n)$ is called *state space*. The curve of points $(t, x_1(t), \ldots, x_n(t))$ in $\mathbb{R}^{n+1}$ is a *time-state curve*

Sometimes it is more convenient to use a compact notation for the differential system in (11) and the IVP in (12). Let's see how we can do this by working first in three dimensions. Choosing three mutually orthogonal unit vectors $\mathbf{i}$, $\mathbf{j}$, and $\mathbf{k}$, we see that every point can be located by its position vector $\mathbf{R} = x\mathbf{i} + y\mathbf{j} + z\mathbf{k}$ for appropriate numbers x, y, and z, the coordinates of the point. If the coordinates are functions of t, we see that the derivative of $\mathbf{R}(t)$, denoted by $\mathbf{R}'(t)$, is given by $\mathbf{R}'(t) = x'(t)\mathbf{i} + y'(t)\mathbf{j} + z'(t)\mathbf{k}$. It is inefficient to keep writing vectors in terms of $\mathbf{i}, \mathbf{j}, \mathbf{k}$. Instead, we will write vectors as a column array with the coordinates listed in descending order. For example,

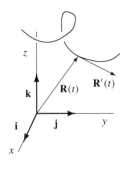

$$
R(t) = \begin{bmatrix} x(t) \\ y(t) \\ z(t) \end{bmatrix} \quad \text{and} \quad R'(t) = \begin{bmatrix} x'(t) \\ y'(t) \\ z'(t) \end{bmatrix}
$$

Notice that $R(t)$ and $R'(t)$ are *not boldfaced* when they denote coordinate column vectors. These *column vectors* are an alternate way of writing vectors using coordinates.

We will use this approach to write system (11) and IVP (12) in abbreviated form. Adapting the discussion above to n state variables, let's denote the *state vector* x, the *initial state vector* x^0 (the superscript 0 is a label that reminds us that this is the initial value of the state vector at t_0) and the *vector rate function* f by the column vectors

$$x = \begin{bmatrix} x_1 \\ \vdots \\ x_n \end{bmatrix}, \qquad x(t_0) = x^0 = \begin{bmatrix} a_1 \\ \vdots \\ a_n \end{bmatrix}, \qquad f = \begin{bmatrix} f_1 \\ \vdots \\ f_n \end{bmatrix}$$

Notice that these column vectors are denoted by the symbols x, x^0, and f. Now when the state vector x varies differentiably with time, we see that we can use the column vector notation

$$x(t) = \begin{bmatrix} x_1(t) \\ \vdots \\ x_n(t) \end{bmatrix}, \qquad x'(t) = \begin{bmatrix} x_1'(t) \\ \vdots \\ x_n'(t) \end{bmatrix}$$

System (11) and IVP (12) can then be written in the compact forms

$$x' = f(t, x); \qquad x' = f(t, x), \quad x(t_0) = x^0$$

We will use this notation for general systems, but with specific systems we usually write out each of the component ODEs.

Let's introduce a class of systems widely used to model natural processes.

Linear Systems

Just as linear ODEs have special properties and uses that have earned them a central place in the theory and applications, linear systems of ODEs have the same importance. Here is how they are defined:

❖ **Linear Differential System.** A *linear differential system* in *normal form* is

$$x_1' = a_{11}x_1 + \cdots + a_{1n}x_n + F_1$$
$$\vdots \qquad\qquad\qquad\qquad (13)$$
$$x_n' = a_{n1}x_1 + \cdots + a_{nn}x_n + F_n$$

where the *coefficients* a_{ij} and the *input* (or *driving*) *terms* $F_1, \ldots, F_n$ are constants or functions of t (but *not* of $x_1, \ldots, x_n$).

☞ These are often simply called *linear systems*

We can illustrate this notation with the coupled spring-mass system of Example 5.1.2.

EXAMPLE 5.1.3

Linear System Notation for Coupled Springs and Masses
Let's take system (10) for a pair of coupled springs and masses, attach some initial conditions, for example, $x_1(0) = 1$, $x_2(0) = 0$, $x_3(0) = 2$, and $x_4(0) = 0$, and write everything in vector form:

$$x' = f(t, x), \qquad x(0) = x^0$$

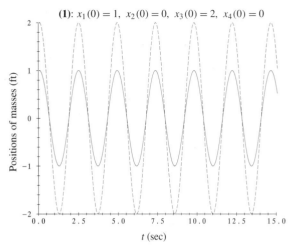

FIGURE 5.1.1 In-phase periodic oscillations of the coupled spring-mass system (Example 5.1.4): mass m_1 (solid), mass m_2 (dashed).

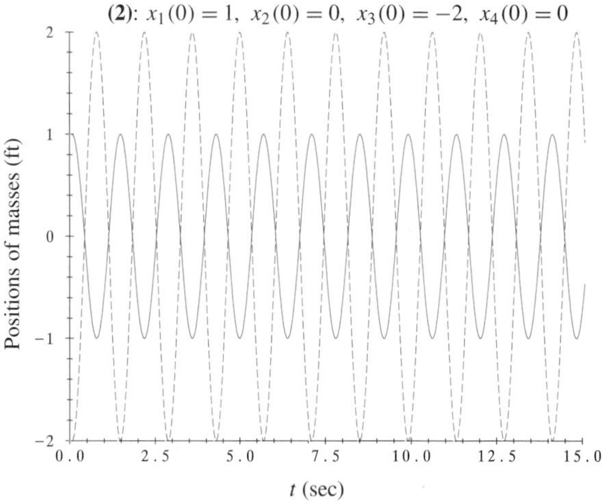

FIGURE 5.1.2 Out-of-phase periodic oscillations of the coupled spring-mass system (Example 5.1.4): mass m_1 (solid), mass m_2 (dashed).

where

$$x(t) = \begin{bmatrix} x_1(t) \\ x_2(t) \\ x_3(t) \\ x_4(t) \end{bmatrix}, \quad f = \begin{bmatrix} x_2 \\ -(k_1+k_2)x_1/m_1 + k_2x_3/m_1 \\ x_4 \\ k_2x_1/m_2 - k_2x_3/m_2 \end{bmatrix}, \quad x(0) = x^0 = \begin{bmatrix} 1 \\ 0 \\ 2 \\ 0 \end{bmatrix} \quad (14)$$

The system of ODEs in (14) is an *undriven* linear system because it has the form (13) with $F_1 = \cdots = F_n = 0$.

We will see a lot of linear systems in Chapter 7, but let's turn now to the geometry of the solutions of systems, linear and nonlinear.

State Space, Orbits, Time-State Curves, Component Curves

A solution of a system of first-order ODEs determines several curves whose behavior highlights various properties of the solution.

☞ Some of these terms were introduced in Sections 1.7 and 3.2.

❖ **Orbits, Time-State Curves, State Space, Component Curves**. Suppose that $x = x(t)$ is a solution of the system $x' = f(t, x)$, over a t-interval I. The point $x(t)$ traces out an *orbit* (or *trajectory*) in the *state* (or *phase*) *space* of the variables $x_1, \ldots, x_n$. A collection of orbits is a *portrait*. The point $(t, x(t))$ traces out a *time-state curve* in the time-state space of the variables $t, x_1, x_2, \ldots, x_n$. The projection of a time-state curve onto the tx_j-plane is the x_j-*component curve*

We will illustrate some of these concepts by using a numerical solver to plot graphs associated with the coupled spring-mass system.

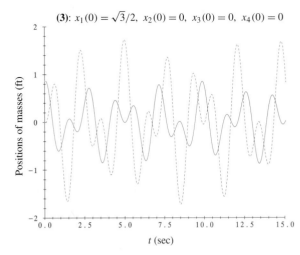

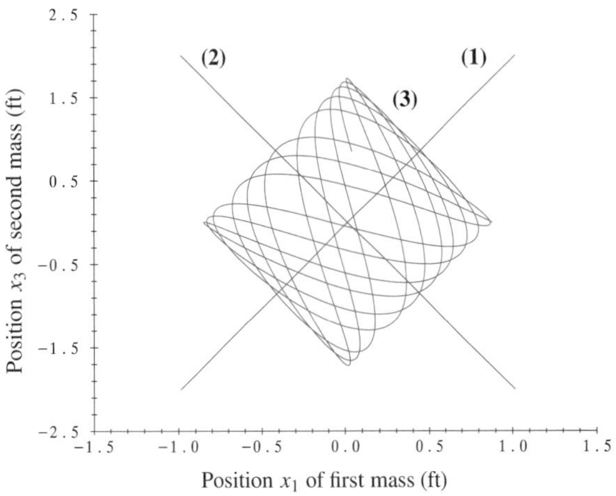

(3): $x_1(0) = \sqrt{3}/2, \ x_2(0) = 0, \ x_3(0) = 0, \ x_4(0) = 0$

FIGURE 5.1.3 Both masses oscillate, but not periodically (Example 5.1.4): mass m_1 (solid), mass m_2 (dashed).

FIGURE 5.1.4 Periodic in-phase oscillations along line **(1)**, out of phase along **(2)**, nonperiodic along the Lissajous curve **(3)** (Example 5.1.4).

EXAMPLE 5.1.4

Oscillating Springs

What happens if we compress the springs of Example 5.1.2 by pushing the first mass one foot to the left $[x_1(0) = 1]$ and the second mass two feet to the left $[x_3(0) = 2]$ and then release them from rest $[x_2(0) = 0$ and $x_4(0) = 0]$? Let's suppose that

☞ Slugs and other units are defined in Appendix B.6.

$$m_1 = 4 \text{ slugs}, \quad m_2 = 1 \text{ slug}, \quad k_1 = 40 \text{ slug/sec}^2, \quad k_2 = \frac{40}{3} \text{ slug/sec}^2$$

From system (10) we see that the modeling IVP with these initial conditions is

$$
\begin{aligned}
x_1' &= x_2, & x_1(0) &= 1 \\
x_2' &= -\frac{40}{3}x_1 + \frac{10}{3}x_3, & x_2(0) &= 0 \\
x_3' &= x_4, & x_3(0) &= 2 \\
x_4' &= \frac{40}{3}x_1 - \frac{40}{3}x_3, & x_4(0) &= 0
\end{aligned}
\tag{15}
$$

Let's use a numerical solver and plot the tx_1- and tx_3-component curves over a 15-sec time interval. The masses oscillate periodically (and in phase) about their equilibrium positions (Figure 5.1.1).

Repeat the process but push the first mass one foot to the left and the second mass two feet to the right of equilibrium [so $x_1(0) = 1$ and $x_3(0) = -2$] and release them from rest. Now we see periodic oscillations that are out of phase (Figure 5.1.2), and the frequency is higher. Finally, move the first mass $\sqrt{3}/2$ feet to the left [$x_1(0) = \sqrt{3}/2$] but keep the second mass in its equilibrium position [$x_3(0) = 0$]. Figure 5.1.3 shows the *nonperiodic* oscillations of the two masses.

Figure 5.1.4 shows the positions of the two masses relative to one another. The straight line segments [**(1)** and **(2)**] correspond to the two periodic oscillations of Figures 5.1.1 and 5.1.2. The nonperiodic motion in Figure 5.1.3 shows up as a curve [see **(3)**] wandering through a parallelogram. This is called a *Lissajous curve*.

Now let's model a very different kind of process. A chemical reaction is a process in which various species combine, interact, and recombine to form other species. A chemical reactor is a vessel in which the reaction takes place. Chemists want to know how the concentrations of the species in the reaction evolve over time and mathematical models often work better than empirical studies for this purpose.

Strange Oscillations in a Chemical Reactor

Let's suppose that a reactant in a chemical reaction changes into intermediate species, and the intermediates into a final product. We can adapt the compartmental model idea and the balance law to model this process. Capital letters denote the various species, lower case letters the corresponding concentrations. In the reaction diagrammed below, reactant W becomes the intermediate X, $W \to X$, at the rate of aw. As the reaction proceeds, X becomes Y, $X \to Y$, at the rate bx, and Y turns into the final product Z, $Y \to Z$, at the rate cy. So let's use a compartment model to describe these reactions, writing the rates over the arrows.

☞ Compartment
models are introduced in
Section 1.8.

$$ \boxed{w} \xrightarrow{\ aw\ } \boxed{x} \xrightarrow{\ bx\ } \boxed{y} \xrightarrow{\ cy\ } \boxed{z} $$

Since the transformation rate of each chemical into another is proportional to its concentration, the reactions are first order, so this gives us a set of four linear rate equations in the concentrations:

$$ w' = -aw, \qquad x' = aw - bx, \qquad y' = bx - cy, \qquad z' = cy \qquad (16) $$

Note that linear system (16) is a linear cascade that can be solved from the top down. In particular, we can start with a concentration w_0 of the reactant and obtain $w(t) = w_0 e^{-at}$. Substituting that expression for w into the linear rate equation for x, and setting $x(0) = 0$, for example, we could solve the second ODE in (16) for $x(t)$, and so on, step-by-step down the line. Typical graphs for $x(t)$ and $y(t)$ are sketched in the margin: these intermediate concentrations reach maximum values, and then diminish exponentially.

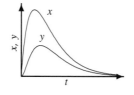

In recent times a particular class of nonlinear reactions called *autocatalytic* reactions have attracted a great deal of attention.[1] In an autocatalytic reaction a chemical species promotes its own production. Here's an example: two units of Y react with one unit of X to produce three units of Y, a net gain of one unit of Y.

---

[1] In 1951, the Russian chemist Boris Pavlovich Belousov (1894–1970) discovered a specific chemical reaction that behaved like the autocatalator, but no one believed him and his work was ignored. In disgust he abandoned his chemical research, and it was only years later that the importance of his work was recognized. Belousov had a turbulent life, starting out as a young revolutionary in the tsarist days. After the Communist Revolution of 1917, he joined the Soviet Army and rose to the rank of Brigade Commander. It was only after retirement from the army that he began his career as a research chemist. A decade after his death his work was awarded the highest civilian award of the Soviet era. For more on the chemistry and the mathematics of these reactions, see P. Gray and S. K. Scott, *Chemical Oscillations and Instabilities* (Oxford: Clarendon Press, 1990), and S. K. Scott, *Chemical Chaos* (Oxford: Clarendon Press, 1991).

In an autocatalytic reaction the rate of transformation from one chemical species into another does not follow a first-order rate law. In order to model reactions that may not be of first order we need to introduce a more general modeling principle called the Chemical Law of Mass Action.

> **Chemical Law of Mass Action.** If n reactant molecules $X_1, \ldots, X_n$ react to produce m product molecules $P_1, \ldots, P_m$ in one step of a reaction, then the rate of decrease of the concentration of each reactant and the rate of increase of the concentration of each product is proportional to the product of the concentrations of the n reactants. This is the *Chemical Law of Mass Action*.

☞ The reactant and product molecules don't need to be distinct.

Let's apply this new principle to the reaction modeled by sytstem (16), but with the $X \to Y$ reaction step augmented by an autocatalytic step $X + 2Y \to 3Y$. By the Chemical Law of Mass Action the rate of decrease of X in the autocatalytic step is $\alpha x y^2$ (α is a positive rate constant), while the rate of increase of Y is $3\alpha x y^2 - 2\alpha x y^2 = \alpha x y^2$. These rates are shown in the augmented compartment model

$$w \xrightarrow{aw} x \xrightarrow[\alpha x y^2]{bx} y \xrightarrow{cy} z$$

Starting with a positive initial concentration of W and zero initial concentrations for the other species, we have the nonlinear initial value problem:

$$
\begin{aligned}
w' &= -aw, & w(0) &= w_0 \\
x' &= aw - bx - \alpha x y^2, & x(0) &= 0 \\
y' &= bx - cy + \alpha x y^2, & y(0) &= 0 \\
z' &= cy, & z(0) &= 0
\end{aligned}
\tag{17}
$$

Let's assume that time and the concentrations have been scaled to dimensionless quantities and that the values of w_0 and the rate constants and w_0 are given by

$$w_0 = 500, \qquad a = 0.002, \qquad b = 0.08, \qquad c = 1, \qquad \alpha = 1 \tag{18}$$

Now let's feed these rate equations and initial data into a numerical solver and see what comes out. We will plot the x- and y-component curves. Figure 5.1.5 shows unusual and unexpected oscillations in the concentrations in the intermediates. These oscillations are internally generated and are *not* due to oscillations in external factors such as temperature or pressure. Notice that after some initial swings the concentrations behave almost normally until around $t = 170$ when the violent oscillations in the concentrations begin. The oscillations stop around $t = 600$ and the concentrations of the intermediates gradually decline.

In the next example we will treat part of the autocatalytic model as an isolated system of two ODEs.

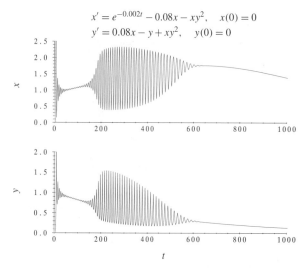

$$x' = e^{-0.002t} - 0.08x - xy^2, \quad x(0) = 0$$
$$y' = 0.08x - y + xy^2, \quad y(0) = 0$$

$$x' = e^{-0.002t} - 0.08x - xy^2, \quad x(0) = 0$$
$$y' = 0.08x - y + xy^2, \quad y(0) = 0$$

FIGURE 5.1.5 Autocatalytic oscillations of the concentrations turn on and then switch off [IVP (19)].

FIGURE 5.1.6 The orbit of the autocatalytic interaction between the intermediates [IVP (19)].

EXAMPLE 5.1.5

The Two-Dimensional Autocatalator

We know from (18) and the first IVP in (17) that $w(t) = w_0 e^{-0.002t} = 500 e^{-0.002t}$, so the second and third rate equations in IVP (17) become

$$x' = 0.002w - 0.08x - xy^2 = e^{-0.002t} - 0.08x - xy^2, \qquad x(0) = 0$$
$$y' = 0.08x - y + xy^2, \qquad\qquad\qquad\qquad\qquad y(0) = 0 \tag{19}$$

We can treat IVP (19) numerically on its own, and that's just what we do. Figure 5.1.5 shows the component curves, and Figure 5.1.6 shows the corresponding orbit in the xy-plane. Note the appearance of the orbital arc when it emerges from the tangle of the oscillations and starts to head back to $x = 0$, $y = 0$ as the reaction nears completion. Finally, Figure 5.1.7 shows the time-state curve for $7 \leq t \leq 700$. The wineglass in the chapter cover figure is the part of this figure limited to the time span $7 \leq t \leq 377$.

We will discuss this nonlinear system further in Example 5.2.1 and Problem 6, Section 9.3.

Figure 5.1.7 and the cover figure show that, by the proper choice of ranges for the variables, you can create curious and unusual graphs that tell a story. This aspect of the presentation of scientific and mathematical graphs is extremely important, because above all else we want the graphs to give us information about the way solutions behave. If the graph is poorly presented, that information won't come across.

Comments

Our aim has been to introduce first-order systems of ODEs, the graphs they generate, and some of the many processes that can be modeled by these systems. In the next section we describe the basic theory of these systems.

$$x' = e^{-0.002t} - 0.08x - xy^2, \quad x(0) = 0$$
$$y' = 0.08x - y + xy^2, \quad y(0) = 0$$

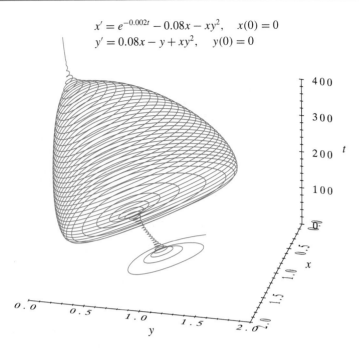

FIGURE 5.1.7 The time-state curve for IVP (19).

PROBLEMS

www **1.** (*From Scalar ODEs to Systems*). Solve each linear ODE; construct and solve an equivalent first-order system of ODEs. For each first-order system of linear ODEs below find an equivalent single ODE, solve it, and then use these solutions to construct all solutions of the system.

(a) $y'' - 4y = 0$ (b) $y'' + 9y = 0$

(c) $y'' + 5y' + 4y = 0$ (d) $x_1' = x_2, \quad x_2' = -2x_1 - 2x_2$

(e) $x_1' = x_2, \quad x_2' = -16x_1$ (f) $y''' + 6y'' + 11y' + 6y = 0$

 2. (*Hooke's Law Springs*). The system $x' = y$, $y' = -bx - ay + A\cos\omega t$ is equivalent to the scalar ODE $x'' + ax' + bx = A\cos\omega t$ that models the motion of a body on a damped and driven Hooke's Law spring. In each case, find $x(t)$ and $y(t)$ explicitly. Then use a graphics package or a numerical ODE solver to plot x- and y-component graphs, the orbit, and the time-state curve. Interpret what you see in terms of the behavior of the spring.

(a) $x' = y, \quad y' = -x - y; \quad x(0) = 1, \ y(0) = 0; \quad 0 \le t \le 15$

(b) $x' = y, \quad y' = -x - 3y; \quad x(0) = 1, \ y(0) = 0; \quad 0 \le t \le 10$

(c) $x' = y, \quad y' = -25x + \cos 5t; \quad x(0) = 0, \ y(0) = 0; \quad 0 \le t \le 10$

(d) $x' = y, \quad y' = -25x + \cos(5.5t); \quad x(0) = 0, \ y(0) = 0; \quad 0 \le t \le 30$

3. (*Cascades*). Linear cascade systems (Section 1.8) can be solved one ODE at a time. Find explicit solutions of the cascade IVPs below. Find the limiting value of each state variable as $t \to +\infty$.

(a) $x_1' = -3x_1, \quad x_2' = x_1, \quad x_3' = 2x_1; \quad x_1(0) = 10, \ x_2(0) = x_3(0) = 0; \quad 0 \le t \le 10$

(b) $x_1' = -x_1, \quad x_2' = x_1 - 3x_2; \quad x_1(0) = 10, \ x_2(0) = 20; \quad 0 \le t \le 6$

(c) $x_1' = -2x_1, \quad x_2' = -3x_2, \quad x_3' = 2x_1 + 3x_2; \quad x_1(0) = 1, \ x_2(0) = 2, \ x_3(0) = 0; \quad 0 \le t \le 5$

4. (*Coupled Springs*). The spring constants k_1 and k_2 and the masses m_1 and m_2 of a coupled spring system have values such that $k_1/m_1 + k_2/m_1 = 1$, $k_2/m_1 = \alpha$, $k_2/m_2 = 1$, where the magnitude of the positive constant α is m_2/m_1. For these values, system (10) becomes $x_1' = x_2$, $x_2' = -x_1 + \alpha x_3$, $x_3' = x_4$, $x_4' = x_1 - x_3$. For each of the values of α given below, plot

- tx_1- and tx_3-component graphs for initial data $(\sqrt{\alpha}, 0, 1, 0)$.
- tx_1- and tx_3-component graphs for initial data $(-\sqrt{\alpha}, 0, 1, 0)$.
- The projections onto $x_1 x_3 x_2$-space of the orbit with initial data $(2\sqrt{\alpha}, 0, 0, 0)$.
- The $x_1 x_3$-graphs for initial data $(\pm\sqrt{\alpha}, 0, 1, 0)$ and $(2\sqrt{\alpha}, 0, 0, 0)$.

Use $0 \le t \le 25$ and interpret what you see in terms of the behavior of the springs.

(a) $\alpha = 0.05$ **(b)** $\alpha = 0.5$ **(c)** $\alpha = 0.95$

5. (*Orbits of Autonomous Systems*). Find formulas for the orbits by solving $dx_2/dx_1 = f_2/f_1$; be careful not to lose orbits for which $f_1 = 0$. Plot some orbits in the rectangle $|x_1| \le 10$, $|x_2| \le 10$.

(a) $x_1' = x_1$, $x_2' = -3x_2$ **(b)** $x_1' = x_1 x_2$, $x_2' = x_1^2 + x_2^2$ [*Hint*: See Example 1.9.1.]

(c) $x_1' = x_2$, $x_2' = -e^{-x_1}$

6. (*Solving a Linear System*). The solutions of the scalar ODE $x' = ax$ have the form $x = x_0 e^{at}$. This suggests that solutions of the system $x_1' = ax_1 + bx_2$, $x_2' = cx_1 + dx_2$, where a, b, c, and d are constants, might have the form $x_1 = \alpha e^{rt} + \beta e^{st}$, $x_2 = \gamma e^{rt} + \delta e^{st}$, where r, s, α, β, γ, and δ must be determined by inserting x_1 and x_2 into the differential system and matching coefficients of like exponentials. Using this method, solve the following systems.

(a) $x_1' = x_1 + 3x_2$, $x_2' = x_1 - x_2$ **(b)** $x_1' = 2x_1 - x_2$, $x_2' = 3x_1 - 2x_2$

7. (*Chemical Rate Equations*). Given the diagrams below for the chemical reaction steps, write out the ODEs for the concentrations. The quantities k_1, k_2, k are positive rate constants. [*Hint*: The rate equation for X in part **(a)** is $x' = -k_1 xy$.]

(a) $X + Y \xrightarrow{k_1} Z \xrightarrow{k_2} W$ **(b)** $X + 2Y \xrightarrow{k} Z$ **(c)** $X + 2Y \xrightarrow{k} 6Y + W$

8. (*The Autocatalator*). The two-dimensional autocatalator orbit of Figure 5.1.5 is the projection of the orbit of IVP (17) onto the xy-plane. The following problems extend IVP (17) by varying the initial concentration $w(0)$ of the reactant w.

(a) Solve IVP (17) after setting $w(0) = 50$, then 100. What unusual features do you see in the x and y component graphs in comparison to what is visible in Figure 5.1.5?

(b) Find the minimal value of $w(0)$ for which you can see oscillations in the component graphs.

(c) Draw graphs like Figure 5.1.7 for $w(0) = 50, 250, 500, 800$. Explain what you see.

9. (*A Chemical Reaction*). Describe the behavior of the solutions of system (17) for various values of the rate constants and various values of $w(0)$. Assume that $x(0) = y(0) = z(0) = 0$.

10. (*Two-Dimensional Autocatalator, Turning Oscillations On and Off*). Look at the rate equations for x and y in IVP (17) with the four parameters $a = 0.002$, $b = 0.08$, $c = 1$, $\alpha = 1$ and with $w(t)$ replaced by $500e^{-at}$. It may be possible to turn the x and y oscillations on or off by changing any one of the four parameters. That is the aim of the project: change a parameter up or down from the value given until the oscillations of the concentrations of the intermediates X and Y disappear. Then explain why you think that happens. Keep the following points in mind.

- Duplicate Figures 5.1.5, 5.1.6, 5.1.7.

- If your solver supports three-dimensional graphs, duplicate the wineglass on the first page of the chapter. What happens if you plot the solution curve for $0 \le t \le 2000$ rather than the truncated curve of the wineglass, $6.5 \le t \le 377$?

- First vary the coefficient b from 0.08 up to 0.14, keeping all other parameters fixed. What happens? Any explanation?

- Vary the coefficients a from 0.002 and c from 1, but keep the other parameters fixed at the values given. Can you turn the oscillations of the intermediates off?

5.2 Properties of Systems

The normal form of an IVP for a first-order differential system is

$$x' = f(t, x), \qquad x(t_0) = x^0 \tag{1}$$

$$x = \begin{bmatrix} x_1 \\ \vdots \\ x_n \end{bmatrix}$$

$$f = \begin{bmatrix} f_1 \\ \vdots \\ f_n \end{bmatrix}$$

$$x^0 = \begin{bmatrix} a_1 \\ \vdots \\ a_n \end{bmatrix}$$

where x, f, and x^0 are n-vectors. The vector x is the *state variable*, the vector f is the *rate function*, and the vector x^0 is the *initial state* at time t_0. Sometimes formulas for solutions can be found, but for a system which models an intricate natural process, that is rarely true. So the focus we take in this chapter is on properties of solutions, not formulas.

Before going on, we need some terminology. A box is a generic name, in any number of dimensions, for what in one dimension is an interval, and in two dimensions is a rectangle. Boxes may extend to infinity (e.g., the first octant in tx_1x_2-space is a box that reaches infinity). Boxes always include the points on their boundaries (i.e., they are *closed*).

The fundamental result below guarantees that under mild conditions on the rate function f, IVP (1) has exactly one maximally extended solution and this solution is a continuous function of the data (by data we mean x^0 and f).

THEOREM 5.2.1

> Fundamental Theorem for Systems.   Consider the IVP
>
> $$x' = f(t, x), \qquad x(t_0) = x^0$$
>
> where all of the functions f_i and $\partial f_i / \partial x_j$ are continuous on a box B in $(n+1)$-dimensional tx-space, and (t_0, x^0) is an interior point of B.
>
> **Existence:** The IVP has a solution on a t-interval containing t_0.
>
> **Uniqueness:** The IVP has at most one solution on any t-interval containing t_0.
>
> **Extension:** The solution can be extended to a maximal t-interval for which the time-state curve lies in B and extends to the boundary of B as t tends to each endpoint of the interval.
>
> **Continuity/Sensitivity:** The solution is continuous in the data x^0 and f.

From now on, we consider only systems for which the Fundamental Theorem 5.2.1 applies. To avoid ambiguity, we always assume that the solution of IVP (1) and the corresponding time-state curve are maximally extended. A maximally extended time-state curve can't disappear inside B. The curve must cross B from boundary to boundary, and this means that the curve may possibly escape to infinity if B itself extends to infinity. The Fundamental Theorem 5.2.1 includes the Fundamental Theorems 2.3.3 and 3.2.1 as special cases.

The two-dimensional system for the chemical intermediates in an autocatalytic reaction (introduced in the previous section) illustrates the Fundamental Theorem.

EXAMPLE 5.2.1

☞ See
Example 5.1.5.

Solutions that Extend to Infinity

The IVP that models the changing concentrations A and B of the intermediates

$$x' = f_1(t, x, y) = e^{-0.002t} - 0.08x - xy^2, \qquad x(0) = 0$$
$$y' = f_2(t, x, y) = 0.08x - y + xy^2, \qquad\qquad y(0) = 0 \tag{2}$$

has rate functions f_1 and f_2 that are continuous functions for all t, x and y. In addition, the various first order partial derivatives with respect to x and y

$$\partial f_1/\partial x = -0.08 - y^2, \qquad \partial f_1/\partial y = -2xy$$
$$\partial f_2/\partial x = 0.08 + y^2, \qquad \partial f_2/\partial y = -1 + 2xy$$

☞ This alone
doesn't mean that all
solutions are defined
on $-\infty < t < \infty$.
Some solutions may
escape to infinity in
finite time.

are continuous for all t, x and y. This means that the maximally extended solution curve of IVP (2) reaches backward and forward in time across txy-space. The interpretation of the system in (2) as a model for the concentration x and y of intermediate species in an autocatalytic reaction falls apart if t is negative or if x and y are too large. No model of autocatalation is valid everywhere in time-state space.

For the remainder of this chapter we will look only at systems whose rate functions don't explicitly depend on time. This allows us to use special techniques that give invaluable information about solution behavior.

Autonomous Systems, Equilibrium Points, Cycles

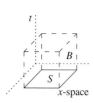

Let's look at the *autonomous IVP* (i.e., the rate function f does *not* depend on t)

$$x' = f(x), \qquad x(t_0) = x^0 \tag{3}$$

where the conditions of the Fundamental Theorem 5.2.1 hold in a box B of tx-space. Since f doesn't depend on t, we may look at a box S in x-space instead of the box B. We may think of S as the projection of B into the x-space along the t-axis. Suppose that x^0 is in S and t_0 is any real number. The four conclusions of Fundamental Theorem 5.2.1 may now be interpreted in terms of S instead of B. For example, if the box S is bounded, then either the orbit of the unique maximally extended solution curve of IVP (3) meets the boundary of S at some finite time beyond the initial time t_0, or else the orbit stays in S as $t \to +\infty$. There is a similar result as t decreases from t_0. This means that some orbits may remain strictly inside S even as $t \to +\infty$ or as $t \to -\infty$. This extends the process described in Section 2.2 for projecting solution curves of a single autonomous ODE in one state variable onto a state line.

☞ We met these
properties earlier when
looking at autonomous
ODEs in Section 2.2 and
at systems generated by
autonomous second-order
ODEs in Chapter 3.

Solutions of autonomous systems have some remarkable properties. First, suppose that $x = x(t)$, $a < t < b$, is a solution of the autonomous system, $x' = f(x)$. Then, for any constant c, the function $x = x(t + c)$, $a - c < t < b - c$, is also a solution, so the two solutions determine exactly the same orbit in state space because the two time-state curves have the same projection onto the state space.

Next, we claim that distinct orbits of an autonomous system never meet.

THEOREM 5.2.2

> **Separation of Orbits.** Suppose that the functions f_i and $\partial f_i / \partial x_j$, $i, j = 1, \ldots, n$, are continuous on a box S in state space. Then the orbits in S of two maximally extended solutions of $x' = f(x)$ either never meet, or else coincide.

To see why this is true, suppose that two orbits do meet at a point. Since the rate functions don't depend on time, we could reset the clock so that both orbits are at the same point at the same time. By the uniqueness property for IVPs, the orbits must coincide.

If a solution $x(t)$ stays constant for all time, it is an *equilibrium solution*. The corresponding time state curve is a straight line in tx-space parallel to the t-axis, and the orbit is a point in state space (an *equilibrium point*). Equilibrium points correspond to the zeros of the rate function $f(x)$. There are computer techniques for finding these zeros, but we mostly determine them by inspection. Here's an example:

EXAMPLE 5.2.2

Equilibrium Points for a Planar System

The equilibrium points for the planar autonomous system

$$\begin{aligned} x' &= x - y + x^2 - xy \\ y' &= -y + x^2 \end{aligned} \tag{4}$$

are all points (x, y) in the xy-plane at which both rate functions are zero:

$$-y + x^2 = 0, \quad \text{and} \quad x - y + x^2 - xy = 0 \tag{5}$$

The first equation in (5) says that $y = x^2$, and when y is replaced by x^2 in the second equation we have

$$x - x^2 + x^2 - x^3 = 0, \quad \text{or} \quad x(1 - x^2) = 0$$

So the x-coordinate of an equilibrium point for system (4) must be either $x = 0$, $x = +1$, or $x = -1$. Using the first equation in (5) again we discover the three equilibrium points $(0, 0)$, $(1, 1)$, $(-1, 1)$.

Periodic solutions of an autonomous system generate closed orbits in state space which are called *cycles*. Sometimes periodic solutions and their cycles can be found by inspection of the system, sometimes by actually constructing solution formulas, and sometimes by more intricate theoretical means (as we shall see in later sections). Frequently, however, our only clue that periodic behavior even exists is on the computer screen after using a numerical solver to graph component curves and orbits. Cycles have the following properties:

- No orbit can touch a cycle (other than the cycle itself), again by the Separation of Orbits Theorem 5.2.2.

- An orbit that intersects itself defines a cycle because at the intersection point we can reset the clock and regenerate the same closed curve over the same time span.

Cycles and equilibrium points play a basic role in the way nonequilibrium orbits behave in state space. They direct nearby orbits to go one way or another. Note that no other orbit can ever touch an equilibrium point because that would violate the Separation of Orbits Theorem.

Let's go into the two dimensional state space of a pair of autonomous ODEs where we can provide visual information about what it all means.

Planar Autonomous Systems, Direction Fields

From this point on in the chapter we will mostly look at the planar autonomous system in normal form:

$$
\begin{aligned}
x' &= f(x, y) \\
y' &= g(x, y)
\end{aligned}
\tag{6}
$$

where the real-valued scalar rate functions f and g and their first order partial derivatives are continuous on a box S in xy-space. Then we know that through each point (x_0, y_0) in S there passes precisely one maximally extended orbit.

The state space of system (6) is the xy-plane, and an orbit is a curve described by the endpoint of the *position vector*

$$
\mathbf{R}(t) = x(t)\mathbf{i} + y(t)\mathbf{j}
$$

where $x(t)$, $y(t)$ is a solution of the system (6), and $\mathbf{i}$, $\mathbf{j}$ are the unit vectors in the positive x and y directions. Now as the endpoint of $\mathbf{R}(t)$ traces out the orbit in time, the velocity $\mathbf{v}(t)$ of the endpoint of $\mathbf{R}(t)$ is

$$
\mathbf{v}(t) = \mathbf{R}'(t) = x'(t)\mathbf{i} + y'(t)\mathbf{j} = f(x, y)\mathbf{i} + g(x, y)\mathbf{j}
\tag{7}
$$

Since $\mathbf{v}(t)$ is tangent to the orbit, equation (7) tells us that the path itself need not be known to find $\mathbf{v}(t)$ at the point (x_0, y_0) on the "flight" path. We see that $\mathbf{v}$ at (x_0, y_0) is $f(x_0, y_0)\mathbf{i} + g(x_0, y_0)\mathbf{j}$, and since the time t that the particle arrives at (x_0, y_0) is irrelevant, we may as well think of $\mathbf{v}$ as a function of x and y, and not of t. The length $\|\mathbf{v}\|$ of $\mathbf{v}$ gives the *speed* of the particle as it traverses an orbit.

Another device for understanding orbital behavior is the direction field: Place a grid on the rectangle $a \leq x \leq b$, $c \leq y \leq d$ where you want to study orbits. Then draw a line segment centered at each point (x, y) of the grid where the projection of the segment on the x-axis has length $|f(x, y)|$ and the projection on the y-axis has length $|g(x, y)|$. Then scale the segments so that no two segments intersect. The resulting field of line segments is a *direction field*. It is a remarkable fact that a continuously differentiable curve in the xy-plane is an orbit of system (6) if the tangent vector at any point (x, y) on the curve is parallel to the *velocity field* vector $f(x, y)\mathbf{i} + g(x, y)\mathbf{j}$. So, a smooth curve in the xy-plane is an orbit of system (6) if it "fits" the direction field generated by system (6). The direction field line centered at the grid point (x, y) is often oriented with an arrowhead which points in the direction of the velocity vector $\mathbf{v}(x, y)$ to show the direction in which orbits are traced out; we use dots instead of arrowheads.

The next example shows a direction field for a planar autonomous system.

☞ Direction fields for first-order ODEs should not be confused with these direction fields.

☞ A sneaky way to find orbits. Useful, too!

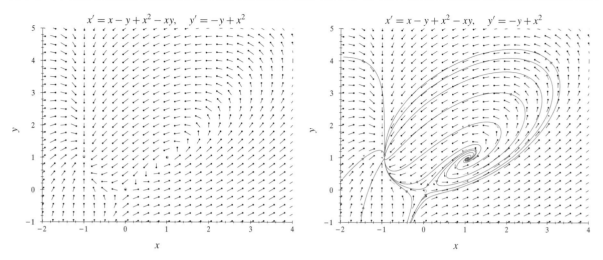

FIGURE 5.2.1 A direction field and three equilibrium points (Example 5.2.3).

FIGURE 5.2.2 Note how the orbits fit the direction field of Figure 5.2.1 (Example 5.2.3).

EXAMPLE 5.2.3

☞ We use dots instead of arrowheads on our direction fields.

An Unusual Direction Field and Some Orbits

Figure 5.2.1 shows a direction field for the system in Example 5.2.2:

$$x' = x - y + x^2 - xy, \qquad y' = -y + x^2$$

The direction field suggests what the orbits of the system look like. This view is especially interesting near the system's three equilibrium points: $(0,0)$, $(1,1)$, and $(-1,1)$. Figure 5.2.2 shows orbits that approach $(0,0)$, but they seem to veer off. Orbits that head toward $(-1,1)$ don't turn away, but they can't reach that equilibrium point in finite time. Orbits seem to emerge (as t increases from $-\infty$) from the equilibrium point $(1,1)$ in a spiral counter-clockwise motion and eventually head off toward the equilibrium point $(-1,1)$.

Certain curves in the state plane help us understand how orbits rise, fall, and change direction, so we now take a look at these curves.

Nullclines

☞ This process is like the *sign analysis* of Section 2.2.

The curves in the xy-plane defined by $f(x,y) = 0$ are called the *x-nullclines* for system (6). The *y-nullclines* are defined by $g(x,y) = 0$. The x-nullclines meet the y-nullclines at the equilibrium points, so sketching the nullclines gives us a way to locate the equilibrium points of system (6). It is important to note that the rate function $f(x,y)$ has fixed sign on each side of an x-nullcline. Similarly, $g(x,y)$ has fixed sign on each side of a y-nullcline. The nullclines divide the xy plane into regions where orbits rise ($g > 0$), fall ($g < 0$), move to the right ($f > 0$), or move to the left ($f < 0$). Knowing where these regions are and the signs of f and g inside them can be of considerable help in visualizing the portrait of the orbits even before the actual orbits are constructed.

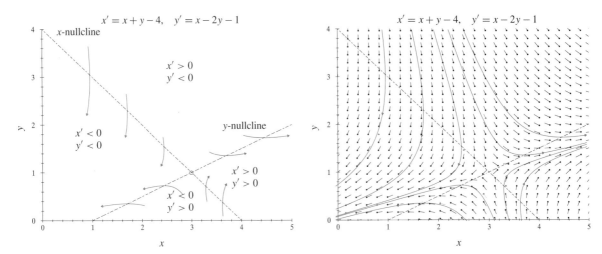

FIGURE 5.2.3 Nullclines, time-directed orbital arcs crossing nullclines (Example 5.2.4).

FIGURE 5.2.4 Orbits, direction field, nullclines (Example 5.2.4).

An example helps us visualize the behavior of orbits of planar autonomous linear systems.

EXAMPLE 5.2.4

Nullclines of a Planar Autonomous Linear System

The nullclines of the system

$$x' = x + y - 4$$
$$y' = x - 2y - 1 \tag{8}$$

are the straight lines defined by

x-nullcline: $x + y - 4 = 0$,   y-nullcline: $x - 2y - 1 = 0$

The two lines cross at the equilibrium point $(3, 1)$ in the xy-plane. The dashed lines in Figure 5.2.3 are the nullclines. We can label each sector formed by x- and y-nullclines according to whether x' and y' are positive or negative in the sector. For example, above the x-nullcline $x + y - 4 = 0$ we must have $x + y - 4 > 0$ (just check the sign at one point in the sector), so sign analysis tells us that an orbit must move to the right as t increases. But above the y-nullcline $x - 2y - 1 = 0$, we must have $x - 2y - 1 < 0$. In the top sector of Figure 5.2.3 orbits move to the right ($x' > 0$) and downward ($y' < 0$). A similar sign analysis can be applied to the other sectors.

With all of this information about the signs of x' and y' at hand, we can hand-draw arcs of orbits as they cross the nullclines. For example the arc at the upper left of Figure 5.2.3 starts in the upper sector near the x-nullcline, moves down and to the right, crosses the x-nullcline vertically (since $x' = 0$ on the nullcline), and then turns to the left since in this new sector $x' < 0$. The nullclines give us a good sense of orbital behavior. The arrowhead indicates the direction of time's increase on each orbital arc. See Figure 5.2.4 for some computer-generated orbits of system (8). Note the good fit between direction field and orbits.

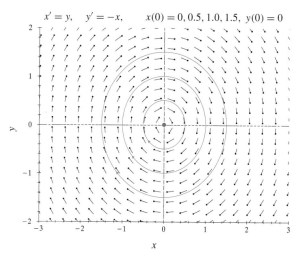

$x' = y, \quad y' = -x, \qquad x(0) = 0, 0.5, 1.0, 1.5, \ y(0) = 0$

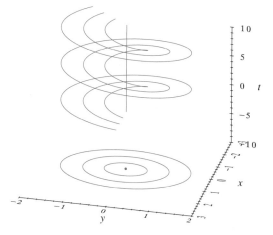

$x' = y, \quad y' = -x, \qquad x(0) = 0, 0.5, 1.0, 1.5, \ y(0) = 0$

FIGURE 5.2.5 Direction field, nullclines (the axes), equilibrium point (origin), cycles for a harmonic oscillator (Example 5.2.5).

FIGURE 5.2.6 Four time-state curves (above) for the harmonic oscillator and their corresponding orbits which are cycles (below) (Example 5.2.5).

Now let's take a system we saw in Section 3.5 as a second-order ODE and give it the full treatment of nullclines, direction field, orbits, and time state curves.

EXAMPLE 5.2.5

The Harmonic Oscillator

The harmonic oscillator system

$$x' = y, \qquad y' = -\omega^2 x \tag{9}$$

where ω is a positive constant, has the nullclines

$$x\text{-nullcline: } y = 0, \qquad y\text{-nullcline: } x = 0$$

and the single equilibrium point $(0, 0)$. As we saw in Example 3.5.1, the nonconstant solutions are all periodic, so the orbits are cycles.

☞ The orbits are plotted in the plane $t = -10$ so that orbits and time-state curves don't appear to meet.

Figure 5.2.5 shows the nullclines crossing at the equilibrium point at the origin; a direction field and three cycles in the case $\omega = 1$. Figure 5.2.6 shows the time-state curves with the orbits below in the xy-plane. Note how the constant time-state curve $x = 0, \ y = 0$ (corresponding to the equilibrium point at the origin in Figure 5.2.5) shows up as a straight line parallel to the t-axis in Figure 5.2.6.

System (9) is linear, and can be solved explicitly (see Example 3.5.1). The next system is neither linear, nor solvable in terms of elementary functions, but the direction field and nullcline approach gives us information about orbital behavior.

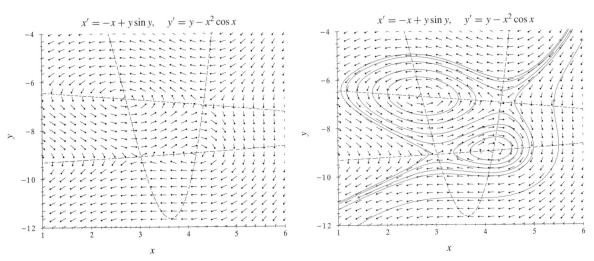

FIGURE 5.2.7 Direction field, nullclines (dashed), and four equilibrium points. The y-nullcline opens upward (Example 5.2.6).

FIGURE 5.2.8 Some orbits are cycles, other orbits enter the computer screen and then leave as t increases (Example 5.2.6).

EXAMPLE 5.2.6

An Intriguing Direction Field: Nullclines, Orbits
The direction field and the nullclines in Figure 5.2.7 for the system

$$x' = -x + y \sin y$$
$$y' = y - x^2 \cos x$$

show something unusual happening near four points (which apparently are equilibrium points). The plotted orbits of Figure 5.2.8 verify our suspicions. Two of the equilibrium points seem to turn away approaching orbits. The other two equilibrium points seem to be surrounded by nested closed orbits (i.e., cycles).

See how orbits change direction as they cross the nullclines. The nullclines divide the rectangle $1 \le x \le 6$, $-12 \le y \le -4$ into eight regions, inside each of which x' and y' have fixed signs. As an orbit moves horizontally across a y-nullcline (or vertically across an x-nullcline) y' (or x') changes its sign. For example, both x' and y' are negative in the region at the upper right, so orbits fall to the left as t increases. The orbit that starts at $x = 6$, $y = -4.5$ moves down vertically across an x-nullcline, enters a region where x' is positive, and turns to the right since x' is positive in the new region.

Planar autonomous systems may be analyzed using direction fields and nullclines, but these techniques are not available for nonautonomous planar systems or for systems with more than two state variables.

Now let's look at the effect of changing the parameters in a system of ODEs.

Continuity/Sensitivity of Pendulum Motion

How do orbits change as some parameter in a system changes? The mathematical model of the damped simple pendulum was introduced in Section 4.1, and some of its orbits were portrayed in the cover figure for Chapter 4. Let's look at the second-order nonlinear ODE that models the motion of an undriven pendulum of length L supporting a bob of mass m and subject to a damping force:

$$mLx'' + cLx' + mg\sin x = 0 \qquad (10)$$

where c is the damping coefficient, x is the angle of the pendulum from the vertical (counterclockwise is the positive direction for x), and x' is the angular velocity. Divide by mL, set $c/m = \alpha$, and (to be specific) set the value of $g/L = 10$. To do a sensitivity study of a system as a parameter changes we can introduce the parameter as a new state variable by adding another differential equation to the system. Let's do this with a system which is equivalent to ODE (10). Our state variables will be x, $y = x'$, and our parameter $\alpha = c/m$. Suppose that initially the pendulum is at $x = 0$, has angular velocity 10 rad/sec, and $\alpha = \alpha_0$. The corresponding IVP is

☞ We used this technique back in Example 2.3.1.

$$
\begin{aligned}
x' &= y, & x(0) &= 0 \\
y' &= -10\sin x - \alpha y, & y(0) &= 10 \\
\alpha' &= 0, & \alpha(0) &= \alpha_0
\end{aligned}
\qquad (11)
$$

We have converted α into a state variable so that we can use our solver to explore the effect of changes in α on the xy-orbits. According to Theorem 5.2.1, the solution of system (11) is a continuous function of the initial data, which means α_0 here because we fix x_0 at 0 and y_0 at 10. Small changes in α_0 mean small changes in the angle $x(t)$ of the pendulum, at least over a short time span (low sensitivity). But a small change in α may cause a big change in the state variables $x(t)$ and $y(t)$ over time (high sensitivity). Let's see how that happens.

EXAMPLE 5.2.7

Changing the Mass Affects the Motion of the Pendulum

Because $\alpha = c/m$, we see that increasing the mass (or lowering the damping constant) in the pendulum model (11) lowers the magnitude of the damping term $-\alpha y$. So a pendulum with a large mass might whirl all the way up and over the pivot several times before the small frictional forces gradually dissipate energy and compel the pendulum to settle down to back-and-forth and decaying oscillations about a downward vertical position. Figure 5.2.9 shows four orbits with initial point $(0, 10)$: $\alpha = 0$ (no damping), $\alpha = 0.3$, $\alpha = 0.5$, and $\alpha = 1.75$. The orbits are graphed over $0 \le t \le 30$. The top orbit shows the pendulum whirling counterclockwise forever, that is, the angle x goes to $+\infty$ as $t \to +\infty$, because there is no friction to stop the over-the-top motion.

The big change occurs when α switches from value 0 to a positive number, that is, when the damping is turned on. The graph with $\alpha = 0.3$ shows how light friction (or a fairly large mass) results in a couple of over-the-top turns followed by decaying oscillations that approach the equilibrium point $(4\pi, 0)$. If $\alpha = 0.5$, the somewhat heavier friction (or smaller mass) produces an orbit that goes over the top once and then settles down to decaying oscillations about the equilibrium point $(2\pi, 0)$. If $\alpha = 1.75$,

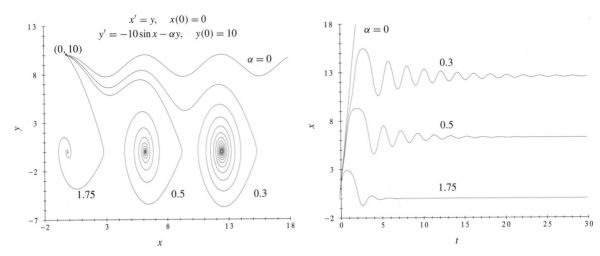

FIGURE 5.2.9 Sensitivity to changes in the parameter $\alpha = c/m$ of the pendulum bob (Example 5.2.7).

FIGURE 5.2.10 The x-component graphs for the orbits of Figure 5.2.9.

the heavy friction (or very small mass) prevents the pendulum from going over the top at all. Figure 5.2.10 shows the graphs of the x-components of these orbits. The components stay close for a short while (low sensitivity to changes in α in the short run), but as time goes on the x-components differ a lot (high sensitivity in the long run).

Finally, we show how to move from rectangular coordinates to the polar coordinates that are particularly useful for some kinds of rotational motion.

Planar Autonomous Systems in Polar Coordinates

The orbits of the planar autonomous system

$$
\begin{aligned}
x' &= f(x, y) \\
y' &= g(x, y)
\end{aligned}
\tag{12}
$$

can be described in polar coordinates r, θ, where

$$
x = r\cos\theta, \qquad y = r\sin\theta
\tag{13}
$$

To transform the ODEs of the system into polar coordinates, we proceed as follows. Suppose that $x = x(t)$, $y = y(t)$ is a solution of the system. Then differentiating each side of the identities in (13) with respect to t and using the ODEs, we have

$$
\begin{aligned}
x' &= r'\cos\theta - r\theta'\sin\theta = f(r\cos\theta, r\sin\theta) \\
y' &= r'\sin\theta + r\theta'\cos\theta = g(r\cos\theta, r\sin\theta)
\end{aligned}
\tag{14}
$$

Solving for r' and θ' in terms of f and g, we have

$$
\begin{aligned}
r' &= (\cos\theta)f(r\cos\theta, r\sin\theta) + (\sin\theta)g(r\cos\theta, r\sin\theta) = F(r, \theta) \\
\theta' &= (1/r)[(\cos\theta)g(r\cos\theta, r\sin\theta) - (\sin\theta)f(r\cos\theta, r\sin\theta)] = G(r, \theta)
\end{aligned}
\tag{15}
$$

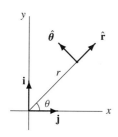

Observe that $x_0 = r_0 \cos \theta_0$, $y_0 = r_0 \sin \theta_0$ with $r_0 \neq 0$ is an equilibrium point for the system (12) if and only if $F(r_0, \theta_0) = 0$, $G(r_0, \theta_0) = 0$ in system (15).

How is the velocity field

$$\mathbf{v} = f(x, y)\mathbf{i} + g(x, y)\mathbf{j}$$

of system (12) related to the functions $F(r, \theta)$ and $G(r, \theta)$ defined in (15)? At each point (r, θ) of the plane, $\hat{\mathbf{r}}$, $\hat{\boldsymbol{\theta}}$ are unit vectors with $\hat{\mathbf{r}}$ and pointing outward along r and $\hat{\boldsymbol{\theta}}$ orthogonal to $\hat{\mathbf{r}}$ and pointing in the direction of increasing θ (i.e., counterclockwise). See the margin figure. A little trigonometry shows that

$$\hat{\mathbf{r}} = \cos \theta \mathbf{i} + \sin \theta \mathbf{j} \quad \text{and} \quad \hat{\boldsymbol{\theta}} = -\sin \theta \mathbf{i} + \cos \theta \mathbf{j}$$

which can be solved for $\mathbf{i}$ and $\mathbf{j}$ to obtain

$$\mathbf{i} = \cos \theta \hat{\mathbf{r}} - \sin \theta \hat{\boldsymbol{\theta}}, \qquad \mathbf{j} = \sin \theta \hat{\mathbf{r}} + \cos \theta \hat{\boldsymbol{\theta}}$$

So we see that the velocity field vector $\mathbf{v} = f\mathbf{i} + g\mathbf{j}$ for the system (12) takes the form

$$\mathbf{v} = f\mathbf{i} + g\mathbf{j} = (\cos \theta \hat{\mathbf{r}} - \sin \theta \hat{\boldsymbol{\theta}}) f + (\sin \theta \hat{\mathbf{r}} + \cos \theta \hat{\boldsymbol{\theta}}) g$$
$$= (f \cos \theta + g \sin \theta) \hat{\mathbf{r}} + (g \cos \theta - f \sin \theta) \hat{\boldsymbol{\theta}}$$
$$= F(r, \theta) \hat{\mathbf{r}} + r G(r, \theta) \hat{\boldsymbol{\theta}}$$

which shows how to express $\mathbf{v}$ in terms of $\hat{\mathbf{r}}$ and $\hat{\boldsymbol{\theta}}$.

Here is an example where the use of polar coordinates leads directly to solution formulas.

EXAMPLE 5.2.8

From Rectangular to Polar Coordinates

Figure 5.2.11 shows a direction field for the autonomous planar system

$$\begin{aligned} x' &= x - 10y \\ y' &= 10x + y \end{aligned} \tag{16}$$

The x-nullcline $x - 10y = 0$ and the y-nullcline $10x + y = 0$ are shown in the figure as tilted straight lines. Since rotational motion of some kind is strongly suggested by the direction field, the outward spiraling orbits of Figure 5.2.12 are no surprise. What is a surprise is that we can find a solution formula for these orbits. To do this, let's rewrite system (16) in terms of polar coordinates as

☞ The orbits of Figure 5.2.12 were formed by solving *backward* in time from the initial points $(-3, -3)$, $(3, 3)$.

$$r' = \cos \theta (r \cos \theta - 10r \sin \theta) + \sin \theta (10r \cos \theta + r \sin \theta) = r$$
$$\theta' = (1/r)[\cos \theta (10r \cos \theta + r \sin \theta) - \sin \theta (r \cos \theta - 10r \sin \theta)] = 10 \tag{17}$$

Since these ODEs decouple, we can solve them separately to find that

$$r = r_0 e^t, \quad r_0 \geq 0, \qquad \theta = 10t + \theta_0 \tag{18}$$

is the general solution of system (17) in terms of t and arbitrary constants r_0 and θ_0.

We see that the velocity field $\mathbf{v}$ for system (16) can be written as

$$\begin{aligned} \mathbf{v} &= (x - 10y)\mathbf{i} + (10x + y)\mathbf{j} \\ &= r\hat{\mathbf{r}} + 10r\hat{\boldsymbol{\theta}} \end{aligned} \tag{19}$$

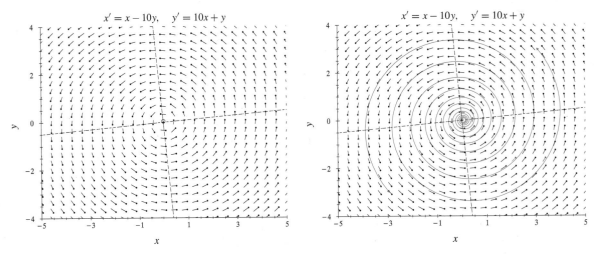

FIGURE 5.2.11 Direction field, nullclines, and equilibrium point (Example 5.2.8).

FIGURE 5.2.12 Outward spiraling orbits cut across nullclines (Example 5.2.8).

So all orbits spiral counterclockwise away from the origin with ever-increasing speed since from (18) and (19),

$$\mathbf{v} = r_0 e^t (\hat{\mathbf{r}} + 10\hat{\boldsymbol{\theta}})$$

and the speed is

$$||\mathbf{v}|| = r_0 e^t ||\hat{\mathbf{r}} + 10\hat{\boldsymbol{\theta}}|| = \sqrt{101}\, r_0 e^t$$

The general solution of system (16) in Cartesian coordinates is obtained from the r, θ solution formulas (18) by using the formulas $x = r\cos\theta$ and $y = r\sin\theta$:

$$x = r_0 e^t \cos(10t + \theta_0), \qquad y = r_0 e^t \sin(10t + \theta_0) \tag{20}$$

where r_0 and θ_0 are arbitrary constants.

Polar coordinates are useful when rotational motion is suspected, which is mainly where you will see them from this point on.

Comments

The basic properties given in the Fundamental Theorem 5.2.1 are usually taken for granted and not specifically mentioned when dealing with a system, which doesn't mean that they are unimportant. Without the assurances given in the Fundamental Theorem 5.2.1, there could be no general theory of systems of ODEs because we would never be sure that an IVP even had a solution, or if it did, that the solution would be unique, or that the solution could be extended forward and backward in time, or that the solution would change continuously with the data.

PROBLEMS

www **1.** Verify that each IVP satisfies the hypotheses of Theorem 5.2.1 for all values of t and the state variables. Find formulas for the solutions. For **(a)–(e)** plot the nullclines and the nine orbits corresponding to all possible combinations of $x(0)$, $y(0) = a$, $b = -1, 0, 2$ in the rectangle $|x| \leq 3$, $|y| \leq 3$. Identify any orbits that are equilibrium points or cycles.

(a) $x' = y$, $y' = -x - 2y$ [*Hint*: Note that $x'' + 2x' + x = 0$.]

(b) $x' = 2x$, $y' = -4y$

(c) $x' = y$, $y' = -9x$

(d) $x' = y^3$, $y' = -x^3$ [*Hint*: Write as $dy/dx = -x^3/y^3$.]

(e) $x' = -x^3$, $y' = -y$

(f) $x' = y$, $y' = -26x - 2y$, $z' = -z/2$; $x(0) = y(0) = z(0) = 1$. Sketch the projection of the orbit in each coordinate plane. Use the ranges $|x| \leq 1$, $-4 \leq y \leq 3$, $1 \leq z \leq 15$, and the time interval $0 \leq t \leq 10$.

 2. Find the equilibrium points and cycles; plot nullclines, direction fields, and several orbits in a rectangle large enough to contain several equilibrium points. Describe what you see.

(a) $x' = x - y^2$, $y' = x - y$ \qquad\qquad **(b)** $x' = y \sin x$, $y' = xy$

(c) $x' = 2 + \sin(x + y)$, $y' = x - y^3 + 27$ \qquad **(d)** $x' = y + 1$, $y' = \sin^2 3x$

(e) $x' = 3(x - y)$, $y' = y - x$

3. Find all solutions. Verify that if $x = x(t)$, $y = y(t)$ is a solution, then so is $x = x(t - T)$, $y = y(t - T)$, where T is any constant. Specify the interval on which each solution is defined.

(a) $x' = 3x$, $y' = -y$ \qquad **(b)** $x' = 1/x$, $y' = -y$

(c) $x' = -x^3$, $y' = 1$ \qquad **(d)** $x' = x^2(1 + y)$, $y' = -y$

4. Consider the planar autonomous system $x' = 1 - y^2$, $y' = 1 - x^2$.

(a) Locate all the equilibrium points of the system.

 (b) Create a portrait of orbits in the rectangle $|x| \leq 3$, $|y| \leq 3$. Use arrowheads on orbits to indicate the direction of increasing time. Are there any cycles?

(c) Use geometric reasoning with the direction field to show that the line $y = x$ is composed of five orbits. Identify these orbits and use arrowheads on the orbits to show the direction of increasing time. [*Hint*: Plotting a direction field first might help.]

(d) For each nonconstant orbit in part **(c)** above, find a function $f(t)$ such that $x = f(t)$, $y = f(t)$ describes the orbit. Show that the orbit that originates at $x(0) = a$, $y(0) = a$, $a < -1$, escapes to infinity in finite time.

5. (*Cycles and Equilibrium Points*). Suppose the autonomous system $x' = f(x)$ satisfies the conditions of the Fundamental Theorem 5.2.1 in a box S in state space. Show the following properties of solution curves, equilibrium points, and cycles in S (converses are true, but not proved here).

(a) Suppose that $x(t)$ is a nonconstant solution for which $x(t) \to P$ as $t \to T$. Show that if P is an equilibrium point, then $|T| = \infty$. [*Hint*: Since $x(t)$ is a continuous function of t, $x(T)$ must be the point P if T is finite. But this violates the uniqueness part of the Fundamental Theorem 5.2.1.]

(b) Show that if $x = x(t)$ is a nonconstant periodic solution, then the corresponding orbit is a simple closed curve (i.e., a cycle). [*Hint*: Suppose that the fundamental period of $x(t)$ is T and that $x(t_1) = x(0)$ for some $t_1, 0 < t_1 \leq T$. Then $y(t) = x(t + t_1)$ is also a solution (why?), and $y(t) = x(t)$ since $y(0) = x(0)$. So $x(t)$ has period t_1. Show that $t_1 = T$.]

6. (*Attraction and Repulsion: Polar Coordinates*). The system below illustrates the attraction and repelling properties of cycles and equilibrium points.

(a) Explain why the unit circle $x^2 + y^2 = 1$ is a cycle of the system

$$x' = x - y - x(x^2 + y^2), \qquad y' = x + y - y(x^2 + y^2) \tag{21}$$

and the origin is the only equilibrium point. [*Hint*: Use (15) to write (21) in polar coordinates.]

(b) Using the polar coordinate form of system (21) above, solve for $r(t)$ and $\theta(t)$. Explain why all nonconstant orbits approach the cycle as time $t \to +\infty$.

(c) Draw a direction field for system (21), plot x- and y-nullclines, and plot orbits inside, on, and outside the unit circle for $|x| \le 3$, $|y| \le 2$. Why could you call the origin a "repeller" and the unit circle an "attractor."

(d) Plot x- and y-component graphs for the orbits plotted in part **(c)**. What is the period of the cycle?

7. (*Here Is the Solution; What Is the System?*). Alex Trebek hands you two vector functions

$$u = \begin{bmatrix} \sin t \\ t \end{bmatrix}, \qquad v = \begin{bmatrix} te^t \\ t^3 \end{bmatrix}$$

and asks you for the continuously differentiable functions $f(x, y, t)$ and $g(x, y, t)$ so that u and v solve the system of ODEs $x' = f$, $y' = g$. Can you do it? Explain why, or why not.

8. (*Sensitivity of Pendulum Motion*). Consider the pendulum system $x' = y$, $y' = -A \sin x - y$. Use a solver, a variety of initial points, and forward and backward solutions (in time) to give a complete analysis of what happens to the motion of the pendulum as the parameter A varies from close to zero to very large values. Justify all your assertions. Now, in a different direction altogether, turn to IVP (11), which involves the parameter α, and try to determine a value α_0 such that the solution curve tends to the equilibrium point $(\pi, 0)$ as $t \to +\infty$. What happens if the value of α is just slightly different from α_0? Interpret everything in terms of the motion of a pendulum.

 Sensitivity to changes in A.

Sensitivity to changes in α.

5.3 Models of Interacting Species

In the dynamics of life, no species is alone. The survival or the flourishing of a species rests on strategies of cooperation, competition, and predation. We can understand a small part of the biological universe by considering the interactions of just two species. We saw an example of competition in the combat models of Section 1.7, but we take a very different approach here. Let's begin with a general discussion of the model-building process for species interactions.[2]

Leah Edelstein-Keshet

---

[2] For additional information on mathematical modeling in the biological area consult the book *Mathematical Models in Biology* (New York: McGraw-Hill, 1988), an excellent introduction to the subject of modeling biological systems. The author Leah Edelstein-Keshet is a contemporary applied mathematician and researcher who is currently a mathematics professor at the University of British Columbia. Her latest research interests include molecular biology (specifically actin filament networks and their dynamics) and population phenomena such as aggregation and swarming in social organisms. Her advice to aspiring young scientists is to take as many mathematics courses as possible, to gain familiarity with computers, and to persevere on the often difficult but ultimately rewarding path to knowledge.

Building the Models

Suppose that $x(t)$ and $y(t)$ denote the population sizes of two species at time t. We also suppose that the two species form a community isolated from all other influences—not likely, but simplification is usually a first step in modeling. The Balance Law applies to each population:

☞ Remember the Balance Law from Section 1.4?

$$\text{Net rate of change of a population } = \text{ Rate in } - \text{ Rate out}$$

Let's suppose at first that migration into and out of the community is negligible, so the net rate of change is just the difference between the birth and death rates. These rates are proportional to the population size, so one model for this is the IVP

$$\begin{aligned} x' &= R_1 x, & x(0) &= x_0 \\ y' &= R_2 y, & y(0) &= y_0 \end{aligned} \tag{1}$$

where R_1 and R_2 are coefficients of proportionality and measure the contribution of the average individual of a species to the overall growth rate of that species. In the simplest case each coefficient is a constant, positive if the birth rate exceeds the death rate, negative the other way around. In this setting the populations grow or decay exponentially [e.g., $x(t) = x_0 e^{R_1 t}$]. Realistically, however, these coefficients aren't constant, but depend on time, population sizes, individual ages, and a host of other factors. We will average out the time variation in the coefficients, so the ODEs in (1) are autonomous.

Different choices for the coefficients R_1 and R_2 model different kinds of interactions. We will assume that R_1 and R_2 are continuously differentiable functions of x and y in a box S that includes the *population quadrant* $x \geq 0$, $y \geq 0$. This means that Theorem 5.2.1 applies, so IVP (1) has a unique solution for each choice of initial point $x_0 \geq 0$, $y_0 \geq 0$. This has an important consequence.

THEOREM 5.3.1

> Isolation of the Population Quadrant. If (x_0, y_0) is in the population quadrant, then the solution $x = x(t)$, $y = y(t)$ of IVP (1) lies entirely in the quadrant.

Here's the explanation. The x- and y-axes are composed of orbits of the ODEs in IVP (1). Reason: $x(t) = 0$, for all t, solves the first ODE of (1), so $x(t)$ and the solution $y(t)$ of the ODE $y' = R_2(0, y)y$ define orbits on the positive y-axis. Similarly, the positive x-axis is also composed of orbits of the system. The Separation of Orbits Theorem 5.2.2 implies that no orbit of IVP (1) can enter or leave the population quadrant, for to do that the orbit would have to cross one of the orbits on the x- or the y-axis. So if $x_0 > 0$, $y_0 > 0$, then the maximally extended solution of IVP (1) remains inside the quadrant.

The observations of real populations are consistent with much of what Theorem 5.3.1 says. However, there is still something to criticize on biological grounds. For example, the model ODEs (1) surely can't be correct for very small populations. At low population levels, real species often become extinct in finite time, although Theorem 5.3.1 implies that extinction, if it occurs at all, takes an infinitely long time.

Based on observations of many populations the following principle has been formulated by population biologists.

☞ This is a lot like the *Chemical Law of Mass Action*.

> Population Law of Mass Action.   The rate of change of one population due to interaction with another is proportional to the product of the two populations.

The constants of proportionality measure the effects of the interaction on the rates of change of the populations. For example, if we have the rate equations for two populations x and y

$$x' = -x + 0.0001xy, \qquad y' = y - 0.5xy$$

then mass-action interactions between x and y increase the rate of change of the x population a little (the term $+0.0001xy$) but are disastrous for the y population (the term $-0.5xy$). We can also use the law for interactions between individuals of the same species, so include terms like $-0.01x^2$ or $0.3x^2$, depending on whether interactions diminish or increase the rate of change.

The Population Law of Mass Action is not as rigorous as, say, Newton's force laws, but it does give us a way to convert species interactions into terms in a mathematical model.

Seven Types of Interaction

We mostly assume that R_1 and R_2 in IVP (1) are first-degree polynomials in x and y. The coefficients $a, b, \alpha, \ldots$ in the models below are nonnegative constants.

- *No Interaction*. The system where R_1 depends only on x, and R_2 only on y

$$\begin{aligned} x' &= R_1(x)x \\ y' &= R_2(y)y \end{aligned} \tag{2}$$

models a community in which the species have no effect on each other.

☞ Predator-prey dynamics are the topic of Section 5.4.

- *Predator-Prey*. The system

$$\begin{aligned} x' &= (-\alpha + by)x = -\alpha x + bxy \\ y' &= (\beta - cx)y = \beta y - cxy \end{aligned} \tag{3}$$

models a community of predators x and their prey y. The positive constants α and β are, respectively, *natural decay* and *natural growth coefficients*. The rate term $-\alpha x$ shows that without prey to eat, the predator population diminishes; the rate term βy shows that without predation, the prey population increases. The number of predator-prey encounters is assumed to be proportional to the product xy of the population of the two species, an assumption based on the Population Law of Mass Action. The coefficients b and $-c$ measure, respectively, predator efficiency in converting food into fertility, and the probability that a predator-prey encounter removes one of the prey.

- *Overcrowding.* Excess population may lower the rate coefficients R_1 and R_2 via predator-predator and prey-prey interactions governed by the Population Law of Mass Action. For example, too large a population may lead to overcrowding effects that lower the growth rate. If the effects of overcrowding on each of the species of a predator-prey model such as system (3) are taken into account via the Population Law of Mass Action, then we have the system

$$\begin{aligned} x' &= (-\alpha - ax + by)x \\ y' &= (\beta - cx - dy)y \end{aligned} \tag{4}$$

Overcrowding is modeled by the mass action terms $-ax^2$ and $-dy^2$.

- *Harvesting* If the x-species in system (1) is harvested at the rate H, the ODE for x becomes

$$x' = R_1 x - H \tag{5}$$

In one form of *seasonal harvesting*, the harvest rate is a periodic on-off function that is on for a time at the start of the year, then off for the rest of the year:

☞ Appendix B.1 has more about on-off functions, like $\text{sqw}(t, d, 1)$.

$$x' = R_1 x - H\,\text{sqw}(t, d, 1)$$

The function $H\,\text{sqw}(t, d, 1)$ is the square wave of period one year and constant amplitude H that is on for the first $d\%$ of each year and then off. For example, if the harvest season lasts for three months out of each year, then $d = 25$ (i.e., 25% of the period). In *constant effort harvesting*, however, the harvest rate is proportional to the population size:

$$x' = R_1 x - Hx \tag{6}$$

where H is a positive constant.

- *Cooperation.* Some biological interactions are of mutual benefit. An example is an ant-aphids community. Watch ants on a rose bush tending their aphids. The aphids feed off of the internal sap of the plant, and the ants dine on a fluid secreted by the aphids. As one part of the plant withers, the ants carry their "cows" to fresh "pasture." This kind of interaction is termed *cooperation* or *mutualism*. One model for cooperation based on the Population Law of Mass Action is

$$\begin{aligned} x' &= (\alpha + by)x \\ y' &= (\beta + cx)y \end{aligned} \tag{7}$$

where each species, if isolated, would have a natural, exponentially growing population (modeled by the terms αx and βy.) The mass action terms $+bxy$ and $+cxy$ show that each species promotes the growth of the other. Systems such as (7) can't model real populations for very long because explosive growth in x and y is unrealistic, but a slight modification may produce bounded populations (take a look at Example 5.3.2).

- *Competition.* Two species may compete for a resource in short supply. One model for competitive interaction with overcrowding is

$$x' = (\alpha - ax - by)x$$
$$y' = (\beta - cx - dy)y \tag{8}$$

The mass action terms $-ax^2$, $-bxy$, $-cxy$, and $-dy^2$ model the negative effects of large populations on growth rates.

☞ This model is analyzed in Section 9.3.

- *Satiation.* The predator's appetite may be satiated by a glut of prey. This phenomenon may be modeled by the following variant of predator-prey system (3):

$$x' = -\alpha x + \frac{b}{m + ky} \cdot xy$$
$$y' = \beta y - \frac{c}{m + ky} \cdot xy \tag{9}$$

The mass action coefficients contain a division by the positive term $m + ky$ and so tend to zero if the prey population y tends to infinity.

Now let's apply the power of numerical solvers to help us understand how the populations evolve. Keep in mind that in all of the models of interaction we have discussed so far, the x- and y-axes are composed of orbits of the system. So orbits which originate in the population quadrant remain in the population quadrant.

Computer Simulations

Concrete examples and their computer-generated orbits put some meat on the bones of abstract models. In each of the following models we begin with two populations that separately would evolve according to one or the other of the models

$$x' = (\alpha - ax)x, \quad \text{or} \quad x' = (-\alpha - ax)x$$
$$y' = (\beta - by)y, \quad \text{or} \quad y' = (-\beta - by)y$$

But coupling terms of the form $\pm cxy$ lead to very different scenarios, which we explore with computer simulations.

Just as in Example 5.2.4, a first step in the sign analysis of each model is to draw the nullclines (the dashed lines in the figures) and then determine the signs of x' and the signs of y' in the regions between the nullclines. This process tells us whether orbits rise ($y' > 0$) or fall ($y' < 0$), move to the left ($x' < 0$) or to the right ($x' > 0$) as time increases. For example, if x' is positive and y' is negative in a region, orbits must move to the right and downward.

EXAMPLE 5.3.1

Cooperation and Explosive Growth
The model ODEs

$$x' = (2 - x + 2y)x = (2 - x)x + 2xy$$
$$y' = (2 + 2x - y)y = (2 - y)y + 2xy \tag{10}$$

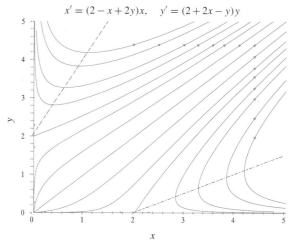

$$x' = (2 - x + 2y)x, \quad y' = (2 + 2x - y)y$$

$$x' = (4 - 2x + y)x, \quad y' = (4 + x - 2y)y$$

FIGURE 5.3.1 Cooperation leading to explosively growing populations (Example 5.3.1).

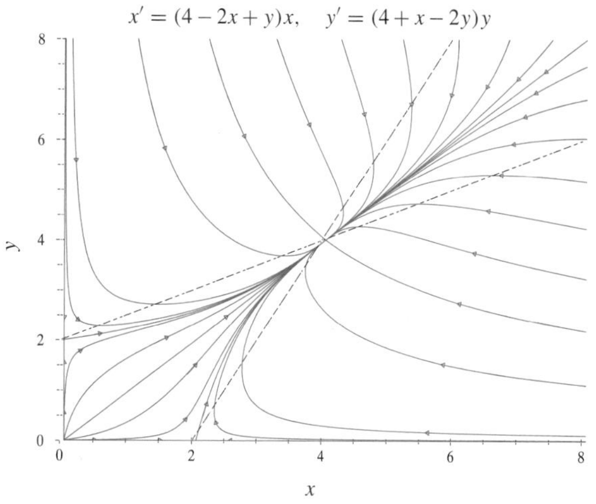

FIGURE 5.3.2 Cooperation leading to stability (Example 5.3.2).

represent mutually beneficial interactions (the terms $+2xy$) and self-limitation (the terms $-x^2$ and $-y^2$).

The nullclines for the x-species are defined by $(2 - x + 2y)x = 0$ and consist of the y-axis (i.e., $x = 0$) and the line through the point $(2, 0)$ of slope $1/2$ (i.e., $2 - x + 2y = 0$). Similarly, the nullclines for the y-species are defined by $(2 + 2x - y)y = 0$ and consist of the x-axis and the line through the point $(0, 2)$ of slope 2. See Figure 5.3.1 for the nullclines (the dashed lines and the coordinate axes). Three of the equilibrium points are on the edge of the population quadrant, $(0, 0)$, $(2, 0)$, $(0, 2)$; but the fourth, $(-2, -2)$ is outside the quadrant.

☞ In this context, a separatrix is an orbit that separates orbits with one kind of behavior as $t \to -\infty$ (or as $t \to +\infty$) from orbits with another kind of behavior.

Figure 5.3.1 shows some of the orbits in the population quadrant $x \geq 0, \quad y \geq 0$. It certainly appears that the populations multiply explosively. Note the separatrix emerging from $(2, 0)$ and another from $(0, 2)$.

System (10) can't be valid for a real species because the populations get too large too fast. So let's improve the model.

EXAMPLE 5.3.2   **Stable Cooperation**

The model and the population can be stabilized by raising the costs of overcrowding and lowering the benefits of cooperation:

$$\begin{aligned} x' &= (4 - 2x + y)x = (4 - 2x)x + xy \\ y' &= (4 + x - 2y)y = (4 - 2y)y + xy \end{aligned} \tag{11}$$

For each species, the increase in the overcrowding coefficient from 1 to 2 and the decrease in the mutualism coefficient from 2 to 1 generate a model that is stable and a lot more realistic than the unstable growth model of system (10). These coefficient changes alter the slopes of the nullclines and move their intersection (which is an equilibrium point) inside the population quadrant. The increase in the natural growth coefficient for each species from the value 2 in system (10) to the value 4 in system (11) shifts the location of the new nullclines (dashed lines in Figure 5.3.2) and pushes the equilibrium further inside the quadrant to the point $(4, 4)$. As in Example 5.3.1, separatrices come out of the equilibrium points $(2, 0)$ and $(0, 2)$ as t increases from $-\infty$, but now the separatrices move toward the equilibrium point $(4, 4)$.

Figure 5.3.2 shows that the internal population orbits are pulled into a pair of wedge-shaped regions (bounded by the dashed nullclines) whose apex is at the internal equilibrium point. The graphs give compelling visual evidence that these cooperating species tend to a stable state of coexistence.

Can populations stabilize and survive if both species compete for resources in limited supply? The next two examples give opposite answers to this question.

EXAMPLE 5.3.3

Stable Competition

Suppose that two species share the same habitat and compete with one another for scarce resources. One kind of competition dynamics is modeled by

$$x' = (2 - 2x - y)x = (2 - 2x)x - xy$$
$$y' = (2 - x - 2y)y = (2 - 2y)y - xy \tag{12}$$

The x-nullclines are the y-axis and the line $2 - 2x - y = 0$, while the y-nullclines are the x-axis and the line $2 - x - 2y = 0$ (dashed lines in Figure 5.3.3). The x-nullclines intersect the y-nullclines in the equilibrium points $(0, 0)$, $(1, 0)$, $(0, 1)$, and $(2/3, 2/3)$. The orbits of Figure 5.3.3 are drawn into wedge-shaped regions bounded by the dashed nullclines, and are attracted by the equilibrium point $(2/3, 2/3)$ at the apex. Separatrices emerge from the equilibrium points $(1, 0)$ and $(0, 1)$ and head toward $(2/3, 2/3)$.

☞ Could competition work better than cooperation?

It appears from Figure 5.3.3 that the species tend to a stable, attracting equilibrium.

Competition isn't always stable. Here's an example where it leads to extinction.

EXAMPLE 5.3.4

Competitive Exclusion

Suppose that the slanting nullclines (dashed lines) in Figure 5.3.3 are exchanged, the x-nullcline becoming a y-nullcline, and vice versa. The new model of competition is

$$x' = (2 - x - 2y)x = (2 - x)x - 2xy$$
$$y' = (2 - 2x - y)y = (2 - y)y - 2xy \tag{13}$$

Sign analysis and the interchanged position of the dashed nullclines suggest a surprising change in the ultimate fate of each species. Referring to Figure 5.3.4, we see that some population orbits are attracted into the wedge between the nullclines at the lower right; these orbits then approach the point $(2, 0)$ of x-species equilibrium and y-species extinction. Other orbits are attracted into the wedge-shaped region at the upper left and approach the point $(0, 2)$ of x-extinction and y-equilibrium.

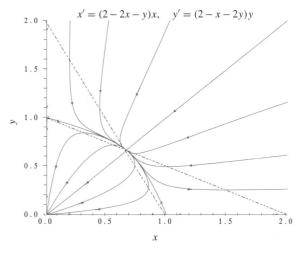

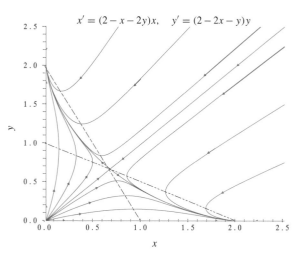

FIGURE 5.3.3 Orbits of stable competition (Example 5.3.3).

FIGURE 5.3.4 Orbits of competitive exclusion (Example 5.3.4).

☞ As in Section 4.1, separatrices divide regions of very different orbital behavior.

It can be shown that exactly two separatrices tend to the internal equilibrium point $(2/3, 2/3)$ as $t \to +\infty$; other orbits may move near the point, but they always turn away and tend to $(2, 0)$ or to $(0, 2)$ as time advances. The two separatrices lie along the line $y = x$ and divide the population quadrant into a region (above the separating line) that is favorable to the y-species but devastating to the x-species, and a complementary region that favors the x-species. This feature is called the *Principle of Competitive Exclusion* because the population quadrant is split into two regions in each of which one of the species thrives but the other dies out. Finally, note that there seem to be separatrices leaving $(2/3, 2/3)$ as time advances; one heads toward $(2, 0)$, the other toward $(0, 2)$.

These computer simulations give us some insight into the effects of interactions on the populations.

Comments

The interaction models introduced in this section should be critically examined:

- Are they good models of the changing populations of real species?
- What conclusions are possible through sign analysis and computer simulations?
- Can these conclusions be verified by mathematical arguments?

We have answered the second question for a few specific model systems. The first and third questions are addressed for predator-prey interactions in Section 5.4. Several examples in Chapters 8 and 9 take up some of the mathematical issues raised here, questions of stability, for example.

None of the models have the explicit predictive character of, say, the models for the motion of a pendulum. However, ever since the models were introduced in the 1920s and 1930s, they have directed the way scientists and mathematicians think about species interactions. Studies of population biology often begin with these models, if only to refute and to replace them by something better.[3]

PROBLEMS

1. Explain the biological dynamics; identify predators, prey, competitors, cooperators; explain population behavior in terms of the rate functions; and identify harvesting and restocking terms and their nature. The constants $\alpha, \beta, \gamma, a, b, c, d, H, k, m$ are all positive.

   (a) $x' = (\alpha - by)x, \quad y' = (\beta - cx)y + H$

   (b) $x' = (\alpha - ax - by)x, \quad y' = (\beta - cx - dy)y$

   (c) $x' = (\alpha + by)x, \quad y' = (-\beta + cx - dy)y$

   (d) $x' = (\alpha - ax - by)x, \quad y' = (-\beta + cx - dy)y + (2 + \cos t)$

   (e) $x' = (\alpha + by)x - H \operatorname{sqw}(t, 25, 1), \quad y' = (\beta + cx - dy)y$

   (f) $x' = (\alpha - ax)x, \quad y' = \left(-\beta + \dfrac{cx}{m + kx}\right)y + H$

   (g) $x' = (\alpha - bz)x, \quad y' = (\beta - my - kz)y, \quad z' = (-\gamma + ax + cy)z - Hz$

2. Find the equations of the x-nullclines and the y-nullclines for each system and sketch their graphs by hand in the population quadrant. Find the equilibrium points. Determine the direction of orbital motion on or across the nullclines as time advances. Identify the nature of the x- and of the y-species (e.g., predator, prey, cooperator, competitor). If initially the populations are at a nonequilibrium point inside the population quadrant, what happens to $x(t)$ and to $y(t)$ as time increases?

   (a) $x' = (5 - x + y)x, \quad y' = (10 + x - 5y)y$ [*Hint*: See Example 5.3.2.]

   (b) $x' = (10 - x + 5y)x, \quad y' = (5 + x - y)y$ [*Hint*: See Example 5.3.1.]

   (c) $x' = (5 - x - y)x, \quad y' = (10 - x - 2y)y$

   (d) $x' = (10 - x - 5y)x, \quad y' = (5 - x - y)y$ [*Hint*: See Example 5.3.4.]

 3. (a)–(d). Use a computer solver to sketch the orbits, direction fields, and nullclines of the systems in Problem 2(a)–(d). Describe what you see in terms of the long-term behavior of each species.

4. (*Models of Cooperation*). In the cooperation model with self-limiting terms $x' = (2 - x + 2y)x$, $y' = (2 + 2ax - y)y$, the positive parameter a can be adjusted to "steer" the system.

   (a) Discuss the meaning of the coefficient $2a$ in population terms.

 (b) Find the critical value a_0 of a that divides the explosive growth model of Example 5.3.1 from a stable model in which both populations tend to an equilibrium inside the population quadrant as time increases. Plot orbits and a direction field in the population quadrant for a value of $a < a_0$. Repeat with $a = a_0$, and then with a value of $a > a_0$. [*Hint*: Find the maximal value a_0 of a such that if $0 < a < a_0$, then there is an equilibrium point inside the population quadrant.]

---

[3]A basic reference in mathematical biology is the encyclopedic volume (and advanced text) by the renowned mathematical biologist J. D. Murray, *Mathematical Biology* (New York: Springer-Verlag, 1989).

www **5.** (*Knocking Out the Competition*). The system $x' = (2 - x - 2y)x + Hx$, $y' = (2 - 2x - y)y$ models competition where the x-species can be restocked ($H > 0$) at a rate proportional to the population. Using a numerical solver to plot solution curves, find the restocking coefficient H of minimal magnitude H_0 so that the y-species becomes extinct regardless of its initial population. Plot orbits for values of H below, at, and above H_0 and explain what you see. [*Hint*: Find the value H_0 of the restocking coefficient that has the property that as H increases through H_0, the equilibrium point inside the population quadrant exits across an axis.]

 6. (*Competitive Exclusion*). In the competition model $x' = (\alpha - ax - by)x$, $y' = (\beta - cx)y$ only the x species has a self-limitation term. Regardless of the values of the positive coefficients, α, β, a, b, c, the Principle of Competitive Exclusion applies (see Example 5.3.4). Explain why. Plot orbits for your choices of values for the coefficients. Explain what you see. [*Hint*: Consider the two cases $\alpha/a > \beta/c$ and $\alpha/a \leq \beta/c$.]

 7. (*A Competition Model: Rescaling*). Suppose that you belong to the x-species and are able to change the rate constants in the x' equation of system (8). Analyze the effects on your competitor of changing each of your rate constants. First, however, rescale the system by letting $x = ku$, $y = mv$, and $t = nT$ and then choosing the positive scaling constants k, m, and n so that the new rate equation for your competitor is $dv/dT = (1 - u - v)v$. Your rescaled rate equation is $du/dT = (\alpha^* - a^*u - b^*v)u$. Explain what happens as you tune the system by changing α^*, a^*, and b^* in turn. What values of these parameters might be both realistic and optimal for your species? Plot and interpret the orbits.

☞ Rescaling cuts the number of system parameters from 6 to 3. More on rescaling in Appendix B.6.

5.4 Predator-Prey Models

A struggle for existence inevitably follows from the high rate at which all organic beings tend to increase. Every being, which during its natural lifetime produces several eggs or seeds, must suffer destruction during some period of its life, and during some season or occasional year; otherwise, on the principle of geometrical increase, its numbers would quickly become so inordinately great that no country could support the product. Hence, as more individuals are produced than can possibly survive, there must in every case be a struggle for existence, either one individual with another of the same species, or with the individuals of distinct species, or with the physical conditions of life. It is the doctrine of Malthus applied with manifold force to the whole animal and vegetable kingdoms; for in this case there can be no artificial increase of food, and no prudential restraint from marriage. Although some species may be now increasing, more or less rapidly, in numbers, all cannot do so, for the world would not hold them. . . .

The amount of food for each species of course gives the extreme limit to which each can increase; but very frequently it is not the obtaining food, but the serving as prey to other animals, which determines the average numbers of a species.[4]

Charles Darwin

In the 1920s and 1930s, Vito Volterra and Alfred Lotka independently reduced Darwin's predator-prey interactions to mathematical models. It is these models that are presented in this section.

---

[4]Charles Darwin, "Struggle for Existence," *The Origin of Species*, new ed., Chap. 3 (from 6th English ed.) (New York: Appleton, 1882).

Predator and Prey

The simplest model of predator and prey association includes only natural growth or decay and the predator-prey interaction itself. All other relationships are assumed to be negligible. We will assume that the prey population grows exponentially in the absence of predation, while the predator population declines exponentially if the prey population is extinct. The predator-prey interaction is modeled by mass action terms proportional to the product of the two populations. The model for the predator and prey populations is the *predator-prey* (or *Lotka-Volterra*) *system*

$$x' = (-a + by)x = -ax + bxy$$
$$y' = (c - dx)y = cy - dxy$$

where x is the predator population and y the prey population, and the rate constants a, b, c, d are positive. The linear rate terms $-ax$ and cy model the natural decay and growth, respectively, of the predator and the prey if each were isolated from the other (so y no longer is the food supply for x). The quadratic terms $+bxy$ and $-dxy$ model the effects of interaction on the rates of change of the two species: food promotes the predator population's growth rate, while serving as food diminishes the prey population's growth rate.

The coordinates of the equilibrium points are found by simultaneously solving the equations for the nullclines, $-ax + bxy = 0$, $cy - dxy = 0$. The equilibrium points are the origin, and the point $(c/d, a/b)$, which is inside the population quadrant. Here is a specific example of a predator-prey system.

EXAMPLE 5.4.1

Predator-Prey Orbits and Component Graphs

Figures 5.4.1 and 5.4.2 show orbits and component graphs of the system

$$x' = -x + xy/10, \quad y' = y - xy/5$$

with equilibrium points $(0, 0)$ and $(5, 10)$. The numerically computed orbits of Figure 5.4.1 through the points $(5, 20)$, $(5, 30)$, and $(5, 40)$ seem to be closed curves (i.e., cycles) that turn clockwise around the equilibrium point $(5, 10)$. The component graphs of Figure 5.4.2 suggest that the solutions are indeed periodic, that predator peaks lag behind prey peaks (is that what you would expect?), and that the period of an orbit increases with the y-amplitude, where the y-amplitude of an orbit is measured upward from the equilibrium point $(5, 10)$.

The computational results are simple, interesting, and curious, but do they have anything to do with the populations of real species?

The Adriatic Fisheries and the First World War

In 1926, Humberto D'Ancona, an Italian biologist, completed a statistical study of the changing populations of various species of fish in the northern reaches of the Adriatic Sea. His estimates of the populations during the years 1910 to 1923 were based on the

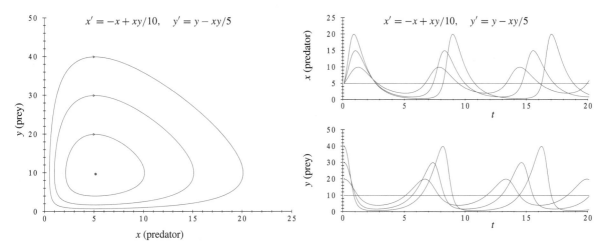

FIGURE 5.4.1 Some closed predator-prey system orbits (Example 5.4.1).

FIGURE 5.4.2 The component graphs for orbits of Figure 5.4.1 (Example 5.4.1).

TABLE 5.4.1 Percentages of Predator Species in the Total Fish Catch

Port	1914	1915	1916	1917	1918	1919	1920	1921	1922	1923
Fiume	12%	21%	22%	21%	36%	27%	16%	16%	15%	11%
Trieste	14%	7%	16%	15%	—	18%	15%	13%	11%	10%

numbers of each species sold on the fish markets of the three ports Trieste, Fiume, and Venice. D'Ancona assumed, as we will, that the numbers of the various species in the markets reflected the relative abundance of the species in the Adriatic. A part of the data is given in Table 5.4.1.

As often happens, the data do not provide overwhelming support for any particular theory of changing fish populations. D'Ancona observed, however, that the percentages of predator species were generally higher during and immediately after World War I (1914–1918). Fishing was drastically curtailed during the war years as fishermen abandoned their nets to fight in the war, and D'Ancona concluded that the decline in fishing caused the change in the proportions of predator to prey. He formulated the hypothesis that during the war the predator-prey community was close to its natural state of a relatively high proportion of predator fish, while the more intensive fishing of the prewar and postwar years disturbed that equilibrium to the advantage of the prey species. Unable to give a reason for the phenomenon, D'Ancona asked his father-in-law, the noted Italian mathematician Vito Volterra (1860–1940), if there was a mathematical model that might cast some light on the matter. Within a few months, Volterra had outlined a series of models for the interactions of two or more species.

System (1) is the simplest of Volterra's models.[5]

Volterra's Laws

Volterra summarized his conclusions about the solutions and orbits of the system

$$x' = -ax + bxy$$
$$y' = cy - dxy$$
(1)

in the form of three laws. We suppose throughout that a, b, c, d are positive constants.

THEOREM 5.4.1

> The Law of the Periodic Cycle. The fluctuations of the populations of the predator and its prey are periodic. The period depends on the values of the rate coefficients of system (1) and on the initial data. The period increases with the amplitude of the corresponding cycle.

Here is how we can find an equation for the cycles. Divide the second rate equation of system (1) by the first and obtain

$$\frac{dy}{dx} = \frac{cy - d^*xy}{-ax + bxy} = \left(\frac{y}{-a + by}\right)\left(\frac{c}{x} - d^*\right)$$
(2)

where we have temporarily replaced the interaction coefficient d by d^* to avoid confusion with the symbol d in the differential dx. Separating the variables in ODE (2) we have that

$$\left(\frac{c}{x} - d^*\right) dx + \left(\frac{a}{y} - b\right) dy = 0$$
(3)

Antidifferentiating each term in (3), we see that an equation for the orbits in the population quadrant is

$$(c \ln x - d^*x) + (a \ln y - by) = C$$
(4)

where C is a constant. If we want the equation of the orbit through the point (x_0, y_0), where x_0 and y_0 are positive, then we evaluate C as follows:

$$(c \ln x_0 - d^*x_0) + (a \ln y_0 - by_0) = C$$
(5)

Vito Volterra

[5]Volterra eventually wrote a book on his theories, *Leçons sur la théorie mathématique de la lutte pour la vie* (Paris: Gauthier-Villars, 1931; reproduced by University Microtexts, Ann Arbor, Mich., 1976). The Soviet biologist G. F. Gause (or Gauze) tested Volterra's theories by carrying out numerous laboratory experiments with various competing and predatory microorganisms. He published his results in *The Struggle for Existence* (Baltimore: Williams & Wilkins, 1934: reissued by Hafner, New York, 1964). D'Ancona defended Volterra's work in a book that again used the memorable phrase of Malthus and Darwin, *The Struggle for Existence* (Leiden: E. J. Brill, 1954). A. J. Lotka, an American biologist and, later in life, an actuary, arrived at many of the same conclusions independently of Volterra; see his book, *Elements of Physical Biology* (Baltimore: Williams & Wilkins, 1925); reprinted as *Elements of Mathematical Biology* (New York: Dover, 1956).

Let's write the equation for the orbits in terms of exponentials instead of logarithms. To do this, exponentiate each side of (4), using the facts that $e^{A+B} = e^A e^B$, $e^{\alpha \ln \beta} = e^{\ln(\beta^\alpha)} = \beta^\alpha$, and C is given in (5) :

$$\left(e^{c \ln x - d^* x}\right)\left(e^{a \ln y - by}\right) = \left(e^{c \ln x_0 - d^* x_0}\right)\left(e^{a \ln y_0 - by_0}\right)$$

$$\left(e^{c \ln x} e^{-d^* x}\right)\left(e^{a \ln y} e^{-by}\right) = \left(e^{c \ln x_0} e^{-d^* x_0}\right)\left(e^{a \ln y_0} e^{-by_0}\right)$$

$$\left(x^c e^{-dx}\right)\left(y^a e^{-by}\right) = \left(x_0^c e^{-dx_0}\right)\left(y_0^a e^{-by_0}\right) \tag{6}$$

where we have replaced d^* by d in the last equation. Formula (6) defines a simple closed curve (i.e., a cycle) for each $(x_0, y_0) \neq (c/d, a/b)$ inside the population quadrant (see Problem 6), and that means that the corresponding solution $x = x(t)$, $y = y(t)$ of system (1) is indeed periodic. The computed orbits of Figure 5.4.1 show some of these cycles.

Remarkably, the average population of each species over any of the cycles is a fixed constant. This is Volterra's second law.

THEOREM 5.4.2

> The Law of Averages. In system (1), the average predator and prey populations over the period of a cycle are, respectively, c/d and a/b.

Let's see why the Law of Averages is true. Suppose that $x = x(t)$, $y = y(t)$ is a nonconstant solution that defines a cycle of period T. The average populations $\bar{x}$ and $\bar{y}$ over one period are defined to be

$$\bar{x} = \frac{1}{T} \int_0^T x(t)\, dt, \qquad \bar{y} = \frac{1}{T} \int_0^T y(t)\, dt \tag{7}$$

Let's show that $\bar{x} = c/d$. First rearrange the terms of the second rate equation of (1) to obtain

$$x(t) = \frac{c}{d} - \frac{1}{d}\frac{y'(t)}{y(t)} \tag{8}$$

Then integrate each side of (8) from 0 to T, divide by T, and use (7):

$$\bar{x} = \frac{1}{T}\int_0^T x(t)\, dt = \frac{1}{T}\int_0^T \frac{c}{d}\, dt - \frac{1}{T}\int_0^T \frac{1}{d}\frac{y'(t)}{y(t)}\, dt$$

$$= \frac{c}{d} - \frac{1}{T}\cdot\frac{\ln y(T) - \ln y(0)}{d} = \frac{c}{d}$$

because $y(T) = y(0)$. A similar argument shows that $\bar{y} = a/b$, and the validity of the Law of Averages is proved.

The Law of Averages brings us closer to the actual data given earlier for the fish catches, because that data consists of averages. But we still haven't put the fishermen into the picture.

Why Harvesting Hurts the Predator and Helps the Prey

Volterra's third law explains what happens when the two species are harvested. The simplest model is that of constant-effort harvesting, in which the amount caught per unit of time is proportional to the population:

$$x' = -ax + bxy - H_1 x = (-a - H_1 + by)x$$
$$y' = cy - dxy - H_2 y = (c - H_2 - dx)y \tag{9}$$

The nonnegative numbers H_1 and H_2 are the harvesting coefficients. When harvesting occurs, the equilibrium point lying inside the population quadrant shifts to the left and upward from $x = c/d$, $y = a/b$ to the point

$$x = (c - H_2)/d, \qquad y = (a + H_1)/b \tag{10}$$

It is assumed that $H_2 < c$. Otherwise, the heavy harvesting of the prey species y doesn't leave enough food for the predator, so the predator species heads toward extinction [modeled by a nonpositive x-coordinate for the equilibrium point given in (10)].

By the Law of Averages the population averages around any cycle are given by the coodinates of the equilibrium point. Since harvesting causes the equilibrium point to move up and to the left of the original position in the population quadrant, harvesting raises the average prey population but lowers the average predator population.

THEOREM 5.4.3

> The Law of Harvesting. Constant-effort harvesting raises the average number of prey per cycle and lowers the average number of predators.

Before the data of Table 5.4.1 can be compared with the predictions of Volterra's Law of Harvesting, that law must be reformulated in terms of percentages, rather than averages.

THEOREM 5.4.4

> The Percentage Law of Harvesting. Constant-effort harvesting raises the average percentage of prey per cycle in the total fish population and lowers the average percentage of predators per cycle.

The verification of the Theorem 5.4.4 is left to the reader (Problem 4).

If the harvesting coefficients in system (9) are too large, the internal equilibrium point $((c - H_2)/d, (a + H_1)/b)$ crosses the positive y-axis, and one (or both) species becomes extinct, as we see in the next example.

EXAMPLE 5.4.2

The Effect of Constant Effort Harvesting

Let's consider the IVP with equal harvesting coefficients $H_1 = H_2 = H$:

$$x' = -x + xy/10 - Hx, \qquad x(0) = 8$$
$$y' = y - xy/5 - Hy, \qquad y(0) = 16 \tag{11}$$

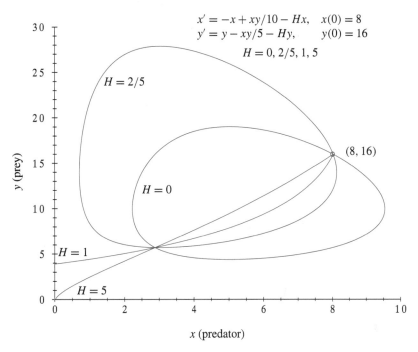

$$x' = -x + xy/10 - Hx, \quad x(0) = 8$$
$$y' = y - xy/5 - Hy, \quad y(0) = 16$$
$$H = 0, 2/5, 1, 5$$

FIGURE 5.4.3 The effects of harvesting (Example 5.4.2).

Figure 5.4.3 shows orbits for IVP (11) with four different values for the harvesting coefficient: $H = 0$ (no harvesting), 2/5 (light harvesting), 1 (critical harvesting), and 5 (heavy harvesting). Two of the orbits in Figure 5.4.3 suggest that with no harvesting or with only light harvesting both species survive. But if the harvesting is critical ($H = 1$), only the prey species survives. If H exceeds 1, then both predator and prey become extinct.

Volterra's Percentage Law of Harvesting answers D'Ancona's question. A *decrease* in the harvesting rate leads to an *increase* in the predator percentage, while an increase in the harvesting rate causes a decrease in the predator percentage. Now look again at Table 5.4.1. As the harvesting rate fell during the war years and then increased during the postwar years, the predator percentage increased and then dropped, just as Volterra's model predicts.

Validity of Volterra's Law of Harvesting

Volterra's models have been both challenged and supported many times in the years since their formulation, but the model continues to be the starting point for most serious attempts to understand just how predator-prey communities evolve with or without harvesting.

One dramatic confirmation of the general validity of Volterra's Law of Harvesting occurred when the insecticide DDT was applied to control the cottony cushion scale insect that infested American citrus orchards. The scale insect had been accidentally introduced from Australia in 1868. Its numbers were controlled (but not eliminated) by importation of the insect's natural predator, a particular kind of ladybird beetle. When DDT was first introduced as the harvesting agent, it was hoped that the scale insect could be completely wiped out. But DDT acts indiscriminately, killing all insects it touches. The consequence was that the numbers of the ladybirds dropped, while the population of the scale insects, freed from extensive ladybird predation, increased.

Comments

Darwin's observation that it is "the serving as prey to other animals which determines the average numbers of a species" is supported by Volterra's model and by the data of the fish catches in the Adriatic. The model has its flaws: No account is taken of the delay in time between an action and its effect on population numbers, the averaging over all categories of age, fertility, and sex is dubious, and the parameters of the model don't change over time. Nonetheless, Occam's Razor applies: "What can be accounted for by fewer assumptions is explained in vain by more."

PROBLEMS

1. (*Lotka-Volterra Systems*). Which is the predator and which is the prey? Find the average predator and prey populations. Do the population cycles turn clockwise around the equilibrium point in the first quadrant? [*Hint*: Use Theorem 5.4.2. To determine cycle orientation, find the sign of x' if $x > 0$ and $y = 0$, and the sign of y' if $y > 0$ and $x = 0$.]

    (a) $x' = -x + xy$, $\quad y' = y - xy$ (b) $x' = 0.2x - 0.02xy$, $\quad y' = -0.01y + 0.001xy$

    (c) $x' = (-1 + 0.09y)x$, $\quad y' = (5 - x)y$

 2. (*Estimating the Periods of Cycles*). (a)–(c) Plot the equilibrium point and the cycles that pass through the points $(5, 10)$ and $(10, 5)$ for the systems of Problem 1. Plot component graphs for the orbits, and use these graphs to estimate periods.

www 3. (*Linearizing to Estimate the Periods of Cycles*). To linearize a planar system $x' = f(x, y)$, $y' = g(x, y)$ about an equilibrium point (x_0, y_0), expand each rate function in a Taylor series about (x_0, y_0) and discard all terms higher than first order. The result is called the *linearization* of the original system at (x_0, y_0). Let's apply this process to the predator-prey rate equations $x' = -ax + bxy$, $y' = cy - dxy$ in order to estimate the periods of the population cycles near the equilibrium point $(c/d, a/b)$ inside the population quadrant.

    (a) Show that the system of linearized rate equations for the predator-prey system is $x' = bcy/d - ac/d$, $y' = -adx/b + ac/b$.

    (b) Show that for certain values of ω and A/B, $x = c/d + A\cos\omega t$, $y = a/b + B\sin\omega t$ solves the linearized system in part (a). What are these values?

    (c) Set $a = b = c = d = 1$ and plot orbits and component graphs of the nonlinear and the linearized systems, using the common initial data $x_0 = 1$, $y_0 = 1, 1.1, 1.3, 1.5, 1.9$. Plot over $0 \le t \le 20$. Explain the graphs and compare the periods of the cycles of the two systems using a common initial point.

4. (*Law of the Percentages*). Verify Theorem 5.4.4 for the predator percentages. [*Hint*: The harvested system is $x' = -ax + bxy - H_1x$, $y' = cy - dxy - H_2y$, with $H_2 < c$. Explain why the predator fraction F of the total average catch is

$$F = \left[1 + \frac{d(a + H_1)}{b(c - H_2)}\right]^{-1}$$

Explain why F decreases as H_1 and H_2 increase.]

5. (*Harvesting a Predator-Prey Community to Extinction*). High harvesting coefficients H_1 and H_2 may lead to extinction of the species. This problem explores what happens.

(a) Explain the graphs in Figure 5.4.3.

(b) Set H_1 at a fixed positive value in system (9). Show that as $H_2 \to c^-$, the equilibrium point approaches the point $(0, (a + H_1)/b)$ on the y-axis and that if $H_2 = c$, all points on the y-axis are equilibrium points of system (9).

(c) Let $H_2 = c$ in system (9). Show that $dy/dx = (dxy)((a + H_1)x - bxy)^{-1}$. Separate the variables, solve, and find x as a function of y and the initial data x_0, y_0, where $x_0 > 0$, $y_0 > 0$.

 (d) What do you think happens to the harvested species as time increases for the case $H_2 = c$? Give an informal explanation, but with reasons for your conclusions. As an aid, set $a = b = c = d = 1$, $H_1 = 0.1$, and plot orbits.

6. (*Orbits of the Lotka-Volterra System Are Cycles*). Follow the steps below to show that for $0 < K \leq K_0 = x_0^c e^{-dx_0} y_0^a e^{-by_0}$ the graph of the orbit defined by equation (6) is a simple closed curve (i.e., a cycle) in the interior of the population quadrant, unless $K = K_0 = a^a(be)^{-a}c^c(de)^{-c}$ when the graph is the equilibrium point $(c/d, a/b)$.

(a) Show that $f(x) = x^c e^{-dx}$ is defined for $x \geq 0$, rises from the value of 0 at $x = 0$ to its maximum value of $M_1 = c^c(de)^{-c}$ at $x = c/d$, falls as x increases beyond c/d, and $f(x) \to 0$ as $x \to +\infty$. Show that the function $g(y) = y^a e^{-by}$ has the same properties, but its maximum value is $M_2 = a^a(be)^{-a}$ attained at $y = a/b$.

(b) Show that no nonnegative values of x and y satisfy equation (6) if $K > K^*$, where $K^* = M_1M_2$. Show that (6) has the unique solution $x = c/d$, $y = a/b$ if $K = K^*$.

(c) Suppose that γ is any positive number, $\gamma < M_1$. Show that $f(x) = \gamma$ has two solutions, x_1 and x_2, where $x_1 < c/d < x_2$. Show that $g(y) = \gamma M_2/f(x)$ has no solution y if $x > x_2$ or $x < x_1$; exactly one solution, $y = a/b$, if $x = x_2$ or if $x = x_1$; and two solutions, $y_1(x) < a/b$ and $y_2(x) > a/b$, if $x_1 < x < x_2$. Show that $y_1(x) \to a/b$ and $y_2(x) \to a/b$ if $x \to x_1$ or $x \to x_2$.

(d) Explain why equation (6) defines a cycle inside the population quadrant if $0 < K < K_0$.

7. (*Harvesting Strategies*). Suppose that system (9) models a predator-prey system and that you are the harvester. Suppose that it is required that $0 < x_m \leq x(t) \leq x_M$ and $0 \leq y_m \leq y(t) \leq y_M$ for all t, where the population bounds x_m, x_M, y_m, y_M are given. Suppose also that $x(0) = c/d$, $y(0) = a/b$. Describe how you would choose the positive coefficients H_1 and H_2 to maintain each species within the prescribed bounds. Justify your arguments.

High harvesting rates can be maintained if the harvesting season is short. Construct and justify your own strategy for maximizing the yield by imposing a limit on the harvest season, while still maintaining the populations within reasonable bounds.

5.5 The Possum Plague: A Model in the Making

The ecological balance in New Zealand has been disturbed by the introduction of the Australian possum, a marsupial the size of a domestic cat with the proper name *Trichosurus vulpecula*. The animal was introduced in the 1830s for a planned fur trade. But there are no natural predators in the forests of New Zealand, and the possum population rapidly increased. Today the estimated population is 70 million, and only a few areas are possum-free. The animals have become a reservoir of a type of tuberculosis with about half of the possum population infected. This poses a threat of transfer of the disease to the livestock that form an important part of the New Zealand economy. An intense effort is under way to understand possum ecology and the disease. This effort includes the construction of mathematical models for the possum/disease dynamics. The simplest of these models is treated here. Our approach will be somewhat telegraphic, and we leave the detailed analysis to the reader.[6]

Modeling the Population/Disease Dynamics

Suppose that P is the population of possums and I is the subpopulation of infected possums, both measured in units of tens of millions. Note that $P \geq I$ and that the population of healthy possums is $P - I$. The simplest dynamical model for $P(t)$ and $I(t)$ is

$$P' = (a - b)P - \alpha I$$
$$I' = \beta I(P - I) - (\alpha + b)I \tag{1}$$

where time t is measured in years, a and b are the respective natural birth and death rate coefficients measured in units of year^{-1}, and α is the disease-induced death rate coefficient in year^{-1} units. The number β [measured in units of $(10^7 \cdot \text{year})^{-1}$] is the mass action coefficient that measures the effectiveness of interactions in transmitting the disease from the population I of the diseased to the population $P - I$ of healthy animals. Once a possum contracts tuberculosis, it never recovers, so there is no subpopulation of "recovered."

This model is only a crude approximation to reality and neglects such factors as patchiness (i.e., spatial variations in the population levels) and differences due to age

---

[6]This section is adapted from the note "Percy Possum Plunders" by Graeme Wake, Professor of Applied Mathematics at the University of Aukland, New Zealand. The note appeared in the newsletter CODEE (Winter 1995). A detailed paper, "Thresholds and Stability Analysis of Models for the Spatial Spread of a Fatal Disease," that includes other mathematical models of the possum population dynamics, was written by Wake and his colleagues K. Louie and M. G. Roberts and appeared in *IMA Journal for Mathematics Applied to Medicine and Biology* **10**, (1993), pp. 207–226. Wake is a distinguished mathematician who enjoys applying mathematics to "real" problems like the possum problem discussed here. He has applied mathematical techniques to help solve the mysteries of why piles of wool being shipped overseas frequently begin to smolder in the boat holds, and why fish-and-chips shops in Australia suddenly catch fire. An excellent general reference on mathematical models for the spread of disease is an article by another distinguished applied mathematician, H. W. Hethcote, "Qualitative Analysis of Communicable Disease Models," *Math. Biosci.* **28** (1976), pp. 335–356.

structure in the possum population. But the model is surprisingly effective in giving a macroscopic picture of the situation and in giving some indication of how much effort is going to be needed to change the long-term outcome.

Nonconstant solutions of system (1) can't be expressed in terms of elementary functions, so we must rely on mathematical theory and computer simulation to understand solution behavior.

Scaling the Variables

System (1) has four parameters, a, b, α, and β. We can scale the populations P and I and time t to reduce the number of parameters before we do computer simulations:

$$P = hx, \qquad I = jy, \qquad t = ks \tag{2}$$

☞ More on scaling in Appendix B.6.

where h, j, and k are positive scaling constants which are to be determined. From system (1), the variable relations given in (2), and the Chain Rule we have

$$\frac{dP}{dt} = \frac{dP}{dx}\frac{dx}{ds}\frac{ds}{dt} = \frac{h}{k}\frac{dx}{ds} = (a-b)hx - \alpha jy$$

$$\frac{dI}{dt} = \frac{dI}{dy}\frac{dy}{ds}\frac{ds}{dt} = \frac{j}{k}\frac{dy}{ds} = \beta jy(hx - jy) - (\alpha + b)jy \tag{3}$$

The scaling constants will be chosen to focus attention on the coefficients of the x-term in the first ODE of system (3) and the y-term in the second. If we set

$$h = j = \alpha/\beta, \quad k = 1/\alpha, \quad c = (a-b)/\alpha, \quad r = (\alpha+b)/\alpha \tag{4}$$

then we have the simplified system in dimensionless variables and parameters:

$$\frac{dx}{ds} = cx - y$$

$$\frac{dy}{ds} = y(x - y) - ry = (x - y - r)y \tag{5}$$

There is a reason for focusing on the parameters c and r. The most likely way to reduce the possum population is to reduce its birth rate or raise its natural death rate, that is, alter a or b in the formulas in (4), and so change c and r.

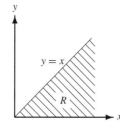

Let's look at the region R, $0 \le y \le x$ because the number of infected animals cannot exceed the entire population. We can ensure that if $0 \le y_0 \le x_0$, then the orbit starting at (x_0, y_0) stays in the region $0 \le y \le x$ as time increases by requiring that $c + r \ge 1$ (see Problem 4).

The equilibrium points of system (5) are found by simultaneously solving the nullcline equations $cx - y = 0$ and $(x - y - r)y = 0$ for x and y. We find that the equilibrium points are the origin and a point whose coordinates are

$$x = r/(1 - c), \qquad y = cr/(1 - c)$$

The latter equilibrium point disappears if $c = 1$. Note also that if $c = 0$, then $(x, 0)$ is an equilibrium point in the region R for every $x \ge 0$.

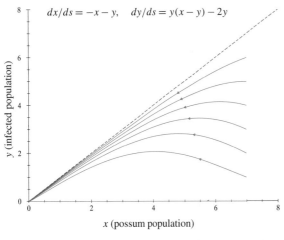

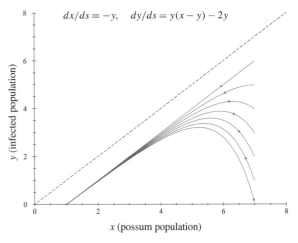

FIGURE 5.5.1 The unlikely case of possum extinction: $c = -1$, $r = 2$.

FIGURE 5.5.2 Another unlikely situation: the infected population dies off if $c = 0$ and $r = 2$.

One goal of the computer simulation will be to determine whether the population orbits $x = x(s)$, $y = y(s)$ of system (5) tend to an equilibrium point or become unbounded as s increases. This can be accomplished by a study of the orbits of system (5) for various values of c and r.

EXAMPLE 5.5.1

From Extinction to Explosion

Set $r = 2$ in system (5) and plot orbits in the region R for $c = -1$, 0, 1.5, and 0.25. Figures 5.5.1–5.5.4 show the results. In Figure 5.5.1 $c = -1$ and all orbits in R tend to the origin; that is, as s advances, the possums die out (and the disease, presumably, with them). Despite appearances, the orbits in Figure 5.5.2 (where $c = 0$) are not all asymptotic to a single equilibrium state $(x, 0)$. Zooming on the region near $(1, 0)$ will show that the seven orbits tend to seven different equilibrium points. Each equilibrium point models a disease-free steady-state possum population.

Figure 5.5.3 ($c = 1.5$) shows a disastrous situation. But Figure 5.5.4 ($c = 0.25$) shows the intriguing scenario where the possum population and its diseased subpopulation approach the equilibrium state E. This may be interpreted in terms of the disease becoming endemic and the healthy and diseased populations at equilibrium.

Figures 5.5.5 and 5.5.6 show an oscillating orbit, its approach to another endemic equilibrium $(22/19, 11/190)$ and the corresponding component graphs in the case $c = 0.05$ and $r = 1.1$ (somewhat more realistic than the values $c = 0.25$, $r = 2$ of Figure 5.5.4). But the time between successive population maxima is so large that it would take years of observation before one could decide whether the data support the validity of a model like this.

This possum model is a model in the making. It isn't yet known whether this or some other model will provide the breakthrough in understanding the dynamics of the situation and how the basic problem can be resolved. Any ideas?

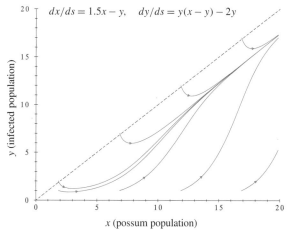

FIGURE 5.5.3 The nightmare of explosive growth: $c = 1.5$, $r = 2$.

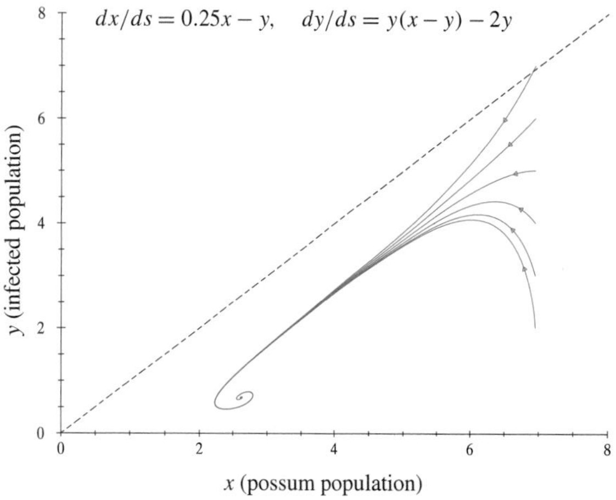

FIGURE 5.5.4 Healthy and infected populations stabilize: $c = 0.25$, $r = 2$.

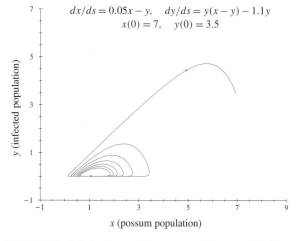

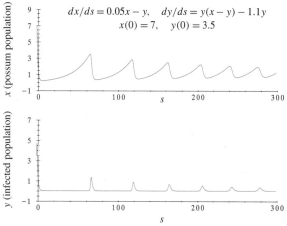

FIGURE 5.5.5 Approach to stable equilibrium where disease is endemic at low population levels: $c = 0.05$, $r = 1.1$.

FIGURE 5.5.6 Population bursts during approach to low population levels $x \approx 1.16$ and $y \approx 0.06$: $c = 0.05$, $r = 1.1$.

Comments

The possum/tuberculosis model is one of a large number of disease models that have been formulated over the last seventy-five years as aids in understanding the dynamics of the spread of disease. These models are widely used, and one of them is outlined in the problem set.

PROBLEMS

www **1.** (*Outbreak of a Measles Epidemic: Kermack–McKendrick SIR Model*). We may model the evolution of an epidemic in a fixed population by dividing the population into three distinct classes:

$$S = \textit{Susceptibles, those who have never had the illness and can catch it}$$

$$I = \textit{Infectives, those who are infected and are contagious}$$

$$R = \textit{Recovered, those who had the illness and have recovered}$$

Assume that the disease is mild (everyone eventually recovers) the disease confers immunity on the recovered, and that the diseased are infective until they recover.

(a) Argue that the evolution of the epidemic may be modeled by the nonlinear system of first-order ODEs $S' = -aSI$, $I' = aSI - bI$, $R' = bI$, where the parameter b represents the reciprocal of the period of infection and a represents the reciprocal of the level of exposure of a typical person. [*Hint*: Compare with models for predator-prey interaction.]

(b) Show that the onset of an epidemic can only occur if the susceptible population is large enough. Specifically, find the threshold value for S above which more people are infected each day than recover.

(c) Suppose that German measles lasts for four days. Suppose also that the typical susceptible person meets 0.3% of the infected population each day and that the disease is transmitted in 1 out of every 6 contacts with an infected person. Find the values of the parameters a and b in the SIR model. How small must the susceptible population be for this illness to fade away without becoming an epidemic? Verify by plotting the component graph for $I(t)$ for your choice of $I(0)$ and for values of $S(0)$ that are 50% above and 50% below the threshold value found in part **(b)**. Plot over $0 \le t \le 30$ and discuss what you see.

(d) Suppose that another illness has parameter values $a = 0.001$ and $b = 0.08$, and suppose that 100 infected individuals are introduced into a population. Investigate how the spread of the infection depends on the size of the population by plotting S-, I-, and R-component graphs for $0 \le t \le 50$, $I(0) = 100$, $R(0) = 0$, and values of $S(0)$ ranging from 0 to 2000 in increments of 500. How does the value of $S(0)$ affect the speed with which the epidemic runs its course?

(e) Using $a = 0.001$ and $b = 0.08$, find I as a function of S, I_0, and S_0, and plot $I(t)$ against $S(t)$ for various values of S_0 and I_0, $0 < S_0 < 1600$, $0 < I_0 < 1250$. Zooming near the origin, look at the long-term behavior. Interpret your graphs in terms of the model. [*Hint*: $dI/dS = -1 + b(aS)^{-1}$.]

2. (*Scaling the Possum Plague*). Suppose that the possum populations P and I are measured in units of a million and that time is measured in years. Show that if the scale constants h and j, k of (2) and (4) are measured, respectively, in units of 10^{-7} and $(\text{years})^{-1}$, then x, y, and s are dimensionless variables.

3. Write the possum model in terms of the susceptibles, $S = P - I$, and the infected, I. Then introduce scaled variables $z = x - y$, y, and s (where x, y, and s are as in the text), write the z, y ODEs, find the equilibria, and plot orbits for $z \ge 0$, $y \ge 0$ with the values of c and r as given in Figures 5.5.1–5.5.5. Interpret your graphs (zoom if necessary to detect long-term behavior).

4. Suppose that R is the region in the first quadrant bounded by the lines $y = x$ and $y = 0$ (R includes its boundaries). Show that if $c + r \ge 1$, then orbits of system (5) that originate in R stay in R as t increases.

5. (*Parameter Study of the Possum Plague*). Make a parameter study of the possum plague model, using parameters of your choice. Explain your results and relate your conclusions to an assessment of the situation in New Zealand.

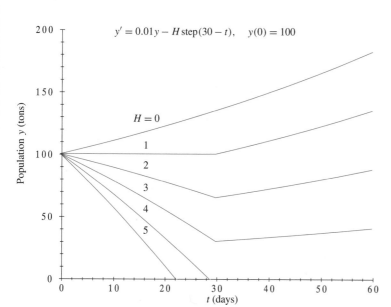

$$y' = 0.01y - H\,\mathrm{step}(30 - t), \quad y(0) = 100$$

Effect of a thirty-day harvest at rate H on an exponentially growing population. How much harvesting is too much (Section 6.2)?

CHAPTER

6

The Laplace Transform

The Laplace transform is a special linear operator that is widely used to construct solution formulas for linear differential equations with a wide variety of driving terms, including on-off functions, and with terms involving a time delay. The operator involves an integral, and transforms problems in linear differential equations into problems with a simpler structure.

6.1 Introduction to the Laplace Transform

When a particular type of equation is used by mathematicians, engineers, and scientists to model and understand a wide variety of natural processes, there is great interest in finding an effective method for solving all equations of that type. One approach is to transform equations of a given type to other equations that are easier to work with, solve the new equations, and then finally reverse the transformation and interpret the solutions in terms of the original equations.

In this chapter the equation types are driven linear ODEs and systems of ODEs with constant coefficients as well as equations with time delays. Laplace[1] showed how to transform equations of these types into algebraic equations that can be "solved" by algebraic means. Our aim is to explore this transformation process and how one reverses the process to obtain a solution of the original problem.

The traffic-flow event outlined below leads to a mathematical model that involves time delays of a type not seen before in this book. We will show in Section 6.3 that the Laplace transform approach is well suited for this model.

A Model for Car Following

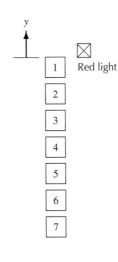

Let's suppose that several cars are stopped at a red light. The light turns green, and the first car accelerates. Ten seconds later the second car rear-ends the first. Why? Let's build a model which we can use to answer questions like this.

Many stimuli may cause the driver in a line of cars to respond with an acceleration or deceleration: a speeding car to the rear, a changing traffic light ahead, a variation in the velocity of the car directly in front. Of course, the driver's response doesn't occur instantaneously. The driver must first detect the stimulus and then decide whether to ease up on, or to depress, the accelerator or the brake pedal. The car itself does not respond instantaneously and contributes to the lag in the response of the car-driver system. We use the word "car" to denote the car-driver system and refer to the delay in the response as the *response time* of the car.

Let's assume that the dominant stimulus to the car is the difference between the velocity of the car ahead and its own velocity. The car's response, we will assume, is an acceleration if the car ahead has a positive relative velocity, and a deceleration if the relative velocity is negative, hence the term *velocity control*. The magnitude of the car's acceleration or deceleration is assumed to be proportional to the magnitude of the relative velocity. The positive constant of proportionality is the *sensitivity* of the car to the stimulus.

Let's measure time from the moment the lead car accelerates after the light turns green. We will assume that the lead car accelerates at α ft/sec^2, that each car is 15 ft long, and that there is a gap of 5 ft between successive cars in the line. The location $y_j(t)$ of the j-th car at time t is measured from the stoplight to its front bumper, so $y_j(0) \leq 0$. Let's look at the first three cars in line and use the modeling principles above to track their positions after the light changes.

Pierre-Simon de Laplace

[1] The French mathematical physicist Marquis Pierre-Simon de Laplace (1749–1827) developed most of his mathematics in applying the Newtonian theory of gravitation to our solar system as a whole. In his *Mécanique Céleste*, 2000 pages long and published in parts between 1799 and 1825, Laplace proved the stability of the solar system under his simplifying assumptions. Laplace was not only a great mathematician, but a prominent politician as well. Having examined and passed the sixteen-year-old Napoleon Bonaparte for entrance to the French Military School in 1785, Laplace survived the French Revolution and the many changes in the French political structure in the years that followed. One wonders what would have happened to the course of European history if Laplace had flunked Napoleon. Talk about sensitivity to data!

Assuming identical car-driver systems, so a common delay time T and sensitivity coefficient λ, we obtain the specific mathematical model involving *differential-delay equations*:

$$
\begin{aligned}
y_j' &= v_j, \qquad j = 1, 2, 3 \\
v_1' &= \alpha \\
v_2'(t+T) &= \lambda[v_1(t) - v_2(t)] \\
v_3'(t+T) &= \lambda[v_2(t) - v_3(t)]
\end{aligned}
\tag{1}
$$

Note the time delay T in the acceleration equations for the second and third cars. We have the following initial conditions

$$
y_1(0) = 0, \quad y_2(0) = -20, \quad y_3(0) = -40, \quad v_j(0) = 0, \quad j = 1, 2, 3 \tag{2}
$$

Our goal is to find the functions $y_j(t)$ and, in particular, to determine if a collision occurs, that is, if for some time t^* and some integer $j = 2, 3$, $y_j(t^*) = y_{j-1}(t^*) - 15$, which corresponds to the j-th and $(j-1)$-st car colliding. We will solve the problem at the end of Section 6.3 after we have developed the Laplace transform that will be the main tool for finding the solution.

The Laplace Transform Operator

☞ Read $\mathcal{L}[f]$ as "ell of f."

The Laplace transform operator $\mathcal{L}$ acts on a function f and returns another function $\mathcal{L}[f]$. Here's how it works.

❖ **The Laplace Transform Operator.** The Laplace transform of a function $f(t)$, $0 \le t < \infty$, is a function $\mathcal{L}[f]$ of a real variable s given by

$$
\mathcal{L}[f](s) = \int_0^\infty e^{-st} f(t)\, dt = \lim_{\tau \to +\infty} \int_0^\tau e^{-st} f(t)\, dt \tag{3}
$$

where the transform is defined for all real s for which the limit in (3) exists.

Let's see what the Laplace transform operator $\mathcal{L}$ does to some elementary functions.

EXAMPLE 6.1.1

The Laplace Transforms of the Functions $f(t) = c, t, e^{at}$
Let's use definition (3) to find some Laplace transforms:

• The transform of the constant function $f(t) = c$, all $t \ge 0$, is

$$
\begin{aligned}
\mathcal{L}[c](s) &= \int_0^\infty e^{-st} \cdot c\, dt = \lim_{\tau \to +\infty} c \int_0^\tau e^{-st}\, dt \\
&= \lim_{\tau \to +\infty} c \left[-\frac{1}{s} e^{-st} \right]_{t=0}^{t=\tau} = \lim_{\tau \to +\infty} c \left[-\frac{1}{s} e^{-s\tau} + \frac{1}{s} e^{-s \cdot 0} \right] \\
&= 0 + \frac{c}{s} = \frac{c}{s}, \qquad s > 0
\end{aligned}
\tag{4}
$$

because, if s is positive, then $e^{-s\tau} \to 0$ as $\tau \to +\infty$. We chose to restrict s to be positive so that the transform exists.

- Here's how to transform $f(t) = t$, $t \geq 0$, by using a table of integrals [e.g., the Table 1.3.1 on page 20]:

$$\mathcal{L}[t](s) = \int_0^\infty e^{-st} \cdot t \, dt = \lim_{\tau \to +\infty} \int_0^\tau e^{-st} t \, dt = \lim_{\tau \to +\infty} \left[-\frac{1}{s^2} e^{-st}(st+1) \right]_{t=0}^{t=\tau}$$

$$= \lim_{\tau \to +\infty} \left[-\frac{s\tau+1}{s^2 e^{s\tau}} \right] + \frac{1}{s^2} e^{-s \cdot 0}(s \cdot 0 + 1)$$

$$= \lim_{\tau \to +\infty} \left[\frac{-s}{s^3 e^{s\tau}} \right] + \frac{1}{s^2}$$

$$= 0 + \frac{1}{s^2} = \frac{1}{s^2}, \qquad s > 0 \tag{5}$$

where at one point we used L'Hopital's Rule [differentiate the numerator $s\tau + 1$ and the denominator $s^2 e^{s\tau}$ with respect to τ, divide, and then find the limit as $\tau \to +\infty$]. Once again, we require that $s > 0$.

- Finally, we calculate $\mathcal{L}[e^{at}]$ in the following way:

$$\mathcal{L}[e^{at}](s) = \int_0^\infty e^{-st} \cdot e^{at} \, dt = \lim_{\tau \to +\infty} \int_0^\tau e^{(a-s)t} \, dt$$

$$= \lim_{\tau \to +\infty} \left[\frac{1}{a-s} e^{(a-s)t} \right]_{t=0}^{t=\tau}$$

$$= \lim_{\tau \to +\infty} \left[\frac{1}{a-s} e^{(a-s)\tau} \right] - \frac{1}{a-s}$$

$$= 0 + \frac{1}{s-a} = \frac{1}{s-a}, \qquad s > a \tag{6}$$

☞ The transform tables at the end of the chapter list transforms of many more functions.

In this case we take $s > a$ so that the transform exists.

Note that $\mathcal{L}[f](s)$ in these examples is not defined for all of s, but only for values of s beyond some number s_0 that varies from one function f to another. This is typical of the Laplace transform and we will see many examples of this feature.

The next example hints at the reason the transform is useful in differential equations. We will need the integration-by-parts formula from calculus: If $u(t)$ and $v(t)$ are continuously differentiable for $a \leq t \leq b$, then

$$\int_a^b u(t)v'(t) \, dt = \left[u(t)v(t) \right]_{t=a}^{t=b} - \int_a^b u'(t)v(t) \, dt$$

Notice that integration-by-parts "moves" the derivative from $v(t)$ to $u(t)$.

EXAMPLE 6.1.2

Transforming a Derivative

Let's suppose $y'(t)$ is continuous for $t \geq 0$ and that for all s larger than some s_0 we have $e^{-s\tau} y(\tau) \to 0$ as $\tau \to +\infty$. Then we have the formula

$$\mathcal{L}[y'](s) = -y(0) + s\mathcal{L}[y](s) \tag{7}$$

Here's an explanation of this fundamental transform formula:

$$\mathcal{L}[y'](s) = \lim_{\tau \to +\infty} \int_0^\tau e^{-st} y'(t)\, dt$$

(integration by parts)
$$= \lim_{\tau \to +\infty} \left\{ \left[e^{-st} y(t) \right]_{t=0}^{t=\tau} - \int_0^\tau (e^{-st})' y(t)\, dt \right\}$$

$$= \lim_{\tau \to +\infty} \left[e^{-s\tau} y(\tau) \right] - e^{-s\cdot 0} y(0) + s \int_0^\infty e^{-st} y(t)\, dt$$

$$\,\dot{=}\, -y(0) + s\mathcal{L}[y](s)$$

where (as noted earlier) we have assumed that $e^{-s\tau} y(\tau) \to 0$ as $\tau \to +\infty$.

Before we can apply the Laplace transform technique in some practical situations we need to check out other properties of the transform.

Properties of the Laplace Transform

If $\mathcal{L}[f](s)$ and $\mathcal{L}[g](s)$ are defined on an interval $s > s_0$, so is $\mathcal{L}[\alpha f + \beta g](s)$ for any real constants α, β. $\mathcal{L}$ is a linear operator since (using properties of integrals)

☞ And so we meet another linear operator.

$$\mathcal{L}[\alpha f + \beta g] = \int_0^\infty e^{-st}(\alpha f(t) + \beta g(t))\, dt$$

$$= \alpha \int_0^\infty e^{-st} f(t)\, dt + \beta \int_0^\infty e^{-st} g(t)\, dt = \alpha \mathcal{L}[f] + \beta \mathcal{L}[g] \qquad (8)$$

Here's how (8) is used in a specific transform problem:

$$\mathcal{L}[7t - e^{2t}] = 7\mathcal{L}[t] - \mathcal{L}[e^{2t}] = \frac{7}{s^2} - \frac{1}{s-2}$$

where we have also used formulas (5) and (6).

In order for the transform to be useful, we need to be able to recover $f(t)$ from $\mathcal{L}[f](s)$. Suppose that $f(t)$ and $g(t)$ are continuous for $t \geq 0$ and that the transforms coincide for $s \geq s_0$:

$$\mathcal{L}[f](s) = \mathcal{L}[g](s), \qquad s \geq s_0$$

Then it can be shown that $f(t) = g(t)$, $t \geq 0$. In other words, a continuous function can be uniquely recovered from its Laplace transform. The operator that does this is linear; it is denoted by $\mathcal{L}^{-1}$ and is called the *inverse Laplace transform*. It is not easy to construct an explicit formula describing the action of $\mathcal{L}^{-1}$, but we can get along without a formula by using tables of transforms such as those at the end of the chapter.[2] For example, if $\mathcal{L}[f](s) = 1/s^2$ for $s > 0$, the only continuous function f with this transform is $f(t) = t$, $t \geq 0$ (see Example 6.1.1).

---

[2]The formula for $\mathcal{L}^{-1}[g]$ involves a contour integral in the complex plane, but we will skip this and rely instead on tables to find inverse transforms.

Initial Value Problems and Transforms

The usefulness of the Laplace transform in solving linear initial value problems is suggested by the following example, which shows that an IVP can be transformed into an algebraic equation that involves no derivatives at all.

EXAMPLE 6.1.3

Solving an Initial Value Problem by Laplace Transforms
Let's look at the IVP

$$y'(t) + ay(t) = f(t), \qquad y(0) = y_0 \tag{9}$$

where a is a constant and f is piecewise continuous on $[0, +\infty)$. Suppose that $L[y']$, $L[y]$, and $L[f]$ are defined on a common interval $s > s_0$. Apply L to each side of the ODE in (9), and use the linearity of L to obtain

$$L[y'](s) + aL[y](s) = L[f](s), \qquad s > s_0 \tag{10}$$

If $\lim_{t\to+\infty} e^{-st}y'(t) = 0$ for each $s > s_0$, using the transform formula (7), formula (10) becomes

$$sL[y](s) - y(0) + aL[y](s) = L[f](s) \tag{11}$$

Solving (11) algebraically for $L[y](s)$, we have

$$L[y](s) = \frac{y(0)}{s+a} + \frac{L[f](s)}{s+a} \tag{12}$$

Note that the single algebraic equation (12) encodes all of the information given in IVP (9).

☞ Now you will need to consult the tables at the end of the chapter.

Let's take $y(0) = 1$ and $f(t) = 4t^3 e^{-at}$. Using entry 3 in Table II we see that

$$L[4t^3 e^{-at}] = \frac{24}{(s+a)^4}$$

So

$$L[y](s) = \frac{1}{s+a} + \frac{24}{(s+a)^5}$$

Using the linearity of L^{-1} and the fact that L^{-1} takes $L[y]$ back to y, we have

$$y(t) = L^{-1}\left[\frac{1}{s+a}\right](t) + L^{-1}\left[\frac{24}{(s+a)^5}\right]$$

☞ "Table entry II.3" means "entry 3 in Table II"

Now using Table entry II.3, but backwards, we have

$$y(t) = e^{-at} + t^4 e^{-at}$$

The integrating-factor technique of Section 1.3 gives the same result.

The advantage of the transform approach is that it reduces the problem of solving an IVP such as (9) to the inspection of a table of transforms.

The Transform of Piecewise-Continuous Functions

Now let's be specific about the kinds of functions that can be transformed. First let's define a new class of functions.

☞ Recall from Section 1.4 that all continuous functions are also peicewise continuous.

❖ **Piecewise Continuous Functions of Exponential Order**. A piecewise continuous function $f(t), 0 \leq t$, is of *exponential order* if there are positive constants M and a such that $|f(t)| \leq Me^{at}$ for all $0 \leq t < \infty$. We denote by **E** the collection of all these functions.

Sums, products, and antiderivatives of functions in **E** are also in **E**. Note that if f is in **E** then for some positive constant b we have $e^{-bt}f(t) \to 0$ as $t \to +\infty$.

EXAMPLE 6.1.4

A Function of Exponential Order

The function $f(t) = -10te^{70t}\cos t$ is in **E** because it is continuous on $[0, \infty)$ and

$$|f(t)| = 10|te^{70t}\cos t| \leq 10te^{70t} \leq 10e^te^{70t} = 10e^{71t}, \qquad t \geq 0$$

Here we used the fact that $t < e^t$ if $t \geq 0$.

EXAMPLE 6.1.5

A Function not of Exponential Order

The function $g(t) = e^{t^2}$ is not in **E**. If it were, then there would be positive constants M and a such that $e^{t^2} \leq Me^{at}$ for $t \geq 0$, so $e^{t(t-a)} \leq M$ for $t \geq 0$. But this is false since $e^{t(t-a)} \to +\infty$ as $t \to +\infty$.

☞ Unless stated otherwise, every function f in this chapter is assumed to be in **E**.

So we see that functions in **E** can become unbounded as $t \to +\infty$ (Example 6.1.4), but they can't "blow up" too fast (Example 6.1.5). It can be shown that if f is in **E** then $\mathcal{L}[f](s)$ exists for all $s \geq s_0$ for some s_0. That's why we restrict attention mostly to functions in **E**. Here's an example of the transform of a piecewise continuous function.

EXAMPLE 6.1.6

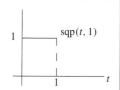

Transform of a Square Pulse

Suppose that $f(t)$ is the square pulse function sqp$(t, 1)$ defined by

$$\text{sqp}(t, 1) = \begin{cases} 1, & 0 \leq t < 1 \\ 0, & t \geq 1 \end{cases}$$

See the margin sketch. Then for all $s > 0$,

$$\mathcal{L}[f](s) = \int_0^\infty e^{-st}f(t)\,dt = \int_0^1 e^{-st}\cdot 1\,dt + \int_1^\infty e^{-st}\cdot 0\,dt = -\frac{1}{s}e^{-st}\Big|_{t=0}^{t=1} = \frac{1}{s}[1 - e^{-s}]$$

Note that the transform is continuous and is differentiable to all orders for $s > 0$.

It can be shown that if f and g are functions in **E** and have identical transforms on an interval $s \geq s_0$, then $f(t) = g(t)$ wherever f and g are both continuous. So if f is continuous and in **E**, then $\mathcal{L}^{-1}\big[\mathcal{L}[f]\big] = f$. At points where f or g is discontinuous, then the values of f and g may differ.

The Laplace transforms of functions in **E** are very well behaved, even though the functions themselves may have discontinuities, as we saw in the examples. In addition, the transforms of functions in **E** decay to 0 as $s \to +\infty$ regardless of the behavior of the functions themselves. The next theorem summarizes these two properties.

THEOREM 6.1.1

Smoothness and Decay Property for $\mathcal{L}[f]$. For f in **E**, the Laplace transform $\mathcal{L}[f](s)$ exists on some half-line, $s > s_0$. Moreover, $\mathcal{L}[f](s)$ has derivatives of every order, and $\mathcal{L}[f](s) \to 0$ as $s \to +\infty$.

So we see that the Laplace transform of a function $f(t)$ in the class **E** is very well-behaved indeed. Not only does $\mathcal{L}[f](s) \to 0$ as $s \to +\infty$, but $\mathcal{L}[f]$ also has derivatives of every order. We won't use this result directly, but it does come in handy as a check against computation errors.

Comments

In Example 6.1.3 we showed that the transform of a derivative does not involve a derivative. This is one of the reasons that the Laplace transform is a useful alternative method for solving linear ODEs, but there are other reasons. For example, Laplace transforms can be used to solve ODEs involving a time delay, which would be hard to do by any other method. The transform method is practical only if a sufficiently long list of transforms is available—that's why we have the tables at the end of the chapter.

PROBLEMS

www **1.** Find the Laplace transform and the largest interval $s > s_0$ on which the transform is defined. The number a is a constant and n is a positive integer.

(a) $3t - 5$ (b) t^2 (c) t^n (d) $\cos at$ (e) te^{at} (f) $t \sin at$

2. Find the Laplace transform and the largest interval $s > s_0$ on which the transform is defined.

(a) $f(t) = \begin{cases} \sin t, & 0 \leq t \leq \pi \\ 0, & t > \pi \end{cases}$ (b) $f(t) = \begin{cases} 0, & 0 \leq t \leq 1 \\ t, & t > 1 \end{cases}$

(c) $f(t) = \begin{cases} t, & 0 \leq t \leq 1 \\ 0, & t > 1 \end{cases}$ (d) $f(t) = \begin{cases} t, & 0 \leq t \leq 1 \\ 2 - t, & t > 1 \end{cases}$

3. Solve each IVP using the Laplace transform. [*Hint*: See Example 6.1.3.]

(a) $y' + 2y = 0$, $y(0) = 1$

(b) $y' + 2y = e^{-3t}$, $y(0) = 5$ [*Hint*: $1/(s+2)(s+3) = 1/(s+2) - 1/(s+3)$.]

(c) $y' + 2y = e^{it}$, $y(0) = 0$ [*Hint*: Treat i as a constant.]

 4. Create a model for a line of three nonidentical cars at a stop light. The cars have distinct response times, sensitivity constants, car lengths, and initial separations.

6.2 Calculus of the Transform

☞ There's the differential calculus, the integral calculus, and now the transform calculus!

Using the definition of the Laplace transform is not always the best way to find the transform of a specific function. To speed up the process, we develop a calculus for transforming a combination of functions from the transforms of the individual functions. We also develop some general rules for calculating inverse Laplace transforms. The partial fraction formulas in Table 6.2.1 are useful for this purpose. For ease of reference, we have placed transform formulas in tables at the end of the chapter.

The Transform Calculus of Derivatives

To solve an ODE with transforms we need to know how derivatives transform. Section 6.1 showed how to transform f'. There is a similar transform of f'' and a general formula for $\mathcal{L}[f^{(n)}]$ (n a positive integer). The following theorem lists these formulas.

THEOREM 6.2.1

Transform of f', f''. Suppose that f and f' are in **E** and f is continuous. Then

$$\mathcal{L}[f'](s) = s\mathcal{L}[f](s) - f(0) \qquad (1)$$

☞ Formula (1) is derived in Example 6.1.2.

Suppose that f' and f'' are in **E** and f' is continuous. Then f is continuous and

$$\mathcal{L}[f''](s) = s^2\mathcal{L}[f](s) - sf(0) - f'(0) \qquad (2)$$

For an integer $n \geq 1$, suppose that $f^{(n-1)}$ and $f^{(n)}$ are in **E**, and that $f^{(n-1)}$ is continuous. Then $f, f', \ldots, f^{(n-2)}$ are continuous, and

$$\mathcal{L}[f^{(n)}] = s^n\mathcal{L}[f] - s^{n-1}f(0) - s^{n-2}f'(0) - \cdots - sf^{(n-2)}(0) - f^{(n-1)}(0) \quad (3)$$

Here's how to get formula (2) from (1).

$$\mathcal{L}[f''] = \mathcal{L}[(f')'] = s\mathcal{L}[f'(s)] - f'(0)$$
$$= s\{s\mathcal{L}[f](s) - f(0)\} - f'(0) = s^2\mathcal{L}[f](s) - sf(0) - f'(0)$$

Look up formula (3) in the long list at the end of the chapter.

There is a companion formula that tells us what transforms into $d^n\mathcal{L}[f]/ds^n$, rather than what $f^{(n)}(t)$ transforms into.

THEOREM 6.2.2

Transform of $t^n f(t)$. Suppose that f is in **E**. Then

$$\mathcal{L}[t^n f(t)] = (-1)^n d^n \mathcal{L}[f]/ds^n \qquad (4)$$

Formula (4) is found by differentiating $\int_0^\infty e^{-st} f(t)\, dt$ with respect to s n times, and bringing the derivative under the integral sign.

Let's work out some examples.

EXAMPLE 6.2.1

Transforms of Cos at, **Sin** at, $t\text{Cos }2t$, t^n

- Using the linearity of L and formula (2) with $f(t) = \cos at$, we see that

$$L[-a^2 \cos at] = s^2 L[\cos at] - s \cdot \cos 0 - a \sin 0$$

$$-a^2 L[\cos at] = s^2 L[\cos at] - s$$

$$L[\cos at] = \frac{s}{s^2 + a^2} \tag{5}$$

- Now let's find the transform of $\sin at$ by using formulas (1) and (5). Since $\sin at = (-a^{-1}\cos at)'$, we have by (1) that

$$L[\sin at] = sL\left[-\frac{1}{a}\cos at\right] - \left(-\frac{1}{a}\cos 0\right)$$

$$= -\frac{s}{a}\frac{s}{s^2 + a^2} + \frac{1}{a} = \frac{a}{s^2 + a^2} \tag{6}$$

- Next we use formulas (4) and (5) to transform $t\cos 2t$:

$$L[t\cos 2t](s) = -\frac{d}{ds}L[\cos 2t](s) = -\frac{d}{ds}\left[\frac{s}{s^2+4}\right] = \frac{s^2-4}{(s^2+4)^2}$$

- Finally, suppose that $f(t) = t^n$; then $f^{(n)}(t) = n!$. Solving formula (3) for $L[f]$,

$$L[f] = \frac{1}{s^n}\{L[f^{(n)}] + s^{n-1}f(0) + s^{n-2}f'(0) + \cdots + f^{(n-1)}(0)\}$$

Because $L[n!] = n!/s$ and $f^{(k)}(0) = 0$, $k = 0, \ldots, n-1$, we have

$$L[t^n] = \frac{1}{s^n}\left\{\frac{n!}{s} + 0 + \cdots + 0\right\} = \frac{n!}{s^{n+1}} \tag{7}$$

Now let's see how to use formula (2) to solve an IVP.

EXAMPLE 6.2.2

The Transform Way to Solve a Second-Order IVP
Let's use transforms to solve the IVP

$$y'' - y = 1, \qquad y(0) = 0, \quad y'(0) = 1$$

We know that the IVP is uniquely solvable and that the solution and its derivatives of all orders are in **E**. Transform both sides of the ODE using formula (2):

$$L[y'' - y] = L[1]$$

$$s^2 L[y] - sy(0) - y'(0) - L[y] = 1/s$$

☞ We used partial fractions for $[s(s-1)]^{-1}$.

$$L[y] = \frac{sy(0) + y'(0) + 1/s}{s^2 - 1} = \frac{1 + 1/s}{s^2 - 1} = \frac{1}{s(s-1)} = \frac{1}{s-1} - \frac{1}{s}, \qquad s > 1$$

Using formula (6) in Example 6.1.1 with $a = 1$, and formula (7) with $n = 0$, we have

$$y(t) = e^t - 1$$

This illustrates again how to use the transform to solve an IVP.

We used the Method of Partial Fractions in Example 6.2.2. See Table 6.2.1 for some other partial fraction formulas.

We could have used the methods of Chapter 3 for the IVP in Example 6.2.2, but the transform technique is often a more efficient way to solve an IVP. Note that the transform method automatically involves the initial data, unlike the situation with earlier methods where the data were used almost as an afterthought to evaluate constants of integration. Whatever method is used, the same solution will result whenever the Existence and Uniqueness Theorem applies. Of course, the transform approach will be successful in producing a formula for the solution of an IVP only if we can actually find a function of t on the interval $0 \le t < +\infty$ whose transform solves the transformed IVP. Since the Laplace transform takes functions defined on a t-interval into functions defined on an s-interval, engineers will often refer to the functions being in the *time domain* or in the *s-domain* depending on whether they are looking at a function defined on $0 \le t < \infty$ or its Laplace transform.

The Transform Calculus of Integrals

We can transform the integral of a function as easily as its derivative.

THEOREM 6.2.3

> **Transform of an Integral.** Suppose that f is in **E** and $a \ge 0$. Then the function $F(t) = \int_a^t f(x)\,dx$, is in **E** and
>
> $$\mathcal{L}\left[\int_a^t f(x)\,dx\right] = \frac{1}{s}\mathcal{L}[f] - \frac{1}{s}\int_0^a f(x)\,dx \qquad (8)$$

As noted in Section 6.1, since f is in **E** the function $\int_a^t f$ is in **E**. Integration by parts yields formula (8):

$$\mathcal{L}\left[\int_a^t f(x)\,dx\right] = \int_0^\infty e^{-st}\left(\int_a^t f(x)\,dx\right) dt$$

$$= \lim_{\tau \to +\infty}\left[-\frac{1}{s}e^{-st}\int_a^t f(x)\,dx\right]_{t=0}^{t=\tau} + \frac{1}{s}\int_0^\infty e^{-st}f(t)\,dt$$

Using the fact that

$$\lim_{\tau \to +\infty}\left[-\frac{1}{s}e^{s\tau}\int_a^\tau f(x)\,dx\right] = 0$$

we see that formula (8) follows.

EXAMPLE 6.2.3

Using Formula (8) to Calculate a Transform

Integrating by parts, we find $\int_0^t xe^x\,dx = te^t - e^t + 1$, so

$$\mathcal{L}\left[\int_0^t xe^x\,dx\right] = \mathcal{L}[te^t] - \mathcal{L}[e^t] + \mathcal{L}[1]$$

On the other hand, from formula (8) with $a = 0$,

$$\mathcal{L}\left[\int_0^t xe^x\,dx\right] = \frac{1}{s}\mathcal{L}[te^t] - \frac{1}{s}\int_0^0 xe^x\,dx = \frac{1}{s}\mathcal{L}[te^t]$$

Equating these two expressions for $\mathcal{L}[\int_0^t xe^x\,dx]$, we can solve for $\mathcal{L}[te^t]$ to find that

$$\mathcal{L}[te^t] = \frac{1}{(s-1)^2}, \qquad s > 1$$

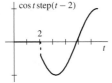

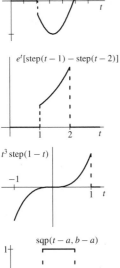

Oliver Heaviside

The power of the Laplace transform approach becomes evident when engineering functions appear in formulas and models. The most important of these functions is the unit step function which was introduced in earlier chapters. Let's see how the step function is used in the transform calculus.

Shifting Theorems: The Step Function

The unit step function

$$\text{step}(t) = \begin{cases} 1, & t \geq 0 \\ 0, & t < 0 \end{cases} \tag{9}$$

plays a central role in the transform calculus. Oliver Heaviside[3] introduced the function, and it is often called the *Heaviside function* in recognition of his contributions to the transform calculus. The step function's main use is to turn other functions "on" or "off." For example, the function $\cos t\,\text{step}(t - 2)$ turns on at $t = 2$, the function $e^t[\text{step}(t - 1) - \text{step}(t - 2)]$ turns on at $t = 1$ and off at $t = 2$, while the function $t^3\,\text{step}(1 - t)$ turns off at $t = 1$.

Many on-off functions can be written in terms of step functions. One example is the *pulse function* [i.e., $\text{sqp}(t - a, b - a)$ based on the square pulse $\text{sqp}(t, a)$ defined in Appendix B.1]:

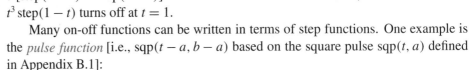

$$\text{sqp}(t - a, b - a) = \text{step}(t - a) - \text{step}(t - b) = \begin{cases} 1, & a \leq t < b \\ 0, & \text{all other } t \end{cases} \tag{10}$$

---

[3]Heaviside (1850–1925) was a self-taught Englishman who invented his own operational calculus to solve differential equations many years before Laplace transforms were introduced into engineering for the same purpose. Heaviside coined the terms "inductance" and "capacitance" and introduced the telegrapher's equation, a partial differential equation for the voltage in a cable as a function of time, location, resistance, inductance, and capacitance. He also predicted the existence of a reflecting ionized region around the earth (the ionosphere). Poverty stricken and scorned in his early years because of his lack of education, his genius was eventually recognized, and he was given many honors and awards.

The next theorem shows the importance of step functions in finding transforms.

THEOREM 6.2.4

The Shifting Theorem. Suppose that f is in **E**. Then $e^{at} f(t)$ is in **E** and

$$\mathcal{L}[e^{at} f(t)](s) = \mathcal{L}[f](s - a) \tag{11}$$

$$\mathcal{L}[f(t - a) \text{step}(t - a)] = e^{-as} \mathcal{L}[f(t)], \qquad a \geq 0 \tag{12}$$

$$\mathcal{L}[f(t) \text{step}(t - a)] = e^{-as} \mathcal{L}[f(t + a)], \qquad a \geq 0 \tag{13}$$

where we define $f(t) = 0$ for $t < 0$.

Formula (11) follows directly from the definition of the Laplace transform operator, so multiplying a function in the time domain by e^{at} shifts the transform variable s in $\mathcal{L}[f](s)$ by the amount a. To show formula (12), note that

$$\mathcal{L}[f(t - a) \text{step}(t - a)] = \int_0^\infty e^{-st} f(t - a) \text{step}(t - a) \, dt = \int_a^\infty e^{-st} f(t - a) \, dt$$

$$= \int_0^\infty e^{-s(x+a)} f(x) \, dx = e^{-as} \mathcal{L}[f]$$

where the variable change $t = x + a$ was used in the last integral.

Formula (12) can be viewed this way: Define f to be 0 for $t < 0$, and shift the graph of f to the right by a. The effect of transforming this translated function is to multiply the transform of the original function by e^{-as}. To derive formula (13), set $g(t) = f(t - a)$, so $g(t + a) = f(t)$ in formula (12); then rename g as f.

The Shifting Theorem is a very useful instrument in our transform calculus arsenal. Let's practice using it.

EXAMPLE 6.2.4

Shifting to Find a Transform and an Inverse Transform
The Shifting Theorem can be used to find both transforms and inverse transforms.

- To find the transform of $e^{-2t} \cos 3t$, use formulas (5) and (11). We see that

$$\mathcal{L}[e^{-2t} \cos 3t] = \frac{s + 2}{(s + 2)^2 + 9}$$

- To find the function $f(t)$ whose transform is $(2s + 3)/(s^2 - 4s + 20)$, write

$$\frac{2s + 3}{s^2 - 4s + 20} = \frac{2(s - 2) + 7}{(s - 2)^2 + 16}$$

$$= 2 \left[\frac{s - 2}{(s - 2)^2 + 16} \right] + \frac{7}{4} \left[\frac{4}{(s - 2)^2 + 16} \right]$$

From formulas (5), (6), and (11) we see that $f(t) = 2e^{2t} \cos 4t + (7/4)e^{2t} \sin 4t$.

We often use the step function to help define piecewise-continuous functions which arise often in Engineering. Now let's see how these functions transform.

EXAMPLE 6.2.5

Transforming Step Functions
According to formula (13) in the Shifting Theorem,

$$\mathcal{L}[\text{step}(t-a)] = \mathcal{L}[1 \cdot \text{step}(t-a)] = e^{-as}\mathcal{L}[1] = \frac{e^{-as}}{s}, \qquad s > 0 \qquad (14)$$

which could also be obtained by direct integration. The transform of $\text{step}(a-t)$ is

$$\mathcal{L}[\text{step}(a-t)] = \int_0^a e^{-st} \cdot 1\, dt + \int_a^\infty e^{-st} \cdot 0\, dt = \frac{1 - e^{-as}}{s}, \qquad s > 0, \quad a > 0 \quad (15)$$

Now formula (14) can be used to transform any function that can be written with step functions, for example a "ramp" function.

EXAMPLE 6.2.6

Transforming a Piecewise Linear Ramp Function
The margin sketch shows the graph of the function

$$f(t) = \begin{cases} 1, & 0 \le t < 1 \\ 2 - t, & 1 \le t < 2 \\ 0, & 2 \le t \end{cases}$$

Using (10), first with $a = 0$ and $b = 1$ and then with $a = 1$ and $b = 2$, we see that

$$f(t) = \{\text{step}(t) - \text{step}(t-1)\} + (2-t)\{\text{step}(t-1) - \text{step}(t-2)\}$$
$$= \text{step}(t) - (t-1)\,\text{step}(t-1) + (t-2)\,\text{step}(t-2)$$

Applying the Laplace transform, we have

$$\mathcal{L}[f] = \mathcal{L}[\text{step}(t)] - \mathcal{L}[(t-1)\,\text{step}(t-1)] + \mathcal{L}[(t-2)\,\text{step}(t-2)]$$

Using (14) and formula (12) in the Shifting Theorem, we find the transform

$$\mathcal{L}[f] = \frac{1}{s} - e^{-s}\frac{1}{s^2} + e^{-2s}\frac{1}{s^2} = \frac{s - e^{-s} + e^{-2s}}{s^2}, \qquad s > 0$$

A function $f(t)$ can be "switched on" at $t = a$ if it is multiplied by the step function $\text{step}(t-a)$ to produce $f(t)\,\text{step}(t-a)$. Here's how we transform such a function.

EXAMPLE 6.2.7

☞ Appendix B.4 has a list of trigonometric identities.

How to Transform a Switched-on Sinusoid
Using formula (13) in the Shifting Theorem and a trigonometric identity we have

$$\mathcal{L}[(\sin t)\,\text{step}(t-a)] = e^{-as}\mathcal{L}[\sin(t+a)] = e^{-as}\mathcal{L}[\sin t \cos a + \cos t \sin a]$$
$$= e^{-as}\left\{\frac{\cos a}{s^2 + 1} + \frac{s \sin a}{s^2 + 1}\right\}$$

which is not the formula you might expect.

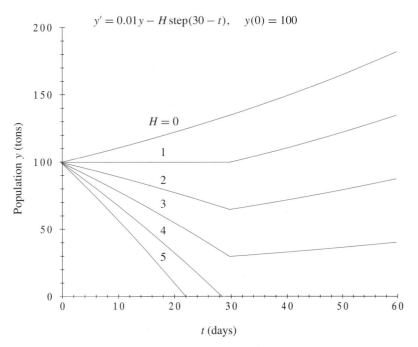

$$y' = 0.01y - H\,\text{step}(30 - t), \quad y(0) = 100$$

FIGURE 6.2.1 A 30-day harvest on an exponential growth population.

Let's take a look at a first-order IVP with a discontinuous driving term.

The Effect of a Thirty-Day Harvest on a Population

Suppose that the growth of a population (think "fish") may be modeled by a law of exponential change. What is the effect of harvesting the population at a constant rate for a fixed time span (think "fishermen" and "fishing season")? Denote by $y(t)$ the size of the population (in tons), measure time in days, and suppose that $y(0) = A$, a positive constant. Let's suppose that the rate coefficient is another positive constant k measured in $(\text{days})^{-1}$, and that harvesting is carried out at the rate H tons per day over the 30 days of the harvest season. This situation is modeled by the IVP

$$y'(t) = ky(t) - H\,\text{step}(30 - t), \qquad y(0) = A \qquad (16)$$

Our goal is to find an expression for $y(t)$ for $t \geq 0$.

Applying the Laplace transform to IVP (16) and using (15), we have for $s > k$

$$s\mathcal{L}[y] - A = k\mathcal{L}[y] - \frac{H}{s}(1 - e^{-30s})$$

$$\mathcal{L}[y] = \frac{A}{s - k} - \frac{H}{s(s - k)}(1 - e^{-30s})$$

$$= \frac{A}{s - k} + \frac{H}{k}\left(\frac{1}{s} - \frac{1}{s - k}\right)(1 - e^{-30s}) \qquad (17)$$

where we have used partial fractions to write

$$\frac{1}{s(s-k)} = -\frac{1}{k}\left(\frac{1}{s} - \frac{1}{s-k}\right)$$

Rearranging the terms on the right of (17) to find the inverse transforms more easily,

$$\mathcal{L}[y] = \frac{A - H/k}{s-k} + \frac{H}{k}\frac{1}{s}(1 - e^{-30s}) + \frac{H}{k}e^{-30s}\frac{1}{s-k}$$

Using $\mathcal{L}[e^{kt}] = (s-k)^{-1}$ and formulas (15) and (12), we move back into the t-domain:

$$y(t) = \left(A - \frac{H}{k}\right)e^{kt} + \frac{H}{k}\operatorname{step}(30 - t) + \frac{H}{k}e^{k(t-30)}\operatorname{step}(t - 30) \qquad (18)$$

Equation (18) represents the changing population only for $y(t) > 0$. So the solution makes sense only for those values of the parameters A, H, and k for which $y(t)$ remains positive. If $H \le Ak$, then $y(t) > 0$ for all $t \ge 0$ If $H > Ak$, it may happen that at some time the population becomes extinct. In fact, if we solve the original ODE over the time interval $0 \le t \le 30$ by the method of integrating factors, then we see that if $H \ge kA(1 - e^{-30k})^{-1}$ the population does indeed become extinct within the thirty day period. The model is not valid after the time of extinction.

In Figure 6.2.1 we have plotted the population curves for several values of the harvesting rate H. The two highest harvesting rates lead to extinction.

Partial Fractions

Because they are useful in finding inverse transforms we have gathered together in one place several partial fraction formulas. The entries in Table 6.2.1 will be all you need to help you solve the problems in this section.

TABLE 6.2.1 Partial fraction formulas. A, B, C, a, b are constants; s is a variable.

- For $a \neq b$: $\dfrac{As + B}{(s-a)(s-b)} = \dfrac{(Aa + B)/(a-b)}{s-a} + \dfrac{(Ab + B)/(b-a)}{s-b}$

- $\dfrac{As + B}{(s-a)^2} = \dfrac{A}{s-a} + \dfrac{Aa + B}{(s-a)^2}$

- For $a \neq b$: $\dfrac{As^2 + Bs + C}{(s-a)(s-b)^2} = \dfrac{D}{s-a} + \dfrac{E}{s-b} + \dfrac{F}{(s-b)^2}$

  where $D = \dfrac{Aa^2 + Ba + C}{(b-a)^2}$, $E = A - D$, $F = \dfrac{Ab^2 + Bb + C}{b-a}$

- $\dfrac{1}{s(s^2 + a^2)} = \dfrac{1/a^2}{s} - \dfrac{s/a^2}{s^2 + a^2}$

Comments

The Shifting Theorem suggests that the transform calculus could be used to solve a *difference equation* like

$$x(t) = ax(t - b) + f(t - b) \tag{19}$$

where a and $b > 0$ are constants and f is a given function. In the difference equation, the current value $x(t)$ of the unknown function depends explicitly on the value at the earlier time $t - b$ (see Problem 10 for an example).

As we will show in Section 6.3, the Shifting Theorem is also the key to solving the differential-delay equations that appeared in the car-following model of Section 6.1. The transform calculus can also be used to solve a linear system (Problem 9).

We have focused our attention mostly on techniques for calculating the transform of a function defined in the time domain (i.e., functions of the t variable). However, this is only half of what is needed when using the Laplace transform to solve an IVP. Once the transform of the solution is found, we still have to invert a function in the s-domain back into a function in the time domain. The partial fraction decompositions in Table 6.2.1 are useful for this purpose.

PROBLEMS

1. Find the Laplace transform of each of the following functions.

   (a) $\sinh at$　　　　　　　　　(b) $\cosh at$　　　　(c) $t^2 e^{at}$

   (d) $(1 + 6t)e^{at}$　　　　　　(e) $te^{2t} f(t)$　　　(f) $(D^2 + 1) f(t)$

   (g) $(t + 1)\,\text{step}(t - 1)$　　　(h) $te^{at}\,\text{step}(t - 1)$　(i) $e^{at}[\text{step}(t - 1) - \text{step}(t - 2)]$

   (j) $(t - 2)[\text{step}(t - 1) - \text{step}(t - 3)]$

2. (*Using Transform Tables*). Use and identify the formulas and examples of this section or the transform tables at the end of the chapter to find $\mathcal{L}[f]$ if f is given and f if $\mathcal{L}[f]$ is given.

   (a) $f(t) = 3\,\text{step}(t - 2) - 3\,\text{step}(t - 5)$　　(b) $f(t) = 2\sin t\,\text{step}(\pi - t)$

   (c) $f(t) = t^{12} e^{5t}$　　　　　　　　　　　　　(d) $f(t) = 6t\sin 3t$

   (e) $\mathcal{L}[f] = \sqrt{s + 2} - \sqrt{s}$　　　　　　　　(f) $\mathcal{L}[f] = \ln(s + 5) - \ln(s - 2)$

3. Use the Laplace transform to solve each IVP. [*Hint*: Use transform tables, partial fraction formulas in Table 6.2.1, and Theorem 6.2.4.]

   (a) $y'' - y' - 6y = 0,$　　　　　　$y(0) = 1,\quad y'(0) = -1$

   (b) $y'' + y = \sin t,$　　　　　　　$y(0) = 0,\quad y'(0) = 1$

   (c) $y'' - 2y' + 2y = 0,$　　　　　$y(0) = 0,\quad y'(0) = 1$

   (d) $y'' + 4y' + 4y = e^t,$　　　　$y(0) = 1,\quad y'(0) = 1$

   (e) $y'' - 2y' + y = \text{step}(t - 1),$　$y(0) = 1,\quad y'(0) = 0$

   (f) $y'' + 2y' - 3y = \text{step}(1 - t),$　$y(0) = 1,\quad y'(0) = 0$

4. Find the Laplace transform of each of the following.

   (a) $\int_0^t (x - 1)e^x\,dx$　(b) $\int_0^t (x^2 - 2x)\,dx$　(c) $\int_0^t \sin(x - \pi/4)e^x\,dx$　(d) $\int_0^t e^{(x-a)} \cos x\,dx$

5. (*Square Pulse*). Show that for $0 < a < b$, $\text{step}(t - a) - \text{step}(t - b) = \text{sqp}(t - a, b - a)$, and find $\mathcal{L}[\text{sqp}(t - 1, 3)]$.

6. (*Radioactive Decay*). A sample of a radioactive element decays at a rate proportional to the amount y present, where k_1 is the proportionality constant. Suppose that at some time $t = a$ more of the element is allowed to enter the sample from outside at a constant rate k_2, and at a later time $t = b$ this flow is stopped.

(a) Show that this system can be modeled by the IVP

$$y' = -k_1 y + k_2[\text{step}(t-a) - \text{step}(t-b)], \qquad y(0) = y_0, \quad 0 < a < b$$

(b) Use the Laplace transform to find $y(t)$. Plot the graph of $y(t)$ with $k_1 = 1$, $k_2 = 0.5$, $a = 1$, $b = 10$, $y_0 = 1$, $0 \le t \le 20$.

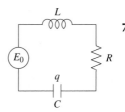

7. (*Circuits*). A basic LC or RLC circuit has a voltage source that can be switched on and off. In the problems below denote the roots of the appropriate characteristic polynomial by r_1 and r_2 (assumed distinct, but not necessarily real). Express your results as functions of t, a, r_1, and r_2.

(a) Find the charge $q(t)$ if $R = 0$ and the switch is turned on at time $t = a > 0$:

$$L\frac{d^2q}{dt^2} + \frac{1}{C}q = E_0\,\text{step}(t-a), \qquad q(0) = 0, \quad q'(0) = 0$$

(b) Find $q(t)$ if $R = 0$ and the switch is turned on at $t = a$ and off at $t = b$, $0 < a < b$:

$$L\frac{d^2q}{dt^2} + \frac{1}{C}q = E_0[\text{step}(t-a) - \text{step}(t-b)], \qquad q(0) = 0, \quad q'(0) = 0$$

(c) Repeat part (a) for the RLC circuit modeled by the IVP

$$L\frac{d^2q}{st^2} + R\frac{dq}{dt} + \frac{1}{C}q = E_0\,\text{step}(t-a), \qquad q(0) = 0, \quad q'(0) = 0$$

(d) Repeat part (b) for the RLC circuit modeled by the IVP

$$L\frac{d^2q}{dt^2} + R\frac{dq}{dt} + \frac{1}{C}q = E_0[\text{step}(t-a) - \text{step}(t-b)], \qquad q(0) = 0, \quad q'(0) = 0$$

8. (*Maintenance of a Game Species*). In the model of the 30-day harvest, find a relation among A, h, and k that will ensure that exactly 330 days after the end of the harvest, the population will once more be at the initial level A.

www 9. (*Laplace Transform for Linear Systems*). Laplace transforms may be used to solve some linear systems. For example, if $x_1' = ax_1 + bx_2 + f_1(t)$, $x_2' = cx_1 + dx_2 + f_2(t)$, with a, b, c, and d constants, then applying the transform to the first ODE, we have $s\mathcal{L}[x_1](s) - x_1(0) = a\mathcal{L}[x_1](s) + b\mathcal{L}[x_2](s) + \mathcal{L}[f_1](s)$. There is a similar equation for $s\mathcal{L}[x_2](s) - x_2(0)$. These two equations may be solved for $\mathcal{L}[x_1]$ and for $\mathcal{L}[x_2]$ in terms of a, b, c, d, $\mathcal{L}[f_1]$, and $\mathcal{L}[f_2]$. Then $x_1(t)$ and $x_2(t)$ are obtained by inverse transforms. Use this approach to solve the following systems.

(a) $x_1' = 3x_1 - 2x_2 + t$, $\quad x_2' = 5x_1 - 3x_2 + 5$, $\quad x_1(0) = x_2(0) = 0$

(b) $x_1' = x_1 + 3x_2 + \sin t$, $\quad x_2' = x_1 - x_2$, $\quad x_1(0) = 0$, $x_2(0) = 1$

10. (*Difference Equation*). Solve the difference equation $3x(t) - 4x(t-1) = 1$, where $x(t) = 0$ if $t \le 0$. [*Hint*: First show that $\mathcal{L}[x(t-1)](s) = e^{-s}\mathcal{L}[x]$. Next show that

$$\mathcal{L}[x] = 1/[s(3 - 4e^{-s})] = (1/3s)(1 + 4/3e^{-s} + \cdots + (4/3e^{-s})^n + \cdots)$$

by using geometric series. Use the transform tables to find a series for $x(t)$.]

6.3 Applications of the Transform: Car Following

The driving force of a natural process is often periodic: a periodic voltage drives an electric circuit, a pulsating magnetic field acts on an iron weight suspended by a spring, medication enters the GI-tract every six hours. Any practical calculus of transforms must be able to handle periodic functions ranging from smoothly contoured sinusoids to a square or triangular wave train. We will find the transform of such functions in connection with a circuit having a square wave input.

We will also use the Laplace transform approach to solve the differential-delay equations for the car-following model introduced in Section 6.1.

Transform of a Periodic Function

Our task will be made simpler if we first look at geometric series. If $|x| < 1$, then the *geometric series* $1 + x + \cdots + x^n + \cdots = \sum_{n=0}^{\infty} x^n$ sums to $1/(1 - x)$. We will use the expansion primarily with $x = \pm e^{-as}$, where $-as < 0$, so $|x| < 1$. We have the series

$$\frac{1}{1 - e^{-as}} = \sum_{n=0}^{\infty} e^{-ans}, \qquad \frac{1}{1 + e^{-as}} = \sum_{n=0}^{\infty} (-1)^n e^{-ans} \qquad (1)$$

These formulas will be useful later on.

Next, we derive the transform formula for a periodic function.

THEOREM 6.3.1

Transforming Periodic Functions. Suppose that f is piecewise continuous and periodic with period T. Then

$$\mathcal{L}[f] = \frac{1}{1 - e^{-Ts}} \int_0^T e^{-su} f(u)\, du \qquad (2)$$

To see this, split $[0, \infty)$ into the period intervals $[0, T], [T, 2T], \ldots$. Then

$$\mathcal{L}[f] = \int_0^{\infty} e^{-st} f(t)\, dt = \int_0^T e^{-st} f(t)\, dt + \int_T^{2T} e^{-st} f(t)\, dt + \cdots \qquad (3)$$

Make the variable change $t = u + nT$ in the integral from nT to $(n+1)T$, and use the fact that by periodicity $f(t + nT) = f(t)$, to obtain for each n:

$$\int_{nT}^{(n+1)T} e^{-st} f(t)\, dt = \int_0^T e^{-s(u+nT)} f(u + nT)\, du = e^{-Tns} \int_0^T e^{-su} f(u)\, du \qquad (4)$$

We see from (3) and (4) that

$$\mathcal{L}[f] = \left\{ \sum_{n=0}^{\infty} e^{-Tns} \right\} \left\{ \int_0^T e^{-su} f(u)\, du \right\} = \frac{\int_0^T e^{-su} f(u)\, du}{1 - e^{-Ts}}$$

because by formula (1), $\sum_{n=0}^{\infty} e^{-Tns} = (1 - e^{-Ts})^{-1}$. Formula (2) follows.

Let's see how the theorem applies to a specific periodic function.

EXAMPLE 6.3.1

Transform of a Sinusoid

Although we have already shown that $\mathcal{L}[\sin at] = a(s^2 + a^2)^{-1}$, we will now derive the formula using Theorem 6.3.1. The period here is $T = 2\pi/a$. By formula (2) and a table of integrals, we have

☞ Table 1.3.1 is helpful in evaluating this integral.

$$\mathcal{L}[\sin at] = \frac{1}{1 - e^{-2\pi s/a}} \int_0^{2\pi/a} e^{-su} \sin au \, du = \frac{a}{s^2 + a^2}$$

☞ More about on-off functions in Appendix B.1.

We turn now to the on-off wave trains that are used in science and engineering.

EXAMPLE 6.3.2

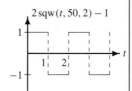

$2\,\mathrm{sqw}(t, 50, 2) - 1$

Transform of a Square Wave

Let's find the transform of the periodic function $f = 2\,\mathrm{sqw}(t, 50, 2) - 1$ by using formula (2). In this case the period is 2, and the integral in formula (2) becomes

$$\int_0^2 e^{-su} f(u) \, du = \int_0^1 e^{-su} \, du - \int_1^2 e^{-su} \, du = \frac{1}{s}(1 - 2e^{-s} + e^{-2s}) = \frac{1}{s}(1 - e^{-s})^2$$

So by formula (2) and the definition of the hyperbolic tangent, we have

☞ Recall that

$$\tanh x = \frac{e^x - e^{-x}}{e^x + e^{-x}}$$

$$\mathcal{L}[f] = \frac{1}{1 - e^{-2s}} \frac{1}{s}(1 - e^{-s})^2 = \frac{(1 - e^{-s})^2}{s(1 - e^{-s})(1 + e^{-s})}$$

$$= \frac{1 - e^{-s}}{s(1 + e^{-s})} = \frac{1}{s} \frac{e^{s/2} - e^{-s/2}}{e^{s/2} + e^{-s/2}} \tag{5}$$

$$= \frac{1}{s} \tanh\left(\frac{s}{2}\right), \qquad s > 0 \tag{6}$$

Formula (6) is Table entry III.2 at the end of the chapter.

A more useful formula for $\mathcal{L}[f]$ can be obtained by applying formula (1) to expand $(1 + e^{-s})^{-1}$ in a geometric series. Doing this we have [from (5)]

☞ Sometimes this is easier than (5) to use in applications.

$$\mathcal{L}[f] = \frac{1 - e^{-s}}{s(1 + e^{-s})} = \frac{1 - e^{-s}}{s} \cdot \frac{1}{1 + e^{-s}} = \frac{1 - e^{-s}}{s} \sum_{n=0}^{\infty} (-1)^n e^{-ns}$$

$$= \frac{1}{s}(1 - e^{-s})(1 - e^{-s} + e^{-2s} - e^{-3s} + \cdots) = \frac{1}{s}\left[1 + 2\sum_{n=1}^{\infty} (-1)^n e^{-ns}\right] \tag{7}$$

which is another way to write $\mathcal{L}[f]$.

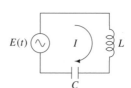

$E(t)$ I L C

A Driven *LC* Circuit

The current I in an *LC* circuit satisfies the ODE $LI'' + I/C = E'$, where E is the input voltage. Divide the ODE by L, set $1/LC = \omega^2$, and suppose that E/L is the triangular wave train $2\,\mathrm{trw}(t, 100, 2) - 1$ and that $I(0) = 0$, $I'(0) = 0$. Notice that the derivative of $2\,\mathrm{trw}(t, 100, 2) - 1$ is $4\,\mathrm{sqw}(t, 50, 2) - 2$ (except at the corner points where there is no derivative). We have the IVP

$$I''(t) + \omega^2 I(t) = \frac{E'(t)}{L} = \frac{2}{L}\left[2\,\mathrm{sqw}(t, 50, 2) - 1\right], \quad I(0) = 0, \quad I'(0) = 0 \quad (8)$$

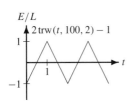

E/L

$2\,\mathrm{trw}(t, 100, 2) - 1$

To be specific, let's assume that I is measured in amperes, the inductance L in henries, C in farads, time in seconds, and E in volts. Sketches of the circuit and the functions E/L and E'/L are given in the margin (for $L = 1$).

Using (5) and the transform formula for $\mathcal{L}[I'']$, we have [since $I(0) = I'(0) = 0$]

$$s^2 \mathcal{L}[I] + \omega^2 \mathcal{L}[I] = \frac{2}{L}\frac{1 - e^{-s}}{s(1 + e^{-s})}, \quad s > 0$$

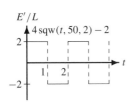

E'/L

$4\,\mathrm{sqw}(t, 50, 2) - 2$

Solving this equation for the transform $\mathcal{L}[I]$ and using formula (7), we have

$$\begin{aligned}
\mathcal{L}[I] &= \frac{2/L}{s(s^2 + \omega^2)}\left(1 + 2\sum_{n=1}^{\infty}(-1)^n e^{-ns}\right) \\
&= \frac{2/L}{s(s^2 + \omega^2)} + \frac{4}{L}\sum_{n=1}^{\infty}(-1)^n \frac{e^{-ns}}{s(s^2 + \omega^2)}
\end{aligned} \quad (9)$$

If we can find $\mathcal{L}^{-1}[1/s(s^2 + \omega^2)]$, we can use Table entry I.3 to determine $I(t)$.

To find this inverse transform, let's decompose $1/s(s^2 + \omega^2)$ via partial fractions (see Table 6.2.1):

$$\frac{1}{s(s^2 + \omega^2)} = \frac{1/\omega^2}{s} - \frac{s/\omega^2}{s^2 + \omega^2} \quad (10)$$

Using (10) and Table entry II.6, we have

$$\begin{aligned}
\mathcal{L}^{-1}\left[\frac{1}{s(s^2 + \omega^2)}\right] &= \frac{1}{\omega^2}\mathcal{L}^{-1}\left[\frac{1}{s}\right] - \frac{1}{\omega^2}\mathcal{L}^{-1}\left[\frac{s}{s^2 + \omega^2}\right] \\
&= \frac{1}{\omega^2} - \frac{1}{\omega^2}\cos \omega t
\end{aligned} \quad (11)$$

Applying $\mathcal{L}^{-1}$ to both sides of (9) and interchanging $\mathcal{L}^{-1}$ and $\sum$, we have that

$$I(t) = \frac{2}{L}\mathcal{L}^{-1}\left[\frac{1}{s(s^2 + \omega^2)}\right] + \frac{4}{L}\sum_{n=1}^{\infty}(-1)\mathcal{L}^{-1}\left[e^{-ns} \cdot \frac{1}{s(s^2 + \omega^2)}\right]$$

Using (11) and formula (12) of the Shifting Theorem 6.2.4, we have that

$$I(t) = \frac{2}{L\omega^2}(1 - \cos \omega t) + \frac{4}{L\omega^2}\sum_{n=1}^{\infty}(-1)^n[1 - \cos \omega(t - n)]\,\mathrm{step}(t - n) \quad (12)$$

This form of the solution looks daunting, but for any positive value of t, there are only a finite number of terms in the sum because the remaining terms are turned off by

$$I'' + I = 4\,\text{sqw}(t, 50, 2) - 2, \quad I(0) = 0, \; I'(0) = 0$$

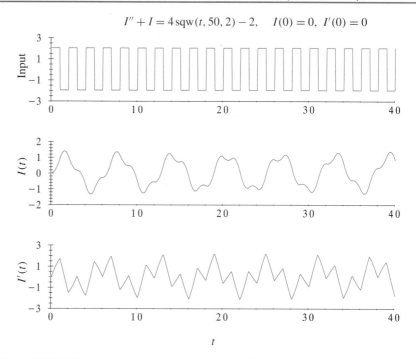

FIGURE 6.3.1 Input-response curves for a driven *LC* circuit [IVP (8)].

the step functions. For example, $I(t)$ is given by

$$\frac{2}{L\omega^2}(1 - \cos \omega t), \qquad\qquad\qquad\qquad\qquad\qquad 0 \le t < 1$$

$$\frac{2}{L\omega^2}(1 - \cos \omega t) - \frac{4}{L\omega^2}(1 - \cos \omega(t - 1)), \qquad\qquad 1 \le t < 2$$

$$\frac{2}{L\omega^2}(1 - \cos \omega t) - \frac{4}{L\omega^2}(1 - \cos \omega(t - 1)) + \frac{4}{L\omega^2}(1 - \cos \omega(t - 2)), \quad 2 \le t < 3$$

See Figure 6.3.1 for the plot of the square wave input E'/L, the solution $I(t)$ of IVP (8), and $I'(t)$, where $\omega^2 = 1$ and $L = 1$.

Now let's solve the car-following model introduced in Section 6.1.

Solution of the Car-Following Model of Section 6.1

In heavy traffic or in constricted passages such as bridges or tunnels, cars follow one another with little or no lane changing or passing. Rear-end collisions are a common occurrence in this setting. Bad driving is the cause of many of these accidents, but some seem to occur for no apparent reason. Mathematical models of car following have been created in an attempt to understand the phenomenon and to help design traffic controls and driving codes that promote safe, efficient traffic flow. We analyze a car-following model, specialized to a line of cars stopped at a red light. What happens when the light changes and the lead car accelerates? What could cause a collision somewhere down the line as each car accelerates in turn?

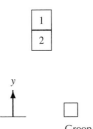

y

Green light

As we showed in Section 6.1, there is a linear differential-delay model for the situation where there are three identical cars in line:

$$\begin{aligned}
y'_j &= v_j, \qquad j = 1, 2, 3 \\
v'_1 &= \alpha \\
v'_2(t+T) &= \lambda[v_1(t) - v_2(t)] \\
v'_3(t+T) &= \lambda[v_2(t) - v_3(t)]
\end{aligned} \tag{13}$$

with initial data given by

$$y_1(0) = 0, \quad y_2(0) = -20, \quad y_3(0) = -40, \quad v_j(0) = 0, \quad j = 1, 2, 3 \tag{14}$$

where $y_j(t)$ is the distance from the light to the bumper of the j-th car, $v_j(t)$ is that car's velocity, $\alpha > 0$ is the constant acceleration of the lead car, T is the common delay (or response) time and λ is the common sensitivity coefficient. The positions of the cars at time t are found by integrating velocities and using initial data:

$$\begin{aligned}
y_1(t) &= \frac{1}{2}\alpha t^2, & t \geq 0 \\
y_j(t) &= -20(j-1) + \int_0^t v_j(s)\, ds, & j = 2, 3, \quad t \geq 0
\end{aligned} \tag{15}$$

We will use the Laplace transform to solve the delay equations in system(13) for the velocities $v_j(t)$. Then an application of (15) will give us the locations $y_j(t)$ of the cars at time t, and we can determine whether the cars have crashed.

Applying the transform to the velocity equations in (13) and using the data in (14), we have

$$\begin{aligned}
\mathcal{L}[v_1] &= \frac{\alpha}{s^2} \\
\mathcal{L}[v'_j(t+T)] &= \lambda\{\mathcal{L}[v_{j-1}] - \mathcal{L}[v_j]\}, \quad j = 2, 3
\end{aligned} \tag{16}$$

Next, we use the fact that $v'_j(t+T) = v'_j(t+T)\,\text{step}(t+T)$ since $t + T \geq 0$, and formula (12) in the Shifting Theorem to evaluate $\mathcal{L}[v'_j(t+T)]$ as lead.tex follows:

$$\mathcal{L}[v'_j(t+T)] = \mathcal{L}[v'_j(t+T)\,\text{step}(t+T)] = e^{Ts}\mathcal{L}[v'_j(t)] = e^{Ts}s\mathcal{L}[v_j] \tag{17}$$

Using formulas (16) and (17) for $j = 2, 3$, we have

$$\begin{aligned}
\mathcal{L}[v_2] &= \frac{\lambda}{\lambda + se^{Ts}}\mathcal{L}[v_1] = \frac{\lambda}{\lambda + se^{Ts}} \cdot \frac{\alpha}{s^2} \\
\mathcal{L}[v_3] &= \frac{\lambda}{\lambda + se^{Ts}}\mathcal{L}[v_2] = \left(\frac{\lambda}{\lambda + se^{Ts}}\right)^2 \frac{\alpha}{s^2}
\end{aligned} \tag{18}$$

The velocity of the j-th car in line may be found by applying $\mathcal{L}^{-1}$ to (18). First, however, we will rewrite $\mathcal{L}[v_j]$ in (18) by using a binomial expansion:

☞ Binomial expansions appear under item 9 in Appendix B.2

$$\begin{aligned}
\mathcal{L}[v_j] &= \frac{\alpha}{s^2}\left(\frac{\lambda}{se^{Ts}}\right)^{j-1}\left(1 + \frac{\lambda}{s}e^{-Ts}\right)^{-(j-1)} \\
&= \alpha\lambda^{j-1}\left\{\frac{e^{-(j-1)Ts}}{s^{j+1}} - \frac{\lambda(j-1)e^{-jTs}}{s^{j+2}} + \frac{\lambda^2(j-1)je^{-(j+1)Ts}}{2!s^{j+3}} - \cdots\right\}
\end{aligned} \tag{19}$$

So from formula (12) of Theorem 6.2.4 and the fact that $\mathcal{L}^{-1}[s^{-n}] = t^{n-1}/(n-1)!$, we have from (19) that for $j = 2, 3$

$$
\begin{aligned}
v_j(t) = \alpha\lambda^{j-1}\Big\{ &\frac{1}{j!}[t-(j-1)T]^j \operatorname{step}[t-(j-1)T] \\
&-\lambda\frac{j-1}{(j+1)!}[t-jT]^{j+1}\operatorname{step}[t-jT] \\
&+\lambda^2\frac{(j-1)j}{2!(j+2)!}[t-(j+1)T]^{j+2}\operatorname{step}[t-(j+1)T]-\cdots\Big\}
\end{aligned}
\tag{20}
$$

Integrating each side of formula (20) from 0 to t and using the formula in (15), we have for $j = 2, 3$

☞ This formula looks useful, but it is not hard to use!

$$
\begin{aligned}
y_j(t) = -20(j-1)+\alpha\lambda^{j-1}\Big\{ &\frac{1}{(j+1)!}[t-(j-1)T]^{j+1}\operatorname{step}[t-(j-1)T] \\
&-\lambda\frac{j-1}{(j+2)!}[t-jT]^{j+2}\operatorname{step}[t-jT] \\
&+\lambda^2\frac{(j-1)j}{2!(j+3)!}[t-(j+1)T]^{j+3}\operatorname{step}[t-(j+1)T]-\cdots\Big\}
\end{aligned}
\tag{21}
$$

The location of the lead car is given by

$$
y_1(t) = \frac{1}{2}\alpha t^2, \qquad t \geq 0
\tag{22}
$$

The series in formula (21) looks formidable, but appearances are deceiving, and it is not hard to use. For a fixed time $t > 0$, there are only a few nonvanishing terms, and they are all at the start of the series, because the remaining terms are switched off by the step functions. For example, for the second car in line (i.e., $j = 2$) we see that its location at time t, $0 \leq t \leq 3T$, is given by

$$
y_2(t) = \begin{cases} -20, & 0 \leq t < T \\ -20+\dfrac{\alpha\lambda}{6}(t-T)^3, & T \leq t < 2T \\ -20+\dfrac{\alpha\lambda}{6}(t-T)^3-\dfrac{\alpha\lambda^2}{24}(t-2T)^4, & 2T \leq t < 3T \end{cases}
\tag{23}
$$

We have reached the stage of model building and testing where we need to select specific values for the parameters and interpret the results. Will there be collisions?

Let's suppose that the lead car accelerates at $\alpha = 6$ ft/sec^2. Then the distance between the first two cars in line can be obtained by using $y_1(t) = 3t^2$, $t \geq 0$, and $y_2(t)$ from (23). So for $0 \leq t \leq 3T$, the separation is

$$
y_1(t)-y_2(t) = \begin{cases} 3t^2+20, & 0 \leq t \leq T \\ 3t^2+20-\lambda(t-T)^3, & T \leq t \leq 2T \\ 3t^2+20-\lambda(t-T)^3+\lambda^2(t-2T)^4/4, & 2T \leq t \leq 3T \end{cases}
\tag{24}
$$

If $y_1(t^*)-y_2(t^*) = 15$, the first and second cars collide at time t^*. Evidently, this can't happen in the first time span of T seconds. A crash will occur during the second span of T seconds, $T \leq t \leq 2T$, if $y_1(t)-y_2(t)$ goes from its value of $3T^2+20$ at

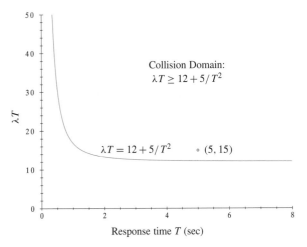

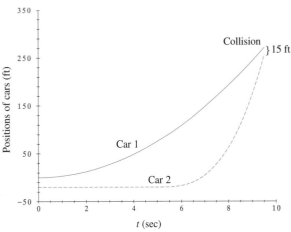

FIGURE 6.3.2 If $(T, \lambda T)$ lies above the curve $\lambda T = 12 + 5/T^2$, a crash occurs in time interval $[T, 2T]$. See equation (25).

FIGURE 6.3.3 Collision between first car and second car at $t = 9.52$ sec: $\lambda = 3$, $T = 5$. The point $T = 5$, $\lambda T = 15$ is in the collision domain.

$t = T$ to a value less than or equal to 15 at $t = 2T$. So a condition for a guaranteed collision in this second time span is that

$$y_1(2T) - y_2(2T) = 12T^2 + 20 - \lambda T^3 \le 15$$

This condition can be written as

$$5 \le T^2(\lambda T - 12), \quad \text{or} \quad \lambda T \ge 12 + 5/T^2 \tag{25}$$

Inequality (25) says that T and the product λT are the important parameters. So let's sketch the curve defined by $\lambda T = 12 + 5/T^2$ using T- and λT-axes (Figure 6.3.2). If the point $(T, \lambda T)$ lies anywhere on or above the curve, the second car hits the first in the second span of T seconds; the point $(T, \lambda T)$ is said to be in the *collision domain*.

Suppose that $T = 5$ sec and that $\lambda = 3$ sec^{-1}. It is straightforward to check that $(T, \lambda T) = (5, 15)$ belongs to the collision domain. So the second car rear-ends the first at some time t, $5 \le t \le 10$. Figure 6.3.2 shows that the collision occurs [i.e., $y_1^*(t) - y_2^*(t) = 15$] at about $t^* = 9.52$ seconds and 250 feet down the road from the stoplight.

A response time of $T = 5$ sec and sensitivity $\lambda = 3$ sec^{-1} (as above) leads to a crash during the first 10 sec, while the same response time but the lower sensitivity of $\lambda = 1$ sec^{-1} does not. One might imagine a lower sensitivity to be inherently more dangerous. In fact, just the opposite is true since a time lag is also involved. The response doesn't occur at the time an observation is made, but T seconds later, at which time the situation may have changed. High sensitivity may lead to a large response to a large velocity difference that no longer exists.

Comments

Mathematical models of traffic flow are relatively new and have not acquired the acceptance of, say, Newton's laws or the laws of electrical circuits. There is too much of the unknown, unknowable, and random in traffic flow, as in most human processes, for there to be a single comprehensive model. There are a good many other models of car-following, models that include other stimuli such as the separation distance between cars or the relative velocity or separation of the car behind as well as the car ahead. Some of these models are given in the Problems.

PROBLEMS

1. Using the material in Sections 6.1–6.3 and in integration tables, find the Laplace transform of each of the following functions. Use the tables at the end of the chapter to check your answers.

    (a) $e^t \sin t$ (b) $\sin^2 t$ (c) $\cos^2 t$ (d) $t \sin at$

    (e) $t^2 \sin at$ (f) $\cos^3 t$ (g) $e^{-3t} \cos(2t + \pi/4)$ (h) $t^2 e^t \cos t$

2. Find the inverse Laplace transform without using tables.

    (a) $\dfrac{1 + e^{-s}}{s}$ (b) $\dfrac{3e^{-2s}}{3s^2 + 1}$ (c) $\dfrac{1}{(s-a)^n}$ (d) $\dfrac{1 - e^{-s}}{s^2}$ (e) $\dfrac{(s-1)e^{-s} + 1}{s^2}$ (f) $\ln \dfrac{s+3}{s+2}$

3. Use partial fractions to find the inverse Laplace transform of each of the following.

    (a) $\dfrac{1}{s(s+1)}$ (b) $\dfrac{1}{s(s+2)^2}$ (c) $\dfrac{1}{(s-a)(s-b)}$, $a \neq b$

    (d) $\dfrac{s^2 + 3}{(s-1)^2(s+1)}$ (e) $\dfrac{3s+1}{(s^2 + 2s + 2)(s-1)}$ (f) $\dfrac{s+1}{(s-2)(s^2+9)}$

4.   Use Theorem 6.2.1, partial fractions (Table 6.2.1), and the tables at the end of the chapter to solve the following initial value problems. Plot the solution curves.

    (a) $y'' + 6y' + 5y = t$, $y(0) = y'(0) = 0$

    (b) $y'' + 2y' + y = e^t$, $y(0) = y'(0) = 0$

    (c) $y'' - 2y' + 2y = \sin t$, $y(0) = y'(0) = 0$

    (d) $y'' - 4y' + 4y = 2e^t + \cos t$, $y(0) = 3/25$, $y'(0) = -4/25$

    (e) $y'' - 2y' + y = e^t \sin t$, $y(0) = y'(0) = 0$

    (f) $y'' + 2y' + y = te^{-t}$, $y(0) = 1$, $y'(0) = -2$

    (g) $y''' + y'' + 4y' + 4y = -2$, $y(0) = 0$, $y'(0) = 1$, $y''(0) = -1$

www   5. (*LC Circuit*). The response $I(t)$ of the IVP $I'' + \omega^2 I = 4 \, \text{sqw}(t, 50, T) - 2$, $I(0) = 0$, $I'(0) = 0$, is sensitive to changes in ω and in T. Plot the input and the response for $0 \le t \le 50$ $[0 \le t \le 100$ in part(**b**)] if ω and T are as given below. Does the response appear to be periodic? If it is, what is the period? If not periodic, describe and explain what the response is doing.

    (a) $\omega^2 = 1$, $T = 5\pi$ (b) $\omega^2 = 20$, $T = 5\pi$ (c) $\omega^2 = 1$, $T = 2\pi$

6. (*Transform of Square Wave*). Let $f(t) = A \, \text{sqw}(t, 50, 2a) - B$, A and B constants. Show that

$$\mathcal{L}[f] = \frac{A}{s}\left[1 + \sum_{1}^{\infty}(-1)^n e^{-nas}\right] - \frac{B}{s}$$

[*Hint*: See Example 6.3.2. Use the linearity of $\mathcal{L}$.]

7. (*Square Wave Inputs*). Solve each IVP below. Plot the input and the output. [*Hint*: Use the expansion given in Problem 6.]

(a) (*Low-Pass Filter*). $y' + y = \text{sqw}(t, 50, T)$, $y(0) = 0$. First let $T = 30$, $0 \le t \le 100$; then let $T = 1$, $0 \le t \le 10$. The IVP models a low-pass communication channel that transmits low-frequency signals with much less distortion than high-frequency signals. See Example 2.2.7.

(b) (*Beats*). $y'' + 4y = \text{sqw}(t, 50, T)$, $y(0) = 0$, $y'(0) = 0$, $0 \le t \le 60$. First let $T = 7$ and then $T = 3$. Plot the input, $y(t)$, and $y'(t)$. Explain why pronounced beats appear when $T = 3$. [*Hint*: The natural period is $\pi = 3.14159\ldots$, while the period of the driving term is 3. See Section 4.2 for a discussion of beats.]

8. (*Transform of Triangular Wave*). Suppose that $f(t) = A \, \text{trw}(t, 100, a) - B$, A and B constants. Using the fact that $\mathcal{L}[\text{trw}(t, 100, a)] = 2(as^2)^{-1} \tanh(as/4)$, show that

$$\mathcal{L}[f] = \frac{2A}{as^2}\left[1 + 2\sum_{n=1}^{\infty}(-1)^n e^{-nas/2}\right] - \frac{B}{s}$$

 9. (*Triangular Wave Input*). Solve the IVP $y' + y = \text{trw}(t, 100, T)$, $y(0) = 0$. First let $T = 30$, $0 \le t \le 10$; then let $T = 1$, $0 \le t \le 10$. Plot the input and the output. If the IVP models a communication channel, which input, $T = 30$ or $T = 1$, is transmitted with less distortion? Explain. [*Hint*: See Problem 8 and Example 2.2.7.]

10. (*Cold Pills and Transforms*). The linear system

$$x' = 12 \, \text{sqw}(t, 25/3, 6) - k_1 x, \qquad x(0) = 0$$
$$y' = k_1 x - k_2 y, \qquad\qquad\qquad y(0) = 0$$

where $k_1 = 0.6931$ and $k_2 = 0.0231$, models the amounts x and y of antihistimine in the GI-tract and in the bloodstream, respectively. The model was derived in Examples 1.8.5 and 1.8.6. Use the methods of this section to find formulas for the solutions $x(t)$ and $y(t)$ valid over the interval $0 \le t \le 48$.

11. (*Velocity Control Model*). Using the velocity control model in the text, which of the following sets of values of T and λT will result in a collision between the first two cars within the second span of T seconds? [*Hint*: See inequality (25).]

(a) $T = 1$, $\lambda T = 20$ (b) $T = 5$, $\lambda T = 12.3$ (c) $T = 2$, $\lambda T = 14$

12. (*Collision Between Second and Third Car*). The velocity control model in the text is used here to give conditions that the second and third cars collide.

(a) Use (21) to show that

$$y_3(t) = \begin{cases} -40, & 0 \le t \le 2T \\ -40 + \dfrac{\alpha\lambda^2}{24}(t - 2T)^4, & 2T \le t < 3T \end{cases}$$

(b) Use (23) and the formula in part (a) above to show that

$$y_2(t) - y_3(t) = \begin{cases} 20, & 0 \le t < T \\ 20 + \dfrac{1}{6}\alpha\lambda(t - T)^3, & T \le t < 2T \\ 20 + \dfrac{1}{6}\alpha\lambda(t - T)^3 - \dfrac{1}{12}\alpha\lambda^2(t - 2T)^4, & 2T \le t < 3T \end{cases}$$

(c) Suppose that $\alpha = 6 \, \text{ft/sec}^2$. Show that the third car rear-ends the second car for some value of t between $2T$ and $3T$ if the first and second cars have not collided and if $10 \le T^2(\lambda T - 16)\lambda T$.

 (d) Show that if the first two cars don't collide during the time span $0 \le t \le 3T$, then $8 + 8(1 + 10/(64T^2))^{1/2} < \lambda T < 12 + 5/T^2$. [*Hint*: Use (24).] Plot the following three curves in the T,

λT-plane: $\lambda T = 8 + 8(1 + 10/(64T^2))^{1/2}$, $\lambda T = 12 + 5/T^2$, and $T^2(\lambda T - 16)\lambda T = 10$ over the time span $0.4 \leq T \leq 0.7$. Shade the region of points $(T, \lambda T)$ where the second and third cars crash during the time span $0 \leq t \leq 3T$, if the first two cars don't. Find three points in the region and discuss the reality of the corresponding physical situation.

13. (*Separation Control Model*). The distance between cars (i.e., their separation) rather than the relative velocity may be taken as the control mechanism for cars at a stoplight. Assume for simplicity that the cars are "points" rather than the 15-ft cars used in the text.

(a) Write out a car-following model in which the stimulus to the $(j + 1)$-st driver is the separation from the car ahead. Assume a response time of T seconds and a sensitivity of λ seconds^{-1}. [*Hint*: Compare with the model in (13), and make the needed changes.]

(b) Repeat part (a) but in a situation where the stimulus is a linear combination of the velocity control model treated in the text and the separation control model of part (a). Use sensitivity coefficients λ_1 and λ_2 for the respective models.

(c) For the model in part (a), assume that the lead car accelerates from a stop with constant acceleration $\alpha = 6$ ft/sec^2 and that the initial separation between cars is 5 ft. Find the motion of the following car. Are there values of λ, given the delay time $T = 2$, that will result in a rear-end collision for some t, $2 \leq t \leq 4$? [*Hint*: Use (17) and (16) to find $\mathcal{L}[v_2'(t + T)]$. As in (19), you will have to use a geometric series expansion before taking the inverse transform to get $v_2(t)$.]

6.4 Convolution

When we use the Laplace transform to solve an initial value problem, the transform of the solution often involves a product of functions. Although the Shifting Theorem (Theorem 6.2.4) allows us to find the inverse transform of certain special products, we don't have a general procedure for inverting a product of functions. The far-reaching notion of the convolution product fills this gap.

The Convolution Product

Let's denote $\mathcal{L}[f](s)$ by $F(s)$ and $\mathcal{L}[g](s)$ by $G(s)$. We will derive an elegant formula for $\mathcal{L}^{-1}[FG]$. First let's define a new kind of product.

❖ **Convolution.** The *convolution product* $f * g$ of f and g is given by

$$(f * g)(t) = \int_0^t f(t - u)g(u)\,du \tag{1}$$

where f and g are assumed to be in **E**.

Here are the basic properties of the convolution product.

THEOREM 6.4.1

> Properties of Convolution. Suppose that f, g, and h are in **E**. Then
>
> 1. (*Closure*) $f * g$ is in **E**.
> 2. (*Commutativity*) $f * g = g * f$.
> 3. (*Associativity*) $(f * g) * h = f * (g * h)$.
> 4. (*Distributivity*) $(f + g) * h = f * h + g * h$.

 In Problem 6 you are asked to show the other properties.

We will verify only property 2. Set $u = t - v$ in the integral in formula (1):

$$(f * g)(t) = \int_0^t f(t - u)g(u)\, du = \int_t^0 f(v)g(t - v)(-dv)$$

$$= \int_0^t f(v)g(t - v)\, dv = \int_0^t g(t - v)f(v)\, dv = (g * f)(t)$$

which shows that the convolution product is commutative.

The convolution product is the sought-for inverse transform of a product.

THEOREM 6.4.2

> Convolution Theorem. Suppose that f and g are in **E** and that F and G are their respective transforms. Then we have the (equivalent) formulas
>
> $$\mathcal{L}^{-1}[FG] = f * g, \qquad \mathcal{L}[f * g] = FG \tag{2}$$

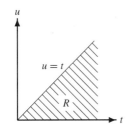

 This is Fubini's Theorem (Theorem B.5.11).

Let's verify the second formula in (2). Suppose that R is the region in the tu-plane defined by $0 \le u \le t$, $0 \le t$ (see the margin figure). Then using (1), we have

$$\mathcal{L}[f * g] = \int_0^\infty e^{-st} \left\{ \int_0^t f(t - u)g(u)\, du \right\} dt$$

$$= \int_0^\infty \left\{ \int_0^t e^{-st} f(t - u)g(u)\, du \right\} dt \tag{3}$$

which may be written as a double integral over the shaded region R; that is,

$$\mathcal{L}[f * g] = \iint_R e^{-st} f(t - u)g(u)\, du$$

Because f and g are in **E**, they are piecewise continuous, so the double integral may be evaluated by an iterated integral in either order, that is, in the order of the iterated integral (3) or in the reverse order. In the reverse order we have

$$\mathcal{L}[f * g] = \int_0^\infty g(u) \left\{ \int_u^\infty e^{-st} f(t - u)\, dt \right\} du \tag{4}$$

Making the change of variable $t = v + u$ in the inner integral in (4), we see that

$$\mathcal{L}[f * g] = \int_0^\infty g(u) \left\{ \int_0^\infty e^{-s(v+u)} f(v)\, dv \right\} du$$

$$= \int_0^\infty g(u)e^{-su}\left\{\int_0^\infty e^{-sv}f(v)\,dv\right\}du$$

$$= \left\{\int_0^\infty e^{-su}g(u)\,du\right\}\left\{\int_0^\infty e^{-sv}f(v)\,dv\right\} = GF = FG$$

So we have established the second formula in (2); the first formula is equivalent.

Note that the inverse transform of a product is *not* the product of the inverse transforms. Here's an example that illustrates how we can use the Convolution Theorem to find the inverse transform of a product if we know the inverse transform of each factor.

EXAMPLE 6.4.1

Using the Convolution Product to Find Inverse Transforms
Since $L[1] = 1/s$ and $L[\sin t] = (s^2+1)^{-1}$, we see that

$$L^{-1}\left[\frac{1}{s(s^2+1)}\right] = L^{-1}\left[\frac{1}{s}\right] * L^{-1}\left[\frac{1}{s^2+1}\right]$$

$$= \int_0^t 1\cdot\sin u\,du = 1 - \cos t$$

which is a formula found earlier by the lengthier method of partial fractions.

Now let's show the real power of the convolution product and use it to generate a solution formula for driven linear ODEs.

IVPs and the Convolution

An example will show just how useful the convolution is in constructing a formula for the solution of an IVP with an arbitrary input function.

EXAMPLE 6.4.2

Solving an IVP by Convolution
Suppose that $f(t)$ is in **E**, and consider the initial value problem

$$y'' + y' - 6y = f(t), \qquad y(0) = y'(0) = 0 \tag{5}$$

Applying the Laplace transform, we see that $(s^2+s-6)L[y] = L[f]$, and so

$$L[y] = L[f]/(s^2+s-6)$$

Taking inverse transforms and using the Convolution Theorem,

$$y = f * L^{-1}\left[\frac{1}{s^2+s-6}\right] \tag{6}$$

Now, by partial fractions

$$\frac{1}{s^2+s-6} = \frac{1}{5}\left[\frac{1}{s-2} - \frac{1}{s+3}\right]$$

and it follows from Table entry II.3 that

$$\mathcal{L}^{-1}\left[\frac{1}{s^2+s-6}\right] = \frac{1}{5}e^{2t} - \frac{1}{5}e^{-3t} \tag{7}$$

Using (7) and the definition of the convolution product, we see that (6) becomes

$$y(t) = \int_0^t \left\{\frac{1}{5}e^{2(t-u)} - \frac{1}{5}e^{-3(t-u)}\right\} f(u)\,du \tag{8}$$

Since the function $y(t)$ in (8) is continuous, we conclude that it must be the unique solution of IVP (5).

Let's use the same approach to solve the general, constant-coefficient linear IVP

$$\begin{aligned} P(D)[y] = (D^2 + aD + b)[y] = f \\ y(0) = y'(0) = 0 \end{aligned} \tag{9}$$

where a and b are real constants. Transforming IVP (9), we have

$$(s^2 + as + b)\mathcal{L}[y] = \mathcal{L}[f]$$

which can be written as

$$P(s)\mathcal{L}[y] = \mathcal{L}[f]$$

Solving for $\mathcal{L}[y]$ we have

$$\mathcal{L}[y] = \frac{\mathcal{L}[f]}{P(s)} \tag{10}$$

If we set $g(t) = \mathcal{L}^{-1}[1/P(s)]$, then the Convolution Theorem gives the solution of IVP (9) as

$$y(t) = \int_0^t g(t-u)f(u)\,du = g * f \tag{11}$$

This justifies the following result:

THEOREM 6.4.3

Solving an IVP with a Convolution Product. If $P(D) = D^2 + aD + b$, where the coefficients are real constants, then the solution of the IVP

$$P(D)[y] = f(t), \qquad y(0) = y'(0) = 0 \tag{12}$$

can be written in terms of $g(t) = \mathcal{L}^{-1}[1/P(s)]$ and the driving term f as

$$y(t) = (g * f)(t) = \int_0^t g(t-u)f(u)\,du \tag{13}$$

Comments

Now we see the power of the convolution product, because it allows us to write the solution of IVP (12) in a form that displays the driving function explicitly. Note that, given $P(D)$, we can construct $g(t) = \mathcal{L}^{-1}[1/P(s)]$ once and for all, regardless of the driving term. This is reminiscent of the Method of Variation of Parameters (Problem 8 in Section 3.7). Although we have restricted attention to the second-order operator $P(D) = D^2 + aD + b$, Theorem 6.4.3 can be extended to any constant coefficient polynomial operator $P(D)$.

PROBLEMS

1. Use convolution to find the Laplace transform of each function. [*Hint*: Each integral is the convolution of two functions.]

   (a) $\int_0^t (t-u)u\,du$ **(b)** $\int_0^t \sin u\,du$

   (c) $\int_0^t (t^2 - 2tu + u^2)\,du$ **(d)** $\int_0^t (\sin t \sin u \cos u - \cos t \sin^2 u)\,du$

2. Find the inverse Laplace transform of each function by using the Convolution Theorem. Write your answers as convolution products, but don't evaluate the integrals. [*Hint*: Write each expression as the product of two functions whose inverse transforms you can find.]

   (a) $\dfrac{s}{(s^2+1)^2}$ **(b)** $\dfrac{s}{(s^2+10)^2}$ **(c)** $\dfrac{1}{s^2(s+1)}$ **(d)** $\dfrac{s}{(s^2+9)^3}$

   (e) $\dfrac{s}{(s+1)(s+2)^3}$ **(f)** $\dfrac{s^2+4s+4}{(s^2+4s+13)^2}$ **(g)** $\dfrac{\mathcal{L}[f]}{s^2+1}$ **(h)** $\dfrac{e^{-3s}\mathcal{L}[f]}{s^3}$

www 3. Write the solution of each IVP as a convolution product (don't evaluate the integral). [*Hint*: Theorem 6.2.4 and partial fraction formulas in Table 6.2.1 may be useful.]

   (a) $y'' + y = t\,\text{step}(t-1),\quad y(0) = y'(0) = 0$ **(b)** $y'' + y = f(t),\quad y(0) = y'(0) = 0$

   (c) $2y'' + y' - y = f(t),\quad y(0) = y'(0) = 0$ **(d)** $y'' + 2y' + y = f(t),\quad y(0) = y'(0) = 0$

   (e) $y'' + 4y = \begin{cases} 0, & 0 \le t < 1 \\ t-1, & 1 \le t < 2, \\ 1, & t \ge 2 \end{cases}\quad y(0) = y'(0) = 0$

4. Use the formula $g(t) = \mathcal{L}^{-1}[1/P(s)]$ to construct $g(t)$ for each of the following differential operators, where $D = d/dt$.

   (a) $D^2 + 6D + 13$ **(b)** $D^2 + (1/3)D + 1/36$ **(c)** $D^3 + 1$ [*Hint*: See Comments.]

5. Use Theorem 6.4.3 to solve the following IVPs.

   (a) $y'' + 6y' + 13y = f(t),\qquad y(0) = y'(0) = 0$

   (b) $y'' + y'/3 + y/36 = f(t),\quad y(0) = y'(0) = 0$

   (c) $y''' + y = t,\qquad\qquad\qquad y(0) = y'(0) = y''(0) = 0$ [*Hint*: See Comments.]

6. (*Convolution Closure, Associativity, Distributivity*). Suppose that f, g, and h are in **E**. Show that $f * g$ is in **E**, that $(f * g) * h = f * (g * h)$, and that $(f + g) * h = f * h + g * h$.

7. Show that $e^{at}(f * g) = (e^{at}f) * (e^{at}g)$.

6.5 Convolution and the Delta Function

The convolution product can be used to find the response of a dynamical system to a sudden force of large amplitude and short duration (like batting a baseball). This can be modeled by the Dirac[4] delta "function." We begin on a hypothetical level, "defining" the delta function by a property we want it to possess [see (1) below]. We then see how far we can develop the theory and applications of the delta function from this shaky beginning.

The Dirac Delta Function and Its Properties

Let's start by extending the domains of all the functions f in $\mathbf{E}$ to the entire real line by taking the value of $f(t)$ to be zero for all negative t. This new collection of functions is denoted by $\mathbf{E}_1$. We can now give the definition of a very strange object, which Dirac called a generalized "function."

> ❖ **Dirac Delta Function.** Suppose that there is an element δ in $\mathbf{E}_1$ such that for every $t_0 \geq 0$ and every function f in $\mathbf{E}_1$ that is continuous at t_0,
>
> $$\int_{-\infty}^{\infty} \delta(t_0 - u) f(u)\, du = f(t_0) \qquad (1)$$
>
> Any such element δ is called a *Dirac delta function.*

Actually there can't be more than one Dirac delta function. To see this, say that there are two functions $\delta(t)$ and $\mu(t)$ in $\mathbf{E}_1$ which satisfy the definition (1). Then for any point t_0 where $\mu(t)$ is continuous,

$$\int_{-\infty}^{\infty} \delta(t_0 - u)\mu(u)\, du = \mu(t_0)$$

On the other hand, if $\delta(t)$ is also continuous at t_0, then the change of variables $v = t_0 - u$ shows that

$$\int_{-\infty}^{\infty} \delta(t_0 - u)\mu(u)\, du = \mu(t_0) = \int_{-\infty}^{\infty} \delta(v)\mu(t_0 - v)\, dv = \delta(t_0)$$

So $\delta = \mu$ because $\delta(t) = \mu(t)$ at all t-values where δ and μ are continuous.

Now let's assume that a Dirac delta function exists. The reader may feel uncomfortable about developing the calculus of a function that may not exist, but we will plunge ahead.

Paul A. Dirac

---

[4]Paul Adrien Maurice Dirac (1902–1984), the British theoretical physicist, was awarded (jointly with Erwin Schrödinger) the 1933 Nobel Prize in physics for his work in quantum mechanics. Among other achievements, Dirac described the motion of an electron by four simultaneous differential equations. Finding negative energy states predicted by this mathematical model, Dirac hypothesized the existence of positrons, or electron antiparticles, whose existence was later experimentally confirmed.

THEOREM 6.5.1

Properties of the Delta Function. We have

$$\int_{-\infty}^{\infty} \delta(t)\,dt = 1, \qquad L[\delta](s) = 1, \qquad L[\delta(t-u)] = e^{-us} \qquad (2)$$

To verify the first property, let $f(t) = \text{step}(t)$, which is the Heaviside function. From (1) we can let $u = t - v$ and show that

$$\text{step}(t) = \int_{-\infty}^{\infty} \delta(t-u)\,\text{step}(u)\,du = \int_{-\infty}^{\infty} \delta(v)\,\text{step}(t-v)\,dv$$
$$= \int_{-\infty}^{t} \delta(v)\,dv \quad [\text{since } \text{step}(t-v) = 0, \text{ if } v > t] \qquad (3)$$

So $\int_{-\infty}^{t} \delta(v)\,dv = 1$ for all $t > 0$, and the first equality in (2) is proved. To show the second property we use the Convolution Theorem (Theorem 6.4.2):

$$L[\delta * \text{step}] = L[\delta]L[\text{step}] \qquad (4)$$

☞ Recall that "step" is the name of the function step(t).

On the other hand, we have that

$$L[\text{step}] = L[1] \qquad (5)$$

and that (after setting $u = t - v$)

$$L[\delta * \text{step}] = L\left[\int_{0}^{t} \delta(t-u)\,\text{step}(u)\,du\right] = L\left[\int_{t}^{0} \delta(v)\,\text{step}(t-v)(-dv)\right]$$
$$= L\left[\int_{0}^{t} \delta(v)\,\text{step}(t-v)\,dv\right] = L\left[\int_{-\infty}^{\infty} \delta(v)\,\text{step}(t-v)\,dv\right] \qquad (6)$$
$$= L[\text{step}] = L[1]$$

where we have used the facts that $\delta(v) = 0$ for $v < 0$, $\text{step}(t-v) = 0$ for $v > t$, and the identity (3). From (4), (5), and (6) we see that $L[\delta] = 1$. The justification of the last identity in (2) is left to the reader (Problem 4).

The delta function is often used to solve IVPs. Suppose that $P(D) = D^2 + aD + b$ for real constants a and b. Then the following result gives a solution formula for an IVP:

THEOREM 6.5.2

The Delta Function and IVPs. The solution of the IVP

$$P(D)[y] = f, \qquad y(0) = y'(0) = 0 \qquad (7)$$

where f lies in $\mathbf{E}_1$, is given by

$$y = \int_{-\infty}^{\infty} G(t,u)f(u)\,du \qquad (8)$$

where for each u the function $z = G(t,u)$ is the solution of the IVP

$$P(D)[z] = \delta(t-u), \qquad z(0) = z'(0) = 0 \qquad (9)$$

To see why formula (8) gives the solution of IVP (7), we reason as follows: By Theorem 6.4.3 if $g(t) = \mathcal{L}^{-1}[1/P(s)]$, then the solution of IVP (7) is given by

$$y(t) = \int_0^t g(t-u)f(u)\,du = \int_{-\infty}^t g(t-u)f(u)\,du$$

$$= \int_{-\infty}^{\infty} \text{step}(t-u)g(t-u)f(u)\,du \tag{10}$$

where we have used the facts that $f(u) = 0$ if $u < 0$ and $\text{step}(t-u) = 0$ for $u > t$. On the other hand, using formula (10) in Section 6.4, we see that the transform of the solution $G(t,u)$ of IVP (9) is given by

$$\mathcal{L}[G(t,u)] = \frac{\mathcal{L}[\delta(t-u)]}{P(s)} = \frac{e^{-us}}{P(s)} \tag{11}$$

where we have also used the third identity in (2). So, by (11) and formula (12) of the Shifting Theorem 6.2.4 we see that

$$G(t,u) = g(t-u)\,\text{step}(t-u) \tag{12}$$

Finally, from (10) and (12), we have the desired formula (8):

$$y(t) = \int_{-\infty}^{\infty} G(t,u)f(u)\,du$$

So we have come up with yet another way to find the unique solution of an IVP. Compare the new formula with that derived in Theorem 6.4.2.

Does the Dirac Delta Function Exist?

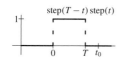

For now, let's suppose that $\delta(t)$ *is* a function in $\mathbf{E}_1$ and see where this takes us. Suppose that $t_0 > 0$ is a point of continuity for δ and that $\delta(t_0) > 0$. Then there is an interval $0 < t_0 - T \le t \le t_0$ on which $\delta(t)$ is positive and continuous. Let's look at the function $f(t) = \text{step}(T-t)\,\text{step}(t)$. Now since $T - t_0 < 0$, $f(t_0) = 0$, so from (1) we have

$$0 = f(t_0) = \int_{-\infty}^{\infty} \delta(t_0 - u)f(u)\,du = \int_0^T \delta(t_0 - u)\,du$$

where we have also used the fact $f(u) = 0$ for $u < 0$ and $u > T$. Making the change of variables $v = t_0 - u$ we see from the choice of t_0 and T that

☞ This is hypothetical reasoning at its best, or worst!!

$$0 = \int_0^T \delta(t_0 - u)\,du = \int_{t_0-T}^{t_0} \delta(v)\,dv > 0$$

since $\delta(t)$ was assumed to be positive for $t_0 - T \le t \le t_0$. This contradiction shows that our $\delta(t_0)$ cannot be positive at any point of continuity. Similarly, $\delta(t_0)$ can't be negative either, so δ vanishes at *every* point where it is continuous. This means that the left-hand side of (1) vanishes for all t, regardless of the function f selected from $\mathbf{E}_1$. But this is absurd, so whatever $\delta(t)$ may be, it is *not* a function in $\mathbf{E}_1$. We already had an inkling from the second formula in (2) that δ can't be a function, since any function in $\mathbf{E}_1$ must have a transform that decays to zero as $s \to +\infty$.

Since Dirac's time, "functions" such as δ have been very important in applications. In advanced treatments of modern applied mathematics, a logically rigorous theory is constructed that contains objects, called *distributions*, *generalized functions*, or *symbolic functions* that behave just like the "delta function" should. So with this in mind, we continue to use δ just as if it were a function like any other.

Impulsive Forces

We can use the δ-function to model a sudden sharp force acting on a physical system.

EXAMPLE 6.5.1

What Happens When You Hit an Oscillating Spring?
Suppose that the weighted end of an oscillating and undamped Hooke's Law spring is struck a hard blow. How does the system respond? If y measures the displacement of the weight from equilibrium, then we can model the system by

$$my''(t) + ky(t) = A\delta(t - T), \qquad y(0) = \alpha, \quad y'(0) = \beta \qquad (13)$$

where m is the mass, k is the spring constant, A is a positive constant, $T > 0$ is the time when the spring is struck, and α and β are given initial values. The "force" $A\delta(t - T)$ is called an *impulsive force*, while its integral over time is the *impulse*. Divide the ODE in (13) by m, set $k/m = \omega^2$, and transform:

$$(s^2 + \omega^2)\mathcal{L}[y] - s\alpha - \beta = Ae^{-Ts}/m$$

$$\mathcal{L}[y] = \frac{s\alpha + \beta}{s^2 + \omega^2} + \frac{Ae^{-Ts}/m}{s^2 + \omega^2} \qquad (14)$$

where the third formula in (2) has been used. Inverting (14), we have

$$y(t) = \alpha \cos \omega t + \frac{\beta}{\omega} \sin \omega t + \frac{A}{m\omega} \sin \omega(t - T)\,\text{step}(t - T) \qquad (15)$$

$$= \left[\alpha - \frac{A}{m\omega} \sin \omega T\,\text{step}(t - T)\right] \cos \omega t + \left[\frac{\beta}{\omega} + \frac{A}{m\omega} \cos \omega T\,\text{step}(t - T)\right] \sin \omega t$$

where we used the trigonometric identity $\sin \omega(t - T) = \sin \omega t \cos \omega T - \cos \omega t \sin \omega T$. The solution $y(t)$ of IVP (13) is continuous, but notice the sharp angle at $t = T$, while $y'(t)$ has a jump discontinuity at $t = T$ (Figure 6.5.1).

The spring in Example 6.5.1 continues to oscillate after the impulsive blow. Can we choose the amplitude and timing so that the blow stops the oscillations altogether?

EXAMPLE 6.5.2

Stopping the Oscillations
If we want to stop the vibrations of the spring in Example 6.5.1, we need to choose A and T so that $y(t) = 0$ for $t > T$. We see from formula (15) that $y(t) = 0$ for $t > T$ if

$$\frac{A}{m\omega} \sin \omega T = \alpha, \qquad \frac{A}{m\omega} \cos \omega T = -\frac{\beta}{\omega} \qquad (16)$$

We see that if $\beta \neq 0$ we must time the blow and choose its amplitude so that

$$\frac{A}{m} = \frac{\alpha\omega}{\sin \omega T}, \qquad \tan \omega T = -\frac{\alpha\omega}{\beta}$$

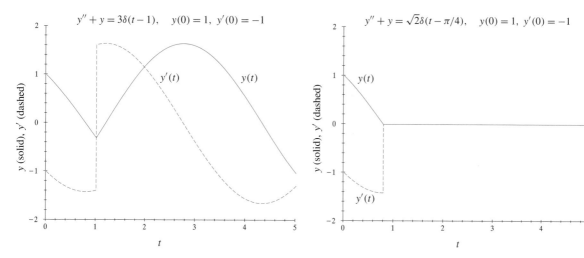

FIGURE 6.5.1 Response of a spring to a blow at time $T = 1$ (Example 6.5.1).

FIGURE 6.5.2 Stopping the oscillations with a well-timed blow at $T = \pi/4$ (Example 6.5.2).

that is,

$$T = \frac{1}{\omega} \arctan\left(\frac{-\alpha\omega}{\beta}\right), \qquad A = \frac{ma\omega}{\sin \omega T}$$

where we assume that $\omega T \neq k\pi$. With this choice of T and A, the vibrations cease at $t = T$. Figure 6.5.2 shows the graphs of $y(t)$ and $y'(t)$ with $m = 1$, $\alpha = 1$, $\beta = -1$, $\omega = 1$, $T = \pi/4$, and $A = \sqrt{2}$.

PROBLEMS

www **1.** Use the Laplace transform to find the solution of an IVP, where $y(0) = y'(0) = 0$.

    (a) $y'' + 2y' + 2y = \delta(t - \pi)$ **(b)** $y'' + 4y = \delta(t - \pi) - \delta(t - 2\pi)$

    (c) $y'' + 3y' + 2y = \sin t + \delta(t - \pi)$ **(d)** $y'' + y = \delta(t - \pi)\cos t$

    (e) $y'' + y = e^t + \delta(t - 1)$

2. (*Driven Spring*). A Hooke's Law spring with spring constant k supports an object of mass 1 and is subject to a force $f(t) = A \sin \omega t$, $\omega^2 \neq k$, for $t \geq 0$. The mass is given a sharp upward blow at $t = 2$ that gives an impulse of 2 units, so the force of the blow is $2\delta(t - 2)$. Determine the motion if the object is at rest at its natural equilibrium at $t = 0$. Plot the solution when $k = 1$, $A = 1$, $\omega = 2$.

3. (*Driven LC Circuit*). Suppose that an LC circuit is subject to a constant voltage F_0. At time $t = 1$ the circuit is dealt a sharp voltage burst $2F_0$ (i.e., $2F_0\delta(t - 1)$). Find the charge as a function of time if $q(0) = q'(0) = 0$. Plot the charge as a function of time if $F_0 = 10$.

4. Assume that $\delta(t)$ exists. Show that $\mathcal{L}[\delta(t - u)](s) = e^{-us}$.

5. Show that $\delta(at) = (1/|a|)\delta(t)$, $a \neq 0$. [*Hint*: Replace t by at in the first formula of (2).]

6. Suppose that $f(t)$ is in $\mathbf{E}_1$. Show that $\delta(t) * f(t) = f(t)$.

Tables of Laplace Transforms
I. General Properties

	$f(t)$	$g(s) = \mathcal{L}[f] = \int_0^\infty e^{-st} f(t)\, dt$
1.	$f(at)$	$\dfrac{g(s/a)}{a}$
2.	$e^{at} f(t)$	$g(s - a)$
3.	$\text{step}(t - a) f(t - a)$	$e^{-as} g(s)$
4.	$f^{(n)}(t)$	$s^n g(s) - s^{(n-1)} f(0) - s^{(n-2)} f'(0) - \cdots - f^{(n-1)}(0)$
5.	$t^n f(t)$	$(-1)^n g^{(n)}(s)$
6.	$\displaystyle\int_0^t f(u)\, du$	$\dfrac{g(s)}{s}$
7.	$\displaystyle\int_0^t (t - u)^{n-1} f(u)\, du$	$(n - 1)! \dfrac{g(s)}{s^n}$
8.	$\displaystyle\int_0^t f_1(t - u) f_2(u)\, du$	$g_1(s) g_2(s)$
9.	$\dfrac{f(t)}{t}$	$\displaystyle\int_s^\infty g(u)\, du$
10.	$f(t)$ is periodic with period p	$\dfrac{1}{1 - e^{-ps}} \displaystyle\int_0^p e^{-su} f(u)\, du$
11.	$\dfrac{1}{\sqrt{\pi t}} \displaystyle\int_0^\infty e^{-u^2/4t} f(u)\, du$	$\dfrac{g(\sqrt{s})}{\sqrt{s}}$
12.[1]	$t^{n/2} \displaystyle\int_0^\infty u^{-n/2} J_n(2\sqrt{ut}) f(u)\, du$	$\dfrac{1}{s^{n+1}} g\left(\dfrac{1}{s}\right), \quad n \geq 0$
13.	$\displaystyle\sum_{k=1}^n \dfrac{P(\alpha_k)}{Q_k(\alpha_k)} e^{\alpha_k t}$	$\dfrac{P(s)}{Q(s)}$

P is a polynomial, degree $< n$;
$Q(s) = (s - \alpha_1) \cdots (s - \alpha_n)$, $\alpha_1, \ldots, \alpha_n$ distinct; $Q_k(\alpha_k) = \prod_{i=1, i \neq k}^n (\alpha_k - \alpha_i)$

[1] J_n is the Bessel function of the first kind of order n. See Section 11.6.

II. Special Laplace Transforms

	$f(t)$	$g(s) = \mathcal{L}[f] = \int_0^\infty e^{-st} f(t)\, dt$
1.	t^{n-1}	$(n-1)!/s^n, \quad n = 1,\, 2,\, 3,\, \ldots$
2.[2]	t^{p-1}	$\Gamma(p)/s^p, \quad p > 0$
3.	$t^{n-1}e^{at}$	$\dfrac{(n-1)!}{(s-a)^n}, \quad n = 1, 2, 3, \ldots$
4.[2]	$t^{p-1}e^{at}$	$\dfrac{\Gamma(p)}{(s-a)^p}, \quad p > 0$
5.	$\sin at$	$a(s^2 + a^2)^{-1}$
6.	$\cos at$	$s(s^2 + a^2)^{-1}$
7.	$\sinh at$	$a(s^2 - a^2)^{-1}$
8.	$\cosh at$	$s(s^2 - a^2)^{-1}$
9.	$e^{bt} - e^{at}$	$\dfrac{b-a}{(s-a)(s-b)}, \quad a \neq b$
10.	$be^{bt} - ae^{at}$	$\dfrac{(b-a)s}{(s-a)(s-b)}, \quad a \neq b$
11.	$\sin at - at\cos at$	$2a^3(s^2 + a^2)^{-2}$
12.	$t\sin at$	$2as(s^2 + a^2)^{-2}$
13.	$\dfrac{e^{-at} - e^{-bt}}{2(\pi t^3)^{1/2}}$	$\sqrt{s+b} - \sqrt{s+a}$
14.[1]	$J_n(t), \quad n > -1$	$(s^2 + 1)^{-1/2}[s + (s^2+1)^{1/2}]^{-n}$
15.	$f(t) = \displaystyle\sum_{k=1}^{[t]} r^{k-1}$ where $[t] =$ greatest integer $\leq t$	$\dfrac{1}{s(e^s - r)} = \dfrac{e^{-s}}{s(1 - re^{-s})}$
16.	$\dfrac{a}{2\sqrt{\pi t^3}} e^{-a^2/4t}$	$e^{-a\sqrt{s}}, \quad a > 0$
17.	$\dfrac{e^{-bt} - e^{-at}}{t}$	$\ln\left(\dfrac{s+a}{s+b}\right)$
18.	$\delta(t-a)$	e^{-as}
19.	$\text{step}(t-a)$	e^{-as}/s

[2] $\Gamma(p)$ is the gamma function. See Section 11.6.

III. Transforms of Graphically Defined Functions

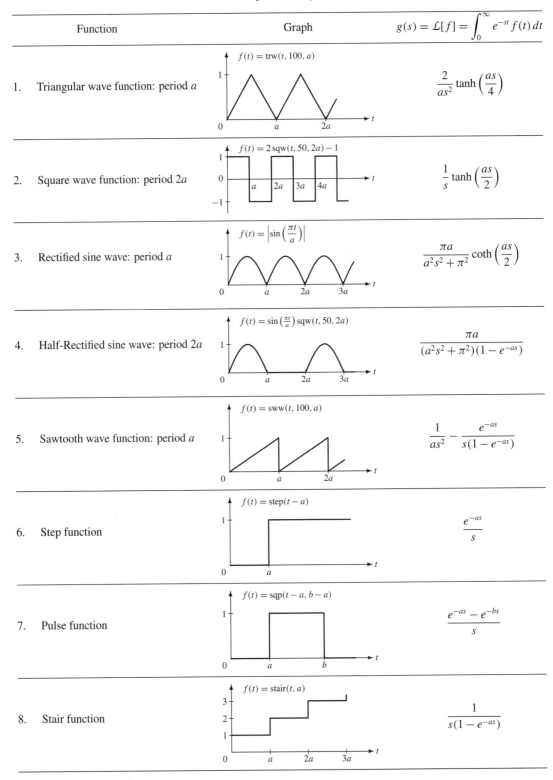

Function	Graph	$g(s) = \mathcal{L}[f] = \int_0^\infty e^{-st} f(t)\, dt$
1. Triangular wave function: period a	$f(t) = \mathrm{trw}(t, 100, a)$	$\dfrac{2}{as^2} \tanh\left(\dfrac{as}{4}\right)$
2. Square wave function: period $2a$	$f(t) = 2\,\mathrm{sqw}(t, 50, 2a) - 1$	$\dfrac{1}{s} \tanh\left(\dfrac{as}{2}\right)$
3. Rectified sine wave: period a	$f(t) = \left|\sin\left(\dfrac{\pi t}{a}\right)\right|$	$\dfrac{\pi a}{a^2 s^2 + \pi^2} \coth\left(\dfrac{as}{2}\right)$
4. Half-Rectified sine wave: period $2a$	$f(t) = \sin\left(\dfrac{\pi t}{a}\right)\mathrm{sqw}(t, 50, 2a)$	$\dfrac{\pi a}{(a^2 s^2 + \pi^2)(1 - e^{-as})}$
5. Sawtooth wave function: period a	$f(t) = \mathrm{sww}(t, 100, a)$	$\dfrac{1}{as^2} - \dfrac{e^{-as}}{s(1 - e^{-as})}$
6. Step function	$f(t) = \mathrm{step}(t - a)$	$\dfrac{e^{-as}}{s}$
7. Pulse function	$f(t) = \mathrm{sqp}(t - a, b - a)$	$\dfrac{e^{-as} - e^{-bs}}{s}$
8. Stair function	$f(t) = \mathrm{stair}(t, a)$	$\dfrac{1}{s(1 - e^{-as})}$

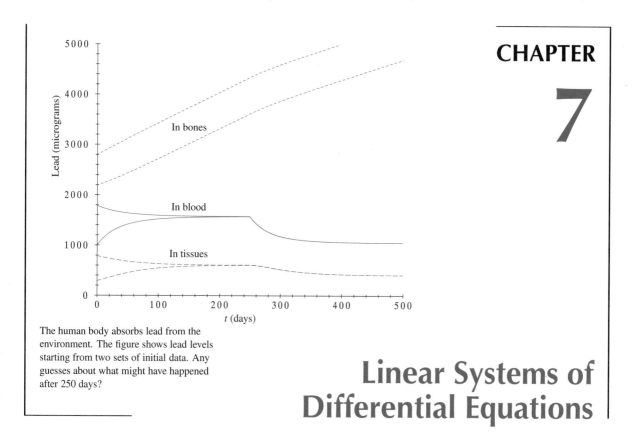

The human body absorbs lead from the environment. The figure shows lead levels starting from two sets of initial data. Any guesses about what might have happened after 250 days?

Linear Systems of Differential Equations

Systems of linear ODEs arise in modeling multiloop electrical circuits, the motion of coupled springs or a double pendulum, the passage of lead through the compartments of the body, and in many other settings. In this chapter we develop techniques that characterize the solutions of a system of linear ODEs.

7.1 Tracking Lead Through the Body

Lead is an ingredient in many objects of everyday life: car batteries, water pipes, glassware, ceramics, paint, and gasoline. But lead is toxic, and high levels in the blood and tissues will impair motor and mental capacities. One way to begin to understand this is to build a model of lead flow in the body.

Lead enters the bloodstream via food, air, and water. It accumulates in the blood, in tissues, and especially in the bones. Some lead is excreted by the kidneys and by hair, nails, and sweat. Let's develop a mathematical model for the flow of lead through each of three body compartments: blood, tissues, and bones.

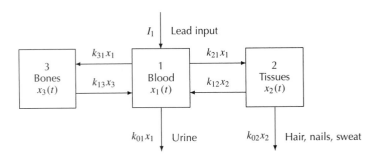

FIGURE 7.1.1 Lead transport in the body (Example 7.1.1).

Numbering the body compartments as 1, 2, and 3, the schematic diagram in Figure 7.1.1 represents the flow of lead through the compartments. The amount of lead in compartment i at time t is denoted by $x_i(t)$. The rate of transfer of lead into compartment j from compartment i is proportional to $x_i(t)$ (a *first-order rate law*), and the constant of proportionality is denoted by the *rate constant* k_{ji}. It is always assumed that $k_{ji} \geq 0$. If there is no transfer of lead into compartment j from compartment i, then $k_{ji} = 0$. A reverse transfer from compartment j into compartment i may also occur, but the rate constant k_{ij} doesn't have to be equal to k_{ji}. The symbols by each arrow give the exit rate of lead from one compartment and the entrance rate into another. Transport out of the compartment system is denoted by a 0 subscript; this outside absorbing region is called a *sink*.

EXAMPLE 7.1.1

Building the Model ODEs

According to the Balance Law,

$$\text{Net rate of change} = \text{Rate in} - \text{Rate out} \tag{1}$$

Applying the Balance Law to the lead flow through the blood, tissue, and bone compartments diagrammed in Figure 7.1.1, we have a system of three rate equations:

$$
\begin{aligned}
\text{(blood)} \quad & x_1' = -(k_{01} + k_{21} + k_{31})x_1 + k_{12}x_2 + k_{13}x_3 + I_1 \\
\text{(tissues)} \quad & x_2' = k_{21}x_1 - (k_{02} + k_{12})x_2 \\
\text{(bone)} \quad & x_3' = k_{31}x_1 - k_{13}x_3
\end{aligned}
\tag{2}
$$

☞ Section 5.2 gives the definition of equilibrium solutions.

The intake rate I_1 of lead into the blood from the GI-tract and the lungs is assumed to be a piecewise continuous function of time. If I_1 and the k_{ij} are positive constants, then system (2) has a unique equilibrium solution (see Problem 1 for an example).

System (2) is a plausible model, but what does it have to do with the specifics of real people ingesting contaminated food and drink and breathing air containing traces of lead from industrial pollution and exhaust fumes? Let's look at a case study.

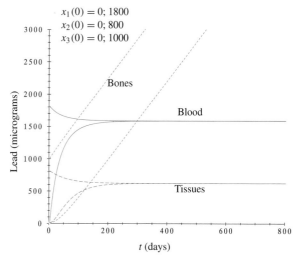

FIGURE 7.1.2 Buildup of lead from two sets of initial data (Example 7.1.2).

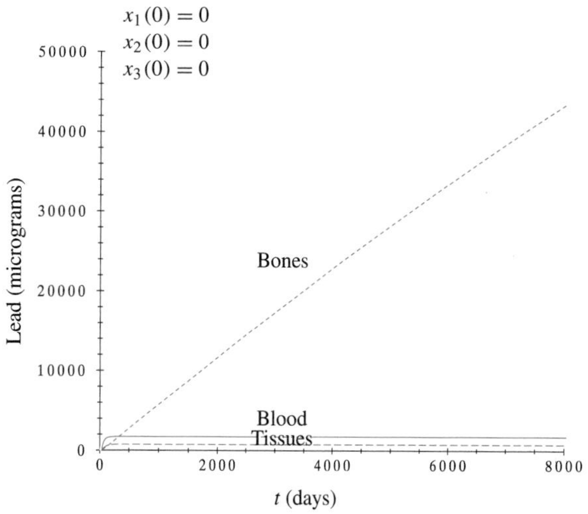

FIGURE 7.1.3 After 8000 days the lead content of the bones is still rising (Example 7.1.2).

EXAMPLE 7.1.2

A Case Study

Michael Rabinowitz, George Wetherill, and Joel Kopple made a controlled study of the lead intake and excretion of a healthy volunteer living in Southern California.[1] The data from this study were used to estimate the values of the lead intake rate I_1 in micrograms/day and the rate constants k_{ji} in (days)$^{-1}$:

$$I_1 = 49.3; \quad k_{01} = 0.0211, \quad k_{21} = 0.0111, \quad k_{31} = 0.0039$$
$$k_{02} = 0.0162, \quad k_{12} = 0.0124, \quad k_{13} = 0.000035 \tag{3}$$

Using system (2) and the data from (3) and taking two different sets of initial values, we have the following IVPs for tracking lead through the body compartments:

$$
\begin{aligned}
x_1' &= -0.0361x_1 + 0.0124x_2 + 0.000035x_3 + 49.3, & x_1(0) &= 0; 1800 \\
x_2' &= 0.0111x_1 - 0.0286x_2, & x_2(0) &= 0; 800 \\
x_3' &= 0.0039x_1 - 0.000035x_3, & x_3(0) &= 0; 1000
\end{aligned}
\tag{4}
$$

☞ Figure 7.1.2 shows a comparison set of component curves using two initial data sets. Realistically, some lead is in the body initially.

Let's apply a numerical solver. Figure 7.1.2 shows the buildup of lead in the body compartments over a period of 800 days.

Lead levels in the bloodstream and the tissues appear to have settled down to equilibrium values after the first 200 days, but the lead level in the bones hasn't begun to level off even after 8000 days (Figure 7.1.3). And that is because the transfer coefficient $k_{13} = 0.000035$ (day)$^{-1}$ of lead from the bones back into the bloodstream is so small that the skeleton accumulates and stores lead like a reservoir.

---

[1]Their work was reported in *Science* **182** (1973), pp. 725–727, and later extended by Batschelet, Brand, and Steiner in *J. Math. Biol.* **8** (1979), pp. 15–23.

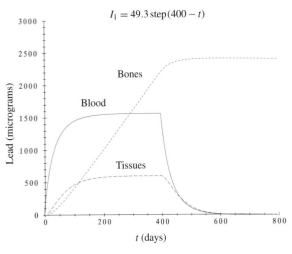

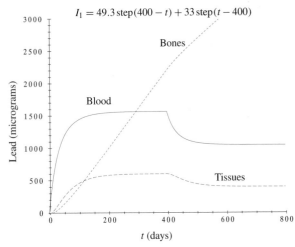

FIGURE 7.1.4 Lead levels drop after all lead is re-moved from the environment (Example 7.1.3).

FIGURE 7.1.5 Removing lead from paint and gasoline lowers lead levels (Example 7.1.3).

Remarkably, no matter what the initial data are, the lead levels approach a unique equilibrium state (although this isn't very clear in Figures 7.1.2 and 7.1.3 for the lead levels in the bones). We will verify this fact in Section 7.10. From this point on we use the initial data $x_1(0) = x_2(0) = x_3(0) = 0$; the second set of initial data listed in IVP (4) leads to similar conclusions.

How can lead levels be lowered? One way would be to clean up the environment by banning lead as an ingredient in paint, water pipes, ceramic glazes, and gasoline. Let's test the effect of a total ban, and then of a more realistic partial ban.

EXAMPLE 7.1.3	**Get the Lead Out of the Environment**

Get the Lead Out of the Environment

After 400 days the subject in the study described in Example 7.1.2 moves into a completely lead-free environment and the intake rate drops from 49.3 micrograms of lead per day to zero. This change can be modeled by replacing the rate term 49.3 in IVP (4) by $49.3 \, \text{step}(400 - t)$. Figure 7.1.4 shows a dramatic drop in lead in the blood and tissues, and some improvement in the lead level in the bones as lead intake stops.

Since a lead-free environment is unrealistic in today's world, let's see what would happen if some (but not all) of the lead were removed from the environment. Figure 7.1.5 shows what happens to the solutions of IVP (4) when the intake rate of 49.3 drops to 33 micrograms of lead per day. This change is modeled by replacing $I_1 = 49.3$ in IVP (4) with $I_1 = 49.3 \, \text{step}(400 - t) + 33 \, \text{step}(t - 400)$. Note the one-third drop in the lead levels for the blood and the tissues in comparison with the levels shown in Figure 7.1.2, but the drop in the rate of increase of the lead in the bones is slight.

☞ More on this model in Section 7.10.

Will this strategy work? Yes, at least it has in Southern California where lead levels in the environment have dropped by almost one-third. But there are still many cases of lead poisoning—children who live in older houses with lead-based paint, workers in battery factories. In individual cases like these, there is a medication that combines

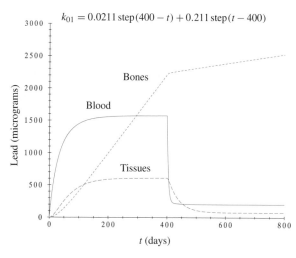

$$k_{01} = 0.0211\,\text{step}(400 - t) + 0.211\,\text{step}(t - 400)$$

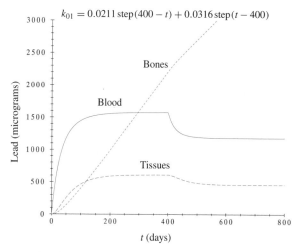

$$k_{01} = 0.0211\,\text{step}(400 - t) + 0.0316\,\text{step}(t - 400)$$

FIGURE 7.1.6 Massive medication from day 400 on causes a sharp drop in the lead levels in the blood and tissues (Example 7.1.4).

FIGURE 7.1.7 A little medication from day 400 on lowers the lead levels in the blood and tissues (Example 7.1.4).

with lead and promotes its passage out of the body. Let's see what happens when this drug is administered. We will model the effects of a huge dose of medication, then of a more realistic dose.

EXAMPLE 7.1.4

Get the Lead Out of the Body

Suppose a massive dose of the medication is given from the 400th day on and the clearance coefficient k_{01} increases tenfold to 0.211. This change is modeled by replacing k_{01} in IVPs (2) and (4) with $k_{01} = 0.0211\,\text{step}(400 - t) + 0.211\,\text{step}(t - 400)$, while retaining the other values given in (3). Figure 7.1.6 shows what happens. The lead level drop in the blood is sudden, but all that medication may have killed the patient. A safer dosage increases the effective value of k_{01} by 50%, to 0.0316. Figure 7.1.7 shows the lowered lead levels with this more realistic dosage.

Case studies, ODE models, and computer simulations all play a role in this account. Mathematical theory does, too, but we postpone that story to Section 7.10 because we need some concepts developed in the next several sections.

Comments

The lead system (2) is an example of the linear compartment models introduced in Section 1.8. When the flow through the compartments is one-way only (as in Section 1.8), the models are called cascades and can be solved one ODE at a time. However, the lead system models back-and-forth flow between the compartments, so the model ODEs are coupled and it is not possible to solve the system one ODE at a time. Other methods (to be given later in this chapter) are needed to explain the behavior of solutions in this more general setting, so we have relied mostly on numerical solvers here.

PROBLEMS

1. (*Lead in the Body: Equilibrium Levels*). Figures 7.1.2 and 7.1.3 show the amounts of lead in the body compartments. The amounts of lead in the blood and tissues quickly approach equilibrium levels. The same is true for lead in the bones, but it takes much longer.

   (a) Find the equilibrium levels by setting the rate functions of IVP (4) equal to zero and solving for x_1, x_2, and x_3.

   (b) Using IVP (4) (with the first set of initial data) and a numerical solver, estimate how long it takes for the lead in the blood to reach 85% of its equilibrium level. Repeat for the tissues. You may try the same problem for the bones, but the time span is incredibly long.

   (c) Replace the lead ingestion rate of 49.3 micrograms per day used in IVP (4) by the positive constant I_1 (unspecified). Now find the equilibrium levels in terms of I_1. Explain why the equilibrium levels are reduced by 50% if I_1 is reduced by 50%.

   (d) For $I_1 = 10, 20, 30, 40, 50$, plot the $x_3(t)$-component curve, and then for each value of I_1, estimate the time it takes the amount of lead in the bones to reach 1000 micrograms.

2. (*Lead in the Body: Lead-Free Environment*). Suppose that the lead-free subject in Example 7.1.2 [i.e., $x_1(0) = x_2(0) = x_3(0) = 0$] is exposed to lead for 400 days, then is removed to a lead-free environment (Figure 7.1.4). Use a computer to estimate how long it takes the amount x_3 of lead in the bones to decline to 50% of $x_3(400)$; repeat for 25% and 10%.

3. (*Getting Some of the Lead Out*). In this problem we see the effect of reducing the lead intake rate by 50% and (at the same time) using medication to double the clearance coefficient k_{01}.

   (a) Replace 49.3 in IVP (4) by $49.3\,\text{step}(400 - t) + 24.65\,\text{step}(t - 400)$. Explain why this models a cut in the intake rate by 50% for $t \geq 400$. Plot the lead levels in blood, tissues, and bones. Do you see a 50% drop in the eventual lead levels in the blood and tissues?

   (b) Leave the term 49.3 in IVP (4) as is, but replace the coefficient 0.0361 in the first ODE by 0.0572 from day 400 onward. This corresponds to doubling the value of k_{01}. Plot the lead levels in blood, tissues, and bones. Is there as much of a drop as observed in part **(a)**? [*Hint*: Replace 0.0361 by $0.0361\,\text{step}(400 - t) + 0.0572\,\text{step}(t - 400)$ in the first ODE of (4).]

   (c) From the 400th day onward, model a partial cleanup of the environment *and* administration of medication as in parts **(a)** and **(b)**. Plot the results and describe the effects of this strategy.

4. (*A Compartment System*). The sketch below shows a compartment system where a substance Q is injected at a rate of I_3 units of substance per unit of time into compartment 3. The substance Q moves out of compartment i into compartment j at a rate proportional to the amount $x_i(t)$ of Q in compartment i; the constant of proportionality is k_{ji}.

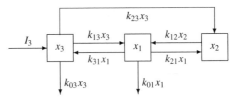

   (a) Find ODEs for $x_1(t)$, $x_2(t)$, and $x_3(t)$ if

   $$I_3 = 1, \quad k_{31} = 3, \quad k_{21} = 1, \quad k_{01} = 16, \quad k_{13} = 1, \quad k_{03} = 5, \quad k_{23} = 2, \quad k_{12} = 4$$

   (b) Find the equilibrium levels of Q in each of the three compartments.

   (c) Plot the three component curves over a long enough time span that each is within 90% of its equilibrium level. Assume that $t \geq 0$ and $x_1(0) = x_2(0) = x_3(0) = 0$.

www **5.** (*The Tracer Inulin*). Water moves one way from the blood into the urinary system, but moves back and forth between the blood (compartment 1) and intercellular areas (compartment 2). This motion is tracked by injecting the tracer inulin into the blood at the rate I_1 gram/hour. Molecules of inulin attach to molecules of water and may be tracked by x-ray.

(a) Draw a boxes-and-arrows diagram like that for the lead system. Find ODEs that model inulin levels in the system. Assume that inulin leaves compartment i and enters compartment j at the rate $k_{ji}x_i(t)$ for some constant $k_{ji} > 0$, where $x_i(t)$ is the amount of inulin in compartment i at time t.

(b) Find the equilibrium levels of inulin in the blood and in the intercellular areas in terms of the rate constants (assume that I_1 is a positive constant).

 (c) Graph the component curves of your system if

$$I_1 = 1, \quad k_{21} = 0.01, \quad k_{12} = 0.02, \quad k_{01} = 0.005, \quad x_1(0) = x_2(0) = 0$$

Use a long enough time span that the levels of inulin in the blood and in the intercellular area have reached at least 90% of their equilibrium values.

 6. The assumption in IVP (4) that $x(0) = 0$ is not realistic. Discuss the long-term effects on $x(t)$ if $x(0) \neq 0$ in IVP (4).

7.2 Overview of Vectors and Matrices

In many branches of science, engineering, and mathematics rectangular arrays of numbers or functions play an important role in organizing and working with systems of all kinds, especially linear systems. The rectangular structure of these arrays (called matrices) makes it easy to define operations for them. Here is how this is done.

Matrices and Matrix Operations

Let's review some basic properties of matrices of numbers, focusing on those ideas that will be useful in the study of systems of linear ODEs in later sections.

❖ **Matrix.** An $m \times n$ *matrix* A is a rectangular array of real or complex numbers a_{ij} with m rows and n columns enclosed in square brackets:

$$A = \begin{bmatrix} a_{11} & a_{12} & \cdots & a_{1n} \\ a_{21} & a_{22} & \cdots & a_{2n} \\ \vdots & \vdots & & \vdots \\ a_{m1} & a_{m2} & \cdots & a_{mn} \end{bmatrix}$$

An entry in a rectangular array is referenced by using a double index. Note that the first index of a_{ij} indicates the row, and the second the column in which the number a_{ij} appears. For example, a_{23} denotes an entry that appears in the second row of the third column. We often abbreviate the notation for A by $[a_{ij}]$. Two $m \times n$ matrices A and B are *equal* if $a_{ij} = b_{ij}$ for all i, j.

A $1 \times n$ row matrix is a *row vector.*

If A is an $m \times 1$ column matrix, it is called a *column vector.* If $m = n$, then A is called a *square matrix* of order n. The entries a_{11}, a_{22}, ..., a_{nn} form the *main*

☞ A *lower triangular matrix* is similarly defined.

diagonal of a square matrix. A square matrix with only zeros below the main diagonal is an *upper triangular matrix*, or simply a *triangular matrix*. In a *diagonal matrix* all entries above and below the main diagonal are zero. Here's how to add matrices.

❖ **Addition of Matrices.** If $A = [a_{ij}]$, $B = [b_{ij}]$ are $m \times n$ matrices, then the sum $A + B$ is the matrix $[a_{ij} + b_{ij}]$.

☞ $\mathbb{R}$ and $\mathbb{C}$ are often used to denote the real and complex numbers, respectively.

Numbers are often referred to as *scalars*; scalars belong to the set $\mathbb{R}$ of real numbers, or the set $\mathbb{C}$ of complex numbers. We can multiply a scalar and a matrix.

❖ **Product of a Scalar with a Matrix.** If $A = [a_{ij}]$ is any $m \times n$ matrix and α any scalar, then $\alpha A = [\alpha a_{ij}]$.

EXAMPLE 7.2.1

Matrix Addition; Product with a Scalar

Suppose that $A = \begin{bmatrix} 2 & -1 \\ 0 & 4 \end{bmatrix}$, $B = \begin{bmatrix} 4 & 7 \\ 9 & 3 \end{bmatrix}$, and $\alpha = 3$. Then

$$A + B = \begin{bmatrix} 2+4 & -1+7 \\ 0+9 & 4+3 \end{bmatrix} = \begin{bmatrix} 6 & 6 \\ 9 & 7 \end{bmatrix}, \quad \alpha A = \begin{bmatrix} 3(2) & 3(-1) \\ 3(0) & 3(4) \end{bmatrix} = \begin{bmatrix} 6 & -3 \\ 0 & 12 \end{bmatrix}$$

and we see that $A + B$ and αA are easy to calculate.

Notice that $A + B = B + A$ and that $(A + B) + C = A + (B + C)$ for any $m \times n$ matrices A, B, and C. Let's now define of the product of two matrices.

❖ **Matrix Products.** Let $A = [a_{ij}]$ be an $m \times n$ matrix and $B = [b_{ij}]$ be an $n \times k$ matrix. Then AB is defined to be the $m \times k$ matrix $AB = C = [c_{ij}]$, where $c_{ij} = \sum_{s=1}^{n} a_{is} b_{sj}$.

In the definition of a matrix product AB, the number of columns in the matrix A must equal the number of rows in matrix B (i.e., the matrices A and B must have compatible dimensions). The element c_{ij} in the product $C = AB$ is the sum of products of elements in the i-th row of A with elements in the j-th column of B. Here's an example.

EXAMPLE 7.2.2

Matrix Product

Let A be the 2×3 matrix $\begin{bmatrix} 2 & 1 & -1 \\ 0 & 3 & 1 \end{bmatrix}$ and B the 3×1 column matrix (i.e., vector) $\begin{bmatrix} -3 \\ 4 \\ 2 \end{bmatrix}$. Then we can calculate the matrix product:

$$AB = \begin{bmatrix} 2 & 1 & -1 \\ 0 & 3 & 1 \end{bmatrix} \begin{bmatrix} -3 \\ 4 \\ 2 \end{bmatrix} = \begin{bmatrix} (2)(-3) + (1)(4) + (-1)(2) \\ (0)(-3) + (3)(4) + (1)(2) \end{bmatrix} = \begin{bmatrix} -4 \\ 14 \end{bmatrix}$$

The product BA is undefined since the dimensions of B and A are 3×1 and 2×3.

For both of the matrix products AB and BA to be defined A and B must be square matrices of the same dimension, but even then it may be that $AB \neq BA$ (i.e., matrix multiplication is *not* always commutative). So in any calculation involving matrix products the order of the factors should not be altered. Here is an example where $AB \neq BA$:

$$\begin{bmatrix} 1 & 2 \\ 3 & 4 \end{bmatrix} \begin{bmatrix} 0 & 1 \\ -1 & 0 \end{bmatrix} = \begin{bmatrix} -2 & 1 \\ -4 & 3 \end{bmatrix}, \qquad \begin{bmatrix} 0 & 1 \\ -1 & 0 \end{bmatrix} \begin{bmatrix} 1 & 2 \\ 3 & 4 \end{bmatrix} = \begin{bmatrix} 3 & 4 \\ -1 & -2 \end{bmatrix}$$

Using the product definition, we readily see that some matrix products can be expressed in a different form, a form that we will often find useful. We'll write out these identities below and refer to them as needed.

Matrix Product Identities

☞ So, for example,

$$x_1 b^1 = \begin{bmatrix} x_1 b_{11} \\ x_1 b_{21} \\ \vdots \\ x_1 b_{n1} \end{bmatrix}$$

Bx is a column vector, AB a matrix of column vectors.

Suppose that B is an $r \times n$ matrix. If we denote the j-th column of B by b^j, then

- For any column vector x with scalar entries $x_1, x_2, \ldots, x_n$ we have

$$Bx = x_1 b^1 + x_2 b^2 + \cdots + x_n b^n \tag{1}$$

- For any $m \times r$ matrix A we have

$$AB = [Ab^1 \ Ab^2 \ \cdots \ Ab^n] \tag{2}$$

These identities are a direct result of the definition of matrix products.

EXAMPLE 7.2.3

Writing Matrix Products in a Different Form

Suppose that

$$B = \begin{bmatrix} 2 & -1 & 1 & 2 \\ 3 & 2 & -4 & 1 \\ 1 & 3 & 1 & 1 \end{bmatrix}$$

Then the columns of B are

$$b^1 = \begin{bmatrix} 2 \\ 3 \\ 1 \end{bmatrix}, \quad b^2 = \begin{bmatrix} -1 \\ 2 \\ 3 \end{bmatrix}, \quad b^3 = \begin{bmatrix} 1 \\ -4 \\ 1 \end{bmatrix}, \quad b^4 = \begin{bmatrix} 2 \\ 1 \\ 1 \end{bmatrix}$$

So for any column vector with entries x_1, x_2, x_3, x_4 we have from formula (1) that

$$B \begin{bmatrix} x_1 \\ x_2 \\ x_3 \\ x_4 \end{bmatrix} = x_1 b^1 + x_2 b^2 + x_3 b^3 + x_4 b^4 = \begin{bmatrix} 2x_1 - x_2 + x_3 + 2x_4 \\ 3x_1 + 2x_2 - 4x_3 + x_4 \\ x_1 + 3x_2 + x_3 + x_4 \end{bmatrix}$$

For the matrix A given below, formula (2) tells us what AB looks like:

$$A = \begin{bmatrix} 1 & 0 & 3 \\ 0 & -1 & 0 \\ 1 & 2 & 1 \end{bmatrix}, \qquad AB = [Ab^1 \ Ab^2 \ Ab^3 \ Ab^4] = \begin{bmatrix} 5 & 8 & 4 & 5 \\ -3 & -2 & 4 & -1 \\ 9 & 6 & -6 & 5 \end{bmatrix}$$

So the j-th column of AB is A multiplied into the j-th column of B.

There are three special types of matrices that come up all the time:

❖ **Zero Matrix, Zero Vector.** The $m \times n$ matrix in which all entries are zero is denoted by 0 (regardless of its dimension) and is called the *zero matrix*. A column vector each of whose entries is zero is called the *zero vector*.

❖ **Block Matrix.** Suppose that A is an $m \times n$ matrix, B is an $i \times j$ matrix, and C is an $r \times s$ matrix. If $A = \begin{bmatrix} B & 0 \\ 0 & C \end{bmatrix}$, where the 0's indicate the zero matrices needed to fill out A, then A is said to be a *block matrix*.

❖ **Identity Matrix.** The $n \times n$ diagonal matrix with $a_{ii} = 1$ for all i and $a_{ij} = 0$ whenever $i \neq j$ is called the *identity matrix* and is denoted by I_n. We usually write it simply as I when its dimension is clear from the context.

☞ We also use *I* as a name for an interval, so watch out!

The symbol 0 is also used for the scalar zero, but context will distinguish it from the zero matrix. Suppose that $A = [a_{ij}]$ is an $m \times n$ matrix. From the definitions it can be directly shown that $0A = 0$ (where the zero on the left is the zero scalar, and the zero on the right is a zero matrix) and $A0 = 0$ (with zero vectors on both sides of the equation). Also, $A + 0 = 0 + A = A$. Finally, for any $m \times n$ matrix A, $I_m A = A I_n = A$.

❖ **Transpose, Symmetric Matrix.**   Given any $m \times n$ matrix $A = [a_{ij}]$, the $n \times m$ matrix obtained from A by interchanging the rows and columns is called the *transpose* of A and is denoted by A^T. Note that the ij-th entry of A^T is the ji-th entry of A. An $n \times n$ matrix A is *symmetric* if $A = A^T$, so $a_{ij} = a_{ji}$.

| EXAMPLE 7.2.4 | **Transpose of a Matrix** |

The transpose of the matrix $A = \begin{bmatrix} 1 & 2 & 3 \\ 4 & 5 & 6 \\ 7 & 8 & 9 \end{bmatrix}$ is the matrix $A^T = \begin{bmatrix} 1 & 4 & 7 \\ 2 & 5 & 8 \\ 3 & 6 & 9 \end{bmatrix}$.

Properties of Matrix Addition, Multiplication, Transposition

For all compatible matrices A, B, and C:
- $A + B = B + A$ (Commutativity of addition)
- $A + (B + C) = (A + B) + C$ (Associativity of addition)
- $A(BC) = (AB)C$ (Associativity of multiplication)
- $A(B + C) = AB + AC$ (Distributivity of multiplication over addition)
- $(A^T)^T = A$, $(A + B)^T = A^T + B^T$, $(AB)^T = B^T A^T$ (Transposition)

Vectors and Vector Spaces

The word "vector" can be used in a very general context, but here we use it to denote a column matrix. The collection of all vectors with n entries is denoted by $\mathbb{R}^n$ if the

$$\begin{bmatrix} 2 \\ -1 \\ 1 \end{bmatrix}$$

Vector in $\mathbb{R}^3$

$$\begin{bmatrix} i \\ 1+i \\ 2 \end{bmatrix}$$

Vector in $\mathbb{C}^3$

entries are real and $\mathbb{C}^n$ if the entries are complex. For $\mathbb{R}^n$ the scalars are real numbers; for $\mathbb{C}^n$ the scalars are complex numbers. Note that $\mathbb{R}^n$ is part of $\mathbb{C}^n$. Both $\mathbb{R}^n$ and $\mathbb{C}^n$, with vector addition and multiplication by a scalar defined as before, are examples of *vector spaces* (also called *linear spaces*).

Here are some inequalities involving vectors and matrices that are useful in proving properties of linear systems that come up in this and later chapters. You won't see these inequalities used very much in the book itself, but they do come up in the proofs in the appendices and in the Resource Manual.

Vector/Matrix Inequalities

☞ $[x_1 \ x_2 \ x_3]^T$ is an "in-line" way to write

$$\begin{bmatrix} x_1 \\ x_2 \\ x_3 \end{bmatrix}$$

- *Triangle Inequality*: If a and b are real numbers, then

$$||a| - |b|| \le |a + b| \le |a| + |b|$$

If $u = [u_1, \ldots, u_n]^T$ and $v = [v_1, \ldots, v_n]^T$ are vectors in $\mathbb{R}^n$, and if we denote the *length* of the vector u by $\|u\| = (u_1^2 + \cdots + u_n^2)^{1/2}$, then

$$|\|u\| - \|v\|| \le \|u + v\| \le \|u\| + \|v\|$$

- *Cauchy-Schwarz Inequality*: If u and v are vectors in $\mathbb{R}^n$, then

☞ So $\|u\|$ and $\|A\|$ measure how "big" the vector u and the matrix A are.

$$|u^T v| \le \|u\| \, \|v\|$$

where $u^T v = u_1 v_1 + \cdots + u_n v_n$.

- *Matrix Estimate*: The *norm* $\|A\|$ of a real matrix $A = [a_{ij}]$ is $\|A\| = \sum_{i,j} |a_{ij}|$. If A and B are matrices for which AB is defined, then $\|AB\| \le \|A\| \cdot \|B\|$.

☞ Vectors are often indexed with superscripts because subscripts are reserved for the entries in a vector.

Suppose that W is a subset of a vector space V (either $\mathbb{R}^n$ or $\mathbb{C}^n$ for now). Then for any finite collection of vectors $v^1, v^2, \ldots, v^m$ in W and any scalars $a_1, a_2, \ldots, a_m$, the sum $a_1 v^1 + a_2 v^2 + \cdots + a_m v^m$ is said to be a *finite linear combination* over W. The *span* of W, denoted by Span(W), is the collection of all vectors in V that are finite linear combinations over W. Span(W) is called the *subspace of V spanned by W*. The set W is said to *span* V if Span(W) = V. In general, a subset W of a vector space V is a *subspace* of V if Span(W) = W. In other words, a subset W of a vector space V is a subspace of V if and only if any linear combination of any two elements of W is a vector in W. Every vector space has the *trivial subspace*, the set consisting of only the zero vector, and, of course, V is a subspace of itself.

There is a simple device for visualizing vectors in $\mathbb{R}^3$: Associate with each vector $[x_1 \ x_2 \ x_3]^T$ the geometric vector $x_1 \mathbf{i} + x_2 \mathbf{j} + x_3 \mathbf{k}$, where $\{\mathbf{i}, \mathbf{j}, \mathbf{k}\}$ is an orthogonal frame in Euclidean 3-space. We'll use x as the name of the vector in either form. Then x_1, x_2, and x_3 are the coordinates of the head of the geometric vector, and the origin is the tail. The sum of two vectors $x = [x_1 \ x_2 \ x_3]^T$ and $y = [y_1 \ y_2 \ y_3]^T$ corresponds with the sum of the associated geometric vectors via the parallelogram law. We can use our geometric intuition in Euclidean 3-space to "see" subspaces of $\mathbb{R}^3$ and learn the following: (1) The span of a nonzero vector x is the line through the origin parallel to x; (2) the span of two nonparallel vectors x and y is the plane through the origin

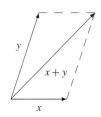

☞ Geometric vectors were reviewed in Section 4.1.

determined by those two vectors; and (3) if three nonzero vectors do not lie in the same plane, then the span of those three vectors must be $\mathbb{R}^3$ itself. So every subspace of $\mathbb{R}^3$ is either a line or a plane through the origin, the whole vector space, or the zero vector itself.

EXAMPLE 7.2.5

Subspaces in $\mathbb{R}^3$

The vectors in $[x_1 \ x_2 \ x_3]^T$ whose components solve the homogeneous linear system

$$3x_1 - x_2 + 2x_3 = 0$$

$$x_1 + 4x_2 - x_3 = 0$$

form a subspace of $\mathbb{R}^3$ because any linear combination of solutions of these equations is again a solution of these same equations. The solution space of these equations must either be the trivial subspace consisting only of the origin, or a line or plane through the origin, or $\mathbb{R}^3$ itself—there are no other possibilities.

Matrices of Functions, Function Vector Spaces

We often use functions of t as the entries in a matrix. The matrix itself then becomes a function. Evaluating the entries of a matrix function at a value of t yields a matrix of constants.

We differentiate matrix functions entrywise. If $A(t) = [a_{ij}(t)]$, then the *matrix derivative* $A'(t) = [a'_{ij}(t)]$. A direct calculation shows that $(A + B)' = A' + B'$ and $(AC)' = A'C + AC'$, where A, B, and C are any matrices of compatible dimensions.

All column vector functions with n entries that are continuously differentiable on an interval can be treated as a set of vectors because they satisfy the same rules as vectors in $\mathbb{R}^n$ or $\mathbb{C}^n$. For each n we call the collection of all such column vector functions a *function vector space*. Solutions of linear differential systems can be described with function vectors.

Linear Differential Systems

The *general linear system* of ODEs in the n state variables $x_1, \ldots, x_n$ has the form

$$x'_1 = a_{11}x_1 + \cdots + a_{1n}x_n + F_1(t)$$

$$\vdots \tag{3}$$

$$x'_n = a_{n1}x_1 + \cdots + a_{nn}x_n + F_n(t)$$

where the *coefficients* a_{ij} and the *driving functions* F_i are constants or functions of t but do not depend on the state variables. If the coefficients a_{ij} are not constants, then we assume they are all defined and continuous (or, perhaps, piecewise continuous) on a common t-interval I.

System (3) can be written in the compact matrix form as

$$x' = Ax + F \tag{4}$$

☞ As noted before, we use I to denote an interval, and sometimes the identity matrix.

where

$$x = \begin{bmatrix} x_1(t) \\ x_2(t) \\ \vdots \\ x_n(t) \end{bmatrix}$$

State Vector

$$A = \begin{bmatrix} a_{11} & \cdots & a_{1n} \\ \vdots & & \vdots \\ a_{n1} & \cdots & a_{nn} \end{bmatrix}, \qquad x = \begin{bmatrix} x_1 \\ \vdots \\ x_n \end{bmatrix}, \qquad F = \begin{bmatrix} F_1 \\ \vdots \\ F_n \end{bmatrix} \qquad (5)$$

The matrix A is the *system matrix*, x is the *state vector*, and F is the *input* or *driving force*. If an initial condition is imposed, then we have the IVP

$$F = \begin{bmatrix} F_1(t) \\ F_2(t) \\ \vdots \\ F_n(t) \end{bmatrix}$$

Input Vector

$$x' = Ax + F, \qquad x(t_0) = x^0 = \begin{bmatrix} x_1^0 \\ \vdots \\ x_n^0 \end{bmatrix} \qquad (6)$$

where $x_1^0, \ldots, x_n^0$ are any given constants and t_0 is any point in I. In most cases A will be assumed to be a constant matrix and I to be the real line. If $F(t)$ is the zero vector for all t, the system is said to be *undriven* (or *homogeneous*, or *free*, or *unforced*).

EXAMPLE 7.2.6

The Lead System is a Linear Differential System

System (2) in Section 7.1 is a linear system in the matrix form (4) with

$$\begin{array}{llll} a_{11} = -(k_{01} + k_{21} + k_{31}) & a_{12} = k_{12} & a_{13} = k_{13} & F_1 = I_1 \\ a_{21} = k_{21} & a_{22} = -(k_{02} + k_{12}) & a_{23} = 0 & F_2 = 0 \\ a_{31} = k_{31} & a_{32} = 0 & a_{33} = -k_{13} & F_3 = 0 \end{array} \qquad (7)$$

For the particular values given in Example 7.1.2, we have

$$A = \begin{bmatrix} -0.0361 & 0.0124 & 0.000035 \\ 0.0111 & -0.0286 & 0 \\ 0.0039 & 0 & -0.000035 \end{bmatrix}, \qquad F = \begin{bmatrix} 49.3 \\ 0 \\ 0 \end{bmatrix} \qquad (8)$$

There is more on the lead system in Section 7.10.

Comments

When the entries of a system matrix A are constants and the entries of the input vector F are zero, we will show in Sections 7.5 and 7.6 that all solutions of the system $x' = Ax$ can be constructed from certain algebraic elements defined by the constant matrix A. In the next few sections we continue to give an overview of these elements and some of their underlying properties.

PROBLEMS

www **1.** Find a 2×2 matrix A (if there is one) that satisfies the equation

$$2A + \begin{bmatrix} 0 & 1 \\ 1 & -1 \end{bmatrix} A = \begin{bmatrix} 1 & -2 \\ 0 & 1 \end{bmatrix}$$

2. Look at an $m \times r$ matrix A and an $r \times n$ matrix B. If b^j is the j-th column of B, show that $AB = [Ab^1 \ Ab^2 \ \cdots \ Ab^n]$. Give some examples.

3. Denote the columns of the $m \times n$ matrix A by $a^1, a^2, \ldots, a^n$. Show that if $x = [x_1 \ x_2 \ \cdots \ x_n]^T$ is any n-vector, then $Ax = x_1 a^1 + x_2 a^2 + \cdots + x_n a^n$. Give some examples.

4. Verify all the *Properties of Matrix Addition, Multiplication, and Transposition* given in the text for the matrices in each part below:

(a) $A = \begin{bmatrix} 2 & -1 \\ 3 & 5 \end{bmatrix}$, $\qquad B = \begin{bmatrix} -2 & 2 \\ 4 & -1 \end{bmatrix}$, $\qquad C = \begin{bmatrix} 0 & 1 \\ -1 & 0 \end{bmatrix}$

(b) $A = \begin{bmatrix} 2 & 0 & 1 \\ -1 & 3 & -1 \\ 0 & -2 & 0 \end{bmatrix}$, $\quad B = \begin{bmatrix} 1 & -1 & 2 \\ 0 & -3 & 1 \\ 2 & 0 & -1 \end{bmatrix}$, $\quad C = \begin{bmatrix} 1 & 2 & 0 \\ 0 & -1 & 1 \\ -2 & 0 & 3 \end{bmatrix}$

5. Suppose that U and W are subspaces of a vector space V (either $\mathbb{R}^n$ or $\mathbb{C}^n$).

(a) Show that the set of all vectors common to U and W is also a subspace of V.

(b) If $V = \mathbb{R}^3$, U is the span of vectors $\{[2 \ \ 1 \ \ 0]^T, \ [1 \ \ 0 \ \ -1]^T\}$, and W is the span of $\{[-1 \ \ 1 \ \ 1]^T, \ [0 \ \ -1 \ \ 1]^T\}$, describe the subspace of all vectors common to U and W.

6. Suppose that A is a 3×3 matrix of real numbers.

(a) Show that $A(\alpha x + \beta y) = \alpha Ax + \beta Ay$, where α and β are reals, and x and y are any 3-vectors.

(b) For a given vector y in $\mathbb{R}^3$, suppose that x^d is a vector such that $Ax^d = y$. Show that every solution of the equation $Ax = y$ is given by $x = x^d + v$, where v is any solution of $Az = 0$.

7. Determine whether or not each differential system below is linear. Write each linear system in the standard matrix form (4).

(a) $x_3' = -2x_2 + x_3 - x_1 + \sin t$, $\quad x_2' = 2x_3 - 5x_1 - x_2' + e^{-t}$, $\quad x_1' = x_1 - 2x_3 + x_2$

(b) $x_2' = x_1 + x_3 - 2x_2 + 1$, $\quad x_1' = x_2 + x_1 x_3$, $\quad x_3' = x_3 + 2x_1 - x_2$

 8. Verify the *Vector/Matrix Inequalities* given in the text with a variety of vectors and matrices. Do the inequalities hold for vectors and matrices with complex entries?

7.3 Systems of Linear Equations

☞ This is *not* a system of ODEs.

The system of n linear equations in n variables $x_1, \ldots, x_n$

$$
\begin{aligned}
a_{11}x_1 + \cdots + a_{1n}x_n &= b_1 \\
&\ \ \vdots \qquad\qquad \vdots \\
a_{n1}x_1 + \cdots + a_{nn}x_n &= b_n
\end{aligned}
\tag{1}
$$

can be written in matrix form as

$$
\begin{bmatrix} a_{11} & \cdots & a_{1n} \\ \vdots & & \vdots \\ a_{n1} & \cdots & a_{nn} \end{bmatrix} \begin{bmatrix} x_1 \\ \vdots \\ x_n \end{bmatrix} = \begin{bmatrix} b_1 \\ \vdots \\ b_n \end{bmatrix}
$$

or more compactly in matrix-vector form as $Ax = b$. Notice that the solution set of linear system (1) is the same as the solution set of an equivalent system obtained from (1) by performing any combination of operations from the list below:

Operations with Linear Equations

1. Interchanging two equations.
2. Multiplying one equation by a nonzero scalar.
3. Adding a multiple of one equation to another.

One easy way to keep track of what happens to linear system (1) when these operations are applied is to form the *augmented matrix* of size $n \times (n+1)$

$$[A|b]$$

and apply any of the following three *elementary row operations*.

Elementary Row Operations for an Augmented Matrix

1. Interchanging two rows.
2. Multiplying one row by a nonzero number.
3. Adding a multiple of one row to a different row.

In Example 7.3.1 we show how to use elementary row operations to find all solutions of linear system (1).

EXAMPLE 7.3.1

Solving a Linear System
Let's find all solutions of the system

$$\begin{aligned} x_1 - x_2 &= 1 \\ -x_1 + 3x_2 + 4x_3 &= 1 \\ x_1 - 3x_2 - 4x_3 &= -1 \end{aligned} \qquad (2)$$

Form the augmented 3×4 matrix

$$[A|b] = \begin{bmatrix} 1 & -1 & 0 & 1 \\ -1 & 3 & 4 & 1 \\ 1 & -3 & -4 & -1 \end{bmatrix}$$

Subtracting the first row from the third, and adding the first row to the second, we have

$$\begin{bmatrix} 1 & -1 & 0 & 1 \\ 0 & 2 & 4 & 2 \\ 0 & -2 & -4 & -2 \end{bmatrix}$$

Adding the second row to the third, we have

$$\begin{bmatrix} 1 & -1 & 0 & 1 \\ 0 & 2 & 4 & 2 \\ 0 & 0 & 0 & 0 \end{bmatrix}$$

which translates to the equations

$$x_1 - x_2 = 1$$
$$2x_2 + 4x_3 = 2 \tag{3}$$
$$0 = 0$$

Since we have three unknowns and only two equations that must be satisfied, we can choose one unknown arbitrarily. Let's say $x_3 = s$, where s is an arbitrary real number; then from the first and second equations in system (3) we have $x_2 = 1 - 2s$ and $x_1 = 2 - 2s$. So the solution set of system (2) is given by

$$\begin{bmatrix} x_1 \\ x_2 \\ x_3 \end{bmatrix} = \begin{bmatrix} 2 - 2s \\ 1 - 2s \\ s \end{bmatrix} = \begin{bmatrix} 2 \\ 1 \\ 0 \end{bmatrix} + s \begin{bmatrix} -2 \\ -2 \\ 1 \end{bmatrix} \quad \text{for arbitrary real } s \tag{4}$$

Geometrically, (4) defines a line through the point $(2, 1, 0)$ parallel to $-2\mathbf{i} - 2\mathbf{j} + \mathbf{k}$.

Linear systems with complex coefficients are solved in the same way, except that we use complex arithmetic throughout.

EXAMPLE 7.3.2

A System with Complex Coefficients

Let's find the solution set of the system

$$ix_1 + x_2 = 1 - i$$
$$-2x_1 + ix_2 = 2i \tag{5}$$

Forming the $[A|b]$ matrix, we have

$$\begin{bmatrix} i & 1 & | & 1 - i \\ -2 & i & | & 2i \end{bmatrix}$$

Adding $-2i$ times the first row to the second row, we have

$$\begin{bmatrix} i & 1 & | & 1 - i \\ 0 & -i & | & -2 \end{bmatrix}$$

So an equivalent system to (5) is

$$-ix_1 + x_2 = 1 - i$$
$$-ix_2 = -2$$

whose unique solution is

$$x_2 = 2/i = -2i$$
$$x_1 = [(1 - i) + 2i]/(-i) = -1 + i$$

Working with complex systems is just like working with real systems, only using complex arithmetic instead.

For any $n \times n$ matrix A with real or complex entries here's the way we find *all* solutions of the linear system $Ax = b$:

General Method for Solving $Ax = b$, **where** A **is an** $n \times n$ **Matrix**

- Form the matrix $[A|b]$ and use elementary row operations to put zeros beneath the main diagonal of A.

- Write out the equivalent system and check to see if it is consistent (i.e., there are no equations of the type *zero = nonzero number*).

- If inconsistent, then stop. The system has no solutions.

- If consistent, discard the p equations $0 = 0$ (if any), assign arbitrary values to p of the unknowns, and solve for the others by back substitution.

Invertible Matrices

In what follows, we will deal exclusively with square matrices, since it is these matrices that occur in the study of linear first-order systems of ODEs.

❖ **Invertibility**. An $n \times n$ matrix A is *invertible* if there is an $n \times n$ matrix B such that $AB = BA = I$. The matrix B is said to be the *inverse of A* and is denoted by A^{-1}. If A^{-1} exists, it is unique. If a matrix A does not have an inverse, it is said to be *singular*.

EXAMPLE 7.3.3   **The Inverse of a Matrix**

Let $A = \begin{bmatrix} 3 & 1 \\ 5 & 2 \end{bmatrix}$ and $B = \begin{bmatrix} 2 & -1 \\ -5 & 3 \end{bmatrix}$. The product $AB = BA = \begin{bmatrix} 1 & 0 \\ 0 & 1 \end{bmatrix} = I$, so A is invertible with $A^{-1} = B$. Also B is invertible with $B^{-1} = A$.

Here is a method for calculating the inverse of a square matrix. If the $n \times n$ matrix A is invertible, then it is possible, using the three elementary row operations, to reduce the matrix A to the identity matrix I. It can be shown that the sequence of elementary row operations that carries the invertible matrix A to I carries the matrix I to A^{-1}. This method is illustrated below in solving $Ax = b$. Note that $[A|b||I]$ denotes the $n \times (2n + 1)$ matrix of A augmented by b and I.

EXAMPLE 7.3.4   **Finding the Inverse of a Matrix**

We wish to solve the system

$$
\begin{aligned}
3x_1 + x_2 &= 3 \\
5x_1 + 2x_2 &= 4
\end{aligned} \tag{6}
$$

which is equivalent to the matrix equation $Ax = b$, where

$$
A = \begin{bmatrix} 3 & 1 \\ 5 & 2 \end{bmatrix}, \qquad x = \begin{bmatrix} x_1 \\ x_2 \end{bmatrix}, \qquad b = \begin{bmatrix} 3 \\ 4 \end{bmatrix}
$$

If A is invertible, we can simultaneously find A^{-1} and solve for x. We have

$$
[A|b||I] = \begin{bmatrix} 3 & 1 & 3 & 1 & 0 \\ 5 & 2 & 4 & 0 & 1 \end{bmatrix}
$$

Applying elementary row operations on $[A|b\|I]$, we have

$$\begin{bmatrix} 6 & 2 & 6 & 2 & 0 \\ 5 & 2 & 4 & 0 & 1 \end{bmatrix} \qquad \begin{matrix} (2 \times \text{Row 1}) \\ (\text{Row 2}) \end{matrix}$$

$$\begin{bmatrix} 1 & 0 & 2 & 2 & -1 \\ 5 & 2 & 4 & 0 & 1 \end{bmatrix} \qquad \begin{matrix} (\text{Row 1} - \text{Row 2}) \\ (\text{Row 2}) \end{matrix}$$

$$\begin{bmatrix} 1 & 0 & 2 & 2 & -1 \\ 0 & 2 & -6 & -10 & 6 \end{bmatrix} \qquad \begin{matrix} (\text{Row 1}) \\ (\text{Row 2} - 5 \times \text{Row 1}) \end{matrix}$$

$$\begin{bmatrix} 1 & 0 & 2 & 2 & -1 \\ 0 & 1 & -3 & -5 & 3 \end{bmatrix} \qquad \begin{matrix} (\text{Row 1}) \\ (1/2 \times \text{Row 2}) \end{matrix}$$

☞ If $Ax = b$ and A is invertible, then $x = A^{-1}b$.

So $A^{-1} = \begin{bmatrix} 2 & -1 \\ -5 & 3 \end{bmatrix}$. The third column of the reduced matrix represents the unique solution $x_1 = 2$, $x_2 = -3$ of system (6). Notice that the solution vector $x = [2 \ -3]^T = A^{-1}[3 \ 4]^T = A^{-1}b$.

For a 2×2 matrix $A = \begin{bmatrix} a & b \\ c & d \end{bmatrix}$, it can be shown by the above process that

$$A^{-1} = \frac{1}{ad - bc} \begin{bmatrix} d & -b \\ -c & a \end{bmatrix} \tag{7}$$

Notice that if $ad - bc = 0$, then A^{-1} does not exist.

Determinants

Every square matrix A has a number associated with it called its *determinant* and denoted by $\det A$. We will define $\det A$ inductively. For a 2×2 matrix,

$$\det \begin{bmatrix} a & b \\ c & d \end{bmatrix} = ad - bc \tag{8}$$

Now assume that $\det A$ has been defined for all square matrices of size $(n-1) \times (n-1)$ and smaller. As we shall see, the definition of $\det A$, where A is an $n \times n$ matrix, is in terms of the determinants of some associated $(n-1) \times (n-1)$ matrices called minors. The ij-th *minor matrix*, denoted by A_{ij}, is the $(n-1) \times (n-1)$ matrix obtained by striking out the i-th row and the j-th column of A (indicated in blue) to obtain

$$A_{ij} = \begin{bmatrix} a_{11} & \cdots & a_{1j} & \cdots & a_{1n} \\ \vdots & & \vdots & & \vdots \\ a_{i1} & \cdots & a_{ij} & \cdots & a_{in} \\ \vdots & & \vdots & & \vdots \\ a_{n1} & \cdots & a_{nj} & \cdots & a_{nn} \end{bmatrix}$$

The number $(-1)^{i+j} \det A_{ij}$ is called the *ij-th cofactor*, denoted by $\text{cof } A_{ij}$.

❖ **Cofactor Expansion.** Suppose that A is an $n \times n$ matrix. Then the *cofactor expansions* on the i-th row and on the j-th column, respectively, are given by

$$\sum_{k=1}^{n} a_{ik} \operatorname{cof} A_{ik}, \qquad \sum_{k=1}^{n} a_{kj} \operatorname{cof} A_{kj}$$

For any matrix A, it can be shown that all cofactor expansions are equal.

Now we have all we need to define the determinant of an $n \times n$ matrix.

❖ **Determinant.** If A is an $n \times n$ matrix, then the *determinant* of A, denoted by det A, is the scalar defined by any one of the cofactor expansions.

This definition is *recursive* (or *inductive*) because it defines the determinant of an $n \times n$ matrix in terms of the determinants of $(n-1) \times (n-1)$ matrices (the cofactor determinants), and so on down to the determinants of 2×2 matrices, which are defined right at the start in formula (8).

Properties of det A (A **any** $n \times n$ **matrix)**

☞ Useful properties, but we won't prove them.

- If A is the block matrix $\begin{bmatrix} B & 0 \\ 0 & C \end{bmatrix}$, then det $A = (\det B)(\det C)$.

☞ You can replace "row" by "column" in bulleted items 2–4.

- If a row of A is multiplied by a scalar a, the determinant of that matrix is $a \det A$. So, if A has a row of zeros, then det $A = 0$.

- Adding a multiple of a row of A to another row gives a matrix B and det $A = \det B$.

- Interchanging two rows of A gives a matrix B, and det $A = -\det B$.

- If A is an upper or lower triangular matrix, then det A is the product $a_{11}a_{22}\cdots a_{nn}$ of the diagonal elements.

- det $A = \det A^T$; A is invertible if and only if det $A \neq 0$; $\det(A^{-1}) = 1/\det A$.

- $\det(AB) = (\det A)(\det B) = \det(BA)$ for any $n \times n$ matrices A, B.

EXAMPLE 7.3.5

Evaluating a Determinant

Expanding $A = \begin{bmatrix} 3 & 6 & 2 \\ 2 & 5 & 1 \\ -1 & 2 & 4 \end{bmatrix}$ by cofactors on the first row, we have

$$\det A = 3 \det \begin{bmatrix} 5 & 1 \\ 2 & 4 \end{bmatrix} - 6 \det \begin{bmatrix} 2 & 1 \\ -1 & 4 \end{bmatrix} + 2 \det \begin{bmatrix} 2 & 5 \\ -1 & 2 \end{bmatrix}$$

$$= 3(20 - 2) - 6(8 + 1) + 2(4 + 5) = 18$$

Performing some row or column operations first would make it easier to evaluate det A.

A good strategy to follow in calculating the determinant of a matrix is to do a cofactor expansion on the row or column with the largest number of zeros.

Determinants and the solvability of $Ax = b$ are intimately related, as the next theorem shows.

THEOREM 7.3.1

> Solvability of Linear Equations. Look at the $n \times n$ linear system $Ax = b$.
>
> - If $\det A \neq 0$, then the system $Ax = b$ has the unique solution $x = A^{-1}b$, so the homogeneous system $Ax = 0$ has only the trivial solution $x = 0$.
>
> - If $\det A = 0$, then A is singular, the homogeneous system $Ax = 0$ has infinitely many solutions, and the system $Ax = b$ has either no solutions, or an infinite number of solutions.
>
> - $Ax = b$ has a unique solution if and only if $\det A \neq 0$, that is, if and only if A is invertible.

Now let's look at three concepts that will be a big help to us when we solve linear differential systems.

Linear Independence, Basis, and Dimension

Suppose that W is a subspace of the vector space V. There are many sets of vectors in the subspace W that span W. If we require that each vector in W be written uniquely as a finite linear combination over a spanning set B, then we must choose B in a special way. Observe that if there are two different finite linear combinations over B that add up to the same vector in W, then by subtracting them we obtain a finite linear combination over B (not all of whose scalar coefficients are zero) that adds up to the zero vector. We refer to a finite linear combination of vectors with at least one nonzero scalar coefficient as *nontrivial*, and *trivial* otherwise.

> ❖ **Linear Independence**. A set B in a vector space is said to be *linearly independent* if and only if there is no nontrivial finite linear combination over B whose sum is the zero vector.

So we see that a set B in a vector space is linearly independent if and only if each vector in Span(B) can be written as a finite linear combination over B in exactly one way. Any subset of an independent set is also independent.

The following result is often useful:

THEOREM 7.3.2

☞ Say v^1 aloud as "v super one".

> Linear Independence and Determinants. A set of n vectors $\{v^1, v^2, \ldots, v^n\}$ in $\mathbb{R}^n$ (or $\mathbb{C}^n$) is linearly independent if and only if $\det[v^1 \ v^2 \ \cdots \ v^n] \neq 0$.

For example, let u^j be the vector in $\mathbb{R}^n$ whose j-th entry is 1 and all other entries 0. Since $\det[u^1 \ u^2 \ \cdots \ u^n] = 1$, the set $\{u^1, u^2, \ldots, u^n\}$ is linearly independent in $\mathbb{R}^n$.

❖ **Basis.** A subset B of a linear space V is a *basis* for V if B is linearly independent and $V = \text{Span}(B)$.

If B is a basis for V, then every vector in V can be written as a finite linear combination over B in exactly one way. We just saw that $B = \{u^1, \ldots, u^n\}$ is an independent set in $\mathbb{R}^n$, and evidently $\mathbb{R}^n = \text{Span}(B)$. So B is a basis for $\mathbb{R}^n$.

☞ The connection between basis and dimension.

It is a fact that every subspace V of $\mathbb{R}^n$ (or $\mathbb{C}^n$) has a basis and that every basis for that subspace contains the same number of elements. This number, denoted by $\dim V$, is called the *dimension* of V. The dimension of the trivial linear space (which consists only of the zero vector) is zero. It can be shown that if W is a subspace of the vector space V, then $\dim W \leq \dim V$; $W = V$ if $\dim W = \dim V < \infty$.

EXAMPLE 7.3.6

Dimensions of the Vector Spaces $\mathbb{R}^n$ and $\mathbb{C}^n$
The vector spaces $\mathbb{R}^n$ and $\mathbb{C}^n$ have dimension n because the set $B = \{u^1, \ldots, u^n\}$, with the u^j as defined above, is a basis for each.

☞ Function vectors are column matrices with functions as entries.

In a function vector space, the concepts of span, subspace, linear independence, basis, and dimension are the same as before, except that now the equality is interpreted to hold for all t-values in an interval I. One consequence of this is that function vector spaces can be infinite-dimensional. The only function vector spaces that interest us here are the *continuity sets* $\mathbf{C}^n(I)$. For each n, $\mathbf{C}^n(I)$ is the collection of all function vectors with the same number of entries, each of which is n-times continuously differentiable over the t-interval I. The context of the problem under consideration reveals the number of entries in function vectors and whether the entries are real- or complex-valued functions.

EXAMPLE 7.3.7

Linear Independence in a Function Vector Space
Here are a few tricks for showing that a set of functions such as $\{\sin t, \cos t, e^t\}$ is linearly independent in the function vector space $\mathbf{C}^2(\mathbb{R})$. Say that there are constants a, b, and c such that

☞ Sin t is a vector function: think of it as the one-rowed column vector $[\sin t]$.

$$a \sin t + b \cos t + c e^t = 0, \quad \text{for all } t \tag{9}$$

We will show that $a = b = c = 0$ by evaluating this identity at the three t-values $t = 0$, $\pi/2$, and π to obtain the linear system $b + c = 0$, $a + ce^{\pi/2} = 0$, and $-b + ce^\pi = 0$. The unique solution of this system is $a = b = c = 0$, so the set $\{\sin t, \cos t, e^t\}$ is linearly independent.

Another approach would be to differentiate the identity (9) twice to produce three identities that when evaluated at $t = 0$ produce the respective equations $b + c = 0$, $a + c = 0$, and $-b + c = 0$; the unique solution is $a = b = c = 0$. Using either approach, we have shown that the set $\{\sin t, \cos t, e^t\}$ is linearly independent in $\mathbf{C}^2(\mathbb{R})$.

EXAMPLE 7.3.8

Linear Independence in $\mathbf{C}^0(\mathbb{R})$

The function vectors v^1, v^2, and v^3 in $\mathbf{C}^0(\mathbb{R})$ given by

$$v^1 = \begin{bmatrix} e^t \\ -e^t \\ 0 \end{bmatrix}, \qquad v^2 = \begin{bmatrix} e^{-t} \\ 0 \\ e^{-t} \end{bmatrix}, \qquad v^3 = \begin{bmatrix} e^{-2t} \\ 2e^{-2t} \\ -e^{-2t} \end{bmatrix}$$

are linearly independent. To show this, suppose that c_1, c_2, and c_3 are real numbers such that $c_1 v^1 + c_2 v^2 + c_3 v^3 = [0 \ \ 0 \ \ 0]^T$, for all t in $\mathbb{R}$. For $t = 0$ we must have $c_1 v^1(0) + c_2 v^2(0) + c_3 v^3(0) = [0 \ \ 0 \ \ 0]^T$, which can be written in matrix form as

$$Ac = \begin{bmatrix} 1 & 1 & 1 \\ -1 & 0 & 2 \\ 0 & 1 & -1 \end{bmatrix} \begin{bmatrix} c_1 \\ c_2 \\ c_3 \end{bmatrix} = \begin{bmatrix} 0 \\ 0 \\ 0 \end{bmatrix}$$

Since the determinant of A is nonzero, it follows from Theorem 7.3.1 that $c_1 = c_2 = c_3 = 0$ is the only solution, so $\{v^1, v^2, v^3\}$ is an independent set.

EXAMPLE 7.3.9

Bases and Linear Independence

☞ The zero polynomial is the polynomial that is zero for all t.

The set $B = \{1, t, t^2, \ldots, t^n, \ldots\}$ of all nonnegative powers of t in the real vector space $\mathbf{C}^0(\mathbb{R})$ has the set of all polynomials with real coefficients as its span. B does not span $\mathbf{C}^0(\mathbb{R})$ because there are continuous functions (e.g., $\sin t$) that are not polynomials. The set B is linearly independent because no nontrivial finite linear combination of the vectors in B can be the zero polynomial. So B is a basis for $\mathrm{Span}(B)$ and $\dim B = \infty$. Since B is a subspace of $\mathbf{C}^0(\mathbb{R})$, then $\mathbf{C}^0(\mathbb{R})$ is also infinite-dimensional, but B is *not* $\mathbf{C}^0(\mathbb{R})$.

EXAMPLE 7.3.10

Solution Set of an ODE as a Function Vector Space

In Section 3.4 we learned how to find a formula for all solutions of the undriven linear ODE $y'' + 2y' + 2y = 0$. The roots of the characteristic polynomial $r^2 + 2r + 2 = 0$ are $r_1 = -1 + i$ and $r_2 = -1 - i$, so the general real-valued solution of the ODE is $y = C_1 e^{-t} \cos t + C_2 e^{-t} \sin t$, where C_1 and C_2 are arbitrary reals. To show that the pair of functions $\{e^{-t} \cos t, e^{-t} \sin t\}$ is linearly independent in the function vector space $\mathbf{C}^2(\mathbb{R})$, we might proceed as follows. Suppose that a and b are constants such that $ae^{-t} \cos t + be^{-t} \sin t = 0$ for all t. Evaluating that identity at $t = 0$ yields that $a = 0$, and evaluating it at $t = \pi/2$ yields that $b = 0$. So the function pair $\{e^{-t} \cos t, e^{-t} \sin t\}$ is linearly independent, and since the span of this pair generates all solutions of the ODE, we conclude that the ODE's solution set is a two-dimensional function vector subspace of $\mathbf{C}^2(\mathbb{R})$.

Comments

The concepts of vector spaces and function vector spaces are extremely useful in characterizing all solutions of linear systems. Key ideas in this connection are linear independence and basis in a vector space. We will come back to these ideas many times in this chapter.

PROBLEMS

1. Find all solutions of each system and characterize them geometrically (see Example 7.3.1).

   (a) $x_1 - 2x_2 = 0$, $\quad 4x_2 - 2x_1 = 0$

   (b) $x_1 - 2x_2 - x_3 = 0$, $\quad x_2 + x_3 = 0$, $\quad x_1 + x_2 + 2x_3 = 0$

   (c) $x_1 + x_2 + 2x_3 = 1$, $\quad 3x_1 + 4x_2 - x_3 = -1$

2. Suppose that the matrix A is given by

$$A = \begin{bmatrix} 1 & -1 & 0 \\ -1 & 3 & -2 \\ 1 & 3 & -4 \end{bmatrix}$$

   (a) Find a solution of the system $Ax = [1 \ -1 \ 1]^T$.

   (b) Find all solutions of the system $Ax = 0$.

   (c) For the matrix A in part (a), what conditions on the vector $y = [a \ b \ c]^T$ guarantee that the system $Ax = y$ has a solution?

3. Find all solutions (if there are any) of the following systems.

   (a) $\begin{array}{rl} 3x_1 - x_2 + x_3 + 2x_4 - 2x_5 = & 1 \\ x_1 + 2x_2 \quad - x_4 + x_5 = & -1 \\ x_2 + 2x_3 - 5x_4 + 2x_5 = & 2 \end{array}$
   (b) $\begin{array}{rl} x_1 - 2x_2 + 4x_3 - x_4 = & -1 \\ 2x_1 - x_2 - x_3 + x_4 = & 1 \\ 5x_1 \quad + 2x_3 + x_4 = & -1 \end{array}$

4. For $A = \begin{bmatrix} 1 & 2 \\ -i & i \end{bmatrix}$ and $y = [-i \ 1]^T$, find all solutions of the system $Ax = y$.

5. Show that a 2×2 upper triangular matrix is invertible if and only if the elements on its main diagonal are all nonzero, and that the inverse (if it exists) must also be upper triangular.

6. Show that if a square matrix A is invertible, so is A^T, and $(A^{-1})^T = (A^T)^{-1}$.

7. Suppose that A and B are invertible $n \times n$ matrices.

   (a) Show that an invertible matrix has a unique inverse.

   (b) Show that AB is invertible and that $(AB)^{-1} = B^{-1}A^{-1}$.

8. Evaluate the following determinants.

   (a) $\det \begin{bmatrix} 3 & 5 & 7 & 2 \\ 2 & 4 & 1 & 1 \\ -2 & 0 & 0 & 0 \\ 1 & 1 & 3 & 4 \end{bmatrix}$
   (b) $\det \begin{bmatrix} 1 & -2 & 3 & -2 & -2 \\ 2 & -1 & 1 & 3 & 2 \\ 1 & 1 & 2 & 1 & 1 \\ 1 & -4 & -3 & -2 & -5 \\ 3 & -2 & 2 & 2 & -2 \end{bmatrix}$

www 9. Are the sets of functions below linearly dependent or independent in $C^0(\mathbb{R})$?

   (a) $\{e^{-t}, -3e^t, \cosh t\}$ $\quad$ (b) $\{e^t, te^t, -t^2e^t\}$

   (c) $\{e^{-t}\cos t, e^{-t}\sin t\}$ $\quad$ (d) $\{1, t-1, 3t^2+t+1, 1-t^2\}$

10. Find each of the subsets of $C^2(\mathbb{R})$ described below. If it is a subspace, state its dimension.

   (a) $\{y : y'' = 0, \ 2y(0) + y'(0) = 0, \ 2y(1) - y'(1) = 0\}$

   (b) $\{y : y'' = 2, \ y(0) = 0, \ y(1) = 0\}$

   (c) $\{y : y'' = 2, \ 2y(0) + y'(0) = 0, \ 2y(1) - y'(1) = 0\}$

11. Which of the following are subspaces of $C^0(\mathbb{R})$? If not a subspace, why not?

   (a) All polynomials of degree 2 $\quad$ (b) All polynomials of degree greater than 3

   (c) All odd functions $[f(-x) = -f(x)]$ $\quad$ (d) All nonnegative functions

12. (*Rank-Nullity Theorem*). Suppose that A is an $n \times m$ matrix and define the operator $L : \mathbb{R}^n \to \mathbb{R}^m$ whose action is given by $L[x] = Ax$, for any vector x in $\mathbb{R}^n$.

(a) Show that L is a linear operator.

(b) The *null space* of L is the set of all vectors v in $\mathbb{R}^n$ such that $L[v] = 0$. Show that $R(L)$, the range of L, is a subspace of $\mathbb{R}^m$ and that $N(L)$, the null space of L, is a subspace of $\mathbb{R}^n$.

(c) The Rank-Nullity Theorem states that $\dim R(L) + \dim N(L) = n$. Verify this assertion for the matrix A given in Problem 2.

13. (*Fredholm Alternative*). Suppose that A is an $n \times m$ matrix. The Fredholm[2] Alternative states that the system $Ax = y$ has a solution if and only if $y^T v = 0$ for all v that solve $A^T x = 0$. Verify this assertion for the matrix A given in Problem 2.

7.4   Eigenvalues and Eigenvectors of Matrices

For an $n \times n$ constant matrix A, there is a great deal of interest in finding a nonzero column vector v for which $Av = \lambda v$ for some real or complex constant λ. The action of multiplying the matrix A into such a vector v is to return v, but multiplied by the constant λ. It is a remarkable fact that any matrix A has at least one such v, λ pair. Vectors and associated scalars with this property play a very important role in solving a linear differential system, $x' = Ax$, where A is a constant matrix.

It's time for some terminology:

❖ **Eigenvalues and Eigenvectors.** A scalar λ is an *eigenvalue* of the $n \times n$ constant matrix A if there is a nonzero vector v such that

$$Av = \lambda v \tag{1}$$

The vector v is called an *eigenvector* of A corresponding to the eigenvalue λ. For any eigenvalue λ of A, the set V_λ of all solutions v of the equation $Av = \lambda v$ is called the *eigenspace* of A corresponding to λ.

The eigenspace V_λ is actually a subspace of $\mathbb{R}^n$ (or $\mathbb{C}^n$, if appropriate), and any nonzero element in V_λ is an eigenvector of A corresponding to λ. Indeed, if v^1 and v^2 are any two elements in the eigenspace V_λ of a matrix A, then $Av^1 = \lambda v^1$ and $Av^2 = \lambda v^2$. So for any constants c_1 and c_2,

$$A(c_1 v^1 + c_2 v^2) = c_1 Av^1 + c_2 Av^2 = \lambda(c_1 v^1 + c_2 v^2)$$

and $c_1 v^1 + c_2 v^2$ is in V_λ, showing that V_λ is a subspace. Note that two eigenspaces V_λ and V_μ of a matrix A have no vector in common (except 0) if $\lambda \neq \mu$.

---

[2]The Swedish mathematician Ivar Fredholm (1866–1927) was one of the founders of the theory of integral equations.

EXAMPLE 7.4.1

Finding Eigenvalues and Eigenvectors

Denoting a 2-vector v by $[a \; b]^T$, let's determine all scalars λ for which the equation

$$A\begin{bmatrix} a \\ b \end{bmatrix} = \lambda \begin{bmatrix} a \\ b \end{bmatrix}, \quad \text{where } A = \begin{bmatrix} 5 & 3 \\ -6 & -4 \end{bmatrix} \tag{2}$$

has a solution $v = [a \; b]^T$, where a and b are *not* both zero.

With the 2×2 identity matrix I, this linear system can be written as

$$(A - \lambda I)\begin{bmatrix} a \\ b \end{bmatrix} = \begin{bmatrix} 0 \\ 0 \end{bmatrix} \tag{3}$$

Now (3) has the solution $a = b = 0$, but this corresponds to the zero vector. From Theorem 7.3.1, system (3) has nontrivial solutions if and only if the determinant of the coefficient matrix vanishes. This means that λ must satisfy the condition

$$\det(A - \lambda I) = \det \begin{bmatrix} 5 - \lambda & 3 \\ -6 & -4 - \lambda \end{bmatrix} = \lambda^2 - \lambda - 2 = 0$$

This happens for just two values of λ: $\lambda_1 = -1$ and $\lambda_2 = 2$. First setting $\lambda = -1$, the system in (2) reduces to the single equation $6a + 3b = 0$. So, for example, $a = 1$ and $b = -2$ will do, and $v = [1 \; -2]^T$ is an eigenvector corresponding to $\lambda = -1$.

Next setting $\lambda = 2$, system (2) reduces to the single equation $3a + 3b = 0$. So $a = 1$ and $b = -1$ will do, and $w = [1 \; -1]^T$ is an eigenvector corresponding to $\lambda = 2$. So we see that

$$A\begin{bmatrix} 1 \\ -2 \end{bmatrix} = (-1)\begin{bmatrix} 1 \\ -2 \end{bmatrix} \quad \text{and} \quad A\begin{bmatrix} 1 \\ -1 \end{bmatrix} = 2\begin{bmatrix} 1 \\ -1 \end{bmatrix} \tag{4}$$

In other words, the matrix A applied to the vectors $[1 \; -2]^T$ and $[1 \; -1]^T$ produces those same vectors again but multiplied by the scalars -1 and 2, respectively.

EXAMPLE 7.4.2

Finding Eigenspaces

From the calculations in Example 7.4.1 we see that A has no eigenvalues other than λ_1 and λ_2, that v is in the eigenspace V_{λ_1} if $v = k[1 \; -2]^T$ for any real k, and that w is in the eigenspace V_{λ_2} if $w = k[1 \; -1]^T$ for any real k. Note that V_{λ_1} is spanned by $[1 \; -2]^T$, that V_{λ_2} is spanned by $[1 \; -1]^T$, and that $\{[1 \; -2]^T, \; [1 \; -1]^T\}$ is a basis for $\mathbb{R}^2$.

Even though the entries of A are real, the eigenvalues and the eigenvectors may be complex. Recall that when dealing with vectors in $\mathbb{C}^n$, the scalars are complex numbers; and when dealing with vectors in $\mathbb{R}^n$, the scalars are real numbers. The complex conjugate of the complex number $\lambda = \alpha + i\beta$ is $\bar{\lambda} = \alpha - i\beta$. For a vector $v = [v_1 \; v_2 \; \cdots \; v_n]^T$ in $\mathbb{C}^n$, the complex conjugate $\bar{v} = [\bar{v}_1 \; \bar{v}_2 \; \cdots \; \bar{v}_n]^T$. If A is real and $Av = \lambda v$ where λ is complex, then v is complex. In this case $\bar{\lambda}$ is also an eigenvalue of A and $\bar{v}$ a corresponding eigenvector since $\bar{\lambda}\bar{v} = \overline{\lambda v} = \overline{Av} = A\bar{v}$. Here's an example of complex eigen-elements.

☞ If you are a bit rusty on working with complex quantities, you might review Appendix B.3.

EXAMPLE 7.4.3

Complex Eigenvalues

To find the eigenvalues and eigenvectors of $A = \begin{bmatrix} 1 & -2 \\ 2 & 1 \end{bmatrix}$, we need to find scalars λ for which there are nonzero vectors v such that

$$(A - \lambda I)v = \begin{bmatrix} 1-\lambda & -2 \\ 2 & 1-\lambda \end{bmatrix} \begin{bmatrix} v_1 \\ v_2 \end{bmatrix} = \begin{bmatrix} 0 \\ 0 \end{bmatrix} \tag{5}$$

From Theorem 7.3.1, system (5) has nonzero solutions v exactly when λ is chosen so that

$$\det \begin{bmatrix} 1-\lambda & -2 \\ 2 & 1-\lambda \end{bmatrix} = \lambda^2 - 2\lambda + 5 = 0$$

The values of λ that satisfy this condition are $\lambda_1 = 1 + 2i$ and $\lambda_2 = \overline{\lambda_1} = 1 - 2i$. The eigenspace V_{λ_1} corresponding to λ_1 is the solution set of the system $(A - \lambda_1 I)v = 0$. Observe that $v = [a\ b]^T$ is a solution of $(A - \lambda_1 I)v = 0$ if and only if

$$(1 - \lambda_1)a - 2b = -2ia - 2b = 0$$

or $a = ib$ (the other equation is a multiple of this one). So $v = [ib\ b]^T$, where b is an arbitrary real or complex number, is in the eigenspace we seek, and $V_{\lambda_1} = \text{Span}([i\ 1]^T)$. Similarly, we obtain that $V_{\lambda_2} = \text{Span}([-i\ 1]^T)$.

Eigenvectors have the following useful property.

THEOREM 7.4.1

> Eigenspace Property. Suppose that V_{λ_j}, $j = 1, \ldots, p$, are eigenspaces of an $n \times n$ matrix A corresponding to the distinct eigenvalues λ_j, $j = 1, \ldots, p$. Suppose that B_j is an independent subset of V_{λ_j}. Then the set B consisting of all the elements of $B_1, \ldots, B_p$ is an independent set.

This can be shown by mathematical induction on p (Problem 8).

☞ The Eigenspace Property is used a lot in solving linear systems of ODEs.

An important consequence of the Eigenspace Property is that if an $n \times n$ matrix A has n distinct eigenvalues $\lambda_1, \ldots, \lambda_n$, then there is a set of n independent eigenvectors. In this case if B is a set of eigenvectors, one from each V_{λ_j}, $j = 1, 2, \ldots, n$, then the Eigenspace Property shows that B is an independent set. The set is actually a basis for $\mathbb{R}^n$ (or $\mathbb{C}^n$). Any basis of $\mathbb{R}^n$ (or $\mathbb{C}^n$) consisting of eigenvectors is called an *eigenbasis*.

EXAMPLE 7.4.4

Eigenbases for $\mathbb{R}^2$ and $\mathbb{C}^2$

The matrix A of Example 7.4.1 has eigenvectors $[1\ -2]^T$ and $[1\ -1]^T$ corresponding, respectively, to the eigenvalues -1 and 2. The set $\{[1\ -2]^T, [1\ -1]^T\}$ is an eigenbasis of $\mathbb{R}^2$. Similarly, the set $\{[1\ i]^T, [1\ -i]^T\}$ of eigenvectors of the matrix of Example 7.4.3 is independent and so is an eigenbasis basis of $\mathbb{C}^2$.

The Characteristic Polynomial of a Matrix

Examples 7.4.1 and 7.4.3 suggest a method for finding the eigenvalues of any matrix. Suppose that A is an $n \times n$ matrix with real or complex entries. If λ_0 is an eigenvalue of A, there must be a column vector $v \neq 0$ such that $Av = \lambda_0 v$. So the matrix equation $(A - \lambda_0 I)v = 0$ has a nontrivial solution v, and the set of all solutions of this equation forms the eigenspace V_{λ_0}. Recall that the linear system $(A - \lambda I)v = 0$ has a nonzero solution v if and only if the determinant of the system $\det(A - \lambda I) = 0$. Let's define a polynomial function $p(\lambda)$ as follows:

$$p(\lambda) = \det(A - \lambda I) = \det \begin{bmatrix} a_{11} - \lambda & a_{12} & \cdots & a_{1n} \\ a_{21} & a_{22} - \lambda & \cdots & a_{2n} \\ \vdots & \vdots & \ddots & \vdots \\ a_{n1} & a_{n2} & \cdots & a_{nn} - \lambda \end{bmatrix} \tag{6}$$

$$= (-1)^n \lambda^n + (-1)^{n-1}(a_{11} + \cdots + a_{nn})\lambda^{n-1} + \cdots + \det A \tag{7}$$

Note that $p(\lambda) = \det(A - \lambda I)$ is a polynomial of degree n in λ. The coefficients of λ^n and λ^{n-1} in formula (7) can be established by expanding the determinant in (6) by cofactors and using induction on n. That the constant term of $p(\lambda)$ is $\det A$ follows from the definition (6) of $p(\lambda)$ if λ is equated to 0. The polynomial p has a name:

❖ **Characteristic Polynomial.** The polynomial $p(\lambda) = \det(A - \lambda I)$ given in (6) is called the *characteristic polynomial* of the square matrix A.

From this definition and the discussion above, we have the following result.

THEOREM 7.4.2

> Characteristic Polynomial Theorem. A scalar λ_0 is an eigenvalue of an $n \times n$ matrix A if and only if λ_0 is a root of the characteristic polynomial $p(\lambda) = \det(A - \lambda I)$.

The calculation of the eigenvalues of an $n \times n$ matrix reduces to finding the roots of a polynomial of degree n. If $n = 2$, the quadratic formula may be used, but for $n > 2$ it may be a difficult problem to find the roots of the corresponding polynomial. Sometimes, factoring techniques of algebra allow the determination of all the eigenvalues even if $n > 2$. If A is the block matrix $\begin{bmatrix} B & 0 \\ 0 & C \end{bmatrix}$, then the eigenvalues of A are precisely those of B and of C, which follows from the fact that $\det(A - \lambda I) = \det(B - \lambda I)\det(C - \lambda I)$. Computer software exists for finding approximations to the eigenvalues of a matrix, and for $n > 2$ this may be the only practical method available. Here's an example where the eigenvalues can be found by factoring and the quadratic formula.

EXAMPLE 7.4.5

Finding the Eigenvalues

The characteristic polynomial of

$$A = \begin{bmatrix} -3 & 1 & -2 \\ 0 & -1 & -1 \\ 2 & 0 & 0 \end{bmatrix}$$

is given by

$$p(\lambda) = \det \begin{bmatrix} -3-\lambda & 1 & -2 \\ 0 & -1-\lambda & -1 \\ 2 & 0 & -\lambda \end{bmatrix} = -\lambda^3 - 4\lambda^2 - 7\lambda - 6$$

Substitution of small integers for λ quickly shows that -2 is a root of $p(\lambda)$. So $p(\lambda) = -(\lambda+2)q(\lambda)$, where the quadratic $q(\lambda)$ is found by division to be $\lambda^2 + 2\lambda + 3$. The roots of $q(\lambda)$ are found by the quadratic formula to be the pair of complex conjugates $-1 + \sqrt{2}i$ and $-1 - \sqrt{2}i$. So the eigenvalues of A are -2, $-1 \pm \sqrt{2}i$.

A polynomial may have a multiple root, so we have the following definition. If k is the largest positive integer for which $(\lambda - \lambda_0)^k$ is a factor of the polynomial $p(\lambda)$, then k is said to be the *multiplicity* of the root λ_0.

❖ **Multiplicity of an Eigenvalue.** The number λ_0 is an eigenvalue of A of *multiplicity* k if λ_0 is a root of multiplicity k of the characteristic polynomial $p(\lambda) = \det(A - \lambda I)$. Eigenvalues of multiplicity 1 are called *simple* eigenvalues.

EXAMPLE 7.4.6

An Eigenvalue of Multiplicity Two

Suppose that the matrix A and its characteristic polynomial are

$$A = \begin{bmatrix} 2 & 2 & 1 \\ 1 & 3 & 1 \\ 1 & 2 & 2 \end{bmatrix}, \qquad \det[A - \lambda I] = -(\lambda - 1)^2 (\lambda - 5)$$

The roots are $\lambda_1 = 1$, $\lambda_2 = 1$, $\lambda_3 = 5$, so $\lambda = 1$ is an eigenvalue of multiplicity 2 and $\lambda = 5$ is a simple eigenvalue. To find the eigenspace V_{λ_1}, we must find all solutions of

$$(A - I)v = \begin{bmatrix} 1 & 2 & 1 \\ 1 & 2 & 1 \\ 1 & 2 & 1 \end{bmatrix} \begin{bmatrix} v_1 \\ v_2 \\ v_3 \end{bmatrix} = \begin{bmatrix} 0 \\ 0 \\ 0 \end{bmatrix}$$

So $v_1 + 2v_2 + v_3 = 0$, which is the equation of a plane through the origin and defines the eigenspace V_{λ_1}. To find a basis of V_{λ_1}, let $v_1 = r$, $v_2 = s$, so $v_3 = -r - 2s$, yielding $v = [r \ s \ -r - 2s]^T = r[1 \ 0 \ -1]^T + s[0 \ 1 \ -2]^T$ as the general vector in V_{λ_1}. So $\{[1 \ 0 \ -1]^T, \ [0 \ 1 \ -2]^T\}$ spans V_{λ_1}. It is an independent set, so it is a basis for V_{λ_1}, which is a two-dimensional subspace of $\mathbb{R}^3$.

It is *not* always the case that the multiplicity of an eigenvalue equals the dimension of its eigenspace. The matrix $A = \begin{bmatrix} 3 & 1 \\ 0 & 3 \end{bmatrix}$ is an example. This matrix has a double eigenvalue 3, while the corresponding eigenspace has dimension 1. We do have the following result, which we will not prove, concerning the dimension of an eigenspace and the multiplicity of the corresponding eigenvalue.

THEOREM 7.4.3

Eigenspace Dimension Theorem. Suppose that A is an $n \times n$ matrix and that λ is an eigenvalue of A of multiplicity m. Then the dimension of the eigenspace V_λ is no larger than m, but at least 1.

If an eigenvalue is simple, then its eigenspace is one-dimensional.

Deficient Eigenspaces, Generalized Eigenvectors

An eigenspace of an $n \times n$ matrix is said to be *nondeficient* if it has the maximum possible dimension (equal to the multiplicity of the corresponding eigenvalue); otherwise the eigenspace is *deficient*. Here's a 3×3 matrix with a deficient eigenspace.

EXAMPLE 7.4.7

A Deficient Eigenspace
The characteristic polynomial of the matrix

$$A = \begin{bmatrix} 1 & -2 & -2 \\ -2 & 2 & 3 \\ 2 & -3 & -4 \end{bmatrix}$$

is $p(\lambda) = (1 + \lambda)^2(1 - \lambda)$, and the eigenvalues are $\lambda_1 = -1$ (double) and $\lambda_2 = 1$ (simple). The eigenspace V_{λ_1} is only one-dimensional, because the only solutions of the system

$$[A - \lambda_1 I]\begin{bmatrix} a \\ b \\ c \end{bmatrix} = \begin{bmatrix} 2 & -2 & -2 \\ -2 & 3 & 3 \\ 2 & -3 & -3 \end{bmatrix}\begin{bmatrix} a \\ b \\ c \end{bmatrix} = \begin{bmatrix} 0 \\ 0 \\ 0 \end{bmatrix}$$

are $a = 0$, $b = k$, and $c = -k$, where k is an arbitrary constant. So V_{λ_1} is spanned by $\{[0 \quad 1 \quad -1]^T\}$ and is one-dimensional, and is therefore deficient. Since the other eigenspace V_{λ_2} is spanned by $\{[1 \quad -1 \quad 1]^T\}$, there does *not* exist an independent set of three eigenvectors of A.

In Example 7.4.7 the eigenspace V_{λ_1} of A corresponding to the eigenvalue $\lambda_1 = -1$ is deficient because the multiplicity of that eigenvalue is 2, but dim $V_{\lambda_1} = 1$. We now define a concept that is useful if an eigenspace is deficient.

❖ **Generalized Eigenvectors.** Suppose that u^1 is an eigenvector for the matrix A corresponding to the eigenvalue λ. The vectors u^1, u^2, u^3, ... form a *chain of generalized eigenvectors* based on u^1 if

$$(A - \lambda I)u^{j+1} = u^j, \qquad j = 1, 2, \ldots$$

Notice that if $(A - \lambda I)u^2 = u^1$ has *no* solution, then the chain of generalized eigenvectors based on u^1 has length 1, and the chain begins and ends with u^1. Although it is not evident, each chain of generalized eigenvectors is linearly independent.

If we construct a chain of generalized eigenvectors from each basis vector for V_λ, then the collection B_λ of these chains is an independent set. Let N denote the number of elements in B_λ. It is always possible to choose a basis for V_λ such that N = multiplicity of λ, but it is never possible for N to exceed that multiplicity.

EXAMPLE 7.4.8

Generalized Eigenvectors for a Deficient Eigenspace
In Example 7.4.7 the eigenspace V_{λ_1} corresponding to the double eigenvalue $\lambda_1 = -1$ is deficient because dim $V_{\lambda_1} = 1$. The eigenvector $u^1 = [0 \ \ 1 \ \ -1]^T$ is a basis for V_{λ_1} and must support a chain of generalized eigenvectors of length 2. Solving the equation $(A + I)u^2 = [0 \ \ 1 \ \ -1]^T$ by row operations, we find that $u^2 = [1 \ \ 1 \ \ 0]^T$ is one such solution but there are many others since the matrix $A + I$ is singular. In fact, $[1 \ \ s \ \ 1 - s]^T$ is a generalized eigenvector for any real value of s. You are free to select for u^2 any one of these solutions. Notice that the equation $(A + I)u^3 = u^2$ has no solution (just try to find one by row operations), so the chain has just two vectors, u^1 and u^2, as expected.

EXAMPLE 7.4.9

A Basis of Generalized Eigenvectors
The characteristic polynomial of the matrix

$$A = \begin{bmatrix} 0 & 1 & 0 \\ -1 & 2 & 0 \\ 0 & 0 & 1 \end{bmatrix}$$

is $p(\lambda) = -(\lambda - 1)^3$, and $\lambda_1 = 1$ is an eigenvalue of multiplicity 3. So by Theorem 7.4.3, the eigenspace V_{λ_1} has dimension not greater than 3. Let's determine V_{λ_1} by finding all solutions of the system $(A - \lambda_1 I)x = 0$. Writing out the system we have

$$-x_1 + x_2 = 0, \qquad -x_1 + x_2 = 0, \qquad 0 = 0$$

Since x_3 doesn't appear in the system, it can have any value. Put $x_3 = s$, where s is an arbitrary real. Now x_1 and x_2 are related by $x_1 = x_2$, so if $x_2 = r$, where r is an arbitrary real number, then also $x_1 = r$. So V_{λ_1} is described by

$$x = \begin{bmatrix} x_1 \\ x_2 \\ x_3 \end{bmatrix} = \begin{bmatrix} r \\ r \\ s \end{bmatrix} = r \begin{bmatrix} 1 \\ 1 \\ 0 \end{bmatrix} + s \begin{bmatrix} 0 \\ 0 \\ 1 \end{bmatrix}, \qquad r, s \text{ arbitrary reals}$$

The vectors $u^1 = [1 \ \ 1 \ \ 0]^T$, $v^1 = [0 \ \ 0 \ \ 1]^T$ are linearly independent and since $V_{\lambda_1} = \text{Span}\{u^1, v^1\}$, we see that dim $V_{\lambda_1} = 2$. So V_{λ_1} is a deficient eigenspace with a basis $\{u^1, v^1\}$.

Now let's find an independent set of three generalized eigenvectors corresponding to $\lambda_1 = 1$. First, we try to do this by generating the chains based at u^1 and at v^1: try to find a vector u^2 such that $(A - I)u^2 = u^1$. We see that $u^2 = [0 \ \ 1 \ \ 0]^T$ is such a vector. There is no vector u^3 such that $(A - I)u^3 = u^2$, so the chain based at u^1 has only two vectors. To find the chain based at v^1, try to find a vector v^2 such that $(A - I)v^2 = v^1$. There is no such vector v^2 (why not?), so this chain has just the one vector v^1 in it.

The set of generalized eigenvectors $B = \{u^1, u^2, v^1\}$ is linearly independent, so our approach has produced a basis for $\mathbb{R}^3$ consisting of generalized eigenvectors, two of which are actually eigenvectors.

☞ We have to figure out just where in V_{λ_1} to start the chain.

If we used the approach in the example above, but with the basis $\tilde{u} = [1 \; 1 \; 1]^T$, $\tilde{v} = [0 \; 0 \; 1]^T$ for the eigenspace V_{λ_1}, we would have failed to produce three generalized eigenvectors from the chains based at $\tilde{u}$ and $\tilde{v}$. In fact, the chains based at $\tilde{u}$ and $\tilde{v}$ have just one element each. We are assured that there is a basis for V_{λ_1} that will generate three generalized eigenvectors. How can we find such a basis? Let's replace $\tilde{u}$ by $w^1 = \tilde{u} + c\tilde{v}$, where c is a nonzero constant, and determine c so that the system $(A - I)w^2 = w^1$ has a solution w^2. If we can do this, then $\{w^1, w^2, \tilde{v}\}$ is a set of generalized eigenvectors corresponding to the eigenvalue $\lambda_1 = 1$. Suppose that $w^2 = [\alpha \; \beta \; \gamma]^T$, and note that the system $(A - I)w^2 = w^1$ becomes

$$-\alpha + \beta = 1$$

$$-\alpha + \beta = 1$$

$$0 = c + 1$$

which has a solution if $c = -1$; for example, $\alpha = 0$, $\beta = 1$, $\gamma = 0$ is a solution. So $w^1 = \tilde{u} - \tilde{v} = [1 \; 1 \; 0]^T$, $w^2 = [0 \; 1 \; 0]^T$, $\tilde{v} = [0 \; 0 \; 1]^T$ form a triple of generalized eigenvectors that we seek. The last part of Problem 10 gives an explanation of just why $w^1 = [1 \; 1 \; 0]^T$ is expected to be the start of a chain.

The next result generalizes Theorem 7.4.1.

THEOREM 7.4.4

☞ In particular, if every eigenspace is nondeficient then $\mathbb{R}^n$ has a basis B of eigenvectors of A.

Properties of Generalized Eigenvectors. Suppose that the $n \times n$ matrix A has distinct eigenvalues $\lambda_1, \lambda_2, \ldots, \lambda_k$ with respective multiplicities $m_1, m_2, \ldots, m_k$, and suppose that V_{λ_j} is the eigenspace corresponding to λ_j. Then

- Each eigenspace V_{λ_j} has a basis such that the collection B_j of generalized eigenvectors generated as chains from that basis has m_j elements.

- The collection B of all generalized eigenvectors in the sets $B_1, \ldots, B_k$ has n elements and is linearly independent. Consequently, B is a basis for $\mathbb{R}^n$ (or $\mathbb{C}^n$, as the case may be).

PROBLEMS

www **1.** Find the eigenvalues of each matrix, their multiplicities, and a basis for each eigenspace.

(a) $\begin{bmatrix} 1 & 0 \\ 2 & 1 \end{bmatrix}$
(b) $\begin{bmatrix} 1 & 1 \\ 1 & 1 \end{bmatrix}$
(c) $\begin{bmatrix} 0 & 3 \\ 3 & 0 \end{bmatrix}$
(d) $\begin{bmatrix} 6 & -7 \\ 1 & -2 \end{bmatrix}$

(e) $\begin{bmatrix} 3 & -5 \\ 5 & 3 \end{bmatrix}$
(f) $\begin{bmatrix} 1 & 0 & 1 \\ 0 & 1 & 0 \\ 1 & 0 & 1 \end{bmatrix}$
(g) $\begin{bmatrix} 5 & -6 & -6 \\ -1 & 4 & 2 \\ 3 & -6 & -4 \end{bmatrix}$

(h) $\begin{bmatrix} \cos\theta & -\sin\theta \\ \sin\theta & \cos\theta \end{bmatrix}$, real θ
(i) $\begin{bmatrix} -1 & 36 & 100 \\ 0 & -1 & 27 \\ 0 & 0 & 5 \end{bmatrix}$
(j) $\begin{bmatrix} a & b \\ -b & a \end{bmatrix}$, real a, b

2.   (*Eigenvalues of a Triangular Matrix*). Show that the eigenvalues of a triangular matrix are the diagonal entries, the multiplicity of each eigenvalue equalling the number of times the eigenvalue appears on the diagonal.

☞ A^k for a positive integer k is the product of k factors of A.

3.   (*Eigenvalues of A^k*). Show that if λ is an eigenvalue of A, then λ^k is an eigenvalue of A^k. If μ is an eigenvalue of A^k, is $\mu^{1/k}$ always an eigenvalue of A?

4.   (*Eigenvalues of A^T*). Show that a square matrix A and its transpose A^T have the same characteristic polynomial, so the same eigenvalues with the same multiplicities.

5.   (*Using Eigenvalues to Find Det A*). Suppose that $\lambda_1, \ldots, \lambda_n$ are the eigenvalues of a matrix A, each repeated according to its multiplicity. Show that $\det A = \lambda_1 \cdots \lambda_n$. [*Hint*: Write $p(\lambda)$ as $(-1)^n(\lambda - \lambda_1) \cdots (\lambda - \lambda_n)$, expand, and compare with the form of $p(\lambda)$ in (6).]

6.   (*Trace*). The *trace* of an $n \times n$ matrix A is the sum of its diagonal entries: $\operatorname{tr} A = a_{11} + \cdots + a_{nn}$.

(a) Show that $\operatorname{tr} A$ is the sum of the eigenvalues of A. [*Hint*: See the hint in Problem 5.]

(b) Use part (a) to show that

$$A = \begin{bmatrix} 1 & 5 & 3 \\ 6 & -7 & 10 \\ 37 & 56 & 2 \end{bmatrix}$$

has an eigenvalue with negative real part.

7.   If A is an $n \times n$ matrix show that A is nonsingular if and only if $\lambda = 0$ is not an eigenvalue.

8.   (*Justification of Theorem 7.4.1*). Suppose that B_j is an independent subset of V_{λ_j}, where V_{λ_j} is an eigenspace of an $n \times n$ matrix A corresponding to eigenvalue λ_j, $j = 1, \ldots, p$. Use induction on p to show that the collection B of all vectors in $B_1, \ldots, B_p$ is an independent set.

(a) First assume that each B_j contains a single vector v^j. Verify that the Eigenspace Property holds for all p by showing (1) that it holds for $p = 1$ (the anchor step), and (2) that if it holds whenever $p \leq k - 1$, then it also holds for $p = k$ (the induction step).

(b) Show the Eigenspace Property for the general case. [*Hint*: Reduce to the case in part (a).]

9.   Find a basis for $\mathbb{R}^3$ that consists only of generalized eigenvectors of the matrix

$$A = \begin{bmatrix} -2 & -2 & 0 \\ 2 & 3 & 0 \\ 0 & 0 & -1 \end{bmatrix}$$

 10.   (*Generalized Eigenspace Property*). Suppose that the $n \times n$ matrix A has k distinct eigenvalues $\lambda_1, \lambda_2, \ldots, \lambda_k$ with respective multiplicities $m_1, m_2, \ldots, m_k$. Furthermore, suppose that $V_{\lambda_1}, V_{\lambda_2}, \ldots, V_{\lambda_k}$ are the eigenspaces of A corresponding to $\lambda_1, \ldots, \lambda_k$. For each j, if V_{λ_j} is nondeficient, suppose that B_j is a basis for V_{λ_j}. If V_{λ_j} is deficient, we know that there is a basis for V_{λ_j} such that the total number of generalized eigenvectors using each element of this basis as a starting element is the multiplicity m_j. Denote the collection of these generalized eigenvectors by B_j. Show the set B consisting of all the elements in all the B_j is a basis for $\mathbb{R}^n$ (or $\mathbb{C}^n$, as the case may be). [*Hint*: Use induction on k. Apply the matrix $(A - \lambda_1 I)$ to show first that B_1 is linearly independent no matter how many chains of generalized eigenvectors it contains. Then show by induction that B is independent. Use $(A - \lambda_i I)$ repeatedly.]

Now use the Fredholm Alternative of Problem 13, Section 7.3 to explain why the eigenvector $[1 \ 1 \ 0]^T$ in Example 7.4.9 must be the start of a chain of generalized eigenvectors of length at least 2.

7.5 Undriven Linear Systems with Constant Coefficients

The general real *linear differential system* in the n state variables $x_1, x_2, \ldots, x_n$ is

$$x_1' = a_{11}x_1 + \cdots + a_{1n}x_n + F_1(t)$$
$$\vdots \tag{1}$$
$$x_n' = a_{n1}x_1 + \cdots + a_{nn}x_n + F_n(t)$$

where the *coefficients* a_{ij} are real constants or real-valued continuous functions on some common t-interval I, and the *input functions* F_i are continuous on I and (usually) real-valued. System (1) can be written in the compact form

$$x' = Ax + F \tag{2}$$

where

$$A = \begin{bmatrix} a_{11} & \cdots & a_{1n} \\ \vdots & & \vdots \\ a_{n1} & \cdots & a_{nn} \end{bmatrix} = [a_{ij}], \qquad x = \begin{bmatrix} x_1 \\ \vdots \\ x_n \end{bmatrix}, \qquad F = F(t) = \begin{bmatrix} F_1 \\ \vdots \\ F_n \end{bmatrix}$$

The column vector x is the state vector, and A is the *system* (or *coefficient*) *matrix* of the differential system. The column vector F is called the *input vector* or *driving term*. The cascade models in Section 1.8, the coupled spring model of Section 5.1, and the lead model in Section 7.1 are all linear systems.

An Operator Formulation of Linear Systems

Let's reformulate system (2) by introducing an operator that acts on column vectors $x(t)$ with continuously differentiable components on an interval I. If we define the operator L like this, $L[x] = x' - Ax$, then system (2) assumes the form $L[x] = F$. The operator L has two very important properties that will help us to find solution formulas for linear systems of ODEs.

THEOREM 7.5.1

Properties of L. The operator L defined by $L[x] = x' - Ax$, where A is an $n \times n$ constant matrix, has the following properties:

Linearity of L: For any constants c_1, c_2, and any differentiable vector functions $x^1(t)$ and $x^2(t)$, we have

$$L[c_1x^1 + c_2x^2] = c_1L[x^1] + c_2L[x^2] \tag{3}$$

Closure Property: For any constants c_1, c_2, and any solutions $x^1(t)$, $x^2(t)$ of the system $L[x] = 0$, the vector function $c_1x^1 + c_2x^2$ is also a solution.

☞ This property is also called the *Principle of Superposition*

The Linearity Property is a consequence of the operations of differentiation and matrix multiplication. The Closure Property follows from the linearity of the operator L because

$$L[c_1x^1 + c_2x^2] = c_1L[x^1] + c_2L[x^2] = 0$$

So $y = c_1x^1 + c_2x^2$ is a solution of $L[x] = 0$.

The Closure Property says that if $\{x^1(t), \ldots, x^m(t)\}$ are solutions of $L[x] = 0$, then the *linear combination* $x = c_1x^1 + \cdots + c_mx^m$ is also a solution. So the solution set of $L[x] = 0$ is a subspace of function vectors; it is called the *null space* of L.

The Linearity Property of L has a consequence that will help us solve the driven system $L[x] = F(t)$. As was done for linear first-order and second-order ODEs, we can characterize all solutions of system (2) as the sum of a particular solution and the set of all solutions of $L[x] = 0$. The collection af all solutions of $L[x] = F$ is called the *general solution* and is here denoted by x_{gen}.

General Solution of the System $L[x] = x' - Ax = F$

☞ x^d is called a *particular solution* of the driven system $L[x] = F$.

Two things are needed to find x_{gen}, the collection of all solutions of $L[x] = F$:

- **Any one solution** $x^d(t)$ of the driven system $L[x] = F$.
- **Every solution** $w(t)$ of the undriven system $L[x] = 0$.

Given this information, the general solution $x_{gen}(t)$ of $L[x] = F$ is given by

$$x_{gen} = x^d(t) + w(t), \quad \text{for every solution } w(t) \text{ of } L[x] = 0 \tag{4}$$

☞ Familiar? We did this in Section 3.3 for the linear operator $P(D)$.

The decomposition (4) is a direct consequence of the linearity of the operator L. Indeed, if x^d is a particular solution of $L[x] = F$, then for any solution x of $L[x] = F$ we have from the linearity of L that $L[x - x^d] = L[x] - L[x^d] = F - F = 0$, so $w = x - x^d$ belongs to the null space of L. The other way around, if x^d is a particular solution of $L[x] = F$ and if w belongs to the null space of L, then

$$L[x^d + w] = L[x^d] + L[w] = F + 0 = F$$

So $x^d + w$ is a solution of $L[x] = F$.

We can break up our search for all solutions of $x' = Ax + F$ into two steps: first, find all solutions of $L[x] = 0$, and second, find a solution of the driven system $L[x] = F$. Let's go to work on finding all solutions w of $L[x] = 0$ and postpone the task of finding a particular solution x^d to a later section. The approach we take is called the *Method of Eigenvectors* for reasons that will become obvious.

Solving $x' = Ax$, When A is a Constant Matrix

First, we might try to guess solutions of $x' = Ax$ when A is a constant matrix and see how far we get. By analogy with the one state variable case, the linear system

$$x' = Ax \tag{5}$$

might be expected to have solutions x of the form $x = ve^{\lambda t}$, for some constant number λ and some constant vector v. Substituting $ve^{\lambda t}$ into system (5) yields

$$\lambda v e^{\lambda t} = A v e^{\lambda t}$$

Cancelling $e^{\lambda t}$, we see that

$$\lambda v = Av \tag{6}$$

So $x = ve^{\lambda t}$ solves system (5) if λ is an eigenvalue of A and v is a corresponding eigenvector. Let's see how we can use this information. Suppose that A has distinct real eigenvalues λ_1 and λ_2 with corresponding eigenvectors v^1 and v^2. Then we see from the above that $x^1 = v^1 e^{\lambda_1 t}$ and $x^2 = v^2 e^{\lambda_2 t}$ are solutions of $x' = Ax$. From the Closure Property in Theorem 7.5.1 we see that $x = C_1 x^1 + C_2 x^2$ gives us a solution of $x' = Ax$ for any values of the constants C_1 and C_2. The question is: *Are these all the solutions, or are there others?* Let's look at an example.

EXAMPLE 7.5.1

Constructing All Solutions of $x' = Ax$

Let's look at the system

$$x' = Ax, \quad \text{with} \quad A = \begin{bmatrix} 5 & 3 \\ -6 & -4 \end{bmatrix} \tag{7}$$

The eigenvalue problem for this matrix was solved in Example 7.4.1, so we have the two eigenvalues $\lambda_1 = -1$, $\lambda_2 = 2$ and corresponding eigenvectors $v^1 = [1 \ -2]^T$, $v^2 = [1 \ -1]^T$. For arbitrary real constants C_1 and C_2 all solutions of $x' = Ax$ look like

$$x = C_1 v^1 e^{-t} + C_2 v^2 e^{2t} \tag{8}$$

☞ The discussion from this point on is only to show that (8) does indeed give all solutions.

Here's how to show that formula (8) captures *all* solutions of ODE (7). Define the matrix

$$V = [v^1 \ \ v^2] = \begin{bmatrix} 1 & 1 \\ -2 & -1 \end{bmatrix}$$

whose columns are the eigenvectors v^1 and v^2. Let's use the change of state variables $x = Vy$. The transform of the system $x' = Ax$ is $Vy' = AVy$. Using the matrix product identities in (1) and (2) of Section 7.2, we see that

$$x' = Vy' = y_1' v^1 + y_2' v^2$$
$$Ax = AVy = A[v^1 \ \ v^2]y = [Av^1 \ \ Av^2]y$$
$$= [(-1)v^1 \ \ 2v^2]y = (-1)y_1 v^1 + 2y_2 v^2$$

So we have the identity

$$y_1' v^1 + y_2' v^2 = (-1)y_1 v^1 + 2y_2 v^2$$

Since $\{v^1, v^2\}$ is a basis for $\mathbb{R}^2$, we can equate the coefficients of v^1 and v^2 on both sides of this equation to obtain the transformed system

$$\begin{aligned} y_1' &= -y_1 \\ y_2' &= 2y_2 \end{aligned} \tag{9}$$

☞ This is the key: y_1 and y_2 give *all* solutions of (9).

System (9) totally decouples and we can solve each ODE by the method of integrating factors to obtain *all* the solutions of the transformed system (9):

$$y_1 = C_1 e^{-t}, \quad y_2 = C_2 e^{2t}, \qquad C_1, C_2 \text{ arbitrary reals}$$

Going back to the x-variables, we see that all solutions of $x' = Ax$ are given by

$$x = Vy = [v^1 \; v^2] \begin{bmatrix} C_1 e^{-t} \\ C_2 e^{2t} \end{bmatrix} = C_1 v^1 e^{-t} + C_2 v^2 e^{2t} \tag{10}$$

Our method of construction shows that formula (8) captured *all* solutions of $x' = Ax$.

Here's an interesting sidelight on the solution formula for system (7) derived in the above example. Put

$$x^1(t) = v^1 e^{-t}, \qquad x^2(t) = v^2 e^{2t}$$

We claim that each solution $x(t)$ of system (7) can be written as a linear combination of x^1 and x^2 in only one way. In fact, say that $x(t) = ax^1 + bx^2$ and also that $x(t) = cx^1 + dx^2$. Then the numbers $a - c$ and $b - d$ satisfy the identity

$$(a - c)x^1(t) + (b - d)x^2(t) = 0, \quad \text{for all } t$$

and from the condition $\det V \neq 0$ it follows (using Theorem 7.3.1) that $a - c = 0$ and $b - d = 0$, establishing our claim.

Let's solve an IVP associated with system (7).

EXAMPLE 7.5.2

Solving an IVP: $x' = Ax, \; x(0) = x^0$

Suppose that A is the matrix of Example 7.5.1. Let's solve the IVP $x' = Ax$, $x(0) = [2 \; 3]^T$. To do this we'll look for constants C_1 and C_2 in formula (10) so that $x(0) = [2 \; 3]^T$; that is,

$$x(0) = \begin{bmatrix} 2 \\ 3 \end{bmatrix} = C_1 \begin{bmatrix} 1 \\ -2 \end{bmatrix} + C_2 \begin{bmatrix} 1 \\ -1 \end{bmatrix} = \begin{bmatrix} C_1 + C_2 \\ -2C_1 - C_2 \end{bmatrix}$$

From the last equality, we see that

$$C_1 + C_2 = 2, \qquad -2C_1 - C_2 = 3$$

and $C_1 = -5$ and $C_2 = 7$. So, from formula (10), the solution of the IVP is

$$x = -5 \begin{bmatrix} 1 \\ -2 \end{bmatrix} e^{-t} + 7 \begin{bmatrix} 1 \\ -1 \end{bmatrix} e^{2t}$$

This method works for any set of initial data.

Examples 7.5.1 and 7.5.2 suggest that all solutions of a constant matrix linear system $x' = Ax$ can be found by purely matrix-algebraic means. We will freely use the terminology of eigenvalues and eigenspaces of matrices throughout what follows, including the notion of generalized eigenvectors. If you are a bit rusty on these topics (or perhaps never studied them at all), then look over the previous sections before proceeding any further. All the relevant matrix-algebraic concepts we will need are developed in those sections.

Keep in mind that even if the constant matrix A has *real* entries, it may have complex eigenvalues and the corresponding eigenvectors may have complex entries. Our approach also works in this case.

Solving $x' = Ax$ When A has Nondeficient Eigenspaces

An eigenspace is nondeficient if its dimension is equal to the multiplicity of its corresponding eigenvalue. If all eigenspaces of a matrix A are nondeficient, it is a straightforward matter to construct the general solution of $x' = Ax$.

THEOREM 7.5.2

☞ The margin comment by Theorem 7.4.4 assures us that $\mathbb{R}^n$ has an eigenbasis

☞ This result is also true if some of the eigenvalues and eigenvectors are complex. In this case we get an eigenbasis for $\mathbb{C}^n$.

Nondeficient Eigenspaces. Suppose that A is an $n \times n$ constant matrix with real entries whose eigenvalues are real and corresponding eigenspaces are nondeficient. Then the general solution of $x' = Ax$ is given by

$$x = C_1 v^1 e^{\lambda_1 t} + \cdots + C_n v^n e^{\lambda_n t} \tag{11}$$

where $C_1, \ldots, C_n$ are arbitrary constants, $\{v^1, \ldots, v^n\}$ is any eigenbasis for $\mathbb{R}^n$, and $\lambda_1, \ldots, \lambda_n$ are the corresponding eigenvalues (repeated as required).

To derive formula (11), first define the matrix $V = [v^1 \; v^2 \cdots v^n]$. Since V is non-singular (because its column vectors are linearly independent), we can change the state variable x in $x' = Ax$ to the state variable y, where $x = Vy$. So $x' = Ax$ transforms to the system $Vy' = AVy$. Evaluating these matrix products we have that

☞ Check out the Matrix Product Identities on page 347.

$$Vy' = [v^1 \; v^2 \; \cdots \; v^n][y_1' \; y_2' \; \cdots \; y_n']^T$$
$$= y_1' v^1 + y_2' v^2 + \cdots + y_n' v^n$$

and

$$AVy = A[v^1 \; v^2 \; \cdots \; v^n]y = [Av^1 \; Av^2 \; \cdots \; Av^n]y$$
$$= [\lambda_1 v^1 \; \lambda_2 v^2 \; \cdots \; \lambda_n v^n]y$$
$$= \lambda_1 y_1 v^1 + \lambda_2 y_2 v^2 + \cdots + \lambda_n y_n v^n$$

Since $\{v^1, v^2, \ldots, v^n\}$ is a basis, we can equate coefficients of the v^j in $Vy' = AVy$ to obtain the transformed system

$$y_j' = \lambda_j y_j, \qquad j = 1, 2, \ldots n \tag{12}$$

All solutions of system (12) are given by $y_j = C_j e^{\lambda_j t}$, $j = 1, 2, \ldots, n$, where C_1, $\ldots, C_n$ are arbitrary constants. So all solutions of the original system are

$$x = Vy = C_1 v^1 e^{\lambda_1 t} + \cdots + C_n v^n e^{\lambda_n t}$$

which is just formula (11).

The next result assures that each IVP has a unique solution.

THEOREM 7.5.3

☞ This result holds even if some of the eigenvalues and eigenspaces are complex.

> Unique Solution for an IVP. Suppose that A is an $n \times n$ matrix of real entries whose eigenvalues $\lambda_1, \ldots, \lambda_n$ are real (not necessarily distinct), and all the eigenspaces are nondeficient. Suppose that $\{v^1, \ldots, v^n\}$ is a corresponding eigenbasis of $\mathbb{R}^n$. Then for each x^0 there is a unique collection of constants $c_1, \ldots, c_n$ such that
>
> $$x = c_1 v^1 e^{\lambda_1 t} + \cdots + c_n v^n e^{\lambda_n t} \tag{13}$$
>
> is the unique solution of the IVP $x' = Ax$, $x(0) = x^0$.

Solution formula (13) follows immediately from the general solution (11) because $\{v^1, \ldots, v^n\}$ is a basis for $\mathbb{R}^n$, so the initial vector x^0 can be written in a unique way as $x^0 = c_1 v^1 + \cdots + c_n v^n$.

EXAMPLE 7.5.3

Finding the General Solution, Solving an IVP
Let's solve the system $x' = Ax$, where

$$A = \begin{bmatrix} 1 & -2 & -2 \\ -2 & 1 & 2 \\ 2 & -2 & -3 \end{bmatrix}$$

The characteristic polynomial of A is $p(\lambda) = (1+\lambda)^2(1-\lambda)$, and the eigenvalues are $\lambda_1 = -1$ (a double eigenvalue) and $\lambda_2 = 1$ (a simple eigenvalue). Direct calculation shows that V_{λ_1} is two-dimensional and is spanned by $\{v^1, v^2\}$, where $v^1 = [1 \ 1 \ 0]^T$, $v^2 = [1 \ 0 \ 1]^T$, while V_{λ_2} is spanned by $\{v^3\}$, where $v^3 = [1 \ -1 \ 1]^T$. The eigenspaces of A are both nondeficient, so $\{v^1, v^2, v^3\}$ is an eigenbasis for $\mathbb{R}^3$. It follows from Theorem 7.5.2 that the general solution of $x' = Ax$ is

$$x = C_1 \begin{bmatrix} 1 \\ 1 \\ 0 \end{bmatrix} e^{-t} + C_2 \begin{bmatrix} 1 \\ 0 \\ 1 \end{bmatrix} e^{-t} + C_3 \begin{bmatrix} 1 \\ -1 \\ 1 \end{bmatrix} e^t \tag{14}$$

☞ Written out, this system is

$$\begin{bmatrix} 1 & 1 & 1 \\ 1 & 0 & -1 \\ 0 & 1 & 1 \end{bmatrix} \begin{bmatrix} C_1 \\ C_2 \\ C_3 \end{bmatrix} = \begin{bmatrix} 1 \\ 0 \\ 0 \end{bmatrix}$$

To solve the IVP $x' = Ax$, $x(0) = [1 \ 0 \ 0]^T$, first insert $t = 0$ into formula (14) and then solve the resulting system $x(0) = C_1 v^1 + C_2 v^2 + C_3 v^3$ for C_1, C_2, and C_3 to obtain $C_1 = 1$, $C_2 = -1$, $C_3 = 1$. So the solution of the IVP above is

$$x = \begin{bmatrix} 1 \\ 1 \\ 0 \end{bmatrix} e^{-t} - \begin{bmatrix} 1 \\ 0 \\ 1 \end{bmatrix} e^{-t} + \begin{bmatrix} 1 \\ -1 \\ 1 \end{bmatrix} e^t$$

Note that if the eigenvalues of A are all real and distinct, then all the eigenvalues of A are simple, so the eigenspaces are nondeficient since they must be one-dimensional. So there is an independent set of n eigenvectors of A in $\mathbb{R}^n$ (an eigenbasis). The only problematic matrices are those with some multiple eigenvalues and some deficient eigenspaces.

Solving $x' = Ax$ When A has Deficient Eigenspaces

How can we find the general solution of the system $x' = Ax$ when the constant $n \times n$ matrix A has real entries, real eigenvalues, and one or more deficient eigenspaces? In that case A does not generate an eigenbasis for $\mathbb{R}^n$, so the approach described in Theorem 7.5.2 doesn't work. Fortunately, Theorem 7.4.4 shows us that this approach will do the job if we use a basis consisting of generalized eigenvectors. Rather than state the result in the general case, we'll focus on one chain of eigenvectors at a time. The process can be repeated for each chain that appears in the selected generalized eigenvector basis. Let's see how this goes.

Generalized Eigenvector Chain of One Vector ──────────────────────────

We already know what to do for a chain that terminates with one vector, v. This vector must be an eigenvector v with a corresponding eigenvalue λ, and we see that

$$x(t) = ve^{\lambda t} \tag{15}$$

is the solution of $x' = Ax$ generated by this 1-vector chain.

Generalized Eigenvector Chain of Two Vectors ──────────────────────────

Now let's see how a chain $\{u^1, u^2\}$ of two generalized eigenvectors of a matrix A create a pair of independent solutions of $x' = Ax$. We have

$$(A - \lambda I)u^1 = 0, \qquad (A - \lambda I)u^2 = u^1 \tag{16}$$

Suppose that V is the matrix whose columns are generalized eigenvector chains that form a basis for $\mathbb{R}^n$. The chain u^1, u^2 appears somewhere in the matrix V like this: $V = [\cdots \; u^1 \; u^2 \; \cdots]$, where u^1 and u^2 are in the j-th and $(j+1)$-st places. Now let's change the state variable x in the system $x' = Ax$ to the new state variable y by the formula $x = Vy$, obtaining the system $Vy' = AVy$. Using the matrix product identities in (1) and (2) of Section 7.2 and equating the u^1-, u^2-components of the new system, we have

$$y'_j u^1 + y_{j+1} u^2 = Au^1 y_j + Au^2 y_{j+1}$$
$$= \lambda u^1 y_j + (\lambda u^2 + u^1) y_{j+1}$$

Equating coefficients of u^1 and u^2, we have the ODEs

$$\begin{aligned} y'_j &= \lambda y_j + y_{j+1} \\ y'_{j+1} &= \lambda y_{j+1} \end{aligned} \tag{17}$$

A formula for all solutions of this system can be found as follows: The last ODE in (17) decouples and can be solved separately to obtain $y_{j+1} = C_{j+1} e^{\lambda t}$, where C_{j+1} is an arbitrary constant. Substituting $y_{j+1}(t)$ into the top ODE of (17) and solving, we obtain $y_j = (C_{j+1}t + C_j)e^{\lambda t}$, where C_j is another arbitrary constant. So we finally have

$$x = Vy = \cdots + (C_{j+1}t + C_j)e^{\lambda t}u^1 + C_{j+1}e^{\lambda t}u^2 + \cdots$$
$$= \cdots + C_j u^1 e^{\lambda t} + C_{j+1}(u^2 + tu^1)e^{\lambda t} + \cdots$$

and we see that

$$x^1 = u^1 e^{\lambda t} \quad \text{and} \quad x^2 = (u^2 + tu^1)e^{\lambda t} \tag{18}$$

are solutions of the system $x' = Ax$ (since the constants C_j, C_{j+1} are arbitrary).

Generalized Eigenvector Chain of m-Vectors

Solution formulas generated by an m-vector chain of generalized eigenvectors are found in the same way. For example, if A is a matrix with a triple eigenvalue λ and a one-dimensional eigenspace, and if u^1, u^2, u^3 is a corresponding chain of generalized eigenvectors, then

$$x = C_1 u^1 e^{\lambda t} + C_2(u^2 + tu^1)e^{\lambda t} + C_3\left(u^3 + tu^2 + \frac{t^2}{2}u^1\right)e^{\lambda t} \tag{19}$$

where C_1, C_2, and C_3 are arbitrary constants, is a solution of $x' = Ax$.

Now we are ready to apply what we have learned above to solve a system $x' = Ax$ with deficient eigenspaces.

EXAMPLE 7.5.4

A Matrix with a Deficient Eigenspace

Look at the system $x' = Ax$, where A is the 3×3 matrix in Example 7.4.7,

$$A = \begin{bmatrix} 1 & -2 & -2 \\ -2 & 2 & 3 \\ 2 & -3 & -4 \end{bmatrix}$$

We saw that A has a double eigenvalue $\lambda_1 = -1$ and a simple eigenvalue $\lambda_2 = 1$. The deficient eigenspace V_{λ_1} is spanned by the single eigenvector $v^1 = [0 \ 1 \ -1]^T$ and the nondeficient eigenspace V_{λ_2} by the eigenvector $[1 \ -1 \ 1]^T$. In Example 7.4.8 we found a chain of generalized eigenvectors based on the eigenvector u^1:

$$u^1 = [0 \ 1 \ -1]^T \quad \text{and} \quad u^2 = [1 \ 1 \ 0]^T$$

Using (15) and (18), we find the three solutions of $x' = Ax$:

$$x^1 = \begin{bmatrix} 0 \\ 1 \\ -1 \end{bmatrix} e^{-t}, \qquad x^2 = \begin{bmatrix} 1 \\ 1 \\ 0 \end{bmatrix} e^{-t} + \begin{bmatrix} 0 \\ 1 \\ -1 \end{bmatrix} te^{-t}, \qquad x^3 = \begin{bmatrix} 1 \\ -1 \\ 1 \end{bmatrix} e^t$$

The general solution of $x' = Ax$ is $x = C_1 x^1 + C_2 x^2 + C_3 x^3$, where C_1, C_2, and C_3 are arbitrary reals.

EXAMPLE 7.5.5

Another Matrix with a Deficient Eigenspace

The 3×3 matrix

$$A = \begin{bmatrix} -1 & 1 & 0 \\ -1 & 0 & 1 \\ 1 & 0 & -2 \end{bmatrix}$$

has $\lambda_1 = -1$ as its only eigenvalue, and its multiplicity m_1 is 3. The corresponding eigenspace $V_{\lambda_1} = \text{Span}\{[1 \ 0 \ 1]^T\}$. Since the deficient eigenspace V_{λ_1} has dimension one, any eigenvector will produce a 3-vector chain of generalized eigenvectors. So a little computation shows that the eigenvector $u^1 = [1 \ 0 \ 1]^T$ generates the chain of generalized eigenvectors

$$u^1 = \begin{bmatrix} 1 \\ 0 \\ 1 \end{bmatrix}, \qquad u^2 = \begin{bmatrix} 1 \\ 1 \\ 0 \end{bmatrix}, \qquad u^3 = \begin{bmatrix} 0 \\ 1 \\ 0 \end{bmatrix}$$

Note that $\{u^1, u^2, u^3\}$ is a basis for $\mathbb{R}^3$, as expected. Using (19), we see that the system $x' = Ax$ has the solutions

$$x^1 = \begin{bmatrix} 1 \\ 0 \\ 1 \end{bmatrix} e^{-t}, \quad x^2 = \left(\begin{bmatrix} 1 \\ 1 \\ 0 \end{bmatrix} + t \begin{bmatrix} 1 \\ 0 \\ 1 \end{bmatrix} \right) e^{-t}, \quad x^3 = \left(\begin{bmatrix} 0 \\ 1 \\ 0 \end{bmatrix} + t \begin{bmatrix} 1 \\ 1 \\ 0 \end{bmatrix} + \frac{t^2}{2} \begin{bmatrix} 1 \\ 0 \\ 1 \end{bmatrix} \right) e^{-t}$$

and that the formula $x = C_1 x^1 + C_2 x^2 + C_3 x^3$, where C_1, C_2, C_3 are arbitrary reals, describes all solutions of the system $x' = Ax$.

Comments

The approach outlined in this section for finding all solutions of $x' = Ax$ also works when the eigenvalues and eigenvectors of the matrix A are complex-valued as we have already noted. The solution formula for $x' = Ax$ looks the same in this case, but the arbitrary constants are now complex numbers and the solutions are complex-valued. If the matrix has real entries, then we would naturally like to know what the real-valued solutions are. We will take up this question in the next section.

In later sections we will look at two models that involve linear differential systems with constant coefficients. The approaches of this and the next section give us the tools we need to examine these models.

PROBLEMS

1. (*General Solutions*). Find the general solution for each of the systems below.

   (a) $x_1' = 5x_1 + 3x_2$; $x_2' = -x_1 + x_2$　　(b) $x_1' = 5x_1 - 2x_2$; $x_2' = -2x_1 + 8x_2$

   (c) $x_1' = x_1 - x_2$; $x_2' = x_1 + 3x_2$

2. (*Solving IVPs*). (a)–(c). Solve the IVP consisting of the rate equations given in 1(a)–(c) and the initial condition $x_1(0) = 1$, $x_2(0) = 1$.

3. (*Nondeficient Eigenspaces*). Find all solutions of $x' = Ax$, where the matrix A is as given. Write your answers in the form given in (11) in the text.

   (a) $\begin{bmatrix} 0 & 1 \\ 8 & -2 \end{bmatrix}$　　(b) $\begin{bmatrix} 3 & -2 \\ 2 & -2 \end{bmatrix}$　　(c) $\begin{bmatrix} 2 & 1 \\ -3 & 6 \end{bmatrix}$　　(d) $\begin{bmatrix} -1 & 4 \\ 2 & 1 \end{bmatrix}$

   (e) $\begin{bmatrix} 3 & 2 & 2 \\ 1 & 4 & 1 \\ -2 & -4 & -1 \end{bmatrix}$　　(f) $\begin{bmatrix} 2 & 2 & 1 \\ 1 & 3 & 1 \\ 1 & 2 & 2 \end{bmatrix}$

4. (*Solving IVPs*). (a)–(f). Solve the IVP $x' = Ax$, $x(0) = [-1 \ 1]^T$, where A is the matrix given in 3(a)–(d). For (e)–(f), $x(0) = [1 \ 1 \ 1]^T$.

www 5. (*Scalar ODE and its Equivalent System*). The "characteristic polynomial" of the n-th-order constant coefficient ODE $y^{(n)} + a_{n-1} y^{(n-1)} + \cdots + a_1 y' + a_0 y = 0$ is the polynomial $r^n + a_{n-1} r^{n-1} + \cdots + a_1 r + a_0$.

(a) Show that if $x_1 = y$, $x_2 = y', \ldots, x_n = y^{(n-1)}$, the ODE becomes $x' = Ax$, where

$$A = \begin{bmatrix} 0 & 1 & 0 & \cdots & \cdots & 0 \\ 0 & 0 & 1 & & & \\ \vdots & & \ddots & & & \vdots \\ \vdots & & & \ddots & & \vdots \\ & & & & 0 & 1 \\ -a_0 & -a_1 & -a_2 & \cdots & -a_{n-2} & -a_{n-1} \end{bmatrix}$$

(b) Show that the characteristic polynomial $p(\lambda)$ of matrix A is $(-1)^n (\lambda^n + a_{n-1} \lambda^{n-1} + \cdots + a_1 \lambda + a_0)$. [*Hint*: Using column operations, transform $A - \lambda I$ into a matrix where the submatrix of the top $(n-1)$ rows starts with a column of zeros followed by an $(n-1) \times (n-1)$ triangular matrix with ones on the diagonal. Then calculate the determinant by cofactors.]

6. (*Deficient Eigenspace*). Find the general solution of each system below.

(a) $x' = \begin{bmatrix} 1 & -2 & -2 \\ -2 & 2 & 3 \\ 2 & -3 & -4 \end{bmatrix} x$ [*Hint*: See Example 7.4.7.]   (b) $x' = \begin{bmatrix} 1 & 0 & -2 \\ 0 & -1 & 1 \\ 0 & 0 & -1 \end{bmatrix} x$

7. (*Deficient Eigenspaces*). Find the general solution of the system

$$x' = Ax = \begin{bmatrix} -1 & 0 & 2 \\ 0 & -1 & 1 \\ 0 & 0 & -1 \end{bmatrix} x$$

by following the outline below:

- Show that the matrix A has only one eigenvalue λ and that the corresponding eigenspace has dimension 2.

- To find an eigenvector that yields a chain of generalized eigenvectors of length 2, proceed as follows: Suppose that v^1 and v^2 are a basis of the eigenspace of A corresponding to λ. Find conditions on the constants c_1 and c_2 such that the system $(A - \lambda I) u^2 = c_1 v^1 + c_2 v^2$ has a nonzero solution.

- Using values for c_1 and c_2 as determined above, show that $u^1 = c_1 v^1 + c_2 v^2$ and a nonzero solution u^2 of $(A - \lambda I) u^2 = u^1$ form a chain of generalized eigenvectors.

- Using v^1, u^1, and u^2, construct the general solution of $x' = Ax$.

8. (*More on Deficient Eigenspaces*). For a 10×10 matrix A with real entries, say it is known that A has three real eigenvalues $\lambda_1 = -1$, $\lambda_2 = 3$, and $\lambda_3 = 2$ with multiplicities $m_1 = 3$, $m_2 = 3$, and $m_3 = 4$. Given the additional information below about the eigenspaces V_{λ_1}, V_{λ_2}, and V_{λ_3} and their bases, find all solutions of the undriven system in each of the two cases:

(a) $\dim V_{\lambda_1} = 1$; V_{λ_1} has a basis $\{u^1\}$,

$\dim V_{\lambda_2} = 2$; V_{λ_2} has a basis $\{v^1, w^1\}$ and v^1 generates a 2-vector chain of generalized eigenvectors,

$\dim V_{\lambda_3} = 2$; V_{λ_3} has a basis $\{p^1, q^1\}$ and q^1 generates a 3-vector chain of generalized eigenvectors.

(b) $\dim V_{\lambda_1} = 3$; V_{λ_1} has a basis $\{u^1, v^1, w^1\}$,

$\dim V_{\lambda_2} = 1$; V_{λ_2} has a basis $\{p^1\}$,

$\dim V_{\lambda_3} = 3$; V_{λ_3} has a basis $\{a^1, b^1, c^1\}$ and a^1 generates a 2-vector chain of generalized eigenvectors.

7.6 Undriven Linear Systems: Complex Eigenvalues

The $n \times n$ constant matrix A in the undriven linear system $x' = Ax$ can have complex entries, or real entries and complex eigenvalues. Either way, if we follow the Method of Eigenvectors described in the previous section, the solutions that result are complex-valued functions. For example, the solution formula (11) in Theorem 7.5.2 is still valid when some eigenvectors have complex entries, but now the arbitrary constants $C_1, C_2, \ldots, C_n$ must be allowed to be complex numbers. If A has real entries, then we would usually rather know what all the real-valued solutions of the system $x' = Ax$ are. In this section we'll learn how to extract all real-valued solutions from the general solution formula, but first we'll review a few things about the arithmetic of complex numbers and complex-valued functions of a real variable.

☞ If you are shaky on working with complex quantities, review Appendix B.3.

Complex Numbers and Complex-Valued Functions

A complex number r has the form $\alpha + i\beta$, where α and β are real numbers called, respectively, the *real part* and the *imaginary part* of r. In symbols, $\alpha = \text{Re}\,[r]$ and $\beta = \text{Im}\,[r]$. So r is a real number if and only if $\text{Im}\,[r] = 0$. Two complex numbers are equal if and only if their respective real and imaginary parts are equal. The arithmetic of complex numbers follows the standard rules of algebra, as long as we remember to replace i^2 by -1. The *complex conjugate* of the complex number $r = \alpha + i\beta$ is defined as $\alpha - i\beta$ and is denoted by $\bar{r}$. Notice that r is a real number if and only if $r = \bar{r}$. The conjugate of a sum, product, or quotient of complex numbers is the sum, product, or quotient of the conjugates. For example, $\overline{r+s} = \bar{r} + \bar{s}$, $\overline{rs} = \bar{r}\bar{s}$, and $\overline{r/s} = \bar{r}/\bar{s}$.

☞ Yes, we know that much of this information has already appeared, but it is handy to have it all in one place.

Complex-valued functions of a real variable can also be resolved into a real part plus i times an imaginary part. For example, as we saw in Section 3.4,

$$e^{(\alpha+i\beta)t} = e^{\alpha t} \cos \beta t + i e^{\alpha t} \sin \beta t$$

If $f(t) = u(t) + iv(t)$, where $u(t)$ and $v(t)$ are real-valued functions, then $\overline{f(t)} = u(t) - iv(t)$. So for any complex number $r = \alpha + i\beta$ we see that

$$\overline{e^{rt}} = \overline{e^{(\alpha+i\beta)t}} = e^{\alpha t} \cos \beta t - i e^{\alpha t} \sin \beta t = e^{(\alpha-i\beta)t} = e^{\bar{r}t}$$

Column vectors with complex entries can be uniquely written as a real part plus i times an imaginary part. For example,

$$\begin{bmatrix} -i \\ 1+i \\ 2 \end{bmatrix} = \begin{bmatrix} 0 \\ 1 \\ 2 \end{bmatrix} + i \begin{bmatrix} -1 \\ 1 \\ 0 \end{bmatrix}, \qquad \begin{bmatrix} t^2 e^{-2it} \\ e^{(1+i)t} \end{bmatrix} = \begin{bmatrix} t^2 \cos 2t \\ e^t \cos t \end{bmatrix} + i \begin{bmatrix} -t^2 \sin 2t \\ e^t \sin t \end{bmatrix}$$

The conjugate of a matrix A is denoted by $\bar{A}$ and is obtained by conjugating the entries. For example,

$$\overline{\begin{bmatrix} 2i & -1+2i \\ -1 & i \end{bmatrix}} = \begin{bmatrix} -2i & -1-2i \\ -1 & -i \end{bmatrix}, \qquad \overline{\begin{bmatrix} e^{rt} \\ 1+i \end{bmatrix}} = \begin{bmatrix} e^{\bar{r}t} \\ 1-i \end{bmatrix}$$

The conjugation rule applies to sums and products of matrices with complex entries and to the product of matrices and complex numbers; that is, $\overline{AB} = \bar{A}\bar{B}$ and $\overline{cA} = \bar{c}\bar{A}$, for matrices A, B and scalars c. For example, note that for any column vectors v and w with complex entries and any complex numbers r and s we have

$$\overline{ve^{rt}} = \bar{v}e^{\bar{r}t}, \qquad \overline{ve^{rt} + we^{st}} = \bar{v}e^{\bar{r}t} + \overline{w}e^{\bar{s}t}$$

An application of conjugation noted earlier is that if matrix A has real entries, then the eigenvalues occur in conjugate pairs and the eigenspace $V_{\bar{\lambda}}$ is just $\overline{V_\lambda}$. This follows by conjugating the eigenvalue-eigenvector equation $Av = \lambda v$ to obtain $A\bar{v} = \bar{\lambda}\bar{v}$. The same holds for a chain of generalized eigenvectors. For example, look at the 3-chain generalized eigenvectors based on the eigenvector u^1 corresponding to the eigenvalue λ:

$$u^1, \quad u^2, \quad u^3, \quad \text{where} \quad (A - \lambda I)u^2 = u^1, \quad (A - \lambda I)u^3 = u^2$$

Conjugating these matrix-vector equations (and remembering that A and I have real entries), we see that $(A - \bar{\lambda}I)\overline{u^3} = \overline{u^2}$, $(A - \bar{\lambda}I)\overline{u^2} = \overline{u^1}$; so $\overline{u^1}, \overline{u^2}, \overline{u^3}$ is a chain of generalized eigenvectors of A based on the eigenvector $\overline{u^1}$ corresponding to the eigenvalue $\bar{\lambda}$.

Now we are ready to tackle some examples.

Finding Real-Valued Solutions

When the square matrix A has real-valued entries we usually prefer to find all the real-valued solutions of the system $x' = Ax$. It's sometimes easier to first find all *complex*-valued solutions and then extract all the real-valued ones from among them.

Our approach uses the fact that the real and imaginary parts of a complex-valued solution of the system $x' = Ax$ are real-valued solutions of the system if the $n \times n$ matrix A has real entries. To see this recall from Theorem 7.5.1 that the operator L defined by $L[x] = x' - Ax$ is a linear operator. Suppose that $x(t)$ is a solution of $x' = Ax$. Then $L[x] = 0$ and if we write $x(t) = u(t) + iv(t)$, where $u(t) = \text{Re}[x(t)]$ and $v(t) = \text{Im}[x(t)]$, then

$$L[x] = L[u + iv] = L[u] + iL[v] = 0 \tag{1}$$

Since A has real entries, we see that $L[u] = u' - Au$ and $L[v] = v' - Av$ are both real-valued, so identity (1) says that $L[u] = 0$ and $L[v] = 0$. In other words, if $x(t)$ is any solution of the system $x' = Ax$, then so are $\text{Re}[x(t)]$ and $\text{Im}[x(t)]$.

This observation suggests a way to find all real-valued solutions of the system $x' = Ax$, where A is any $n \times n$ constant matrix with real entries. The first step is to use the Method of Eigenvectors to find a formula for n solutions $x^1(t), x^2(t), \ldots, x^n(t)$ of $x' = Ax$ (some possibly complex-valued). This requires us to find all eigenvalues of A and enough eigenvectors (and generalized eigenvectors, if necessary) to form a basis for $\mathbb{C}^n$. We can identify all the real-valued solutions of $x' = Ax$ by using the constructed solutions $x^1(t), x^2(t), \ldots, x^n(t)$ in the following way: Say the n constructed solutions are indexed so that $x^1(t), x^2(t), \ldots, x^k(t)$ are real-valued and the remaining $n - k$ ones are complex-valued. Since A has real entries, the complex eigenvalues of A occur in conjugate pairs (so $n - k$ is an even number). Discard the solutions that arise from an eigenvalue λ with $\text{Im}[\lambda] < 0$, and that should leave us with $(n - k)/2 = m$ complex-valued solutions. Say that these solutions are indexed as $x^{k+1}(t), \ldots, x^{k+m}(t)$. Suppose that

$$u^{k+j}(t) = \text{Re}[x^{k+j}(t)], \quad v^{k+j}(t) = \text{Im}[x^{k+j}(t)], \qquad \text{for each } j = 1, \ldots, m$$

Now because $L[x] = x' - Ax$ defines a linear operator, we see that any linear combination of the solutions $x^1, \ldots, x^k, u^{k+1}, \ldots, u^{k+m}, v^{k+1}, \ldots, v^{k+m}$ is again a solution of $x' = Ax$. In other words,

☞ Since $n - k = 2m$, note that this formula has n summands.

$$x = a_1 x^1 + \cdots + a_k x^k + b_1 u^{k+1} + \cdots + b_m u^{k+m} + c_1 v^{k+1} + \cdots + c_m v^{k+m} \qquad (2)$$

for arbitrary reals $a_1, \ldots, a_k, b_1, \ldots, b_m, c_1, \ldots, c_m$ is always a solution of $x' = Ax$. It is a remarkable fact that (2) gives *all* the real-valued solutions of $x' = Ax$ in this case, but we won't show that here.

☞ In this book we keep $n \le 4$, so how bad can things get?

In practice, however, things are not as tough as it appears to arrive at a formula for all real-valued solutions of a system with complex eigenvalues. Knowing how things will work out in the end, we can skip over the intermediate steps to obtain the final result. The procedure below summarizes this process.

Formula for All Real-Valued Solutions of $x' = Ax$

Suppose that A is an $n \times n$ constant matrix with real entries. Then all real-valued solutions of $x' = Ax$ look like

$$x(t) = C_1 y^1(t) + C_2 y^2(t) + \cdots + C_n y^n(t) \qquad (3)$$

where the C_j are arbitrary real constants and the special real-valued solutions $y^j(t)$ are determined as follows:

- Find all eigenvalues of A. If two eigenvalues are complex conjugates, discard the one with negative imaginary part.

- For each of the retained eigenvalues find a basis for the corresponding eigenspace such that the number of generalized eigenvectors in the chains generated by that basis is equal to the multiplicity of that eigenvalue.

- Use the Method of Eigenvalues to find the real-valued solutions $x^1(t), \ldots, x^k(t)$, generated by the real eigenvalues, and the complex-valued solutions $x^{k+1}(t), \ldots, x^{k+m}(t)$ generated by the retained complex eigenvalues (note that $k + 2m = n$).

- The $y^j(t)$ in (3) are defined as follows: take $y^j(t) = x^j(t)$, $j = 1, 2, \ldots, k$, and the remaining $y^j(t)$ by $\text{Re}[x^{k+1}(t)], \text{Im}[x^{k+1}(t)], \ldots, \text{Re}[x^{k+m}(t)], \text{Im}[x^{k+m}(t)]$.

Let's look at some examples.

EXAMPLE 7.6.1	**Real-Valued Solutions from Complex-Valued Solutions** Let's look at the system

$$x' = Ax, \quad \text{where } A = \begin{bmatrix} 1 & -2 \\ 2 & 1 \end{bmatrix} \qquad (4)$$

☞ Example 7.4.3
shows how to find
these eigenspaces.

The roots of the characteristic polynomial $p(\lambda) = \det(A - \lambda I) = \lambda^2 - 2\lambda + 5$ are the eigenvalues of A. The two roots are $\lambda_1 = 1 + 2i$ and $\lambda_2 = 1 - 2i$. Let's select the basis $[i \ 1]^T$ for V_{λ_1} and $[-i \ 1]^T$ for V_{λ_2}. Notice that $\{[i \ 1]^T, [-i \ 1]^T\}$ is an eigenbasis for $\mathbb{C}^2$, and from Theorem 7.5.2 we have the general solution formula

$$x(t) = K_1 \begin{bmatrix} i \\ 1 \end{bmatrix} e^{(1+2i)t} + K_2 \begin{bmatrix} -i \\ 1 \end{bmatrix} e^{(1-2i)t} \qquad (5)$$

where K_1, K_2 are arbitrary complex numbers.

Let's follow the above procedure to recognize all the real-valued solutions hidden among the complex-valued solutions of formula (5). To simplify our task notationally, let's put

$$x^1(t) = \begin{bmatrix} i \\ 1 \end{bmatrix} e^{(1+2i)t}, \qquad x^2(t) = \begin{bmatrix} -i \\ 1 \end{bmatrix} e^{(1-2i)t} \qquad (6)$$

Now let's discard $x^2(t)$ because it arises from the eigenvalue $\lambda_2 = 1 - 2i$ with negative real part. Next, let's resolve $x^1(t)$ into its real and imaginary parts as follows:

$$x^1(t) = \left(\begin{bmatrix} 0 \\ 1 \end{bmatrix} + i \begin{bmatrix} 1 \\ 0 \end{bmatrix} \right) \left(e^t \cos 2t + i e^t \sin 2t \right)$$

$$= \left(\begin{bmatrix} 0 \\ 1 \end{bmatrix} e^t \cos 2t - \begin{bmatrix} 1 \\ 0 \end{bmatrix} e^t \sin 2t \right) + i \left(\begin{bmatrix} 1 \\ 0 \end{bmatrix} e^t \cos 2t - \begin{bmatrix} 0 \\ 1 \end{bmatrix} e^t \sin 2t \right)$$

$$= u^1(t) + i v^1(t)$$

where $u^1(t) = \text{Re}[x^1(t)]$ and $v^1(t) = \text{Im}[x^1(t)]$:

$$u^1(t) = \begin{bmatrix} 0 \\ 1 \end{bmatrix} e^t \cos 2t - \begin{bmatrix} 1 \\ 0 \end{bmatrix} e^t \sin 2t, \qquad v^1(t) = \begin{bmatrix} 1 \\ 0 \end{bmatrix} e^t \cos 2t + \begin{bmatrix} 0 \\ 1 \end{bmatrix} e^t \sin 2t \qquad (7)$$

So all real-valued solutions of system (4) are given by

$$x(t) = C_1 u^1(t) + C_2 v^1(t), \quad \text{for any reals } C_1 \text{ and } C_2$$

where $u(t)$ and $v(t)$ are as given in (7).

EXAMPLE 7.6.2	**Solution Formula for a Linear System** Let's find a formula for all the real-valued solutions of the system

$$x' = \begin{bmatrix} -2 & -1 & 3 \\ 5 & 0 & -5 \\ 0 & 0 & 1 \end{bmatrix} x \qquad (8)$$

We will follow the procedure above and use the Method of Eigenvectors to find three real-valued solutions $y^1(t)$, $y^2(t)$, and $y^3(t)$ of system (8) such that x_{gen} is given by formula (3). Denoting the system matrix by A, the eigenvalue-eigenvector equation $(A - \lambda I)w = 0$ becomes

$$\begin{bmatrix} -2-\lambda & -1 & 3 \\ 5 & -\lambda & -5 \\ 0 & 0 & 1-\lambda \end{bmatrix} \begin{bmatrix} w_1 \\ w_2 \\ w_3 \end{bmatrix} = \begin{bmatrix} 0 \\ 0 \\ 0 \end{bmatrix}$$

Since $\det(A - \lambda I) = (1 - \lambda)(\lambda^2 + 2\lambda + 5)$, we see that A has three eigenvalues $\lambda_1 = 1$, $\lambda_2 = -1 + 2i$, $\lambda_3 = -1 - 2i$. Let's drop λ_3 (since $\text{Im}[\lambda_3] < 0$) and find the eigenspaces V_{λ_1} and V_{λ_2} corresponding to λ_1 and λ_2. First we find V_{λ_1} by solving the system $(A - I)w = 0$. Using row operations, this system reduces to $w_2 = 0$, $w_1 - w_2 - w_3 = 0$. So $[1 \ 0 \ 1]^T$ is an eigenvector for λ_1 and spans the eigenspace V_1. In the notation of the procedure

$$y^1 = \begin{bmatrix} 1 \\ 0 \\ 1 \end{bmatrix} e^t$$

is a real-valued solution of system (8). Next, let's find V_{λ_2} by solving the system $(A - (-1 + 2i)I)w = 0$. Row operations reduce this system to $5w_1 + (1 - 2i)w_2 = 0$ and $w_3 = 0$. So $[1-2i \ -5 \ 0]^T$ is an eigenvector for λ_2 and spans the eigenspace V_{λ_2}. In the notation of the procedure, $x^2(t) = [1-2i \ -5 \ 0]^T e^{(-1+2i)t}$ and is a complex-valued solution of system (8). Let's find the real and imaginary parts of $x^2(t)$.

$$x^2(t) = \left(\begin{bmatrix} 1 \\ -5 \\ 0 \end{bmatrix} + i \begin{bmatrix} -2 \\ 0 \\ 0 \end{bmatrix} \right) e^{-t}(\cos 2t + i \sin 2t) = y^2(t) + iy^3(t)$$

where $y^2(t) = \text{Re}[x^2(t)]$ and $y^3(t) = \text{Im}[x^2(t)]$. We have that

$$y^2 = \begin{bmatrix} 1 \\ -5 \\ 0 \end{bmatrix} e^{-t} \cos 2t - \begin{bmatrix} -2 \\ 0 \\ 0 \end{bmatrix} e^{-t} \sin 2t, \quad y^3 = \begin{bmatrix} -2 \\ 0 \\ 0 \end{bmatrix} e^{-t} \cos 2t + \begin{bmatrix} 1 \\ -5 \\ 0 \end{bmatrix} e^{-t} \sin 2t$$

are real-valued solutions of system (8). So we see that the general solution of system (8) is

$$x_{gen} = C_1 y^1 + C_2 y^2 + C_3 y^3$$

where C_1, C_2, and C_3 are arbitrary reals.

Another Approach to Finding Solution Formulas

☞ This approach will come in handy in the sections that follow.

For constant $n \times n$ matrices A, the Method of Eigenvectors constructs all solutions of the system $x' = Ax$ by first calculating eigenvalues, eigenvectors, and generalized eigenvectors of the matrix A. There is another way to come up with a general solution formula for $x' = Ax$ without directly mentioning eigenvectors. It uses the following result (which is interesting in its own right).

THEOREM 7.6.1

Existence and Uniqueness Theorem. For the $n \times n$ constant matrix A, and any constant vector x^0, the IVP

$$x' = Ax, \qquad x(0) = x^0$$

has a unique solution defined on the whole t-axis. If $x^0 = 0$, then the solution of the IVP is $x = 0$, for all t.

☞ Proving this result from scratch takes a little doing.

This theorem is basically the Fundamental Theorem (Theorem 5.2.1), but with the added fact that the unique solution lives on the whole t-axis.

Theorem 7.6.1 gives us another approach to finding a formula for all the solutions of $x' = Ax$. Say that $y^1(t), \ldots, y^n(t)$ are n solutions of $x' = Ax$ with the property that $\det[y^1(0) \;\; \cdots \;\; y^n(0)] \neq 0$. We will show that all solutions of $x' = Ax$ are given by

$$x(t) = c_1 y^1(t) + c_2 y^2(t) + \cdots + c_n y^n(t)$$

where $c_1, \ldots, c_n$ are arbitrary constants. To show this say that $y(t)$ is any solution of $x' = Ax$; then from Theorem 7.3.1 there are unique constants $k_1, k_2, \ldots, k_n$ such that

$$k_1 y^1(0) + \cdots + k_n y^n(0) = y(0)$$

Now, put $z(t) = y(t) - k_1 y^1(t) - \cdots - k_n y^n(t)$. Note that $z' = Az$ because

$$Az = Ay - k_1 Ay^1 - \cdots - k_n Ay^n = y' - k_1(y^1)' - \cdots - k_n(y^n)' = z'$$

But since $z(0) = 0$, it follows from Theorem 7.6.1 that $z(t) = 0$, for all t, and we must have that $y(t) = k_1 y^1(t) + \cdots + k_n y^n(t)$. So every solution of $x' = Ax$ is a linear combination of the solution set $\{y^1(t), \ldots, y^n(t)\}$.

This motivates the following definition.

❖ **Basic Solution Set**. If A is an $n \times n$ constant matrix, then a set of n solutions $\{y^1, \ldots, y^n\}$ of the system $x' = Ax$ is called a *basic solution set* if every solution $y(t)$ can be written as $y(t) = c_1 y^1 + \cdots + c_n y^n$ for some constants $c_1, \ldots, c_n$.

We just saw that n solutions $y^1, y^2, \ldots, y^n$ form a basic solution set for $x' = Ax$ if $\det[y^1(0) \;\; y^2(0) \;\; \cdots \;\; y^n(0)] \neq 0$. For example, if $\mathbb{R}^n$ (or $\mathbb{C}^n$, if appropriate) has a basis for $\{v^1, \ldots, v^n\}$ consisting of eigenvectors of A, then

$$\{y^j = v^j e^{\lambda_j t}, \;\; j = 1, 2, \ldots, n, \;\; \lambda_j \text{ is the eigenvalue corresponding to } v^j\}$$

is a basic solution set.

More generally, it can be shown that for any $n \times n$ constant matrix A, and for any n solutions $y^1, y^2, \ldots, y^n$ of the system $x' = Ax$, the determinant $\det[y^1(t) \;\; \cdots \;\; y^n(t)]$ is either always zero for all t, or never zero for any t. So we have the following test.

THEOREM 7.6.2

> **Test for a Basic Solution Set.** Suppose that A is an $n \times n$ constant matrix. The n solutions $y^1, \dots, y^n$ of the system $x' = Ax$ comprise a basic solution set if and only if $\det[y^1(t) \; \cdots \; y^n(t)] \neq 0$, for all t.

Basic solution sets are often convenient to use in describing the general solution of the system $x' = Ax$, as we will see in later sections. Summing up, we have the following result.

THEOREM 7.6.3

> **Existence of a Basic Solution Set.** For any $n \times n$ constant matrix A, the system $x' = Ax$ has a basic solution set $\{x^1(t), \dots, x^n(t)\}$, so all solutions of the system are given by
>
> $$x = c_1 x^1 + c_2 x^2 + \cdots + c_n x^n \tag{9}$$
>
> where the coefficients c_j are arbitrary constants.

The Method of Eigenvectors provides a basic solution set by constructing it from the eigenvalues, eigenvectors and, if necessssary, the generalized eigenvectors of the constant matrix A. Basic solution sets are not unique; any one will do for the general solution formula (9). In Section 7.8 we return to this point.

Our next result characterizes the form of all entries of any real-valued solution of the system $x' = Ax$, where A is an $n \times n$ constant matrix with real entries.

Components of Solutions of $x' = Ax$

☞ The power of t in $t^k e^{\lambda t}$ depends on the "depth" of the deficiency of the eigenspace V_λ.

We have seen in Examples 7.5.4 and 7.5.5 that if there are deficient eigenspaces, then some solutions have components containing polynomial-exponential functions such as $te^{\lambda t}$ or $t^2 e^{\lambda t}$. If λ had been a complex number $\alpha + i\beta$ with $\beta \neq 0$, this would mean terms such as $te^{\alpha t} \cos \beta t$ or $te^{\alpha t} \sin \beta t$, and so on. In fact, the following result gives the form of the components of all real-valued solutions of $x' = Ax$.

THEOREM 7.6.4

☞ So every component is a polynomial-exponential function.

> **Components of Solutions of $x' = Ax$.** Any component of any real-valued solution of $x' = Ax$, where A is an $n \times n$ matrix of real constants, is a linear combination of terms of the form $t^k e^{\alpha t} \cos \beta t$ and $t^k e^{\alpha t} \sin \beta t$, where $\lambda = \alpha + i\beta$ is an eigenvalue of A and k is any nonnegative integer less than the multiplicity of λ.

Comments

Sections 7.5 and 7.6 tell us all we need to know about finding a formula for all the real-valued solutions of the linear system $x' = Ax$, where A is a constant $n \times n$ matrix with real entries. The key to finding such a formula is finding eigenvalues and eigenvectors (and perhaps generalized eigenvectors) of the matrix A. In theory this is always possible, but it can be difficult for large n.

PROBLEMS

www **1.**   (*General Solutions*). Find the general solution for the system $x_1' = 2x_1 - x_2,\ \ x_2' = x_1 + 2x_2$.

2.   (*Solving an IVP*). Solve the IVP consisting of the rate equations given in Problem 1 and the initial condition $x_1(0) = 1,\ \ x_2(0) = 1$.

3.   (*Nondeficient Eigenspaces*). Find all solutions of $x' = Ax$, where the matrix A is as given. Write your answers in the form given in (11) of Section 7.5.

     (a) $\begin{bmatrix} 0 & 1 \\ -64 & 0 \end{bmatrix}$ **(b)** $\begin{bmatrix} 2 & -1 \\ 8 & -2 \end{bmatrix}$

4.   (*Solving IVPs*). **(a)–(b)**. Solve the IVP $x' = Ax,\ x(0) = [-1 \ \ 1]^T$, for A given in 3**(a)**, **(b)**.

5.   (*Complex Eigenvalues of a Real Matrix*). Let A be a constant matrix with real entries and suppose that λ is a nonreal eigenvalue with corresponding eigenvector v.

     (a) Show that v must have at least one nonreal component and that $\bar{v}$ is an eigenvector of A with $\bar{\lambda}$ its corresponding eigenvalue.

     (b) Show that $\{ve^{\lambda t}, \bar{v}e^{\bar{\lambda}t}\}$ is an independent set of complex-valued solutions of $x' = Ax$.

     (c) Show that $\{\text{Re}[ve^{\lambda t}], \text{Im}[ve^{\lambda t}]\}$ is an independent set of real-valued solutions of $x' = Ax$. [*Hint*: $\text{Re}[z] = (z + \bar{z})/2,\ \text{Im}[z] = (-i/2)(z - \bar{z})$.]

     (d) Find the general real-valued solutions of each system $x' = Ax$, where

$$A = \begin{bmatrix} 1 & 1 \\ -5 & 3 \end{bmatrix}, \qquad A = \begin{bmatrix} 1 & 2 \\ -2 & 1 \end{bmatrix}, \qquad A = \begin{bmatrix} -3 & 1 & -2 \\ 0 & -1 & -1 \\ 2 & 0 & 0 \end{bmatrix}$$

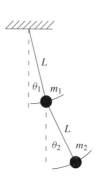

6.   (*The Double Pendulum*). Suppose that one pendulum is suspended from another (see the margin sketch). It can be shown (after an appropriate rescaling of time) that the following linear system models the small-amplitude oscillations about the equilibrium position $x_1 = x_2 = x_3 = x_4 = 0$:

$$x_1' = x_2$$
$$x_2' = -x_1 + \alpha x_3$$
$$x_3' = x_4$$
$$x_4' = x_1 - x_3$$

where $\alpha = (m_2/m_1)(1 + m_2/m_1)^{-1}$ is the *reduced mass*, x_1 and x_3 are the angles θ_1 and θ_2, and x_2 and x_4 are the angular velocities, θ_1' and θ_2'. Observe that $0 < \alpha < 1$.

     (a) Build the general real-valued solution of this system using the methods of this section.

     (b) Put $\alpha = 0.3$. Does the system have any periodic solutions? If so, what are they?

     (c) For $\alpha = 0.3$, find the unique solution of the system for each set of initial conditions:
         (1) $x_1(0) = (0.15)\sqrt{0.3},\quad x_2(0) = 0,\quad x_3(0) = -0.15,\quad x_4(0) = 0$
         (2) $x_1(0) = (0.15)\sqrt{0.3},\quad x_2(0) = 0,\quad x_3(0) = 0.15,\quad\ \ \ x_4(0) = 0$
         (3) $x_1(0) = (0.3)\sqrt{0.3},\quad\ \ x_2(0) = 0,\quad x_3(0) = 0,\quad\ \ \ \ \ \ x_4(0) = 0$
         [*Hint*: Note that the initial data in **(3)** are the sum of the initial data in **(1)** and **(2)**.]

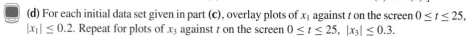

     (d) For each initial data set given in part **(c)**, overlay plots of x_1 against t on the screen $0 \le t \le 25$, $|x_1| \le 0.2$. Repeat for plots of x_3 against t on the screen $0 \le t \le 25$, $|x_3| \le 0.3$.

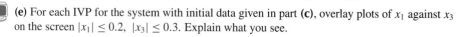

     (e) For each IVP for the system with initial data given in part **(c)**, overlay plots of x_1 against x_3 on the screen $|x_1| \le 0.2,\ |x_3| \le 0.3$. Explain what you see.

7.7 Orbital Portraits

For a constant 2×2 matrix A the orbital portraits of the system $x' = Ax$ fall into distinct types. We will develop a complete classification of orbital portrait types when $\lambda = 0$ is not an eigenvalue of A. We show by an example that a planar system's orbital portrait type may change drastically even if the entries in the system matrix change only a tiny bit. Finally, we model the vibrations of a pair of coupled springs.

Orbital Portrait Gallery

Let's classify by type all possible orbital portraits for the planar system

$$\begin{bmatrix} x \\ y \end{bmatrix}' = A \begin{bmatrix} x \\ y \end{bmatrix}, \qquad A = \begin{bmatrix} a & b \\ c & d \end{bmatrix} \tag{1}$$

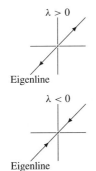

λ > 0

Eigenline

λ < 0

Eigenline

where A is a matrix of real constants. The six cases cover all of the possibilities for a real 2×2 matrix A with nonzero eigenvalues. They are listed in Table 7.7.1 along with the nature of the eigenvalues of A and a basic set of solutions. The names that are given in the table define the classification type for the equilibrium point at the origin.

An *eigenline* is a line through the origin parallel to a real eigenvector. So if λ is a real eigenvalue, v a corresponding eigenvector, then the family of solution curves $x = cve^{\lambda t}$, for any real constant c, trace out an eigenline through the origin. If λ is

TABLE 7.7.1 *Classification of Orbital Portraits for the Planar System $x' = Ax$*. The eigenvectors u, v of A correspond, respectively, to eigenvalues λ_1, λ_2; the vector w in the basic set for the deficient node is a generalized eigenvector based on u; it is assumed that $\lambda = 0$ is not an eigenvalue of A.

Classification	Basic Solution Set
Improper node λ_1 and λ_2 real, distinct, and same sign	$\{ue^{\lambda_1 t}, ve^{\lambda_2 t}\}$
Deficient node $\lambda_1 = \lambda_2$; one-dimensional eigenspace	$\{ue^{\lambda_1 t}, (w + tu)e^{\lambda_1 t}\}$
Star (or **proper**) **node** ($\{u, v\}$ any basis of $\mathbb{R}^2$) $\lambda_1 = \lambda_2$; two-dimensional eigenspace	$\{ue^{\lambda_1 t}, ve^{\lambda_1 t}\}$
Saddle point $\lambda_1 < 0 < \lambda_2$	$\{ue^{\lambda_1 t}, ve^{\lambda_2 t}\}$
Center (or **vortex**) $\lambda_1 = i\beta = \bar{\lambda}_2$, β real and nonzero	$\{\text{Re}[ue^{i\beta t}], \text{Im}[ue^{i\beta t}]\}$
Spiral point (or **focus**) $\lambda_1 = \alpha + i\beta = \bar{\lambda}_2$, α, β real and nonzero	$\{\text{Re}[ue^{(\alpha+i\beta)t}], \text{Im}[ue^{(\alpha+i\beta)t}]\}$

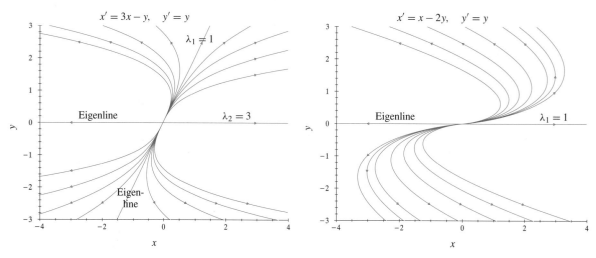

FIGURE 7.7.1 Improper node (Example 7.7.1).   **FIGURE 7.7.2** Deficient node (Example 7.7.2).

positive, orbits move outward from the origin along the eigenline as t increases; if λ is negative, orbits move inward along the eigenline. Each eigenline divides the plane into sectors, and no solution curve can cross from one sector into another. Improper nodes and saddles have two eigenlines, deficient nodes only one. For star nodes, all lines through the origin are eigenlines, while centers and spiral points have no eigenlines.

The orbital portraits of the six following examples are representative.

EXAMPLE 7.7.1

Improper Node

The matrix of the system $x' = 3x - y$, $y' = y$ has eigenvalues $\lambda_1 = 1$ and $\lambda_2 = 3$ with respective eigenvectors $[1 \quad 2]^T$ and $[1 \quad 0]^T$. The general solution is

$$\begin{bmatrix} x \\ y \end{bmatrix} = c_1 \begin{bmatrix} 1 \\ 2 \end{bmatrix} e^t + c_2 \begin{bmatrix} 1 \\ 0 \end{bmatrix} e^{3t} = \left(c_1 \begin{bmatrix} 1 \\ 2 \end{bmatrix} + c_2 \begin{bmatrix} 1 \\ 0 \end{bmatrix} e^{2t} \right) e^t \tag{2}$$

The eigenlines $y = 2x$ and $y = 0$ are generated by the eigenvectors $[1 \quad 2]^T$ and $[1 \quad 0]^T$, respectively. All solution curves with $c_1 \neq 0$ emerge from the origin (as t increases) tangent to the eigenline $y = 2x$ because the term in parentheses in solution formula (2) tends to $c_1 [1 \quad 2]^T$ as $t \to -\infty$ (Figure 7.7.1). To see what happpens as $t \to +\infty$ let's write solution formula (2) as

$$\begin{bmatrix} x \\ y \end{bmatrix} = e^{3t} \left(c_1 \begin{bmatrix} 1 \\ 2 \end{bmatrix} e^{-2t} + c_2 \begin{bmatrix} 1 \\ 0 \end{bmatrix} \right)$$

So as $t \to +\infty$, the state vector $[x \quad y]^T$ looks more and more like $c_2 [1 \quad 0]^T e^{3t}$, if $c_2 \neq 0$.

The term "node" is used in geometry for a point where a collection of curves intersects with all possible tangents. The node here is "improper" because there are only two tangents. The node of the next example has one tangent line at the origin.

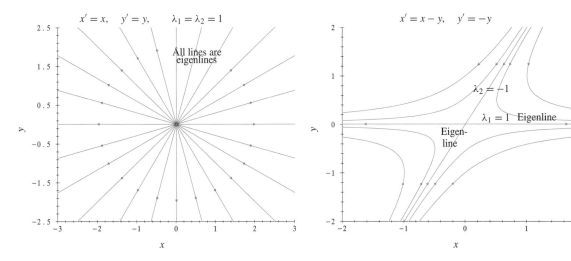

FIGURE 7.7.3 Star node (Example 7.7.3).

FIGURE 7.7.4 Saddle point (Example 7.7.4).

EXAMPLE 7.7.2

Deficient Node

The matrix of the system $x' = x - 2y$, $y' = y$ has the double eigenvalue $\lambda_1 = 1$. The eigenspace is deficient and the eigenvector $v = [1 \quad 0]^T$ is a basis for the eigenspace. After finding a generalized eigenvector $w = [0 \quad -1/2]^T$ by solving $(A - I)w = v$ for w, we see that the general solution is

$$\begin{bmatrix} x \\ y \end{bmatrix} = c_1 \begin{bmatrix} 1 \\ 0 \end{bmatrix} e^t + c_2 \left(\begin{bmatrix} 0 \\ -1/2 \end{bmatrix} + t \begin{bmatrix} 1 \\ 0 \end{bmatrix} \right) e^t \tag{3}$$

All orbits in Figure 7.7.2 emerge from the origin tangent to the eigenline $y = 0$ as t increases from $-\infty$, because for large negative values of t the dominant terms in formula (3) are $c_1[1 \quad 0]^T e^t$ (if $c_2 = 0$) and $c_2 t[1 \quad 0]^T e^t$ (if $c_2 \neq 0$). Writing solution formula (3) as

$$\begin{bmatrix} x \\ y \end{bmatrix} = \left(c_1 \begin{bmatrix} 1 \\ 0 \end{bmatrix} + c_2 \begin{bmatrix} 0 \\ -1/2 \end{bmatrix} + c_2 t \begin{bmatrix} 1 \\ 0 \end{bmatrix} \right) e^t$$

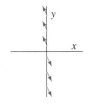

☞ Remember: orbits are traced out in the direction indicated by the direction field.

we see that the state vector $[x \quad y]^T$ looks a lot like $c_2 t[1 \quad 0]^T e^t$ as $t \to +\infty$. So as orbits emerge from the origin they eventually reverse direction and head the other way. But without a computer how could we tell if the state vector rotates clockwise or counterclockwise? Just find the direction field vectors on the y-axis!

EXAMPLE 7.7.3

Star Node

The matrix $A = I$ of the system $x' = x$, $y' = y$ has the double eigenvalue $\lambda = 1$, and all nonzero vectors are eigenvectors. Solutions have the form ve^t for any vector v (Figure 7.7.3).

The term "star node" is aptly chosen: the node is "proper" because every direction is a direction of tangency at the origin.

EXAMPLE 7.7.4

Saddle Point

The eigenvalues of the matrix of the system $x' = x - y$, $y' = -y$ are $\lambda_1 = 1$ and $\lambda_2 = -1$ with corresponding eigenvectors $[1\ 0]^T$ and $[1\ 2]^T$. The general solution is

$$\begin{bmatrix} x \\ y \end{bmatrix} = c_1 \begin{bmatrix} 1 \\ 0 \end{bmatrix} e^t + c_2 \begin{bmatrix} 1 \\ 2 \end{bmatrix} e^{-t}$$

The eigenlines are $y = 0$ and $y = 2x$. The orbits of Figure 7.7.4 resemble branches of hyperbolas; each branch is asymptotic to one eigenline as $t \to -\infty$ and to the other as $t \to +\infty$.

    Curves in Figure 7.7.4 may be visualized as contour curves on a topographic map with two peaks in opposite corners and a mountain pass between them.

EXAMPLE 7.7.5

Center

The eigenvalues of the matrix of the system $x' = 3x - 2y$, $y' = 5x - 3y$ are $\lambda_1 = i$, and $\lambda_2 = -i$. The eigenspace corresponding to λ_1 is spanned by the eigenvector $v = [2\ \ 3 - i]^T$, while the eigenspace corresponding to $\lambda_2 = \bar{\lambda}_1$ is spanned by the eigenvector $\bar{v}$. Since the eigenvectors are complex, the eigenlines are not visible in the real plane of Figure 7.7.5. The general, real-valued solution is given by linear combinations of

$$\text{Re}\left[\begin{bmatrix} 2 \\ 3 - i \end{bmatrix} e^{it}\right] = \text{Re}\left[\begin{bmatrix} 2\cos t + 2i\sin t \\ 3\cos t + \sin t + i(3\sin t - \cos t) \end{bmatrix}\right] = \begin{bmatrix} 2\cos t \\ 3\cos t + \sin t \end{bmatrix}$$

and

$$\text{Im}\left[\begin{bmatrix} 2 \\ 3 - i \end{bmatrix} e^{it}\right] = \begin{bmatrix} 2\sin t \\ 3\sin t - \cos t \end{bmatrix}$$

The orbits are tilted ellipses (Figure 7.7.5), which are periodic with period 2π and oriented counterclockwise by increasing time.

    A center (or vortex) is a point that is surrounded by whirling masses (e.g., the eye of a hurricane), so the name is suitable in this context. The easiest way to determine the direction of rotation is to look at the direction field on the x- or y-axis. In Example 7.7.5 the field vector at a point $(x, 0)$ is $3x\mathbf{i} + 5x\mathbf{j}$, which points into the first quadrant from the point $x\mathbf{i}$ for x positive and into the third quadrant for x negative, so orbital motion is counterclockwise.

EXAMPLE 7.7.6

Spiral Point

The matrix of the system $x' = x - 4y$, $y' = 5x - 3y$ has eigenvalues $\lambda_1 = -1 + 4i$, and $\lambda_2 = \bar{\lambda}_1$. The eigenspaces corresponding to λ_1 and λ_2 are spanned, respectively, by the eigenvector $v = [2\ \ 1 - 2i]^T$ and by $\bar{v}$. These vectors are complex and can't be seen in the plane of Figure 7.7.6. The general, real-valued solution is given by linear combinations of

$$\text{Re}\left[\begin{bmatrix} 2 \\ 1 - 2i \end{bmatrix} e^{(-1+4i)t}\right] = e^{-t}\begin{bmatrix} 2\cos 4t \\ \cos 4t + 2\sin 4t \end{bmatrix}$$

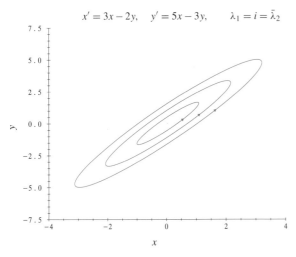

$$x' = 3x - 2y, \quad y' = 5x - 3y, \quad \lambda_1 = i = \bar{\lambda}_2$$

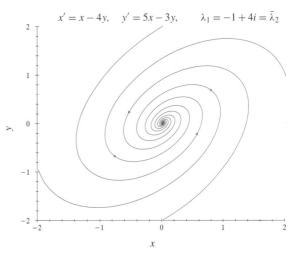

$$x' = x - 4y, \quad y' = 5x - 3y, \quad \lambda_1 = -1 + 4i = \bar{\lambda}_2$$

FIGURE 7.7.5 Center or vortex (no eigenlines exist) (Example 7.7.5).

FIGURE 7.7.6 Spiral point (no eigenlines exist) (Example 7.7.6).

and

$$\text{Im}\left[\begin{bmatrix} 2 \\ 1 - 2i \end{bmatrix} e^{(-1+4i)t}\right] = e^{-t}\begin{bmatrix} 2\sin 4t \\ \sin 4t - 2\cos 4t \end{bmatrix}$$

The orbits spiral toward the origin (because the real part of λ_1 is negative) and counterclockwise (as we see by inspecting the direction of the field vectors $x\mathbf{i} + 5x\mathbf{j}$ on the x-axis). Figure 7.7.6 shows this motion.

Orbits of any planar linear system with eigenvalues $\lambda = \alpha \pm i\beta$, $\alpha \neq 0$, $\beta \neq 0$, behave in the same way as the orbits in Figure 7.7.6, but the spiraling may be clockwise or counterclockwise and inward or outward.

Sensitivity

If entries in the system matrix of a linear system are changed, the orbits sometimes change in drastic ways. The next example shows this.

EXAMPLE 7.7.7

How to Transform a Saddle into an Improper Node

The planar autonomous system

$$\begin{aligned} x' &= \alpha x + y \\ y' &= -x - y \end{aligned} \tag{4}$$

has an adjustable real parameter α in the system matrix. The characteristic polynomial of the system matrix is $p(\lambda) = \lambda^2 + (1 - \alpha)\lambda + 1 - \alpha$, and the eigenvalues are

$$\lambda_1, \lambda_2 = -\frac{1}{2}(1 - \alpha) \pm \frac{1}{2}\sqrt{(1 - \alpha)^2 - 4 + 4\alpha} \tag{5}$$

By analyzing the eigenvalues as functions of α (Problem 1), we see that

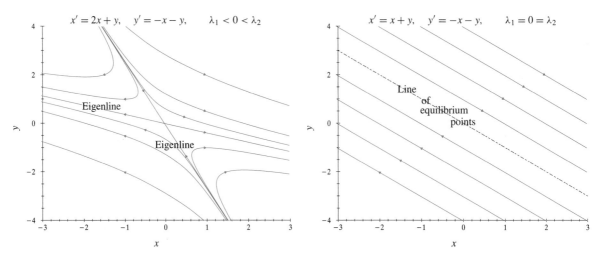

FIGURE 7.7.7 Saddle point (two eigenlines): $\alpha = 2$ (Example 7.7.7).

FIGURE 7.7.8 Eigenline is a line of equilibrium points: $\alpha = 1$ (Example 7.7.7).

- λ_1, λ_2 have opposite signs if $\alpha > 1$. So we have a saddle with two oppositely directed eigenlines (Figure 7.7.7).
- $\lambda_1 = \lambda_2 = 0$ if $\alpha = 1$. The line $x + y = 0$ of equilibrium points is bracketed by oppositely directed straight-line orbits parallel to $x + y = 0$ (Figure 7.7.8).
- λ_1, λ_2 are complex conjugates with negative real parts if $-3 < \alpha < 1$. Now we have spirals (Figures 7.7.9 and 7.7.10).
- $\lambda_1 = \lambda_2 = -2$ if $\alpha = -3$. Here's a deficient node with eigenline $x - y = 0$ (Figure 7.7.11).
- λ_1, λ_2 are distinct and negative if $\alpha < -3$. We finally see an improper node with two eigenlines (Figure 7.7.12).

Figures 7.7.7–7.7.12 show how the orbital portrait changes from saddle, to parallel lines (one of which is a line of equilibrium points), to spiral, to deficient node, and finally to improper node as α changes. Note that as α transits $+1$ and -3 drastic changes occur in the structure of the orbits.

Now that we have shown the two-dimensional orbital portraits, what next? We could draw three-dimensional orbital portraits for a system of three first-order autonomous linear ODEs, but we leave that to Problem 10, and venture instead in a different direction.

Second-Order Systems: Coupled Springs

We have used the eigenvalues and eigenvectors of the matrix A to construct solutions of the first-order linear system $z' = Az$. Let's use this "eigen-element" approach to find the solutions of the *second-order linear system*

$$z'' = Az \tag{6}$$

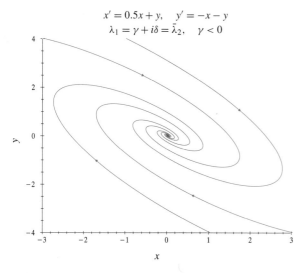

$$x' = 0.5x + y, \quad y' = -x - y$$
$$\lambda_1 = \gamma + i\delta = \bar{\lambda}_2, \quad \gamma < 0$$

FIGURE 7.7.9 Spiral point (no eigenlines): $\alpha = 0.5$ (Example 7.7.7).

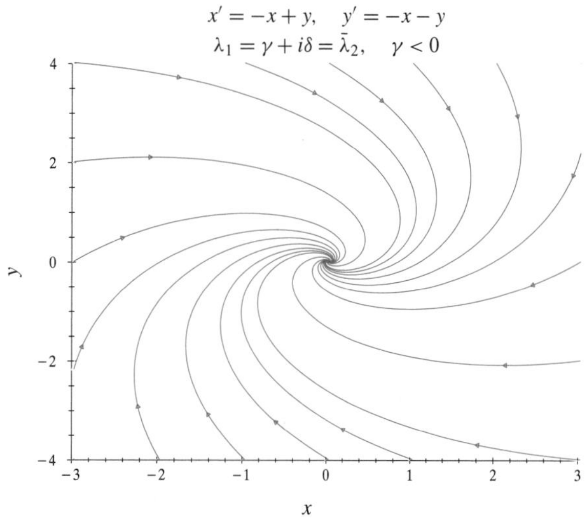

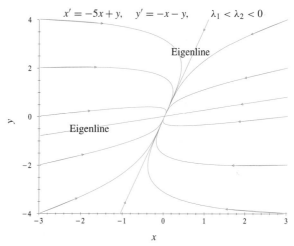

$$x' = -x + y, \quad y' = -x - y$$
$$\lambda_1 = \gamma + i\delta = \bar{\lambda}_2, \quad \gamma < 0$$

FIGURE 7.7.10 Spiral point (no eigenlines): $\alpha = -1$ (Example 7.7.7).

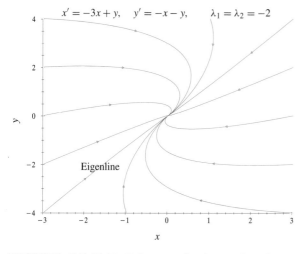

$$x' = -3x + y, \quad y' = -x - y, \quad \lambda_1 = \lambda_2 = -2$$

Eigenline

FIGURE 7.7.11 Deficient node (one eigenline): $\alpha = -3$ (Example 7.7.7).

$$x' = -5x + y, \quad y' = -x - y, \quad \lambda_1 < \lambda_2 < 0$$

Eigenline

Eigenline

FIGURE 7.7.12 Improper node (two eigenlines): $\alpha = -5$ (Example 7.7.7).

where z is a column n-vector and A is an $n \times n$ matrix of real constants.

THEOREM 7.7.1

☞ If any $\lambda_k < 0$, then r_k and the c_j are all complex numbers.

General Solution of $z'' = Az$. Suppose that the constant $n \times n$ matrix A with real entries has the n distinct and nonzero eigenvalues $\lambda_1, \ldots, \lambda_n$ and corresponding eigenvectors $v^1, \ldots, v^n$. Then the general solution of $z'' = Az$ is

$$z = v^1(c_1 e^{r_1 t} + c_2 e^{-r_1 t}) + \cdots + v^n(c_{2n-1} e^{r_n t} + c_{2n} e^{-r_n t}) \tag{7}$$

where $c_1, c_2, \ldots, c_{2n}$ are arbitrary constants and $r_1 = \lambda_1^{1/2}, \ldots, r_n = \lambda_n^{1/2}$.

Here's why you would expect Theorem 7.7.1 to be true. The function vector $z = ve^{rt}$ is a solution of $z'' = Az$ if $r^2 = \lambda$ is a nonzero eigenvalue of A and v is a corresponding eigenvector because

$$z'' = r^2 ve^{rt} = \lambda ve^{rt}$$

$$Az = Ave^{rt} = \lambda ve^{rt}$$

☞ We deal with $z'' = Az$ directly, rather than with the equivalent first-order system $z' = y$, $y' = Az$.

The same argument shows that $z = ve^{-rt}$ is also a solution, so we have a pair of independent solutions $\{ve^{rt}, ve^{-rt}\}$ for each eigenvalue λ and corresponding eigenvector v. It should come as no surprise that the basic solution set $\{v^1 e^{r_1 t}, v^1 e^{-r_1 t}, \ldots, v^n e^{-r_n t}\}$ has $2n$ elements because $z'' = Az$ has the $2n$ state variables $z_1, z_1', \ldots, z_n, z_n'$.

Now let's apply all this to the second-order system introduced in Section 5.1 to model the motion of a coupled spring-mass system. In the process, we are led to some interesting two-dimensional orbital portraits.

EXAMPLE 7.7.8

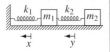

☞ Look back at Example 5.1.2 for the modeling that leads to system (8).

Coupled Springs and Normal Modes

The vibrations of the coupled spring-mass system sketched in the margin are modeled by the second-order system

$$\begin{aligned} x'' &= -(a+b)x + by \\ y'' &= cx - cy \end{aligned} \tag{8}$$

where the positive constants a, b, and c are given by

$$a = k_1/m_1, \qquad b = k_2/m_1, \qquad c = k_2/m_2 \tag{9}$$

We can write (8) as $z'' = Az$, where $z = [x \ y]^T$, and the matrix A is $\begin{bmatrix} -a-b & b \\ c & -c \end{bmatrix}$.

Its eigenvalues are the roots λ_1, λ_2 of the characteristic polynomial

$$p(\lambda) = \lambda^2 + (a+b+c)\lambda + ac$$

$$\lambda_1, \lambda_2 = -\frac{1}{2}(a+b+c) \pm \frac{1}{2}\left((a+b+c)^2 - 4ac\right)^{1/2}$$

☞ The key to showing this inequality is that $(a-c)^2 \geq 0$ for all a and c.

Since a, b, and c are positive and $(a+b+c)^2 > 4ac$, both eigenvalues λ_1 and λ_2 are negative, and their square roots are the pure imaginary numbers $\pm i\omega_1$, $\pm i\omega_2$ where $\omega_1 = \sqrt{-\lambda_1}$ and $\omega_2 = \sqrt{-\lambda_2}$. In the notation of Theorem 7.7.1 with v^1 and v^2 the eigenvectors of A corresponding to λ_1 and λ_2, the general complex-valued solution of the second-order system (8) is

$$z = \begin{bmatrix} x \\ y \end{bmatrix} = v^1 \left(c_1 e^{i\omega_1 t} + c_2 e^{-i\omega_1 t}\right) + v^2 \left(c_3 e^{i\omega_2 t} + c_4 e^{-i\omega_2 t}\right) \tag{10}$$

where c_1, c_2, c_3, and c_4 are arbitrary complex constants. From (10) we can extract the general real-valued solution:

$$z = v^1(C_1 \cos \omega_1 t + C_2 \sin \omega_1 t) + v^2(C_3 \cos \omega_2 t + C_4 \sin \omega_2 t) \tag{11}$$

☞ Example 3.5.1
shows how to do this.

where C_1, C_2, C_3, C_4 are arbitrary real constants. Furthermore, the general solution (11) can be written in the amplitude phase-shift form

$$z = K_1 v^1 \cos(\omega_1 t + \theta_1) + K_2 v^2 \cos(\omega_2 t + \theta_2) \tag{12}$$

where K_1, K_2, θ_1, θ_2 are arbitrary real constants.

The vectors v^1 and v^2 in the xy-plane are *normal mode vectors* and ω_1 and ω_2 are the respective *normal mode circular frequencies*. From formula (12), the general solution is an arbitrary linear combination of *normal mode oscillations*, $v^1 \cos(\omega_1 t + \theta_1)$ and $v^2 \cos(\omega_2 t + \theta_2)$.

Now let's put numbers into the picture and plot (in the xy-plane) specific normal mode oscillations and some of their linear combinations.

EXAMPLE 7.7.9

Orbital Portraits of the Motions of Coupled Springs

In system (8) let's take these values for the masses and the spring constants:

$$m_1 = 4 \text{ slugs}, \quad m_2 = 1 \text{ slug}, \quad k_1 = 40 \text{ slugs/sec}^2, \quad k_2 = (40/3) \text{ slugs/sec}^2$$

Then the ratios in (9) become

$$a = k_1/m_1 = 10, \quad b = k_2/m_1 = 10/3, \quad c = k_2/m_2 = 40/3$$

The coefficient matrix in (8) and its characteristic polynomial are

$$A = \begin{bmatrix} -40/3 & 10/3 \\ 40/3 & -40/3 \end{bmatrix}, \qquad p(\lambda) = \lambda^2 + 80\lambda/3 + 400/3$$

The eigenvalues and normal mode circular frequencies are

$$\lambda_1 = -20/3, \quad \lambda_2 = -20, \quad \omega_1 = \sqrt{20/3}, \quad \omega_2 = \sqrt{20}$$

The normal modes, two normal mode oscillations, and the general solution z are

$$v^1 = \begin{bmatrix} 1 \\ 2 \end{bmatrix}, \quad v^2 = \begin{bmatrix} 1 \\ -2 \end{bmatrix}, \quad \begin{bmatrix} 1 \\ 2 \end{bmatrix} \cos(\sqrt{20/3}t), \quad \begin{bmatrix} 1 \\ -2 \end{bmatrix} \cos(\sqrt{20}t)$$

$$z = \begin{bmatrix} x \\ y \end{bmatrix} = K_1 \begin{bmatrix} 1 \\ 2 \end{bmatrix} \cos(\sqrt{20/3}t + \theta_1) + K_2 \begin{bmatrix} 1 \\ -2 \end{bmatrix} \cos(\sqrt{20}t + \theta_2)$$

☞ Look back at
Section 5.1 for a
discussion of tx- and
ty-graphs.

Figure 7.7.13 shows two normal mode oscillations: **(1)** $K_1 = 1$, $\theta_1 = 0$, $K_2 = 0$, and **(2)** $K_1 = 0$, $K_2 = 1$, $\theta_2 = 0$. The figure also shows a linear combination of normal mode oscillations: **(3)** $K_1 = K_2 = 1/2$, $\theta_1 = \theta_2 = 0$. The latter is called a Lissajous curve; it is *not* periodic because the ratio of the two normal mode frequencies is $\omega_1/\omega_2 = 1/\sqrt{3}$, which is irrational. The Lissajous curve never returns to its starting point as it wanders through a rectangle whose sides are parallel to the normal modes.

We can interpret the curves and lines in Figure 7.7.13 as showing us how the positions x and y of the two weights on the springs play off against each other, given different sets of initial data. The oscillations along the normal modes reflect the fact that if the initial positions of the weights are in the ratio $1 : 2$ or $-1 : 2$ and the initial

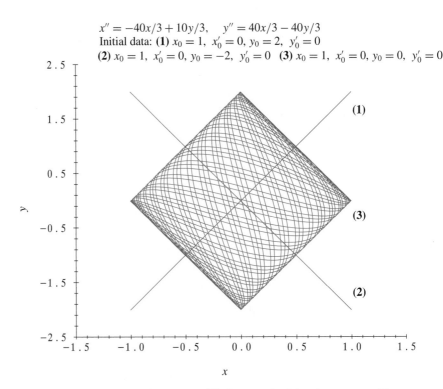

$$x'' = -40x/3 + 10y/3, \quad y'' = 40x/3 - 40y/3$$
Initial data: (1) $x_0 = 1$, $x_0' = 0$, $y_0 = 2$, $y_0' = 0$
(2) $x_0 = 1$, $x_0' = 0$, $y_0 = -2$, $y_0' = 0$ (3) $x_0 = 1$, $x_0' = 0$, $y_0 = 0$, $y_0' = 0$

FIGURE 7.7.13 Normal mode oscillations and a Lissajous curve (Example 7.7.9).

velocities are 0, then the weights oscillate periodically and always maintain the initial ratio. But if the initial ratio of the positions is anything other than $1 : 2$ or $-1 : 2$, a wandering Lissajous curve must result. The curves in Figure 7.7.13 intersect each other, but this does not violate the property that distinct orbits of an autonomous system can't meet. The reason is that the state space of the second-order system (8) is four-dimensional because there are four state variables (x, x', y, y'). What we see in Figure 7.7.13 is the projection of three orbits from this four-dimensional state space into the two-dimensional space of the xy-plane.

☞ Example 5.1.4 treats the equivalent system of four first-order ODEs.

PROBLEMS

www **1.** The system for the following problems is $x' = \alpha x + y$, $y' = -x - y$ (see Example 7.7.7).

(a) Explain why the equilibrium point at the origin is a saddle if $\alpha > 1$.

(b) Explain why the origin is a spiral point if $-3 < \alpha < 1$.

(c) Why is the origin an improper node if $\alpha < -3$? Explain why the system matrix has a pair of independent real eigenvectors in this case.

 (d) Plot orbital portraits in the frame $|x| \le 3$, $|y| \le 4$ for the three cases $\alpha = -4, 0, 4$.

2. Find the characteristic polynomial of the system matrix, the eigenvalues, a set of two independent eigenvectors (or generalized eigenvectors, if necessary), and the general real-valued solution of each system. Then classify each as a saddle, center, spiral, or one of the nodal types.

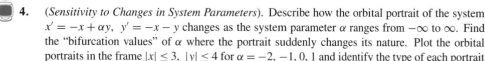

    (a) $x' = x + y$, $y' = 4x - 2y$ **(b)** $x' = 6x - 8y$, $y' = 2x - 2y$

    (c) $x' = 7x + 6y$, $y' = 2x + 6y$ **(d)** $x' = 4y$, $y' = -x$

    (e) $x' = -x - 4y$, $y' = x - y$ **(f)** $x' = -2x$, $y' = -2y$

 3. **(a)–(f)** Plot orbital portraits in the frame $|x| \le 3$, $|y| \le 4$ for each system in Problem 2.

4. (*Sensitivity to Changes in System Parameters*). Describe how the orbital portrait of the system $x' = -x + \alpha y$, $y' = -x - y$ changes as the system parameter α ranges from $-\infty$ to ∞. Find the "bifurcation values" of α where the portrait suddenly changes its nature. Plot the orbital portraits in the frame $|x| \le 3$, $|y| \le 4$ for $\alpha = -2, -1, 0, 1$ and identify the type of each portrait (e.g., a spiral point if $\alpha = 1$). [*Hint:* Read Example 7.7.7 for the general approach.]

5. (*Planar Affine Systems*). The system $x' = ax + by + r$, $y' = cx + dy + s$, where a, b, r, c, d, s are real constants, is said to be an *affine system*. Suppose that $ad - bc \ne 0$.

    (a) Show that the affine system has a unique equilibrium point.

    (b) For an equilibrium point (x_0, y_0), show that the change of variables from x and y to u and v given by $x = u + x_0$, $y = v + y_0$ yields the undriven system $u' = au + bv$, $v' = cu + dv$.

    (c) Show that if $u = u(t)$, $v = v(t)$ is the general solution of the u', v' system in part **(b)**, then the general solution of the affine system is $x(t) = u(t) + x_0$, $y(t) = v(t) + y_0$.

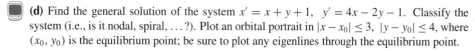   **(d)** Find the general solution of the system $x' = x + y + 1$, $y' = 4x - 2y - 1$. Classify the system (i.e., is it nodal, spiral, ...?). Plot an orbital portrait in $|x - x_0| \le 3$, $|y - y_0| \le 4$, where (x_0, y_0) is the equilibrium point; be sure to plot any eigenlines through the equilibrium point.

6. (*Second-Order ODE*). The scalar ODE $y'' + 2ay' + by = 0$, where a and b are real constants, is equivalent to the system $x_1' = x_2$, $x_2' = -bx_1 - 2ax_2$.

    (a) Explain why the system has a saddle point if $a = 1$, $b < 0$, while if $a = 1$, $0 < b < 1$, the system has an improper node.

    (b) Explain why if $a = 1$, $b = 1$, the system has a deficient node, while if $a = 1$, $b > 1$, the system has a spiral point.

    (c) Find all values for a and b for which the system has a center.

    (d) Show that no values of a and b lead to a system that has a star node.

7. (*An Inverse Problem*). An undriven 2×2 linear system with constant coefficients has the form $x' = Ax$ for a constant matrix A, where $x(t)$ is a column vector with two components $x_1(t)$ and $x_2(t)$. It is known that $x_1 = 2\sin(2t - \pi)$, $x_2 = \cos(2t - \pi)$ is one solution of the system.

    (a) Find the general, real-valued solution of the system.

     (b) Plot some representative orbits of the system near the origin.

 8. (*Coupled Springs*). Suppose that $m_1 = 1$, $k_1 = k_2 = 1$ in Example 7.7.8, but the second mass m_2 may take on various values. Calculate the normal modes and frequencies for $m_2 = 1/2, 1, 3/2$. In each case plot a pair of normal mode oscillations and a Lissajous curve to obtain a figure resembling Figure 7.7.13.

  9. (*Zero Eigenvalues*). Except for system (4) with $\alpha = 1$, we have avoided considering the case of a planar autonomous system where the real system matrix A has at least one zero eigenvalue. What happens if A has a zero eigenvalue? Explain why either there is one straight line consisting solely of equilibrium points or else all points are equilibrium points if at least one eigenvalue is zero. Plot orbital portraits of the various distinct cases where at least one of the eigenvalues is zero, and explain why your "gallery" of portraits represents all the possible cases.

10. (*Orbital Portraits in Three Dimensions*). Consider the system $x' = Ax$, where $x = [x_1 \; x_2 \; x_3]^T$ and A is a 3×3 matrix of real constants with nonzero eigenvalues. Construct a gallery of representative orbital portraits in $x_1 x_2 x_3$ state space that covers all of the cases. For example, if λ_1, λ_2, and λ_3 denote the eigenvalues of A, you will need to consider the case where the three eigenvalues are real, distinct, and have a common sign, and plot orbits in three dimensions of a system such as $x_1' = -x_1$, $x_2' = -2x_2$, $x_3' = -3x_3$. Another one of the several cases you should consider is $\lambda_1 = \alpha + i\beta$, $\lambda_2 = \alpha - i\beta$, α, β, and λ_3 real and nonzero; the representative system in this case might be $x_1' = -x_1 + 5x_2$, $x_2' = -5x_1 - x_2$, $x_3' = -x_3$. Since any system with eigenvalues λ_1, λ_2, and λ_3 behaves as $t \to \infty$ exactly the same as the system with eigenvalues $-\lambda_1$, $-\lambda_2$, and $-\lambda_3$ as $t \to -\infty$, their orbital portraits are identical, so don't treat them as distinct cases. For example, consider the case with three distinct positive real eigenvalues and the case with three distinct negative real eigenvalues as the same. Your gallery should have more than 10 distinct orbital portraits. [*Hint*: To construct your example system, start with the six planar cases of Examples 7.7.1–7.7.6 and add a third ODE such as $x_3' = \alpha x_3$.]

7.8 Driven Systems and the Matrix Exponential

In this section we develop a strategy for finding a formula for all solutions of the driven system

$$x' = Ax + F(t) \tag{1}$$

where A is an $n \times n$ constant matrix with real entries and the vector function $F(t)$ is continuous on some t-interval containing the origin. We have seen in Section 7.5 that all solutions of the driven system (1) are given by $x^d(t) + w(t)$, where $x^d(t)$ is a particular solution of the driven system (any one will do) and $w(t)$ is any solution of the undriven system $x' = Ax$. In Sections 7.5 and 7.6 we learned how to derive a formula for all solutions of $x' = Ax$, so what remains is to find a formula for $x^d(t)$.

EXAMPLE 7.8.1

General Solution for a Driven System
Let's look at the driven system

$$x' = Ax + b \tag{2}$$

where A is an $n \times n$ nonsingular constant matrix with real entries and b is a constant vector. For example, the lead system (4) in Section 7.1 has this form. We can guess a particular solution $x^d(t)$ to system (2). The constant function $x^d(t) = -A^{-1}b$ does the job (try it and see), so

$$x_{gen} = -A^{-1}b + w, \qquad w \text{ any solution of } x' = Ax$$

is the general solution of system (2).

Let's go back and look at the undriven system $x' = Ax$ again and define some terms that will be useful in solving the driven system $x' = Ax + F$.

Solution Matrices for $x' = Ax$

The solutions of the undriven system

$$x' = Ax \qquad (3)$$

are column vector functions. If A is an $n \times n$ constant matrix, a *solution matrix* is an $n \times n$ matrix $X(t)$ whose columns $x^j(t)$, $j = 1, \ldots, n$, are solutions of system (3). A solution matrix $X(t)$ satisfies the linear *matrix differential equation*

$$X' = AX \qquad (4)$$

since $(x^j)' = Ax^j$ for each column x^j of X. There is a corresponding matrix IVP

$$X' = AX, \qquad X(0) = B \qquad (5)$$

where B is an $n \times n$ matrix of constants. From Theorem 7.6.1 we see that IVP (5) has a unique solution matrix $X = X(t)$.

EXAMPLE 7.8.2

Constructing a Solution Matrix

The constant system matrix for the system

$$x' = 5x + 3y, \qquad y' = -6x - 4y \qquad (6)$$

has eigenvalues -1 and 2 and respective eigenvectors $[1 \;\; -2]^T$ and $[1 \;\; -1]^T$. The pair of solutions $[1 \;\; -2]^T e^{-t}$ and $[1 \;\; -1]^T e^{2t}$ form a basic solution set for the system. The following are all solution matrices:

$$\begin{bmatrix} e^{-t} & e^{2t} \\ -2e^{-t} & -e^{2t} \end{bmatrix}, \quad \begin{bmatrix} e^{-t} & 2e^{-t} \\ -2e^{-t} & -4e^{-t} \end{bmatrix}, \quad \begin{bmatrix} e^{2t} & e^{-t} + e^{2t} \\ -e^{2t} & -2e^{-t} - e^{2t} \end{bmatrix} \qquad (7)$$

Theorem 7.6.3 says that there is a basic solution set of n solutions $\{x^1, x^2, \ldots, x^n\}$ of system (3) such that every solution of the system is just a linear combination of these n solutions. System (3) does not have a unique basic solution set, but we saw in Sections 7.5 and 7.6 that the Method of Eigenvectors will produce a basic solution set.

If the columns of a solution matrix $X(t)$ form a basic solution set for $x' = Ax$, then $X(t)$ is a *basic solution matrix* (also called a *fundamental matrix*). If $\det B \neq 0$, then the solution matrix for IVP (5) is a basic solution matrix for system (3). Each basic solution matrix $X(t)$ is a nonsingular matrix for all t in the interval I since $\det X(t) \neq 0$; the columns of $X(t)$ form a basic solution set for $x' = Ax$. Notice that the first and third matrices in (7) are basic solution matrices for system (6) because their determinants are nonzero at $t = 0$.

If C is any $n \times n$ nonsingular matrix of constants and $X(t)$ is a basic solution matrix, then $X(t)C$ is also a basic solution matrix. Reason: columns of $X(t)C$ are linear combinations of the columns of $X(t)$, and $X(t)C$ is nonsingular since $\det(X(t)C) = (\det X(t))(\det C) \neq 0$. One type of basic solution matrix is of central importance.

☞ Since the columns of $X(t)$ are a basic solution set of $x' = Ax$, they form and independent set, so $\det X(t) \neq 0$.

☞ If the notation e^{tA} seems weird to you, use $\exp(tA)$. Whatever the notation, e^{tA} is just another matrix.

❖ **Matrix Exponential** e^{tA}. The *matrix exponential* for the system $x' = Ax$ is the unique basic solution matrix $X(t)$ for which $X' = AX$ and $X(0) = I_n$, the $n \times n$ identity matrix. This solution matrix is denoted by e^{tA} and is called the *matrix exponential* for A.

Suppose that u^j is the column vector with 1 in the j-th row, 0's elsewhere. The columns y^j of e^{tA} are the unique solutions of the n distinct IVPs $x' = Ax$, $x(0) = u^j$, $j = 1, 2, \ldots, n$. We can also derive e^{tA} from any basic solution matrix $X(t)$ as follows: define the matrix $Y(t)$ by

$$Y(t) = X(t)X^{-1}(0) \tag{8}$$

Since the columns of $Y(t)$ solve the system $x' = Ax$, it is a solution matrix. In fact it must be e^{tA} because $Y(0) = X(0)X^{-1}(0) = I_n$.

EXAMPLE 7.8.3

A Matrix Exponential
The matrix exponential for system (6) in Example 7.8.2 can be constructed by using formula (8) because the first matrix in (7) is a basic solution matrix. We have

$$e^{tA} = \begin{bmatrix} e^{-t} & e^{2t} \\ -2e^{-t} & -e^{2t} \end{bmatrix} \begin{bmatrix} 1 & 1 \\ -2 & -1 \end{bmatrix}^{-1}$$

☞ Exponentiating each entry in tA does *not* produce e^{tA}.

$$= \begin{bmatrix} e^{-t} & e^{2t} \\ -2e^{-t} & -e^{2t} \end{bmatrix} \begin{bmatrix} -1 & -1 \\ 2 & 1 \end{bmatrix} \tag{9}$$

$$= \begin{bmatrix} -e^{-t} + 2e^{2t} & -e^{-t} + e^{2t} \\ 2e^{-t} - 2e^{2t} & 2e^{-t} - e^{2t} \end{bmatrix}$$

Notice that $e^{0A} = I_2$.

Because of the way e^{tA} was defined we see that $x(t) = e^{tA}c$ is a solution of the system $x' = Ax$ for any value of the constant vector c. Notice that $x(0) = c$, so we have the following result.

THEOREM 7.8.1

> Solution of the IVP $x' = Ax$, $x(0) = x^0$. The solution of the IVP $x' = Ax$, $x(0) = x^0$, for any constant vector x^0 is given by
> $$x = e^{tA}x^0 \tag{10}$$

Theorem 7.8.1 tells us how the state vector $x(t)$ of the dynamical system modeled by the differential system $x' = Ax$ evolves in time. To find the state vector at time t, we just multiply the initial state vector by the exponential matrix e^{tA}.

EXAMPLE 7.8.4

Solving $x' = Ax$
We saw in Example 7.8.3 that formula (9) gives the matrix exponential e^{tA} of the planar system

$$x' = 5x + 3y, \qquad y' = -6x - 4y \tag{11}$$

We see from formula (10) that the solution for which $x(0) = a$, $y(0) = b$, is

$$\begin{bmatrix} x(t) \\ y(t) \end{bmatrix} = e^{tA} \begin{bmatrix} a \\ b \end{bmatrix} = \begin{bmatrix} -e^{-t} + 2e^{2t} & -e^{-t} + e^{2t} \\ 2e^{-t} - 2e^{2t} & 2e^{-t} - e^{2t} \end{bmatrix} \begin{bmatrix} a \\ b \end{bmatrix} \tag{12}$$

With this formula in hand, we can solve any IVP for system (11).

The matrix exponential has a number of other important properties.

THEOREM 7.8.2

Properties of e^{tA}. Suppose that A is a matrix of real constants. Then

Derivative: $(e^{tA})' = Ae^{tA} = e^{tA}A$, all t.

Transition: $e^{(t-t_0)A}$ solves the matrix IVP $X' = AX$, $x(t_0) = I$.

Product: $e^{tA}e^{sA} = e^{(t+s)A}$, all t and s.

Inverse: $e^{tA}e^{-tA} = I$, so $(e^{tA})^{-1} = e^{-tA}$, all t.

Series: $e^{tA} = I + tA + t^2A^2/2! + \cdots + t^kA^k/k! + \cdots$, all t.

Here is how we can justify these properties:

Derivative: The first equality is immediate since e^{tA} is a solution of the matrix system $X' = AX$. To show the second equality, note that $Y = e^{tA}A$ and $Z = Ae^{tA}$ are both solutions of the matrix IVP $X' = AX$, $X(0) = I$. So, we have $Y' = (Ae^{tA})A = A(e^{tA}A) = AY$, $Z' = A(Ae^{tA}) = AZ$, and $Y(0) = Z(0) = I$. By the uniqueness of the solution of an IVP, $Y = Z$.

Transition: Since the system $x' = Ax$ is autonomous, if $x(t)$ is a solution, then so is $x(t - t_0)$. So $e^{(t-t_0)A}$ solves the matrix system $X' = AX$; $e^{(t-t_0)A}$ evaluates to I at $t = t_0$.

Product: Since the system $x' = Ax$ is autonomous, we see that for any constant s, $X_1 = e^{t+s}A$ is a solution matrix for the system $X' = AX$. Next, note that for any constant s, the matrix e^{sA} is nonsingular; so $X_2 = e^{tA}e^{sA}$ is also a solution matrix for $X' = AX$. Now $X_1(0) = e^{sA} = X_2(0)$, so from Theorem 7.6.1, $X_1 = X_2$ for all t and s.

Inverse: Using the product formula, we see that $e^{tA}e^{-tA} = e^{0A} = I$.

Series: The power series $S(t)$ of matrices on the right side of the equality converges absolutely for all t and for all matrices A (i.e., each element of the matrix sum converges absolutely). Under these circumstances, we note that $S(0) = I$ and that $S'(t)$ may be obtained by differentiating the given power series term by term:

$$S'(t) = A + tA^2 + \cdots + \frac{t^{(k-1)}}{(k-1)!}A^k + \cdots$$

$$= A\left[I + tA + \cdots + \frac{t^{k-1}}{(k-1)!}A^{k-1} + \cdots\right] \tag{13}$$

$$= AS(t)$$

So the solution matrices e^{tA} and $S(t)$ both solve the matrix IVP $X' = AX$, $X(0) = I$. By Theorem 7.6.1, $e^{tA} = S(t)$.

Here are some examples on how to find e^{tA}.

EXAMPLE 7.8.5

Diagonal Matrices

Let's find e^{tD} if $D = \begin{bmatrix} \lambda_1 & 0 \\ 0 & \lambda_2 \end{bmatrix}$. By definition, the first column of e^{tD} is the unique solution of the IVP

$$x' = \lambda_1 x, \qquad x(0) = 1$$
$$y' = \lambda_2 y, \qquad y(0) = 0$$

The ODEs in this system decouple and can be solved separately to find $x = e^{\lambda_1 t}$, $y = 0$. The second column of e^{tD} is the unique solution of the IVP

$$x' = \lambda_1 x, \qquad x(0) = 0$$
$$y' = \lambda_2 y, \qquad y(0) = 1$$

Again the ODEs decouple and are solved separately to find $x = 0$, $y = e^{\lambda_1 t}$. So

$$e^{tD} = \begin{bmatrix} e^{\lambda_1 t} & 0 \\ 0 & e^{\lambda_2 t} \end{bmatrix}$$

so it is easy to find e^{tD} in this case.

In general, if D is an $n \times n$ *diagonal matrix* with $d_{jj} = \lambda_j$, $j = 1, \ldots, n$, and $d_{ij} = 0$, $i \neq j$, then e^{tD} is the diagonal matrix with diagonal entries $e^{t\lambda_1}, \ldots, e^{t\lambda_n}$, and zero for all off-diagonal entries. It is not usually so easy to calculate a matrix exponential.

EXAMPLE 7.8.6

Triangular Matrix

Let's find e^{tA} if $A = \begin{bmatrix} a & b \\ 0 & a \end{bmatrix}$. By definition, the first column of e^{tA} is the unique solution of the IVP

$$x' = ax + by, \qquad x(0) = 1$$
$$y' = ay, \qquad\qquad y(0) = 0$$

Solve the second ODE first to find $y = 0$, substitute this into the first ODE, and solve to find $x = e^{at}$. The second column in e^{tA} is the unique solution of the IVP

$$x' = ax + by, \qquad x(0) = 0$$
$$y' = ay, \qquad\qquad y(0) = 1$$

Again solving the second ODE first we find that $y = e^{at}$, and after substituting into the first ODE we find $x = bte^{at}$. So we have

$$e^{tA} = \begin{bmatrix} e^{at} & bte^{at} \\ 0 & e^{at} \end{bmatrix}$$

The upper right entry reflects the fact that a is an eigenvalue with a deficient eigenspace. Note that e^{tA} is triangular since A is triangular.

EXAMPLE 7.8.7

Nilpotent Matrices

A matrix A is *nilpotent* if some power A^k is the zero matrix. Then e^{tA} can be calculated easily from the expression in the series property of Theorem 7.8.2 because the series stops with the power A^{k-1}. That is, we have $A^k = A^{k+1} = \cdots = 0$, so

$$e^{tA} = I + tA + \frac{t^2}{2!}A^2 + \cdots + \frac{t^{k-1}}{k!}A^{k-1}$$

For example, with the matrix A given below we have

$$A = \begin{bmatrix} 0 & 1 & 0 \\ 0 & 0 & 1 \\ 0 & 0 & 0 \end{bmatrix}, \qquad A^2 = \begin{bmatrix} 0 & 0 & 1 \\ 0 & 0 & 0 \\ 0 & 0 & 0 \end{bmatrix}, \qquad A^3 = \begin{bmatrix} 0 & 0 & 0 \\ 0 & 0 & 0 \\ 0 & 0 & 0 \end{bmatrix}$$

So it is not hard to find e^{tA}:

$$e^{tA} = I + tA + \frac{t^2}{2!}A^2 = \begin{bmatrix} 1 & t & t^2/2! \\ 0 & 1 & t \\ 0 & 0 & 1 \end{bmatrix}$$

Since A is triangular we could also have found e^{tA} as in Example 7.8.6.

Variation of Parameters

According to our plan for solving the driven IVP $x' = Ax + F$, $x(0) = x^0$, the next step is to find a particular solution $x^d(t)$ of the driven system $x' = Ax + F(t)$. The matrix exponential constructed above can be used to carry out the construction.

THEOREM 7.8.3

Variation of Parameters for Driven Systems. Suppose that e^{tA} is the matrix exponential for $x' = Ax$, where A is an $n \times n$ constant matrix. Then for any vector function $F(t)$ whose components are continuous on some t-interval I containing the origin, the vector function

$$x^d(t) = e^{tA} \int_0^t e^{-sA} F(s)\, ds = \int_0^t e^{(t-s)A} F(s)\, ds \qquad (14)$$

lives on the interval I and is the unique solution of the IVP

$$x' = Ax + F(t), \qquad x(0) = 0 \qquad (15)$$

The integral in (14) acts on vector functions component by component.

Note that the second equality in formula (14) is a consequence of the product property of matrix exponentials: $e^{tA}e^{-sA} = e^{(t-s)A}$. To verify that formula (14) defines a solution, we proceed as follows:

$$(x^d)' = \frac{d}{dt}\left\{e^{tA}\int_0^t e^{-sA}F(s)\,ds\right\}$$

$$= \left\{(e^{tA})'\int_0^t e^{-sA}F(s)\,ds\right\} + e^{tA}e^{-tA}F(t)$$

(by the product rule and the Fundamental Theorem of Calculus). So

$$(x^d)' = \left\{Ae^{tA}\int_0^t e^{sA}F(s)\,ds\right\} + F(t) = Ax^d(t) + F(t)$$

where we used several properties of the matrix exponential in Theorem 7.8.2.

The origin of the title "Variation of Parameters" is outlined in Problem 9, Section 7.11. We can now use Theorems 7.8.1 and 7.8.3 to solve the central problem for systems of linear ODEs.

THEOREM 7.8.4

Solution of a Linear System. Suppose that e^{tA} is the matrix exponential for $x' = Ax$, where A is an $n \times n$ constant matrix, and suppose that $F(t)$ is a continuous vector function on a t-interval I containing the origin. Then the general solution of $x' = Ax + F(t)$ is

$$x(t) = e^{tA}c + e^{tA}\int_0^t e^{-sA}F(s)\,ds, \qquad t \text{ in } I \qquad (16)$$

where c is any constant vector. The solution of the initial value problem $x' = Ax + F(t)$, $x(0) = x^0$, t in I, is

$$x(t) = e^{tA}x^0 + e^{tA}\int_0^t e^{-sA}F(s)\,ds \qquad (17)$$

$$\text{Total response} = \text{Response to initial data} + \text{Response to input } F$$

☞ If this looks familiar, turn back to Theorem 1.4.1.

In formulas (16) and (17) the matrix factor e^{tA} can be brought inside the integral and combined with e^{-sA} to obtain $e^{(t-s)A}$. Formulas (16) and (17) require the prior construction of the matrix exponential (not always an easy task) and then integration of the components of the vector $e^{-sA}F(s)$ (also not always easy). Nevertheless, solution formulas like (16) and (17) for whole classes of systems of ODEs are rare, and whenever they exist they provide the foundation for much theoretical and practical work. Note that the term $e^{tA}c$ in formula (16) can be replaced by $X(t)c$, where $X(t)$ is any basic solution matrix for $x' = Ax$.

EXAMPLE 7.8.8

Total Response

Formula (17) represents the total response of a driven linear system with given initial data as the sum of the responses to the initial data and to the driving force. Figure 7.8.1 illustrates this principle for the IVP

$$x' = -x + 10y + 5\cos t, \qquad x(0) = 0$$
$$y' = -10x - y + 5\cos \pi t, \qquad y(0) = 0.5 \qquad (18)$$

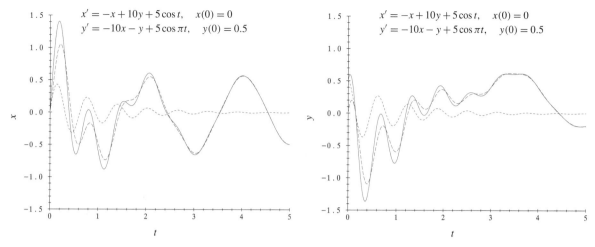

FIGURE 7.8.1 Total x-response (solid) is the sum of x-responses to initial data (short dashes) and to driving force (long dashes) (Example 7.8.8).

FIGURE 7.8.2 Total y-response (solid) is the sum of y-responses to initial data (short dashes) and to driving force (long dashes) (Example 7.8.8).

The short-dashed curve in Figure 7.8.2 is the y-component of the response of the un-driven system to the initial data, the long-dashed curve is the y-component of the response to the driving cosine terms (the initial data here are $x(0) = y(0) = 0$), and the solid curve is the y-component of the sum of the two responses.

EXAMPLE 7.8.9

Solving $x' = Ax + F(t)$

Using formula (17) for the IVP $x' = 5x + 3y + F_1(t)$, $y' = -6x - 4y + F_2(t)$, $x(0) = a$, $y(0) = b$, where the matrix exponential e^{tA} for this system is given by (9) in Example 7.8.3, we have

$$\begin{bmatrix} x(t) \\ y(t) \end{bmatrix} = e^{tA} \begin{bmatrix} a \\ b \end{bmatrix} + \int_0^t e^{(t-s)A} \begin{bmatrix} F_1(s) \\ F_2(s) \end{bmatrix} ds \tag{19}$$

This formula can be used with any initial data and driving force.

Comments

If A is any constant matrix, then the matrix exponential e^{tA} can *always* be calculated from any basic matrix $X(t)$ for $x' = Ax$ by using formula (8): $e^{tA} = X(t)X^{-1}(0)$. The results of this section apply even if some of the components of the input vector $F(t)$ are only piecewise continuous, although solution curves, component curves, and orbits may have "corners" where $F(t)$ is not continuous.

The results of this section can be generalized to the system $x' = A(t)x$, where the system matrix $A(t)$ is not constant. However, we postpone that discussion to the last section in this chapter.

PROBLEMS

www 1. Find e^{tA}, where A is given. Then solve the IVP $x' = Ax$, $x(0) = [1 \ 2]^T$. [*Hint:* Find the eigenvalues of A and a basis for each eigenspace. Use them to construct a basic matrix $X(t)$ as in Example 7.8.3. Then $e^{tA} = X(t)X^{-1}(0)$. In part **(c)**, solve the system directly.]

(a) $\begin{bmatrix} 1 & 3 \\ 1 & -1 \end{bmatrix}$ **(b)** $\begin{bmatrix} 0 & 1 \\ -9 & 0 \end{bmatrix}$ **(c)** $\begin{bmatrix} -1 & 1 \\ 0 & -1 \end{bmatrix}$ **(d)** $\begin{bmatrix} 1 & 1 \\ -1 & 1 \end{bmatrix}$

2. Find e^{tA}. Then solve the IVP $x' = Ax$, $x(0) = [1 \ 2 \ 3]^T$. [*Hint:* See Examples 7.8.5, 7.8.7.]

(a) $\begin{bmatrix} 0 & 1 & 1 \\ 0 & 0 & 1 \\ 0 & 0 & 0 \end{bmatrix}$ **(b)** $\begin{bmatrix} 2 & 0 & 0 \\ 0 & -3 & 0 \\ 0 & 0 & 7 \end{bmatrix}$ **(c)** $\begin{bmatrix} 0 & 0 & 0 \\ 2 & 0 & 0 \\ 3 & 4 & 0 \end{bmatrix}$

3. (*Block Matrices*).

(a) Show that if A is the block matrix $\begin{bmatrix} B & 0 \\ 0 & C \end{bmatrix}$, then e^{tA} is the block matrix $\begin{bmatrix} e^{tB} & 0 \\ 0 & e^{tC} \end{bmatrix}$. [*Hint:* The system $x' = Ax$ "decouples" into two subsystems.]

(b) Use part **(a)** to find e^{tA} if

$$A = \begin{bmatrix} 1 & 3 & 0 \\ 1 & -1 & 0 \\ 0 & 0 & 2 \end{bmatrix} \quad \text{and} \quad A = \begin{bmatrix} 1 & 3 & 0 & 0 \\ 1 & -1 & 0 & 0 \\ 0 & 0 & 0 & 1 \\ 0 & 0 & 0 & 0 \end{bmatrix}$$

4. Solve the following driven systems. Leave your answers in the form of formula (17); calculate e^{tA}, $e^{tA}x^0$, $e^{-sA}F(s)$, but don't evaluate the integrals.

(a) $x' = \begin{bmatrix} 0 & 2 \\ -2 & 0 \end{bmatrix} x + \begin{bmatrix} 1 \\ 0 \end{bmatrix}$, $x(0) = \begin{bmatrix} a \\ b \end{bmatrix}$

(b) $x' = \begin{bmatrix} 2 & -1 \\ 3 & -2 \end{bmatrix} x + \begin{bmatrix} 3e^t \\ t \end{bmatrix}$, $x(0) = \begin{bmatrix} 1 \\ 2 \end{bmatrix}$.   Plot the component curves.

(c) $x' = \begin{bmatrix} 2 & -5 \\ 1 & -2 \end{bmatrix} x + \begin{bmatrix} \cos t \\ 0 \end{bmatrix}$, $x(0) = \begin{bmatrix} a \\ b \end{bmatrix}$

(d) $x' = \begin{bmatrix} -1 & -4 \\ 1 & -1 \end{bmatrix} x + \begin{bmatrix} e^{-3t} \\ 1 \end{bmatrix}$, $x(0) = \begin{bmatrix} 0 \\ 0 \end{bmatrix}$.   Plot the component curves.

(e) $x' = \begin{bmatrix} 3 & -1 & 1 \\ 2 & 0 & 1 \\ 1 & -1 & 2 \end{bmatrix} x + \begin{bmatrix} f_1(t) \\ f_2(t) \\ f_3(t) \end{bmatrix}$, $x(0) = \begin{bmatrix} a \\ b \\ c \end{bmatrix}$

5. (*Undetermined Coefficients*). Sometimes a particular solution $x^d(t)$ of $x' = Ax + F$ can be found by a method of undetermined coefficients. Find all solutions for each system below.

(a) $x' = \begin{bmatrix} 2 & -1 \\ 5 & -2 \end{bmatrix} x + \begin{bmatrix} e^t \\ 1 \end{bmatrix}$. [*Hint:* Assume x^d has the form $x_1 = ae^t + b$, $x_2 = ce^t + d$ and find the coefficients by inserting x_1 and x_2 into the ODEs and matching coefficients of corresponding terms. Then add x^d to the general solution of the undriven system.]

(b) $x' = \begin{bmatrix} 1 & 3 \\ 1 & -1 \end{bmatrix} x + \begin{bmatrix} \cos t \\ 2t \end{bmatrix}$. [*Hint:* Assume x^d has the form $x_1 = a_1 \cos t + b_1 \sin t + c_1 + d_1 t$, $x_2 = a_2 \cos t + b_2 \sin t + c_2 + d_2 t$.]

6. (*Exponential Matrices May Not Commute*). Here are some unexpected properties of e^{tA}.

(a) Show that $e^{tA}e^{tB}$ doesn't have to be either $e^{t(A+B)}$ or $e^{tB}e^{tA}$ by calculating all three where

$$A = \begin{bmatrix} 0 & 1 \\ 0 & 0 \end{bmatrix} \quad \text{and} \quad B = \begin{bmatrix} 1 & 0 \\ 0 & 0 \end{bmatrix}$$

(b) Suppose that $AB = BA$. Show that $e^{t(A+B)} = e^{tA}e^{tB} = e^{tB}e^{tA}$ for all t. [*Hint:* Show that if

$P(t) = e^{t(A+B)}e^{-tA}e^{-tB}$, then $P'(t) = 0$ for all t. Since $P(0) = I$, we must have $P(t) = I$.]

7. (*Laplace Transforms for Systems*). The following steps outline the method of Laplace transforms for solving the IVP

$$x' = Ax + F, \qquad x(0) = x^0$$

where A is an $n \times n$ matrix of constants. The Laplace transform $\mathcal{L}[x(t)]$ of a vector function is the vector of the Laplace transform of the individual components.

(a) Apply the operator $\mathcal{L}$ to the above system and obtain $(sI - A)\mathcal{L}[x] = x^0 + \mathcal{L}[F]$.

(b) Show that the matrix $sI - A$ is invertible for all real s that are large enough. [*Hint*: $sI - A$ is singular only when s is an eigenvalue of A, and A has a finite number of eigenvalues.]

(c) Show that the solution of the IVP above is

$$x = \mathcal{L}^{-1}[(sI - A)^{-1}x^0] + \mathcal{L}^{-1}\left[(sI - A)^{-1}\mathcal{L}[F]\right]$$

Compare with (17) and show that

$$\mathcal{L}^{-1}[(sI - A)^{-1}] = e^{tA}, \qquad \mathcal{L}^{-1}\left[(sI - A)^{-1}\mathcal{L}[F]\right] = e^{tA} * F(t)$$

which is the convolution of e^{tA} and $F(t)$.

(d) Use the Laplace transform to solve the IVP of Problem 1(b).

 8. (*Laplace Transforming the Coupled Springs IVPs*). Use the Laplace transform to solve the system of linear second-order ODEs $x'' = -40x/3 + 10y/3$, $y'' = 40x/3 - 40y/3$, where $x(0) = 1$, $x'(0) = 0$, $y(0) = 2$, $y'(0) = 0$. Repeat with $x(0) = 1$, $x'(0) = 0$, $y(0) = 0$, $y'(0) = 0$. [*Hint*: Compare with the results in Examples 7.7.8 and 7.7.9.]

7.9 Steady States of Driven Linear Systems

A well-designed physical system resists disturbances. A shock may knock the system out of its steady state, but not far away if the disturbance is small. Once the disturbance ends, the system returns to the steady state. In this section we formulate the mathematical equivalent of these ideas for the linear system

$$x' = Ax + F(t) \tag{1}$$

where A is an $n \times n$ matrix of real constants, x is an n-vector, and the driving force F is either constant or periodic.

We distinguish two kinds of steady states:

❖ **Constant Steady State**. A constant solution x^s of system (1) is a *steady state* if all solutions of system (1) tend to x^s as $t \to +\infty$.

❖ **Periodic Steady State**. A periodic solution $x^s(t)$ of system (1) is a *steady state* if all solutions of system (1) tend to $x^s(t)$ as $t \to +\infty$.

These are commonsense engineering definitions that involve a constant equilibrium solution or a periodic solution. In both cases the steady state attracts all solutions as time advances, so in the long term only the steady state is visible.

Constant Steady State

Suppose that A is a nonsingular matrix and that F_0 is a constant vector. Direct substitution shows that

$$x(t) = -A^{-1}F_0 \tag{2}$$

is a constant solution of the system

$$x' = Ax + F_0 \tag{3}$$

Since A is nonsingular, system (3) has no other constant solutions.

THEOREM 7.9.1

☞ So we see why eigenvalues with negative real parts are so important.

> Constant Steady State. Suppose that all eigenvalues of the constant matrix A have negative real parts and that F_0 is a constant vector. Then the system $x' = Ax + F_0$ has the unique steady state $x^s = -A^{-1}F_0$.

We can verify this result in the following way. Because the eigenvalues of A have negative real parts, 0 is not an eigenvalue, so A is nonsingular and A^{-1} exists. The general solution of $x' = Ax + F_0$ is then

$$x = e^{tA}c - A^{-1}F_0$$

☞ We noted this polynomial-exponential feature in Theorem 7.6.4.

where c is any constant vector. Every entry in the matrix e^{tA} is a real polynomial-exponential function, every term of which has a factor of the form $e^{\alpha t}$, where α is the real part of an eigenvalue of A. Since every α is negative, all of these terms tend to 0 as $t \to +\infty$, so $e^{tA}c$ tends to the zero vector as $t \to +\infty$, and we are finished.

EXAMPLE 7.9.1

A Constant Steady State

The system matrix of the system

$$\begin{aligned} x' &= -x + 4y + 14 \\ y' &= -3x - 2y + 28 \end{aligned} \tag{4}$$

has eigenvalues $-3/2 \pm i\sqrt{47}/2$, while $x^s = -A^{-1}F_0 = [10 \quad -1]^T$. From Theorem 7.9.1, all solutions of system (4) tend to the unique equilibrium solution and steady state x^s. Figures 7.9.1 and 7.9.2 display orbits and component curves.

Periodic Steady State, Periodic Forced Oscillations

Now suppose that the driving vector $F(t)$ is periodic with period T. A *periodic forced oscillation* of the system

$$x' = Ax + F(t) \tag{5}$$

is a periodic solution of period T. We have the following result (see Problem 13).

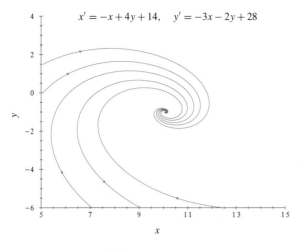

$$x' = -x + 4y + 14, \quad y' = -3x - 2y + 28$$

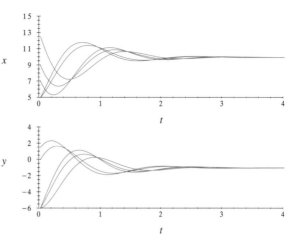

FIGURE 7.9.1 Orbits approach constant steady state at $(10, -1)$ (Example 7.9.1).

FIGURE 7.9.2 Component curves for the orbits of Figure 7.9.1.

THEOREM 7.9.2

Periodic Forced Oscillation. Suppose that A is a constant matrix in system (5) and that the driving term $F(t)$ is continuous (or piecewise continuous) and periodic with period T. If $2\pi i / T$ is *not* an eigenvalue of A, then system (5) has a unique periodic forced oscillation of period T.

We require that $2\pi i / T$ not be an eigenvalue because if it were an eigenvalue, then the undriven system $x' = Ax$ would have solutions containing terms such as $\cos(2\pi t / T)$ or $\sin(2\pi t / T)$ of period T. Since $F(t)$ also has period T, this could lead to resonance and unbounded oscillations (see Section 4.2).

The conditions below guarantee that a periodic forced oscillation attracts all solutions.

THEOREM 7.9.3

Periodic Steady State. Suppose all eigenvalues of the constant matrix A have negative real parts and that $F(t)$ is periodic with period T. Then the system $x' = Ax + F(t)$ has a unique steady state, which is periodic of period T.

Here's why this theorem is true. If the eigenvalues of A have negative real parts, then $2\pi i / T$ is not an eigenvalue of A. The conditions of Theorem 7.9.2 hold, and the system $x' = Ax + F(t)$ has a periodic forced oscillation of period T; call it $x^s(t)$. Since all solutions of the system have the form

$$x(t) = e^{tA}c + x^s(t)$$

it follows that all solutions tend to $x^s(t)$ as $t \to +\infty$, because $e^{tA}c$ tends to the zero vector as $t \to +\infty$, and we are done.

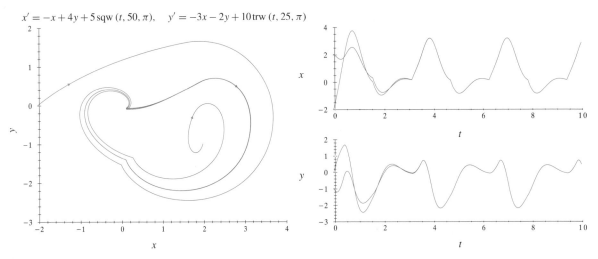

$$x' = -x + 4y + 5\,\text{sqw}(t, 50, \pi), \quad y' = -3x - 2y + 10\,\text{trw}(t, 25, \pi)$$

FIGURE 7.9.3 Orbits approach periodic steady-state orbit of period π (Example 7.9.2).

FIGURE 7.9.4 Component curves of the two orbits of Figure 7.9.3.

EXAMPLE 7.9.2

Periodic Steady State

The linear system shown here is driven by engineering functions of period π:

$$x' = -x + 4y + 5\,\text{sqw}(t, 50, \pi)$$

$$y' = -3x - 2y + 10\,\text{trw}(t, 25, \pi)$$

The system matrix has eigenvalues $-3/2 \pm i\sqrt{47}/2$. Figure 7.9.3 shows the orbits that start at $(-2, 0)$ and $(2, -1)$ as they approach the unique periodic steady state, and Figure 7.9.4 shows the corresponding component curves. Inspection of the component curves verifies that the periodic steady state has period π. Can you identify the points on the orbits and component curves where the square wave input is turned on or off?

In Examples 7.9.1 and 7.9.2 all that can eventually be seen is the steady state.

Decay, BIBO

We might guess that all solutions of $x' = Ax + F(t)$ decay exponentially to a single solution if all eigenvalues of the constant matrix A have negative real parts. That is just what happens, as the following theorem implies.

THEOREM 7.9.4

> Decay Estimate. Suppose that the eigenvalues of A have real parts less than the negative constant a. Then there is a positive constant M such that the solution $x = x(t, x^0)$ of the IVP $x' = Ax$, $x(0) = x^0$, satisfies
>
> $$\|x(t, x^0)\| = \|e^{tA}c\| \le Me^{at}\|x^0\|$$
>
> for all $t \ge 0$ and all initial states x^0.

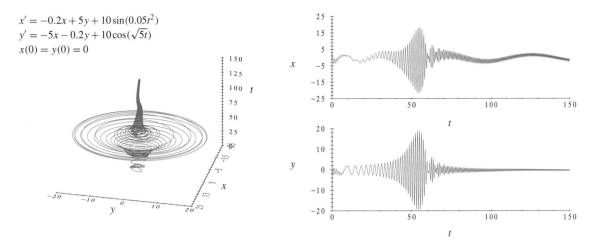

$$x' = -0.2x + 5y + 10\sin(0.05t^2)$$
$$y' = -5x - 0.2y + 10\cos(\sqrt{5}t)$$
$$x(0) = y(0) = 0$$

FIGURE 7.9.5 Time-state curve (Example 7.9.3). An inverted tornado?

FIGURE 7.9.6 Bounded components of the time-state curve of Figure 7.9.5.

Suppose, finally, that we know little about the input vector $F(t)$ except that it is bounded for $t \geq 0$, that is, for some positive constant M, $\|F(t)\| \leq M$ for all $t \geq 0$.

THEOREM 7.9.5

☞ Theorem 2.3.1 gives a scalar version of BIBO.

> **Bounded Input, Bounded Output (BIBO).** Suppose that the eigenvalues of the matrix A have negative real parts and that the input vector $F(t)$ is bounded for $t \geq 0$. Then every solution $x(t)$ of $x' = Ax + F(t)$ is bounded for $t \geq 0$.

The verification of BIBO follows from Theorem 7.9.4.

This result is of considerable importance in science and engineering. It implies that as long as the eigenvalues of A have negative real parts, then a persistent disturbance, that is, the bounded input vector $F(t)$, is not likely to destroy the system because there can be no unbounded solutions.

EXAMPLE 7.9.3

BIBO

The system below has the driving vector $10[\sin(0.05t^2) \quad \cos(\sqrt{5}t)]^T$:

$$x' = -0.2x + 5y + 10\sin(0.05t^2)$$
$$y' = -5x - 0.2y + 10\cos(\sqrt{5}t) \tag{6}$$

The system matrix has eigenvalues $-0.2 \pm 5i$, and the magnitude of the driving vector is bounded from above by $10\sqrt{2}$. All conditions of Theorem 7.9.5 are met, so all solutions are bounded for $t \geq 0$. See Figures 7.9.5 and 7.9.6 for the strange time-state curve and components of the bounded solution with initial state $x(0) = y(0) = 0$.

We have restricted the term "steady state" to denote an attracting constant solution or an attracting periodic forced solution of a linear system with a constant or periodic input. If we try to define a steady state for the system $x' = Ax + F(t)$ for an input F that is neither constant nor periodic, then we run into trouble. Even if all solutions of the system are bounded for $t \geq 0$ and even if all the eigenvalues of A have negative real parts, it is not clear just how to define the steady state. The reason is that if $x^1(t)$ and $x^2(t)$ are any two solutions of the system, then $x^1 - x^2$ solves $x' = Ax$. So by Theorem 7.9.4 and by the solution formula for a driven linear system we have that for some constant vector c, some positive constant M, and some negative constant a

$$\lim_{t\to\infty} \|x^1(t) - x^2(t)\| = \lim_{t\to\infty} \|e^{tA}c\| \leq \lim_{t\to\infty} Me^{at}\|c\| = 0$$

All solutions of the driven system tend to one another, and any solution could be called the steady state. And that is why we have restricted the term "steady state" to the specific situations where F is either constant or periodic because then there is a unique constant or periodic solution that attracts all solutions.

Where are the Eigenvalues?

The existence of steady states and the BIBO principle depend on all eigenvalues of the system matrix having negative real parts. The eigenvalues can be found directly if the matrix is 2×2, but for most larger matrices other methods are needed. Many numerical solvers have programs for approximating eigenvalues, which is one way to determine whether the conditions of the theorems of this section are met. We list below several simple tests that you can carry out with pencil and paper. These tests will give you information about the location of the eigenvalues, but not their precise values.

The first two tests give information about the roots of a polynomial and can be applied to the characteristic polynomial of a matrix.

THEOREM 7.9.6

☞ The Student Resource Manual has the verification.

> Coefficient Test. Suppose that $P(r) = r^n + a_{n-1}r^{n-1} + \cdots + a_0$, where the coefficients are real. If any coefficient of $P(r)$ is either zero or negative, then at least one root has a nonnegative real part.

EXAMPLE 7.9.4

Applying the Coefficient Test
The Coefficient Test implies that the polynomials $r^3 + 3r + 10$ and $r^3 - r^2 + 5r + 7$ each have at least one root with nonnegative real part. The test does *not* apply to the polynomial $r^3 + r^2 + r + 1$ because all coefficients are positive.

The Coefficient Test is simple to use, but it has limited application. What we want is a test that applies to polynomials with positive coefficients. The Routh Test is just what is needed.[3] We will only give the test as it applies to polynomials of degree no more than four, although the test can be extended to any polynomial.

---

[3]Edward John Routh (1831–1907) was born in Canada, but spent his adult life at the University of Cambridge in England. His books on dynamics and the stability of moving bodies became nineteenth-century classics.

THEOREM 7.9.7

☞ The Student Resource Manual extends the Routh Test to a polynomial of any degree.

> The Routh Test. All of the roots of the indicated polynomials have negative real parts precisely when the given conditions are met.
>
> - $r^2 + a_1 r + a_0$: all coefficients are positive.
> - $r^3 + a_2 r^2 + a_1 r + a_0$: all coefficients are positive and $a_2 a_1 > a_0$.
> - $r^4 + a_3 r^3 + a_2 r^2 + a_1 r + a_0$: all coefficients are positive, $a_3 a_2 > a_1$, and $a_3 a_2 a_1 > a_3^2 a_0 + a_1^2$.

EXAMPLE 7.9.5

Applying the Routh Test

The polynomial $r^3 + 2r^2 + 3r + 5$ has positive coefficients, and since $a_2 = 2$, $a_1 = 3$, and $a_0 = 5$, the polynomial meets the Routh condition

$$6 = a_2 a_1 > a_0 = 5$$

So all roots of this cubic have negative real parts.

The Routh Test and the Coefficient Test give us information about the location of the eigenvalues of a matrix, but we can only use the tests after we have found the characteristic polynomial of the matrix. The next tests are weaker, but they have the big advantage of applying directly to the matrix, so there is no need to find the characteristic polynomial first.

THEOREM 7.9.8

☞ See Problem 6 in Section 7.4 for why this is so.

> The Trace Test. **(a)** Suppose that A is an $n \times n$ matrix of real constants and that trace $A = a_{11} + \cdots + a_{nn}$ is negative [positive]. Then at least one eigenvalue of A has a negative [positive] real part. **(b)** If the trace is zero, then either all eigenvalues have zero real parts, or there is a pair of eigenvalues whose real parts have opposite signs.

EXAMPLE 7.9.6

Using the Trace Test

The matrix A_1 below must have an eigenvalue with positive real part because the trace is $+1$. The trace of the matrix A_2 below is negative (tr $= -2$), but $+9$ is an eigenvalue of A_2. This does not contradict the Trace Test because the test asserts that *if* the trace is negative, then there is at least one eigenvalue whose real part is negative, not that all eigenvalues must have negative real parts.

$$A_1 = \begin{bmatrix} -165 & 13 & 360 \\ -200 & 15 & 267 \\ -150 & -8 & 151 \end{bmatrix}, \qquad A_2 = \begin{bmatrix} -10 & 50 & 0 \\ 300 & -1 & 0 \\ 0 & 0 & 9 \end{bmatrix}$$

Finally, we present a method which was discovered by the Russian mathematician S. Gerschgorin in the 1930s. Suppose that A is an $n \times n$ matrix of constants. The *Gerschgorin row disk* R_i in the complex plane is defined by

$$R_i = \text{ the set of all complex numbers } z \text{ where } |z - a_{ii}| \leq r_i \text{ and } r_i = \sum_{\substack{k=1 \\ k \neq i}}^{n} |a_{ik}|$$

So R_i is the circular disk of radius r_i centered at the point a_{ii} in the complex plane. Note that r_i is the sum of the magnitudes of all entries in the i-th row of A, except for $|a_{ii}|$. Similarly, the *Gerschgorin column disk* C_i is defined by

$$C_i = \text{ the set of all } z \text{ where } |z - a_{ii}| \leq c_i \text{ and } c_i = \sum_{\substack{k=1 \\ k \neq i}}^{n} |a_{ki}|$$

Here is Gerschgorin's surprising theorem.

THEOREM 7.9.9

☞ The Student Resource Manual outlines a proof.

> Eigenvalues and the Gerschgorin Disks. Suppose that A is an $n \times n$ matrix of constants and that R_i and C_i, $i = 1, \ldots, n$, are the respective Gerschgorin row and column disks of A. Then all eigenvalues of A lie in the union of all the row disks, and also in the union of all the column disks. If a union of k of the row disks [column disks] is disjoint from the other row disks [column disks], then the union contains exactly k of the eigenvalues, counting multiplicities.

EXAMPLE 7.9.7

Gerschgorin Disks

The Gerschgorin row and column disks of $A = \begin{bmatrix} -1+i & 1.5 & 1 \\ 0.5 & -2 & -1 \\ 0.25 & 0 & -7 \end{bmatrix}$ are given by

Row: $|z + 1 - i| \leq 2.5$, $|z + 2| \leq 1.5$, $|z + 7| \leq 0.25$

Column: $|z + 1 - i| \leq 0.75$, $|z + 2| \leq 1.5$, $|z + 7| \leq 2$

The disks are sketched in Figures 7.9.7 and 7.9.8. Since the column disks lie entirely inside the left half of the complex plane, the eigenvalues of A must all have negative real parts. From the location of the row disks and the last part of Theorem 7.9.9, we conclude that A has an eigenvalue very close to the number -7.

Comments

The Routh Test and the Gershgorin Disk Test are both used to locate the roots of a polynomial in the complex plane. It is important to know when the roots have negative real parts, since then all the solutions of the undriven system are transients (i.e., die out as $t \to +\infty$). This guarantees that a periodic forced oscillation is a steady state. Although there are several good numerical root finders and eigenvalue solvers, the techniques just outlined are so easy that they should be used first.

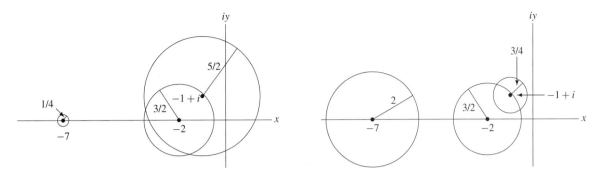

FIGURE 7.9.7 Row disks (Example 7.9.7).   **FIGURE 7.9.8** Column disks (Example 7.9.7).

PROBLEMS

1. (*Constant Steady States*). Is the constant solution $x = 0$ a steady state for $x' = Ax$ if A is as given? Give reasons. [*Hint*: Look at the eigenvalues of A.]

    (a) $\begin{bmatrix} -10 & 1 & 2 \\ -3 & 4 & 5 \\ -7 & -6 & 7 \end{bmatrix}$ (b) $\begin{bmatrix} -10 & 1 & 2 \\ -3 & 4 & -2 \\ -7 & -6 & -40 \end{bmatrix}$ (c) $\begin{bmatrix} -10 & 1 & 2 \\ 0 & -1 & 100 \\ 0 & 0 & -1 \end{bmatrix}$

2. (*Tuning a System to Create a Steady State*). Find the values of the real constant α for which $x = 0$ is the steady state of $x' = Ax$. [*Hint*: Look at the eigenvalues of A. In (b) and (c), examine the equivalent single third-order ODE.]

    (a) $\begin{bmatrix} \alpha & 2 \\ 3 & -4 \end{bmatrix}$ (b) $\begin{bmatrix} 0 & 1 & 0 \\ 0 & 0 & 1 \\ \alpha & -1 & -1 \end{bmatrix}$ (c) $\begin{bmatrix} 0 & 1 & 0 \\ 0 & 0 & 1 \\ -2 & -3 & \alpha \end{bmatrix}$

3. (*BIBO*). The system $x' = y$, $y' = -bx - ay + f(t)$ is equivalent to the single ODE $x'' + ax' + bx = f(t)$. Show that every solution of the system is bounded for $t \geq 0$ if $a > 0$, $b > 0$, and $f(t)$ is continuous and bounded for $t \geq 0$.

4. (*Unbounded Solutions*). Show that the system $x' = y$, $y' = -x + \cos t$ has unbounded solutions as $t \to +\infty$, but that this does not contradict Theorem 7.9.5.

5. (*Unbounded Solutions*). Show that the system $x' = -x + e^{3t}y$, $y' = -y$ has unbounded solutions as $t \to +\infty$ even though the eigenvalues of the nonconstant system matrix are negative constants. Why is this not a contradiction of Theorem 7.9.5?

6. (*Unbounded Solutions*). Suppose that the matrix A has an eigenvalue with nonnegative real part. Then there always exists a bounded driving function $F(t)$ such that $x' = Ax + F$ has unbounded solutions. In other words, bounded input may *not* result in bounded output in this setting. Parts (a) and (b) show why this is so.

    (a) Show that $x' = Ax$ has unbounded solutions if A has an eigenvalue with positive real part.

    (b) Suppose that the matrix A has an eigenvalue λ with zero real part. Show that if v is an eigenvector corresponding to λ, the system $x' = Ax + e^{\lambda t}v$ has the unbounded solution $x = te^{\lambda t}v$, $t \geq 0$.

7. (*Applying the Routh Test*). Which polynomials below have all roots with negative real parts?

    (a) $z^3 + z^2 + 2z + 1$ (b) $z^4 + z^3 + 2z^2 + 2z + 3$ (c) $z^4 + z + 1$

    (d) $6z^3 + 11z^2 + 6z + 6$ (e) $12z^3 + 12z^2 + 11z + 6$

8. (*Applying the Routh Test*). Show that the roots of $z^4 + 2z^3 + 3z^2 + z + a$ have negative real parts if and only if $0 < a < 1.25$.

www **9.** (*Steady State*). Explain why each system below has a steady state. Then find a formula for it, and plot component graphs of the steady state and of several other solutions.

(a) $x' = y + 5$, $y' = -2x - 3y + 10$

(b) $x' = y$, $y' = -x - 2y + \cos t$. [*Hint*: Write as an equivalent second-order ODE.]

(c) $x' = -x + y + \sin t$, $y' = -x - y$. [*Hint*: Assume a particular solution of the form $x = A \cos t + B \sin t$, $y = C \cos t + D \sin t$ and match coefficients.]

10. (*Periodic Steady State*). Explain why each system below has a periodic steady state. Then plot the components of several solutions over a long enough time period that the periodic steady state is clearly visible. Highlight the periodic steady state on the hard copy of your graph.

(a) $x' = -x + y/2 + \cos^2 t$, $y' = -x - 2y + \sin t$

(b) $x' = -2x + 5y + \text{sqw}(t, 50, 2)$, $y' = -5x - 2y + \cos \pi t$

(c) $x' = y + \text{trw}(t, 50, 1)$, $y' = -x - y + \text{sqw}(t, 50, 2)$

(d) $x' = y + (\sin t)\,\text{sqw}(t, 50, 2\pi)$, $y' = -4x - y + 1$

11. (*Unbounded Solutions*). Show that if all eigenvalues of A have negative real parts and if one solution of $x' = Ax + F(t)$ is unbounded as $t \to +\infty$, then all solutions are unbounded as $t \to +\infty$.

12. (*Gerschgorin Disks*). Use Gerschgorin disks and explain why all eigenvalues of the matrix in part **(a)** have negative real parts, but the matrix in part **(b)** has an eigenvalue with positive real part.

(a) $A = \begin{bmatrix} -10 & 1 & 8 \\ 1 & -3 & 2 \\ 8 & 1 & -11 \end{bmatrix}$ (b) $A = \begin{bmatrix} 10 & 1 & 10 \\ 1 & -5 & 8 \\ 1 & 1 & -20 \end{bmatrix}$

13. (*Periodic Forced Oscillations*). A periodic forced oscillation of the system $x' = Ax + F(t)$ is a nonconstant solution $x(t)$ of period T, where A is an $n \times n$ matrix of real constants and $F(t)$ has period T. Show that the system has a unique periodic forced oscillation if $2\pi i/T$ is *not* an eigenvalue of A. [*Hint*: $x(t) = e^{tA}x^0 + e^{tA}\int_0^t e^{-sA}F(s)\,ds$ is periodic with period T if and only if $x(0) = x(T) = e^{TA}x^0 + e^{TA}\int_0^T e^{-sA}F(s)ds$.]

7.10 Lead Flow, Noise Filter: Steady States

In this section we'll apply the results of Section 7.9 to show that the equilibrium solution of the lead model is a steady-state solution as advertised in Section 7.1. Next, we'll model a two-loop electrical circuit that returns a periodic steady-state output (i.e., an attracting periodic forced oscillation) for any periodic input, again building on Section 7.9. We'll see that the circuit filters out high-frequency signals (noise) but passes low-frequency signals.

Sensitivity of Lead Levels to Antilead Medication

In Section 7.1 we applied the Balance Law to a compartment model and found a system of ODEs that models the flow of lead through the blood, tissues, and bones. Using the data given in Section 7.1, we have the model system

$$x_1' = -(0.0150 + k_{01})x_1 + 0.0124x_2 + 0.000035x_3 + I_1$$
$$x_2' = 0.0111x_1 - 0.0286x_2 \tag{1}$$
$$x_3' = 0.0039x_1 - 0.000035x_3$$

where $x_1(t)$, $x_2(t)$, and $x_3(t)$ are the respective amounts of lead (in micrograms) in the blood, tissues, and bones at time t (measured in days); k_{01} is a rate constant that measures the rate at which the kidneys clear lead from the blood; and I_1 is the input rate of lead into the blood.

Let's show that for all positive constants k_{01} and I_1, the equilibrium solution of system (1) is, in fact, a steady-state solution, that is, it attracts all solutions as $t \to +\infty$. This explains why the solution curves in the figures of Section 7.1 behave as they do.

The Gerschgorin disks introduced in Section 7.9 help to show that all eigenvalues of the system matrix

☞ A compartment model with a constant or periodic input must have a steady state if every compartment is linked to an external absorbing compartment (or sink) by a direct chain of flowrate arrows (as is the case here). More on this in the Student Resource Manual.

$$A = \begin{bmatrix} -(0.0150 + k_{01}) & 0.0124 & 0.000035 \\ 0.0111 & -0.0286 & 0 \\ 0.0039 & 0 & -0.000035 \end{bmatrix}$$

have negative real parts, so the equilibrium solution is a steady state. The three row disks in the complex plane are defined by

$$|z + 0.0150 + k_{01}| \le 0.012435, \quad |z + 0.0286| \le 0.0111, \quad |z + 0.000035| \le 0.0039$$

and the three column disks are defined by

$$|z + 0.0150 + k_{01}| \le 0.0150, \quad |z + 0.0286| \le 0.0124, \quad |z + 0.000035| \le 0.000035$$

The third row disk overlaps the right half of the complex plane, so we can't conclude from examination of the row disks that the eigenvalues lie inside the left half-plane. What about the column disks? If k_{01} is positive, then the first two column disks lie inside the left half-plane, but the third column disk just touches the origin and so raises the possibility that 0 is an eigenvalue. But $\lambda = 0$ is an eigenvalue of the matrix A if and only if $\det A = 0$, and since by direct computation

$$\det A = -0.000035[0.0286k_{01} + (0.0111)(0.0189)]$$

we see that $\det A \ne 0$ for any $k_{01} \ge 0$. So with the help of the Gerschgorin column disk test, we see that all eigenvalues of A actually have negative real parts.

So now we know that the results of Section 7.9 can be applied to system (1) for the transport of lead through the body compartments. For example, if $k_{01} = 0.0211$ and $I_1 = 49.3$, the constant steady state is given by

$$x^s = -A^{-1} \begin{bmatrix} 49.3 \\ 0 \\ 0 \end{bmatrix} \approx \begin{bmatrix} 1800 \\ 699 \\ 200\,583 \end{bmatrix}$$

and all solutions of IVP (1) approach x^s as $t \to +\infty$ (check this out with Figure 7.1.3).

Now let's look at a system with a periodic input and investigate the existence of steady-state solutions and their sensitivity to changes in the input frequency.

A Two-Loop Low-Pass Filter

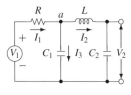

The electrical circuit sketched in the margin is a common component in electronic equipment. The input is a known voltage $V_1(t)$, and the circuit responds with the output voltage $V_2(t)$ across the second capacitor. Candidates for state variables are the currents into and out of the junction a and the respective voltages across the resistor, the inductor, and the capacitors. However, several of these quantities are interrelated, so we don't need all of them. For example, by Kirchhoff's Current Law applied at junction a, we have

$$I_3 = I_1 - I_2$$

☞ Section 4.4 has more information about circuits.

so we don't need I_3 if we keep the currents I_1 and I_2 as state variables. As it happens, V_2, I_1, and I_2 will serve as state variables for the circuit. Let's find ODEs that relate these three quantities.

The derivative form of Coulomb's Law implies that the rate of change of the voltage across the second capacitor is

$$V_2' = I_2/C_2 \tag{2}$$

Now let's apply Kirchhoff's Voltage Law to the outer loop through the input, resistor, inductor, and second capacitor:

$$V_1 = RI_1 + LI_2' + V_2 \tag{3}$$

where RI_1 is the voltage across the resistor (Ohm's Law) and LI_2' is the voltage across the inductor (Faraday's Law). So ODEs (2) and (3) involve V_2' and I_2', but we need a third ODE, one that involves I_1'.

Coulomb's Law in derivative form implies that the voltage drop across the second capacitor is I_3/C_1, which can be written as $(I_1 - I_2)/C_1$. Then Kirchhoff's Voltage Law (in derivative form) for the left-hand loop through the input, resistor, and first capacitor implies that

$$V_1' = RI_1' + (I_1 - I_2)/C_1 \tag{4}$$

which is an ODE with I_1'.

So now we have the linear system of three ODEs in the three state variables V_2, I_1, and I_2:

$$x' = Ax + F(t) \tag{5}$$

$$
x = \begin{bmatrix} V_2 \\ I_1 \\ I_2 \end{bmatrix}, \qquad
A = \begin{bmatrix} 0 & 0 & 1/C_2 \\ 0 & -1/RC_1 & 1/RC_1 \\ -1/L & -R/L & 0 \end{bmatrix}, \qquad
F = \begin{bmatrix} 0 \\ V_1'/R \\ V_1/L \end{bmatrix} \tag{6}
$$

We want to see how system (5) responds to a periodic input voltage $V_1(t)$. In particular, we expect a steady-state periodic response, that is, a periodic forced oscillation. But to back up our expectations, we first need to verify that all eigenvalues of A have negative real parts so that we can use the steady-state results of Section 7.9. The Gerschgorin disks aren't any help here because two of the diagonal elements of A are 0, so the corresponding disks overlap both the right and the left half of the complex plane. Instead, we try another approach.

The characteristic polynomial $p(\lambda)$ of A is

$$p(\lambda) = -\left[\lambda^3 + \frac{1}{RC_1}\lambda^2 + \frac{1}{L}\left(\frac{1}{C_1} + \frac{1}{C_2}\right)\lambda + \frac{1}{LRC_1C_2}\right] \tag{7}$$

Since the coefficients of $p(\lambda)$ are all positive, the eigenvalues of A all have negative real parts by the Routh Test (Theorem 7.9.7) if and only if

$$\frac{1}{RC_1} \cdot \frac{1}{L}\left(\frac{1}{C_1} + \frac{1}{C_2}\right) > \frac{1}{LRC_1C_2}$$

which is certainly so since all of the circuit parameters are positive. So we know now that if the input voltage is periodic with period T, then there is a unique periodic steady-state output that has the same period T.

Let's take as input the function $V_1 = a_0 e^{i\omega t}$ of period $2\pi/\omega$. Then

$$F = \begin{bmatrix} 0 \\ V_1'/R \\ V_1/L \end{bmatrix} = \alpha e^{i\omega t}, \quad \text{where } \alpha = \begin{bmatrix} 0 \\ a_0 i\omega/R \\ a_0/L \end{bmatrix} \tag{8}$$

Let's find the steady-state output. Since the eigenvalues of A have negative real parts, Theorem 7.9.3 implies that there is a unique periodic steady state $x^s(t)$ of period $2\pi/\omega$. We expect the steady-state response to have the form $x^s = \beta e^{i\omega t}$ for some constant vector β. To find the vector β, insert x^s into system (5):

$$i\omega e^{i\omega t}\beta = e^{i\omega t}A\beta + e^{i\omega t}\alpha \tag{9}$$

Solving (9) for β, we have

$$\beta = [i\omega I - A]^{-1}\alpha, \quad \text{so} \quad x^s(t) = [i\omega I - A]^{-1}\alpha e^{i\omega t} \tag{10}$$

The matrix inverse in (10) exists because every eigenvalue of A has a negative real part, so the pure imaginary number $i\omega$ can't be an eigenvalue of A.

The voltage $V_2(t)$ [i.e., the first component of $x^s(t)$] is considered to be the steady-state output voltage. A long calculation using formulas (8), (10) shows that the response V_2 to the input $V_1 = a_0 e^{i\omega t}$ is

$$V_2(t) = \frac{a_0}{p(i\omega)LRC_1C_2}e^{i\omega t} \tag{11}$$

where p is the characteristic polynomial (7). The ratio of the amplitude of the steady-state output voltage V_2 to the amplitude of the input voltage V_1 is

$$\left|\frac{V_2}{V_1}\right| = \frac{1}{|p(i\omega)LRC_1C_2|}$$

$$= \left|1 - LC_2\omega^2 + iR\omega(C_1 + C_2 - \omega^2 LC_1C_2)\right|^{-1} \tag{12}$$

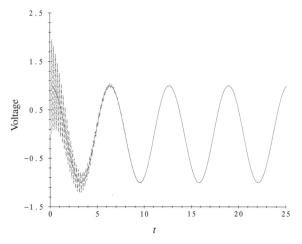
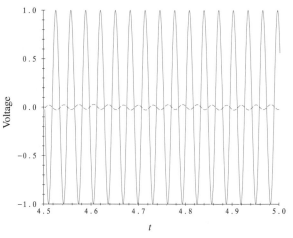

FIGURE 7.10.1 Input voltage $V_1 = \cos t$ (solid), output voltage V_2 (dashed) (Example 7.10.1).

FIGURE 7.10.2 Input voltage $V_1 = \cos 200t$ (solid), output voltage V_2 (dashed) (Example 7.10.1).

If ω is small, the ratio $|V_2/V_1| \approx 1$ and the circuit passes the input voltage with little change in its amplitude. However, if ω is large, then $|V_2/V_1|$ is small and the circuit essentially stops the input voltage. For these reasons, the circuit is said to be a *low-pass filter*. Note that we considered another kind of circuit in Section 2.2 that also acted as a low-pass filter.

EXAMPLE 7.10.1

A Low-Pass Filter: Sensitivity to Frequency
Suppose that in system (5)

$$R = 1, \quad L = 1, \quad C_1 = C_2 = 0.001$$

$$V_1(t) = \cos \omega t$$

$$V_2(0) = 0, \quad I_1(0) = I_2(0) = 0$$

Figures 7.10.1 and 7.10.2 show that a voltage of low circular frequency $\omega = 1$ passes through the circuit with almost no distortion (after an initial period when the transient voltages are still significant), while an input voltage of circular frequency 200 is severely attenuated after 4.5 units of time have elapsed. So we have a very specific example of the sensitivity of the amplitude of the steady-state output voltage to changes in the frequency of the input voltage.

The filtering property of the circuit applies not only to a periodic input of the form $F = \alpha e^{i\omega t}$, but to any periodic input. See Problem 4 for the case of a triangular wave input voltage.

Suppose that a low-frequency input voltage is contaminated by high-frequency noise. The low-pass circuit acts as a filter, effectively stops the noise, and returns a voltage virtually identical to the noiseless input voltage.

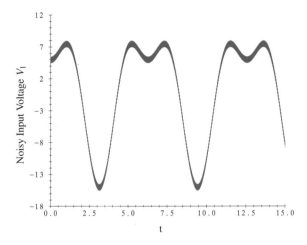

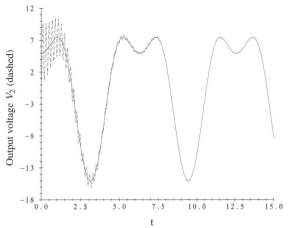

FIGURE 7.10.3 The input voltage is contaminated by high frequency noise: $V_1 = 10\cos t - 5\cos 2t + 0.5\cos 200t$. Will the circuit filter out the noise term $0.5\cos 200t$ (Example 7.10.2)?

FIGURE 7.10.4 The circuit filters high frequency noise $0.5\cos 200t$ from the input. The output voltage V_2 (dashed) approaches the "ideal" noiseless voltage $10\cos t - 5\cos 2t$ (solid) (Example 7.10.2).

EXAMPLE 7.10.2

Filtering Out Noise

Suppose that the input voltage V_1 is $10\cos t - 5\cos 2t + 0.5\cos 200t$, where the high frequency term $0.5\cos 200t$ represents noise (see Figure 7.10.3). The dashed curve in Figure 7.10.4 is the output voltage $V_2(t)$, given the actual noisy input V_1 for system (5), (6) with $R = 1$, $L = 1$, $C_1 = C_2 = 0.001$, $V_2(0) = 0$, and $I_1(0) = I_2(0) = 0$. We see that the high frequency noise in the input is filtered out in a very short time, so that all that remains in the output are the low frequency terms $10\cos t - 5\cos 2t$ (see Figure 7.10.4).

The circuit acts as a low-pass filter for all positive values of the circuit parameters R, L, C_1, and C_2, but the amount of attenuation of the output voltage will vary with the values of the parameters. An interesting computer experiment would involve varying the resistance R, the inductance L, or the capacitances C_1 and C_2 to see what effect this would have on the output voltage V_2.

Comments

We have shown how to find the steady-state solutions of the flow of lead in the human body and of the linear mathematical models of a low-pass filter. We didn't construct the exact solution formulas for the model systems. Instead, we used indirect techniques such as the Routh Test and Gerschgorin disks to check that the eigenvalues of the system matrices have negative real parts, which in turn implies that all solutions are "attracted" to the steady-state solution.

PROBLEMS

1.   (*Low-Pass Filter*). Explain why the circuit modeled in the text is a low-pass filter for all positive values of L, R, C_1, and C_2 if the input F is given by (8). [*Hint*: Show that for $|\omega|$ small, $|V_2(t)|$ is approximately $|a_0|$, while if $|\omega|$ is large, $|V_2(t)|$ is approximately $|a_0|/\omega^3 LRC_1C_2$.]

2.   (*Low-Pass Filter: Computer Simulations*). The system for the circuit discussed in the text is

$$x_1' = x_3/C_2$$

$$x_2' = -x_2/(RC_1) + x_3/(RC_1) + V_1'/R$$

$$x_3' = -x_1/L - Rx_2/L + V_1/L$$

Let $V_1 = \cos \omega t$. We will say that the circuit *passes* [amplifies] the input voltage V_1 if the output voltage $x_1 = V_2(t)$ eventually has amplitude of at least 90% [110%] of the amplitude of V_1. We say that the circuit *stops* the input voltage if the amplitude of the output voltage is eventually less than 10% of the amplitude of V_1. Otherwise, we say that the circuit is a *partial filter* for the input. In each case, use a numerical solver to plot a component graph of $x_1(t) = V_2(t)$ and to decide whether the circuit amplifies, passes, stops, or acts as a partial filter for $V_1 = \cos \omega t$. Use $x_1(0) = x_2(0) = x_3(0) = 0$ as initial data; solve over a sufficiently long time span that the solution is essentially in the steady state at the end of the time interval. See also Problem 3.

(a) $L = 1$, $R = 1$, $C_1 = C_2 = 1$; $\omega = 1, 2, 50$

(b) $L = 1$, $R = 0.1$, $C_1 = C_2 = 1$; $\omega = 1, 1.5, 10$. The circuit amplifies one of these input frequencies. Which one, and by how much?

(c) $L = 1$, $R = 100$, $C_1 = C_2 = 0.01$; $\omega = 0.1, 1.5, 10$

3.   (*Low-Pass Filter: Amplitude Analysis*). In the text it is shown that if the input voltage $V_1(t)$ for the system of Problem 2 is $a_0 e^{i\omega t}$, then the ratio of the output voltage $V_2(t)$ to $V_1(t)$ is given by formula (12). Use this formula to justify the following claims.

(a) If $L = 1$, $R = 1$, $C_1 = C_2 = 1$, then the frequencies $\omega = 1, 2, 50$ are, respectively, passed, partially filtered, and stopped.

(b) If $L = 1$, $R = 0.1$, $C_1 = C_2 = 1$, then the frequencies $\omega = 1, 1.5, 10$ are, respectively, passed and amplified, passed, and stopped.

(c) If $L = 1$, $R = 100$, $C_1 = C_2 = 0.01$, then the frequencies $\omega = 1, 2, 50$ are, respectively, passed, partially filtered, and stopped.

4.   (*Triangular Wave Input*). In the circuit system of Problem 2, set $L = 1$, $R = 1$, $C_1 = C_2 = 1000$, $V_1(t) = 2\,\mathrm{trw}(t, 100, T) - 1$.

(a) Explain why $V'(t) = (8/T)\,\mathrm{sqw}(t, 50, T) - 4/T$.

(b) Plot $V_1(t)$ and $x_1 = V_2(t)$ in the two cases $T = 2\pi, \pi/100$. From the graphics decide whether the input V_1 is passed and amplified, passed but not amplified, partially filtered, or stopped.

www 5.   (*Sensitivity of Lead System to Lead Intake Levels*). The intake rate of lead into the blood is 49.3 micrograms per day in system (1). If the lead levels in food, air, and water are lowered (reducing the intake rate below 49.3), the levels in the blood, tissues, and bones diminish.

(a) Explain why if the intake rate is lowered from I_1 to αI_1, where α is a constant between 0 and 1, then the steady-state level in each of the compartments is lowered by the same proportion α. [*Hint*: According to Section 7.9, the steady-state vector of $x' = Ax + F_0$, where F_0 is constant and all eigenvalues of A have negative real parts, is $-A^{-1}F_0$.]

(b) Plot graphs of the amount of lead in the blood, tissues, and bones over a 1200-day period if $x_1(0) = x_2(0) = x_3(0) = 0$, if the system matrix A is that of system (1), and if the intake rate is reduced from 49.3 micrograms per day to (49.3)/2 per day from the 400th day on. Then repeat if the intake rate is cut in half again from the 800th day on to (49.3)/4 micrograms per day. Compare the computed results with predictions based on part (a).

(c) Suppose that the intake rate I_1 of lead into the blood diminishes from the 400th day:

$$I_1 = \begin{cases} 49.3, & 0 \le t \le 400 \\ 49.3\exp(a(400-t)), & t > 400 \end{cases}$$

where a is a positive parameter. Use a solver/plotter package to solve system (1) with this intake function for various positive values of a, $0 < a < 0.1$. Estimate the smallest value of a that will ensure that the lead level in the bones will never exceed 3000 micrograms.

 6. (*Seasonally Varying Lead Levels*). Replace the intake rate $I_1 = 49.3$ in system (1) by the time-dependent intake rate $I_1 = 49.3[1 + 9\sin^2(2\pi t/365)]/10$. Explain why this models seasonal variations in lead intake rates. Then plot $x_3(t)$ over a long enough time span that you can approximate the maximal lead level in the bones. What is this level and when is it first reached?

7.11 The Theory of General Linear Systems

☞ It would be a good idea to scan Section 3.7 before tackling this section.

The linear systems considered so far in this chapter have all had constant coefficients, and our approach to finding solution formulas was constructive. We didn't make use of a Fundamental Theorem in the process because we didn't need it. We found all solutions of $x' = Ax$ for the constant $n \times n$ matrix A by first finding a basis for $\mathbb{R}^n$ (or $\mathbb{C}^n$) consisting of eigenvectors and generalized eigenvectors of A. Our choice to look at constant-coefficient linear systems first was not accidental. Almost all the linear systems we solve in this text have constant coefficients. But now it's time to see what we can do in the way of a solution formula for $x' = Ax$ when the entries of A are not constants.

We end this chapter with a brief summary of the theory of general linear systems. Our description will be somewhat telegraphic because it parallels the approach for $x' = Ax$ presented at the end of Section 7.6, and in Section 7.8. Our starting point is a Fundamental Theorem for the driven system $x' = Ax + F$.

The Fundamental Theorem for Linear Systems

The Fundamental Theorem for systems (Theorem 5.2.1) can be sharpened for linear systems.

THEOREM 7.11.1

Fundamental Theorem for Linear Systems. For an $n \times n$ matrix $A(t) = [a_{ij}(t)]$ and an n-function vector $F(t) = [F_1(t) \ \ F_2(t) \ \ \cdots \ \ F_n(t)]^T$ consider the IVP

$$x' = A(t)x + F(t), \qquad x(t_0) = x^0 \tag{1}$$

where the coefficients $a_{ij}(t)$ and the entries $F_j(t)$ are all continuous on a common t-interval I, t_0 is any point in I, and x^0 is any constant n-vector. Then IVP (1) has a unique solution that is defined at least for all t in the interval I and is a continuous function of t_0 and x^0.

Existence and uniqueness follows from Theorem 5.2.1 since the vector function $f = A(t)x + F(t)$ and all the first partial derivatives $\partial f_i / \partial x_j = a_{ij}(t)$, $i, j = 1, \ldots, n$, are continuous for all t in I and x in $\mathbb{R}^n$ by assumption. We omit showing that the solution $x(t)$ can be extended to I and varies continuously with t_0 and x^0.

Theorem 7.11.1 does not give a formula or process for finding the solution of IVP (1). Our goal is to derive a representation for the solution of IVP (1). This representation has more theoretical than practical value.

Overview of Our Approach

Note that the operator L that produces $x' - A(t)x = L[x]$ is linear just as we saw in Section 7.8 for the constant matrix A case, so the Linearity Property and the Closure Property hold. Just as before, the Linearity Property divides our task of finding all solutions of the system $L[x] = F$ into two parts: (1) find a particular solution x^d of the driven system $L[x] = F$, and (2) find *all* solutions of the undriven system $L[x] = 0$.

Let's first look at the undriven system

$$x' = A(t)x \tag{2}$$

where the $n \times n$ matrix A has entries that are continuous on a t-interval I. For any n solutions $x^1(t), x^2(t), \ldots, x^n(t)$ of system (2), the Closure Property says that $x(t) = c_1 x^1 + c_2 x^2 + \cdots + c_n x^n$ is also a solution for any constants $c_1, \ldots, c_n$. Are there more solutions? Suppose there is a value t_0 in I such that $\det[x^1(t_0) \ \ x^2(t_0) \ \ \cdots \ \ x^n(t_0)] \neq 0$. Now suppose that $z(t)$ is any solution of system (2). Then, according to Theorem 7.3.1, there are unique constants $k_1, k_2, \ldots, k_n$ such that

$$z(t_0) = k_1 x^1(t_0) + k_2 x^2(t_0) + \cdots + k_n x^n(t_0)$$

So $z(t)$ and $k_1 x^1(t) + k_2 x^2(t) + \cdots + k_n x^n(t)$ both solve the IVP

$$x' = A(t)x, \qquad x(t_0) = z(t_0)$$

and by Theorem 7.11.1 they must be one and the same solution. In this case the solution set $\{x^1, x^2, \ldots, x^n\}$ captures all solutions of system (2). This motivates the following definition.

☞ We first saw basic solution sets in Section 7.6.

❖ **Basic Solution Set.** The set of n solutions $\{x^1, x^2, \ldots, x^n\}$ of the system $x' = A(t)x$, where $A(t)$ is an $n \times n$ matrix whose entries $a_{ij}(t)$ are continuous on a t-interval I, is said to be a *basic solution set* if there is a t_0 in I such that $\det[x^1(t_0) \ \ x^2(t_0) \ \ \cdots \ \ x^n(t_0)] \neq 0$.

Actually, the determinant condition in this definition comes up often enough that it has acquired a name of its own:

☞ We first met the Wronskian in Section 3.7, but it had a different form.

❖ **Wronskian.** Suppose that $x^1, x^2, \ldots, x^n$ are n solutions of the system $x' = Ax$, where A is an $n \times n$ matrix. The determinant

$$W[x^1, x^2, \ldots, x^n] = \det[x^1(t) \; x^2(t) \; \cdots \; x^n(t)]$$

is called the *Wronskian* of the n solutions $x^1(t), \ldots, x^n(t)$.

☞ Only need to check that $W \neq 0$ at one point.

If A is an $n \times n$ matrix with entries continuous on some common t-interval, it can be shown that the Wronskian of n solutions of the system $x' = Ax$ is either always zero or never zero. This is a consequence of the following result.

THEOREM 7.11.2

> **Wronskian Property for $x' = Ax$.** If the entries $a_{ij}(t)$ in the $n \times n$ matrix $A(t)$ are continuous on some interval I, and if $x^1, x^2, \ldots, x^n$ are solutions of the system $x' = A(t)x$, then the Wronskian $W = W[x^1, \ldots, x^n]$ satisfies the ODE
>
> $$W' = [a_{11}(t) + a_{22}(t) + \cdots + a_{nn}(t)]\,W, \qquad t \text{ in } I \qquad (3)$$

We show this for $n = 2$. Suppose that $y(t) = [y_1 \; y_2]^T$ and $z(t) = [z_1 \; z_2]^T$ are two solutions of $x' = Ax$. Then we see that

$$\begin{cases} y_1' = a_{11}y_1 + a_{12}y_2 \\ y_2' = a_{21}y_1 + a_{22}y_2 \end{cases} \quad \text{and} \quad \begin{cases} z_1' = a_{11}z_1 + a_{12}z_2 \\ z_2' = a_{21}z_1 + a_{22}z_2 \end{cases}$$

and since $W[y, z] = y_1 z_2 - y_2 z_1$, we see that

$$W' = \frac{d}{dt}W[y, z] = y_1' z_2 + y_1 z_2' - y_2' z_1 - y_2 z_1'$$

$$= (a_{11}y_1 + a_{12}y_2)z_2 + y_1(a_{21}z_1 + a_{22}z_2) - (a_{21}y_1 + a_{22}y_2)z_1 - y_2(a_{11}z_1 + a_{12}z_2)$$

$$= (a_{11} + a_{22})(y_1 z_2 - y_2 z_1) = (a_{11} + a_{22})W$$

which is ODE (3) for $n = 2$. To show this for $n > 2$, the properties of determinants given in Section 7.3 must be used.

So it appears that to find the general solution for $x' = A(t)x$ we only need to find a basic solution set. Problem 1 shows the existence of a basic solution set—the argument uses Theorem 7.11.1. Summarizing, we have the following result.

THEOREM 7.11.3

> **Basic Solution Sets and IVPs.** Suppose that $\{x^1, \ldots, x^n\}$ is a basic solution set for $x' = A(t)x$, where the entries of the $n \times n$ matrix $A(t)$ are continuous on a t-interval I. Then for each t_0 in I and any x^0 in $\mathbb{R}^n$ (or $\mathbb{C}^n$), the IVP
>
> $$x' = A(t)x, \qquad x(t_0) = x^0$$
>
> has the solution $x = c_1 x^1 + \cdots + c_n x^n$ for a unique set of constants $c_1, \ldots, c_n$.

Theorem 7.11.3 generalizes the similar result in Theorem 7.6.3 that was for linear undriven systems with constant coefficients.

Solution Matrices

There is nothing special about the functions $x^1(t)$, ..., $x^n(t)$ used to form a basic solution set for the system $x' = A(t)x$. Any set of n solutions of $x' = A(t)x$ will do, as long as the Wronskian of the set is nonzero on I. The examples of Section 7.5 and 7.6 give ways to construct the basic solution sets when the system matrix A is constant.

From this point on, our development so closely parallels the material in Section 7.8 that we only sketch its outlines. In particular, if the columns of a matrix $X(t)$ form a basic solution set for the system $x' = A(t)x$, then $X(t)$ is a *basic solution matrix*. A basic solution matrix $X(t)$ is nonsingular for all t in I since $\det X(t)$ is the Wronskian, so $\det X(t) \neq 0$ on I. The *transition matrix* $\Phi(t, t_0)$, t, t_0 in I, is the unique basic solution matrix for which $\Phi(t_0, t_0) = I_n$.

☞ The exponential matrix e^{tA} is the transition matrix for $x' = Ax$ if A is a matrix of constants and $t_0 = 0$.

If C is any $n \times n$ nonsingular matrix of constants and $X(t)$ is a basic solution matrix, we can derive Φ from any basic solution matrix $X(t)$ as follows:

$$\Phi = X(t)X^{-1}(t_0), \quad t, t_0 \text{ in } I \tag{4}$$

In the important case where the system matrix A has constant entries, the methods of Sections 7.5 and 7.6 suffice for finding $\Phi(t, t_0)$, which in fact is $e^{(t-t_0)A}$. For a linear system with a time-varying matrix $A(t)$, it is usually hard to find an explicit formula for Φ, but here is an example where we can construct Φ.

EXAMPLE 7.11.1

Transition Matrix if $A(t)$ Is Nonconstant
The nonautonomous, undriven linear system (defined for $|t| < 1$)

$$x' = \frac{1}{1-t^2}(-tx + y)$$
$$y' = \frac{1}{1-t^2}(x - ty) \tag{5}$$

has a solution, $x = 1$, $y = t$, and a second solution, $x = t$, $y = 1$, as a direct calculation shows. The two solutions have a nonvanishing Wronskian and so form a basic solution set. From formula (4) the transition matrix for system (5) is

$$\Phi(t, t_0) = \begin{bmatrix} 1 & t \\ t & 1 \end{bmatrix}\begin{bmatrix} 1 & t_0 \\ t_0 & 1 \end{bmatrix}^{-1} = \frac{1}{1-t_0^2}\begin{bmatrix} 1 - tt_0 & t - t_0 \\ t - t_0 & 1 - tt_0 \end{bmatrix} \tag{6}$$

☞ Why do the orbits appear to be tangent to a curve?

where t and t_0 lie in the interval $(-1, 1)$. Notice that $\Phi(t_0, t_0) = I_2$. See Figures 7.11.1 and 7.11.2 for graphs of orbits and component curves of system (5).

$$x' = (-tx + y)/(1 - t^2), \quad y' = (x - ty)/(1 - t^2)$$
$$x(0) = 5, \quad |y(0)| \le 8$$

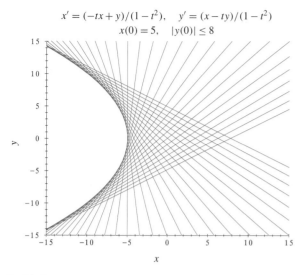

FIGURE 7.11.1 The envelope of straight-line orbits looks like a parabola (Example 7.11.1).

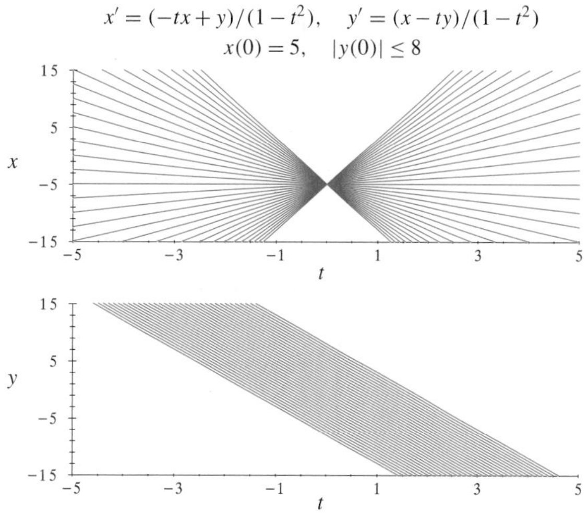

$$x' = (-tx + y)/(1 - t^2), \quad y' = (x - ty)/(1 - t^2)$$
$$x(0) = 5, \quad |y(0)| \le 8$$

FIGURE 7.11.2 Straight line solution curves don't "see" singularities at $t = \pm 1$ (Example 7.11.1).

The transition matrix has a number of important properties that are described in the following theorem:

THEOREM 7.11.4

> **Properties of the Transition Matrix.** Suppose that $\Phi(t, t_0)$ is the transition matrix for $x' = A(t)x$, where the entries of the $n \times n$ matrix A are continuous on some common t-interval I, and t, t_0 are any points in I. Then:
>
> **Identity:** $\Phi(t_0, t_0) = I_n$, the identity matrix.
>
> **Derivative:** $\Phi'(t, t_0) = A(t)\Phi(t, t_0)$.
>
> **Basic Matrices:** $\Phi(t, t_0)B$ is a basic solution matrix for any $n \times n$ nonsingular constant matrix B.
>
> **Transition from x^0 to $x(t)$:** The IVP $x' = A(t)x$, $x(t_0) = x^0$ has the unique solution given by $x(t) = \Phi(t, t_0)x^0$.
>
> **Product:** $\Phi(t, t_0) = \Phi(t, t_1)\Phi(t_1, t_0)$, for all t, t_0, t_1 in I.
>
> **Inverse:** $\Phi^{-1}(t, s) = \Phi(s, t)$, for all s, t in I.

☞ The transition matrix takes the initial state along the solution to the state at time t.

The first four properties follow immediately from the definition of Φ. The other properties are shown as follows:

Product: $X_1 = \Phi(t, t_0)$ and $X_2 = \Phi(t, t_1)\Phi(t_1, t_0)$ solve the same matrix IVP, $X' = AX$, $X(t_1) = \Phi(t_1, t_0)$. So by uniqueness, $X_1 = X_2$.

Inverse: Use the product and identity properties to show that $\Phi(t, s)\Phi(s, t)$ is the identity matrix for all s, t in the interval I.

Variation of Parameters

According to our plan for solving the driven IVP $x' = A(t)x + F(t)$, $x(t_0) = x^0$, the next step is to find a particular solution $x^d(t)$ of the driven system. Any solution will do, and the transition matrix can be used to carry out the construction in exactly the same way as was done in Theorem 7.8.3.

THEOREM 7.11.5

Variation of Parameters for Driven Systems. Suppose that $\Phi(t, t_0)$ is the transition matrix for $x' = A(t)x$, for t, t_0 in the interval I, where the entries of $A(t)$ are continuous on I. Then for any vector function $F(t)$ whose components are continuous on I, the vector function

$$x^d(t) = \Phi(t, t_0) \int_{t_0}^{t} \Phi(t_0, s) F(s)\, ds = \int_{t_0}^{t} \Phi(t, s) F(s)\, ds \qquad (7)$$

is the unique solution of the IVP

$$x' = A(t)x + F(t), \qquad x(t_0) = 0 \qquad (8)$$

The integrals in (7) act on vector functions component by component.

Note that the second equality in formula (7) is a consequence of the product property of transition matrices: $\Phi(t, t_0)\Phi(t_0, s) = \Phi(t, s)$.

We can now solve the central problem for systems of linear ODEs in exactly the same way as was done in Theorem 7.8.4.

THEOREM 7.11.6

Solution of a Linear System. Suppose that $\Phi(t, t_0)$ is the transition matrix for $x' = A(t)x$, where the entries of $A(t)$ are continuous on an interval I, and t, t_0 are in I. Then the general solution of $x' = A(t)x + F(t)$, t in I, is

$$x(t) = \Phi(t, t_0)c + \Phi(t, t_0) \int_{t_0}^{t} \Phi(t_0, s) F(s)\, ds \qquad (9)$$

where c is any constant vector and t_0 is any point in I. The solution of the initial value problem $x' = A(t)x + F(t)$, $x(t_0) = x^0$, t_0 in I, is

$$
\begin{array}{ccccc}
x(t) & = & \Phi(t, t_0)x^0 & + & \Phi(t, t_0) \displaystyle\int_{t_0}^{t} \Phi(t_0, s) F(s)\, ds \\
\text{Total} & = & \text{Response to} & + & \text{Response to} \\
\text{response} & & \text{initial data} & & \text{driving force}
\end{array} \qquad (10)
$$

Formulas (9) and (10) require the prior construction of the transition matrix Φ (not always easy for a general linear system) and then integration of the components of the vector ΦF (also not always easy).

EXAMPLE 7.11.2

Solving $x' = A(t)x + F(t)$

Referring to Example 7.11.1, we see from formula (10) that the solution of the IVP

$$x' = \frac{1}{1-t^2}(-tx+y) + F_1(t), \qquad x(0) = a$$

$$y' = \frac{1}{1-t^2}(x-ty) + F_2(t) \qquad y(0) = b$$

over the interval I, $|t| < 1$, is

$$\begin{bmatrix} x(t) \\ y(t) \end{bmatrix} = \Phi(t,0)\begin{bmatrix} a \\ b \end{bmatrix} + \int_0^t \Phi(t,s)\begin{bmatrix} F_1(s) \\ F_2(s) \end{bmatrix} ds$$

where by formula (6)

$$\Phi(t,0) = \begin{bmatrix} 1 & t \\ t & 1 \end{bmatrix}, \qquad \Phi(t,s) = \frac{1}{1-s^2}\begin{bmatrix} 1-ts & t-s \\ t-s & 1-ts \end{bmatrix}$$

Here is one case where formula (9) is not too formidable.

PROBLEMS

1. (*Basic Solution Set*). Show that there is a basic solution set for the linear system of ODEs $x' = A(t)x$, where the entries in A are continuous on a t-interval I. [*Hint*: Suppose that u^j is the column n-vector with 1 in the j-th place and 0's elsewhere. Use Theorem 7.11.1 to show that each of the n IVPs $x' = Ax$, $x(t_0) = u^j$, $j = 1, 2, \ldots, n$, has a unique solution. Call these solutions $x^1, \ldots, x^n$ and consider $W[x^1, \ldots, x^n](t_0)$.]

2. Look at the system

$$x' = \begin{bmatrix} 0 & 1 \\ -2t^{-2} & 2t^{-1} \end{bmatrix} x, \qquad t > 0$$

   (a) Show that $M(t) = \begin{bmatrix} t^2 & t \\ 2t & 1 \end{bmatrix}$ is a basic solution matrix for the system. [*Hint*: Verify directly that each column vector of $M(t)$ is a solution of the system.]

   (b) Find the transition matrix $\Phi(t, t_0)$, t, $t_0 > 0$. [*Hint*: Use $M(t)$ and equation (4).] Verify that $\Phi(t_0, t_0) = I_2$ and that $\Phi^{-1}(t, t_0) = \Phi(t_0, t)$ if t, $t_0 > 0$.

   (c) Show that the system in part (a) is equivalent to the Euler equation: $t^2 y'' - 2ty' + 2y = 0$.

☞ More on Euler ODEs in Problem 4, Section 3.7.

3. Look at the system

$$x' = \begin{bmatrix} t^{-1} & 0 & -1 \\ 0 & 2t^{-1} & -t^{-1} \\ 0 & 0 & t^{-1} \end{bmatrix} x, \qquad t > 0$$

   (a) Show that $M(t) = \begin{bmatrix} t & t & -t^2 \\ 0 & t^2 & t \\ 0 & 0 & t \end{bmatrix}$ is a basic solution matrix for the system.

   (b) Find the transition matrix $\Phi(t, t_0)$, t, $t_0 > 0$. [*Hint*: Use $M(t)$ and equation (4).]

4. Solve the following initial value problems:

(a) $x' = \begin{bmatrix} 0 & 2 \\ -2 & 0 \end{bmatrix} x + \begin{bmatrix} \cos \omega t \\ \sin \omega t \end{bmatrix}$, $\quad x(0) = \begin{bmatrix} a \\ b \end{bmatrix}$, $\quad \omega \neq 0$. Are there any values of ω for which the response of the system is unbounded?

(b) $x' = \begin{bmatrix} 0 & 1 \\ -2t^{-2} & 2t^{-1} \end{bmatrix} x + \begin{bmatrix} t^3 \\ t^4 \end{bmatrix}$, $\quad x(1) = \begin{bmatrix} 0 \\ 0 \end{bmatrix}$, $\quad t > 0$. [*Hint*: $[t^2 \ 2t]^T$ and $[t \ 1]^T$ are solutions of the undriven system.]

(c) $x' = \begin{bmatrix} t^{-1} & 0 & 0 \\ -t^{-2} & t^{-1} & 0 \\ t^{-3} & -t^{-2} & t^{-1} \end{bmatrix} x + \begin{bmatrix} t^3 \\ t^2 \\ t^3 \end{bmatrix}$, $\quad x(1) = \begin{bmatrix} 1 \\ 0 \\ 1 \end{bmatrix}$, $\quad t > 0$. [*Hint*: $[t \ 1 \ 0]^T$, $[0 \ t \ 1]^T$, and $[0 \ 0 \ t]^T$ are solutions of the undriven system.]

5. Suppose that $A(t)$ is a real $n \times n$ matrix.

(a) If $x'(t) = A(t)x(t)$ and $(z^T(t))' = -z^T(t)A(t)$, show that $\sum_{i=1}^n x_i(t)z_i(t) = z^T(t)x(t)$ is constant for all t. [*Hint*: $(z^T x)' = (z^T)'x + z^T x' = 0$.]

(b) Suppose that A is *skew symmetric* [i.e., $A(t) = -A^T(t)$]. Show that $\|x(t)\|$ is constant for all t if $x'(t) = A(t)x(t)$; that is, show that each orbit of this differential equation lies on a "sphere" of radius $\|x(0)\|$ in state space centered at the origin.

6. Suppose that entries in the matrix A have common period T. Show that if $\Phi(t, t_0)$ is the transition matrix for $x' = A(t)x$, then $\Phi(t + T, t_0) = \Phi(t, t_0)C$ for some constant vector C.

www 7. (*Nonconstant System Matrix*). Consider the system $x' = y$, $y' = -2x/t^2 + 2y/t$, $\quad t > 0$.

(a) Show that $X(t) = \begin{bmatrix} t^2 & t \\ 2t & 1 \end{bmatrix}$ is a basic solution matrix.

(b) Find $\Phi(t, t_0)$, and verify that $\Phi(t_0, t_0) = I$, $\Phi^{-1}(t, t_0) = \Phi(t_0, t)$, $\quad t_0 > 0$.

(c) Plot the component curves and the orbits for the system in part **(a)**, given the initial data $x(1) = 1$, $y(1) = -5, -4, \ldots, 4, 5$.

8. Show that the method outlined in Problem 8 in Section 3.7 is a special case of Theorem 7.8.3. [*Hint*: Transform the second-order scalar ODE $y'' + a(t)y' + b(t)y = f$ to the system $u' = v$, $v' = -bu - av + f$, where $u = y$, $v = y'$.]

9. (*Another Approach to Variation of Parameters*). Prove Theorem 7.8.3 by assuming that $x' = A(t)x + F$, $x(t_0) = 0$, has a solution of the form $x = \Phi(t, t_0)c(t)$, where Φ is the transition matrix for $x' = A(t)x$ and $c(t)$ is the unknown "varied parameter" vector. [*Hint*: Insert Φc for x in the ODE, use properties of Φ, and show that $c' = \Phi(t_0, t)F(t)$, from which $c(t) = \int_{t_0}^t \Phi(t_0, s)F(s) \, ds$ if $c(t_0) = 0$.]

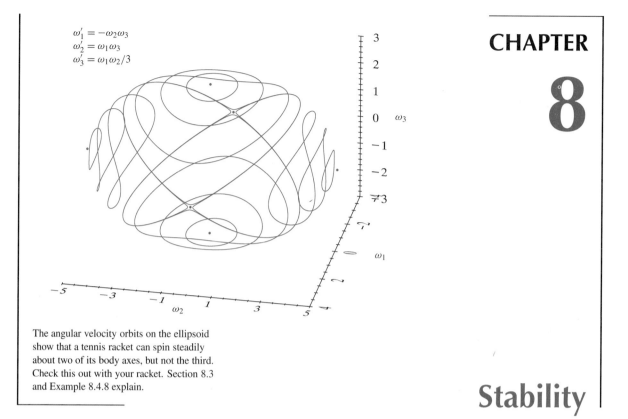

$$\omega_1' = -\omega_2\omega_3$$
$$\omega_2' = \omega_1\omega_3$$
$$\omega_3' = \omega_1\omega_2/3$$

The angular velocity orbits on the ellipsoid
show that a tennis racket can spin steadily
about two of its body axes, but not the third.
Check this out with your racket. Section 8.3
and Example 8.4.8 explain.

CHAPTER

8

Stability

*We expect the systems and structures of everyday life to be stable. Push a
stable system a little bit away from its equilibrium, and it will stay nearby
or even return to the equilibrium state. We will translate this informal idea
of stability into mathematical terms and give tests for stability that can be
applied to natural processes modeled by autonomous systems of ODEs.*

8.1 Stability and Linear Systems

Suppose that a damped simple pendulum initially hangs straight down at rest. The
bob will remain in that so-called equilibrium position forever. Tap the bob slightly
and watch its decaying amplitudes as it oscillates around and approaches its original
position. That position is said to be asymptotically stable. Now balance the bob at
rest vertically upward. The slightest disturbance sends the bob downward and away,
so we call the upward equilibrium position unstable. Finally, grease the support point,
put the pendulum structure in a tall jar, and pump out the air so that we have a simple
pendulum without any damping forces. Release the bob from rest at an angle of $\pi/3$
radians and the bob oscillates periodically without decay about the downward equilibrium position. In this case, the downward equilibrium is said to be neutrally stable.

☞ The stability
properties of a system tell
us about the long-term
behavior of some
solutions.

In this section we translate these informal notions of stability and instability into
mathematical terms, state the criteria for asymptotic stability, neutral stability, and
instability of an equilibrium state of a linear autonomous system, and then give some
examples. In the next section we will find stability criteria for an equilibrium of a
nearly linear autonomous system and return to the pendulum model.

Stability

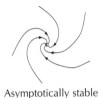

Asymptotically stable

Neutrally stable

Let's suppose that the *n*-vector function $f(x)$ is continuously differentiable in $\mathbb{R}^n$.
According to the Fundamental Theorem (Theorem 5.2.1), the IVP

$$x' = f(x), \qquad x(0) = x^0, \quad x^0 \text{ in } \mathbb{R}^n \tag{1}$$

has a unique, maximally extended solution that we denote by $x(t, x^0)$. From this point
on we suppose that solutions of $x' = f(x)$ are always maximally extended and defined
for $t \geq 0$. The simplest solutions of system (1) are the *equilibrium solutions*, $x = p$ for
all t, where $f(p) = 0$. The orbit of an equilibrium solution is an *equilibrium point* in
state space.

Let's explain what we mean when we say that a system is stable at an equilibrium
point. First, though, let's recall that $\|x - p\|$ denotes the *distance* between the points
$x = (x_1, \ldots, x_n)$ and $p = (p_1, \ldots, p_n)$ in state space, so

$$\|x - p\| = [(x_1 - p_1)^2 + \cdots + (x_n - p_n)^2]^{1/2}$$

Here's the definition that is central to everything we do in this chapter.

Unstable

Unstable

❖ **Stability.**   The system $x' = f(x)$ is *stable* at an equilibrium point p if for
each positive number ε there is a positive number δ such that:

If $\|x^0 - p\| < \delta$,   then $\|x(t, x^0) - p\| < \varepsilon$,   for all $t \geq 0$

The system is *asymptotically stable* at p if it is stable at p and $\|x(t, x^0) - p\| \to 0$
[i.e., $x(t, x^0) \to p$] as $t \to +\infty$ for all points x^0 near p. The system is *neutrally
stable* at p if it is stable at p, but not asymptotically stable. The system is *unstable* at p if it is not stable at p.

So stability means that if an orbit starts within distance δ of the equilibrium point p,
from then on the orbit stays within distance ε of p.

An equilibrium point p is said to be an *attractor* if $\|x(t, x^0) - p\| \to 0$ as $t \to +\infty$
for all x^0 near p. So the system is asymptotically stable at p if it is stable at p and
p is an attractor. The system is neutrally stable at p if it is stable at p, but p is *not*
an attractor. Instability means that for some positive number ε_0 there are points q
arbitrarily close to p with the property that at some future time T the point $x(T, q)$ is
farther away from p than distance ε_0. The sketches in the margin illustrate these ideas.

Sometimes we say simply that "the equilibrium point p is asymptotically stable
[or neutrally stable or unstable]" without referring directly to the system $x' = f(x)$.
Alternatively, we might say that "the system is stable [or unstable]," but we have to

☞ **Warning!** Some
attractors are unstable.
Take a look at Problem 6.

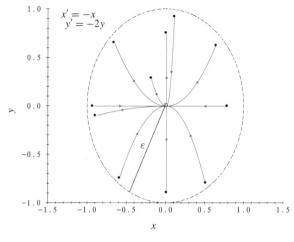

FIGURE 8.1.1 Asymptotic stability: orbits that start closer to the origin than ε stay closer than ε and tend to the origin as $t \to +\infty$ (Example 8.1.1).

FIGURE 8.1.2 Neutral stability: orbits that start (at the solid dots) closer to $(0, 0)$ than δ stay closer than ε, but they do not approach $(0, 0)$ (Example 8.1.2).

be careful about using expressions like this because the system might be stable at one equilibrium point and unstable at another.

Examples

Here are several examples that illustrate the properties of stability and instability.

EXAMPLE 8.1.1

An Asymptotically Stable Linear System

Let's verify the conditions for asymptotic stability for the linear system

$$x' = -x, \qquad y' = -2y \tag{2}$$

at the equilibrium point $(0, 0)$. The general solution is

$$x(t) = C_1 e^{-t}, \qquad y(t) = C_2 e^{-2t}, \qquad C_1, \ C_2 \text{ any constants} \tag{3}$$

The distance r from the orbital point $(x(t), y(t))$ to the origin is

$$r(t) = \left[x^2(t) + y^2(t)\right]^{1/2} = \left[C_1^2 e^{-2t} + C_2^2 e^{-4t}\right]^{1/2} \tag{4}$$

Because the exponential terms in (4) decrease to 0 as $t \to +\infty$, so does $r(t)$. This means that for every positive number ε we can take $\delta = \varepsilon$ in the definition of stability:

If $\left[x^2(0) + y^2(0)\right]^{1/2} < \delta = \varepsilon$, then $[x^2(t) + y^2(t)]^{1/2} < \varepsilon$, for all $t \geq 0$

and system (2) is stable at the origin. Figure 8.1.1 shows the circle of radius ε centered at the origin. Orbits with initial points inside the circle stay inside as time increases. The origin is also an attractor since $r(t) \to 0$ as $t \to +\infty$, so system (2) is actually asymptotically stable at the origin.

Here's a system that is neutrally stable.

EXAMPLE 8.1.2

The Neutral Stability of an Undamped Hooke's Law Spring
The undriven motion of a body of mass m suspended by a Hooke's Law spring, with positive spring constant k and no damping, is described by the ODE

$$mx'' + kx = 0$$

The equivalent first-order linear system is

$$x' = y, \qquad y' = -(k/m)x \tag{5}$$

Let's use a method introduced in Section 1.7 to find the equations of the orbits of system (5). Dividing y' by x', we obtain the separable ODE $dy/dx = -kx/(my)$. Separating the variables and solving, we obtain the equations for the orbits

$$kx^2 + my^2 = C \tag{6}$$

where C is any positive constant. So the orbits form a family of ellipses centered at the origin, which is the equilibrium point of system (5).

We will use these ellipses to show that system (5) is stable at the equilibrium. Suppose that ε is any positive number and construct the circle of radius ε about the origin. Given values for k and m, there is a value of C such that the elliptical orbit defined by (6) touches the ε-circle from the inside. Now choose the positive number δ so that the circle about the origin of radius δ touches that ellipse from the inside. From the way the ellipse and the δ-circle are defined, we see that any orbit with initial point inside the δ-circle stays inside the ellipse for all t and so inside the ε-circle.

Figure 8.1.2 shows ε- and δ-circles and the ellipse constructed above for $k = 4$ and $m = 1$. The two smaller elliptical orbits in the figure start inside the δ-circle ($\delta = 1/2$) at the points $(0.15, 0)$ and $(0.35, 0)$ and stay within the ε-circle ($\varepsilon = 1$) for all time.

EXAMPLE 8.1.3

An Unstable Linear System
Look at the system $x' = x$, $y' = -2y$. Some solutions of this system are of the form $x = ae^t$, $y = 0$, where a is any positive constant. They "blow up" as $t \to +\infty$. This means that no matter how small a circle you take centered at the origin, there are orbits that start inside that circle at $t = 0$ but become unbounded as $t \to +\infty$, so the system is unstable (Figure 8.1.3).

The previous examples are planar and linear, but the ideas of stability apply in any dimension and to nonlinear systems as well. Here's an example.

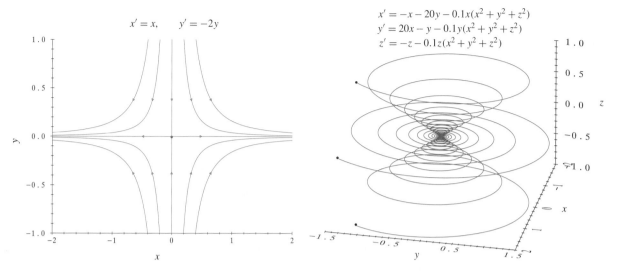

$x' = x, \qquad y' = -2y$

$$x' = -x - 20y - 0.1x(x^2 + y^2 + z^2)$$
$$y' = 20x - y - 0.1y(x^2 + y^2 + z^2)$$
$$z' = -z - 0.1z(x^2 + y^2 + z^2)$$

FIGURE 8.1.3 Only two orbits of this unstable system are attracted to the origin (Example 8.1.3).

FIGURE 8.1.4 Orbits of asymptotically stable system spiral inward from initial points (Example 8.1.4).

EXAMPLE 8.1.4

An Asymptotically Stable Nonlinear System with Three State Variables
The origin is an equilibrium point for the nonlinear system

$$x' = -x - 20y - 0.1x(x^2 + y^2 + z^2)$$
$$y' = 20x - y - 0.1y(x^2 + y^2 + z^2) \qquad (7)$$
$$z' = -z - 0.1z(x^2 + y^2 + z^2)$$

and the system is asymptotically stable at the origin. Here's why.

We have no solution formula for system (7), so we take an indirect approach to determine stability at the origin. Let $\rho(t)$ denote the distance $\left[x^2(t) + y^2(t) + z^2(t)\right]^{1/2}$ from an orbital point $(x(t), y(t), z(t))$ to the origin. If we differentiate $\rho(t)$ using the Chain Rule and the rate functions in (7), we have

$$\rho'(t) = \frac{1}{2}\left[x^2 + y^2 + z^2\right]^{-1/2}[2xx' + 2yy' + 2zz']$$

$$= \frac{1}{\rho}\left[-x^2 - y^2 - z^2 - 0.1(x^2 + y^2 + z^2)^2\right]$$

$$= -\frac{1}{\rho}\left(\rho^2 + 0.1\rho^4\right) = -\rho - 0.1\rho^3$$

Since $\rho \geq 0$, we have

$$\rho' + \rho = -0.1\rho^3 \leq 0$$

Multiplying each side of this inequality by the "integrating factor" e^t, we see that

$$(\rho e^t)' \leq 0$$

This implies that $\rho(t)e^t$ is a nonincreasing function as t increases, so

$$\rho(t)e^t \leq \rho(0)e^0, \quad \text{for } t \geq 0$$

or, solving for $\rho(t)$,

$$\rho(t) \leq \rho(0)e^{-t}$$

☞ So we have shown asymptotic stability without using solution formulas.

This last inequality shows that the distance from the orbital point $(x(t), y(t), z(t))$ to the origin decreases exponentially to 0 as $t \to +\infty$. So for any positive number ε we can set $\delta = \varepsilon$ and meet the criterion for stability.

The origin is also an attractor since $\rho(t) \to 0$ as $t \to +\infty$. This means that system (7) is asymptotically stable at the origin (Figure 8.1.4).

The *basin of attraction* of an asymptotically stable equilibrium point p is the set of all points x^0 such that $\|x(t, x^0) - p\| \to 0$ as $t \to +\infty$. The asymptotic stability at p is *global* if the basin of attraction is all of state space, and is *local* otherwise. The origin in Example 8.1.4 is globally asymptotically stable; its basin of attraction is $\mathbb{R}^3$.

A system may have several equilibrium points, and the system may be stable at some of these points and unstable at others. Here is a one-dimensional example.

EXAMPLE 8.1.5

An ODE with Stable and Unstable Equilibrium Points
The IVP in one state variable,

$$x' = (1 - x)x, \qquad x(0) = x_0 \tag{8}$$

is a logistic IVP with the x-axis as the state space. So orbits of the ODE in (8) are traced out on the x-axis. The points $p = 0$ and $p = 1$ are equilibrium points. The Sign Analysis techniques of Section 2.2 show that as $t \to +\infty$, the solution $x(t, x_0)$ of IVP (8) approaches 1 if $x_0 > 0$. The ODE is locally asymptotically stable at $p = 1$, with basin of attraction the interval $x > 0$. Similarly, if $x_0 < 0$, then $x(t, x_0)$ falls away from 0 as t increases. The ODE is unstable at $p = 0$. The state line in the margin illustrates all these properties in a very compact way.

x-axis

So far, our approach to testing for stability has been case by case. Luckily, there are stability tests that apply to broad categories of systems. The first of these takes care of all autonomous undriven linear systems.

Linear Systems and Stability

It is no accident that several of our examples are constant coefficient linear systems. These systems provide the simplest examples of stability and instability and arise most frequently in the applications. The following stability test for linear systems is a consequence of the analysis in Section 7.9.

THEOREM 8.1.1

Stability Properties of a Linear System. Suppose that A is an $n \times n$ matrix of real constants. For each eigenvalue λ of A, suppose that m_λ denotes the multiplicity of λ and d_λ the dimension of λ's eigenspace. Then:

1. The system $x' = Ax$ is **globally asymptotically stable** at the origin if and only if every eigenvalue of A has a negative real part.

2. The system $x' = Ax$ is **neutrally stable** at the origin if and only if
   - Every eigenvalue of A has a nonpositive real part, and
   - At least one eigenvalue has a zero real part, and $d_\lambda = m_\lambda$ for every eigenvalue λ with a zero real part.

3. The system $x' = Ax$ is **unstable** at the origin if and only if
   - Some eigenvalue of A has a positive real part,

   or
   - There is an eigenvalue λ with a zero real part and $d_\lambda < m_\lambda$.

☞ The eigenvalues of A tell the whole stability and long-term behavior story for the ODE $x' = Ax$.

Let's take a look at **1**. If all eigenvalues of A have negative real parts, then by the Decay Estimate (Theorem 7.9.4) there is a *negative* constant a and a positive constant M such that for all $t \geq 0$,

$$\|x(t, x^0)\| \leq M\|x^0\|e^{at}$$

for every solution $x(t, x^0)$ of $x' = Ax$. In the definition of stability, take $\delta = \varepsilon/M$ for each $\varepsilon > 0$. Then we see that if $\|x^0\| < \varepsilon/M$, then

$$\|x(t, x^0)\| \leq M\|x^0\|e^{at} < M(\varepsilon/M)e^{at} \leq \varepsilon, \quad \text{for } t \geq 0$$

and the origin is stable. The origin is a global attractor since $e^{at} \to 0$ as $t \to +\infty$; so the system is globally asymptotically stable, and we have verified the "if" part of **1**. The origin's basin of attraction is all of state space.

Now let's look at **3**. Suppose that A has an eigenvalue $\lambda = \alpha + i\beta$ with positive real part α. If v is a corresponding eigenvector, then $x = ce^{\alpha t}\text{Re}\left[e^{i\beta t}v\right]$ is a real-valued solution for any real constant c. Since $\|x(t)\| = |c|e^{\alpha t}\|v\| \to \infty$ as $t \to +\infty$, the system $x' = Ax$ is unstable in this case. Now suppose that $\alpha = 0$ and $d_\lambda < m_\lambda$ for some eigenvalue λ; by Theorem 7.6.4 there are solutions of $x' = Ax$ that include terms of the form $t^k \text{Re}[e^{i\beta t}cv]$ for some positive integer k, where c is any real constant. Since $|c|t^k\|v\| \to \infty$ as $t \to +\infty$, the system has solutions that are unbounded as $t \to +\infty$, so the system is unstable. That verifies the "if" part of **3**.

The verification of the other parts of the theorem is omitted.

Theorem 8.1.1 applies to Examples 8.1.1–8.1.3. Because of the nature of the eigenvalues of the system matrices in these examples, the origin is, respectively, a globally asymptotically stable improper node (eigenvalues -2, -1), a neutrally stable center (eigenvalues $\pm i\sqrt{k/m}$), and an unstable saddle (eigenvalues -2, 1).

Sometimes we can look at a matrix A and figure out the stability properties of the system $x' = Ax$ with very little work. Here are some examples.

EXAMPLE 8.1.6

Ways to Determine Stability Without Much Effort

Let's determine the stability properties of these autonomous linear systems at the equilibrium point at the origin:

$$\textbf{(a)} \ x' = \begin{bmatrix} -10 & 300 & 1 \\ -5 & 20 & 11 \\ -4 & 11 & -9 \end{bmatrix} x \qquad \textbf{(b)} \ x' = \begin{bmatrix} -10 & 4 & -5 \\ -3 & -15 & 10 \\ -1 & 8 & -10 \end{bmatrix} x$$

$$\textbf{(c)} \ x' = \begin{bmatrix} 0 & 2 & 0 \\ -2 & 0 & 0 \\ 0 & 0 & -3 \end{bmatrix} x \qquad \textbf{(d)} \ x' = \begin{bmatrix} 0 & 1 & 0 & 0 \\ -1 & 0 & 0 & 0 \\ 0 & 0 & 0 & 1 \\ 0 & 0 & -1 & 0 \end{bmatrix} x$$

System **(a)** is unstable because the trace of the matrix is $-10 + 20 - 9 = +1$, so at least one eigenvalue has a positive real part (Theorem 7.9.8). System **(b)** is globally asymptotically stable because the Gerschgorin row disks for the matrix (Theorem 7.9.9) are entirely inside the left half of the complex plane, so all eigenvalues have negative real parts. System **(c)** is neutrally stable because the block structure of the matrix tells us that its eigenvalues are $\pm 2i$ and -3 (see page 365). System **(d)** is also neutrally stable. The block structure tells us that i and $-i$ are each double eigenvalues and each eigenspace has full dimension, that is, $m_\lambda = d_\lambda = 2$.

☞ Block matrices were defined in Section 7.2.

Comments

The conclusions of Theorem 8.1.1 also hold for the autonomous linear system $z' = A(z - p)$, which has an equilibrium point at p; just translate p to the origin by the variable change $x = z - p$, and apply Theorem 8.1.1 to $x' = Ax$.

PROBLEMS _____

1. (*Visual Stability Properties*). Decide whether the systems whose orbits are shown are asymptotically stable, neutrally stable, or unstable at each of the equilibrium points in the figures. Give informal explanations (i.e., you don't have to use δ's and ε's).

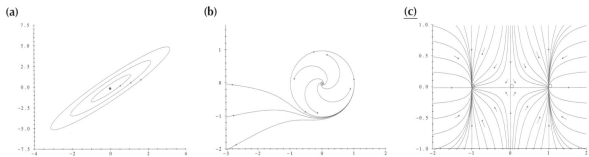

(a)   **(b)**   **(c)**

2. (*Neutral Stability*). Show that each of the following systems is neutrally stable at the origin. Plot portraits of orbits near the origin.

(a) $x' = 2y$, $y' = -8x$ (b) $x' = 0$, $y' = -y^3$ (c) $x' = y$, $y' = -x$, $z' = -z^3$

3. (*Asymptotic Stability*). Explain why each system is asymptotically stable at the origin. Is the asymptotic stability global? Plot portraits (in state space) of the orbits near the origin. [*Hint*: The graphs in parts (a)–(c) will be state lines.]

(a) $x' = -x$ (b) $x' = -x^3$

(c) $x' = -\sin x$ (d) $x' = -4x$, $y' = -3y$

(e) $x' = x - 3y$, $y' = 4x - 6y$ (f) $x' = -x + 4y$, $y' = -3x - 2y$

(g) $x' = -2x^3$, $y' = -5y$ (h) $x' = -x + y + z$, $y' = -2y$, $z' = -3z$

(i) $x' = -x^3$, $y' = -y^3$ (j) $x' = -x - y$, $y' = x - y$, $z' = -z^3$

4. (*A Stable Attractor*). Use polar coordinates to show that the system $x' = -x - 10y - x(x^2 + y^2)$, $y' = 10x - y - y(x^2 + y^2)$ is globally asymptotically stable at the origin.

www **5.** (*Instability*). The following systems are unstable at the origin. Explain why. Plot orbits (in state space) near the origin. Insert arrowheads on orbits to show the direction of motion as time increases. [*Hint*: The graphs in parts (a)–(c) will be state lines].

(a) $x' = x^2$ (b) $x' = \sin x$ (c) $x' = |x|$

(d) $x' = 3x - 2y$, $y' = 4x - y$ (e) $x' = 3x - 2y$, $y' = 2x - 2y$

(f) $x' = x + y$, $y' = -x - y$ (g) $x' = x^3$, $y' = -3y$

6. (*An Unstable Attractor*). Parts (a)–(c) outline the steps needed to show that the planar system $x' = x - y - x^3 - xy^2 + xy(x^2 + y^2)^{-1/2}$, $y' = x + y - x^2y - y^3 - x^2(x^2 + y^2)^{-1/2}$ has an unstable attractor at the equilibrium point $x = 1$, $y = 0$.

(a) Show that this system is equivalent to the system $r' = r(1 - r^2)$, $\theta' = 1 - \cos\theta$, where r, θ are polar coordinates in the xy-plane. Solve the rate equations for r and θ.

(b) Show that the point $x = 1$, $y = 0$ is an unstable attractor. [*Hint*: Use sign analysis to show that the unit circle consists of the equilibrium point $x = 1$, $y = 0$ and an orbit that leaves the point as t increases from $-\infty$ but returns to the point as $t \to +\infty$.]

 (c) Plot a portrait of orbits of the xy-system in the rectangle $|x| \le 2$, $|y| \le 1.5$.

7. Determine the stability properties of $x' = Ax$, where A is the given matrix.[*Hint*: Take a look at some of the tests given in Section 7.9.]

(a) $\begin{bmatrix} -1 & 2 \\ -2 & -1 \end{bmatrix}$ (b) $\begin{bmatrix} -3 & 1 \\ 2 & -4 \end{bmatrix}$ (c) $\begin{bmatrix} 1 & -1 \\ 2 & -2 \end{bmatrix}$

(d) $\begin{bmatrix} 0 & 1 \\ 0 & 0 \end{bmatrix}$ (e) $\begin{bmatrix} 5 & 1 \\ 1 & -2 \end{bmatrix}$ (f) $\begin{bmatrix} 3 & -2 \\ 5 & -3 \end{bmatrix}$

(g) $\begin{bmatrix} -10 & 3 & -4 \\ 1 & -12 & 9 \\ -1 & -2 & -5 \end{bmatrix}$ (h) $\begin{bmatrix} -5 & 0 & 0 \\ 0 & 0 & -3 \\ 0 & 3 & 0 \end{bmatrix}$ (i) $\begin{bmatrix} 0 & 1 & 1 \\ 0 & 0 & 1 \\ 0 & 0 & 0 \end{bmatrix}$

(j) $\begin{bmatrix} 0 & 0 & 1 & 0 \\ 0 & 0 & 0 & 0 \\ 0 & 0 & 0 & 0 \\ 0 & 0 & 0 & 0 \end{bmatrix}$ (k) $\begin{bmatrix} 3 & -2 & 0 & 0 \\ 5 & -3 & 0 & 0 \\ 0 & 0 & 3 & -2 \\ 0 & 0 & 5 & -3 \end{bmatrix}$ (l) $\begin{bmatrix} 3 & -2 & 1 & 0 \\ 5 & -3 & 0 & 1 \\ 0 & 0 & 3 & -2 \\ 0 & 0 & 5 & -3 \end{bmatrix}$

8.2   Stability of a Nearly Linear System

The stability of an autonomous linear system is completely determined by the nature of the eigenvalues of the system matrix (Theorem 8.1.1). But what happens if the system has some "small" nonlinearities? What is the stability story in this case? The first example below suggests that the inclusion of small nonlinearities may not have much effect on an asymptotically stable linear system.

EXAMPLE 8.2.1

Linear and Nonlinear Springs
The two planar systems

$$x' = y, \qquad y' = -101x - y \tag{1}$$

$$x' = y, \qquad y' = -101x - y + x^2 \tag{2}$$

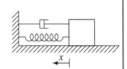

model, respectively, the motion of a viscously damped Hooke's Law spring and a viscously damped soft spring, each attached to a wall and moving back and forth on a horizontal platform. Each system has an equilibrium point at the origin. System (1) is linear, so we can use the methods of Sections 7.7 and 8.1 to describe the nature of its orbits and to determine its stability properties. The matrix of coefficients has characteristic polynomial $\lambda^2 + 2\lambda + 101$ and eigenvalues $-1 \pm 10i$, so the origin is a globally asymptotically stable spiral point.

System (2) is the same as system (1) except for the nonlinear term x^2. For small $|x|$, we expect the x^2 term to have only a slight effect on solutions in comparison to the effect of the term $101x$. Figures 8.2.1 and 8.2.2 bear this out. Although the term x^2 does not seem to change orbits and solution curves very much when $|x|$ is small, the story may be completely different when $|x|$ is large.

The terms y and $101x$ in systems (1) and (2) have "order 1," while x^2 has "order 2." The first order of business in this section is to define just what "the order of a function" means, and to do that we need to define the term "open ball." Then we can explain what we mean by a "nearly linear system" and determine stability criteria for such a system.

Open Ball, Order of a Function

The inside of an interval on the real line, the inside of a circle in $\mathbb{R}^2$, and the inside of a sphere in $\mathbb{R}^3$ are examples of an "open ball," which is defined as follows:

☞ $B_r(q)$ is called a ball because that's what it looks like in $\mathbb{R}^3$. It is *open* because its boundary points aren't included.

❖ **Open Ball in $\mathbb{R}^n$.** Suppose that q is a point in $\mathbb{R}^n$ and r is a positive number. Then the set of all points x in $\mathbb{R}^n$ closer to q than r [all x with $\|x - q\| < r$] is called the *open ball* in $\mathbb{R}^n$ of radius r centered at q; it is denoted by $B_r(q)$.

We use the term "open ball" a lot from now on, but often without reference to $\mathbb{R}^n$. Now let's look into the notion of the order of a function or function vector at a point.

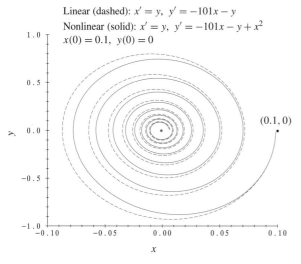

Linear (dashed): $x' = y$, $y' = -101x - y$
Nonlinear (solid): $x' = y$, $y' = -101x - y + x^2$
$x(0) = 0.1$, $y(0) = 0$

Linear (dashed): $x' = y$, $y' = -101x - y$
Nonlinear (solid): $x' = y$, $y' = -101x - y + x^2$
$x(0) = 0.1$, $y(0) = 0$

FIGURE 8.2.1 Orbits of the linear and nonlinear spring systems stay close together and spiral toward the asymptotically stable origin (Example 8.2.1).

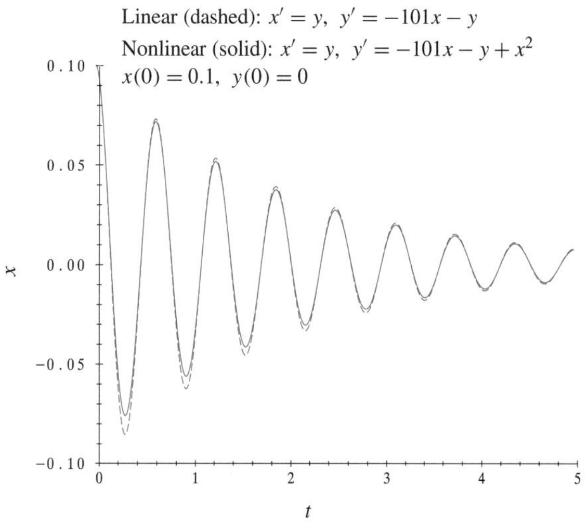

FIGURE 8.2.2 The x-components of the linear and the nonlinear spring systems stay close together as they tend to 0 (Example 8.2.1).

☞ We use $\| \cdot \|$ to denote distance because P, x, and p may be vectors.

❖ **Order of a Function.** Suppose that $P(x)$ is a function for which $P(p) = 0$. $P(x)$ has *order* k, $k \geq 0$, at p if there is a positive constant c such that

$$\| P(x) \| \leq c\|x - p\|^k \tag{3}$$

for all x in some open ball centered at p.

Note that if P has order k, then P also has order m for all $0 < m < k$. Here are three examples of functions and their orders at particular points.

EXAMPLE 8.2.2

Two Function Vectors of Order 2

The function vector $P(x, y) = [0 \;\; x^2]^T$ has order 2 at the origin $x = 0$, $y = 0$ because

$$\| P(x, y) \| = \left(0^2 + \left(x^2 \right)^2 \right)^{1/2} = x^2$$

☞ For single terms such as $3x^2$, $-100y^3$, or $20x^2y^2$, "order" is the same as "degree," which is, respectively, 2, 3, 4.

The column vector function $P(x, y) = [x^2 + xy \;\; 3y^2 - x^2]^T$ also has order 2 at the origin because (using polar coordinates r and θ)

$$\| P(x, y) \| = [(x^2 + xy)^2 + (3y^2 - x^2)^2]^{1/2}$$
$$= [r^4 (\cos^2 \theta + \cos \theta \sin \theta)^2 + r^4 (3 \sin^2 \theta - \cos^2 \theta)^2]^{1/2}$$
$$\leq r^2 [(1 + 1)^2 + 3^2]^{1/2} = \sqrt{13}\, r^2 = \sqrt{13}(x^2 + y^2) = \sqrt{13}\, \|(x, y)\|^2$$

where we have used the Triangle Inequality and the fact that $|\cos \theta| \leq 1$, $|\sin \theta| \leq 1$. In the same way, we see that the function vector $P(x, y) = [0 \;\; -5x^2y^3]^T$ has order 5 at the origin.

Extending the techniques of Example 8.2.2, we see that if the components of P are all polynomials with zero constant terms and if N is the lowest degree of the terms appearing in the polynomial, then P has order N at the origin. So for example the function vector $P = [3xy - y^7 \quad 2x^3y^3 + 5y^4]$ has order 2 at the origin because of the term $3xy$. Now let's look at the order of a linear system.

EXAMPLE 8.2.3

Order of $A(x - p)$

Suppose that $x - p$ is a column n-vector and A is a nonzero $n \times n$ constant matrix. Then the column vector function $A(x - p)$ is first-order at p. This is a consequence of a property of matrix norms listed in Section 7.2:

$$\text{If} \quad \|A\| = \sum_{i,j} |a_{ij}|, \quad \text{then } \|A(x - p)\| \leq \|A\| \cdot \|x - p\|$$

Referring to inequality (3), we see that we can set $c = \|A\|$, $k = 1$ in the definition of order, so the vector $A(x - p)$ has order 1 at p.

Now we can define nearly linear systems and characterize their stability properties.

Nearly Linear Systems and Their Stability

Let's suppose that A is an $n \times n$ matrix of real constants and consider the two systems

$$x' = A(x - p)$$
$$x' = A(x - p) + P(x)$$

Here's the definition of "nearly linear."

❖ **Nearly Linear System**. Suppose that $P(x)$ is a continuously differentiable function vector, that $P(p) = 0$, and that $P(x)$ has order at least 2 at p. Then the system $x' = A(x - p) + P(x)$ is said to be *nearly linear* at p, while the linear system $x' = A(x - p)$ is said to be the *linearization at* p. $P(x)$ is a *higher-order perturbation* of the linearized system.

We can illustrate these concepts by using the results of Examples 8.2.1 and 8.2.2.

EXAMPLE 8.2.4

A Nearly Linear System and Its Linearization

The second system in Example 8.2.1 is

$$\begin{align} x' &= y \\ y' &= -101x - y + x^2 \end{align} \tag{4}$$

The higher-order perturbation $P(x, y) = [0 \quad x^2]^T$ has order 2 at the origin $(0, 0)$, so system (4) is nearly linear at the origin. The linearization of system (4) at the origin is obtained by throwing out the term x^2 from system (4) to obtain the linearized system $x' = y$, $y' = -101x - y$.

Figures 8.2.1 and 8.2.2 suggest that the stability properties at the origin of system (4) and the corresponding linearized system are the same. The following theorem implies that this is indeed the case.

THEOREM 8.2.1

Stability of a Nearly Linear System. Suppose that A is an $n \times n$ matrix of real constants. Suppose also that $P(x)$ is a function vector that is continuously differentiable in an open ball $B_r(p)$ in $\mathbb{R}^n$, that $P(p) = 0$, and that $P(x)$ has order at least 2 at p. Then the nearly linear system

$$x' = A(x - p) + P(x) \qquad (5)$$

has the following stability properties:

1. The system is **asymptotically stable** at p if all eigenvalues of A have negative real parts.

2. The system is **unstable** at p if A has an eigenvalue with a positive real part.

☞ The proof of part **1** of this theorem is given in the Student Resource Manual in the notes for Section 8.4; the proof uses material from that section.

This means that we have a test for the stability properties of system (5) at any equilibrium point p. Here are the steps of the test:

- Determine the signs of the real parts of the eigenvalues of A by:
  - Direct calculation.
  - Using a computer program.
  - Using one of the methods described in Section 7.9.

☞ So now we know a lot about the long-term behavior of a nearly linear system

- Conclude that:
  - The system is **asymptotically stable** at p if the signs are all negative.
  - The system is **unstable** at p if at least one of the signs is positive.
  - More information is needed if some eigenvalue has a zero real part, but none has a positive real part.

The "more information needed" alternative comes about because there is a gap in Theorem 8.2.1: It does not treat the situation where one or more of the eigenvalues of A has a zero real part and no eigenvalue has a positive real part. Indeed, as indicated in Problems 4 and 11, the higher-order terms in the system determine the stability properties in this case. We will take up a general approach to stability in Section 8.4, which will (usually) handle these cases of eigenvalues with zero real parts.

The rest of this section shows how we can use Theorem 8.2.1 to discuss and (usually) determine stability properties of nearly linear systems.

Stability and the Jacobian Matrix

Theorem 8.2.1 is a powerful theorem, but how can we use it if we have a system that doesn't have the form (5) of a nearly linear system? Taylor expansions and Jacobian matrices come to the rescue and show how we can write the n-vector system

$$x' = f(x), \quad \text{where } f(p) = 0$$

in the form of a nearly linear system at p.

☞ Look at
Theorem B.5.16 for
Taylor's Theorem for
function vectors.

Let's suppose that $f(x)$ is at least three times continuously differentiable in the ball $B_r(p)$. Then by Taylor's Theorem we have

$$f(x) = f(p) + J(p)(x - p) + P(x) \tag{6}$$

where $J(p)$ is the $n \times n$ matrix $[\partial f_i / \partial x_j]_p$ of all the first-order partial derivatives of the components of $f(x)$ evaluated at $x = p$ and is called the *Jacobian matrix* of f at p. The remainder term $P(x)$ in the Taylor expansion is continuously differentiable in the ball $B_r(p)$ and has order at least 2 at p: So we can write $x' = f(x)$ in the form of a nearly linear system at p.

$$x' = f(x) = J(p)(x - p) + P(x) \tag{7}$$

We can apply Theorem 8.2.1 to determine the stability properties of system (7) at p.

Let's look at the damped simple pendulum system as our first illustration of how we can use Theorem 8.2.1 and Jacobian matrices.

EXAMPLE 8.2.5

☞ Take a look
back at Section 4.1 for
a discussion of
pendulum motion.

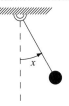

Stability of the Damped Simple Pendulum

The system that models a damped simple pendulum is

$$\begin{aligned} x' &= f(x, y) = y \\ y' &= g(x, y) = -b \sin x - ay \end{aligned} \tag{8}$$

where x is the angle from the downward vertical and a and b are positive constants determined by the mass of the bob, the length of the rod, and the viscous damping coefficient. We are interested in the stability properties at the equilibrium points $(0, 0)$ (pendulum hanging straight down at rest) and $(\pi, 0)$ (pendulum balanced straight up).

Let's look at the equilibrium point at the origin. The Jacobian matrix of the right side of system (8) evaluated at the origin is

$$J(0, 0) = \begin{bmatrix} \partial f / \partial x & \partial f / \partial y \\ \partial g / \partial x & \partial g / \partial y \end{bmatrix}_{(0,0)} = \begin{bmatrix} 0 & 1 \\ -b \cos x & -a \end{bmatrix}_{(0,0)} = \begin{bmatrix} 0 & 1 \\ -b & -a \end{bmatrix}$$

The characteristic polynomial of J is $\lambda^2 + a\lambda + b$ and its eigenvalues are

$$\lambda_{1,2} = -\frac{a}{2} \pm \frac{1}{2}(a^2 - 4b)^{1/2}$$

Since a and b are positive, both eigenvalues have negative real parts. For most pendulums the damping constant a is small in comparison to b, so $(a^2 - 4b)^{1/2}$ is imaginary. This means that λ_1 and λ_2 are complex conjugates with real part the negative number $-a/2$. We will assume this from now on.

Next we need to find the higher-order perturbation by doing a Taylor expansion of the right-hand side of system (8) about $x_0 = 0$, $y_0 = 0$. The infinitely differentiable function $-b \sin x$ is the only nonlinear term. Now the Taylor expansion of $\sin x$ about $x_0 = 0$ is

$$\begin{aligned} \sin x &= \sin 0 + (\cos 0)x - \frac{\sin 0}{2!}x^2 + P(x) \\ &= x + P(x) \end{aligned}$$

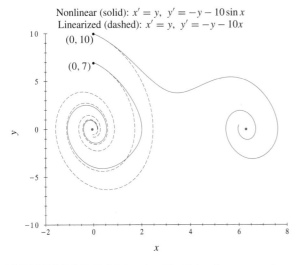

Nonlinear (solid): $x' = y$, $y' = -y - 10\sin x$
Linearized (dashed): $x' = y$, $y' = -y - 10x$

Nonlinear (solid): $x' = (4 - 2x + y)x$, $y' = (4 + x - 2y)y$
Linearized (dashed): $x' = -8(x - 4) + 4(y - 4)$,
$y' = 4(x - 4) - 8(y - 4)$

FIGURE 8.2.3 Initial point $(0, 7)$ is close enough to $(0, 0)$ that nonlinear orbit (solid) and linearized orbit (dashed) have the same behavior; not so for orbits through $(0, 10)$ (Example 8.2.5).

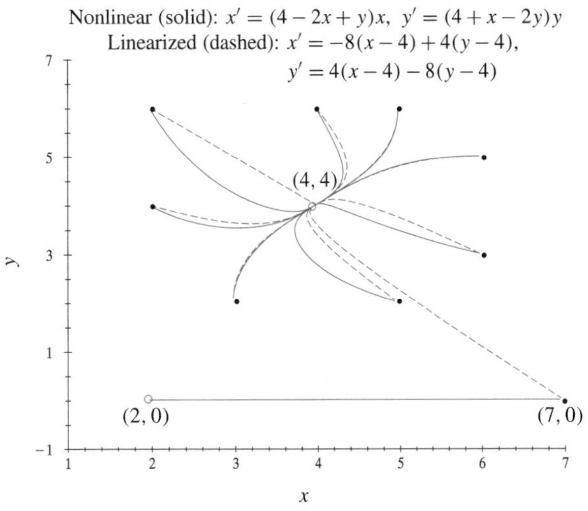

FIGURE 8.2.4 Some orbits of nonlinear system (11) (solid) and of linearized system (12) (dashed). All dashed orbits tend to $(4, 4)$ as $t \to +\infty$, but one of the solid orbits tends to $(2, 0)$ (Example 8.2.6).

The remainder function $P(x)$ is certainly continuously differentiable in x (since $\sin x$ possesses all derivatives) and $P(x)$ has order 3 at $x = 0$. So we can write system (8) as a nearly linear system about $(0, 0)$:

$$\begin{bmatrix} x \\ y \end{bmatrix}' = J(0, 0) \begin{bmatrix} x \\ y \end{bmatrix} + \begin{bmatrix} 0 \\ -bP(x) \end{bmatrix}, \qquad J(0, 0) = \begin{bmatrix} 0 & 1 \\ -b & -a \end{bmatrix} \qquad (9)$$

The linearized system at $(0, 0)$ is

$$\begin{bmatrix} x \\ y \end{bmatrix}' = J(0, 0) \begin{bmatrix} x \\ y \end{bmatrix} \qquad (10)$$

Since the eigenvalues of J have negative real parts, system (8) is asymptotically stable at $(0, 0)$ by Theorem 8.2.1, and so is its linearized system (10).

We can say much more. Not only are the stability properties of systems (8) and (10) the same at the origin, but the orbits behave similarly. Both sets of orbits starting out near the origin execute decaying spirals around the origin; orbits of (8) and (10) starting at the same point remain very close together as t increases (Figure 8.2.3). But all is not well if orbits of (8) and (10) start from a point at some distance from the origin. For example, the linearized orbit from $(0, 10)$ spirals into $(0, 0)$ as $t \to +\infty$ just as all orbits of system (10) must, but the nonlinear orbit starting at $(0, 10)$ has so much kinetic energy that it goes over the top and doesn't approach the origin at all (Figure 8.2.3).

These stability methods can be applied to the nonlinear hard and soft spring models of Section 3.1 and to many of the interacting species models of Chapter 5. Let's take another look at interacting species.

EXAMPLE 8.2.6

Stability Properties of a Cooperation Model
  Cooperative interaction between two self-limiting species is modeled by the system

☞ This system was
introduced in
Example 5.3.2

$$x' = (4 - 2x + y)x$$
$$y' = (4 + x - 2y)y \tag{11}$$

The orbits shown in Figure 5.3.2 suggest that the system is unstable at the equilibrium points $(0, 0)$, $(2, 0)$, and $(0, 2)$, but asymptotically stable at the equilibrium point $(4, 4)$. To verify these stability properties we will find the eigenvalues of the Jacobian matrix of system (11) at each of the equilibrium points and use Theorem 8.2.1. When (11) is written in nearly linear form at any one of its equilibrium points, the higher-order perturbations always have degree 2 at the equilibrium point.

  The Jacobian matrix at a general point (x, y) is

$$J(x, y) = \begin{bmatrix} 4 - 4x + y & x \\ y & 4 - 4y + x \end{bmatrix}$$

The Jacobian matrices at the equilibrium points $(0, 0)$, $(2, 0)$, $(0, 2)$, and $(4, 4)$ are

$$\begin{bmatrix} 4 & 0 \\ 0 & 4 \end{bmatrix}, \quad \begin{bmatrix} -4 & 2 \\ 0 & 6 \end{bmatrix}, \quad \begin{bmatrix} 6 & 0 \\ 2 & -4 \end{bmatrix}, \quad \begin{bmatrix} -8 & 4 \\ 4 & -8 \end{bmatrix}$$

and the respective eigenvalues are

$$\lambda_{1,2} = 4, 4; \qquad \lambda_{1,2} = -4, 6; \qquad \lambda_{1,2} = 6, -4; \qquad \lambda_{1,2} = -4, -12$$

  So we see from Theorem 8.2.1 and by the signs of the eigenvalues that system (11) is unstable at the first three equilibrium points, but asymptotically stable at the fourth point $(4, 4)$. Let's check out the behavior of orbits near the equilibrium point $(4, 4)$.

  Figure 8.2.4 shows some orbits (solid) of the nonlinear system as they tend with increasing time to the asymptotically stable point $(4, 4)$. Also shown is an orbit (solid) of the nonlinear system that approaches the equilibrium point $(2, 0)$. The dashed curves in Figure 8.2.4 are the orbits of the linearized system at $(4, 4)$

$$\begin{bmatrix} x' \\ y' \end{bmatrix} = \begin{bmatrix} -8 & 4 \\ 4 & -8 \end{bmatrix} \begin{bmatrix} x - 4 \\ y - 4 \end{bmatrix} \tag{12}$$

from the same set of initial points. Generally speaking, the two sets of orbits are quite close, but there are some differences—look at the orbits that start at $x_0 = 7$, $y_0 = 0$. All orbits of the linearized system (12) tend to $(4, 4)$ as $t \to +\infty$, but the orbits of the nonlinear system (11) may behave quite differently than the orbits of the linearized system if both start out at a distance from the equilibrium point under study [e.g., the orbits starting at $(7, 0)$ in Figure 8.2.4].

  Linear system (12) has an asymptotically stable improper node at $(4, 4)$ because the eigenvalues of $J(4, 4)$ are distinct and negative (-4 and -12). Figure 8.2.4 shows that the orbits of nonlinear system (11) near $(4, 4)$ also approach the point in a nodal way and are generally close to the linearized orbits through the same points, if these starting points are close enough to $(4, 4)$.

Linear and Nonlinear Nodes, Spirals, and Saddles

The table following this section shows orbital portraits near the origin of the system

$$\begin{bmatrix} x \\ y \end{bmatrix}' = A \begin{bmatrix} x \\ y \end{bmatrix} + \begin{bmatrix} P(x, y) \\ Q(x, y) \end{bmatrix}, \qquad A = \begin{bmatrix} a & b \\ c & d \end{bmatrix} \tag{13}$$

☞ This is just what
Theorem 8.2.1 says.

where $a, b, c,$ and d are real constants, P and Q are continuously differentiable and are of order at least 2 at the origin, and the eigenvalues of A are nonzero. The perturbations P and Q can't alter the stability properties of the linearized system

$$\begin{bmatrix} x \\ y \end{bmatrix}' = A \begin{bmatrix} x \\ y \end{bmatrix} \tag{14}$$

☞ Nodes, saddles,
and spiral points are
described in Section 7.7.

at the origin (as long as the real parts of the eigenvalues are nonzero). But even more is true: The nodal-, spiral-, or saddle-like behavior of the linearized system is preserved by the nonlinear system (13) near the origin if the eigenvalues are distinct and have nonzero real parts, although the perturbed orbits aren't quite the same. For this reason, the terms "nodes," "saddles," and "spiral points" are used even if the system is nonlinear; all of these equilibrium points are said to be *elementary*.

PROBLEMS

1. (*Nearly Linear Systems and Stability*). Determine the stability properties of each system at the origin. [*Hint:* Use Theorem 8.2.1. Use a Jacobian matrix for part **(d)**.]

    (**a**) $x' = -x + y^2, \quad y' = -8y + x^2$ (**b**) $x' = 2x + y + x^4, \quad y' = x - 2y + x^3 y$

    (**c**) $x' = 2x - y + x^2 - y^2, \quad y' = x - y$ (**d**) $x' = e^{x+y} - \cos(x - y), \quad y' = -\sin x$

2. (*Jacobian Matrices and Stability*). Discuss the stability properties of each system at each of its equilibrium points.

    (**a**) $x' = y, \quad y' = -6x - y - 3x^2$ (**b**) $x' = y^2 - x, \quad y' = x^2 - y$

    (**c**) $x' = -x - x^3, \quad y' = y + x^2 + y^2$ (**d**) $x' = -y - x(x^2 + y^2), \quad y' = x - y(x^2 + y^2)$

    (**e**) $x' = x + xy^2, \quad y' = x$ (**f**) $x' = -x + y^2, \quad y' = x + y$

3. Plot several orbits of each of the systems in Problems 2**(a)**–**(f)** in the neighborhood of each equilibrium point. Relying solely on the visual appearance of these orbits, what are your conclusions about the nature of the eigenvalues of the Jacobian matrix at each equilibrium point?

4. (*The Troubling Case of the Zero Eigenvalue*). Show that the linear approximation at the origin to the system $x' = -x, \ y' = ay^3$ has eigenvalues -1 and 0. Show that the system is asymptotically stable at the origin if a is a negative constant, unstable if a is a positive constant. What happens if $a = 0$? Plot the orbits in all three cases. Why does Theorem 8.2.1 not apply? What do you conclude about basing a stability analysis entirely on the eigenvalues of the Jacobian matrix?

☞ This system was
introduced in
Example 5.2.3.

5. (*Planar Portrait*). Look at the system $x' = x - y + x^2 - xy, \ y' = -y + x^2$.

    (**a**) Find the three equilibrium points.

    (**b**) Using Jacobian matrices and their eigenvalues, determine the stability properties of the system at each equilibrium point.

    (**c**) Create the first figure below for the given system: First plot orbits in the rectangle $|x| \le 0.5$, $|y| \le 0.5$ (see solid arcs). Then find the linearized system at the origin and plot some of its orbits in the same rectangle (see dashed arcs).

(d) Create the second orbital portrait shown below from the given system: Locate and mark the equilibrium points. Finally, use other orbits to create a plot like the third figure shown below.

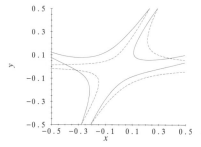 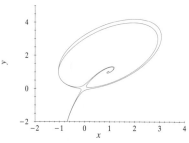 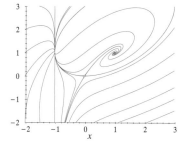

6. (*Basin of Attraction*). Let's consider the system $x' = y$, $y' = -2x - y - 3x^3$.

(a) Show that the system is asymptotically stable at the origin but not globally.

 (b) Use computer graphics to find the approximate basin of attraction of the origin. Use the plotting region $-3 \le x \le 2$, $-3 \le y \le 5$.

7. (*A Perturbed RLC Circuit*). The ODE $Lx'' + Rx' + x/C + g(x, x') = 0$ governs the behavior of a simple *RLC* electric circuit, where x represents the charge on the capacitor, x' is the current in the loop, and L, R, and C are the positive circuit parameters inductance, resistance, and capacitance. The twice continuously differentiable function $g(x, x')$ represents nonlinearities in the circuit. Suppose that $g(x, x')$ has order at least 2 in an open ball centered at $(0, 0)$ in the xx'-plane. Show that the equivalent system $x' = y$, $y' = -x/LC - Ry/L - g(x, y)/L$ is asymptotically stable at the origin.

8. For each system below, find the equilibrium points and the Jacobian matrix and its eigenvalues at each equilibrium point. Use this information to determine the stability properties of the system at each equilibrium point.

(a) $x' = -x - x^3$, $y' = -y$ **(b)** $x' = \alpha x - y + x^2$, $y' = x + \alpha y + x^2$, constant $\alpha > -1$

www 9. (*Stabilizing a System by Tuning a Parameter*). Find all values of the constant α such that the system $x' = z + x^2 y$, $y' = x - 4y + xz^2$, $z' = \alpha x + 2y - z + x^2$ is asymptotically stable at the origin. [*Hint:* Use the Routh Test of Section 7.9.]

 10. (*Interacting Species Models*). Consider the interacting species model $x' = (2 + ax + by)x$, $y' = (2 + cx + dy)y$ in the cases indicated below. Give a complete stability analysis and model interpretations of each system at each equilibrium point. Plot orbits in the population quadrant.

- Cooperation leading to explosive growth: $a = d = -1$ and $b = c = 2$. See Example 5.3.1.

- Stable competition: $a = d = -2$ and $b = c = -1$. See Example 5.3.3.

- Competitive exclusion: $a = d = -1$ and $b = c = -2$. See Example 5.3.4.

- Fix the parameters $b = c = -1$ and take nonzero values for a and d; analyze the resulting model.

 11. (*Another Awkward Case*). Theorem 8.2.1 doesn't cover the case where all eigenvalues have nonpositive real parts and some have zero real parts. Problem 4 showed the difficulties that can occur if there is a zero eigenvalue. Construct planar systems with equilibrium point at the origin where the Jacobian matrix at the origin has some pure imaginary eigenvalues and the perturbed system is asymptotically stable, neutrally stable, or unstable, depending on the nature of the higher-order perturbation term.

 12. (*Upended Pendulum*). Give a complete analysis of the stability properties of the damped simple pendulum system at the equilibrium point $(\pi, 0)$. Use your numerical solver and plot orbits of the system and its linearization near $(\pi, 0)$.

Stability of Perturbed Planar Systems

The following table compares the stability characteristics and orbital portraits of the nonlinear system (13) and its linear approximation (14). Only the asymptotically stable, neutrally stable, and saddle cases for A are portrayed. In the cases where the real parts of both eigenvalues are positive, reverse the arrowheads and label the portraits "unstable."

Eigenvalues of A	$\begin{bmatrix} x \\ y \end{bmatrix}' = A \begin{bmatrix} x \\ y \end{bmatrix}$	$\begin{bmatrix} x \\ y \end{bmatrix}' = A \begin{bmatrix} x \\ y \end{bmatrix} + \begin{bmatrix} P \\ Q \end{bmatrix}$
$\lambda_1 = \alpha + i\beta = \bar{\lambda}_2$ $\alpha < 0,\ \beta \neq 0$ If $\alpha > 0$, reverse arrows.	Asymptotically stable spiral point	Asymptotically stable spiral point
$\lambda_1 = \lambda_2 < 0$ If $\lambda_1 = \lambda_2 > 0$, reverse arrows.	Asymptotically stable star node (nondeficient eigenspace) Improper node (deficient eigenspace)	Asymptotically stable node or spiral point
$\lambda_1 < \lambda_2 < 0$ If $\lambda_1, \lambda_2 > 0$, reverse arrows.	Asymptotically stable improper node	Asymptotically stable improper node
$\lambda_1 = i\beta = \bar{\lambda}_2$ $\beta \neq 0$	Neutrally stable center	Stability depends on P and Q Center, or spiral point, or center-spiral[1]
$\lambda_1 < 0 < \lambda_2$	Unstable saddle	Unstable saddle

[1] See Problem 3 in Section 9.1 for an example of a system of ODEs with a center-spiral.

8.3 Conservative Systems

Consider these scenarios: An undamped spring oscillates periodically about its equilibrium position forever; an undamped simple pendulum whirls unceasingly about its pivot. Something is conserved here. What is it? It turns out that the conserved quantity is a function of position and velocity called the total energy of the system. The total energy of each of the motions never changes, so it's conserved. Start the spring or the pendulum motion out with a little less or a little more energy, and the new level of energy will also be conserved on the new orbit. A *conservative system* of ODEs is an autonomous system for which some real-valued function (e.g., energy) of the state variables remains constant on each orbit. In this section we will examine the orbital behavior and stability properties of conservative systems.

Some Examples of Conservation for Planar Systems

Let's start by looking at planar autonomous systems of the form

$$\begin{aligned} x' &= g(y) \\ y' &= f(x) \end{aligned} \tag{1}$$

> ☞ Planar systems like (1) were treated in Section 1.7, but in the separable form $g(y)dy/dx - f(x) = 0$.

Suppose that $G(y)$ and $F(x)$ are any antiderivatives of $g(y)$ and $f(x)$, respectively. Then the value of the function K defined by

$$K(x, y) = G(y) - F(x) \tag{2}$$

is "conserved" along each orbit $x = x(t),\ y = y(t)$ of system (1). To see this, use the Chain Rule and (1) to show that the rate of change of $K(x(t), y(t))$ is zero:

$$\frac{d}{dt}K(x(t), y(t)) = \frac{\partial K}{\partial x}\frac{dx}{dt} + \frac{\partial K}{\partial y}\frac{dy}{dt} = -f(x)g(y) + f(x)g(y) = 0$$

So the value of K stays constant along the orbit $x(t),\ y(t)$, although the value may change from one orbit to another.

The function $K(x, y)$ is called an *integral* of system (1) because it is defined by integrating g and f. Each orbit of system (1) lies on a *level set* (or *contour curve*) of K: $K(x, y) = C$ for some constant C. Note that if $K(x, y)$ is conserved on each orbit of (1), then so is $\alpha K + \beta$ for any constants $\alpha \neq 0$ and β.

Usually in applications $g(y) = y$ [so $G(y) = y^2/2$], and then K or $\alpha K + \beta$ is interpreted as "energy." Examples taken from Sections 3.1 and 4.1 illustrate this approach.

EXAMPLE 8.3.1

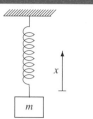

Neutrally Stable Spring System: Conservation of Energy

The linear system

$$\begin{aligned} x' &= y \\ y' &= -kx/m \end{aligned} \tag{3}$$

models the motion of an undamped Hooke's Law spring-mass system about the equilibrium point $x = 0,\ y = 0$. The body of mass m is supported by the spring with

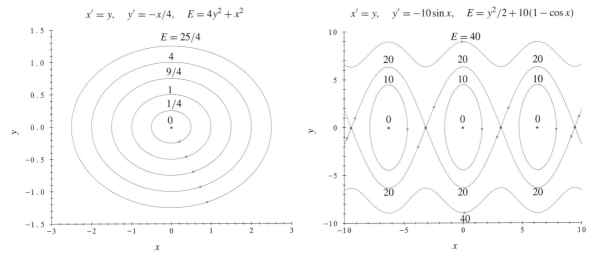

$$x' = y, \quad y' = -x/4, \quad E = 4y^2 + x^2 \qquad\qquad x' = y, \quad y' = -10\sin x, \quad E = y^2/2 + 10(1-\cos x)$$

FIGURE 8.3.1 Constant energy elliptical orbits for an undamped spring-mass system surround the neutrally stable equilibrium point (Example 8.3.1).

FIGURE 8.3.2 Constant energy orbits and neutrally stable and unstable equilibrium points of the undamped simple pendulum (Example 8.3.2).

spring constant k. Since the eigenvalues $\pm i\sqrt{k/m}$ of the coefficient matrix are simple, the system is neutrally stable (Theorem 8.1.1).

In this case $g(y) = y$, $G(y) = y^2/2$, $f(x) = -kx/m$, and $F(x) = -kx^2/(2m)$. So by formula (2), $K(x, y) = G(y) - F(x) = y^2/2 + kx^2/(2m)$ is an integral of system (3). Scientists define the *total energy* E to be mK:

$$E(x, y) = \frac{1}{2}my^2 + \frac{1}{2}kx^2$$

The term $my^2/2$ is called the *kinetic energy* of motion of the body of mass m and $kx^2/2$ is the *potential energy* due to the spring force $-kx$. The level sets of constant energy

$$E = my^2/2 + kx^2/2 = \text{positive constant}$$

are the elliptical orbits of system (3). Figure 8.3.1 shows some of the ellipses surrounding the neutrally stable rest position at the origin for the values $m = 8$ and $k = 2$.

Let's look again at the undamped simple pendulum system from Section 4.1.

EXAMPLE 8.3.2

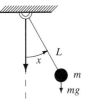

Undamped Simple Pendulum System, Conservation of Energy
The model system for the motion of the undamped simple pendulum is

$$\begin{aligned} x' &= y \\ y' &= -a\sin x \end{aligned} \tag{4}$$

where $a = g/L$ and x is measured in radians. In this case $g(y) = y$, $G(y) = y^2/2$, $f(x) = -a\sin x$, and $F(x) = a\cos x$. By (2), $K = y^2/2 - a\cos x$ is an integral of system (4). In this case scientists define the total energy to be $mK + ma$:

$$E(x, y) = \frac{1}{2}my^2 + ma(1 - \cos x) \tag{5}$$

☞ Take a look at Figure 4.1.1 and the cover of the book for more orbits of this system.

where (as in Example 8.3.1) $my^2/2$ is the *kinetic energy* of motion of the bob of mass m. The *potential energy* $ma(1 - \cos x)$ is due to the gravitational force mg on the pendulum bob. Note that $E(x, y)$ takes on its minimal value of 0 at the equilibrium points $(2n\pi, 0)$ of system (4) corresponding to the downward rest positions of the pendulum. Note also that at an upward position $((2n + 1)\pi, 0)$, the potential energy takes on its maximal value $2ma$.

Some level sets of E are shown in Figure 8.3.2, where we have taken $a = g/L = 10$ and $m = 1$. Note that if the value E_0 is between 0 and 20, then the graph of

$$E = \frac{1}{2}y^2 + 10(1 - \cos x) = E_0$$

is the collection of closed ovals, one oval around each equilibrium point $(2n\pi, 0)$ (shown in Figure 8.3.2 for $E = 10$). If we sweep E_0 over the range $0 < E_0 < 20$, we get families of closed ovals filling up the region about $(2n\pi, 0)$ for each $n = 0, \pm 1, \ldots$. These low-energy ovals model the familiar back-and-forth swings of the pendulum. This implies that each of the equilibrium points $(2n\pi, 0)$ is neutrally stable.

☞ Neutral stability follows by an argument like that given in Example 8.1.2.

In contrast, by taking the value of E_0 to be 20 we get the level set defined by

$$E = \frac{1}{2}y^2 + 10(1 - \cos x) = 20$$

Solving for y in terms of x, we get

$$y = \pm\sqrt{20}\sqrt{1 + \cos x}$$

whose graph is the collection of separatrices joining the unstable equilibrium points $((2n + 1)\pi, 0)$ to one another.

Each of the unstable points is a nonlinear saddle. We can see this by looking at the Jacobian matrix J at the point $((2n + 1)\pi, 0)$ of system (4) with $a = 10$.

$$J = \begin{bmatrix} 0 & 1 \\ -10\cos((2n + 1)\pi) & 0 \end{bmatrix} = \begin{bmatrix} 0 & 1 \\ 10 & 0 \end{bmatrix}$$

The characteristic polynomial of J and its eigenvalues are

$$\lambda^2 - 10 = 0, \qquad \lambda = \pm\sqrt{10}$$

According to the table following Section 8.2, there is a nonlinear saddle at each point $((2n + 1)\pi, 0)$.

The high-energy over-the-top orbits $E > 20$ are shown by the wavy lines above and below the separatrices in Figure 8.3.2.

Note that there is nothing special about Examples 8.3.1 and 8.3.2 except that both systems are special cases of the system

$$x' = y, \qquad y' = f(x)$$

We see that if $F(x)$ is any antiderivative of $f(x)$, then the quantity

$$K = \frac{1}{2}y^2 - F(x)$$

is conserved along each orbit. Because of the interpretation of $\alpha K + \beta$ as energy (α and β are appropriately chosen constants), K itself is sometimes called an energy function. This leads to a wide-ranging definition of conservation that is at the heart of many engineering and science applications.

Conservative Systems and Integrals

We can extend these ideas to the n-vector autonomous system

$$x' = f(x) \tag{6}$$

☞ To make our definition easier, we assume f is defined everywhere.

where f is any continuously differentiable, n-component function vector defined for all x. Suppose that $K(x)$ is a real-valued and continuously differentiable function defined for all x and that $x = x(t)$ is an orbit of system (6). The derivative $dK(x(t))/dt$ is called the *derivative of K following the motion*, and its values give information about how K changes as the orbit $x = x(t)$ is traversed.

> ❖ **Integral.** The real-valued and continuously differentiable function $K(x)$ is an *integral* of system (6) if:
>
> - The derivative of $K(x)$ following the motion of $x' = f(x)$ is zero for all orbits $x = x(t)$ of system (6).

☞ The second condition avoids trivial integrals.

> - K is nonconstant on every ball in $\mathbb{R}^n$.

The function K defined above is called an integral because it is often obtained by integration [e.g., see how we got (2) by integrating the functions in system (1)]. If an integral K of system (6) exists, it is not unique since $\alpha K + \beta$ (for any constants $\alpha \neq 0$ and β) is also an integral. Now we can define a conservative system.

> ❖ **Conservative System.** The system $x' = f(x)$ is *conservative* if the system has an integral $K(x)$. If C is a constant, the set of all x for which $K(x)$ has the value C is called a *level set*.

Let's look at the consequences of this definition.

THEOREM 8.3.1

☞ The exact ODEs of Problem 10, Section 1.6 always have integrals.

> Integrals and Orbits. Suppose that the system $x' = f(x)$ has an integral $K(x)$. Then each orbit of the system lies on a level set of K, and every level set of K is a union of orbits.

To see this, let's select any point x^0, denote the value of $K(x^0)$ by K_0, and denote the orbit of $x'(x) = f(x)$ that passes through x^0 by $x(t)$. By the definition of an integral, $K(x(t)) = K_0$ for all t for which the orbit $x = x(t)$ is defined, so the orbit of $x = x(t)$ lies in the level set $K(x) = K_0$.

Next, suppose that q is any point in the level set $K(x) = K_0$. By the argument above, $K(z(t)) = K_0$ for all t for which the orbit $x = z(t)$ passing through q is defined. But since q is any point in the level set, it follows that $K(x) = K_0$ is a union of orbits.

One important consequence of Theorem 8.3.1 is that if any orbit meets an integral surface, it must stay on the surface because of the uniqueness of orbits through the point of contact.

The planar systems in Examples 8.3.1 and 8.3.2 meet the conditions for conservation. Let's look at an example in 3-space where the integral's level sets are surfaces.

EXAMPLE 8.3.3

A Neutrally Stable and Conservative System in 3-Space
Let's show that $K(x, y, z) = x^2 + y^2 + z^2$ is an integral of the system

$$\begin{aligned} x' &= xz - 10yz \\ y' &= 10xz + yz \\ z' &= -x^2 - y^2 \end{aligned} \tag{7}$$

The derivative of K following the motion of the system is zero because

$$\frac{d}{dt}(K(x(t), y(t))) = 2xx' + 2yy' + 2zz'$$

$$= 2x(xz - 10yz) + 2y(10xz + yz) + 2z(-x^2 - y^2) = 0$$

Although K stays constant on each orbit, K is nonconstant on every open ball of $\mathbb{R}^3$. So system (7) is conservative, and K is an integral. Because the value of K can't change along an orbit, each orbit stays on a level set of K, that is, on a sphere $x^2 + y^2 + z^2 = $ constant. See Figure 8.3.3 for a view of the sphere $K = 4$ and the orbit on the sphere with initial point $x_0 = y_0 = z_0 = \sqrt{4/3}$.

As t increases from 0, this orbit spirals away from the initial point toward the equilibrium point $(0, 0, -4)$. As t decreases from 0, the orbit spirals toward $(0, 0, 4)$. Despite appearances, the point $(0, 0, -4)$ is not an attractor because nearby orbits *not* on the sphere $K = 4$ remain on their own spheres, so can't approach $(0, 0, -4)$ as $t \to +\infty$. For the same reason $(0, 0, 4)$ is not a repeller.

Note that the z-axis is a line of equilibrium points. The Jacobian matrix of system (7) at the equilibrium point $(0, 0, \alpha)$, $\alpha > 0$, and its eigenvalues are

$$\begin{bmatrix} \alpha & -10\alpha & 0 \\ 10\alpha & \alpha & 0 \\ 0 & 0 & 0 \end{bmatrix}, \qquad \lambda = \alpha(1 \pm 10i),\ 0$$

☞ The table following the last section suggests that the orbits behave like this.

Because of the complex conjugate eigenvalues $\alpha(1 \pm 10i)$ with positive real part α, we might expect that the orbits on the sphere $K = \alpha^2$ spiral away from the equilibrium point $(0, 0, \alpha)$ and then spiral toward the equilibrium point $(0, 0, -\alpha)$. In any event, the equilibrium point $(0, 0, \alpha)$ is unstable and the antipodal point $(0, 0, -\alpha)$ is neutrally stable.

Finally, note that the origin is at the center of the family of spherical integral surfaces $K = $ constant, so by an argument like that made in Example 8.1.1 the origin is also neutrally stable.

Surprisingly, conservation and attractors can't coexist.

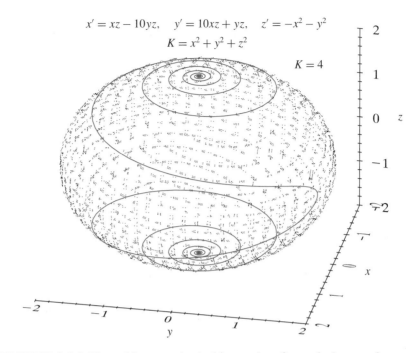

$$x' = xz - 10yz, \quad y' = 10xz + yz, \quad z' = -x^2 - y^2$$
$$K = x^2 + y^2 + z^2$$

$K = 4$

FIGURE 8.3.3 The orbit on a spherical integral surface spirals away from the north pole and toward the south pole as t increases (Example 8.3.3).

THEOREM 8.3.2

> Conservative Systems Have No Attractors. Suppose that the system $x' = f(x)$ is conservative. Then the system has no attractors and no asymptotically stable equilibrium points.

Here's why this is true. Suppose that $K(x)$ is an integral of the system $x' = f(x)$. Suppose, contrary to the assertion in the theorem, that the system has an attracting equilibrium p. So for some positive constant r, the solution $x(t)$ with initial point x^0, where x^0 lies in the open ball $B_r(p)$, has the property that $x(t) \to p$ as $t \to +\infty$. Since $K(x)$ is continuous, $K(x(t))$ must approach $K(p)$ as $x(t) \to p$, but since K must remain constant on the orbit, this implies that $K(x(t)) = K(p)$ for all t. This reasoning applies to every orbit whose initial point lies in $B_r(p)$, so $K(x)$ must have the constant value $K(p)$ throughout the open ball. This violates the condition of the nonconstancy of an integral on any open ball, so p can't be an attractor after all. Consequently, the system can't have any asymptotically stable equilibrium points.

A similar argument implies that a conservative system can't have any repelling equilibrium points. The existence of an integral is a counterindicator for asymptotic stability, but is consistent with instability or neutral stability, as we have seen in Examples 8.3.1–8.3.3.

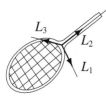

Spin a Tennis Racket, Spin a Book

Throw a tennis racket up into the air and watch its gyrations. Wrap a rubber band around a book (not this one!), toss it into the air, and look at its spinning behavior. Now try to get the racket or the book to spin steadily about each of three perpendicular body axes (see the margin sketch). Not so hard to do about two of the axes—but nearly impossible about the third. Why is that?

We are interested in the body's angular motions while aloft, not its vertical motion, so let's confine our attention to the angular velocity vector ω and its three components ω_1, ω_2, ω_3 along the respective body axes, L_1, L_2, L_3. Let's ignore air resistance. The key parameters that influence the angular velocity are the principal inertias I_1, I_2, I_3 about the respective body axes L_1, L_2, L_3. A principal inertia measures the response of the body to attempts to get the body to spin about the corresponding axis and depends not only on the body's mass but also on its shape. In terms of these inertias, the angular velocity analogues of Newton's force laws are known to be $I_1\omega_1' = (I_2 - I_3)\omega_2\omega_3$, $I_2\omega_2' = (I_3 - I_1)\omega_1\omega_3$, $I_3\omega_3' = (I_1 - I_2)\omega_1\omega_2$.

☞ A complete derivation can be found in *Differential Equations Laboratory Workbook*, John Wiley & Sons, Experiment 6.9.

Dividing by the principal inertias, we have the system

$$\omega_1' = \frac{I_2 - I_3}{I_1}\omega_2\omega_3$$

$$\omega_2' = \frac{I_3 - I_1}{I_2}\omega_1\omega_3 \tag{8}$$

$$\omega_3' = \frac{I_1 - I_2}{I_3}\omega_1\omega_2$$

The constant coefficients in system (8) are unitless. Let's measure angles in radians and time in seconds, so that each ω_i has units of radians per second.

First, we note that for any constant $\alpha \neq 0$, the equilibrium point $\omega = (\alpha, 0, 0)$ of system (8) represents a pure steady rotation (or spinning motion) about the first body axis L_1 with angular velocity α. The equilibrium point $(-\alpha, 0, 0)$ represents steady rotation about L_1 in the opposite direction. Similar statements are true for the equilibrium points $\omega = (0, \alpha, 0)$ and $(0, 0, \alpha)$.

Now the *kinetic energy of angular rotation* is given by

$$KE(\omega_1, \omega_2, \omega_3) = \frac{1}{2}\left(I_1\omega_1^2 + I_2\omega_2^2 + I_3\omega_3^2\right)$$

The function KE is an integral of system (8) since

$$\frac{d(KE)}{dt} = I_1\omega_1\omega_1' + I_2\omega_2\omega_2' + I_3\omega_3\omega_3'$$

$$= (I_2 - I_3)\omega_1\omega_2\omega_3 + (I_3 - I_1)\omega_1\omega_2\omega_3 + (I_1 - I_2)\omega_1\omega_2\omega_3 = 0$$

So system (8) is conservative. The ellipsoidal integral surface $KE = C$, where C is a positive constant, is called an *inertial ellipsoid* for system (8).

System (8) has other integrals. For example,

$$K = \frac{I_3 - I_1}{I_2}\omega_1^2 - \frac{I_2 - I_3}{I_1}\omega_2^2$$

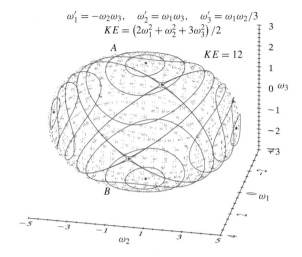

$\omega_1' = -\omega_2\omega_3, \quad \omega_2' = \omega_1\omega_3, \quad \omega_3' = \omega_1\omega_2/3$
$KE = \left(2\omega_1^2 + \omega_2^2 + 3\omega_3^2\right)/2$

$KE = 12$

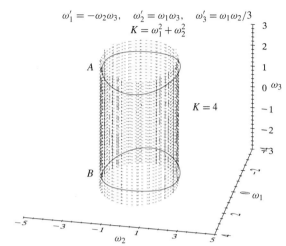

$\omega_1' = -\omega_2\omega_3, \quad \omega_2' = \omega_1\omega_3, \quad \omega_3' = \omega_1\omega_2/3$
$K = \omega_1^2 + \omega_2^2$

$K = 4$

FIGURE 8.3.4 Orbits lie on an inertial ellipsoid. Orbits A and B are also shown in Figure 8.3.5 (Example 8.3.4).

FIGURE 8.3.5 Orbits A and B are on the cylindrical integral surface $K = 4$ and on the inertial ellipsoid $KE = 12$ of Figure 8.3.4 (Example 8.3.4).

is an integral of system (8) because

$$K' = 2\frac{I_3 - I_1}{I_2}\omega_1\omega_1' - 2\frac{I_2 - I_3}{I_1}\omega_2\omega_2'$$

$$= 2\frac{I_3 - I_1}{I_2} \cdot \frac{I_2 - I_3}{I_1}\omega_1\omega_2\omega_3 - 2\frac{I_2 - I_3}{I_1} \cdot \frac{I_3 - I_1}{I_2}\omega_1\omega_2\omega_3 = 0$$

Let's put in some numbers for I_1, I_2, and I_3 and see what happens.

EXAMPLE 8.3.4

Orbits on Ellipsoids
Set $I_1 = 2$, $I_2 = 1$, $I_3 = 3$. Then from (8) we have

$$\omega_1' = -\omega_2\omega_3$$
$$\omega_2' = \omega_1\omega_3 \quad\quad\quad (9)$$
$$\omega_3' = \frac{1}{3}\omega_1\omega_2$$

With the given values for I_1, I_2, I_3 we have the integrals

$$KE = \frac{1}{2}\left(2\omega_1^2 + \omega_2^2 + 3\omega_3^2\right)$$

$$K = \omega_1^2 + \omega_2^2$$

Figure 8.3.4 shows the inertial ellipsoid $KE = 12$ and several orbits on the surface. The Chapter 8 cover figure shows the same orbits, but without the surface.

☞ Finding intersecting integral surfaces is one way to find orbits.

Figure 8.3.5 shows the cylindrical integral surface $K = 4$ Since any orbit that touches an integral surface must lie entirely on that surface, we see that an orbit that lies on two integral surfaces must lie in their intersection. The orbits A and B through the points $(2, 4, 0)$ and $(2, -4, 0)$ in Figures 8.3.4 and 8.3.5 lie on the inertial ellipsoid $KE = 12$ *and* on the cylinder $K = 4$.

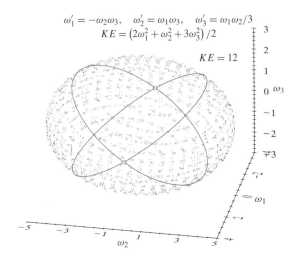

$$\omega_1' = -\omega_2\omega_3, \quad \omega_2' = \omega_1\omega_3, \quad \omega_3' = \omega_1\omega_2/3$$
$$KE = \left(2\omega_1^2 + \omega_2^2 + 3\omega_3^2\right)/2$$

$KE = 12$

$$\omega_1' = -\omega_2\omega_3, \quad \omega_2' = \omega_1\omega_3, \quad \omega_3' = \omega_1\omega_2/3$$

FIGURE 8.3.6 Periodic orbits connect small neighborhoods of two antipodal unstable equilibrium points (Example 8.3.4).

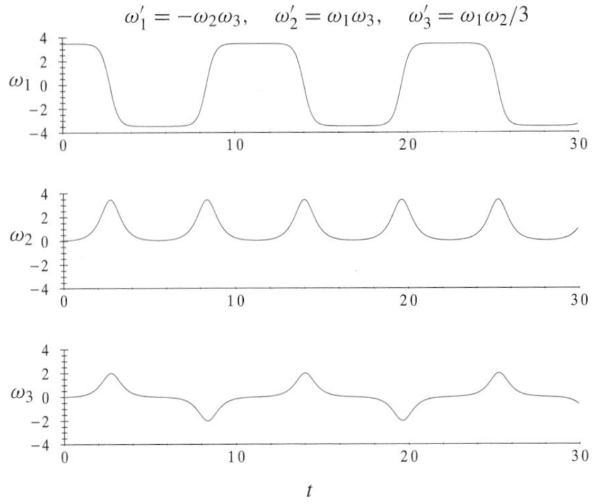

FIGURE 8.3.7 Component curves of one of the periodic orbits of Figure 8.3.6. What is the gyrating tennis racket doing?

What about the easy-to-achieve spinning motions about the L_2 and L_3 body axes? Look at Figure 8.3.4 again. These spinning motions correspond to the nearly circular orbits along the long (L_3) body axis and to orbits encircling the short (L_2) body axis.

But about the body axis of middle length (the L_1-axis) the story is quite different. Figure 8.3.6 shows two orbits that start out very close to the unstable equilibrium point $(\sqrt{12}, 0, 0)$ on the ellipsoid. Note that each orbit goes back near the unstable antipodal equilibrium point $(-\sqrt{12}, 0, 0)$, and then returns in an endlessly repeating periodic path. Figure 8.3.7 shows the corresponding component curves of one of these orbits. Note how ω_1 stays nearly constant for quite a while (i.e., the racket maintains a nearly constant angular velocity about the L_1-axis), then quickly reverses its sign (as the body reverses its direction of rotation about the L_1-axis): strange gyrations here!

The eigenvalues of the Jacobian matrix in (9) at points on the L_2 and L_3 axes are conjugate pure imaginaries and 0; on the L_1 axis the eigenvalues are real numbers of opposite signs and 0. So the table following Section 8.2 suggests the orbital behavior seen in Figures 8.3.4 and 8.3.6.

☞ Look ahead at Example 8.4.8 for information about the equilibrium points on the axes.

Comments

Most physical systems aren't conservative since some sort of damping usually dissipates energy. Still, many systems are nearly conservative over short time spans. So a conservative system of ODEs is often a useful approximation to a physical system's behavior.

Orbits that start off a given integral surface never reach it, and orbits that start on the surface never escape. So "life" on the surface is essentially two-dimensional, which explains why we seem to see the spiral points, saddles, and centers of a planar autonomous system on the integral surfaces of Figures 8.3.3 and 8.3.4.

PROBLEMS

 1. (*Integrals*). Find an integral and plot some orbits. What are the stability properties of the equilibrium points?

(a) $x' = 3x$, $y' = -y$ (b) $x' = -y$, $y' = 25x$

(c) $x' = y^3$, $y' = -x^3$ (d) $x' = x$, $y' = y + x^2 \cos x$ [*Hint*: See Example 2.1.4.]

2. (*A Conservative Attractor?*). Show that the system $x' = -x$, $y' = -2y$, is asymptotically stable at the origin and that $K = yx^{-2}$ is constant on each orbit not touching the y-axis. Why doesn't this contradict Theorem 8.3.2?

www **3.** Show that the system $x' = g(y)$, $y' = f(x)$, $z' = 0$ is conservative. Any attractors?

 4. (*Springs*). Integrals can help us understand the behavior of undamped springs.

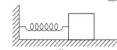

(a) (*Hard Spring*). Find an integral and sketch the orbits of $x' = y$, $y' = -10x - x^3$. What are the stability properties of the system at the origin?

(b) (*Soft Spring*). Repeat part (a) for $x' = y$, $y' = -10x + x^3$ and discuss the stability properties at each of the three equilibrium points. [*Hint*: See Section 3.1.]

5. (*Skew Symmetry*). Consider the system $x' = Ax$ where the real $n \times n$ matrix A is *skew-symmetric*, that is, $A^T = -A$.

(a) Show that the system is conservative on $\mathbb{R}^n$ and neutrally stable at the origin. [*Hint*: Look at the function $K = x^T x = ||x||^2$.]

(b) Suppose that each eigenvalue of A is zero or pure imaginary. Show that the surface $||x|| =$ constant is an integral surface of equilibrium points and nonconstant orbits defined by solutions of $x' = Ax$ whose components are linear combinations of periodic functions.

6. (*Hamiltonian Systems*). Parts (a)–(d) below refer to the *Hamiltonian system*[2]

$$x_i' = \frac{\partial H}{\partial y_i}, \qquad y_i' = -\frac{\partial H}{\partial x_i}, \qquad i = 1, \dots, k$$

where the *Hamiltonian* $H(x_1, \dots, x_k, y_1, \dots, y_k)$ is a differentiable, real-valued scalar function on $\mathbb{R}^{2k}$ and H is nonconstant on every open ball in $\mathbb{R}^{2k}$.

(a) Show that H is an integral of the Hamiltonian system.

(b) Explain why a Hamiltonian system can't be asymptotically stable at an equilibrium point.

(c) Explain why the system $x' = y$, $y' = -f(x)$ has no asymptotically stable equilibrium points. [*Hint*: The system has Hamiltonian $H = y^2/2 + \int_0^x f(s)\, ds$, the "total energy" of the system.]

(d) What does the result in part (c) imply about the possibility of constructing a "restoring force" $-f(x)$ such that all orbits of the system $x' = y$, $y' = -f(x)$ tend toward an equilibrium state as $t \to +\infty$?

 7. (*Exact Planar Systems*). The planar system $x' = N(x, y)$, $y' = -M(x, y)$ is *exact* in a rectangle R of the xy-plane if the functions N and M belong to $\mathbf{C}^1(R)$ and $\partial N/\partial x = \partial M/\partial y$ on R. Show that if the system is exact, then it is a Hamiltonian system (see Problem 6 for the definition). Conversely, show that every planar Hamiltonian system is exact. As a first example show that $x' = y^2 + e^x \cos y + 2\cos x$, $y' = -e^x \sin y + 2y \sin x$ is exact in $\mathbb{R}^2$ and find a Hamiltonian. Plot orbits in the region $-20 \le x \le 10$, $-4 \le y \le 6$. Why is it hard to use a numerical solver for orbits in the region $x > 5$? Now show that the *teddy bear* system of Example 1.7.2 is exact, find a Hamiltonian, and plot some of the bears. Finally, make up your own Hamiltonian system with as weird orbits as possible.

📖 Take a look at Problem 10 in Section 1.6.

---

[2]These systems are named in honor of the Irish mathematician and astronomer William Rowan Hamilton (1805–1865).

8. (*Tennis Racket*). Carry out a general treatment of the tennis racket model

$$\omega_1' = \frac{I_2 - I_3}{I_1}\omega_2\omega_3, \qquad \omega_2' = \frac{I_3 - I_1}{I_2}\omega_1\omega_3, \qquad \omega_3' = \frac{I_1 - I_2}{I_3}\omega_1\omega_2$$

where I_1, I_2, and I_3 are positive constants. Address the following points:

- Assume that $I_2 < I_1 < I_3$ and find integrals whose level sets are cylinders, hyperboloids of one sheet, hyperboloids of two sheets, and cones.
- Assume that $I_1 = I_2 < I_3$ and give a full analysis of the system in this case.
- Explain why all nonconstant orbits on an inertial ellipsoid are periodic (except for the separatrices).
- Use your solver and grapher to plot surfaces, orbits, and solution curves.
- Toss a racket into the air and match what you see with the model predictions.

8.4 Lyapunov Functions

How can a system by tested for asymptotic stability? We can use the methods of Sections 8.1 and 8.2 and (often) determine stability properties by inspection of the signs of the real parts of eigenvalues. But if some eigenvalue has a zero real part, this approach may not work. If the system has an integral, so that it is conservative, the most we can hope for is neutral stability, as we see from Theorem 8.3.2. In this section the wide-ranging method of Lyapunov functions is introduced to test for asymptotic stability even when other methods are inconclusive. Lyapunov[3] defined classes of scalar functions to test for the stability, asymptotic stability, or instability of any system, linear or nonlinear, autonomous or nonautonomous. We will focus on autonomous systems in $\mathbb{R}^2$ or $\mathbb{R}^3$, although Lyapunov's methods are presented for state spaces and systems of any dimension. The approach will be somewhat telegraphic.[4]

Strong Lyapunov Functions: Asymptotic Stability

Suppose that x and $f(x)$ are n-vectors, that $f(x)$ is continuously differentiable, and that $f(0) = 0$; then 0 is an equilibrium point of the system

$$x' = f(x) \tag{1}$$

---

[3] Aleksandr Mikhailovich Lyapunov(1857–1918) introduced the functions in his doctoral dissertation for the University of Moscow. The dissertation was published in 1892 in the research journal *Comm. Soc. Math. Khar'kov* and has been reprinted many times in the last 50 years; see *Stability of Motion* (New York: Academic Press, 1966) for an English translation. At first, little attention was paid to Lyapunov's work, but in the 1930s mathematicians, physicists, and engineers of the former Soviet Union suddenly took up the question of stability. The work of these researchers, based as it was on Lyapunov's ideas, dominated the field for decades. Lyapunov's dissertation and the memoirs of Henri Poincaré are the foundation of most of the contemporary developments in the theory and applications of differential equations and dynamical systems.

[4] A more detailed treatment of the material in this section can be found in the *Differential Equations Laboratory Workbook*, R. L. Borrelli, C./,S. Coleman, and W. B. Boyce, (New York: John Wiley & Sons, Inc., 1992).

If $f(p) = 0$ for $p \neq 0$, we can translate p to the origin, so there is no harm in assuming that the equilibrium point is always at the origin. We will give Lyapunov's test for the asymptotic stability of system (1) at the origin, but first we need some definitions.

☞ The ambiguity of symbols! When we write $V(0) = 0$, the first "0" stands for the origin in $\mathbb{R}^n$, the second for the real number zero.

A real-valued function $V(x)$ is *positive definite* on an open ball B centered at the origin in $\mathbb{R}^n$ if it has only positive values on B except at the origin, where $V(0) = 0$; *negative definite* is defined similarly. A continuously differentiable function $V(x)$ is a *strong Lyapunov function* for system (1) on B if V is positive definite and the derivative V' following the motion is negative definite on B. We can now state a stability theorem.

THEOREM 8.4.1

> Lyapunov's First Theorem. Suppose that there is a strong Lyapunov function $V(x)$ for system (1) on an open ball centered at the origin. Then system (1) is asymptotically stable at the origin.

The conclusion of this theorem is plausible, since the values of the strong Lyapunov function $V(x(t))$ must continually diminish along each orbit $x = x(t)$ as t increases (since V' is negative definite). This means that the orbit $x = x(t)$ must cut across level sets $V = C$ with ever smaller values of C. In fact, $\lim_{t \to \infty} V(x(t)) = 0$, which implies that $x(t) \to 0$ as $t \to \infty$ since $x(t)$ and $V(x)$ are continuous and $V(x)$ has value zero only at the origin [that's where the positive definiteness of $V(x)$ comes into play]. Note that the points in the ball B (where V is a strong Lyapunov function) belong to the basin of attraction of the origin.

$V = C$

Let's try to visualize what Lyapunov's First Theorem is saying geometrically. The gradient vector[5] (denoted by ∇V) at a point x on a level set $V = C$ points outward in the direction of increasing values of V. The gradient vector of a strong Lyapunov asymptotic stability function V and the velocity vector f of the system $x' = f(x)$ must form an obtuse angle θ at each point x near the origin. The reason for this is that $V'(x)$ is negative definite, and

☞ Take a look at Section 4.1 for this formula for the dot product.

$$0 > V' = \sum_i \frac{\partial V}{\partial x_i} f_i = \nabla V \cdot f = \|\nabla V\| \, \|f\| \cos \theta$$

This inequality means that $\cos \theta$ must be negative, and the angle θ obtuse. The obtuseness of the angle implies that the velocity vector f must point inward toward lower values of V. So the orbit must also move inward toward the origin, where the value of V is minimal.

Here are two examples that illustrate Lyapunov's First Theorem. The stability properties are easily found by looking at the eigenvalues of the system matrix of each linear system. But we want to illustrate how V functions can be used, so we will not use any eigen-information.

---

[5]In Cartesian coordinates $\nabla V(x, y) = \partial V / \partial x \, \mathbf{i} + \partial V / \partial y \, \mathbf{j}$, $\nabla V(x, y, z) = \partial V / \partial x \, \mathbf{i} + \partial V / \partial y \, \mathbf{j} + \partial V / \partial z \, \mathbf{k}$. The gradient vector at a point on a level set $V = C$ is perpendicular to the level set.

EXAMPLE 8.4.1   **Decreasing Distances**

The square of the distance from an orbital point $(x(t), y(t))$ to the equilibrium point $(0, 0)$ of the linear system

$$x' = -x, \qquad y' = -2y \qquad (2)$$

is the positive definite function $V = x^2(t) + y^2(t)$. We have

$$V' = \frac{d}{dt}V(x(t), y(t)) = 2x\frac{dx}{dt} + 2y\frac{dy}{dt} = 2x(-x) + 2y(-2y) = -2x^2 - 4y^2$$

which is negative definite. So, without using a solution formula, we know from Lyapunov's First Theorem that the system is asymptotically stable at the origin. See Figure 8.4.1 for the graphs of two level sets of V (dashed) and several orbits crossing these level sets and moving inward toward the origin as time increases.

We used the square of the Euclidean distance between points as the strong Lyapunov function in this example, but the next example makes use of a different V-function.

EXAMPLE 8.4.2   **Another Measure of Distance: Damped Hooke's Law Spring**

Suppose that $(x(t), y(t))$ is an orbital point of the damped Hooke's Law spring system

$$x' = y, \quad y' = -x - y \qquad (3)$$

Although $V_1 = x^2 + y^2$ is positive definite, it's derivative following the motion is no longer negative definite:

$$V_1' = 2xx' + 2yy' = 2xy + 2y(-x - y) = -2y^2$$

which has value zero not only at the equilibrium point $(0, 0)$, but on the entire x-axis (i.e., $y = 0$). So the function V_1 is *not* a strong Lyapunov function, and we can't apply Theorem 8.4.1 with this V_1.

But here is a function that will work as desired:

$$V = 3x^2 + 2xy + 2y^2 = \frac{5}{3}(x + y)^2 + \frac{1}{3}(2x - y)^2$$

☞ The function V seems to have appeared by "magic." We explain all in Example 8.4.6.

where the last equality was obtained by completing the squares in x and y. The function V has positive values everywhere, except at $x = 0$, $y = 0$ where its value is zero. How does $V(x(t), y(t))$ change as $(x(t), y(t))$ moves along an orbit of system (3) with advancing time? Here's the derivative of V following the motion of system (3):

$$V' = \frac{d}{dt}V(x(t), y(t)) = 6x\frac{dx}{dt} + 2y\frac{dx}{dt} + 2x\frac{dy}{dt} + 4y\frac{dy}{dt}$$

$$= (6x + 2y)y + (2x + 4y)(-x - y) = -2x^2 - 2y^2$$

so V' is negative definite. Lyapunov's First Theorem applies and system (3) is asymptotically stable at the origin.

The level sets $V =$ positive constants are tilted ellipses. Figure 8.4.2 shows orbits of (3) cutting across the ellipses as the orbits move inward toward the origin.

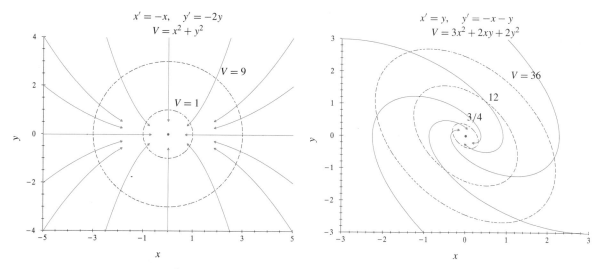

FIGURE 8.4.1 Orbits move inward toward the equilibrium point and across the circular level sets of V (dashed) (Example 8.4.1).

FIGURE 8.4.2 Orbits move inward toward the equilibrium point and across the elliptical level sets of V (dashed) (Example 8.4.2).

The next example begins to show the power of Lyapunov's First Theorem; none of the methods given earlier in the chapter applies.

EXAMPLE 8.4.3

Asymptotic Stability of a Fuzzy System
Sometimes we only have partial information about a system. Consider

$$x' = y - F(x, y)$$
$$y' = -x - G(x, y) \tag{4}$$

Note that the eigenvalues of the linear system $x' = y$, $y' = -x$ are $\pm i$, so Theorem 8.2.1 can't apply even if F and G have order at least 2. Suppose that all we know about the functions F and G is that $F(0, 0) = G(0, 0) = 0$, that F and G are continuously differentiable, and that $xF(x, y) > 0$ for $x \neq 0$ and $yG(x, y) > 0$ for $y \neq 0$ (i.e., F has the sign of x and G has the sign of y). Then $V = x^2 + y^2$ is a strong Lyapunov function because it is positive definite and V' is negative definite:

$$V' = 2xx' + 2yy' = 2x(y - F) + 2y(-x - G) = -xF - yG$$

which is negative definite by the sign properties of F and G.

We have determined that system (4) is asymptotically stable even though we don't know much about F and G (i.e., they are somewhat "fuzzy"). Note in particular that we do *not* assume that F and G have order at least 2. In fact $F = ax$ and $G = by$, where a and b are positive constants, satisfy the given conditions. Figure 8.4.3 shows orbits of a different example of system (4) (i.e., $F = x^3$ and $G = y^3$). The orbits cut inward across level sets of $V = x^2 + y^2$.

Now let's modify the functions and F and G in Example 8.4.3 so that the sign properties of F and G hold near the origin, but not at a distance.

EXAMPLE 8.4.4

Local Asymptotic Stability
Let's look at the system

$$x' = y - x^3[1 - x\sin(x + y)]$$
$$y' = -x - y^3[1 - xy]$$

The functions $F = x^3[1 - x\sin(x + y)]$ and $G = y^3[1 - xy]$ have the same sign properties near $(0, 0)$ as x^3 and y^3, but at a distance from $(0, 0)$ the signs of these functions can change rather rapidly. So we expect to see *local* asymptotic stability, but not necessarily global. Figure 8.4.4 bears this out. In this case we say that $V = x^2 + y^2$ is a *strong local Lyapunov function*.

Before we go any further, let's define three other kinds of "definiteness" and then give some criteria for the various kinds of definiteness of quadratic forms.

Definiteness and Quadratic Forms

☞ Note that the various kinds of definiteness also apply at any point p where $V(p) = 0$.

A real-valued function $V(x)$ for which $V(0) = 0$ is *positive semidefinite* if there is an open ball B centered at 0 such that $V(x) \geq 0$ for all x in B; a *negative semidefinite* V is defined similarly. Also, V is *indefinite* if every open ball B centered at 0 contains points q_1 and q_2 where $V(q_1) > 0$ and $V(q_2) < 0$, but (as always) $V(0) = 0$.

Here is the definiteness test for planar quadratic forms $V(x, y)$.

THEOREM 8.4.2

Planar Quadratic Forms. Suppose that $V(x, y)$ is the *quadratic form* $ax^2 + 2bxy + cy^2$, where a, b, and c are real numbers. Then $V(x, y)$ is:

- Positive definite if and only if $a > 0$ and $b^2 < ac$.
- Positive semidefinite if and only if $a \geq 0$ and $b^2 \leq ac$.
- Negative definite if and only if $a < 0$ and $b^2 < ac$.
- Negative semidefinite if and only if $a \leq 0$ and $b^2 \leq ac$.

Otherwise, $V(x, y)$ is indefinite.

The verification of these tests uses the quadratic formula (Problem 8). Note that we can interchange a and c and obtain equivalent tests.

EXAMPLE 8.4.5

Testing Quadratic Forms for Definiteness
For $V = x^2 + 8y^2$ we have $a = 1$, $b = 0$, and $c = 0$, so $a > 0$, $b^2 < ac$, and V is positive definite. For $V = x^2 + xy + y^2$ we have $a = c = 1$ and $b = 1/2$, so $a > 0$, $b^2 < ac$, and V is positive definite. Similarly, $V = x^2 - y^2$ is indefinite and $V(x, y) = -y^2$ is negative semidefinite.

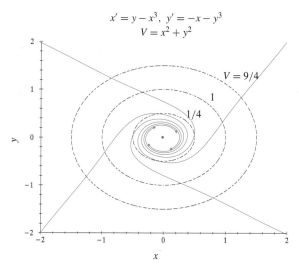

$$x' = y - x^3, \ y' = -x - y^3$$
$$V = x^2 + y^2$$

$$x' = y - x^3[1 - x\sin(x+y)], \ y' = -x - y^3[1 - xy]$$
$$V = x^2 + y^2$$

FIGURE 8.4.3 Orbits move inward across all level sets of the strong Lyapunov function V: global asymptotic stability (Example 8.4.3).

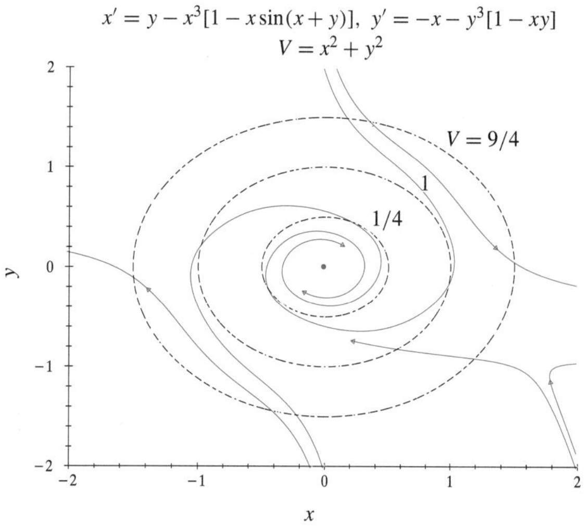

FIGURE 8.4.4 The system is (at least locally) asymptotically stable because V is a strong local (but not global) Lyapunov function (Example 8.4.4).

EXAMPLE 8.4.6

Another Look at Example 8.4.2

Let's explain the strange choice of the strong Lyapunov function $V = 3x^2 + 2xy + 2y^2$ for the system $x' = y, \ y' = -x - y$. As we saw in Example 8.4.2, $V_1 = x^2 + y^2$ can't be used in Theorem 8.4.1 because V' is only negative semidefinite. So let's set $V = ax^2 + 2bxy + cy^2$ and try to find a choice of the constants a, b, and c so that V is positive definite and V' is negative definite. By Theorem 8.4.2 we want a and c to be positive and $b^2 < ac$. We have

$$V' = (2ax + 2by)x' + (2bx + 2cy)y' = (2ax + 2by)y + (2bx + 2cy)(-x - y)$$
$$= -2bx^2 + 2(a - b - c)xy + 2(b - c)y^2$$

So let's get rid of the mixed term xy by setting $a = b + c$ and then choose b so that $0 < b < c$ and $b^2 < ac$. For example, $a = 3$, $b = 1$, and $c = 2$ will work:

$$V = 3x^2 + 2xy + 2y^2$$
$$V' = -2x^2 - 2y^2$$

and we have accomplished what we set out to do. Note that there is nothing unique about the values of a, b, and c; $a = 5$, $b = 2$, and $c = 3$ will also give a positive definite V and a negative definite V'.

Now let's look at "weak" Lyapunov functions and Lyapunov's tests for stability (without distinguishing between neutral and asymptotic) and for instability.

Lyapunov's Tests for Stability and Instability

First define a new kind of Lyapunov function. Suppose that $V(x)$ is a continuously differentiable real-valued function. It is a *weak Lyapunov function* for the system

$$x' = f(x), \qquad f(0) = 0 \tag{5}$$

if V is positive definite, but $V'(x)$ is only negative semidefinite [so $V'(x) \leq 0$ near the origin]. Then we have a test for stability, but the test won't distinguish between neutral and asymptotic stability.

THEOREM 8.4.3

> Lyapunov's Second Theorem. Suppose that there is a weak Lyapunov function V for system (5). Then system (5) is stable at the origin.

The result seems reasonable since the negative semidefiniteness of V' keeps an orbit that starts near the origin close to the origin as t increases. But orbits don't have to cut across level sets of V. In fact, if $V' = 0$ for all x, then $V(x) = $ constant may be an integral surface and we have neutral stability.

EXAMPLE 8.4.7

A Weak Lyapunov Function and Neutral Stability
Let's look at the system

$$x' = y^3$$
$$y' = -x^3$$

whose orbits are given by $x^4 + y^4 = $ constant. In fact, let $V = x^4 + y^4$; then $V' = 4x^3 x' + 4y^3 y' = 4x^3 y^3 - 4y^3 x^3 = 0$. So V is positive definite and V' is negative semidefinite (trivially), and the system is stable at the origin. In fact the system is neutrally stable and conservative; the orbits are the level sets of V.

Finally, we have Lyapunov's test for instability.

THEOREM 8.4.4

> Lyapunov's Third Theorem. The system $x' = f(x)$, where $f(0) = 0$, is unstable at the origin if there is a continuously differentiable indefinite (or positive definite) function $V(x)$ on a ball centered at the origin and $V'(x)$ is positive definite.

This result is plausible. If an orbit starts near the origin at a point q where $V(q)$ is positive, then the positive definiteness of V' implies that as time advances the value of V along the orbit increases. The orbit must move farther and farther away from the origin.

We will end this section by applying Lyapunov functions to the rotating tennis racket problem of Section 8.3.

EXAMPLE 8.4.8

The Spinning Racket

In the last section we modeled the angular motions of a spinning tennis racket. In Example 8.3.4 we put in numbers for the principal inertias and looked at the specific system for the angular velocity of the spinning racket:

$$\omega_1' = -\omega_2\omega_3$$
$$\omega_2' = \omega_1\omega_3 \qquad\qquad (6)$$
$$\omega_3' = \frac{1}{3}\omega_1\omega_2$$

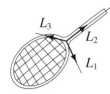

Note that all points on the ω- coordinate axes are equilibrium points of the system; they represent steady rotations about the corresponding body axis of the tennis racket. Let's check out the stability of these points. For simplicity, we will look only at the points $(1, 0, 0)$, $(0, 1, 0)$, and $(0, 0, 1)$, but the approach we take extends to all the equilibrium points.

☞ V_1 was called K in Section 8.3.

First we note that $V_1 = \omega_1^2 + \omega_2^2$ and $V_2 = \omega_1^2 + 3\omega_3^2$ are integrals of system (6). To verify this, check that $V_1' = 0$ and $V_2' = 0$ following the motion of system (6). Now here are the stability properties of system (6):

- **Neutrally stable at** $(0, 1, 0)$. To see why this is so, set $V = (V_1 - 1)^2 + V_2$. The function V is positive definite, vanishing only at the point $(0, 1, 0)$. But we have that $V' = 2(V_1 - 1)V_1' + V_2'$ is zero for all $\omega_1, \omega_2, \omega_3$, so V' is negative semidefinite (trivially), and V is a weak Lyapunov function. Theorem 8.4.3 applies and (6) is stable at $(0, 1, 0)$. Since conservative systems can't have asymptotically stable equilibrium points, we must have neutral stability.

- **Neutrally stable at** $(0, 0, 1)$. Use the approach above, but with the weak Lyapunov function $V = V_1 + (V_2 - 1)^2$.

- **Unstable at** $(1, 0, 0)$. We see that $V = \omega_2\omega_3$ is indefinite at $(1, 0, 0)$ and

$$(\omega_2\omega_3)' = \omega_2\omega_3' + \omega_3\omega_2' = \omega_2\left(\frac{1}{3}\omega_1\omega_2\right) + \omega_3(\omega_1\omega_3) = \omega_1\left(\frac{1}{3}\omega_2^2 + \omega_3^2\right)$$

which is positive definite on the ball $B_1((1, 0, 0))$. So Theorem 8.4.4 applies and we have instability.

The examples suggest that Lyapunov functions are widely used by engineers in testing models of complex systems for stable operation because V-functions can be custom "built" to check out the stability properties of a proposed model. Lyapunov tests coupled with computer simulations are now central to most design processes.

PROBLEMS

1. (*Using Lyapunov Functions to Test Stability*). Test stability at the origin by using $V = ax^2 + cy^2$ or $ax^2 + 2bxy + cy^2$. Distinguish between neutral and asymptotic stability, if possible.

    (**a**) $x' = -4y - x^3$, $y' = 3x - y^3$ (**b**) $x' = y$, $y' = -9x$

    (**c**) $x' = -x + 3y$, $y' = -3x - y$ (**d**) $x' = 2x + 2y$, $y' = 5x + y$

    (**e**) $x' = (-x + y)(x^2 + y^2)$, $y' = -(x + y)(x^2 + y^2)$ (**f**) $x' = -2x - xe^{xy}$, $y' = -y - ye^{xy}$

    (**g**) $x' = -x^3 + x^3y - x^5$, $y' = y + y^3 + x^4$ (**h**) $x' = y$, $y' = -2x - 3y$

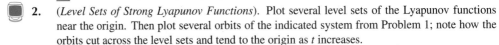

2. (*Level Sets of Strong Lyapunov Functions*). Plot several level sets of the Lyapunov functions near the origin. Then plot several orbits of the indicated system from Problem 1; note how the orbits cut across the level sets and tend to the origin as t increases.

    (**a**) $V = 3x^2 + 4y^2$;   Problem 1(**a**) system   (**b**) $V = x^2 + y^2$;   Problem 1(**c**) system

    (**c**) $V = x^2 + y^2$;   Problem 1(**e**) system   (**d**) $V = x^2 + y^2$;   Problem 1(**f**) system

    (**e**) $V = 5x^2 + 2xy + y^2$;   Problem 1(**h**) system

www 3. (*Testing for Stability*). Use a Lyapunov function of the form $V = ax^2 + cy^2$ and determine whether the system is asymptotically stable or unstable at the origin.

    (**a**) $x' = -x - x^3$, $y' = -y$ (**b**) $x' = x + x^2$, $y' = -y$

    (**c**) $x' = \varepsilon x - y + x^2$, $y' = x + \varepsilon y + x^2$, where ε is a positive constant.

4. (*More Lyapunov Functions*). Use $V = ax^{2m} + by^{2n}$, m and n positive integers, to determine stability properties at the origin. Plot some level sets of V and some orbits of the systems. [*Hint*: Calculate V', and then pick strategic values for a, b, m, n.]

    (**a**) $x' = -x^3 + y^3$, $y' = -x^3 - y^3$ (**b**) $x' = -2y^3$, $y' = 2x - y^3$

5. (*Neutral Stability*). Show that $x' = (x - y)^2(-x + y)$, $y' = (x - y)^2(-x - y)$ is neutrally stable at the origin. [*Hint*: Use $V = x^2 + y^2$. Note the line of equilibrium points.]

6. (*V-functions and the Stability of a Damped Pendulum*). The assertions below fully describe the stability properties of the system $x' = y$, $y' = -(g/L)\sin x - cy/m$ for a damped pendulum.

    (**a**) Show that this system is asymptotically stable at every point $(2k\pi, 0)$. [*Hint*: Put $V(x, y) = (4g/L)[1 - \cos(x - 2k\pi)] + 2y^2 + (2c/m)(x - 2k\pi)y + (c^2/m^2)(x - 2k\pi)^2$.]

    (**b**) Show that the system $x' = y$, $y' = -(g/L)\sin x - (c/m)y$ is unstable at every point $(x_0, 0)$, where $x_0 = (2k + 1)\pi$. [*Hint*: Use $V = -y\sin(x - x_0) + (c/m)(1 - \cos(x - x_0))$.]

7. (*Lotka-Volterra Model*). The system $x' = (a - by)x$, $y' = (-c + dx)y$, where the coefficients are positive, models the interactions of a predator-prey community. Show that the system is neutrally stable at the equilibrium point $(c/d, a/b)$ by using the function $Q(x, y) = (y^a e^{-by})(x^c e^{-dx})$. [*Hint*: Let $V = Q(c/d, a/b) - Q(x, y)$. See Problem 6(**a**) of Section 5.4.]

8. Show the validity of Theorem 8.4.2 for the planar quadratic form $V = ax^2 + 2bxy + cy^2$.

9. Consider the system $x' = y - xf(x, y)$, $y' = -x - yf(x, y)$.

    (**a**) Show that the system is stable at the origin if $f(x, y)$ is positive semidefinite.

    (**b**) Show that the system is asymptotically stable at the origin if $f(x, y)$ is positive definite.

    (**c**) Show that the system is unstable at the origin if $f(x, y)$ is negative definite.

    (**d**) Illustrate parts (**b**) and (**c**) above by using the respective functions $f = |x| + |y|$ and $f = -\cos(x^2 + y^2)$. Plot orbits of each system in a region containing the origin.

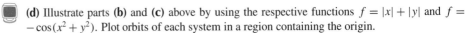

10. (*Tennis Anyone?*). Determine the stability of all equilibrium points in system (8) of Section 8.3. Zoom in on orbits near each equilibrium point on an inertial ellipsoid and explain what you see. What are the corresponding motions of the tennis racket?

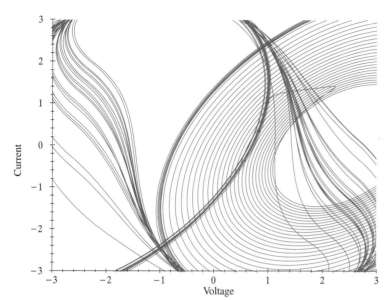

Chaotic current and voltage in a nonlinear circuit? Take a look at the scroll circuit in Section 9.3.

CHAPTER
9

Cycles, Bifurcations, and Chaos

In this chapter we look at the long-term behavior of physical processes and their modeling systems of ODEs (usually autonomous, always nonlinear, often planar). We will see orbits attracted to equilibrium states, periodic states, and regions of chaotic wandering. Unusual behavior suddenly appears as a parameter is pushed beyond a bifurcation point. The material of this chapter takes us to the edge of understanding of dynamical behavior.

9.1 Cycles

Although Balthazar van der Pol is best known for his work with electric circuits, he recognized a wide variety of systems that exhibit oscillations:

> *...the aeolian harp, a pneumatic hammer, the scratching noise of a knife on a plate, the waving of a flag in the wind, the humming noise sometimes made by a water-tap, the squeaking of a door ..., the periodic recurrence of epidemics and of economic crises, the periodic density of an even number of species of animals living together and the one species serving as food for the other ..., and finally, the beating of a heart.*

In this quotation, van der Pol[1] grouped varied phenomena into a single category of systems that exhibit periodic oscillations. According to van der Pol, most of these systems can be modeled by a nonlinear autonomous system $x' = F(x)$, where x is a real n-vector and $F(x)$ is a continuously differentiable n-vector function. So the Fundamental Theorem (Theorem 5.2.1) holds. Distinct orbits can't meet because the system is autonomous.

A periodic oscillation's orbit in state space is a *cycle*. We have already seen many of them—for example, the linear cycles of a harmonic oscillator in Section 3.5 and the nonlinear cycles of the Lotka-Volterra predator-prey system in Section 5.4. But our focus now is on a special kind of cycle, a so-called limit cycle.

❖ **Limit Cycle.** A cycle of an autonomous system is a *limit cycle* if some non-periodic orbit tends to it as $t \to +\infty$ or as $t \to -\infty$. A limit cycle is an *attractor* if every nearby orbit approaches it as $t \to +\infty$, a *repeller* if every nearby orbit approaches it in reverse time, that is, as $t \to -\infty$.

Note that the cycles of the harmonic oscillator (Section 3.5) and of the Lotka-Volterra system (Section 5.4) are *not* limit cycles because all orbits near each of these cycles are also cycles, so they can't approach one another as $t \to +\infty$ or as $t \to -\infty$. In this section and the next we will search for limit cycles. For simplicity we will look only at planar autonomous systems $x' = f(x, y), \quad y' = g(x, y)$, where f and g are continuously differentiable.

EXAMPLE 9.1.1

Using Polar Coordinates to Locate a Cycle

Here's a system with an attracting limit cycle:

$$x' = x - y - x(x^2 + y^2), \qquad y' = x + y - y(x^2 + y^2) \qquad (1)$$

One way to see this is to describe the orbits using polar coordinates

$$x = r\cos\theta, \quad y = r\sin\theta, \quad r^2 = x^2 + y^2, \quad \tan\theta = y/x \qquad (2)$$

☞ Refer to Section 5.2, formula (15) for another way to transform an ODE from rectangular to polar coordinates.

Differentiating the last two equations in (2) with respect to t and using (1) and the Chain Rule, we obtain an ODE in r and another in θ:

$$2rr' = 2xx' + 2yy' = 2x[x - y - x(x^2 + y^2)] + 2y[x + y - y(x^2 + y^2)]$$

$$= 2(x^2 + y^2) - 2(x^2 + y^2)^2 = 2r^2 - 2r^4$$

$$(\sec^2\theta)\theta' = \frac{xy' - yx'}{x^2} = \frac{1}{x}\left[x + y - y(x^2 + y^2)\right] - \frac{y}{x^2}\left[x - y - x(x^2 + y^2)\right]$$

$$= 1 + \left(\frac{y}{x}\right)^2 = \frac{x^2 + y^2}{x^2} = \sec^2\theta$$

---

[1] The Dutch physicist and engineer Balthazar van der Pol (1889–1959) did pioneering work on the first commercially available radios in the 1920s. His mathematical models for the behavior of a radio's internal voltages and currents are still in use, and we will look at one of them later in this section.

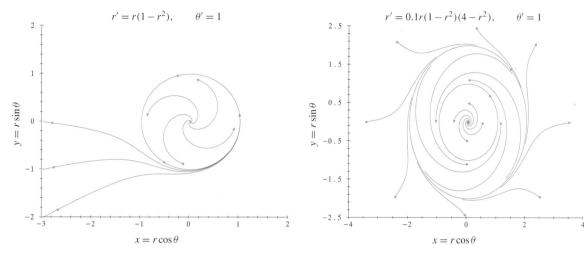

FIGURE 9.1.1 Attractor limit cycle at $r = 1$ encloses an unstable equilibrium point (Example 9.1.1).

FIGURE 9.1.2 Attractor and repeller limit cycles at $r = 1, 2$ enclose an unstable origin (Example 9.1.2).

So we obtain the equivalent system in polar coordinates:

$$r' = r(1 - r^2), \qquad \theta' = 1 \tag{3}$$

We see from (3) that $\theta(t) = t + C$, so $\theta(t + 2\pi) = \theta(t) + 2\pi$. This tells us that orbits move counterclockwise around the equilibrium point of system (1) at the origin in a full rotation every 2π units of time. From the ODE for r in (3) we see that $r = 0$ and $r = 1$ are solutions. The solution $r = 0$ corresponds to the equilibrium point at the origin, while $r = 1$ corresponds to the circular cycle $x^2 + y^2 = 1$ for system (1). Using Sign Analysis techniques for $r' = r(1 - r^2)$, we see that if $0 < r_0 < 1$, then $r(t)$ decreases to 0 as $t \to -\infty$ and increases to 1 as $t \to +\infty$. But if $r_0 > 1$, then $r(t)$ decreases to 1 as $t \to +\infty$ and increases to $+\infty$ as t decreases (see the r state line in the margin). So the unit circle is a limit cycle of period 2π, and it attracts all nonconstant orbits of system (1). See Figure 9.1.1.

In the next example, we don't even write out the planar system in xy-coordinates, but go directly to the equivalent system in polar coordinates and use Sign Analysis.

EXAMPLE 9.1.2

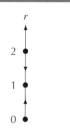

Two Limit Cycles: One Attracts, the Other Repels

The system in polar coordinates

$$r' = 0.1r(1 - r^2)(4 - r^2), \qquad \theta' = 1$$

has an attracting limit cycle ($r = 1$) and a repelling limit cycle ($r = 2$). These results are suggested by the changes of sign of r' at the values 1 and 2 (see the state line in the margin). As in the last example, the period of each cycle is 2π, and the rotation is counterclockwise because θ' is positive (Figure 9.1.2). The factor of 0.1 in the rate function for r was inserted to make the graphs in Figure 9.1.2 attractive. What would the graphs look like if we replaced 0.1 by 10 or by 0.001?

Now let's turn to the kind of planar system that van der Pol used for modeling radio circuits.

Van der Pol Circuits and Limit Cycles

Suppose that we are to design an electrical circuit with a periodic output current of prescribed amplitude and period. We also require that the current must return to the periodic state quickly after a disturbance. Van der Pol's circuit is an *RLC* loop in which the passive resistor is replaced by an active element. First we will review the equations of the passive *RLC* circuit from a state-variable and system point of view.

Passive *RLC* Circuit

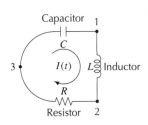

Suppose that a source of constant voltage is attached to the circuit shown in the margin and then withdrawn. How do the current $I(t)$ and the voltages across the circuit elements change with the passage of time? According to Kirchhoff's Voltage Law (Section 4.4), the voltages across the three circuit elements satisfy

$$V_{13} = V_{12} + V_{23} \tag{4}$$

where V_{ij} denotes the voltage drop from node i to node j. The separate voltages are related to the corresponding circuit elements by the laws

$$\begin{aligned} V_{12} &= LI' \\ V_{23} &= RI \\ V_{13}' &= -\frac{1}{C}I \end{aligned} \tag{5}$$

☞ Look back at Section 4.4 for these laws.

where the minus sign in the last equation follows from the relation $V_{13} = -V_{31}$. The relations given in (5) are, respectively, Faraday's Law, Ohm's Law, and the differential form of Coulomb's Law.

From (4) and (5) we see that the two state variables I and $V = V_{13}$ are enough to characterize the dynamics of the circuit via the linear system

☞ From solutions $I(t)$ and $V(t) = V_{13}(t)$ of (6), we can get $V_{23}(t)$ from (5) and $V_{12}(t)$ from (4).

$$\begin{aligned} I' &= \frac{1}{L}V_{12} = \frac{1}{L}(-V_{23} + V) = \frac{1}{L}(-RI + V) \\ V' &= -\frac{1}{C}I \end{aligned} \tag{6}$$

The system matrix has characteristic polynomial $\lambda^2 + R\lambda/L + 1/LC$, so its eigenvalues have negative real parts. As in the results of Example 7.7.6, the current $I(t)$ and the voltage $V(t)$ tend to 0 as $t \to +\infty$, and there are no periodic solutions, so there are no limit cycles. The physical reason for this decay to the zero-state is that the resistor dissipates electrical energy.

Active RLC Circuit

The circuit is active if energy is pumped into the circuit whenever the amplitude of the current falls below some level. One way to do this is to replace the resistor by an active element that acts as a "negative resistor" at low current levels but dissipates energy at high levels. In van der Pol's time there were vacuum tubes and now there are semiconductor devices that do just this.

☞ A negative resistor pumps energy into the circuit.

The rate equations for the active circuit are obtained from the passive circuit system (6) by replacing the Ohm's Law voltage drop RI by $F(I)$:

$$I' = \frac{1}{L}(-F(I) + V)$$
$$V' = -\frac{1}{C}I \tag{7}$$

If time, current I, and voltage V across the capacitor are suitably rescaled to dimensionless form, system (7) takes on the form of a *van der Pol system*

$$\frac{dx}{d\tau} = y - \mu f(x)$$
$$\frac{dy}{d\tau} = -x \tag{8}$$

where

$$\tau = t/\sqrt{LC}, \qquad x = I/I_0, \qquad y = V\sqrt{C/L}/I_0$$

$$\mu = \sqrt{C/L}/I_0, \qquad f(x) = F(I_0 x), \qquad I_0 \text{ is a positive reference current}$$

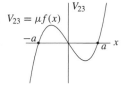

For convenience, we continue to refer to τ as time, x as current, and y as voltage across the capacitor. The margin sketches show an active *van der Pol circuit* and the nonlinear voltage-current relationship of a typical semiconductor device. In particular, note that voltage y and current x have opposite signs if $|x| < a$. The nature of the semiconducting "resistor" is enough to change completely the long-term behavior of the current in the circuit and the voltage across the capacitor.

Van der Pol and other physicists, engineers, and mathematicians have come up with the following theorem.

THEOREM 9.1.1

☞ The proof is sketched in the Student Resource Manual.

> The Van der Pol Cycle. Suppose that the continuous and piecewise smooth function $f(x)$ is defined for all x and has the properties (a) $f(-x) = -f(x)$; (b) for some positive constant a, $f(x) < 0$ if $0 < x < a$ and $f(x) > 0$ if $x > a$; and (c) $f(x) \to +\infty$ as $x \to +\infty$. Then for each positive value of μ the van der Pol system $x' = y - \mu f(x)$, $y' = -x$ has a unique limit cycle that encloses the equilibrium point $x = 0$, $y = 0$. This cycle attracts all nonconstant orbits.

Let's see what the limit cycles look like in a particular case.

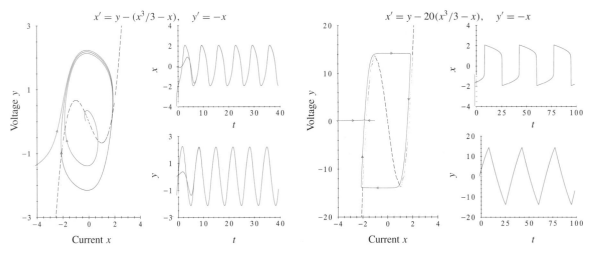

FIGURE 9.1.3 Two orbits approach the limit cycle: $\mu = 1$ (Example 9.1.3).

FIGURE 9.1.4 Two orbits approach the limit cycle: $\mu = 20$ (Example 9.1.3).

EXAMPLE 9.1.3

A Particular Van der Pol System

Van der Pol originally used the function

$$f(x) = \frac{1}{3}x^3 - x \qquad (9)$$

which meets the conditions of Theorem 9.1.1 with $a = \sqrt{3}$. Figures 9.1.3 and 9.1.4 show the graph of $y = \mu(x^3/3 - x)$ for $\mu = 1, 20$ (dashed curves). Also shown are orbits of the corresponding van der Pol system that wind onto the limit cycle, one from the inside and one from the outside, and the corresponding x- and y-component graphs. For large enough values of μ, the limit cycle has slanting sides nearly along arcs of the graph of $y = \mu(x^3/3 - x)$. For these values of μ the "top" and the "bottom" of the cycle correspond to rapid changes in the magnitude of the current x as the voltage stays constant. This is followed by slow changes along the sides of the cycle as the current "relaxes" and the voltage reverses its sign. For this reason, the limit cycle of a van der Pol system for large values of μ is called a *relaxation oscillation*.

☞ The period 6.5 for $\mu = 1$ is close to 2π. That's not surprising since all orbits have period 2π when $\mu = 0$.

We can treat $\mu = \sqrt{C/L}/I_0$ as a tuning parameter for adjusting the period of the current/voltage cycle while leaving the amplitude of the current approximately constant. For example, we see from the graphs in Figures 9.1.3 and 9.1.4 that the period is approximately 6.5 units of dimensionless time τ if $\mu = 1$ and about 33 units if $\mu = 20$. In each graph the dimensionless current has amplitude 2. Note that (in contrast to the current) the amplitude of the dimensionless voltage y across the semiconductor increases with μ.

Finally, note that the Jacobian matrix at the origin for the van der Pol system $x' = y - \mu(x^3/3 - x)$, $y' = -x$ is $J = \begin{bmatrix} \mu & 1 \\ -1 & 0 \end{bmatrix}$ with eigenvalues $[\mu \pm (\mu^2 - 4)^{1/2}]/2$. Since μ is positive, the eigenvalues are positive (if $\mu \geq 2$) or are complex conjugates with positive real parts if $0 < \mu < 2$. By the results of Sections 7.7 and 8.2 we know that the origin is a repelling and unstable equilibrium point, so it's no surprise that nearby orbits move outward from the origin as time advances.

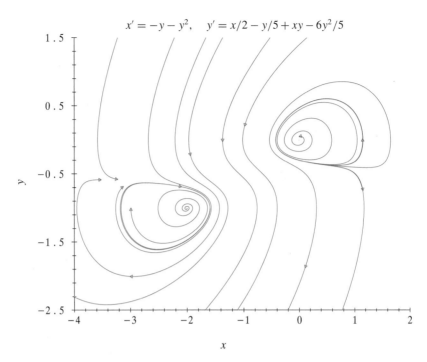

$$x' = -y - y^2, \quad y' = x/2 - y/5 + xy - 6y^2/5$$

FIGURE 9.1.5 Two limit cycles for a quadratic system (Example 9.1.4).

The limit cycles of this section's examples are orbits of planar systems whose rate functions are polynomials (of third or fifth degree) in the xy-state variables. Here is an example where the rate functions are only quadratic and there are two limit cycles.

EXAMPLE 9.1.4

A Quadratic System with Two Limit Cycles
There is no known formula for the solutions of the quadratic system

$$x' = -y - y^2, \qquad y' = \frac{1}{2}x - \frac{1}{5}y + xy - \frac{6}{5}y^2 \qquad (10)$$

so we use a solver to find limit cycles. First we find equilibrium points. The x' ODE shows us that any equilibrium point's y-coordinate is 0 or -1. Using $y = 0, -1$ in the y' ODE, we see that the equilibrium points are $(0, 0)$ and $(-2, -1)$. Figure 9.1.5 shows a repelling limit cycle around $(0, 0)$ and an attracting limit cycle around $(-2, -1)$.

The Chinese mathematician Tung Chin Chu discovered system (10) in the late 1950s and succeeded in proving that it did indeed have a pair of limit cycles.[2]

David Hilbert

[2] A conjecture about planar systems with quadratic rate functions is that they either have no more than four attracting and repelling limit cycles or else infinitely many. The Chinese mathematician Shi Song Li found a system with four in 1979. The Russian mathematician N. N. Il'yaschenko recently resolved part of the conjecture by showing that any such system has only finitely many attracting and repelling cycles; not much else is known. David Hilbert (1860–1943), the best known German mathematician of the late nineteenth and early twentieth century, posed this question about the number and placement of cycles as #16b in his famous list of 23 problems presented to the International Congress of Mathematicians in Paris in 1900. Most of Hilbert's problems have been solved, but not 16b. Will a similar list of open problems be presented at the International Congress in 2000?

Comments

Van der Pol's limit cycles and the limit cycles of the examples are internally generated by the dynamics of the corresponding autonomous systems and are *not* the response to some external driving force that is periodic in time. It is not accidental that these systems are nonlinear. Indeed, an autonomous linear system $x' = Ax$ has no attracting or repelling limit cycles. That is, if $x = x(t)$ is a periodic solution of $x' = Ax$, then by linearity so is $x = cx(t)$ for all real constants c. This means that any cycle of a linear system is part of a family of cycles, so it can't attract (or repel) all nearby orbits.

Cycles in the xy-state plane have two distinctive properties:

- Each cycle encloses one or more equilibrium points.

- Each cycle divides the plane into two regions (i.e., the interior and exterior regions of the cycle), and neither region is accessible from the other via an orbit.

The first property tells us that if we are searching for cycles, we would do well to locate the equilibrium points first. The second property implies that we can regard the dynamics and orbital behavior in the interior region bounded by a cycle as relatively independent of what goes on in the exterior region.

PROBLEMS

www **1.** (*Cycles and Limit Cycles*). Find the equilibrium points and the cycles of the following systems written in polar coordinates. Determine whether the equilibrium point at the origin, $r = 0$, is asymptotically stable, neutrally stable, or unstable. Determine whether each cycle is a limit cycle, and if it is, whether it is attracting or repelling. Sketch the cycles and other orbits in the xy-plane by hand. Use arrowheads to show the direction of increasing time.

(a) $r' = 4r(4 - r)(5 - r)$, $\theta' = 1$

(b) $r' = r(r - 1)(2 - r^2)(3 - r^2)$, $\theta' = -3$

(c) $r' = r(1 - r^2)(4 - r^2)$, $\theta' = 1 - r^2$ [*Hint*: r' and θ' are 0 if $r = 1$.]

(d) $r' = r(1 - r^2)(9 - r^2)$, $\theta' = 4 - r^2$ [*Hint*: $\theta' = 0$ but $r' \neq 0$ if $r = 2$.]

(e) $r' = r\cos \pi r$, $\theta' = 1$

(f) $r' = r\sin(\pi/r)$, $\theta' = -2$, where r' is defined to be 0 if $r = 0$. [*Hint*: Use the definition of stability in Section 8.1 to show that the system is stable at the origin.]

2. (*Limit Cycles*). Find all limit cycles and identify each as attracting or repelling. [*Hint*: Write the systems in parts (a)–(c) in polar coordinates.]

(a) $x' = y - x(x^2 + y^2)$, $y' = -x - y(x^2 + y^2)$

(b) $x' = x + y - x(x^2 + y^2)$, $y' = -x + y - y(x^2 + y^2)$

(c) $x' = 2x - y - x(3 - x^2 - y^2)$, $y' = x + 2y - y(3 - x^2 - y^2)$

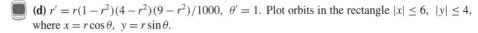 (d) $r' = r(1 - r^2)(4 - r^2)(9 - r^2)/1000$, $\theta' = 1$. Plot orbits in the rectangle $|x| \leq 6$, $|y| \leq 4$, where $x = r\cos \theta$, $y = r\sin \theta$.

3. (*Center-Spiral*). Consider the system in polar coordinates $r' = r^3 \sin(1/r)$, $\theta' = 1$, where r' is defined to be 0 at $r = 0$. Show that the corresponding xy-system has infinitely many circular limit cycles around the equilibrium point at the origin, these cycles "converge" onto the origin, they are alternately attracting and repelling, and the other orbits are spirals that spiral away from one cycle and onto another as t increases. The origin of the xy-system is said to be a *center-spiral*. Explain the name. [*Hint*: See also Problem 1(f).]

4. (*Nonisolated, Nonlinear Cycles*). Show that the nonconstant orbits of the system $x' = y^3$, $y' = -x^3$ are cycles that enclose the equilibrium point at the origin and fill the xy-plane. Plot the orbits and component curves corresponding to initial points $(0, 0)$, $(0.5, 0)$, $(1, 0)$, $(2, 0)$. Are the periods of distinct cycles the same?

5. (*Semistable Cycles*). Some cycles repel on one side and attract on the other; they are one kind of *semistable* cycle. Find all cycles of $r' = r(1 - r^2)^2(4 - r^2)(9 - r^2)$, $\theta' = 1$ and identify each as attracting, repelling, or semistable. Sketch the orbits.

6. (*A Strange Cycle*). Explain why the system $r' = r(r - 1)^2 \sin[\pi/(r - 1)]$, $\theta' = 1$, where r' is defined to be 0 if $r = 1$, has a cycle $r = 1$, that is not a limit cycle, and every neighborhood of which contains other cycles as well as spirals between the successive cycles [*Hint*: As $r \to 1$, $(r - 1)^2 \sin[\pi/(r - 1)] \to 0$; $r' = 0$ at $r = 1$ and $r = 1 \pm 1/n$.]

7. (*Cycles in $\mathbb{R}^3$*). Determine the behavior of orbits near the cycle $r = 1$, $z = 0$ of the system in cylindrical coordinates $r' = r(1 - r^2)$, $\theta' = 25$, $z' = \alpha z$, where α is a constant. Consider separately the three cases $\alpha < 0$, $\alpha = 0$, $\alpha > 0$. Plot orbits in xyz-space for $\alpha = -0.5, 0, 1$.

8. (*Van der Pol Systems*). Verify that each system satisfies the conditions of the Van der Pol Cycle Theorem (Theorem 9.1.1). For each value of μ plot the limit cycle and some orbits that are attracted to it. Estimate the period and the x- and the y-amplitude of each cycle; verify that the larger the value of μ, the longer the period and the larger the y-amplitude of the cycle but that the x-amplitude doesn't change much.

   (a) $x' = y - \mu(x^3 - 10x)$, $y' = -x$; $\mu = 0.1, 2$

   (b) $x' = y - \mu x(|x| - 1)$, $y' = -x$; $\mu = 0.5, 5, 50$

   (c) $x' = y - \mu x(x^4 + x^2 - 1)/10$, $y' = -x$; $\mu = 0.1, 1$

   (d) $x' = y - \mu x(2x^2 - \sin^2 \pi x - 2)$, $y' = -x$; $\mu = 0.5, 5$

   (e) $x' = y - \mu(x - |x + 1| + |x - 1|)$, $y' = -x$; $\mu = 0.5, 5, 50$

9. (*Van der Pol Cycles*). Plot an orbit and its component graphs for solutions near the cycle of the van der Pol system $x' = y - \mu f(x)$, $y' = -x$ for a range of values of μ, $1/10 \leq \mu \leq 10$, for each function f given in (a) and (b) below. Use the x- or y-component graph to estimate the period of the cycle for each value of μ that you used. How does the period change as μ increases? How do the maximal amplitudes of the x- and y-components change as μ increases?

   (a) $f(x) = x^3 - x$ (b) $f(x) = |x|x^3 - x$

10. (*Liénard Equation*). The nonlinear, second-order ODE $x'' + f(x)x' + g(x) = 0$, where $f(x)$ and $g(x)$ are continuous and piecewise smooth, is called the *Liénard equation* after the French mathematician and applied physicist, Alfred Liénard (1869–1958).

    (a) Show that if $y = x' + F(x)$, where $F(x) = \int_0^x f(s)\, ds$, then the Liénard equation can be written in system form as $x' = y - F(x)$, $y' = -g(x)$.

    (b) Plot the orbit of the periodic solution of $x'' + (x^2 - 1)x' + x = 0$ both in the xx'-plane and in the *Liénard xy-plane*.

    (c) (*Rayleigh Equation*). The *Rayleigh equation* is $z'' + \mu[(z')^2 - 1]z' + z = 0$. Differentiate the Rayleigh equation with respect to t, then set $x = \sqrt{3}z'$ and show that $x'' + \mu(x^2 - 1)x' + x = 0$. Show that the ODE in x reduces to a form of system (8) if we set $y = x' + \mu(x^3/3 - x)$. Show that the Rayleigh equation has a unique attracting limit cycle for each $\mu > 0$. Plot the Rayleigh cycle and plot orbits through initial points $(1.5, 1.5)$ and $(0.5, 0.5)$ in the zz'-plane for each of the following values of μ: $\mu = 0.1, 1, 5, 10$.

9.2 Long-Term Behavior

The maximally extended orbits of a planar autonomous system are curves in the state plane. If the rate functions of the system are continuously differentiable, then the orbits completely fill up the plane. By uniqueness no two orbits ever cross or touch. Even though orbits can't touch, they can wriggle around in an incredible variety of ways. But if an orbit is bounded, then we will see that its long-term behavior is strictly limited to one of only three alternatives. Poincaré[3] saw to the heart of this behavior, and his ideas underlie much of what we have to say in this section.

Cycle-Graphs

Orbits of a damped Hooke's Law spring tend to an equilibrium point as $t \to +\infty$. Orbits of a van der Pol system spiral toward a periodic cycle as $t \to +\infty$. In the example below we will see that as $t \to +\infty$ orbits approach a cycle-graph, which is a strange hybrid of equilibrium point and cycle that is neither constant nor periodic.

EXAMPLE 9.2.1

Let's consider the planar autonomous system

$$x' = x(1 - x - 3.75y + 2xy + y^2)$$
$$y' = y(-1 + y + 3.75x - 2x^2 - xy) \tag{1}$$

The x- and y-axes are composed of orbits of system (1), as is the line $y = 1 - x$. The latter is shown by replacing y by $1 - x$ on the right sides of the rate equations of system (1), and then observing that $y' = -x'$; we omit the calculations. The axes and the line $y = 1 - x$ intersect at the equilibrium points $(0, 0)$, $(1, 0)$, and $(0, 1)$. Figure 9.2.1 shows the direction of motion along the sides of the orbital triangle that connects these points. The triangle is a directed graph whose counterclockwise orientation is determined by the motion of system (1) along its edges as time increases. Each edge orbit tends to an equilibrium point at a vertex as $t \to +\infty$ and as $t \to -\infty$. The triangle can't be traversed in finite time since it takes infinitely long to trace out each side, so the triangle is *not* a periodic orbit of system (1). Visual evidence produced by a numerical solver suggests that every orbit inside the triangle tends to the equilibrium

Jules Henri Poincaré

[3]The French mathematician Jules Henri Poincaré (1854–1912) is regarded as the "last universalist," the last person to understand all of the mathematics and much of the physics of his era. His professional life was spent mostly at the University of Paris and the École Polytechnique, where he was professor of mathematical physics, probability, celestial mechanics, and astronomy. Every year he lectured on a different subject, his students taking notes that were later published. His research covered most of the areas of the mathematics of his time: groups, number theory, theory of functions, algebraic geometry, algebraic topology (which he developed), mathematical physics, celestial mechanics, partial differential equations, and ordinary differential equations. His first research paper (1878) and his last (1912) were on ODEs. Poincaré and A. M. Lyapunov created the modern approach to ODEs with its emphasis on the general behavior of orbits and solutions, rather than on solution formulas. Poincaré had a talent for good exposition, wrote many popular books on mathematics and science, and was the first (and, so far, the only) mathematician elected to the literary section of the French Institute.

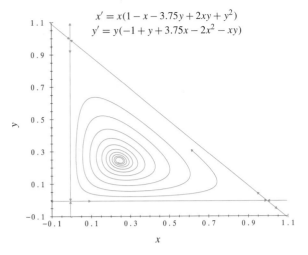

$$x' = x(1 - x - 3.75y + 2xy + y^2)$$
$$y' = y(-1 + y + 3.75x - 2x^2 - xy)$$

FIGURE 9.2.1 A triangular cycle-graph approached by an outward spiraling orbit (Example 9.2.1). The cycle-graph consists of three vertex equilibrium points and three orbits along the edges.

FIGURE 9.2.2 Component graphs of the orbit of Figure 9.2.1 that tends to a triangular cycle-graph: fast near the sides and slow around the corners (Example 9.2.1).

point $(0.25, 0.25)$ as $t \to -\infty$ and spirals outward toward the triangle as $t \to +\infty$ (Figure 9.2.1). Figure 9.2.2 shows the component graphs of the spiraling orbit of Figure 9.2.1. The time intervals where both component graphs are horizontal correspond to the slow motion near the vertices. The nearly vertical segments of the component graphs correspond to rapid changes in x or y away from the vertices.

The triangle in Figure 9.2.1 is an example of a cycle-graph for a system. Here is the general definition.

❖ **Cycle-Graph**. A *cycle-graph* (or *polycycle*) of a planar autonomous system is a graph in state space that closes on itself and consists of N vertices ($N \geq 1$) and at least N edges. The vertices are equilibrium points of the system and the edges are orbits that tend to vertices as $t \to -\infty$ and as $t \to +\infty$. The graph is coherently oriented along the edges by time's increase (i.e., the graph can be traversed by following the arrows).

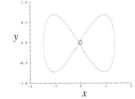

The cycle-graph of Example 9.2.1 has three vertices and three edges and is oriented counterclockwise. Cycle-graphs may have more edges than vertices as the "lazy-eight" cycle-graph in the margin shows (see Problem 10).

Long-Term Behavior of Orbits of a Planar Autonomous System

So far in this chapter we have seen orbits of planar autonomous systems approach an equilibrium point, a limit cycle, and a cycle-graph as $t \to +\infty$. What other types of long-term behavior are possible? It turns out that there are no other alternatives for bounded orbits. In this context an orbit is said to be *bounded* if its graph is enclosed in some rectangle in the state plane, so the orbit is defined for all t and does *not* approach infinity as $t \to -\infty$ or as $t \to +\infty$.

Poincaré and the Swedish mathematician Ivar Bendixson (1861–1935) proved the following fundamental result about long-term behavior.

THEOREM 9.2.1

Long-Term Behavior in the Plane. Suppose that Γ is a maximally extended and bounded orbit of the system $x' = f(x, y)$, $y' = g(x, y)$, where f and g are continuously differentiable. Suppose also that Γ lies in a rectangle that contains only a finite number of equilibrium points. Then, as $t \to +\infty$, Γ must tend to exactly one of the following:

- An equilibrium point
- A cycle
- A cycle-graph

The same alternatives hold as $t \to -\infty$.

These alternatives give us all the possible "histories" and "futures" of an orbit Γ. For example, Γ might tend to an equilibrium point as $t \to -\infty$ and to a cycle as $t \to +\infty$, or maybe tend to a cycle as $t \to -\infty$ and to a cycle-graph as $t \to +\infty$. If Γ is itself an equilibrium point or a cycle, then it simply tends to itself as $t \to -\infty$ and as $t \to +\infty$. If Γ is a nonconstant orbital "edge" in a cycle-graph, then Γ approaches an equilibrium point of the cycle-graph as $t \to -\infty$ and an equilibrium point of the cycle-graph (possibly the same point) as $t \to +\infty$. What Γ can't do as $t \to +\infty$, for example, is tend to a pair of equilibrium points or to a single nonconstant and nonperiodic orbit. Nor can Γ wander around inside its boundary rectangle without any apparent "destination"; as time advances (or regresses) Γ must head toward one of the three Poincaré-Bendixson alternative sets. It is known that if Γ approaches a cycle or a cycle-graph, it must do so in a spiraling fashion that is consistent with the time orientation of the cycle or cycle-graph. See, for example, Figures 9.1.1–9.1.5 and 9.2.1.

☞ So there is no chaos for orbits of a planar autonomous system.

The set of points an orbit Γ approaches as $t \to +\infty$ is called its *positive limit set* and is denoted by $\omega(\Gamma)$. The *negative limit set* $\alpha(\Gamma)$ is defined similarly, except that $t \to -\infty$. The letters alpha and omega are the first and the last letters of the Greek alphabet, so $\alpha(\Gamma)$ tells us how and where the orbit Γ is "born," and $\omega(\Gamma)$ tells how and where it "dies." Theorem 9.2.1 gives us three alternatives for $\alpha(\Gamma)$ and three for $\omega(\Gamma)$.

☞ The positive limit set of the spiralling orbit in Figure 9.2.1 is the triangular cycle-graph and the negative limit set is the equilibrium point $(0.25, 0.25)$.

What can we say about the long-term behavior of Γ if it isn't bounded? If, for example, Γ stays in no rectangle as $t \to -\infty$, but does remain in some rectangle for all $t \geq t_0$ for some t_0, then the alternatives of Theorem 9.2.1 apply only to $\omega(\Gamma)$. We saw this behavior in Example 9.1.1 where orbits outside the limit cycle $r = 1$ become unbounded as t decreases but spiral toward the limit cycle as $t \to +\infty$. It's the other way around in Example 9.1.2 where the orbits outside the limit cycle $r = 2$ are "born" at the cycle but become unbounded as t increases.

The alternatives of Theorem 9.2.1 are simple to state, but there are profound implications.

THEOREM 9.2.2

☞ As always, we assume that f and g are continuously differentiable.

> Unbounded Orbits. Suppose that the system $x' = f(x, y)$, $y' = g(x, y)$ has no equilibrium points. Then all orbits become unbounded as time increases and as time decreases.

☞ Without equilibrium points in the system, orbits are born at infinity and return there to die.

Suppose, to the contrary, that an orbit Γ remains in a rectangle, say for all $t \geq t_0$, for some t_0. Then by Theorem 9.2.1 $\omega(\Gamma)$ must be a cycle because the other two alternatives involve equilibrium points. It was noted in the "Comments" at the end of Section 9.1 that the interior region of a cycle must contain at least one equilibrium point, but by hypothesis in this case there aren't any. This contradiction implies that Γ can't, after all, stay inside any rectangle as t increases. Similarly, Γ can't stay inside any rectangle as t decreases. So Γ becomes unbounded for increasing time and for decreasing time.

The next result also follows from Theorem 9.2.1.

THEOREM 9.2.3

> Bounded Orbits. Every bounded orbit of the system $x' = f(x, y)$, $y' = g(x, y)$ has an equilibrium point or a cycle in its negative and in its positive limit set. If a limit set contains no equilibrium points, then the limit set is a cycle.

This follows from the fact that if a limit set is not a cycle, then it must either be an equilibrium point or a cycle-graph, and cycle-graphs contain equilibrium points.

If we want to see what a bounded orbit does as time tends to $+\infty$ or to $-\infty$, we should first locate all of the equilibrium points, cycles, and cycle-graphs. Finding the equilibrium points is the algebraic problem of solving the equations $f(x, y) = 0$, $g(x, y) = 0$ simultaneously. Many computer solvers use Newton's Method to approximate the coordinates of the equilibrium points. Finding cycles and cycle-graphs (or showing that there aren't any) is a harder task, and it is time now to outline some techniques for doing just that.

Are There Any Cycles or Cycle-Graphs?

☞ Look back at the "Comments" for Section 9.1.

Here is one way to locate a cycle. Suppose that S is a ringlike region without any equilibrium points and all orbits that intersect the inner and outer edges of S move into S as time increases. If such a ring S does exist, then there must be at least one cycle in S that also encloses the "hole." Because the cycle must also enclose one or more equilibrium points, but S doesn't have any, there must be at least one equilibrium point in the hole.

EXAMPLE 9.2.2

👉 It takes some algebra to show that there is only one equilibrium point.

Locating a Cycle

The system

$$x' = y + x\left(1 - 2x^2 - \frac{1}{2}y^2\right)$$
$$y' = -x + y\left(1 - 2x^2 - \frac{1}{2}y^2\right)$$

(2)

has a single equilibrium point at the origin. If there is a cycle, then the cycle must enclose the origin. Let's construct a ring around the origin with the property that orbits cross its perimeters moving into the ring with the advance of time. Because there are no equilibrium points in the ring itself, every orbit Γ entering the ring has a cycle as its positive limit set.

Here is how a ring with the desired properties may be constructed. Set $V = x^2 + y^2$, that is, the square of the distance from an orbital point $x(t)$, $y(t)$ to the origin. The derivative of V following the motion of system (2) is

$$V' = 2xx' + 2yy' = 2(x^2 + y^2)(1 - 2x^2 - y^2/2)$$

Observe that V' is positive if $x^2 + y^2 \leq 1/3$ and V' is negative if $x^2 + y^2 \geq 3$ because

👉 These inequalities are *not* obvious!

$$1 - 2x^2 - y^2/2 > 1 - 3(x^2 + y^2) \geq 0 \quad \text{if} \quad x^2 + y^2 \leq 1/3$$
$$1 - 2x^2 - y^2/2 < 1 - (x^2 + y^2)/3 \leq 0 \quad \text{if} \quad x^2 + y^2 \geq 3$$

These are rough estimates, but they do give us the circular inner and outer perimeters (i.e., $x^2 + y^2 = 1/3$ and $x^2 + y^2 = 3$) of a ring with the property that orbits touching either perimeter must move into the ring as time increases because of the signs of V' (Figure 9.2.3). So there is a limit cycle inside the ring. Figure 9.2.3 suggests that the cycle is unique and attracting.

The next example has an interpretation in terms of competing species.

EXAMPLE 9.2.3

Competing Species

The system

$$x' = x(1 - 0.1x - 0.1y), \qquad y' = y(2 - 0.05x - 0.025y)$$

(3)

models the populations of two competing species (Section 5.3). The equilibrium points $(0, 0)$, $(0, 80)$, $(10, 0)$, and $(70, -60)$ lie on the boundary of or outside the population quadrant $x \geq 0$, $y \geq 0$. So there can be no cycles inside the population quadrant and no cycles that intersect that quadrant. The first assertion follows from the fact that a cycle inside the quadrant must enclose an equilibrium point, but there are no equilibrium points inside the quadrant. The second follows because the x- and y-axes are unions of orbits, so orbits can't enter the first quadrant from any other quadrant because they would have to intersect orbits on the axes.

Bendixson discovered a criterion that is widely used for the *absence* of cycles and cycle-graphs in a simply connected region of the xy-plane.

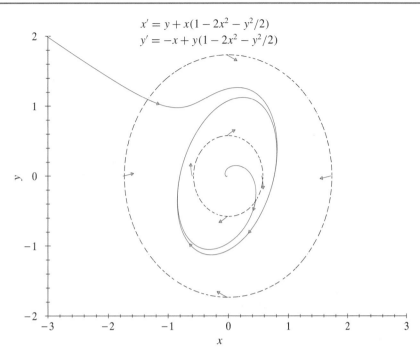

$$x' = y + x(1 - 2x^2 - y^2/2)$$
$$y' = -x + y(1 - 2x^2 - y^2/2)$$

FIGURE 9.2.3 An attracting limit cycle inside a ring (Example 9.2.2).

THEOREM 9.2.4

☞ The Student Resource Manual has the proof of this theorem.

> **Bendixson's Negative Criterion.** Suppose that the function $\partial f/\partial x + \partial g/\partial y$ has a fixed sign in a simply connected[4] region R of the xy-plane. Then the system $x' = f(x, y),\ y' = g(x, y)$ has no cycle or cycle-graph in R.

Here is a practical application of Bendixson's Negative Criterion.

EXAMPLE 9.2.4

Damping Means No Cycles or Cycle-Graphs

The ODE $x'' = g(x, x')$ is a generalization of the ODEs for the motion of a simple pendulum, a body on the end of a spring, and the current in an RLC circuit. The equivalent system is

$$x' = y, \qquad y' = g(x, y)$$

In this case $f = y$, so $\partial f/\partial x + \partial g/\partial y$ reduces to $\partial g/\partial y$. If $\partial g/\partial y$ has fixed sign in the xy-plane, there can be no cycles or cycle-graphs. For example, the system that models a damped simple pendulum is

$$x' = y, \qquad y' = -a\sin x - by$$

---

[4]A connected region in the plane is said to be *simply connected* if it has no holes. So the rectangle described by $|x| \leq 1,\ |y| \leq 2$ is simply connected, but the ring of Example 9.2.2 is not.

☞ Friction dissipates energy, so physically that's why there aren't any cycles or cycle-graphs.

where a and b are positive constants. The term $-by$ models the effect of friction on the motion of the pendulum, and the function $\partial g/\partial y$ is just the negative constant $-b$. So by Theorem 9.2.4 there are no cycles or cycle-graphs.

Comments

The tests and alternatives of this section are qualitative, not quantitative. The long-term behavior of the orbits of a planar system is described in words and graphically, not by solution formulas. Results such as these are appropriate at the early stages of an analysis of a complex system that models some physical phenomenon. These approaches are also useful in the design of a system, if for no other purpose than to tell us what can't be done. For example, if the aim is to construct a planar autonomous system with a cycle or a cycle-graph, we have to allow for an equilibrium point. In analyzing the orbital structure of a specific system, it is often a good idea to use a numerical solver. Attracting or repelling cycles, cycle-graphs, and equilibrium points are (usually) visible on the screen, and they give us useful information about possible long-term orbital behavior.

☞ This section has been about behavior in a state plane. What about behavior on a state line or in three-dimensional state space?

We note that long-term behavior for a single scalar ODE, $x' = f(x)$, has already been described by the Sign Analysis and state-line techniques of Section 2.2. Every nonconstant and bounded orbit on a state line tends to an equilibrium point as $t \to -\infty$ and to another equilibrium point as $t \to +\infty$; there is no room for cycles or cycle-graphs. What happens in three-dimensional state space? That's the subject of the next two sections—expect to see strange behavior!

PROBLEMS

1. (*Limit Sets*). Using either analytical or graphical techniques, find the positive and the negative limit sets of the orbits through the initial points indicated. Sketch orbits for **(c)**, **(d)**, and **(f)**.

(a) $x' = y, \; y' = -x; \; (0, 0), (1, 1)$ (b) $x' = y, \; y' = x; \; (1, 1), (1, -1), (1, 0)$

(c) (*Van der Pol*). $x' = y - (x^3/3 - x), \; y' = -x; \; (0.1, 0), (3, 0)$ [*Hint*: See Section 9.1.]

(d) (*Simple Pendulum*). $x' = y, \; y' = -\sin x; \; (0, 1), (0, \sqrt{2}), (0, 2)$

(e) (*Polar Form*). $r' = r(1 - r^2)(4 - r^2), \; \theta' = 5; \; r_0 = 1/2, 3/2, 5/2; \; \theta_0 = 0$

(f) $x' = [x(1 - x^2 - y^2) - y][(x^2 - 1)^2 + y^2], \; y' = [y(1 - x^2 - y^2) + x][(x^2 - 1)^2 + y^2];$
$(1/2, 0), (0, 1), (3/2, 0)$ [*Hint*: Use a numerical solver, and note that the unit circle seems to be a cycle-graph consisting of two equilibrium points and two joining arcs.]

2. (*Bendixson Negative Criterion*). Show that each ODE has no periodic solutions and no cycle-graphs. [*Hint*: Convert each ODE to a system.]

(a) (*Damped Pendulum*). $mLx'' + cLx' + mg \sin x = 0; \; m, L, c,$ and g are positive constants.

(b) (*Damped Nonlinear Spring*). $mx'' + ax' + bx + cx^3 = 0$, where $m, a,$ and b are positive constants and c is any constant.

(c) $x'' + (2 + \sin x)x' + g(x) = 0$, where $g(x)$ is any continuously differentiable function.

(d) $x'' + f(x)x' + g(x) = 0; \; f$ and g are continuously differentiable and $f(x)$ has fixed sign.

3. (*Bendixson Negative Criterion*). Show that each system has no cycle or cycle-graph in the region indicated.

(a) $x' = 2x - y + 36x^3 - 15y^2, \quad y' = x + 2y + x^2y + y^5; \qquad xy$-plane

(b) $x' = 12x + 10y + x^2y + y\sin y - x^3, \quad y' = x + 14y - xy^2 - y^3; \qquad$ the disk $x^2 + y^2 \le 8$

(c) $x' = x - xy^2 + y\sin y, \quad y' = 3y - x^2y + e^x\sin x; \qquad$ the open disk $x^2 + y^2 < 4$

4. (*No Limit Sets*). Explain why none of the orbits of the following systems have positive and negative limit sets. Plot some of the orbits and observe that as t increases or decreases orbits exit the screen. [*Hint*: Use Theorem 9.2.2.]

(a) $x' = x - y + 10, \quad y' = x^2 + y^2 - 1$

(b) $x' = y, \quad y' = -\sin x - y + 2$

(c) $x' = x^2 + 2y^2 - 4, \quad y' = 2x^2 + y^2 - 16$

5. (*Any Cycles?*). Determine whether the following systems have cycles. Find all cycles (graphically or analytically), if any exist. If there are no cycles, explain why not.

(a) $x' = e^x + y^2, \; y' = xy$ [*Hint*: Does x ever decrease?]

(b) $x' = 2x^3y^4 + 5, \; y' = 2ye^x + x^3$ [*Hint*: Use Bendixson's Negative Criterion.]

(c) (*Polar Coordinates*). $r' = r\sin(r^2), \; \theta' = 1$

6. (*Using Green's Theorem*). Prove that the average value of the function $F(x, y) = \partial f/\partial x + \partial g/\partial y$ on the region R inside a cycle Γ of the system $x' = f(x, y), \; y' = g(x, y)$ must be zero. [*Hint*: The average value of F on R is defined to be $\int\int_R F(x, y)\,dx\,dy/A$, where A is the area of R. Use Green's Theorem (Theorem B.5.12).]

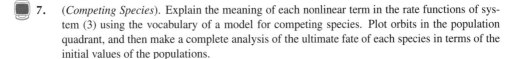

7. (*Competing Species*). Explain the meaning of each nonlinear term in the rate functions of system (3) using the vocabulary of a model for competing species. Plot orbits in the population quadrant, and then make a complete analysis of the ultimate fate of each species in terms of the initial values of the populations.

8. (*Contradicting Bendixson?*). The quantity $\partial f/\partial x + \partial g/\partial y$ for the system $x' = x - 10y - x(x^2 + y^2) = f, \; y' = 10x + y - y(x^2 + y^2) = g$ is $2 - 4x^2 - 4y^2$, which is negative in the region R defined by $x^2 + y^2 > 0.5$. Show that the unit circle is a cycle that lies entirely in R. Why is this not a contradiction of Bendixson's Negative Criterion?

www **9.** (*Triangle Cycle-Graph*). System (1) has a triangular cycle-graph that is the positive limit set of orbits inside the triangle. Let's explore the regions *outside* the triangle.

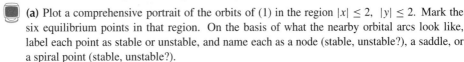

(a) Plot a comprehensive portrait of the orbits of (1) in the region $|x| \le 2, \; |y| \le 2$. Mark the six equilibrium points in that region. On the basis of what the nearby orbital arcs look like, label each point as stable or unstable, and name each as a node (stable, unstable?), a saddle, or a spiral point (stable, unstable?).

(b) Use the Jacobian matrix technique of Section 8.2 to verify that the equilibrium points $(0, 0)$, $(1/4, 1/4)$, $(4/3, 4/3)$ have the stability characteristics claimed in part (a).

(c) (*Cycle-Graph Sensitivity*). Replace the term $-3.75y$ in the first equation of system (1) by cy, but leave the second equation as is. For $c = -4, -3.5$, plot a portrait of the orbits. On the basis of your graphs, what do you think happens to the orbits as c increases through the "critical" value $c = -3.75$?

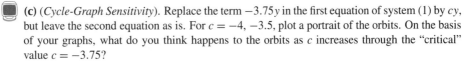

10. (*Lazy Eight Cycle-Graph*). Show that the system $x' = y + x(1 - x^2)(y^2 - x^2 + x^4/2), \; y' = x - x^3 - y(y^2 - x^2 + x^4/2)$ has equilibrium points at $(0, 0)$ and $(\pm 1, 0)$. Use a numerical solver and plot the orbits through the point $(-3, 2)$ forward in time. Repeat with the orbits through $(\pm 0.5, 2)$, but carry these orbits forward and backward in time. Explain what you see. [*Hint*: Look at the graph in the margin on page 481.]

11. (*Scaling Time*). Suppose that $x(t)$, $y(t)$ is a solution of the system $x' = f(x, y)$, $y' = g(x, y)$, and scale the time variable by $s = s(t)$ by setting $dt/ds = h(x(t), y(t))$ for a given function h that has a fixed sign. The scaled system has the form $dx/ds = fh$, $dy/ds = gh$.

(**a**) If $h(x, y)$ is positive everywhere, explain why the two systems have identical orbits with the same orientation induced by time's advance. What happens if $h(x, y)$ is everywhere negative?

(**b**) Suppose that $h(x, y)$ is positive everywhere except that h is zero at a point p that lies on a nonconstant orbit Γ of the unscaled system. What can you say about the corresponding orbit of the scaled system?

(**c**) Suppose that $f = x - 10y - x(x^2 + y^2)$ and $g = 10x + y - y(x^2 + y^2)$, while the factor $h = 1 - \exp[-10(x - 1)^2 - 10y^2]$. Explain why the unscaled system has an attracting limit cycle and the scaled system has an attracting cycle-graph on the unit circle. Plot the orbit through the point $(0.5, 0)$ and the corresponding component curves for the original system. Repeat for the scaled system. Explain what you see. [*Hint*: See the graphs (**d**), (**e**), and (**f**).]

(**d**)   (**e**)   (**f**)

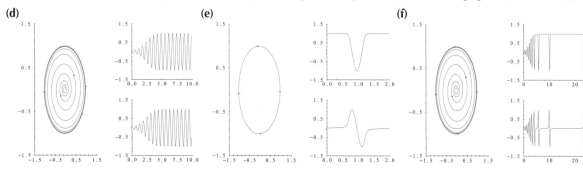

12. (*Dulac's Criterion*). Suppose that R is a simply connected region in the plane. Suppose that f, g, and K are continuously differentiable in R. Suppose that $\partial(Kf)/\partial x + \partial(Kg)/\partial y$ has a fixed sign in R. Show that the system $x' = f$, $y' = g$ can't have a cycle in R. [*Hint*: Look at the proof of Bendixson's Negative Criterion in the Student Resource Manual. Then adapt the proof to $\partial(Kf)/\partial x + \partial(Kg)/\partial y$.]

13. (*Ragozin's Negative Criterion for Interacting Species*). Suppose that a two-species interaction is governed by the model system $x' = xF(x, y)$, $y' = yG(x, y)$. The two species are said to be *self-regulating* if $\partial F/\partial x$ and $\partial G/\partial y$ are each negative throughout the population quadrant. If the x-species, say, is self-regulating, then the negativity of $\partial F/\partial x$ implies that the per unit growth rate F diminishes as x increases. Use Dulac's Criterion (Problem 12) to show that the system of ODEs of a pair of self-regulating species has no cycles in the population quadrant. [*Hint*: Let $K = (xy)^{-1}$ for $x > 0$, $y > 0$.]

  14. Use ODEs to draw the face of a cat.

9.3 Bifurcations

Cycles and equilibrium points are distinctive features of orbital portraits of planar autonomous systems. If the rate functions of the system are changed slightly, one might expect the new portrait to resemble the old; a cycle might contract a little, an equilibrium point shift somewhat, or a spiral tighten or loosen, but the dominant features of the portrait would remain. This seems reasonable, but it is not necessarily true. Actually, small changes in a coefficient in a rate function may mean the disappearance of one feature and the sudden appearance of another that is altogether different. As a parameter changes, an equilibrium point may suddenly appear and spawn a second

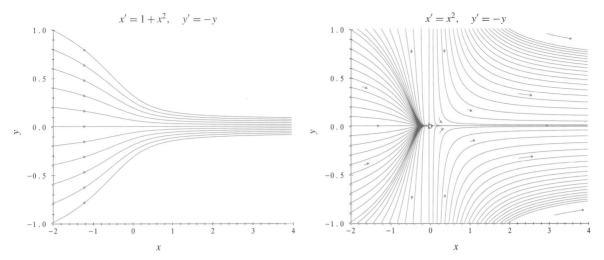

FIGURE 9.3.1 No equilibrium point before the saddle-node bifurcation: $c = 1$ (Example 9.3.1).

FIGURE 9.3.2 Saddle-node equilibrium point at bifurcation: $c = 0$ (Example 9.3.1).

☞ Bifurcations are also treated in Section 2.4.

equilibrium point, or a stable equilibrium point may destabilize and eject an attracting limit cycle. These are examples of *bifurcations*. In the context of ODEs the word "bifurcation" has come to mean any marked change in the structure of the orbits of a system (usually nonlinear) as a parameter passes through a critical value. In this section and the problems we describe several kinds of bifurcation that can occur in a planar autonomous system. We also look at the unusual scroll circuit system with three state variables; this system passes through several bifurcations and then becomes chaotic as a system parameter is changed.

Saddle-Node Bifurcation

The simplest bifurcations involve the appearance, disappearance, or splitting of an equilibrium point as a parameter c changes. One example is given here; other types appear in Problem 1. Some of these bifurcations are direct extensions to planar systems of the scalar bifurcations described in Section 2.4 and its problem set.

EXAMPLE 9.3.1

Saddle-Node Bifurcation

The system

$$x' = c + x^2, \qquad y' = -y \tag{1}$$

has no equilibrium points if the parameter c is positive. See Figure 9.3.1 for orbits if $c = 1$. If $c = 0$, a strange hybrid of saddle point and stable node suddenly appears at the origin, resembling a node on the left, a saddle on the right (Figure 9.3.2); it is an example of a *saddle-node* equilibrium point. It is not an elementary equilibrium point (Section 8.2) because 0 is an eigenvalue of the Jacobian matrix of (1) at the origin. As c decreases below 0, the origin splits into two equilibrium points $(\pm\sqrt{-c}, 0)$, one a saddle point and the other an asymptotically stable node. These points move in opposite directions away from the origin as c decreases (Figure 9.3.3).

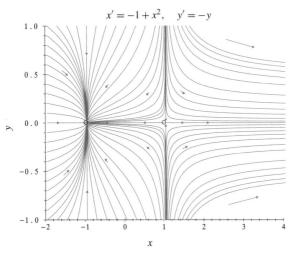

FIGURE 9.3.3 Saddle and node after the saddle-node bifurcation: $c = -1$ (Example 9.3.1).

FIGURE 9.3.4 The saddle-node bifurcation diagram (Example 9.3.1).

Figure 9.3.4 shows the bifurcation diagram for system (1). The diagram shows the x-coordinates of the equilibrium points as functions of c; the solid curve denotes the x-coordinate of the stable equilibrium point and the dashed curve denotes the x-coordinate of the unstable equilibrium point. This *bifurcation diagram* describes the evolution of a system's orbits as a parameter changes. The saddle-node bifurcation is an example of a tangent bifurcation (see Section 2.4). See Problem 1 for other examples.

Now let's take a look at a very different kind of bifurcation.

The Hopf Bifurcation

In the next example an equilibrium point "expands" into a limit cycle as a parameter changes.

EXAMPLE 9.3.2

Bifurcation to a Limit Cycle

For all values of the parameter c the origin is an equilibrium point of the system

$$x' = cx + 5y - x(x^2 + y^2)$$
$$y' = -5x + cy - y(x^2 + y^2)$$ (2)

☞ Linearizations are described in Section 8.2.

The linearization at the origin for system (2) is the system

$$x' = cx + 5y$$
$$y' = -5x + cy$$ (3)

The eigenvalues of the system matrix of (3) are $c \pm 5i$. We see that as the parameter c increases through 0, the equilibrium point of system (3) at the origin changes from an asymptotically stable spiral point ($c < 0$), to a neutrally stable center ($c = 0$), and then to an unstable spiral point ($c > 0$).

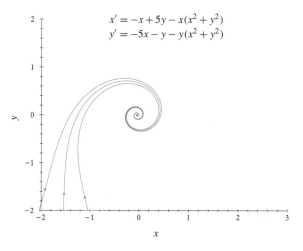

$$x' = -x + 5y - x(x^2 + y^2)$$
$$y' = -5x - y - y(x^2 + y^2)$$

FIGURE 9.3.5 Asymptotic stability before the Hopf bifurcation: $c = -1$ (Example 9.3.2).

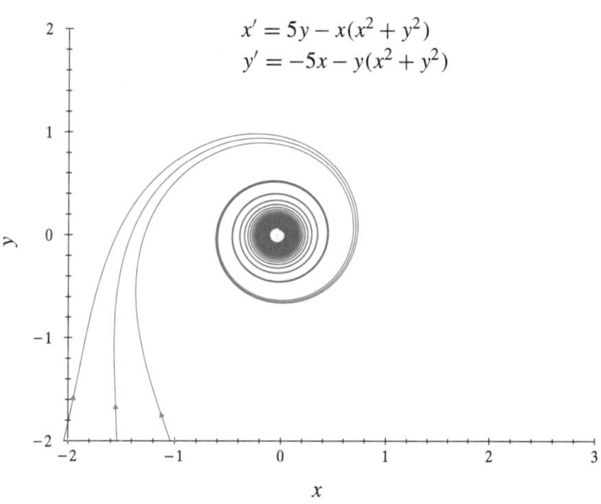

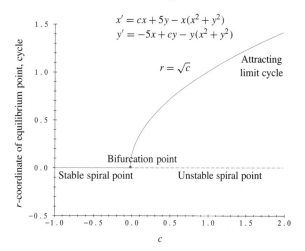

$$x' = 5y - x(x^2 + y^2)$$
$$y' = -5x - y(x^2 + y^2)$$

FIGURE 9.3.6 Asymptotic stability at the Hopf bifurcation value: $c = 0$ (Example 9.3.2).

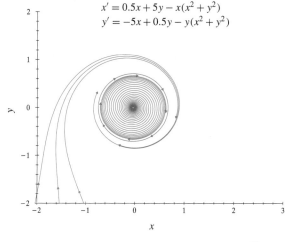

$$x' = 0.5x + 5y - x(x^2 + y^2)$$
$$y' = -5x + 0.5y - y(x^2 + y^2)$$

FIGURE 9.3.7 Attracting limit cycle $r = \sqrt{c}$ after the Hopf bifurcation: $c = 0.5$ (Example 9.3.2).

$$x' = cx + 5y - x(x^2 + y^2)$$
$$y' = -5x + cy - y(x^2 + y^2)$$

$r = \sqrt{c}$

Attracting limit cycle

Bifurcation point

Stable spiral point   Unstable spiral point

r-coordinate of equilibrium point, cycle

FIGURE 9.3.8 Hopf bifurcation diagram (Example 9.3.2).

From Section 8.2 we see that for the nonlinear system (2) the origin is also an asymptotically stable spiral point if $c < 0$ (Figure 9.3.5) and unstable if $c > 0$ (Figure 9.3.7). Something quite different happens at $c = 0$: the origin is still an asymptotically stable spiral point, but the inward motion of the spiral is very slow (Figure 9.3.6). As c becomes positive, the origin destabilizes and emits an attracting circular limit cycle of radius $\sqrt{c}$ (Figure 9.3.7). These results are easier to understand if you write system (2) in polar coordinates:

$$r' = r(c - r^2), \qquad \theta' = -5 \qquad (4)$$

From (4) we see that the circle $r = \sqrt{c}$ is indeed an orbit if $c > 0$. The circle is traversed clockwise (since θ' is negative) and attracts all other nonconstant orbits since r' is positive if $0 < r < \sqrt{c}$ and negative if $r > \sqrt{c}$.

Figure 9.3.8 shows a *bifurcation diagram* for system (2) and the equivalent system (4). The horizontal c-axis is first solid for the asymptotically stable equilibrium point at the origin and then dashed because the origin is unstable for positive c. The solid curve $r = \sqrt{c}$ depicts the amplitude of the attracting limit cycle.

The phenomenon illustrated above is an example of a general mechanism for creating limit cycles discovered by the German mathematician Eberhard Hopf. Take a look at the system

$$x' = \alpha(c)x + \beta(c)y + P(x, y, c)$$
$$y' = -\beta(c)x + \alpha(c)y + Q(x, y, c) \tag{5}$$

where P and Q are at least second order in x, y and twice continuously differentiable in x, y, and c, and $\alpha(c)$ and $\beta(c)$ are continuously differentiable functions of c. The eigenvalues of the linear system matrix of (5) at the origin [i.e., the Jacobian matrix of system (5) at the origin] are $\alpha(c) \pm i\beta(c)$. Here is Hopf's theorem.

THEOREM 9.3.1

> Hopf Bifurcation. Suppose that $\alpha(0) = 0$, $\alpha'(0) > 0$, and $\beta(0) \neq 0$ and that system (5) is asymptotically stable at the origin for $c = 0$. Then as c increases through 0, the origin destabilizes and ejects an attracting limit cycle of amplitude $K(c)$, where $K(0) = 0$ and $K(c)$ is a continuous and increasing function of c if $c \geq 0$ is sufficiently small. The period of the cycle is approximately $2\pi/|\beta|$ for small $c \geq 0$.

This transition from an asymptotically stable equilibrium point to an unstable equilibrium point enclosed by an attracting limit cycle is called a *supercritical Hopf bifurcation*. The proof of Theorem 9.3.1 is omitted.

In the statement of Theorem 9.3.1 the equilibrium point stays fixed at the origin as the parameter c changes, and the Jacobian matrix J of the vector rate function at the origin has the special form given in the margin. But all that is necessary is that **(1)** for a range of values of c the system has an equilibrium point somewhere (the equilibrium point may even move along a curve in the plane as c changes); **(2)** the Jacobian matrix J at the equilibrium point has the complex conjugate eigenvalues $\alpha(c) \pm i\beta(c)$; **(3)** at some "bifurcation value" c_0 we have $\alpha(c_0) = 0$, $\alpha'(c_0) > 0$, and $\beta(c_0) \neq 0$; and **(4)** the system is asymptotically stable at the equilibrium point if $c = c_0$. Under these conditions, as c increases through c_0 a bifurcation to an attracting limit cycle must occur.

☞ The matrix J is
$$J = \begin{bmatrix} \alpha(c) & \beta(c) \\ -\beta(c) & \alpha(c) \end{bmatrix}$$

EXAMPLE 9.3.3

A Hopf Bifurcation

System (2) of Example 9.3.2 is a special case of system (5) with $\alpha(c) = c$, $\beta(c) = 5$, $P = -x(x^2 + y^2)$, and $Q = -y(x^2 + y^2)$. P and Q have order 3 and are twice continuously differentiable; the other conditions, $\alpha(0) = 0$, $\alpha'(0) > 0$, and $\beta(0) \neq 0$, for a Hopf bifurcation are also satisfied. So the attracting cycle in Figure 9.3.7 is a Hopf limit cycle.

In a *subcritical Hopf bifurcation* an unstable spiral point stabilizes and spawns a

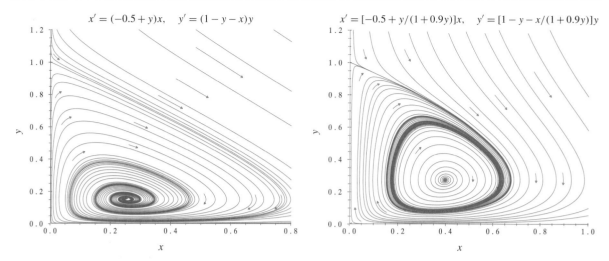

$x' = (-0.5 + y)x, \quad y' = (1 - y - x)y$

$x' = [-0.5 + y/(1 + 0.9y)]x, \quad y' = [1 - y - x/(1 + 0.9y)]y$

FIGURE 9.3.9 Populations approach an equilibrium point: $k = 0$ (Example 9.3.4).

FIGURE 9.3.10 Populations approach a limit cycle after Hopf bifurcation: $k = 0.9$ (Example 9.3.4).

repelling Hopf limit cycle as the parameter c changes [here $\alpha'(0) < 0$]. There are other types of Hopf bifurcations in which the equilibrium point moves as c changes and the form of the system is not as specific as (5). Computer programs are available that test a system for the presence of a Hopf bifurcation, but we will be satisfied with the visual evidence provided by numerical solvers. The next example illustrates this process.

EXAMPLE 9.3.4

Satiable Predation and Hopf Bifurcation: Figures 9.3.9–9.3.12

When food is plentiful, a predator's appetite is soon satiated, so an increase in the prey population has little effect on the interaction terms in the rate equations. Following the discussion in Section 5.3, we see that one model for the interactions of a satiable predator population x and a prey population y susceptible to overcrowding is

$$x' = [-a + by/(c + ky)]x$$
$$y' = [d - ey - fx/(c + ky)]y \tag{6}$$

where $a, b, \ldots, k$ are positive constants. Hopf bifurcations occur for certain ranges of values of the coefficients. In Figures 9.3.9–9.3.11, the coefficient k is taken to be the bifurcation parameter and the values of the other constants are taken to be $a = 0.5$, $b = d = e = f = 1$, and $c = 0.3$. The larger the value of k, the more rapidly the predator's appetite satiates as y increases, so k can be interpreted as the satiation coefficient.

Figures 9.3.9–9.3.11 show how the population orbits behave for three values of k: $k = 0$ (no satiation effect), $k = 0.9$, and $k = 1.35$. In the first and third cases, all orbits inside the population quadrant spiral toward an asymptotically stable equilibrium point. In the second case, orbits are pulled toward an attracting limit cycle. Although we do not show it, as k increases through 0.5 (approximately) a supercritical Hopf bifurcation takes place as an equilibrium point destabilizes and spawns an attracting limit cycle. As k increases, the cycle's amplitude first increases, but around $k = 0.85$

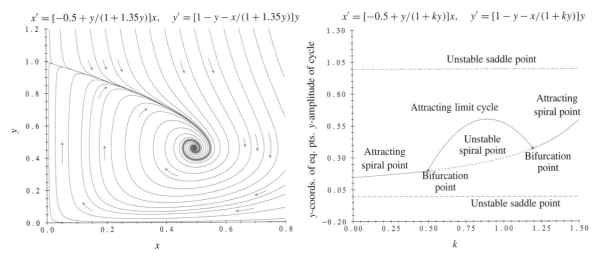

$$x' = [-0.5 + y/(1 + 1.35y)]x, \quad y' = [1 - y - x/(1 + 1.35y)]y$$

$$x' = [-0.5 + y/(1 + ky)]x, \quad y' = [1 - y - x/(1 + ky)]y$$

FIGURE 9.3.11 Approach to equilibrium point after reverse Hopf bifurcation: $k = 1.35$ (Example 9.3.4).

FIGURE 9.3.12 Hopf bifurcation diagram for satiable predation (Example 9.3.4).

the amplitude begins to shrink. By $k = 1.2$ (approximately) the equilibrium point has restabilized and absorbed the cycle in a "reverse" supercritical Hopf bifurcation. As k increases beyond 1.2, only an asymptotically stable equilibrium point remains inside the quadrant.

Figure 9.3.12 shows a bifurcation diagram for system (6), where the values of the constants $a, b, \ldots, f$ are as given above. The vertical axis represents the y-coordinate of each of the three equilibrium points of system (6) and the y-amplitude of cycles as measured from the enclosed equilibrium point. The equilibrium points $(0, 0)$ and $(0, 1)$ are always unstable, so they are represented by dashed lines. The equilibrium point inside the first quadrant has y-coordinate given by $y = 0.15/(1 - 0.5k)$. As k increases from 0, this y-coordinate increases. For $0 \le k \le 0.5$ and $1.2 \le k < 2$ the internal equilibrium point is asymptotically stable (solid curve), but for $0.5 < k < 1.2$ the point is unstable (dashed curve) and there is an attracting limit cycle represented in Figure 9.3.12 by the hump (solid line) of the extreme y-value on the cycle.

☞ Since the Jacobian matrices at $(0, 0)$ and $(0, 1)$ have real eigenvalues with opposite signs, each of these points is a saddle point.

Now let's take a look at a three-dimensional model where a sequence of bifurcations leads to chaotic wandering.

The Scroll Circuit

The diagram in the margin shows a model schematic of the *scroll circuit*. It has two capacitors (one with a "negative capacitance"), a nonlinear resistor, and an inductor. Figure 9.3.15 shows why the circuit has its strange name. The circuit is modeled by the system of ODEs (where the state variables and the function f are defined below)

$$
\begin{aligned}
x' &= -cf(y - x) \\
y' &= -f(y - x) - z \\
z' &= ky
\end{aligned}
\tag{7}
$$

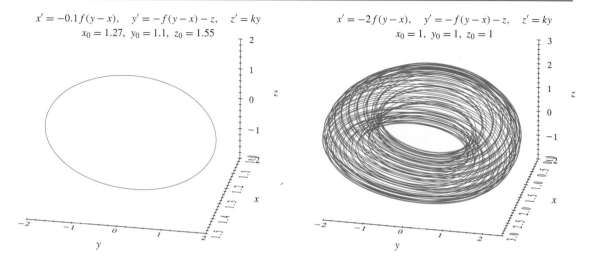

$$x' = -0.1f(y-x), \quad y' = -f(y-x)-z, \quad z' = ky$$
$$x_0 = 1.27, \ y_0 = 1.1, \ z_0 = 1.55$$

$$x' = -2f(y-x), \quad y' = -f(y-x)-z, \quad z' = ky$$
$$x_0 = 1, \ y_0 = 1, \ z_0 = 1$$

FIGURE 9.3.13 An attracting limit cycle ($c = 0.1$): $0 \le t \le 720$, plotted over $700 \le t \le 720$ [system (7)].

FIGURE 9.3.14 Limit cycle of Figure 9.3.13 has bifurcated to an attracting torus ($c = 2.0$): $0 \le t \le 500$.

The state variables x, y, and z are scaled and dimensionless versions, respectively, of the voltages V_1 and V_2 across the capacitors and the current I through the inductor:

$$x = V_1/E_1, \quad y = V_2/E_1, \quad z = I/(C_2 E_1), \quad c = C_2/C_1, \quad k = 1/(LC_2)$$

The scaled current/voltage characteristic $f(V)$ of the nonlinear resistor is given by

$$f(V) = -m_0(V)/C_2 + 0.5(m_0 + m_1)(|V+1| - |V-1|)/C_2 \qquad (8)$$

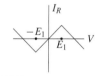

☞ As in the van der Pol circuit, the nonlinear resistor can pump energy into the circuit.

The constants $-m_0$ and m_1 are the slopes of the falling and the rising segments of the voltage/current characteristic shown in the margin; the current I_R reaches its local extreme values when the voltage difference $V_2 - V_1$ has the values $\pm E_1$.

Let's take $k = 1$, $m_0/C_2 = 0.07$, and $m_1/C_2 = 0.1$ and "tune" the circuit by varying the parameter $c = C_2/C_1$. Several Hopf bifurcations occur as c is changed from 0.1 to 33, as well as a number of other bifurcations of a more mysterious nature. In particular, in a *toroidal Hopf bifurcation* an attracting limit cycle for $c = 0.1$ (Figure 9.3.13) "expands" into an attracting torus of orbits for $c = 2.0$. Figure 9.3.14 shows one orbit on this torus.

Figures 9.3.15–9.3.20 show projections of the orbits starting at $x_0 = y_0 = z_0 = 1$, where the parameter c takes on the respective values 0.1, 1, 6, 15, 20, 33. The projections are into the xz-plane of the scaled voltage x and the scaled current y. Compare Figure 9.3.20 with the figure on the opening page of this chapter. **Warning**: for values of c above 13 some solutions of system (7) are very sensitive to changes in initial data or parameters, so your solver may produce orbits different from those shown. What has happened seems to be that the attracting torus has broken up and chaotic wandering has begun. Although much remains unknown and unproven about the scroll circuit and its model system, the pictures give visual evidence of the unusual variations in the circuit's voltages and currents as the ratio c of the capacitances is changed.

$$x' = 0.1f(y - x), \quad y' = -f(y - x), \quad z' = y$$
$$x_0 = 1, \ y_0 = 1, \ z_0 = 1$$

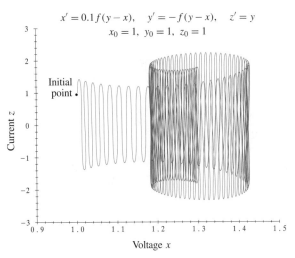

FIGURE 9.3.15 The projected scroll orbit ($c = 0.1$): $0 \le t \le 500$ [system (7)].

$$x' = -f(y - x), \quad y' = -f(y - x), \quad z' = y$$
$$x_0 = 1, \ y_0 = 1, \ z_0 = 1$$

FIGURE 9.3.16 Projected orbit ($c = 1$): $0 \le t \le 250$ [system (7)].

$$x' = -6f(y - x), \quad y' = -f(y - x), \quad z' = y$$
$$x_0 = 1, \ y_0 = 1, \ z_0 = 1$$

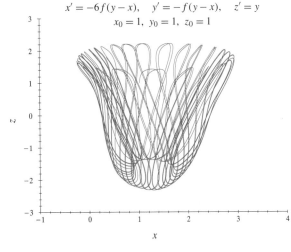

FIGURE 9.3.17 Orbital vase ($c = 6$): $0 \le t \le 325$ [system (7)].

$$x' = -15f(y - x), \quad y' = -f(y - x), \quad z' = y$$
$$x_0 = 1, \ y_0 = 1, \ z_0 = 1$$

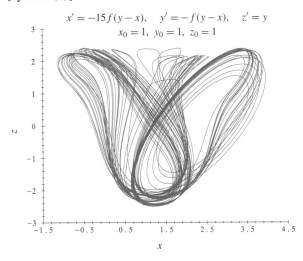

FIGURE 9.3.18 his orbit looks chaotic ($c = 15$): $0 \le t \le 400$ [system (7)].

The electrical engineer L. O. Chua and his colleagues have initiated a series of studies of the scroll and related circuits, their mathematical models, and computer simulations.[5] The circuits are not hard to build and the actual voltages and currents are amazingly close to those seen in the computer simulations of the model ODEs.

[5]T. Matsumoto, L. O. Chua, and R. Tokunaga, "Chaos via Torus Breakdown," *IEEE Trans. Circuits Syst. CAS-34*(3) (March 1987), pp. 240–254. T. S. Parker and L. O. Chua, *Practical Numerical Algorithms for Chaotic Systems* (New York: Springer-Verlag, 1989).

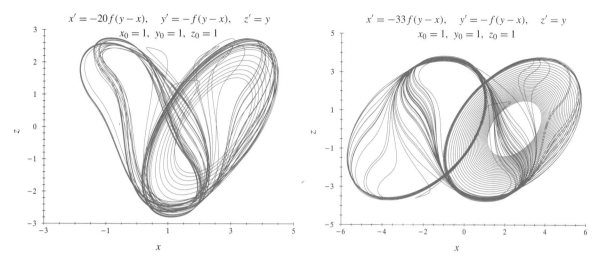

$x' = -20f(y-x), \quad y' = -f(y-x), \quad z' = y$
$x_0 = 1, \ y_0 = 1, \ z_0 = 1$

$x' = -33f(y-x), \quad y' = -f(y-x), \quad z' = y$
$x_0 = 1, \ y_0 = 1, \ z_0 = 1$

FIGURE 9.3.19 More chaos ($c = 20$): $0 \leq t \leq 425$ [system (7)].

FIGURE 9.3.20 The organized chaos of the chapter cover figure ($c = 33$): $0 \leq t \leq 450$ [system (7)].

Comments

Contemporary research suggests that mysterious oscillations and behavioral changes in physical systems can often be explained in terms of a bifurcation. Oscillations in the concentrations of chemicals in a chemical reactor, the life cycles of a cell, and the van der Pol limit cycle in a electrical circuit seem to be triggered as a system parameter is changed and a Hopf or some other kind of bifurcation takes place.

It is not always a simple matter to verify the conditions for a bifurcation. Once a possible bifurcation parameter in the system has been identified, computer graphics can be used to display visual evidence that a bifurcation event has occurred.

PROBLEMS

www **1.** (*Saddle-node, Transcritical, Pitchfork Bifurcations*). Describe the bifurcations in each system. Find the values of c at which a bifurcation occurs, and identify the bifurcation as of saddle-node, transcritical, or pitchfork type. Saddle-node bifurcations are discussed in the text. As a parameter is changed in a *transcritical bifurcation* an asymptotically stable node and a saddle point move toward each other, merge, and then emerge with the node now a saddle, and the saddle a node. As a parameter is changed in a *pitchfork bifurcation*, an asymptotically stable equilibrium point suddenly splits into three equilibrium points, two of which are asymptotically stable and the third unstable. Plot graphs of orbits for values of c before, at, and after the bifurcation. Sketch bifurcation diagrams. [*Hint*: Use values of c near 0. See Section 2.4.]

(a) $x' = c + 10x^2, \ y' = x - 5y$ (b) $x' = cx - x^2, \ y' = -2y$

(c) $x' = cx + 10x^2, \ y' = x - 2y$ (d) $x' = cx - 10x^3, \ y' = -5y$

(e) $x' = cx + x^5, \ y' = -y$

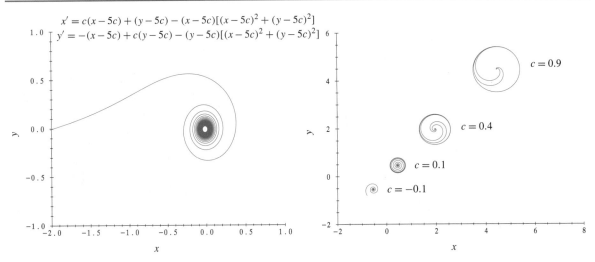

FIGURE 9.3.21 Orbit at the Hopf bifurcation: $c = 0$ [Problem 5(**b**)].

FIGURE 9.3.22 Orbits before and after the Hopf bifurcation [Problem 5(**c**)].

2. (*Hopf Bifurcation*). Show that each system experiences a Hopf bifurcation at $c = 0$. Plot orbits before, at, and after the bifurcation. Draw bifurcation diagrams for (**a**) and (**b**) showing the y-amplitude of the limit cycle and the y-coordinate of each equilibrium point as functions of c. [*Hint*: Write the ODEs in polar coordinates.]

    (**a**) $x' = cx + 2y - x(x^2 + y^2)$, $\ y' = -2x + cy - y(x^2 + y^2)$

    (**b**) $x' = cx - 3y - x(x^2 + y^2)^3$, $\ y' = 3x + cy - y(x^2 + y^2)^3$

    (**c**) $x' = y - x^3$, $\ y' = -x + cy - y^3$

    (**d**) (*Rayleigh's Equation*). The Rayleigh ODE $z'' + c[(z')^2 - 1]z' + z = 0$ is equivalent to the system $x' = y$, $\ y' = -x + cy - y^3$ if we set $z = x$, $z' = y$. [*Hint*: The Rayleigh ODE is shown to be equivalent to a van der Pol system in Problem 10, Section 9.1.]

3. (*Subcritical Hopf Bifurcation*). Show that the origin for $x' = -cx + y + x(x^2 + y^2)$, $\ y' = -x - cy + y(x^2 + y^2)$ bifurcates as c increases through 0 from an unstable spiral point to a stable spiral point surrounded by a repelling limit cycle. Plot orbits and a bifurcation diagram.

4. (*Satiable Predation*). This problem continues the exploration of system (6) given in Example 9.3.4. Let $a = 0.5$, $d = e = f = 1$, $c = 0.3$, $k = 0.9$, and let b be the bifurcation parameter. Explore what happens as b increases from 0.75 to 3. Any Hopf bifurcations? Explain what you see.

5. (*Hopf Bifurcation: Moving Equilibrium Point*). The system

    $$x' = c(x - 5c) + (y - 5c) - (x - 5c)[(x - 5c)^2 + (y - 5c)^2]$$
    $$y' = -(x - 5c) + c(y - 5c) - (y - 5c)[(x - 5c)^2 + (y - 5c)^2]$$

    has the unique equilibrium point $P(5c, 5c)$ for each value of c.

    (**a**) Rewrite the system in terms of coordinates centered at P: $u = x - 5c$, $v = y - 5c$. Then write the new system in polar coordinates based at $u = 0$, $v = 0$.

    (**b**) Using (**a**), show that the system has a supercritical Hopf bifurcation at $c = 0$. Show that for $c > 0$, the circle of radius $\sqrt{c}$ centered at $P(5c, 5c)$ [in x-, y-coordinates] is an orbit. See Figures 9.3.21 and 9.3.22.

    (**c**) Plot orbits for $c = -0.1, 0.1, 0.4, 0.9$ and describe what is happening. [*Hint*: See Figure 9.3.22 where orbits for each of these values of c are plotted on a single graph. This is done by creating a third ODE, $c' = 0$, and using the given values of c as part of the initial data.]

 6. (*The Autocatalator*). The autocatalator (see Section 5.1) models a chemical reaction in which the concentration of a precursor W decays exponentially, generating a new species X in the process. Species X decays to Y and at the same time reacts autocatalytically with Y, the latter reaction creating more Y than is consumed. Y decays in turn to Z. The rate equations for the corresponding concentrations w, x, y, z are

$$w' = -aw, \quad x' = aw - bx - \alpha xy^2, \quad y' = b + \alpha xy^2 - cy, \quad z' = cy$$

where the rate coefficients a, b, c, and α are positive parameters and the independent variable is time. In what appears to be a kind of Hopf bifurcation, the concentrations of the chemical intermediates X and Y undergo violent oscillations for certain ranges of the parameters and certain levels of species W. The goal of this project is to understand the behavior of the reaction, given various sets of data and parameters. First read Example 5.1.5, and **Experiments** in Section 5.1 of the Student Resource Manual; then address the following points:

- The system can be treated as a nonautonomous planar system by setting $w = w(0)e^{-at}$ and ignoring the rate equation for z (see Example 5.1.5). This reduction may help with your solver graphics.

- Why is it reasonable to set $x(0) = y(0) = z(0) = 0$?

- After some initial oscillations in x and y, all four concentrations behave as expected as time increases if $w(0) = 500$, $a = 0.002$, $b = 0.14$, $\alpha = c = 1.0$, $0 \le t \le 1500$. What levels do the four concentrations approach as t gets large?

- Repeat with $b = 0.08$ instead of 0.14, and $w(0) = 5000$. What is happening here?

- Repeat with $b = 0.02$ and $a = 0.001, 0.003, 0.008$, $\alpha = c = 1.0$; set $w(0) = 5000$. Now what is happening?

- In all of the above, plot component curves. If possible with your solver graphics, plot projected orbits in xyz-space or in any other space of three variables selected from t, w, x, y, z. Explain each graph.

- Plot orbits in xy-space.

- Set $a = 0.002$, $b = 0.08$, $\alpha = c = 1.0$, $w(0) = 500$. For various values of b above and below 0.08, plot $x(t)$ against t and $x(t)$ against $y(t)$. Try to explain in terms of a Hopf bifurcation for the system $x' = C - bx - \alpha xy^2$, $y' = bx + \alpha xy^2 - cy$, where C is some positive constant.

 7. (*The Scroll Circuit*). In carrying out this project you will need to solve system (7) over a long time span. Address the following points:

- Using the basic circuit laws of Section 4.4, derive the equations

$$C_1 V_1' = -g(V_2 - V_1)$$
$$C_2 V_2' = -g(V_2 - V_1) - I$$
$$L I' = V_2$$

where $g(V_2 - V_1) = -m_0(V_2 - V_1) + 0.5(m_0 + m_1)(|V_2 - V_1 + E_1| - |V_2 - V_1 - E_1|)$. Then scale time and the state variables V_1, V_2, and I appropriately to obtain the scroll system (7).

- Move the nonlinear resistor over into the right-hand circuit loop and discard the left capacitor entirely. Compare the resulting system with a van der Pol circuit [especially that of Problem 8(**e**) of Section 9.1].

- Duplicate Figures 9.3.15–9.3.20. Then plot other views of the solutions for the given values of c, for example, orbits in xyz-space, projections in xy- and yz-planes, and component curves. [See **(a)**, **(b)**, **(c)** below for orbits through the point $(1, 1, 1)$ with $c = 0.1, 1.0$, and 33.0, respectively].

(a)   **(b)**   **(c)**

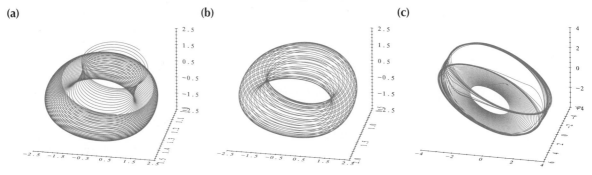

- Let $m_0/C_2 = 0.07$, $m_1/C_2 = 0.1$ in the formula in (8) for $f(y - x)$; let $k = 1$. Solve the scroll system numerically over the time interval $[0, 1000]$, but plot over the time interval $[500, 1000]$ to eliminate transient initial behavior. Use $x_0 = 1$, $y_0 = 1$, $z_0 = 0$, and the following values for the parameter c: $0.1, 0.5, 2, 8.8, 9.6, 10.8, 12, 13, 13.4, 13.45, 13.52, 15, 33$. Graph the projections of the orbits on the xz-plane and in xyz-space (if possible). Then graph the components against t to pick out periodic behavior. The Chapter 9 cover figure shows part of the projection onto the xz-plane of the orbit corresponding to $c = 33$. Explain what you see.

- With $c = 0.1$, $x_0 = 1$, $y_0 = 1$, and $z_0 = 0$, decrease time from $t = 0$ and see if you pick up a repelling toroidal limit set of orbits.

9.4 Chaos

In 1963 the American mathematician and meteorologist, E. N. Lorenz, published an account of the peculiarities he had observed among the computed solutions of a system of three autonomous, first-order, nonlinear ODEs that model thermal variations in an air cell below a thunderhead.[6] For certain values of the parameters in the rate equations of the system, solutions behave in an almost random way. The orbits remain bounded as time advances, but they do not tend to an equilibrium point, or to a cycle, or to a cycle-graph, but instead wander between regions of oscillatory behavior. Small changes in a parameter or in the initial data evolve into large and apparently unpredictable differences in subsequent states. Lorenz could not find an explanation for this strange behavior. Although the problem is still not completely understood, it has led

Edward N. Lorenz

[6]Even as a young boy, Edward Lorenz was fascinated by meteorology and mathematics. He graduated from Dartmouth College in 1938 and worked as a weather forecaster in the Army Air Corps during World War II. In his 1963 paper, "Deterministic Nonperiodic Flow," *J. Atmos. Sci.* **20** (1963), pp. 130–141, Lorenz showed that for certain values of the parameters σ, r, and b the solutions of system (1) are very sensitive to even small changes in the initial data, a property now believed to be characteristic of chaotic systems. His discovery has won him numerous awards such as the Kyoto Prize in 1991 and the Louis J. Battan Author's Award in 1995.

to many studies, both theoretical and applied, by scientists, engineers, and mathematicians trying to formulate a good definition and theory for chaotic but deterministic dynamics: *chaotic* because of the difficulty of predicting future states of the system from information about the present state, and *deterministic* because the Fundamental Theorem for Systems (Theorem 5.2.1) guarantees that each set of initial values determines a unique solution. In this section we will describe some of the curious features of the Lorenz system's long term behavior and its solutions and orbits.

The Lorenz System and Its Equilibrium Points

The *Lorenz system* in dimensionless state variables is

$$x' = -\sigma x + \sigma y$$
$$y' = rx - y - xz \tag{1}$$
$$z' = -bz + xy$$

where σ, r, and b are positive parameters that denote physical characteristics of air flow, x is the amplitude of the convective currents in the air cell, y is the temperature difference between rising and falling currents, and z is the deviation from the normal temperatures in the cell. The equations are derived by applying Fourier series techniques to the nonlinear PDEs of turbulent flow and keeping only the lowest order terms.

For all positive values of the parameters σ, r, and b the Lorenz system has an equilibrium point at the origin; for $0 < r < 1$ there are no other equilibrium points. The eigenvalues of the Jacobian matrix of the system at the origin are given by

$$\lambda_1, \lambda_2 = \frac{1}{2}[-\sigma - 1 \pm ((\sigma - 1)^2 + 4\sigma r)^{1/2}], \qquad \lambda_3 = -b \tag{2}$$

We omit the lengthy algebraic calculations that lead to these formulas. For values of r between 0 and 1 all three of these eigenvalues are negative, so the system is asymptotically stable at the origin (in fact, the origin is a global attractor). For $r > 1$ the first eigenvalue is positive, and the origin has destabilized.

As the parameter r is increased through the value 1, the origin ejects two new equilibrium points:

$$P_1 = (-\alpha, -\alpha, r - 1), \qquad P_2 = (\alpha, \alpha, r - 1) \tag{3}$$

where $\alpha = \sqrt{br - b}$. The eigenvalues of the Jacobian matrix of the Lorenz system at P_1 and P_2 are the roots of the cubic

$$\lambda^3 + (1 + b + \sigma)\lambda^2 + b(\sigma + r)\lambda + 2\sigma b(r - 1) \tag{4}$$

☞ Here and elsewhere in this section we omit specific reference to the (often lengthy) algebraic calculations that underlie the formulas.

If r is any number larger than 1, at least one of these eigenvalues is negative because a cubic polynomial with positive coefficients must have a negative real root. An application of the Routh Test of Section 7.9 to the cubic in (4) shows that the other roots are also negative (or have negative real parts) if r lies between 1 and the *critical value* r_c:

$$r_c = \frac{\sigma(\sigma + b + 3)}{\sigma - b - 1} \tag{5}$$

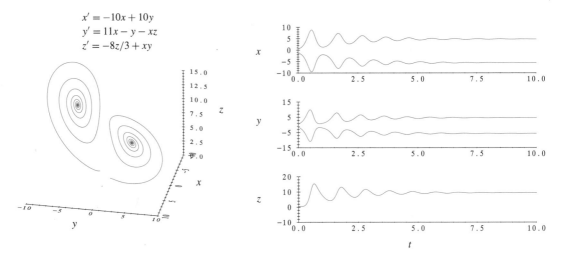

$$x' = -10x + 10y$$
$$y' = 11x - y - xz$$
$$z' = -8z/3 + xy$$

FIGURE 9.4.1 Orbits of the Lorenz system through $(1, 1, 1)$ and $(-1, -1, 1)$: $r = 11$, $\sigma = 10$, $b = 8/3$.

FIGURE 9.4.2 Component curves of the two orbits of Figure 9.4.1.

So for values of r between 1 and r_c, the equilibrium points P_1 and P_2 are locally asymptotically stable. Figures 9.4.1 and 9.4.2 show the orbits and component curves defined by the solutions that start at the points $(1, 1, 1)$ and $(-1, -1, 1)$, where we have set $\sigma = 10$, $b = 8/3$, and $r = 11$. Note the symmetries in the x- and y-component graphs; the z-component graphs are the same for the two orbits. For the given values of σ and b, we have $r_c = 470/19 \approx 24.737$, so the value $r = 11$ is subcritical. Solution behavior is regular and uncomplicated.

Now let's take another look at what happens as the value of the parameter r increases through 1. In fact, a pitchfork bifurcation occurs. The solid curves in the bifurcation diagram of Figure 9.4.3 show the x-coordinate of the asymptotically stable equilibrium points P_1 and P_2, while the dashed line represents the destabilized origin. As r is incremented through the critical value r_c, a subcritical Hopf bifurcation takes place (not shown in Figure 9.4.3), and P_1 and P_2 lose their stability by "swallowing" unstable limit cycles and taking on the unstable behavior of the cycles. These and many other properties of the Lorenz system are treated by Sparrow. [7]

The Squeeze, the Attractor, and the Limit Sets

Let's suppose that S is a closed region such as a solid ellipsoid or a solid spherical ball in the xyz-state space of the Lorenz system. Follow each point q of S forward for t units of time along the orbit which starts at q. This carries q to the point q_t on the orbit. The set S_t of points q_t for all q in S may no longer resemble S very much if t is large,

[7]Colin Sparrow, *The Lorenz Equations: Bifurcations, Chaos, and Strange Attractors*, vol. 41 of the series *Applied Mathematical Sciences* (New York: Springer-Verlag, 1982).

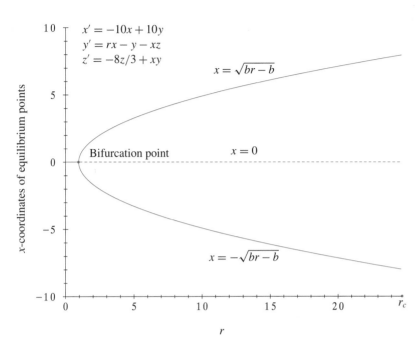

FIGURE 9.4.3 Pitchfork bifurcation diagram for the Lorenz system with $\sigma = 10$, $b = 8/3$, $0 \leq r \leq r_c \approx 24.737$; asymptotically stable points (solid curve), unstable points (dashed line).

but one would expect S_t to be very close to S if t is close to 0 since $S_0 = S$. The point transformation $T_t : S \to S_t$ that carries each point q to q_t is continuous since solutions of the Lorenz system are continuous in the initial data. T_t is called the *orbit map* of the Lorenz system. For the Lorenz system the orbit map T_t is defined for all $t \geq 0$ since it is known that every orbit of a Lorenz system is defined for all $t \geq 0$. For fixed positive σ, r, b, the transformation T_t, $t > 0$, squeezes any region S with positive volume into a region S_t with smaller volume. It can be shown that as $t \to +\infty$ the volume of S_t tends to zero. We call this phenomenon of contracting volumes the *Lorenz squeeze*. This does not necessarily mean that the region S_t squeezes down to a single point as $t \to +\infty$, but it does mean that S_t somehow thins out with increasing time and tends toward a set whose volume is zero. The positive limit setof any orbit in S must lie in this set of zero volume, so the limit set can't be "thick."

☞ The Student Resource Manual has more information on the Lorenz squeeze.

☞ As in Section 9.2, the positive limit set of an orbit is the collection of points the orbit approaches as $t \to +\infty$.

It is known that for each set of positive values for σ, r, and b there is a solid ellipsoid E defined by

$$rx^2 + \sigma y^2 + \sigma(z - 2r)^2 \leq C$$

where C is a large, positive constant with the property that every orbit of the Lorenz system eventually enters and remains inside E as t increases. By the squeeze property, the volume of the set E_t tends to 0 as $t \to +\infty$. Since every orbit enters E, it makes sense to define the *Lorenz attractor* Z to be the intersection of all regions E_t, $t \geq 0$, or more simply,

$$Z = E \cap E_1 \cap E_2 \cap \cdots$$

The Lorenz attractor is a nonempty, closed, bounded, invariant set[8] of zero volume. For values of r between 0 and 1, Z contains only the origin. For values of $r > 1$ there is only partial information about the nature of the orbits inside Z. All orbits tend to Z with advancing time (i.e., Z is a *global attractor*), but the nature of that approach may be quite strange. Z contains the equilibrium points and the cycles of the Lorenz system and all positive limit sets of orbits. The exact shape of Z is not known, although parts of Z can be seen in orbital portraits.

The Lorenz Portraits

First, note that any orbit beginning on the z-axis remains there for all t, that is,

$$x = 0, \qquad y = 0, \qquad z = z_0 e^{-bt}$$

is a solution of the Lorenz system for all values of z_0. If x and y are replaced by $-x$ and $-y$, respectively, in the Lorenz system, the rate equations do not change. This means that if $x(t)$, $y(t)$, $z(t)$ is a solution, so is $-x(t)$, $-y(t)$, $z(t)$, and orbits are symmetric through the z-axis. This explains why there is only one z-component curve in Figure 9.4.2.

It is known that for values of r larger than 1 there are exactly two orbits of the system that exit from the origin as time advances. They are called the *unstable manifolds* from the origin, and one is the reflection through the z-axis of the other. As time increases, an unstable manifold from the origin 0 may tend to either equilibrium point P_1 or P_2, or may jump back and forth from a neighborhood of one to a neighborhood of the other.

The Lorenz attractor Z may contain some unstable cycles. In fact, Z may contain infinitely many, depending on the values of r, σ, and b. It is widely believed that the presence of infinitely many unstable cycles of arbitrarily large periods is a sign of chaos. One of the unsettling characteristics of the Lorenz model is that orbital structure may change considerably with only a small change in the initial point (or in r). As time advances, an orbit may bear little resemblance to orbits that start nearby in state space. This behavior indicates that Lorenz orbits display sensitivity to changes in the data.

The parameters σ and b are fixed at the values

$$\sigma = 10, \qquad b = 8/3$$

in all of the portraits in this section, but r takes on various values. Each orbit is computed over a finite time interval, and its asymptotic behavior as $t \to +\infty$ can only be

---

[8]That is, Z has at least one point, Z contains all of its boundary points, Z lies inside a bounded box in $\mathbb{R}^3$, and Z has the property that any orbit intersecting Z stays in Z for all of t.

$$x' = -10x + 10y, \quad y' = 25x - y - xz, \quad z' = -8z/3 + xy$$
$$x_0 = 1, \; y_0 = 1, \; z_0 = 1$$

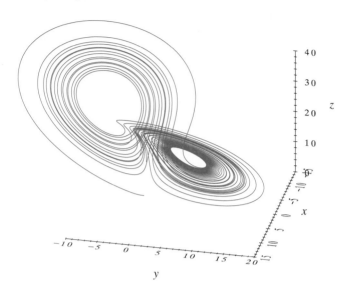

FIGURE 9.4.4 The orbit quickly approaches its positive limit set inside the Lorenz attractor: $r = 25$, above critical value $r_c = 24.737$; $0 \le t \le 40$.

inferred. The graphs in Figures 9.4.4–9.4.8 depict orbits, component curves, and orbital projections near the Lorenz attractor Z. Let's discuss these figures and what they seem to imply.

For values of r larger than the critical value $r_c \approx 24.737$ given in formula (5) where $\sigma = 10$, $b = 8/3$, orbits head rapidly to the Lorenz attractor. What is visible in Figure 9.4.4 can be interpreted as a part of the attractor that consists of two almost planar pieces at an angle to one another, each piece containing oscillatory arcs of orbits near the equilibrium points P_1, P_2. The orbit wanders from one piece to the other, but unpredictably, as the component graphs in Figure 9.4.7 show. In these figures, $r = 25$, which is just above the critical value of $r_c \approx 24.737$.

Figure 9.4.8 shows a distinctive feature of chaotic dynamics: extreme sensitivity to small changes in the initial data. The solutions with initial points $(-1, -1, 1)$ and $(-1.01, -1, 1)$ stay together for the first 20 units of time, but after that they seem to become increasingly uncorrelated. This means that it is not possible to make accurate long-term predictions about the state of the system even if the values of the parameters σ, r, b and the initial values x_0, y_0, z_0 are known quite accurately.

The Lorenz system appears to have attracting periodic orbits (i.e., attracting limit cycles) for certain values of r. Figure 9.4.9 shows part of the x-component graph of a periodic orbit for $r = 230$. Figures 9.4.10, 9.4.11, and 9.4.12 show what happens when r diminishes from 230 to $r = 220, 217$, and finally 210. These graphs of parts of the x-component curves of the periodic orbits indicate that the corresponding period doubles, then doubles again, and finally the solution becomes an aperiodic, seemingly

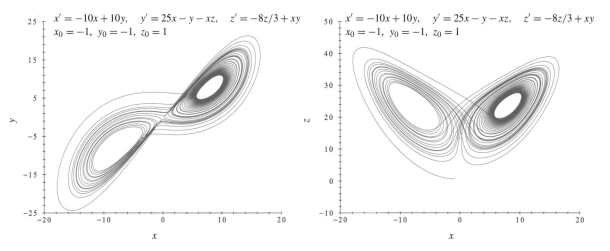

FIGURE 9.4.5 Projection of orbit of Figure 9.4.4 onto the xy-plane.

FIGURE 9.4.6 Projection of orbit of Figure 9.4.4 onto the xz-plane.

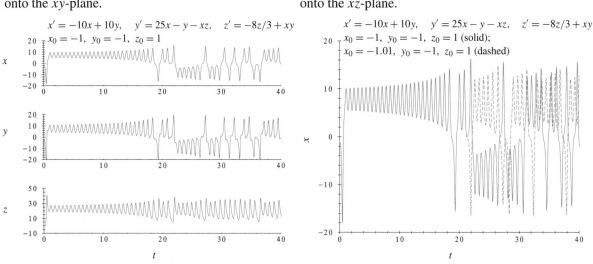

FIGURE 9.4.7 The x-, y-, z-component curves of the orbit in Figure 9.4.4.

FIGURE 9.4.8 Sensitivity to small changes in initial data.

chaotic oscillation. Figures 9.4.9–9.4.12 are obtained by selecting an initial point near a periodic solution, waiting until the corresponding orbit is nearly on the attracting periodic orbit, and then plotting the x-component graph; so the system is solved over $0 \le t \le 50$ and plotted over $40 \le t \le 50$.

It is widely believed that there are many sequences of values of r that correspond to a period-doubling sequence that ends in chaos. The period-doubling sequence above starts at $r = 230$. The reader may find other period-doubling sequences that start at the same value of r. Such sequences are called *period-doubling bifurcations* or *routes to chaos*

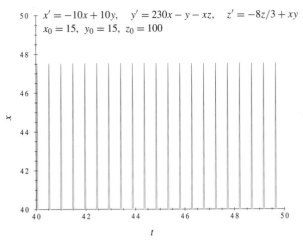

FIGURE 9.4.9 Part of the x-component of a periodic solution of period ≈ 0.48: $r = 230$.

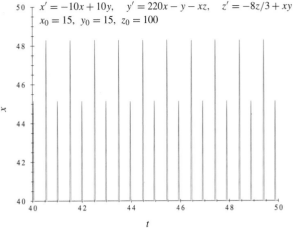

FIGURE 9.4.10 The period of the periodic solution doubles to about 0.96: $r = 220$.

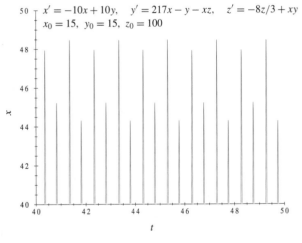

FIGURE 9.4.11 The period doubles again to approximately 1.92: $r = 217$.

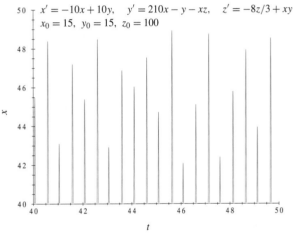

FIGURE 9.4.12 And finally we see chaotic motion: $r = 210$.

Chaos and Strange Attractors

The Lorenz attractor appears to be one of the key elements in understanding the asymptotic behavior of the orbits of a Lorenz system. With σ and b fixed positive constants, as the parameter r is stepped up from the value 0 through the value r_c the attractor becomes increasingly complex. It is thought that with the advance of the parameter r, unstable periodic orbits are created and then disappear inside the attractor and that there will be ranges of r for which the attractor contains infinitely many unstable periodic orbits with distinct periods. Such chaotic behavior within the attractor is thought to induce the turbulent behavior of nearby orbits as they approach the attractor with

$x' = -10x + 10y, \quad y' = 200x - y - xz, \quad z' = -8z/3 + xy$
$x_0 = 15, \quad y_0 = 15, \quad z_0 = 100$

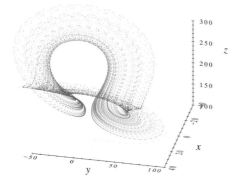

$x' = -10x + 10y, \quad y' = 200x - y - xz, \quad z' = -8z/3 + xy$
$x_0 = 15, \quad y_0 = 15, \quad z_0 = 100$

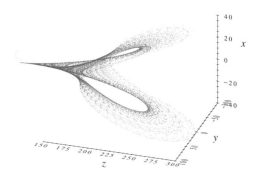

FIGURE 9.4.13 Poincaré time section of an orbit near the Lorenz attractor: $0 \le t \le 80, \quad \Delta t = 0.005$.

FIGURE 9.4.14 Another view of the Poincaré time section of the orbit of Figure 9.4.13.

the advance of time.

Several definitions of chaos have been proposed, but so far there is no standard definition. Here is an informal definition that captures most of the features most researchers believe are characteristic of chaos: A nonempty, closed, and bounded invariant set A for an autonomous system of ODEs is *chaotic* if it contains a *dense orbit* (i.e., an orbit that passes arbitrarily close to every point of A as time advances), if it contains at least two orbits (otherwise a single cycle or equilibrium point would fit the definition), and if orbits in A are sensitive to changes in initial data. An attractor set A is said to be a *strange attractor* if A is chaotic and if A has no proper subsets that are both attracting and invariant. In this definition chaos is a property of a set, not of a single orbit, and not of orbits outside the set. The problem with this definition (and with others known to the authors) is that it is nearly impossible to prove that any particular system (such as the Lorenz system) actually possesses a strange attractor.

Uncertainties abound. As we saw in Section 9.2, a bounded orbit of a planar autonomous system can do only one of three things as $t \to \infty$. But there is room in three dimensions for much more complicated behavior, and so far there is little theoretical guidance about exactly what happens to an orbit as $t \to +\infty$ and the orbit approaches Z or a strange attractor situated somewhere inside Z. Computed orbits suggest that for some values of r the Lorenz attractor Z does indeed contain a strange attractor A, but so far no one has been able to prove it.

Figures 9.4.13–9.4.16 show a different way to view orbital tangles. In a *Poincaré time section* of an orbit the orbital points $(x(t), y(t), z(t))$ are shown only for a discrete set of values of t, in this case at time intervals with $\Delta t = 0.005$ starting at $t = 0$ and ending at $t = 80$. It is as if a strobe light flashes every Δt seconds as the orbit is being traced, and the resulting dots appear on a photographic plate.

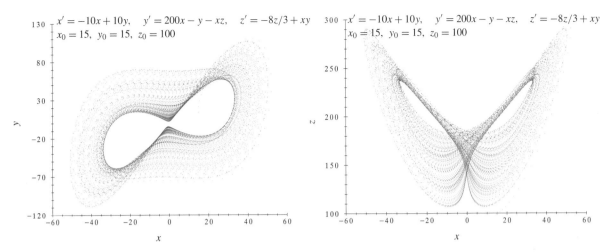

FIGURE 9.4.15 The xy-projection of the time section of Figure 9.4.13.

FIGURE 9.4.16 The xz-projection of the time section of Figure 9.4.13.

Computational Sensitivity

Computers truncate, chop, round-off, and otherwise reduce infinite strings of digits (such as decimal expansions) to finite strings. The effect on computed solutions of ODEs over short time intervals is usually negligible. But in cases of extreme sensitivity to changes in data, computer truncations may result in a graph that is very different from the "true" graph. Figure 9.4.17 shows the graph of the x-component curve of an orbit of a Lorenz system. Now look at Figure 9.4.18 where the same data are used, but the absolute and relative local error bounds used by the authors' solver are slightly tightened. For values of t beyond 26, the x-component curve is quite different from that in Figure 9.4.17. So don't be surprised if your numerical solver produces Lorenz orbits and component curves that don't look like what you see here. It makes one wonder if chaos is only a computational artifact!

However, there is strong (but indirect) evidence that the sensitivity in the Lorenz system is real and not just computational. It is now known that very close to each numerical solution computed over a finite time span there lies a true solution. So, although the computed solution of an IVP may eventually diverge quite a bit from the true solution, close to the computed solution there is a true solution, but perhaps with slightly different initial data.[9]

---

[9]This is called the *shadowing property*. See, for example, S. M. Hammel, J. A. Yorke, and C. Grebogi, "Numerical Orbits of Chaotic Processes Represent True Orbits," *Bull. Amer. Math. Soc.* **19** (1988), pp. 465–469.

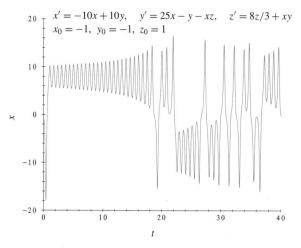

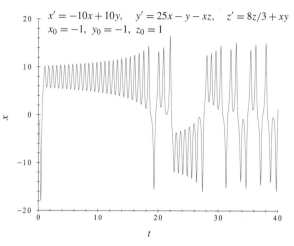

FIGURE 9.4.17 The x-component curves of an orbit: $r = 25$. Our solver error bounds were set at absolute error 10^{-10} and relative error 10^{-7}.

FIGURE 9.4.18 The solver's error bounds are tightened slightly (from 10^{-10} to 10^{-11} and from 10^{-7} to 10^{-8}), and the x-component eventually changes a lot.

Comments

The Lorenz system has other curious features. There appear to be many bifurcations to unstable limit cycles as r is varied in the neighborhood of the value 13.24, even before the critical value $r_c \approx 24.737$. In this case, the unstable manifolds that emerge from the origin shoot out toward the equilibrium points P_1 or P_2 and then fall back toward the origin as time increases. Another odd feature (but believed to be characteristic of chaos) is that for many values of r the dimension of the Lorenz attractor Z and of the strange attractor A inside Z is not 1 or 2, but seems to be some real number a little larger than 2. Dimension must be defined, and in this setting it is what is called the Hausdorff dimension. The Hausdorff dimension of a smooth surface is 2 and of a smooth curve is 1, but a noninteger Hausdorff dimension can be assigned to many other "crinkly" sets.[10] In fact, the nonintegral dimension of the strange attractor may turn out to be typical of a chaotic system.

Chaos requires an autonomous system with at least three state variables or a nonautonomous system with at least two state variables in order to give the orbits enough dimensions to do their strange meanderings. That's why chaos doesn't happen for autonomous systems with one or two state variables. Take a look at the Duffing system in Problem 4 and the driven pendulum system in Problem 5 for examples of chaotic nonautonomous systems with two state variables.

Lorenz proposed his system as a mathematical model of the thermal turbulence under a thunderhead. Although it is now known that the Lorenz system is not a very

---

[10]For the Hausdorff dimension see the book by M. Martelli, *Discrete Dynamical Systems and Chaos* (New York: Longman/Wiley, 1992).

good model for turbulence, it is an excellent model for the general phenomenon of chaotic but deterministic behavior. Engineers, scientists, mathematicians, social scientists, economists, even novelists now use the Lorenz system as a paradigm for chaos. We are only at the beginning of a new theory, and that theory will take many twists and turns before it is put on a solid foundation.

Let's close out this chapter with a tribute to Dame Mary Cartwright, an English mathematician who is now recognized as a pioneer in chaotic dynamics.[11] Her work centered on ODEs much like the van der Pol system of Section 9.1, but with a driving function that is periodic in time. These systems are also similar to the Duffing system of Problem 4.

PROBLEMS

www **1.** (*Lorenz System*). In the following problems you are asked to verify assertions made in this section about the equilibrium points of the Lorenz system (1).

(a) Show that the origin and the points P_1, P_2 given in (3) are the only equilibrium points (the latter two only for $r > 1$).

(b) Construct the Jacobian matrix of the Lorenz system at the origin and show that the eigenvalues are as given in (2). Prove that all eigenvalues of this Jacobian matrix are negative if $0 < r < 1$ but that one eigenvalue is positive if $r > 1$.

(c) Construct the Jacobian matrix of the Lorenz system at P_1 and at P_2 for $r > 1$. Show that the eigenvalues of the matrix are roots of the cubic given in (4). Explain why one root is negative for all $r > 1$, while the other two roots are negative or have negative real parts if and only if r satisfies (5). [*Hint*: Use the Routh Test (Theorem 7.9.7).]

(d) Show that a cubic $\lambda^3 + A\lambda^2 + B\lambda + C$, $B > 0$, has a pair of pure imaginary roots if $C = AB$. Apply this result to the Lorenz system with $\sigma = 10$, $b = 8/3$ to find the value of r for which the Jacobian matrix at P_1 and P_2 has pure imaginary eigenvalues.

[11]Mary Lucy Cartwright (1900–) was one of the early researchers in what has developed into modern dynamical systems theory. In 1938, the British Department of Scientific and Industrial Research requested "really expert guidance" from mathematicians about "certain types of non-linear differential equations involved in the technique of radio engineering." British engineers were involved in a secret project to develop radar before the expected war (World War II) with Nazi Germany broke out. They were modeling the voltages and currents in the non-linear radar circuits with sinusoidally driven van der Pol systems, but couldn't understand the strange behavior the models predicted. This request was the source of many interesting mathematical problems for Cartwright to attack. Cartwright collaborated with J. E. Littlewood, publishing several joint papers that (we now know) give an excellent overview of chaotic dynamics, although those terms were never used. Although a pure mathematician at heart, Cartwright worked on many applied problems. Explaining this apparent contradiction, Cartwright commented that "the [applied] mathematical problems which have interested me most are apt to turn into topological problems or problems of topological dynamics." Cartwright and her work are discussed in "Mr. Littlewood and I: The Mathematical Collaboration of M. L. Cartwright and J. Littlewood," by Shawnee McMurran and James Tattersall, *American Mathematical Monthly* **103**, December 1996, pp. 833–845.

Mary Cartwright

2. (*Lorenz System*). The aim of this group project is to generate a gallery of portraits of orbits, projections, and their component curves for several values of r (set $\sigma = 10$, $b = 8/3$). Explain what you see. Look particularly for periodic orbits, period-doubling sequences, and strange attractors. Be sure to check sensitivity to changes in initial data. Find a copy of Colin Sparrow's book (see footnote 7) and try to duplicate some of his pictures. If your ODE solver is not adequate, replace the Lorenz system by the discrete approximation that uses Euler's Method:

$$x_{n+1} = x_n + h(-\sigma x_n + \sigma y_n)$$

$$y_{n+1} = y_n + h(r x_n - y_n - x_n z_n)$$

$$z_{n+1} = z_n + h(-b z_n + x_n y_n)$$

where h is the step size. The Euler approximation is surprisingly revealing for $h = 0.01$. Try to duplicate some of the figures in the text using your own solver. We guarantee that, although your graphs start out looking like ours, after a while they will be very different. This is a reflection of the sensitivity of the solutions of the Lorenz system to the slightest change, whether in initial data, parameters, or in the numerical method itself. It is not that your solver is more accurate than ours (or the reverse), it is only that it is different.

3. (*Rössler System*). The Rössler system[12] is

$$x' = -y - z, \qquad y' = x + ay, \qquad z' = b - cz + xz$$

where a, b, and c are positive parameters. O. E. Rössler wanted to build the "simplest possible" system that would model the phenomena displayed by the slightly more complicated Lorenz system. The system, now named after the inventor, is the result. In the parameter region of interest, this "model of a model" has two equilibrium points and one nonlinear term rather than the three equilibrium points and two nonlinear terms of the Lorenz system. As in the Lorenz system, there appears to be a folded strange attractor, spiraling, chaotic wandering, and period doubling. There are sequences of values of a that (for $b = 2$ and $c = 4$) correspond to period doubling and apparently terminate with the creation of a strange attractor. The orbits and component curves of the Rössler system are to be studied as one of the three parameters is varied. For most of the study, fix the parameters b and c at the respective values 2 and 4, and vary a. Look for period-doubling sequences, perhaps interspersed with values of a for which there are orbits of other periods. Check for chaotic spiraling, and especially for the folded strange attractor. If possible, plot orbits in xyz-space from different points of view. Plot orbital projections on various state-variable planes and use component plots to detect periodic orbits. If possible, plot only over an end segment of the time interval for which the system is actually solved (e.g., if the system is solved for $0 \le t \le 1000$, plot for $900 \le t \le 1000$). This way any transient behavior is likely to have disappeared and the orbit will be nearly in the strange attractor. Other suggestions:

- Plot orbits for $x(0) = z(0) = 0$, $y(0) = 2$, $b = 2$, $c = 4$, $a = 0.410, 0.400, 0.395$. Explain what you see. The IVP is solved for $0 \le t \le 1000$, but the orbits shown are plotted only for $800 \le t \le 1000$, so initial transients have died out.

[12]The description of the Rössler system, the scroll circuit (Section 9.3), and Duffing's equation (Problem 4) are adapted by permission from R. L. Borrelli, C. Coleman, and W. E. Boyce, *Differential Equations Laboratory Workbook* (New York: Wiley, 1992).

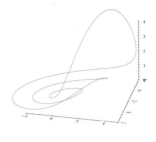

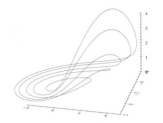

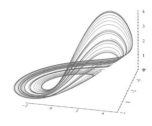

$a = 0.410$ $a = 0.400$ $a = 0.395$

- Plot orbits and component curves for the sequence of values of $a = 0.3, 0.35, 0.375$, $0.386, 0.3909, 0.398, 0.4, 0.411$. Use $x_0 = 0$, $y_0 = 2$, and $z_0 = 0$ and identify periodic orbits and their periods, period doubling, chaotic spiraling, and any other significant phenomena. What happens if different initial data are used?

- Find the equilibrium points of the Rössler system if a, b, and c are positive constants. For $b = 2$, $c = 4$, and the values of a given above, study orbital behavior near each equilibrium point.

- Explain why, if $|z|$ is very small, the subsystem $x' = -y$, $y' = x + ay$ has an unstable spiral point at the origin if $0 < a < 2$. Explain why this leads to a spreading apart of nearby orbits (a central part of chaotic wandering). This spreading is not unbounded, however. Explain from the third equation of the Rössler system how if $x < c$, the coefficient of z is negative and the z-system stabilizes near $b/(c - x)$. If $x > c$, the z-system "diverges" (assuming $b > 0$). Now look at the x-system and explain why orbits alternately lie in a plane parallel to the xy-plane, then are thrown upward, and in turn are folded back and reinserted into the plane, but closer to the origin.

- Vary b and c as well as a and analyze the effect on the orbits. Any period doubling, folding, chaotic wandering? Explain.

4. (*Duffing's Equation*).   Chaos in autonomous differential systems needs at least three dimensions, and the autonomous Lorenz, Rössler, and scroll systems all have three state variables. However, a somewhat different kind of chaotic wandering may appear with just two state variables as long as the system is nonautonomous; time itself may be considered to be the third dimension. Duffing's ODE and system and the ODE and system of the driven pendulum (Problem 5) provide examples of apparently chaotic behavior in two state dimensions with a sinusoidal driving term that is periodic in time.

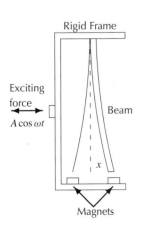

Rigid Frame

Exciting force

$A \cos \omega t$

Beam

x

Magnets

The motion of a long, thin, elastic steel beam with its top end embedded in a pulsating frame and the bottom end hanging just above two magnets (see margin sketch) can be modeled by the *driven damped Duffing's equation*:

$$x'' + cx' - ax + bx^3 = A \cos \omega t \qquad (6)$$

where a, b, c, A, and ω are nonnegative constants. The equivalent *driven damped Duffing's system* is

$$x' = y, \qquad y' = ax - bx^3 - cy + A \cos \omega t$$

Orbits of the system display a remarkable variety of behaviors, ranging from oscillatory, to periodic, to chaotic. For some sets of values of the parameters and some sets of initial data, orbits appear to approach a strange attractor. Warning: numerical solutions of chaotic systems over long time spans are almost certain to be inaccurate. Nevertheless, the shadowing property mentioned in footnote 9 gives us some confidence that what the numerical computations reveal actually do approximate reality. The sources cited below give additional background.[13] In carrying out the Duffing project, follow the outline below:

- Analyze and graph the orbits of an undriven, undamped Duffing system: $x' = y$, $y' = x - x^3$. Find the three equilibrium points. Find an orbital figure eight cycle-graph that consists of the origin and two other orbits. Why are all but five orbits periodic? See Figure **(a)**. Explain why the period of the orbit inside a lobe tends to $\sqrt{2\pi}$ as the amplitude of the orbit tends to 0 while the period tends to $+\infty$ as the orbit tends to the boundary of the lobe. Analyze the periods of the orbits outside the lobes.

- Add damping and obtain the system $x' = y$, $y' = x - x^3 - 0.15y$. Orbits are graphed for $x_0 = 1$, $0.8 \le y_0 \le 1.0$ in Figure **(b)**. Some orbits tend to $(-1, 0)$, others to $(1, 0)$ as $t \to +\infty$. Fix x_0 at 1 and plot orbits for several values of y_0 from 0 to 5.0. For each y_0 solve over the t-interval $0 \le t \le 200$. Do you see any pattern? Construct a table having the values of y_0 in the first column and the number of times the orbit starting at $(1, y_0)$ crosses the y-axis as t increases from 0 to 200 in the second column. From the table, estimate the length of each y_0 interval that generates orbits with a given number of crossings. Explain why all orbits starting inside any one of these y_0 intervals approach one of the equilibrium points $(-1, 0)$, $(1, 0)$, while the orbits starting in adjacent intervals approach the other equilibrium point. What happens to the lengths of these intervals as y_0 increases?

- Now add a periodic driving force: $x' = y$, $y' = x - x^3 - 0.15y + 0.3\cos t$. Solve the system with initial point $x_0 = -1$, $y_0 = 1$ over the interval $0 \le t \le 1000$, but plot over $500 \le t \le 1000$. Describe the chaotic tangle [see Figure **(c)**]. Any periodic orbits? [*Hint:* Try an initial point $x(0) \approx -0.2$, $y(0) \approx 1.4$.]

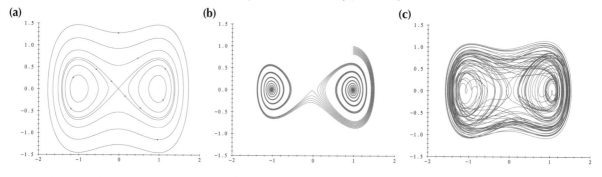

(a)   **(b)**   **(c)**

5. (*Driven Pendulum*). The system for a sinusoidally driven pendulum with a constant torque has the form

$$x' = y, \qquad y' = -ay - b\sin x + c\cos(\omega t) + d$$

where a, b, c, d, and ω are nonnegative parameters. Explain the model. Search for chaotic wandering, and plot and discuss your results.

[13]J. Guckenheimer and P. Holmes, *Nonlinear Oscillations, Dynamical Systems, and Bifurcations of Vector Fields* (New York: Springer-Verlag, 1983); P. Holmes and F. C. Moon, *J. Appl. Mech.* **108** (1983), pp. 1021–1032; J. M. T. Thompson and H. B. Stewart, *Nonlinear Dynamics and Chaos* (New York: Wiley, 1986).

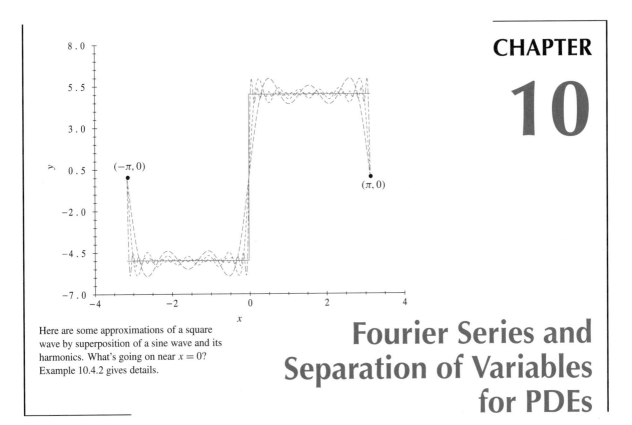

Here are some approximations of a square wave by superposition of a sine wave and its harmonics. What's going on near $x = 0$? Example 10.4.2 gives details.

Fourier Series and Separation of Variables for PDEs

A partial differential equation involves partial derivatives of sought-for functions of more than one independent variable. In this chapter we look at a separation of variables method that provides a solution formula for certain partial differential equations. These solution formulas are infinite series and were first used decisively by Joseph Fourier in the early 1800s to solve a heat conduction problem. We will look at Fourier's method in some detail and apply it to important special cases.

10.1 Vibrations of a String

Periodic disturbances and oscillations play an important role in many diverse areas in science and engineering. Examples are the motion of a pendulum in a gravitational field, the motion of celestial bodies, and oscillations in an electrical circuit. Oscillations that travel through space are known under the collective title of "wave motion." Some familiar examples of wave motion are the motion of ripples on the surface of a pond, sound waves, and the transverse vibrations of a string.

Let's model a basic natural process that involves wave motion: the transverse vibrations of a taut, flexible string. First, a few definitions that will come in handy during our derivation:

☙ **Interior and Boundary points, Region**. A point p is an *interior point* of a set S in $\mathbb{R}^n$ if S contains a ball with center at p. A *region R* in $\mathbb{R}^n$ is a connected (i.e., not in disjoint pieces) set consisting only of interior points. The boundary of a region R is denoted by ∂R. A *closed region* is a region R together with ∂R, and is denoted by R^*.

 The symbol ∂ in ∂R is *not* related to partial derivatives.

☙ **Continuity Sets**. A function u defined on an open (or closed) region S in $\mathbb{R}^n$ is in the *continuity set* $\mathbf{C}^k(S)$, for some integer $k \geq 0$, if all partial derivatives of u of order less than or equal to k exist as continuous functions on S.

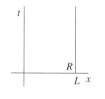 $\mathbf{C}^k(S)$ first appeared in Section 3.7. By convention, $\mathbf{C}^0(S)$ consists of all continuous functions on S.

Model for the Vibrating String

Suppose that a string is stretched between the points $x = 0$ and $x = L$ on the x-axis. Let's build a model for the string's motion under the action of the tension. To simplify the problem, we assume that the string is *flexible*; that is, the tension vector $\boldsymbol{\tau}$ at any point the string always acts tangentially to the string, so the string offers no resistance to bending. The static equilibrium position of the string under the action of tension alone is the interval $0 \leq x \leq L$. We can describe the motion of a point on the string by three functions of x and t which represent the spatial coordinates at time t of the point x on the string. On physical grounds, the motion depends not only on the initial deflection and velocity of the string but also on how the string is restrained at the boundary points $x = 0$ and $x = L$. We discuss boundary conditions later in this section.

The following simplifying assumptions are needed to obtain a simple partial differential equation (PDE) modeling the motion of the string:

- The string moves in a fixed plane, and points on the string move only in a direction transverse to the string. So the deflection of the string may be described by a single function $u(x, t)$ defined in the region $R = \{(x, t) : 0 < x < L, \ t > 0\}$.
- The motion u belongs to $\mathbf{C}^2(R)$ and $\mathbf{C}^1(R^*)$, and only "small" deflections from equilibrium are allowed in the sense that higher powers of $\partial u/\partial x$ (denoted by u_x) can be ignored when compared to the first-order term u_x.[1]
- The string has constant linear density ρ and the tension $\tau(x, t)$ is in $\mathbf{C}^1(R^*)$.
- Any external forces, including the weight of the string, are small compared to the tension and can be neglected.

---

[1] As we saw in Section 8.2, a function $h(z)$ has *order n* in z at $z = 0$ if there is a positive constant c such that for all $|z|$ small enough, $|h(z)| \leq c|z|^n$ [denoted by $h = O(|z|^n)$ as $z \to 0$]. For small $|z|$ the lower-order terms in a sum of nonnegative powers of z are dominant, so higher-order terms can be dropped.

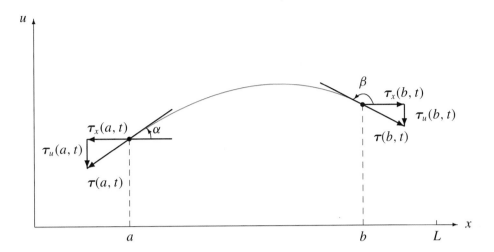

FIGURE 10.1.1 Profile on $[a, b]$ of the string at time t.

To model the motion of the string, let's look at the forces acting on the segment $[a, b]$; by assumption, the only force is tension. Resolving the tension acting at points b and a into transverse and parallel components $\boldsymbol{\tau}_u$ and $\boldsymbol{\tau}_x$, we see that (Figure 10.1.1)

☞ τ denotes the length of the vector $\boldsymbol{\tau}$. Ditto for τ_x and τ_u.

$$\tau_u(b, t) = \tau(b, t)|\sin\beta|; \qquad \tau_x(b, t) = \tau(b, t)|\cos\beta|;$$
$$\tau_u(a, t) = \tau(a, t)|\sin\alpha|; \qquad \tau_x(a, t) = \tau(a, t)|\cos\alpha|$$

where $\tau(a, t) = \|\boldsymbol{\tau}(a, t)\|$, $\tau_u(a, t) = \|\boldsymbol{\tau}_u(a, t)\|$, $\tau_x = \|\boldsymbol{\tau}_x(a, t)\|$, etc. Since

☞ These equations come from the fact that the slope ($\tan\beta$) of the profile at $x = b$ is $u_x(b, t)$.

$$|\sin\beta| = \frac{|u_x(b, t)|}{\sqrt{1 + |u_x(b, t)|^2}}, \qquad |\cos\beta| = \frac{1}{\sqrt{1 + |u_x(b, t)|^2}}$$

and similar formulas hold for $\sin\alpha$, $\cos\alpha$, we see from the second assumption above that

$$|\sin\beta| = |u_x(b, t)|, \qquad |\sin\alpha| = |u_x(a, t)|, \qquad |\cos\beta| = 1, \qquad |\cos\alpha| = 1$$

☞ The binomial series is given in Appendix B.2, item 9.

where second- and higher-order terms in u_x have been ignored. In other words, expand $(1 + |u_x|^2)^{-1/2}$ into a binomial series, $1 - \frac{1}{2}|u_x|^2 + \frac{3}{8}|u_x|^4 - \cdots$, and keep only the first term, 1. With the proper orientations, the transverse and parallel components for the net tension forces acting on the segment $[a, b]$ are

$$\tau(b, t)u_x(b, t) - \tau(a, t)u_x(a, t) \qquad \text{(transverse)} \qquad (1)$$

$$\tau(b, t) - \tau(a, t) \qquad \text{(parallel)} \qquad (2)$$

Since we assumed that only transverse motions of the string are possible, the net force acting parallel to the x-axis must be zero, so from (2) we conclude that $\tau(b, t) = \tau(a, t)$. Since a and b are arbitrary we conclude that the tension $\tau(x, t)$ is independent of x in $[0, L]$.

Now applying Newton's Second Law to the segment $[x, x + \Delta x]$, which has mass

$\rho \cdot (\Delta x)$ and is acted on by the transverse force (1), we have

$$\rho \cdot (\Delta x) u_{tt}(x^*, t) = \tau[u_x(x + \Delta x, t) - u_x(x, t)] \tag{3}$$

where x^* is the x-coordinate of the center of mass of the string segment between x and $x + \Delta x$. Dividing (3) through by $\rho \Delta x$, denoting τ/ρ by c^2, and taking limits as $\Delta x \to 0$, we obtain the *wave equation in one space dimension*,

☞ The subscript notation for partial derivatives will be used a lot from this point on: u_{tt} for $\partial^2 u/\partial t^2$, u_{xx} for $\partial^2 u/\partial x^2$, and so on.

$$u_{tt} = c^2 u_{xx} \tag{4}$$

which must be satisfied for all points (x, t) in R. For small amplitude vibrations the tension τ changes very little over time, so we assume that c^2 is constant. Note that PDE (4) is a second-order and linear partial differential equation and that c has units of length/time.

EXAMPLE 10.1.1

Solutions of the Wave Equation

The wave equation (4) has many solutions. For example, if F is any $\mathbf{C}^2$-function of one variable, then $u = F(x - ct)$ is a solution. This can be shown by repeated application of the Chain Rule to calculate partial derivatives. For example, if r denotes $x - ct$, and we set $u = F(x - ct) = F(r)$, then

$$u_t = \frac{dF}{dr}\frac{\partial r}{\partial t} = F'(r)(-c) \quad \text{and} \quad u_{tt} = -c\frac{d}{dr}[F'(r)]\frac{\partial r}{\partial t} = c^2 F''(r)$$

Similarly, we have

$$u_x = \frac{df}{dr}\frac{\partial r}{\partial x} = F'(r) \quad \text{and} \quad u_{xx} = \frac{d}{dr}[F'(r)]\frac{\partial r}{\partial x} = F''(r)$$

So we see that $u_{tt} = c^2 u_{xx}$ if $u = F(x - ct)$.

Physical intuition suggests that *initial conditions* together with *boundary conditions* are needed to obtain a unique and physically plausible solution. Let's now consider the effect of various conditions at the boundaries of a finite string.

Boundary/Initial Value Problems for the Wave Equation

It is assumed that the string has length L (i.e., $0 \le x \le L$) and that time is measured forward from the initial time $t = 0$.

Conditions at the Ends of the String: Boundary Operators

Only transverse motions of the string are considered. If the endpoints of the string (i.e., $x = 0$, $x = L$) are allowed to move, we can imagine the string to be looped around transverse rods at these two ends, but otherwise free to move (other interpretations are possible). We are interested in three types of boundary conditions. The simplest is when the transverse displacement of an endpoint is a prescribed function of time. For example, at $x = 0$ we may require that

$$u(0, t) = \alpha(t), \qquad t \ge 0 \tag{5}$$

where $\alpha(t)$ is a given function. If $\alpha(t) = 0$ for all t, then the string is fastened securely to the x-axis at $x = 0$. The special case $u(0, t) = 0$, all $t \geq 0$, is a *fixed boundary condition*. A similar condition could be imposed at $x = L$.

Now suppose instead that an endpoint is acted on by a transverse force, say $F = \gamma(t)$. Taking that endpoint to be $x = 0$, let's apply Newton's Second Law using the transverse components of forces acting on the string segment $0 \leq x \leq h$. As we saw earlier, the transverse force acting at $x = h$ is $\tau u_x(h, t)$, so we have that $\rho h u_{tt}(x^*, t) = \tau u_x(h, t) + \gamma(t)$, where ρ is the density of the string, x^* is the center of mass of the string segment $[0, h]$, and τ is the (constant) tension in the string. Since $u_{tt}(x, t)$ is bounded in x for each $t \geq 0$, the left-hand side $\rho h u_{tt}(x^*, t)$ tends to 0 as $h \to 0$. So it follows that

$$\tau u_x(0, t) + \gamma(t) = 0, \qquad t \geq 0 \tag{6}$$

If $\gamma(t) = 0$ for all t, then $u_x(0, t) = 0$, all $t \geq 0$, is a *free boundary condition*. A similar condition can be imposed at the endpoint $x = L$.

Finally, suppose that the left endpoint of the string is connected to the point $x = 0$ on the x-axis by a Hooke's Law spring with spring constant k. Then the function $\gamma(t)$ in (6) must be $-ku(0, t)$, and we have an *elastic boundary condition*:

$$\tau u_x(0, t) - ku(0, t) = 0, \qquad t \geq 0 \tag{7}$$

A similar condition could be imposed at the other endpoint.

The boundary conditions above can be described in terms of *boundary operators*. For example, the condition $2u_x(0, t) - 4u(0, t) = 0$ can be written as $B[u(x, t)] = 0$, where the operator B is defined by $B[u(x, t)] = 2u_x(0, t) - 4u(0, t)$. Note that B is a linear operator because

$$B[au + bv] = 2(au + bv)_x(0, t) - 4(au + bv)(0, t)$$

$$= a[2u_x(0, t) - 4u(0, t)] + b[2v_x(0, t) - 4v(0, t)]$$

$$= aB[u] + bB[v]$$

Operators associated with the other boundary conditions are also linear.

❖ **Boundary Conditions Expressed with Operators**. The *fixed*, *free*, and *elastic boundary conditions* at $x = 0$ are written as $B_0[u] = 0$, where the boundary operator B_0 is defined respectively by:

- Fixed Boundary: $B_0[u(x, t)] = u(0, t)$
- Free Boundary: $B_0[u(x, t)] = u_x(0, t)$
- Elastic Boundary: $B_0[u(x, t)] = \tau u_x(0, t) - ku(0, t)$

Similar definitions hold for boundary operators B_L at $x = L$.

Since the boundary operators B_0 and B_L are linear operators, it follows that the null space of any one of them is closed under linear combinations. This fact will be of some importance later. The operators above can also be used to express driven boundary conditions; for example, the boundary condition (6) implies that $B_0[u(x, t)] = -\gamma(t)/\tau$, where the action of B_0 is given by $B_0[u] = u_x(0, t)$.

The Mixed Initial/Boundary Value Problem for the Vibrating String

As we saw above, the deflection $u(x, t)$ of a vibrating string must satisfy the *wave equation* (4) in the region $R = \{(x, t) : 0 < x < L, \ t > 0\}$, provided that the motion of the string satisfies some simplifying assumptions. We saw that if u is in $\mathbf{C}^1(R^*)$, it is appropriate to demand that u and its first partial derivative with respect to x satisfy certain conditions at the boundaries $x = 0$ and $x = L$ for all $t \geq 0$ (depending on the character of the restraint of the string at the endpoints). As expected, the initial deflection $u(x, 0)$ and the initial velocity $u_t(x, 0)$ of the string may also be specified. So, in mathematical terms, the motion of a vibrating string with given initial and endpoint conditions will be a function u in $\mathbf{C}^2(R)$ and $\mathbf{C}^1(R^*)$ that satisfies all of the equations below inside the region R or on its boundaries [(IC) = initial conditions, (BC) = boundary conditions]:

$$
\begin{aligned}
\text{(PDE)} \quad & u_{tt} - c^2 u_{xx} = 0, & & 0 < x < L, \quad t > 0 \\
\text{(BC)} \quad & B_0[u] = \phi(t), & & t \geq 0 \\
& B_L[u] = \mu(t), & & t \geq 0 \\
\text{(IC)} \quad & u(x, 0) = f(x), & & 0 \leq x \leq L \\
& u_t(x, 0) = g(x), & & 0 \leq x \leq L
\end{aligned} \tag{8}
$$

☞ A PDE with boundary and/or initial conditions is called a *problem*. One PDE may generate several problems.

where $f(x)$, $g(x)$, $\phi(t)$, and $\mu(t)$ are given functions, and the boundary operators B_0 and B_L each can be any of the three types of boundary operators mentioned above. The *mixed initial value/boundary value problem* (8) with *initial conditions* (IC) and *boundary conditions* (BC) can be shown to have no more than one solution for any appropriately smooth *initial data* [$f(x)$ and $g(x)$] and *boundary data* [$\phi(t)$ and $\mu(t)$] (see Problem 3).

Standing Waves and Separated Solutions

There are solutions for the wave equation (4) that have considerable significance in the theory of acoustics. These are the *standing waves* $X(x)T(t)$, where $T(t)$ is a periodic function of time. More accurately, $X(x)$ is the standing wave *with amplitude modulated* by $T(t)$. Substituting $X(x)T(t)$ into PDE (4) and separating variables, we see that if $X(x) \neq 0$ and $T(t) \neq 0$, then

$$
\frac{X''(x)}{X(x)} = \frac{1}{c^2} \frac{T''(t)}{T(t)} \tag{9}
$$

holds for all $0 < x < L$ and $t > 0$. Since x and t are independent variables, the only way identity (9) could hold is for each side to be equal to the same constant. Calling this constant λ, the identity (9) can be separated to yield the ODEs

$$
X''(x) = \lambda X(x), \qquad T''(t) = c^2 \lambda T(t) \tag{10}
$$

Actually, if any value of λ is used in (10) and $X(x)$ and $T(t)$ are any solutions of the corresponding ODEs, then $X(x)T(t)$ is a solution of the wave equation (4). Such solutions are generally called *separated solutions* for PDE (4). In acoustics, standing waves are separated solutions where $T(t)$ is periodic. Thus for standing waves we must have $\lambda < 0$ for $T(t)$ to be periodic.

☞ Recall from Example 3.5.1 that an ODE such as $T'' = c^2 \lambda T$ has periodic solutions if and only if $\lambda < 0$.

Let's consider a guitar string of length L with constant tension τ, constant density ρ, and $c^2 = \tau/\rho$. For small motions, the deflection $u(x, t)$ solves the problem

☞ System (11) models the motion of a string clamped at each end, given an initial deflection f, and then released from rest.

$$
\begin{aligned}
u_{tt} &= c^2 u_{xx} & 0 < x < L, \quad t > 0 \\
u(0, t) &= 0, \quad u(L, t) = 0, & t \geq 0 \\
u(x, 0) &= f(x), & 0 \leq x \leq L \\
u_t(x, 0) &= 0, & 0 \leq x \leq L
\end{aligned}
\tag{11}
$$

where $f(x)$ is the initial deflection of the string, and the boundary operators B_0 and B_L have the action $B_0[u] = u(0, t)$ and $B_L[u] = u(L, t)$. The first thing we do is to find all separated solutions $X(x)T(t)$ of the PDE $u_{tt} = c^2 u_{xx}$ that satisfy the boundary conditions $u(0, t) = u(L, t) = 0$, for all t. Since the string is clamped at both ends, any separated solution must satisfy the boundary conditions $X(0) = X(L) = 0$. So to find all such separated solutions, we see from (10) that we must first determine all constants λ such that the boundary value problem

$$
X''(x) = \lambda X(x), \qquad X(0) = X(L) = 0
\tag{12}
$$

has a nontrivial C^2-solution $X(x)$. This is an example of a *Sturm-Liouville problem* (more on such problems in Section 10.6). Notice that if $X(x)$ is any C^2-function on $[0, L]$ such that $X(0) = X(L) = 0$, integration by parts shows that

$$
\int_0^L X(x) X''(x) \, dx = X(x) X'(x) \Big|_0^L - \int_0^L \big[X' \big]^2 \, dx = - \int_0^L \big[X' \big]^2 \, dx \leq 0
\tag{13}
$$

Now suppose that X is a non-trivial solution of (12); multiply each side of the ODE in (12) by X and use (13) to obtain that

$$
\lambda \int_0^L [X(x)]^2 \, dx = \int_0^L X(x) X''(x) \, dx \leq 0
$$

So it appears that we must take $\lambda \leq 0$. Now (12) does not have a nontrivial solution for $\lambda = 0$, so let's try $\lambda = -k^2$ for some $k > 0$. The general solution of $X'' + k^2 X = 0$ is $X(x) = A \cos kx + B \sin kx$ for arbitrary constants A and B. The condition $X(0) = 0$ implies that $A = 0$, and the condition $X(L) = 0$ implies that $B \sin kL = 0$. To avoid the trivial solution, we must assume that $B \neq 0$, so we must take $kL = n\pi$, where n is a nonnegative integer. For any value $\lambda_n = -(n\pi/L)^2$, $n = 1, 2, \ldots$, the Sturm-Liouville problem (12) has the corresponding solution (up to an arbitrary multiplicative constant).

$$
X_n(x) = \sin(n\pi x/L)
$$

So apparently all separated solutions of the wave equation which satisfy fixed boundary conditions are actually standing waves. Let's find a formula for these standing waves. Now since $\lambda = \lambda_n = -(n\pi/L)^2$, the second ODE in (10) shows that

$$
T_n(t) = A_n \cos \frac{n\pi c}{L} t + B_n \sin \frac{n\pi c}{L} t
$$

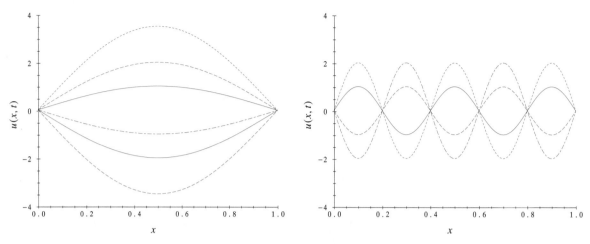

FIGURE 10.1.2 Six time profiles of guitar string standing wave: $n = 1$.

FIGURE 10.1.3 Four time profiles of guitar string standing wave with nodes: $n = 5$.

so all standing waves for the guitar string have the form

$$u_n(x, t) = \sin\frac{n\pi x}{L}\left(A_n\cos\frac{n\pi ct}{L} + B_n\sin\frac{n\pi ct}{L}\right), \qquad n = 1, 2, 3, \ldots \quad (14)$$

where A_n and B_n are arbitrary constants. If time is measured in seconds, the standing wave u_n has the *period* $2L/nc$ seconds (per cycle), and recovers its initial sinusoidal profile $u_n(x, 0) = A_n\sin(n\pi x/L)$ at $t = k(2L/nc)$ seconds for every integer k. The *frequency* of u_n is the reciprocal of the period: $f_n = nc/2L$ cycles/second (or hertz). The numbers f_n are said to be the *natural frequencies* supported by the string. The points $x = 0, L/n, \ldots, (n - 1)L/n$, and L, are rest points, or *nodes*, of the standing wave u_n. Figures 10.1.2 and 10.1.3 show standing waves for $n = 1$ and $n = 5$ and several values of t.

☞ In Section 10.7 we superimpose the standing waves for particular values of A_n and B_n, and solve problem (11) with initial conditions $u(x, 0) = f(x)$, $u_t(x, 0) = g(x)$.

The standing wave u_n is a solution of problem (11) with fixed boundary conditions, $u(0, t) = u(L, t) = 0$, and initial conditions

$$u(x, 0) = A_n\sin\frac{n\pi x}{L}, \qquad u_t(x, 0) = \frac{n\pi c}{L}B_n\sin\frac{n\pi x}{L}$$

A Musical Interlude

The standing waves in (14) and their frequencies have acquired musical names because of their association with musical tones and instruments. The standing wave

$$u_1(x, t) = \sin\frac{\pi x}{L}\left(A_1\cos\frac{\pi ct}{L} + B_1\sin\frac{\pi ct}{L}\right)$$

is called the *fundamental* or *first harmonic*, while the standing waves of higher frequency are *higher harmonics*. For example, if $n = 3$, then $A_3 = -2$, $B_3 = 0$ and

$$u_3(x, t) = -2\sin\frac{3\pi x}{L}\cos\frac{3\pi ct}{L}$$

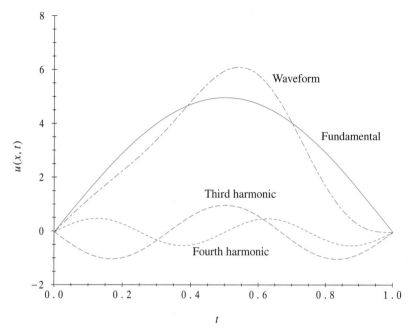

FIGURE 10.1.4 The waveform is a superposition of standing waves.

is a third harmonic. The waveform sketched in Figure 10.1.4 is the profile of a fundamental plus a third and a fourth harmonic. Each stringed instrument produces its own combinations of fundamentals and overtones; the variety of these combinations distinguishes the sound of guitar from a harp, or even one guitar from another.

Are the physical vibrations of the guitar string, violin string, or piano wire what we actually hear? What really happens is that these mechanical vibrations of the wire cause corresponding variations in the air pressure, variations that we hear as different musical sounds. What our ears detect are air-pressure waves set in motion by the vibrations of a string, a larynx, or some other source of sonic energy. It can be shown that air pressure itself is a solution of a wave equation, where the constant c^2 is a ratio of specific heats and of standard air pressures and densities.

Partial Differential Equations

The internal temperatures of a solid body whose boundary temperatures are controlled, voltage along a transmission line, the displacement of a vibrating string—these natural processes and many more can be modeled by partial differential equations with appropriate boundary and initial conditions. Using the "variable subscript" notation for the partial derivatives of the real-valued function $u(x, y, z, t)$, the basic PDEs that occur most frequently in the applications are listed below:

$$
\begin{aligned}
&\text{Laplace (or Potential) Equation} && u_{xx} + u_{yy} + u_{zz} = 0 \\
&\text{Wave Equation} && u_{tt} = c^2\{u_{xx} + u_{yy} + u_{zz}\} \qquad (15)\\
&\text{Heat (or Diffusion) Equation} && u_t = K\{u_{xx} + u_{yy} + u_{zz}\}
\end{aligned}
$$

where c and K are positive constants whose values are determined by the physical properties of the materials being modeled. The variables x, y, and z are space variables and t is time. The wave equation arises in models involving wave propagation (hence the name), while Laplace's equation appears in models of steady-state phenomena, and the heat equation is connected with diffusion processes such as heat flow. A convenient operator notation ∂_t, ∂_x, ∂_{xx}, and so on, is frequently used to denote the operations of differentiation $\partial/\partial t$, $\partial/\partial x$, $\partial^2/\partial x^2$, and so on. So if we define the *Laplacian* operator $\nabla^2 = \partial_{xx} + \partial_{yy} + \partial_{zz}$, we see that the PDEs in (15) can be written as

$$\nabla^2 u = 0, \qquad (\partial_{tt} - c^2 \nabla^2)u = 0, \qquad (\partial_t - K\nabla^2)u = 0 \qquad (16)$$

The *order* of a PDE is the order of the highest derivative appearing in the equation. All the PDEs in (15) are second order. Notice that the Laplacian ∇^2 is a linear operator since $\nabla^2[\alpha u + \beta v] = \alpha \nabla^2 u + \beta \nabla^2 v$ for any constants α, β and any functions u, v that are twice continuously differentiable. The operators ∂_t and ∂_{tt} are linear as well. So the basic PDEs in (16) are *linear* because they have the form $L[y] = f$, where L is a linear operator. Since the *driving term* f is zero for all the PDEs in (16), they are said to be *undriven*. It follows that the PDEs in (16) all have the form $L[u] = 0$, so a linear combination of solutions to any one of them is again a solution of that same equation. This simple but important fact will be used over and over again in this chapter; it is called the *Principle of Superposition of Solutions*

Comments

It is a remarkable fact that problem (11) can be solved as a superposition of the standing waves $u_n(x, t)$ defined in (14), where the constants A_n, B_n are determined appropriately for all $n = 1, 2, \ldots$. We carry out the details in the next section. This technique can be generalized to an important method for solving boundary/initial value problems for PDEs. Although the method is rather limited in its scope, it is very important since it provides explicit solution formulas for a wide class of commonly occurring boundary/initial value problems. This approach is called the Method of Separation of Variables, and we will discuss it in more detail in Section 10.7 after we have developed the necessary background in Fourier methods.

PROBLEMS

www **1.**   (*Standing Waves*). Consider a vibrating string of length L.

   (a) Find all standing waves with free boundary conditions at the endpoints.

   (b) Find all standing waves with a fixed boundary condition at $x = 0$ and a free boundary condition at $x = L$.

2.   Verify directly that the Sturm-Liouville problem (12) has no nontrivial solutions when either $\lambda = 0$, or $\lambda > 0$.

3. (*Uniqueness Theorem*). Suppose that R is the region $0 < x < L$, $t > 0$, and suppose that $F(x, t)$ is continuous on R^*, that $\alpha(t)$, $\beta(t)$ are continuous on $t \geq 0$, and that $\gamma(x)$, $\delta(x)$ are continuous on $0 \leq x \leq L$. Following the outline below, show that the problem below can't have more than one solution $u(x, t)$ that is in $\mathbf{C}^2(R)$ and $\mathbf{C}^1(R^*)$:

$$
\begin{aligned}
&\text{(PDE)} && u_{tt} = c^2 u_{xx} + F(x, t) && \text{in } R \\
&\text{(BC)} && u(0, t) = \alpha(t), && u(L, t) = \beta(t), && t \geq 0 \\
&\text{(IC)} && u(x, 0) = \gamma(x), && u_t(x, 0) = \delta(x), && 0 \leq x \leq L
\end{aligned}
$$

- Suppose the given problem has two solutions u and v. Put $U = u - v$ and show that U satisfies the same problem, but with the data F, α, β, γ, and δ all set equal to zero.

- Define $W(t) = \frac{1}{2} \int_0^L (U_t^2 + c^2 U_x^2) \, dx$ for any $t \geq 0$. Use the identity $(U_x U_t)_x = U_{xx} U_t + U_x U_{tx}$ to show that

$$
W'(t) = U_x(L, t)U_t(L, t) - U_x(0, t)U_t(0, t)
$$

- Show that $U_t(L, t) = 0$ and $U_t(0, t) = 0$ for $t \geq 0$, so $W'(t) = 0$ for $t \geq 0$. [*Hint*: Recall the definition of partial derivatives.]

- Show that $U_t(x, t) = 0$ and $U_x(x, t) = 0$ for all (x, t) in R^*, so $U = $ constant on R.

- Show that actually $U = 0$ on R and conclude that $u = v$ on R.

10.2 Orthogonal Functions

The last section ended with the surprising statement that a formula for the solution of the plucked guitar string problem

$$
\begin{aligned}
&u_{tt} = c^2 u_{xx} && 0 < x < L, \quad t > 0 \\
&u(0, t) = 0, \quad u(L, t) = 0, && t \geq 0 \\
&u(x, 0) = f(x), \quad u_t(x, 0) = 0, && 0 \leq x \leq L
\end{aligned}
\tag{1}
$$

can be constructed with a superposition of the standing waves

$$
u_n(x, t) = \sin \frac{n\pi x}{L} \left(A_n \cos \frac{n\pi ct}{L} + B_n \sin \frac{n\pi ct}{L} \right), \qquad n = 1, 2, 3, \ldots
$$

no matter what the initial deflection $f(x)$ of the string is. Let's see how to do this.

Suppose that a solution $u(x, t)$ of the boundary/initial value problem (1) has the form $u(x, t) = \sum_{n=1}^{\infty} u_n(x, t)$, where the constants A_n, B_n are such that the series converges at each point in the closed region $R = \{(x, t) : 0 \leq x \leq L, \ t \geq 0\}$. We now show how to find the values that the coefficients A_n, B_n must have in order that the series $\sum_{n=1}^{\infty} u_n(x, t)$ converges to a solution of problem (1). First note that each u_n satisfies the endpoint conditions, as does the sum $\sum u_n$. If the partial derivatives u_{xx} and u_{tt} can be computed by differentiation of the series $\sum u_n$ term by term, then $u = \sum u_n$ satisfies the wave equation because each u_n does. The constants A_n, B_n must be determined so that the series $\sum u_n$ satisfies the initial conditions. This implies that

$$
f(x) = u(x, 0) = \sum_{n=1}^{\infty} A_n \sin \frac{n\pi x}{L}, \qquad 0 \leq x \leq L
\tag{2}
$$

$$
0 = u_t(x, 0) = \sum_{n=1}^{\infty} \frac{n\pi c}{L} B_n \sin \frac{n\pi x}{L}, \qquad 0 \leq x \leq L
\tag{3}
$$

Now comes a step that is of crucial importance in determining the constants A_n, B_n in (2) and (3). For all integers m and n we have (see Problem 1):

$$\int_0^L \sin\frac{n\pi x}{L}\sin\frac{m\pi x}{L}\,dx = \begin{cases} 0, & m\neq n,\ \text{or } m=n=0 \\ L/2, & m=n\neq 0 \end{cases}$$

$$\int_0^L \cos\frac{n\pi x}{L}\cos\frac{m\pi x}{L}\,dx = \begin{cases} 0, & m\neq n \\ L, & m=n=0 \\ L/2, & m=n\neq 0 \end{cases} \tag{4}$$

These cosine integrals will be useful later.

Multiplying both sides of (2) and (3) by $\sin(m\pi x/L)$ and integrating over $0\leq x\leq L$ (assuming that integration and infinite summation can be interchanged) and using (4), we have

$$\int_0^L f(x)\sin\frac{m\pi x}{L}\,dx = A_m\left(\frac{L}{2}\right), \qquad m=1,2,\dots$$

$$0 = \frac{m\pi c}{L}B_m\left(\frac{L}{2}\right), \qquad m=1,2,\dots$$

which uniquely determines A_m and B_m (in fact, $B_m=0$). So if our assumptions are valid, we are led to the characterization of the solution of problem (1) as the series

$$u(x,t) = \sum_{n=1}^\infty A_n\sin\frac{n\pi x}{L}\cos\frac{n\pi ct}{L} \tag{5}$$

where

$$A_n = \frac{2}{L}\int_0^L f(x)\sin\frac{n\pi x}{L}\,dx, \qquad n=1,2,\dots \tag{6}$$

We will show later in Section 10.7 that if the data function $f(x)$ is "nice" enough, then formula (5) with coefficients given by (6) does indeed solve problem (1). If we knew in advance that problem (1) has no more than one solution (see Problem 3 in Section 10.1), then (5), (6), must be *the* solution of problem (1).

The integral formulas in (4) are called *orthogonality relations* and are at the heart of deriving the solution formula (5), (6). In Section 4.1 we saw that the idea of orthogonality generated by the dot (or scalar) product for vectors in $\mathbb{R}^3$ brought not only a certain geometric sense to the construction of models but significant computational simplifications as well. In this section we extend the notion of orthogonality to function vector spaces (defined in Section 7.2). Then we will show how orthogonality provides a valuable new tool for the approximation of functions. In Section 10.7 we show how to use these concepts to solve boundary/initial value problems for PDEs.

Linear Spaces

The Student Resource Manual has more on linear spaces.

In Section 7.2 we used the term *linear space* or *vector space* for a collection of objects (called *vectors*) that behave like $\mathbb{R}^n$ (or $\mathbb{C}^n$), the column vectors with n real entries (or n complex entries). In other words, there is an operation of "addition for vectors," denoted by "+," and an operation of "multiplication of a vector by a scalar" that satisfy the same rules as for $\mathbb{R}^n$ (or $\mathbb{C}^n$). If the vectors are *real*-valued objects, then the scalars are $\mathbb{R}$, and if the vectors are *complex*-valued objects, then the scalars are $\mathbb{C}$.

In Section 7.2 we also mentioned *function vector spaces* whose objects are column vectors with entries that are real-valued (respectively, complex-valued) functions. In this chapter our function vectors will have just one entry, and, as before, the scalars are $\mathbb{R}$ if the functions are real-valued, and $\mathbb{C}$ if the functions are complex-valued.

| EXAMPLE 10.2.1 | **Linear Spaces of Functions** |

For any interval I and any integer $k \geq 0$ the continuity set $\mathbf{C}^k(I)$ of *real*-valued functions is a *real* linear space if *real* scalars are used, and the operations of addition and scalar multiplication are given by

$$(f + g)(t) = f(t) + g(t), \qquad (\alpha f)(t) = \alpha f(t), \quad \text{all } t \text{ in } I \qquad (7)$$

If $\mathbf{C}^k(I)$ consists of *complex*-valued functions, then $\mathbf{C}^k(I)$ is a *complex* linear space when *complex* scalars and the operations in (7) are used.

Scalar Products

It turns out that the basic properties of the scalar product on $\mathbb{R}^3$ are *symmetry*, *bilinearity*, and *positive definiteness* (see Problem 1, Section 4.1). This leads us to ask whether or not on *any* given linear space we can find a function that returns a scalar for each pair of vectors so that these three properties are satisfied. When this is possible for a linear space, we can imitate what was done for the dot product on $\mathbb{R}^n$ and develop the ideas of "orthogonality," "distance," and "convergence of a sequence" in that space.

☞ The dots in $\langle \cdot, \cdot \rangle$ are just placeholders waiting for vectors to be inserted.

❖ **Scalar Product, Euclidean Space.** Suppose that V is a linear space, and suppose that $\langle \cdot, \cdot \rangle$ is a function that associates with every pair of vectors u, v the scalar $\langle u, v \rangle$. Then $\langle \cdot, \cdot \rangle$ is a *scalar product* if the following properties are satisfied for all vectors u, v, w and all scalars α, β:

$$(Symmetry) \qquad \langle v, u \rangle = \overline{\langle u, v \rangle}, \text{ the complex conjugate of } \langle u, v \rangle \qquad (8a)$$

$$(Bilinearity) \qquad \langle \alpha u + \beta v, w \rangle = \alpha \langle u, w \rangle + \beta \langle v, w \rangle \qquad (8b)$$

$$(Positive\ definiteness) \qquad \langle u, u \rangle \geq 0 \text{ and } = 0 \text{ if and only if } u = 0 \qquad (8c)$$

A linear space V with a scalar product is called a *Euclidean space*.

There are many scalar products for the same linear space, as we will see later. It is not hard to check that $\langle x, y \rangle = \sum_1^n x_k y_k$ is a scalar product on $\mathbb{R}^n$; this *standard scalar product* makes $\mathbb{R}^n$ into a real Euclidean space and generalizes the dot product on $\mathbb{R}^3$. Similarly, $\langle x, y \rangle = \sum_1^n x_k \overline{y}_k$ makes $\mathbb{C}^n$ into a complex Euclidean space and is the *standard scalar product* on $\mathbb{C}^n$.

Since the scalar product of any vector with itself is always a nonnegative number (even for complex linear spaces), we can define the length of a vector in the same way as for the dot product.

❖ **Norm**. In a Eulidean space E with scalar product $\langle \cdot, \cdot \rangle$, we define the *norm* (or *length* or *magnitude*) of a vector u by the nonnegative quantity

$$\|u\| = (\langle u, u \rangle)^{1/2} \qquad (9)$$

This norm is called *the norm induced by the scalar product* (other norms are possible). Vectors of length 1 are called *unit vectors*.

Dividing any nonzero vector by its length produces a unit vector. This process is called *normalization*.

❖ **Distance Between Two Vectors**. In a Euclidean space E with scalar product $\langle \cdot, \cdot \rangle$, we define the *distance* between any two vectors u and v in E to be $\|u - v\|$, where $\| \cdot \|$ is the norm induced by $\langle \cdot, \cdot \rangle$.

Recall that a function $f(x)$ on a closed interval $[a, b]$ is said to be *piecewise continuous* if there are finitely many points $c_1, c_2, \ldots, c_k$ with $a < c_1 < c_2 < \cdots < c_k < b$, such that f is continuous on each open interval $(a, c_1), (c_1, c_2), \ldots, (c_{k-1}, c_k), (c_k, b)$, and the one-sided limits of f exist at all the points $x = a, c_1, \ldots, c_k, b$. Examples are pulse and step functions over any interval $[a, b]$. [*Note*: The actual values of f at the points $x = a, c_1, \ldots, c_k, b$ play no role in the definition. In fact, f need not even be defined at these points.] The set of all piecewise continuous functions on $[a, b]$ is denoted by PC$[a, b]$. Using the usual notion of addition of functions and multiplication of a function by a scalar, we see that PC$[a, b]$ is a linear space (a *real* linear space if the functions are real-valued and the scalars are real numbers).

EXAMPLE 10.2.2

The Euclidean Space PC$[a, b]$
For the linear space PC$[a, b]$ we see that

$$\langle f, g \rangle = \int_a^b f(x)\overline{g(x)}\,dx, \qquad f, g \text{ in PC}[a, b] \qquad (10)$$

☞ For example, the function $f(x)$ that is zero everywhere except at one point is *not* the zero function and yet $\langle f, f \rangle = 0$.

is a scalar product if we identify any two functions in PC$[a, b]$ that differ at only a finite set of points [otherwise (8c) fails to hold]. We will use the same symbol PC$[a, b]$ for this Euclidean space, but simply remember that the statement $f = 0$ means that f vanishes in $[a, b]$ except at most on a finite set. The scalar product (10) is called the *standard scalar product* on PC$[a, b]$ (there are many others). Note that $\mathbf{C}^n[a, b]$ is contained in PC$[a, b]$ for any nonnegative integer n, and that (10) is also a scalar product on $\mathbf{C}^n[a, b]$. The norm induced by the scalar product (10) has the form

$$\|f\| = \left[\int_a^b |f|^2\,dx \right]^{1/2}, \qquad f \text{ in PC}[a, b] \qquad (11)$$

Any scalar product $\langle \cdot, \cdot \rangle$ and its induced norm $\| \cdot \|$ have some useful properties:

THEOREM 10.2.1

> *Properties of Scalar Product.* Let E be a Euclidean space with scalar product $\langle \cdot, \cdot \rangle$ and induced norm $\| \cdot \|$. For all u, v in E, and scalars α, the following hold:
>
> (*Cauchy-Schwarz Inequality*) $\quad |\langle u, v \rangle| \leq \|u\| \|v\|$
> (*Positive definiteness*) $\quad \|u\| \geq 0$, and $\|u\| = 0$ if and only if $u = 0$
> (*Homogeneity*) $\quad \|\alpha u\| = |\alpha| \|u\|$
> (*Triangle Inequality*) $\quad \left| \|u\| - \|v\| \right| \leq \|u + v\| \leq \|u\| + \|v\|$

These properties are very useful in calculations and in producing error estimates in Euclidean spaces, as we will soon see.

Orthogonality

A scalar product is used to define the orthogonality of vectors.

❖ **Orthogonality.** Two vectors u, v in a Euclidean space E are said to be *orthogonal* if and only if $\langle u, v \rangle = 0$. A subset S in E not containing the zero vector is an *orthogonal set* if $\langle u, v \rangle = 0$ for any two distinct vectors u, v in S.

The zero vector is orthogonal to all vectors of a Euclidean space E, but it is the only such vector. Orthogonality extends the familiar idea of perpendicularity in the standard Euclidean spaces $\mathbb{R}^3$ and $\mathbb{R}^2$. For instance, $(a, b) \neq (0, 0)$ in $\mathbb{R}^2$ is orthogonal to $(-b, a)$, and this coincides with the familiar fact that a pair of lines intersects orthogonally if their slopes are negative reciprocals. The notion of orthogonality in a function vector space has no such simple geometric interpretation.

EXAMPLE 10.2.3

An Orthogonal Set in PC$[-\pi, \pi]$
The set $\Phi = \{1, \cos x, \sin x, \ldots, \cos nx, \sin nx, \ldots\}$ is an orthogonal set in PC$[-\pi, \pi]$ under the standard scalar product. To see this we use a trigonometric identity to show for example that for any integers m and n,

☞ The trig identities in Appendix B.4 come in handy here.

$$\langle \sin nx, \cos mx \rangle = \int_{-\pi}^{\pi} \sin nx \cos mx \, dx$$

$$= \int_{-\pi}^{\pi} \frac{1}{2}[\sin(n + m)x + \sin(n - m)x] \, dx = 0$$

The other orthogonality relations follow in a similar fashion.

EXAMPLE 10.2.4

An Orthogonal Set in the Complex Euclidean Space PC$[-T, T]$
Suppose PC$[-T, T]$ consists of complex-valued functions. The set $\Psi = \{e^{ik\pi/T} : k = 0, \pm 1, \pm 2, \ldots\}$ is an orthogonal set. Using the standard scalar product in PC$[-T, T]$, we have for $k \neq m$,

$$\langle e^{ik\pi x/T}, e^{im\pi x/T} \rangle = \int_{-T}^{T} e^{ik\pi x/T} e^{-im\pi x/T} \, dx = \frac{T e^{i(k-m)\pi x/T}}{i(k-m)\pi} \Bigg|_{-T}^{T} = 0$$

For any integer k we see that

$$\|e^{ik\pi x/T}\|^2 = \langle e^{ik\pi x/T}, e^{ik\pi x/T} \rangle = 2T$$

So the "vector" $e^{ik\pi x/T}/\sqrt{2T}$ has unit length.

Comments

There are other orthogonal sets in PC[0, L] besides the ones defined in this section. In Section 10.6 we will describe a "machine" that will generate them as needed to come up with a formula for the solution of a boundary/initial value problem.

PROBLEMS

1. Use the two trigonometric identities

    $$\sin\alpha\sin\beta = (1/2)[\cos(\alpha-\beta) - \cos(\alpha+\beta)]$$
    $$\cos\alpha\cos\beta = (1/2)[\cos(\alpha+\beta) + \cos(\alpha-\beta)]$$

    to verify the integration formulas (4).

 2. (*Plucked Guitar String*). A guitar string of length π initially has the shape

    $$f(x) = \begin{cases} x, & 0 \le x \le \pi/2 \\ \pi - x, & \pi/2 \le x \le \pi \end{cases}$$

    and is released from rest. Graph the profile of the string at $t = 1$, $t = 5$, and $t = 10$. [*Hint*: Use a computer algebra solver (CAS) package to calculate enough coefficients A_n [given by (6)] in solution formula (5) to give a reasonably good approximation. Then graph the truncated series for the various t-values using a CAS package again.]

www 3. (*Guitar String*). The guitar string modeled by system (1) has length $L = \pi$, tension T, and density ρ. Find the solution in the form of the series in (5) with specific values for the coefficients A_n if the initial profile $f(x)$ is as given.

    (a) $f(x) = \sin x$ **(b)** $f(x) = 0.1 \sin 2x$ **(c)** $f(x) = \sum_{n=1}^{10}(\sin nx)/n$

    (d) Explain in mathematical terms why the tauter the string, the higher the pitch. Now replace the string by a denser string and pluck it again. In mathematical terms, what do you hear now?

4. (*Subspaces*). A subset A of a linear space V is called a *subspace* of V if for any scalar α and any pair of vectors u, v in A, we have that αu and $u + v$ is in A. For any scalars $\alpha_1, \alpha_2, \ldots, \alpha_n$ and any vectors $u_1, u_2, \ldots, u_n$ in A, the sum $\alpha_1 u_1 + \cdots + \alpha_n u_n$ is said to be a *finite linear combination* over A. The *span* of A, denoted by Span(A), is the collection of all finite linear combinations over A.

    (a) For any linear space V, show that the set of all vectors in V and the set consisting of the zero vector alone are both subspaces of V.

    (b) Show that a set of vectors A in a linear space V is a subspace if and only if Span(A) = A.

    (c) For a closed interval $I = [a, b]$ in $\mathbb{R}$ and any nonnegative integers m, n, suppose that the continuity sets $\mathbf{C}^n(I)$ and $\mathbf{C}^m(I)$ are both real or both complex linear spaces of functions. Show that $\mathbf{C}^n(I)$ is a subspace of $\mathbf{C}^m(I)$ if $n \ge m$.

    (d) Suppose that $\mathbf{C}^\infty(I)$ is the set of all real-valued functions on an interval I having derivatives of all orders on I. Show that $\mathbf{C}^\infty(I)$ is a real linear space.

5. (*A Weighted Scalar Product*). For two functions f and g in the real space $\mathbf{C}^0[0, 1]$ define $\langle f, g \rangle = \int_0^1 f(x)g(x)e^x \, dx$. The function e^x is said to be a *weight* or *density* factor.

   (a) Show that $\langle \cdot, \cdot \rangle$ defines a scalar product.

   (b) Calculate the scalar product of each function pair: $f = 1 - 2x$, $g = e^{-x}$; $f = x^2$, $g = e^x$.

   (c) Calculate the scalar product of each function pair: $f = x$, $g = 1 - x$; $f = e^{-x/2}\sin(\pi x/2)$, $g = e^{-x/2}\sin(3\pi x/2)$; $f = \cos(\pi x/2)$, $g = 1$.

6. For any closed interval $I = [a, b]$ and any nonnegative integer n, show that

$$\langle f, g \rangle = \int_a^b f(x)\overline{g(x)} \, dx$$

   is a scalar product on the function vector space $\mathbf{C}^n(I)$ of complex-valued functions.

7. (*Properties of a Scalar Product*). Suppose that $\langle \cdot, \cdot \rangle$ is a scalar product on a real linear space V, and let $\| \cdot \|$ be the norm induced by the scalar product.

   (a) Show that $\|u + v\|^2 + \|u - v\|^2 = 2\|u\|^2 + 2\|v\|^2$ for all u, v in V.

   (b) (*Cauchy-Schwarz Inequality*). Show that $|\langle u, v \rangle| \le \|u\| \cdot \|v\|$ for all u, v in V. [*Hint*: Use the fact that $\|\alpha u - \beta v\|^2 \ge 0$ for *all* scalars α and β.]

   (c) Show that $\langle u, v \rangle = (\|u + v\|^2 - \|u - v\|^2)/4$ for all u, v in V.

☞ There is also "big ell two," but we won't worry about that here.

8. ("*Little Ell Two*": l^2). Suppose that $\mathbb{R}^\infty$ is the set of all sequences of real numbers, $(x_n)_{n=1}^\infty$, abbreviated simply as (x_n). Define addition and multiplication by reals in $\mathbb{R}^\infty$ termwise; that is, $(x_n) + (y_n) = (x_n + y_n)$, $\alpha \cdot (x_n) = (\alpha x_n)$. Suppose that l^2 is the collection of all sequences (x_n) in l^2 for which $\sum_{n=1}^\infty |x_n|^2$ converges; l^2 is called "little ell two."

   (a) Show that for any x, y in l^2, $\sum_{i=1}^\infty x_i y_i$ converges.

   (b) Show that $\langle x, y \rangle = \sum_{i=1}^\infty x_i y_i$ is a scalar product on l^2.

9. (*Orthogonal Projection*). Suppose E is a Euclidean space and S a subspace of E spanned by the finite orthogonal set $\{u^1, u^2, \ldots, u^n\}$. For u in E the vector

$$\text{proj}_S(u) = \sum_{i=1}^n [\langle u, u^i \rangle / \|u^i\|^2] u^i$$

   is called the *orthogonal projection* of u onto S. [*Note*: $\text{proj}_S(u)$ does not depend on the orthogonal spanning set of S chosen.]

   (a) Show that for each u in E we can uniquely write $u = v + w$, with v in S and w orthogonal to S.

   (b) Show that for any u in E, $\|u - \text{proj}_S(u)\| \le \|u - v\|$ for all v in S, and equality holds if and only if $v = \text{proj}_S(u)$.

   (c) In $\mathbb{R}^3$, find the distance from the point $(1, 1, 1)$ to the plane: $x + 2y - z = 0$.

10.3 Fourier Series and Mean Approximation

There is a new kind of convergence (*convergence in the mean*) that works in Euclidean spaces and is extremely helpful in finding a formula for the solution of boundary/initial value problems for linear PDEs. This convergence is easily described in terms of the scalar product on that space and this type of convergence has some far-reaching consequences. Then we will define the Fourier series of a function with respect to an orthogonal set.

Mean Approximation

Suppose that f is a given vector in a Euclidean space E. If g in E is considered as an approximation to f, then $\|f - g\|$ is known as the *error in the sense of the mean*, or simply *error in the mean*. Suppose that $\Phi = \{\phi_1, \phi_2, \ldots\}$ is a finite or infinite orthogonal set in the Euclidean space E. Then for a given element f in E we can ask the following question: What linear combination of the first n elements of Φ is "closest" to f in the sense of the distance measure on E (i.e., *in the mean*)? In symbols: For what values of the constants $c_1, c_2, \ldots, c_n$ is $\|f - \sum_{k=1}^{n} c_k \phi_k\|$ a minimum? The answer to this question follows.

THEOREM 10.3.1

> **Mean Approximation Theorem.** For any orthogonal set $\Phi = \{\phi_1, \phi_2, \ldots\}$ in a Euclidean space E, any f in E, and any positive integer n, the problem
>
> $$\left\| f - \sum_{k=1}^{n} c_k \phi_k \right\| = \text{minimum}$$
>
> has the unique solution
>
> $$(c_k)_{\min} = \langle f, \phi_k \rangle / \|\phi_k\|^2, \qquad k = 1, 2, \ldots, n \tag{1}$$
>
> Moreover,
>
> $$\left\| f - \sum_{k=1}^{n} (c_k)_{\min} \phi_k \right\|^2 = \|f\|^2 - \sum_{k=1}^{n} |\langle f, \phi_k \rangle|^2 / \|\phi_k\|^2 \tag{2}$$

To show this, assume for simplicity that E is a *real* Euclidean space. Now notice that by the bilinearity of the scalar product $\langle \cdot, \cdot \rangle$ in E,

$$\left\| f - \sum_{k=1}^{n} c_k \phi_k \right\|^2 = \left\langle f - \sum_{k=1}^{n} c_k \phi_k, \; f - \sum_{k=1}^{n} c_k \phi_k \right\rangle$$

$$= \|f\|^2 - 2 \sum_{k=1}^{n} c_k \langle f, \phi_k \rangle + \sum_{k=1}^{n} c_k^2 \|\phi_k\|^2 \tag{3}$$

Completing the square on the quadratic polynomial in c_k in (3), we see that

$$\|\phi_k\|^2 c_k^2 - 2\langle f, \phi_k \rangle c_k = \|\phi_k\|^2 \left(c_k - \frac{\langle f, \phi_k \rangle}{\|\phi_k\|^2} \right)^2 - \frac{|\langle f, \phi_k \rangle|^2}{\|\phi_k\|^2}$$

so (3) becomes

$$\left\| f - \sum_{k=1}^{n} c_k \phi_k \right\|^2 = \|f\|^2 + \sum_{k=1}^{n} \|\phi_k\|^2 \left(c_k - \frac{\langle f, \phi_k \rangle}{\|\phi_k\|^2} \right)^2 - \sum_{k=1}^{n} \frac{|\langle f, \phi_k \rangle|^2}{\|\phi_k\|^2} \tag{4}$$

To minimize the right-hand side of (4), the c_k must be chosen as in (1), and when this is done, formula (2) results from (4).

EXAMPLE 10.3.1   **Mean Approximation**

Direct calculation verifies that $\Phi = \{\sin(k\pi x/L) : k = 1, 2, \ldots\}$ is an orthogonal set in PC[0, L], for any positive constant L, and that $\|\sin(k\pi x/L)\|^2 = L/2$, for any $k = 1, 2, \ldots$.

Define a function f as follows:

$$f(x) = \begin{cases} x, & 0 \leq x \leq L/2 \\ 0, & L/2 \leq x \leq L \end{cases}$$

We will find the best mean-square approximation to f in the form $c_1 \sin(\pi x/L) + c_2 \sin(2\pi x/L) + c_3 \sin(3\pi x/L)$. Using (1), we see that for any $k > 0$:

☞ Use integration by parts or a table to evaluate the integral.

$$(c_k)_{\min} = \frac{2}{L} \int_0^{L/2} x \sin \frac{k\pi x}{L} \, dx = \frac{2L}{k\pi} \left(\frac{1}{k\pi} \sin \frac{k\pi}{2} - \frac{1}{2} \cos \frac{k\pi}{2} \right)$$

So $(c_1)_{\min} = 2L/\pi^2$, $(c_2)_{\min} = L/2\pi$, $(c_3)_{\min} = -2L/9\pi^2$, and

$$g_{\min} = \frac{2L}{\pi^2} \sin \frac{\pi x}{L} + \frac{L}{2\pi} \sin \frac{2\pi x}{L} - \frac{2L}{9\pi^2} \sin \frac{3\pi x}{L}$$

is the best approximation in the mean to f of the desired form. The square of the mean error (or the *mean-square error*) of this approximation is given by (1) and (2) as

☞ A direct calculation shows that $\|f\|^2 = L^3/24$.

$$\|f - g_{\min}\|^2 = \|f\|^2 - \sum_{k=1}^{3} (c_k)_{\min}^2 \left\| \sin \frac{k\pi x}{L} \right\|^2 = \frac{L^3}{24} - \frac{L}{2} \left(\frac{4L^2}{\pi^4} + \frac{L^2}{4\pi^2} + \frac{4L^2}{81\pi^4} \right)$$

Convergence in the Mean

As we saw earlier, the induced norm $\| \cdot \|$ on a Euclidean space E measures the "distance" between two vectors u, v in E as $\|u - v\|$. We can use this distance measure to define convergence of infinite sequences and series in E.

❖ **Convergence in the Mean.** Suppose that $\| \cdot \|$ is the induced norm on a Euclidean space E. We say that the sequence $\{f_n\}$, $n = 1, 2, \ldots$, in E *converges in the mean* if and only if there exists a vector f in E such that $\|f_n - f\| \to 0$ as $n \to \infty$. In this case we write $f_n \to f$ as $n \to \infty$.

For g_k in E, $k = 1, 2, \ldots$, the series $\sum_{k=1}^{\infty} g_k$, *converges in the mean* if and only if the sequence of partial sums $f_n = \sum_{k=1}^{n} g_k$, $n = 1, 2, \ldots$, converges in the mean. If $\sum_{k=1}^{n} g_k \to f$ as $n \to \infty$, then we write $f = \sum_{k=1}^{\infty} g_k$.

As in ordinary convergence, a mean convergent sequence $\{f_n\}$ can have only one limit.

THEOREM 10.3.2

> Mean Convergence Theorem. For any sequence $\{f_n\}$ in a Euclidean space E, there is at most one vector f in E such that $\|f_n - f\| \to 0$ as $n \to \infty$.

To show this, suppose, to the contrary, that $f_n \to g$ and $f_n \to h$ as $n \to \infty$, where $g \neq h$. Now $\|g - h\| > 0$, and from the Triangle Inequality we have

$$0 < \|g - h\| = \|(g - f_n) + (f_n - h)\| \leq \|f_n - g\| + \|f_n - h\|, \quad \text{for all } n$$

So it is impossible that both $\|f_n - g\| \to 0$ and $\|f_n - h\| \to 0$, unless $g = h$, establishing our claim.

Mean convergence has another important property (to show this, see Problem 4):

THEOREM 10.3.3

> Continuity of the Scalar Product. Suppose that in a Euclidean space,
>
> $$f_n \to f, \text{ and } g_m \to g \text{ as } n, m \to \infty$$
>
> Then $\langle f_n, g_m \rangle \to \langle f, g \rangle$, as $n, m \to \infty$.

Mean, Pointwise, and Uniform Convergence

Mean convergence for sequences in PC[a, b] is different from pointwise and uniform convergence. Let's review pointwise and uniform convergence for a sequence $\{f_n(x)\}$,

☞ Appendix A.2 has more on pointwise and uniform convergence.

$n = 1, 2, \ldots$, in PC[a, b]. The sequence $\{f_n(x)\}$ is said to *converge pointwise* at x_0 in [a, b] if there is a value y_0 such that $f_n(x_0) \to y_0$ as $n \to \infty$ in the ordinary sense of number sequences. Restated, this means that for any interval of length $2\epsilon > 0$ centered at y_0, the values of $f_n(x_0)$ "eventually" all remain in this interval, that is, for all $n \geq N$, where N is some positive integer (see Figure 10.3.1).

Uniform convergence of the function sequence $\{f_n(x)\}$ on a closed interval [a, b] is *more* than just saying that $\{f(x)\}$ converges pointwise at *each* x_0 in [a, b]. *Uniform convergence* of $\{f_n(x)\}$ to a function $g(x)$ on [a, b] means that for any "tube" with center-spine $g(x)$ and vertical diameter $2\epsilon > 0$, the graphs of $f_n(x)$ are "eventually" all contained in this tube for all $n \geq N$, where N is some positive integer (see Figure 10.3.2).

Note that if the sequence $\{g_k\}$ in PC[a, b] converges uniformly to g on [a, b], then $g_k \to g$ in the mean (over PC[a, b]). Also, uniform convergence over a bounded interval implies mean convergence. This summarizes the general relationship between the three modes of convergence in PC[a, b] considered here.

Orthogonal Series: Bases

Suppose that $\Phi = \{\phi_1, \phi_2, \ldots\}$ is an orthogonal set in a Euclidean space E. The series given by $\sum_{k=1}^{\infty} c_k \phi_k$ for any scalars $c_1, c_2, \ldots$ is called an *orthogonal series* (whether or not it converges). By definition, the series $\sum_{k=1}^{\infty} c_k \phi_k$ converges (in the mean) to the vector f in E if and only if $\|\sum_{k=1}^{n} c_k \phi_k - f\| \to 0$ as $n \to \infty$.

In finite-dimensional linear spaces we have seen how useful it is to have a set with the property that every vector in the space can be uniquely written as a linear

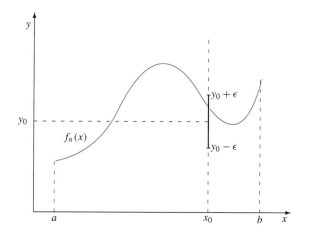

FIGURE 10.3.1 Pointwise convergence geometry.

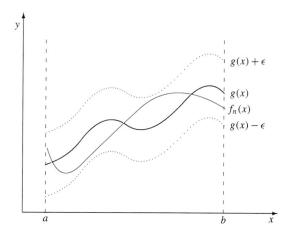

FIGURE 10.3.2 Uniform convergence geometry.

combination over this set. There is an analog of this concept in infinite-dimensional Euclidean spaces like $PC[a, b]$ that makes use of the convergence property to attach a meaning to the sum of an orthogonal series: An orthogonal set $\Phi = \{\phi_k : k = 1, 2, \ldots\}$ in a Euclidean space E is called a *basis* for E if and only if for every f in E there exists a unique set of scalars $\{c_k\}_1^\infty$ such that $f = \sum_{k=1}^\infty c_k \phi_k$.

The continuity of the scalar product (Theorem 10.3.3) says that if $f = \sum_{k=1}^\infty c_k \phi_k$, then $c_n = \langle f, \phi_n \rangle / \|\phi_n\|^2$ for each $n = 1, 2, \ldots$.

THEOREM 10.3.4

> Fourier-Euler Theorem. If for an orthogonal set $\Phi = \{\phi : k = 1, 2, \ldots\}$ in the Euclidean space E we have that $f = \sum_{k=1}^\infty c_k \phi_k$, then the coefficients are given by the *Fourier-Euler Formulas*
>
> $$c_n = \frac{\langle f, \phi_n \rangle}{\|\phi_n\|^2} \quad \text{for each } n = 1, 2, \ldots \qquad (5)$$

Since $\sum_{k=1}^N c_k \phi_k \to f$ as $N \to \infty$, it follows [from the continuity of the scalar product (Theorem 10.3.3)] that $\langle \sum_{k=1}^N c_k \phi_k, \phi_n \rangle \to \langle f, \phi_n \rangle$ for each (fixed) n, as $N \to \infty$. But by orthogonality, $\langle \sum_{k=1}^N c_k \phi_k, \phi_n \rangle = c_n \langle \phi_n, \phi_n \rangle$ if $N > n$, and the assertion is established.

Formula (5) is the reason for the following definition.

❖ **Fourier Coefficient.** The constant $c_k = \langle f, \phi_k \rangle / \|\phi_k\|^2$ is called the k-th *Fourier* (or *Fourier-Euler*) *coefficient* of f over the orthogonal set Φ.

Theorem 10.3.4 and the Mean Approximation Theorem 10.3.1 give the following equivalent definition of a basis.

❖ **Basis.** An orthogonal set $\Phi = \{\phi_k : k = 1, 2, \ldots\}$ is a basis for the Euclidean space E if and only if for each f in E,

$$\sum_{k=1}^{\infty} \frac{\langle f, \phi_k \rangle}{\|\phi_k\|^2} \phi_k = f \tag{6}$$

in the sense of mean convergence. The condition (6) is not a particularly easy test to apply because it requires showing that an orthogonal series converges in the mean to the sum on the right-hand side. There are other basis tests that are more convenient, as we will see in Theorem 10.3.6.

An important property of bases is given by the following result, which follows immediately from the condition in (6).

THEOREM 10.3.5

> Totality Theorem. Suppose that $\Phi = \{\phi_k : k = 1, 2, \ldots\}$ is a basis for the Euclidean space E. If f in E is such that $\langle f, \phi_k \rangle = 0$ for all $k = 1, 2, \ldots$, then $f = 0$.

So we see that the only element in a Euclidean space that is orthogonal to a basis is the zero vector. Doesn't this look familiar? We know this result is valid in the finite-dimensional spaces $\mathbb{R}^n$ and $\mathbb{C}^n$, but it is rather comforting to know that the result remains valid in Euclidean spaces that are function vector spaces.

Now it's time to see what can be done with orthogonal series in a Euclidian space. We will see that they are very useful in devising solution formulas for PDEs.

Fourier Series

Orthogonal series in a Euclidean space constructed using coefficients given by formula (5) have a special name.

❖ **Fourier Series.** Suppose that $\Phi = \{\phi_k : k = 1, 2, \ldots\}$ is an orthogonal set in the Euclidean space E. Then the orthogonal series

👉 If you want to, you can think of FS as an operator that, acting on f, produces the orthogonal series in (7).

$$\mathrm{FS}[f] = \sum_{k=1}^{\infty} \frac{\langle f, \phi_k \rangle}{\|\phi_k\|^2} \phi_k, \quad \text{for any } f \text{ in } E \tag{7}$$

is called *the Fourier series of f over Φ.* The symbol FS[f] stands for the formal series in (7); *no implication is made about convergence.*

The same function may have any number of Fourier series, depending on the particular choice of an orthogonal set Φ in E. Let's combine (6) and (7) into a basis criterion: An orthogonal set Φ in a Euclidean space E is a basis for E if and only if every f in E is the sum of its Fourier series over Φ. The results below will be used later.

THEOREM 10.3.6

Three Theorems. **Suppose** that $\Phi = \{\phi_k : k = 1, 2, \ldots\}$ is an orthogonal set in the Euclidean space E, but not necessarily a basis. Then for any f in E, the series $\sum_{k=1}^{\infty} |\langle f, \phi_k \rangle|^2 / \|\phi_k\|^2$ converges in the mean, and the following properties hold.

Bessel Inequality: We have

$$\|f\|^2 \geq \sum_{k=1}^{\infty} \frac{|\langle f, \phi_k \rangle|^2}{\|\phi_k\|^2} \quad \text{for all } f \text{ in } E \tag{8}$$

Decay of Coefficients: If $c_k = \langle f, \phi_k \rangle / \|\phi_k\|^2$, the k-th Fourier coefficient of f, and if Φ is nonfinite, then

$$\|\phi_k\| \cdot |c_k| \to 0 \quad \text{as } k \to \infty \tag{9}$$

Parseval Relation: Φ is a basis for E if and only if

$$\|f\|^2 = \sum_{k=1}^{\infty} \frac{|\langle f, \phi_k \rangle|^2}{\|\phi_k\|^2} \quad \text{for all } f \text{ in } E \tag{10}$$

The identity (2) in the Mean Approximation Theorem holds for all positive integers n. Since the left-hand side of (2) can never be negative, it follows that

$$\|f\|^2 \geq \sum_{k=1}^{n} \frac{|\langle f, \phi_k \rangle|^2}{\|\phi_k\|^2} \quad \text{for all } n \tag{11}$$

The sequence $\left\{ \sum_{k=1}^{n} |\langle f, \phi_k \rangle|^2 / \|\phi_k\|^2 : n = 1, 2, \ldots \right\}$ is bounded from above and nondecreasing, so the infinite series $\sum_{k=1}^{\infty} |\langle f, \phi_k \rangle|^2 / \|\phi_k\|^2$ is convergent. Inequality (8) now follows immediately from (11) by taking the limit as $n \to \infty$. Because the infinite series in (8) is convergent, the k-th term converges to zero. Property (9) follows from the fact that $|\langle f, \phi_k \rangle|^2 / \|\phi_k\|^2 = \|\phi_k\|^2 |c_k|^2$, where c_k is the Fourier coefficient given by (5). Finally, the Parseval Relation follows immediately from (1) and (2) of the Mean Approximation Theorem.

EXAMPLE 10.3.2

Fourier Series in PC$[-\pi, \pi]$

Consider the orthogonal set Φ in PC$[-\pi, \pi]$ given in Example 10.2.3. We now use the definition (7) to calculate the Fourier series of the function $f(x) = x$ on $[-\pi, \pi]$. Notice first that $\int_{-\pi}^{\pi} \sin^2 nx\, dx = \int_{-\pi}^{\pi} \cos^2 nx\, dx = \pi$ for all $n \geq 1$ and that $\int_{-\pi}^{\pi} 1^2\, dx = 2\pi$. Now FS$[f]$ has the form

$$\text{FS}[f] = A_0 + \sum_{k=1}^{\infty} (A_k \cos kx + B_k \sin kx) \tag{12}$$

☞ Use a table of integrals or integration-by-parts to evaluate B_k.

where $A_0 = (1/2\pi) \int_{-\pi}^{\pi} 1 \cdot x\, dx = 0$, $A_k = (1/\pi) \int_{-\pi}^{\pi} x \cos kx\, dx = 0$ (since $x \cos kx$ is an odd function on $[-\pi, \pi]$), and $B_k = (1/\pi) \int_{-\pi}^{\pi} x \sin kx\, dx = 2(-1)^{k+1}/k$. So we have

$$\text{FS}[f] = 2 \sum_{k=1}^{\infty} \frac{(-1)^{k+1}}{k} \sin kx \tag{13}$$

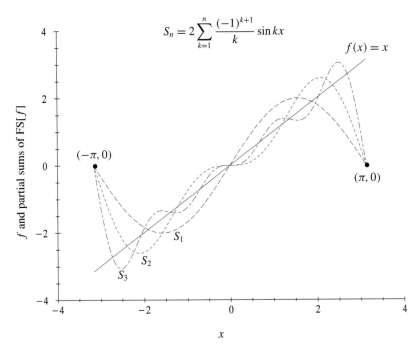

$$S_n = 2 \sum_{k=1}^{n} \frac{(-1)^{k+1}}{k} \sin kx$$

FIGURE 10.3.3 Graphs of three partial sums of the Fourier series (13).
See Example 10.3.2.

Figure 10.3.3 shows the graphs of $f(x) = x$ (solid) and the first three partial sums S_1, S_2, S_3 of the Fourier series in (13) for $|x| \le \pi$.

PROBLEMS

1. (*Best Approximation*). In the Euclidean space PC$[-\pi, \pi]$, consider the subspace S_N spanned by the set $\Phi_N = \{1, \cos x, \sin x, \ldots, \cos Nx, \sin Nx\}$. Find the element in S_N that is closest to the element $f(x) = x$, $|x| < \pi$, in PC$[-\pi, \pi]$.

2. (*Fourier Series*). Consider the orthogonal set $\Phi = \{\sin kx : k = 1, 2, \ldots\}$ in PC$[0, \pi]$. Find the Fourier series of the following functions in PC$[0, \pi]$ with respect to Φ.

   (a) $f(x) = x$ **(b)** $f(x) = 1$ **(c)** $f(x) = 1 + x/\pi$

www 3. (*Calculating Fourier Series*). $\Phi = \{1, \cos x, \sin x, \ldots, \cos nx, \sin nx, \ldots\}$ is an orthogonal subset, with the standard scalar product, of the Euclidean space PC$[-\pi, \pi]$. Find the Fourier Series of the following functions in PC$[-\pi, \pi]$ with respect to Φ. See margin sketches for graphs.

   (a) (*Sawtooth*).   **(b)** (*Triangular*).

$$f(x) = \begin{cases} -A(1 + x/\pi), & -\pi \le x < 0 \\ A(1 - x/\pi), & 0 < x < \pi \end{cases} \qquad f(x) = \begin{cases} A(1 + x/\pi), & -\pi \le x \le 0 \\ A(1 - x/\pi), & 0 < x < \pi \end{cases}$$

   (c) (*Inverted Triangular*).
$$f(x) = |x|, \quad |x| \le \pi$$

4. (*Scalar Product Continuity*). Show the continuity of the scalar product (Theorem 10.3.3). [*Hint*: Write $\langle f_n, g_m \rangle - \langle f, g \rangle = \langle f_n, g_m - g \rangle + \langle f_n - f, g \rangle$ and use the Cauchy-Schwarz inequality.]

5. (*Uniqueness*). Suppose that Φ is a basis for a Euclidean space E. Show that if two vectors f and g in E have the same Fourier series over Φ, then $f = g$.

3(a)

(0, A)

(−π, 0)

(π, 0)

(0, −A)

3(b)

(0, A)

(−π, 0)   (π, 0)

3(c)

π

−π   π

6. (*A Euclidean Space of Functions*). The collection of real-valued functions on $\mathbb{R}$ defined by $E = \{f \text{ in } \mathbf{C}^0(\mathbb{R}) : \int_{-\infty}^{\infty} |f|^2 \, dx < \infty\}$ is a linear space.

(a) Show that $\langle f, g \rangle = \int_{-\infty}^{\infty} fg \, dx$ is a scalar product on E.

(b) Consider the sequence

$$f_n(x) = n^{-1/2} e^{-x^2/n^2}, \qquad n = 1, 2, \ldots$$

Show that $\{f_n\}$ converges uniformly to the zero function of $\mathbb{R}$.

(c) Show that $\{f_n\}$ does not converge in the mean in E. [*Hint:* If $\{f_n\}$ is mean convergent, the limit must be the zero function (why?), but Theorem 10.3.3 contradicts this.]

10.4 Fourier Trigonometric Series

Some techniques of applied analysis crucially depend on the possibility of writing a given real-valued function as the sum of specialized trigonometric series over some interval. In fact, this situation arises so often that there's a huge body of literature on this topic. In this section we look at this idea in a systematic fashion.

Using some trigonometric identities (see Appendix B.4) before calculating the integrals, we can verify directly that

$$\Phi_T = \left\{ 1, \cos\frac{\pi x}{T}, \sin\frac{\pi x}{T}, \ldots, \cos\frac{n\pi x}{T}, \sin\frac{n\pi x}{T}, \ldots \right\} \tag{1}$$

is an orthogonal set in the Euclidean space $PC[-T, T]$ under the standard scalar product $\langle f, g \rangle = \int_{-T}^{T} fg \, dx$. Notice that

$$\left\| \cos\frac{k\pi x}{T} \right\|^2 = \int_{-T}^{T} \cos^2\frac{k\pi x}{T} \, dx = \begin{cases} 2T & \text{if } k = 0 \\ T & \text{if } k > 0 \end{cases}$$

$$\left\| \sin\frac{k\pi x}{T} \right\|^2 = \int_{-T}^{T} \sin^2\frac{k\pi x}{T} \, dx = T, \qquad k > 0$$

So for any f in $PC[-\pi, \pi]$, the Fourier series of f over Φ_T is

☞ This is a special case of (7) in Section 10.3.

$$\text{FS}[f] = A_0 + \sum_{k=1}^{\infty} \left(A_k \cos\frac{k\pi x}{T} + B_k \sin\frac{k\pi x}{T} \right) \tag{2}$$

where the coefficients are given by the Fourier-Euler formulas,

☞ Some people write $A_0/2$ as the first term in FS[f] just so A_0 can be computed using the general formula for A_k.

$$A_0 = \frac{1}{2T} \int_{-T}^{T} f(x) \, dx, \qquad A_k = \frac{1}{T} \int_{-T}^{T} f(x) \cos\frac{k\pi x}{T} \, dx, \qquad k > 0$$

$$B_k = \frac{1}{T} \int_{-T}^{T} f(x) \sin\frac{k\pi x}{T} \, dx, \qquad k > 0 \tag{3}$$

This particular orthogonal series in (2) is called a *Fourier trigonometric series* on the interval $[-T, T]$ and consists of a superposition of a constant term and sinusoids whose circular frequencies are multiples (harmonics) of a basic (fundamental) circular frequency of π/T. There are many other trigonometrical series that come up in applied mathematics, but for now we'll stick to Fourier trigonometrical series. By popular usage, however, series (2) with coefficients defined by (3) is commonly called the

Fourier series[2] of f, and not its more complete name. We will adopt this popular name whenever no confusion can arise.

Although it is not evident, the orthogonal set Φ_T is a basis for $PC[-T, T]$, so we have the following result.

THEOREM 10.4.1

> Fourier Trigonometric Basis Theorem. The set Φ_T given by (1) is a basis for $PC[-T, T]$, and $FS[f]$ converges in the mean to f for any f in $PC[-T, T]$.

Calculation of Fourier Trigonometric Series

For a function f in $PC[-T, T]$, the process of setting up the series in (2) with the coefficients in (3) is known as *"expanding f into a Fourier trigonometric series,"* and the values of A_k and B_k in (3) are *"the Fourier coefficients of f."* Calculating the Fourier coefficients of a function f can be a lengthy exercise in integration. Note

☞ Remember from calculus that the value of an integral $\int_a^b f(x)\,dx$ doesn't change if the value of $f(x)$ is changed at only a finite set of points.

that two functions f and g in $PC[-T, T]$ whose values differ at only a finite number of points in $[-T, T]$ have the *same* Fourier coefficients. So changing the values of a function at a finite number of points in $[-T, T]$ does not change its Fourier coefficients.

The process of calculating Fourier coefficients can be simplified if the function has certain symmetry properties. A function f in $PC[-T, T]$ is *even* if $f(-x) = f(x)$ for $|x| < T$ (except possibly a finite number of points), while f is *odd* if $f(-x) = -f(x)$ for $|x| < T$. It is easy to show that $f(x)g(x)$ is even if f and g are both even or both odd, and $f(x)g(x)$ is odd if one factor is even and the other odd. Note that $\int_{-a}^a f(x)\,dx = 0$ if f is odd, and $\int_{-a}^a f(x)\,dx = 2\int_0^a f(x)\,dx$ if f is even. So if f in $PC[-T, T]$ is odd, the Fourier coefficients $A_k = 0$, since $f(x)\cos(k\pi x/T)$ is odd for each $k = 0, 1, 2, \dots$. Similarly, $B_k = 0$ for $k = 1, 2, \dots$ if f is even. The problems have more shortcuts of this kind.

EXAMPLE 10.4.1

Fourier Series of an Even Function
Let's find the Fourier series of the function $f(x) = |x|$, $-T \le x \le T$. Since f is even, all the coefficients B_k vanish. Now we have

☞ Use integration by parts or a table to do the second integration.

$$A_0 = \frac{1}{T}\int_0^T x\,dx = \frac{T}{2}, \quad A_k = \frac{2}{T}\int_0^T x\cos\frac{k\pi x}{T}\,dx = \frac{2T}{\pi^2 k^2}(\cos k\pi - 1), \quad \text{for } k > 0$$

---

[2]The trigonometric series encountered in this chapter are called Fourier series in honor of the French physicist Jean Baptiste Joseph Fourier [1768–1830] who used them in the study of heat flow. From about 1740 onward, physicists and mathematicians such as Bernoulli, D'Alembert, Lagrange, and Euler had engaged in heated discussions about the possibility of representing an arbitrary periodic function as a sum of a trigonometric series. In fact, in 1777 Euler had discovered the integral formulas for what are now called the Fourier coefficients. Fourier added his thoughts on the subject at the Paris Academy of Sciences in 1807 and 1811. In the papers on heat conduction he presented there, he suggested that a completely arbitrary function could be represented as the sum of a trigonometric series. The referees criticized him for a lack of rigor. Lagrange, who was at that time one of the referees, denied vehemently Fourier's contention, and these papers were never published. Fourier continued his work, however, and in 1822 he published *La Théorie Analytique de la Chaleur*, throughout which he made use of Fourier series in his analysis of heat flow. Besides its widespread use in the study of the partial differential equations modeling wave motion, heat flow, and electrostatics, the theory of Fourier series has directly affected such diverse areas of mathematics as modern algebra and Cantor's work on set theory.

Jean Baptiste Joseph
Fourier

Note that since $\cos k\pi = (-1)^k$, $A_k = 0$ when k is a positive even integer. Setting $k = 2n+1$, $n = 0, 1, \ldots$, we see that $A_{2n+1} = -4T/[\pi^2(2n+1)^2]$. So

$$\text{FS}[f] = \frac{T}{2} + \sum_{n=0}^{\infty} \frac{-4T}{\pi^2(2n+1)^2} \cos\left(\frac{(2n+1)\pi x}{T}\right)$$

Convergence Properties

Since Φ_T in (1) is a basis for PC$[-T, T]$, we know that for any f in PC$[-T, T]$, the Fourier series in (2) with coefficients given by (3) converges in the mean to f; that is,

$$\left\| A_0 + \sum_{k=1}^{n} (A_k \cos kx + B_k \sin kx) - f \right\| \to 0, \quad \text{as } n \to \infty$$

But what about the pointwise or uniform convergence properties of Fourier series?

EXAMPLE 10.4.2

Loss of Uniform Convergence

The function

$$f(x) = \begin{cases} 5, & 0 < x < \pi \\ -5, & -\pi < x < 0 \end{cases}$$

is in PC$[-\pi, \pi]$ and since it is an odd function, its Fourier series contains no cosine terms (or constant term). A straightforward calculation shows that

☞ Use the Fourier-Euler formulas (3).

$$\text{FS}[f] = \frac{20}{\pi} \sum_{k=0}^{\infty} \frac{\sin(2k+1)x}{2k+1} \tag{4}$$

In Figure 10.4.1 (and the chapter cover figure), the graphs of the partial sums of series (4) using 3, 9, and 15 terms are compared with the graph of f. Apparently, we have pointwise convergence of FS$[f]$ to f except at the endpoints $x = \pm\pi$, and at the "jump" discontinuity at $x = 0$. The convergence also appears to be uniform on any closed interval not containing any of these three exceptional points.

The behavior of the partial sums of the Fourier series (4) near $x = 0$ and $x = \pm\pi$ is somewhat mysterious. Apparently, the "hump" in the graph of a partial sum near these points does not diminish in size but merely "moves" over closer to the exceptional points $x = 0$ and $x = \pm\pi$ as more terms are included in the partial sum. This is the *Gibbs phenomenon* and always occurs at jump discontinuities.

Let's examine the precise pointwise convergence properties of the Fourier trigonometric series. First we need some additional definitions and facts.

❖ **Piecewise Smoothness, PS$[-T, T]$.** A piecewise continuous function f on $[-T, T]$ is *piecewise smooth* if it is differentiable at all but at most a finite set of points and f' is piecewise continuous on $[-T, T]$. The set of all piecewise smooth functions on $[-T, T]$ is denoted by PS$[-T, T]$.

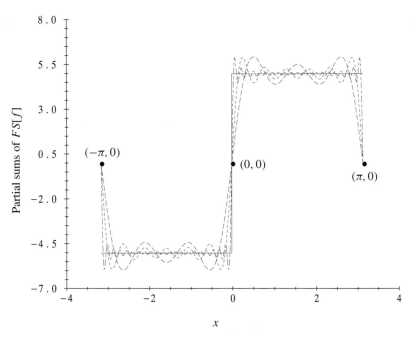

FIGURE 10.4.1 Partial sums (dashed lines) of the Fourier series in (4) for the square wave (solid): 3 terms, 9 terms, 15 terms (Example 10.4.2).

So f is in PS$[-T, T]$ if and only if both f and f' are in PC$[-T, T]$.

THEOREM 10.4.2

Properties of Piecewise Continuous Functions.

Decay of Coefficients: Suppose that f is in PC$[-T, T]$ and that FS$[f]$ is the Fourier series of f given by (2) and (3). Then $A_k \to 0$, $B_k \to 0$ as $k \to \infty$.

Integration of Periodic Functions: Suppose that f is a periodic function on $\mathbb{R}$ with period $2T$ and f is piecewise continuous over the interval $[-T, T]$. Then

☞ A similar formula appears in the verification of Theorem 6.3.1.

$$\int_{-T}^{T} f(x)\, dx = \int_{a}^{a+2T} f(s)\, ds, \quad \text{for any real } a \tag{5}$$

To show the first assertion, note that $\| \cos(k\pi x/t) \|^2 = \| \sin(k\pi x/T) \|^2 = T$ for $k > 0$, and from the Decay of Coefficients Theorem (Theorem 10.3.6) we see that $A_k \to 0$, $B_k \to 0$ as $k \to \infty$. The integration formula in (5) is treated in Problem 7.

The *periodic extension* $\tilde{f}$ of a function f in PC$[-T, T]$ to the whole real line is a periodic function of period $2T$ that has the same values as the given function f on the interval $-T < x < T$.

☞ The values of $\tilde{f}$ at odd multiples of T don't matter.

Note that $\tilde{f}$ is piecewise continuous on $\mathbb{R}$. No harm is done in leaving $\tilde{f}$ undefined at odd multiples of T since integrals involving $\tilde{f}$ are not affected by the values of $\tilde{f}$ at these points. Also, the limits $\tilde{f}(x_0^+)$ and $\tilde{f}(x_0^-)$ are not affected by the value of $\tilde{f}$ at x_0. Now we are ready to state a major result of this section.

THEOREM 10.4.3

☞ The Student Resource Manual has a proof of this result.

> Pointwise Convergence Theorem for Fourier Series. Suppose that f is piecewise smooth on $[-\pi, \pi]$ with $\tilde{f}$ the periodic extension of f into $\mathbb{R}$ with period 2π. Then for each x_0 in $\mathbb{R}$, FS[f] evaluated at x_0 converges to $\frac{1}{2}[\tilde{f}(x_0^+) + \tilde{f}(x_0^-)]$.

So Fourier series of piecewise smooth functions converge pointwise for all x in $\mathbb{R}$, and the sum at a point x_0 where $\tilde{f}$ is continuous is just $\tilde{f}(x_0)$, whereas at both endpoints $x = \pm\pi$, the sum has the same value, $\frac{1}{2}[f(\pi^-) + f(-\pi^+)]$.

EXAMPLE 10.4.3

Pointwise Convergence of Fourier Series

Let's apply the Pointwise Convergence Theorem to the function in Example 10.4.2. For $-\pi < x < \pi$ we calculate that

$$\frac{1}{2}[\tilde{f}(x_0^+) + \tilde{f}(x_0^-)] = \begin{cases} 5, & 0 < x_0 < \pi \\ 0, & x_0 = 0 \\ -5, & -\pi < x_0 < 0 \end{cases}$$

But what about the endpoints $x_0 = \pm\pi$? We see that $\tilde{f}(\pi^+) = f(-\pi^+) = -5$ and $\tilde{f}(\pi^-) = f(\pi^-) = 5$, and that $\tilde{f}(-\pi^+) = f(-\pi^+) = -5$ and $\tilde{f}(-\pi^-) = f(\pi^-) = 5$. So in our case, as it should be,

$$\frac{1}{2}[\tilde{f}(x_0^+) + \tilde{f}(x_0^-)] = 0, \quad \text{for } x_0 = \pm\pi$$

which agrees with Theorem 10.4.3 (see the endpoints in Figure 10.4.1).

Uniform Convergence of Fourier Series

If we assume a bit more smoothness for f, FS[f] will converge uniformly and not just pointwise. The function in Example 10.4.2 shows that something more than merely piecewise smoothness is required. First, however, we need an effective test for uniform convergence. The simplest is the following test.

THEOREM 10.4.4

> Weierstrass M-Test. Suppose that the functions $g_k(x)$, $k = 1, 2, \ldots$, are continuous on the interval I, and suppose that the positive constants M_k are such that $|g_k(x)| \leq M_k$, all x in I, and $\sum_{k=1}^{\infty} M_k$ converges. Then there is a continuous function $g(x)$ on I such that the series $\sum_{k=1}^{\infty} g_k(x)$ converges uniformly to g on I.

Now we have the following result.

THEOREM 10.4.5

> Differentiation of Fourier Series Uniform Convergence. **Suppose that** f **is in** the Euclidean space $PS[-\pi, \pi]$ and that $\tilde{f}$ is in $C^0(\mathbb{R})$, where $\tilde{f}$ is the periodic extension of f into $\mathbb{R}$. Then $FS[f']$ is the term-by-term derivative of $FS[f]$, and $FS[f]$ converges uniformly to $\tilde{f}$ on $\mathbb{R}$.

To show this, first note that $\tilde{f}$ will be in $C^0(\mathbb{R})$ if f is in $C^0[-\pi, \pi]$ and $f(-\pi) = f(\pi)$. The Cauchy-Schwarz Inequality of Theorem 10.2.1 applied to $A_k \cos kx + B_k \sin kx$, the k-th term in $FS[f]$, yields the inequality

$$|A_k \cos kx + B_k \sin kx| \leq (A_k^2 + B_k^2)^{1/2}, \quad \text{for all } x \text{ in } \mathbb{R}$$

So if $M_k = (A_k^2 + B_k^2)^{1/2}$ and if $\sum_{k=1}^{\infty} M_k$ converges, the Weierstrass M-Test gives the desired result. Now let's show that $\sum_{k=1}^{\infty} M_k$ converges. Since both f and f' are in $PC[-\pi, \pi]$, we can write

$$FS[f] = \frac{A_0}{2} + \sum_{k=1}^{\infty} (A_k \cos kx + B_k \sin kx)$$

☞ A'_k and B'_k are just names of the coefficients in $FS[f']$, *not* derivatives!

$$FS[f'] = \frac{A'_0}{2} + \sum_{k=1}^{\infty} (A'_k \cos kx + B'_k \sin kx)$$

where the coefficients A_k, B_k, A'_k, B'_k are given by the Euler formulas in (3). Integration by parts in the Euler formulas for A'_k and B'_k implies that

$$A'_0 = 0, \quad A'_k = kB_k, \quad B'_k = -kA_k, \quad k = 1, 2, \ldots \quad (6)$$

The Bessel inequality applied to $FS[f']$ implies that $\sum k^2(A_k^2 + B_k^2)$ converges. Using this and the Cauchy-Schwarz inequality, we obtain

$$\sum_{k=1}^{n} \frac{1}{k}[k^2 A_k^2 + k^2 B_k^2]^{1/2} \leq \left[\sum_{k=1}^{\infty} \frac{1}{k^2}\right]^{1/2} \left[\sum_{k=1}^{\infty} k^2(A_k^2 + B_k^2)\right]^{1/2}$$

The series $\sum[A_k^2 + B_k^2]^{1/2}$ converges because its partial sums form a nondecreasing sequence bounded from above. The claim that $FS[f']$ is the term by term derivative of $FS[f]$ follows from (6), and so we are done.

Decay Estimates

There is an estimate on how fast the coefficients in $FS[f]$ decay, but first a definition is needed. A function $A(k)$ is said to be $O(1/k)$ as $k \to \infty$ [read as *A is big O of* $1/k$] if there are positive constants c, K such that $|A(k)| \leq c/k$ for all $k \geq K$.

THEOREM 10.4.6

☞ Example 10.4.2 illustrates this theorem.

> Decay of Fourier Coefficients of Piecewise Smooth Functions. **Let** g **be in** $PS[-\pi, \pi]$. Then the coefficients in $FS[g]$ are $O(1/k)$ as $k \to \infty$. If $\tilde{g}$, the periodic extension of g, is not in $C^0(\mathbb{R})$, then this estimate can't be improved.

To show this, first note that g is in $PS[-\pi, \pi]$, so we can partition $[-\pi, \pi]$ with points $-\pi = x_0 < x_1 < x_2 < \cdots < x_N < x_{N+1} = \pi$ such that g is continuously differentiable on each subinterval $[x_n, x_{n+1}]$. The coefficients A_k and B_k of $FS[g]$ are given by the Fourier-Euler formulas in (3) (replacing f by g, and T by π, in those formulas). Now we turn to $FS[g']$. We can evaluate the coefficients of the sine terms in $FS[g']$ by integrating by parts as follows:

$$\frac{1}{\pi} \int_{-\pi}^{\pi} g'(x) \sin kx \, dx = \frac{1}{\pi} \sum_{n=0}^{N} \int_{x_n}^{x_{n+1}} g'(x) \sin kx \, dx$$

$$= \frac{1}{\pi} \sum_{n=0}^{N} g(x) \sin kx \Big|_{x_n^+}^{x_{n+1}^-} - \frac{k}{\pi} \sum_{n=0}^{N} \int_{x_n}^{x_{n+1}} g(x) \cos kx \, dx \qquad (7)$$

$$= G - k \cdot \frac{1}{\pi} \int_{-\pi}^{\pi} g(x) \cos kx \, dx = G - kA_k$$

Since G is bounded and the coefficients of $FS[g']$ decay to zero as $k \to \infty$, we see that the coefficients A_k of the cosine terms in $FS[g]$ are $O(1/k)$ as $k \to \infty$. A similar computation holds for the coefficients of the sine terms in $FS[g]$. We skip showing that this decay rate can't be improved.

Now let's assume that f in $PC[-\pi, \pi]$ is such that $\tilde{f}$ is in $\mathbf{C}^m(\mathbb{R})$ and that $f^{(m)}$ is in $PS[-\pi, \pi]$. Set

$$FS[f^{(j)}] = A_0^j + \sum_{k=1}^{\infty} (A_k^j \cos kx + B_k^j \sin kx), \quad \text{for } j = 0, 1, \ldots, m+1$$

where j in A_k^j and B_k^j is a superscript and *not* a power. Using the definition of the coefficients A_k^j, B_k^j in (3) and integration by parts we have the recursion relation

$$A_k^{j+1} = kB_k^j, \qquad B_k^{j+1} = -kA_k^j, \qquad k > 0, \quad j = 0, \ldots, m \qquad (8)$$

Now let's assume in addition just a bit more smoothness, namely, that $f^{(m+1)}$ is *piecewise smooth*. Taking $g = f^{(m+1)}$ we conclude from the previous result that the coefficients in $FS[f^{(m+1)}]$ are $O(1/k)$ as $k \to \infty$. Putting this fact together with (8) produces the following estimate.

THEOREM 10.4.7

☞ These estimates are a good check when calculating Fourier coefficients. Take a look at Example 10.4.1, where $f = |x|$, $m = 0$.

Decay Theorem for Fourier Trigonometric Coefficients. Let f belong to $PC[-\pi, \pi]$, $\tilde{f}$ belong to $\mathbf{C}^m(\mathbb{R})$, and $f^{(m+1)}$ belong to $PS[-\pi, \pi]$ for some $m = 0, 1, 2, \ldots$. If A_k, B_k are the Fourier coefficients of f given by (3), then

$$A_k = O(1/k^{m+2}), \qquad B_k = O(1/k^{m+2}), \quad \text{as } k \to \infty \qquad (9)$$

Also, if $\tilde{f}$ is *not* in $\mathbf{C}^{m+1}(\mathbb{R})$, the estimates in (9) can't be improved.

The decay estimates in (9) follow from the formulas given in (8) because

$$A_k^{j+1} = kB_k^j = -k^2 A_k^{j-1} = \cdots = -(k)^{j+2} A_k, \quad j \text{ odd}$$

Similar formulas can be obtained if j is even. The coefficients of B_k can be handled in the same way. We skip showing that these are the best possible estimates.

EXAMPLE 10.4.4

Decay of Fourier Coefficients for a Function in $PC[-\pi, \pi]$
The "square wave" of Example 10.4.2 does not come under Theorem 10.4.7 since its periodic extension is not continuous. The square wave *is* in $PS[-\pi, \pi]$, and from the first decay theorem (Theorem 10.4.6) we see that the Fourier coefficients decay precisely like $O(1/k)$, which we see from (4) is the case.

PROBLEMS

1. Use trigonometric identities, common sense, or previous results to find the Fourier series on $[-\pi, \pi]$ of each function below *without* using the Euler formulas in (3).

    (a) $5 - 4\cos 6x - 7\sin 3x - \sin x$ **(b)** $\cos^2 7x + (\sin \frac{1}{2}x)(\cos \frac{1}{2}x) - 2\sin x \cos 2x$

    (c) $\sin^3 x$

2. Calculate $FS[f]$ for the functions below, each assumed to have domain $[-\pi, \pi]$.

    (a) $2x - 2$ **(b)** x^2 **(c)** $a + bx + cx^2$

    (d) $\sin \pi x$ **(e)** $|x| + e^x$ **(f)** $|x(x^2 - 1)|$

    (g) *(Rectangular Pulse).* For a positive number $B \leq \pi$,

    $$f(x) = \begin{cases} A, & |x| < B \\ 0, & B \leq |x| \leq \pi \end{cases}$$

    (h) *(Alternating Pulse).* For a positive number $B \leq \pi$,

    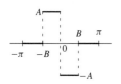

    $$f(x) = \begin{cases} 0, & B \leq |x| \leq \pi \\ A, & -B < x < 0 \\ -A, & 0 < x < B \end{cases}$$

www 3. Compute the Fourier series of each of the following functions on the indicated interval. As you will see, the Fourier series of a fixed function is highly dependent on the interval of the definition of the function.

    (a) $f(x) = \sin x$, $-\pi \leq x \leq \pi$

    (b) $f(x) = \sin x$, $-\pi/2 \leq x \leq \pi/2$ [*Hint:* Use (2), (3) with $T = \pi/2$.]

    (c) $f(x) = \sin x$, $-3\pi/2 \leq x \leq 3\pi/2$ [*Hint:* Use (2), (3) with $T = 3\pi/2$.]

4. Let $f(x)$ be the function $|x|$ on $[-\pi, \pi]$.

    (a) Using $FS[f]$ and Theorem 10.4.3 with $x_0 = 0$, show that

    $$\frac{\pi^2}{8} = \sum_{k=0}^{\infty} \frac{1}{(2k+1)^2}$$

    (b) Discuss the convergence properties of $FS[f]$ on all of $\mathbb{R}$.

5. (*Not All Trigonometric Series Are Fourier Series*). Show that

$$\sum_{1}^{\infty} \frac{1}{n^{1/4}} \cos nx, \qquad \sum_{1}^{\infty} (\sin n) \sin nx, \qquad \sum_{2}^{\infty} \frac{1}{\ln(n)} \sin nx$$

are *not* the Fourier series of any functions in PC$[-\pi, \pi]$. [*Hint*: Use the Parseval Relation (Theorem 10.3.6), where Φ is given by (1) with $T = \pi$.]

6. Evaluate $\lim_{n \to \infty} \int_{-\pi}^{\pi} e^{\sin x} (x^5 - 7x + 1)^{52} \cos nx \, dx$. [*Hint*: Relate the integral to a Fourier coefficient of a portion of the integrand.]

7. (*Integrating a Periodic Function*). Show that $\int_{-T}^{T} f(x) \, dx = \int_{a}^{a+2T} f(s) \, ds$ for all real numbers a if f is periodic with period $2T$ and piecewise continuous on $[-T, T]$. [*Hint*: Break up the integral on the right and change the integration variable.]

8. (*Lagrange's Identity*). Show that

$$\frac{1}{2} + \sum_{k=1}^{n} \cos ks = \frac{\sin(n + \frac{1}{2})s}{2 \sin(s/2)}$$

[*Hint*: Use L'Hôpital's Rule if $s = 2m\pi$, m an integer. For other s, use $2 \sin(s/2) \cos ks = \sin[k + (1/2)]s - \sin[k - (1/2)s]$.]

9. (*Bases for PC$[-\pi, \pi]$*).

  (a) Show that $\{1, \cos x, \ldots, \cos nx, \ldots\}$ is *not* a basis of PC$[-\pi, \pi]$.

  (b) Show that $\{1, \cos x, \ldots, \cos nx, \ldots\}$ *is* a basis of the subspace of all even functions in PC$[-\pi, \pi]$.

  (c) Formulate and prove similar statements for $\{\sin x, \ldots, \sin nx, \ldots\}$.

10. Show that $\|f - S_n\| \le \|f - S_m\|$ if $n \ge m$, where S_n and S_m are the corresponding partial sums of FS$[f]$.

10.5 Half-Range and Exponential Fourier Series

Half-Range Expansions

There are some standard ways of constructing orthogonal sets (and even bases) for the Euclidean space PC$[0, L]$ with the standard scalar product. We begin by finding bases for PC$[0, L]$ that consist only of sine functions or only of cosine functions.

Suppose that f in PC$[0, L]$ is given and put

$$f_{\text{even}}(x) = \begin{cases} f(x), & 0 \le x \le L \\ \\ f(-x), & -L \le x \le 0 \end{cases}$$

☞ Even and odd functions were defined on page 540.

$$f_{\text{odd}}(x) = \begin{cases} f(x), & 0 < x \le L \\ \\ -f(-x), & -L < x < 0 \end{cases}$$

Observe that f_{odd} is odd about $x = 0$, f_{even} is even about $x = 0$, and f_{odd} and f_{even} are in PC$[-L, L]$. The functions f_{odd} and f_{even} are called the *odd* and *even extensions*,[3] respectively, of f to the interval $[-L, L]$. Observe that FS$[f_{\text{odd}}]$ contains only sine functions, and FS$[f_{\text{even}}]$ contains only cosine functions. We will define the *Fourier sine series* and *Fourier cosine series of f*, denoted by FSS$[f]$ and FCS$[f]$:

$$\text{FSS}[f] = \text{FS}[f_{\text{odd}}], \qquad \text{FCS}[f] = \text{FS}[f_{\text{even}}] \tag{1}$$

So, for example, for f in PC$[0, L]$, we have

$$\text{FSS}[f] = \sum_{k=1}^{\infty} b_k \sin \frac{k\pi x}{L}, \qquad b_k = \frac{2}{L} \int_0^L f(x) \sin \frac{k\pi x}{L}\, dx, \qquad k = 1, 2, \ldots \tag{2}$$

Our knowledge of the convergence properties of Fourier series tells us much about Fourier sine series. In particular, the functions

$$\Phi_s = \{\sin(k\pi x/L) : k = 1, 2, \ldots\}$$

form an orthogonal set in PC$[0, L]$ since for all $k \neq m$,

$$0 = \int_{-L}^{L} \sin \frac{k\pi x}{L} \sin \frac{m\pi x}{L}\, dx = 2 \int_0^L \sin \frac{k\pi x}{L} \sin \frac{m\pi x}{L}\, dx$$

Actually, Φ_s is a basis of PC$[0, L]$ since the Fourier sine series of any f in PC$[0, L]$ converges in the mean to f. The convergence properties of FSS$[f]$ can be traced back to those criteria for the Fourier series FS$[f_{\text{odd}}]$. Similarly, the Fourier cosine series of a function f in PC$[0, L]$ is given by

☞ The term $a_0/2$ is used in FCS$[f]$ instead of a_0 just so the coefficient formula is valid for $k = 0$ as well as $k > 0$.

$$\text{FCS}[f] = \frac{a_0}{2} + \sum_{k=1}^{\infty} a_k \cos \frac{k\pi x}{L}$$

$$a_k = \frac{2}{L} \int_0^L f(x) \cos \frac{k\pi x}{L}\, dx, \qquad k = 0, 1, \ldots \tag{3}$$

Again, the convergence properties of FCS$[f]$ are known from the behavior of FS$[f_{\text{even}}]$. In particular, we can verify that the set

$$\Phi_C = \{\cos(k\pi x/L) : k = 0, 1, 2, \ldots\}$$

is an orthogonal set in PC$[0, L]$ and is a basis for that space.

The series FSS$[f]$ and FCS$[f]$ in (2) for f in PC$[0, L]$ are frequently referred to as *half-range expansions* in the applications.

EXAMPLE 10.5.1

Half-Range Versus Full-Range Expansions

Consider the four functions f, g, f_{odd}, and f_{even} defined below:

$$f(x) = e^x, \quad 0 \leq x \leq \pi \qquad\qquad g(x) = e^x, \quad -\pi \leq x \leq \pi$$

$$f_{\text{odd}}(x) = \begin{cases} -e^{-x}, & -\pi \leq x < 0 \\ e^x, & 0 < x \leq \pi \end{cases} \qquad f_{\text{even}}(x) = \begin{cases} e^{-x}, & -\pi \leq x \leq 0 \\ e^x, & 0 \leq x \leq \pi \end{cases}$$

---

[3]Note that f_{odd} is not defined at $x = 0$. This will have no effect at all on what follows. In general, the actual value of a piecewise continuous function at a discontinuity is immaterial when calculating Fourier coefficients.

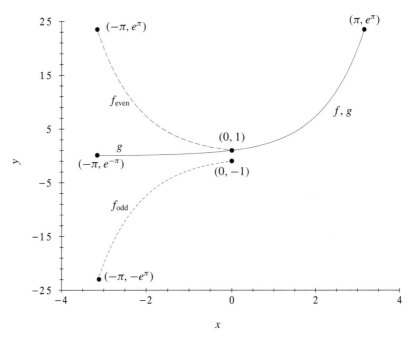

FIGURE 10.5.1 Even and odd extensions (Example 10.5.1). The solid line is e^x. The long-dashed and the short-dashed lines are, respectively, the even and odd extensions of $f(x) = e^x$, $0 \le x \le \pi$.

Note that f_{odd} is the *odd extension* and f_{even} is the *even extension* of f to $(-\pi, \pi)$. See Figure 10.5.1 for a sketch of the graphs.

Notice that the four functions have the same values on $(0, \pi)$. Carrying out the necessary calculations to find the Fourier series of g, f_{odd}, and f_{even}, we obtain

☞ These calculations are omitted.

$$\text{FS}[g] = \frac{2\sinh\pi}{\pi} \sum_{0}^{\infty} (-1)^k \frac{\cos kx - k\sin kx}{1+k^2}$$

$$\text{FS}[f_{\text{odd}}] = \frac{2}{\pi} \sum_{1}^{\infty} \frac{k[1-(-1)^k e^\pi]}{1+k^2} \sin kx$$

$$\text{FS}[f_{\text{even}}] = \frac{2}{\pi} \sum_{0}^{\infty} \frac{e^\pi[(-1)^k - 1]}{1+k^2} \cos kx$$

Now from the definition of half-range expansions in (1) we know what $\text{FSS}[f]$ and $\text{FCS}[f]$ converge to on the entire real line. From the convergence theorem (Theorem 10.4.3) we see that $\text{FCS}[f]$ and $\text{FSS}[f]$ both converge to $f(x) = e^x$ on $0 < x < \pi$, but on $-\pi < x < 0$, $\text{FCS}[f]$ and $\text{FSS}[f]$ converge to distinct functions, neither of which is e^x. Indeed $\text{FCS}[f]$ converges to e^{-x} on $-\pi \le x \le 0$, while $\text{FSS}[f]$ converges to $-e^{-x}$ on $-\pi < x < 0$. Looking at the periodic extensions $\tilde{f}_{\text{odd}}$ and $\tilde{f}_{\text{even}}$ we discover (again from Theorem 10.4.3) that

$$\text{FSS}[f](0) = \text{FSS}[f](\pi) = \text{FSS}[f](-\pi) = 0$$

$$\text{FCS}[f](x) = f_{\text{even}}(x), \qquad |x| \le \pi$$

Fourier Exponential Series

If the functions in PC$[-\pi, \pi]$ are complex-valued, we can use the orthogonal set

$$\Psi = \{e^{ikx} : k = 0, \pm 1, \pm 2, \ldots\}$$

Recall that the scalar product in PC$[-\pi, \pi]$ is $\langle f, g \rangle = \int_{-\pi}^{\pi} f\overline{g}\,dx$, where $\overline{g}$ denotes the complex conjugate of g. Note that $\|e^{ikx}\|^2 = \int_{-\pi}^{\pi} e^{ikx}e^{-ikx}\,dx = 2\pi$, for all k. The Fourier series of f in PC$[-\pi, \pi]$ with respect to Ψ is the *Fourier exponential series*

$$\mathrm{FS}[f] = \sum_{n=-\infty}^{\infty} c_n e^{inx}$$

$$c_n = \frac{\langle f, e^{inx} \rangle}{\|e^{inx}\|^2} = \frac{1}{2\pi} \int_{-\pi}^{\pi} f(x)e^{-inx}\,dx, \qquad n = 0, \pm 1, \pm 2, \ldots \tag{4}$$

Leaving aside the question of ordering, there are advantages to grouping the n-th and the $(-n)$-th terms together to obtain

$$\mathrm{FS}[f] = c_0 + \sum_{n=1}^{\infty} (c_{-n}e^{-inx} + c_n e^{inx}) \tag{5}$$

but it is customary to write FS$[f]$ in the two-sided form (4).

Note that if f in PC$[-\pi, \pi]$ happens to be real-valued, it still can be expanded in a Fourier exponential series. In that case we have

$$c_{-n} = \overline{c}_n \quad \text{and} \quad c_n = \frac{1}{2}(A_n - iB_n) \quad \text{for all } n > 0 \tag{6}$$

where A_n and B_n are the respective Fourier cosine and sine coefficients in the real Fourier series of f. So the series (5) reduces to the Fourier trigonometric series FS$[f]$ given in the last section. We have intentionally used the same symbol, FS$[f]$, to denote both the real and complex forms of Fourier series. There is little risk of confusion in this practice because of the special relation between the orthogonal sets Φ and Ψ used in these Fourier series.

EXAMPLE 10.5.2

A Fourier Exponential Series

Let's find the exponential form of FS$[f]$ when $f(x) = x$, $|x| \le \pi$. We have $c_0 = 0$, and for $n \ne 0$, $c_n = (2\pi)^{-1} \int_{-\pi}^{\pi} xe^{-inx}\,dx = (i/n)\cos n\pi = i(-1)^n/n$. So

$$\mathrm{FS}[f] = i\sum_{n=1}^{\infty} \left[\frac{(-1)^n}{n}e^{inx} - \frac{(-1)^n}{n}e^{-inx}\right]$$

Recalling that $e^{i\theta} = \cos\theta + i\sin\theta$, we also see that

$$\mathrm{FS}[f] = 2\sum_{n=1}^{\infty} \frac{(-1)^{n+1}}{n}\sin nx$$

which we could have obtained by expanding f directly into a Fourier series over the trigonometric set Φ (see Example 10.3.2).

EXAMPLE 10.5.3

Another Fourier Exponential Series

Let's compute FS[f] for $f(x) = e^{i\omega x}$, $|x| \leq \pi$, where ω is *not* an integer:

$$c_n = \frac{1}{2\pi} \int_{-\pi}^{\pi} e^{i\omega x} e^{-inx}\, dx = \frac{1}{2\pi} \frac{1}{i(\omega - n)} e^{i(\omega - n)x}\Big|_{-\pi}^{+\pi} = \frac{\sin(\omega - n)\pi}{(\omega - n)\pi}$$

So we have

$$\text{FS}[f] = \frac{\sin \omega\pi}{\omega\pi} + \sum_{n=1}^{\infty} \left[\frac{\sin(\omega - n)\pi}{(\omega - n)\pi} e^{inx} + \frac{\sin(\omega + n)\pi}{(\omega + n)\pi} e^{-inx} \right]$$

Using the properties of Fourier series developed in Sections 10.2 and 10.4, we can show that Ψ is a basis for the complex linear space PC$[-\pi, \pi]$, and that the Pointwise and Uniform Convergence Theorems (Theorems 10.4.3 and 10.4.5) hold for FS[f] when f is complex-valued.

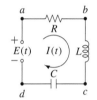

☞ This ODE was derived in Section 4.4.

Application to an *RLC* Circuit

Consider the simple *RLC* circuit shown in the margin where R, L, and C are positive constants. The charge $q(t)$ on the capacitor satisfies the ODE

$$Lq''(t) + Rq'(t) + \frac{1}{C}q(t) = E(t) \tag{7}$$

Although we have already solved such ODEs for arbitrary piecewise continuous input voltages $E(t)$, the Fourier series approach is more natural if $E(t)$ is periodic and we are searching for a periodic solution. Suppose, then, that $E(t)$ has period 2π and that the restriction of $E(t)$ to $[-\pi, \pi]$ is piecewise continuous.

We want to find all functions $q(t)$ of period 2π for which

$$q \text{ is in } C^1(\mathbb{R}), \qquad q'' \text{ is in PC}[-\pi, \pi] \tag{8a}$$

$$Lq'' + Rq' + \frac{1}{C}q - E = 0 \qquad \begin{cases} \text{on } [-\pi, \pi], \text{ except possibly} \\ \text{for a finite number of points} \end{cases} \tag{8b}$$

Suppose that ODE (7) has a solution q meeting these requirements. First put

$$\text{FS}[q] = \sum_{k=-\infty}^{\infty} c_k e^{ikt} \quad \text{and} \quad \text{FS}[E] = \sum_{k=-\infty}^{\infty} b_k e^{ikt} \tag{9}$$

where the c_k are to be found and the b_k are known since $E(t)$ is given. Using the smoothness conditions of (8a), (8b) and term-by-term differentiation, we have

$$\text{FS}[q'] = \sum ikc_k e^{ikt}, \qquad \text{FS}[q''] = \sum (ik)^2 c_k e^{ikt} \tag{10}$$

Inserting the Fourier series for q, E, q', and q'' from (9) and (10) into (8b), we have

$$\sum [P(ik)c_k - b_k]e^{ikt} = 0, \quad \text{where} \quad P(x) = Lx^2 + Rx + \frac{1}{C} \tag{11}$$

☞ $P(ik) \neq 0$ because
$P(x)$ does *not* have pure
imaginary roots.

Since $\{e^{ikt} : k = 0, \pm 1, \pm 2, \ldots\}$ is a basis, it follows from (11) that $P(ik)c_k - b_k = 0$, $k = 0, \pm 1, \ldots$. But $P(ik) \neq 0$, $k = 0, \pm 1, \ldots$, so $c_k = b_k/P(ik)$ for all k. We have

$$\text{FS}[q] = \sum_{k=-\infty}^{\infty} \frac{1}{P(ik)} b_k e^{ikt} \tag{12}$$

We omit showing that the series in (12) converges uniformly to a function $q(t)$ satisfying (8a). By construction, the Fourier series of $Lq'' + Rq' + (1/c)q - E$ is the same as the Fourier series of the zero function. By the Totality Theorem (Theorem 10.3.5), we see that $Lq'' + Rq' + (1/C)q - E = 0$ on $[-\pi, \pi]$, except possibly at a finite number of points. So (8b) holds and we are done.

☞ Actually, this was
first shown in Section 4.3
for sinusoidal inputs, and
in Problem 9 of
Section 4.3 for general
periodic inputs.

We have shown that the *RLC* circuit has a unique periodic response $q(t)$ to a periodic input voltage $E(t)$. The response $q(t)$ is known as a *periodic forced oscillation*. So *every* solution of ODE (7) is the superposition of damped exponentials (the *transients*) and the steady-state periodic solution just constructed.

PROBLEMS

1. (*Fourier Sine and Cosine Series*). Find the Fourier sine series, FSS[f], of each of the following functions. Then find the Fourier cosine series, FCS[f].

   (a) $f(x) = 1, \quad 0 < x < \pi$ (b) $f(x) = \sin x, \quad 0 \leq x \leq \pi$

2. (*Fourier Exponential Series*). Find the Fourier exponential series of $f(x)$, $|x| \leq \pi$.

   (a) $f(x) = \cos x$ (b) $f(x) = x^2$

   (c) $f(x) = |x| + ix$ [*Hint*: Show that FS$[|x|] = (\pi/2) - \pi^{-1} \sum_{k=1}^{\infty} (e^{i2kx} + e^{-i2kx})/(2k^2)$.]

www 3. (*Another Kind of Fourier Sine Series*). Suppose that f is in PC[0, c], $f(x) = f(c - x)$ for all x, $c/2 < x < c$ (i.e., f is even about $x = c/2$).

   (a) Verify that FSS[f] contains only terms of the form $\sin[(2k + 1)\pi x/c]$.

   (b) Reflect the function $\sin x$, $0 \leq x \leq \pi/4$, about $x = \pi/4$ to obtain a function f defined on $[0, \pi/2]$ that is even about $x = \pi/4$. Use part (a) to find FSS[f].

4. (*Another Kind of Fourier Cosine Series*). Suppose that f is in PC[0, c], $f(x) = -f(c - x)$, $c/2 < x < c$ (i.e., f is "odd" with respect to $x = c/2$).

   (a) Verify that FCS[f] contains only terms of the form $\cos[(2k + 1)\pi x/c]$.

   (b) Use part (a) to find FCS[f] for the function f defined by reflecting x^2, $0 \leq x < 1$, "oddly" through $x = 1$.

5. (*Electrical Circuit*). Find the steady-state charge in the capacitor in an *RLC* loop if $R = 10 \, \Omega$, $L = 0.5$ H, $C = 10^{-4}$ F, and $E(t) = \pi \, \text{trw}(t, 100, 2\pi)$ measured in volts.

6. (*Driven Hooke's Law Spring*). In a spring-mass system, let the spring constant be 1.01 newtons/meter and the weight of the mass be 1 kilogram. Suppose that while in motion, the mass is acted on by viscous damping with coefficient equal to 0.2, and also suppose that the mass is driven by the force (measured in newtons) described by sqw$(t, 50, 2\pi)$. Find the steady-state motion of the system.

10.6 Sturm-Liouville Problems

The results below are important for the Method of Separation of Variables for solving boundary/initial value problems for PDEs. Orthogonality and orthogonal series play a key role throughout.

Suppose that M is the operator defined on $\mathbf{C}^2[a, b]$ by

$$M[y] = \frac{1}{\rho(x)} \left\{ \frac{d}{dx} \left[p(x) \frac{dy}{dx} \right] + q(x)y \right\} \tag{1}$$

where

$$\rho \text{ is in } \mathbf{C}^0[a, b], \qquad \rho \neq 0 \text{ on } (a, b), \qquad q \text{ is in } \mathbf{C}^0[a, b], \tag{2}$$
$$p \neq 0 \text{ on } (a, b), \qquad p \text{ is in } \mathbf{C}^1[a, b]$$

The operator M in (1) is not as special as it seems. In fact, the general second-order linear differential operator $P(D) = a_2(x)D^2 + a_1(x)D + a_0(x)$, where $a_i(x)$ is in $\mathbf{C}^0[a, b]$, $i = 0, 1, 2$, and $a_2(x) \neq 0$ on $[a, b]$, can be written in the form (1) if we let

$$p(x) = e^{\alpha(x)}, \qquad q(x) = \frac{a_0(x)}{a_2(x)} e^{\alpha(x)}, \qquad \rho(x) = \frac{e^{\alpha(x)}}{a_2(x)} \tag{3}$$

where $\alpha(x) = \int^x [a_1(x)/a_2(x)] \, dx$.

Using the *weighted scalar product on* $\mathbf{C}^0[a, b]$ *with weight factor* ρ, we see that $\langle f, g \rangle = \int_a^b \rho(x) f(x) \overline{g}(x) \, dx$. Then integration by parts yields the relation

$$\langle Mu, v \rangle = p(u'v - uv') \Big|_a^b + \langle u, Mv \rangle, \quad \text{for all } u, v \text{ in } \mathbf{C}^2[a, b] \tag{4}$$

☞ Recall that
$\langle Mu, v \rangle =$
$\int_a^b \rho \cdot Mu \cdot v \, dx$.

Now let's define an operator L that acts like M in (1) but whose domain Dom (L) is a restriction of $\mathbf{C}^2[a, b]$ defined by conditions at the endpoints $x = a$ and $x = b$. We are interested in symmetric operators.

❖ **Symmetric Operators.** An operator L defined on a subspace of a Euclidean space E and taking values in E is said to be *symmetric* if

$$\langle Lu, v \rangle = \langle u, Lv \rangle, \quad \text{for all } u, v \text{ in Dom } (L)$$

☞ The terminology of operators was introduced in Section 3.7.

Now let's look more carefully at appropriate boundary conditions for our differential operator M.

❖ **Boundary Conditions.** For u in $\mathbf{C}^2[a, b]$, the conditions

$$\alpha u'(a) + \beta u(a) = 0, \qquad \gamma u'(b) + \delta u(b) = 0 \tag{5}$$

are called *separated* (or *unmixed*) *boundary conditions* at $x = a$ and $x = b$, respectively, where the constants α, β, γ, and δ are such that $\alpha^2 + \beta^2 \neq 0$ and $\gamma^2 + \delta^2 \neq 0$. The conditions

$$u(a) = u(b), \qquad u'(a) = u'(b) \tag{6}$$

are called *periodic boundary conditions*.

The following results are implied by identity (4).

THEOREM 10.6.1

> Symmetric Operator Theorem. Suppose that L is an operator with action given by M in (1) where the domain Dom (L) is one of two types:
>
> - Dom $(L) = \{u$ in $\mathbf{C}^2[a, b] : u$ satisfies a separated boundary condition at each endpoint where $p \neq 0\}$.
> - Dom $(L) = \{u$ in $\mathbf{C}^2[a, b] : u(a) = u(b), u'(a) = u'(b)$ and $p(a) = p(b)\}$.
>
> Then L is a *symmetric operator* in $\mathbf{C}^0[a, b]$ under the weighted scalar product $\langle f, g \rangle = \int_a^b \rho(x) f(x) \overline{g(x)}\, dx$.

Symmetric linear operators such as L in the theorem above are defined on a subspace Dom (L) of $\mathbf{C}^0[a, b]$ and take values in $\mathbf{C}^0[a, b]$. The eigenvalues and eigenspaces of such operators play an important role in applied mathematics. The task of finding all the eigenvalues and the corresponding eigenspaces of such operators L is called a *Sturm-Liouville problem*.[4] If the conditions in (2) on the operator M are satisfied, then the Sturm-Liouville problem for L is said to be *regular*. We have the following definition for the eigen-elements of any operator L:

Jacques Charles François
Sturm

❖ **Eigenvalues, Eigenspaces**. Suppose that S is a subspace of a linear space V and that $L : S \to V$ is a linear operator. A scalar λ is an *eigenvalue* for L if there is a nonzero v in S such that $Lv = \lambda v$. The vector v is an *eigenvector* of L corresponding to the eigenvalue λ. For any eigenvalue λ of L, the set V_λ of all vectors in V that satisfy the equation $Lv = \lambda v$ is the *eigenspace* of L corresponding to λ. [Eigenvalues and eigenvectors can be complex-valued if V is a complex linear space.]

A similar definition for operators on $\mathbb{R}^n$ was given in Section 7.4. Since our operators act on functions in $\mathbf{C}^2[a, b]$, their eigenvectors are often called *eigenfunctions*. Notice that for any eigenvalue λ, the eigenspace V_λ contains the eigenvectors for L corresponding to λ. (The zero vector can never be an eigenvector of any linear operator.) V_λ is comprised of all the eigenvectors of L corresponding to λ with the zero vector thrown in. Two eigenspaces, V_λ and V_μ, of a linear operator with $\lambda \neq \mu$ can only have the zero vector in common.

Eigenspaces and eigenvalues for our symmetric differential operators have several important properties, which we list below.

Joseph Liouville

---

[4] The Swiss mathematician Jacques Charles François Sturm (1803–1855) and the French mathematician Joseph Liouville (1809–1882) jointly introduced these ideas to resolve problems concerning the solutions of the partial differential equations of wave motion and thermal diffusion.

THEOREM 10.6.2

Eigenspace Properties. Suppose that L is any differential operator of the type defined in the Symmetric Operator Theorem (Theorem 10.6.1). Then the following properties hold:

Orthogonality of Eigenspaces: Any two eigenspaces of L, V_λ, and V_μ corresponding to two distinct eigenvalues λ and μ must be orthogonal; that is, any element of V_λ is orthogonal to any element of V_μ under the scalar product $\langle f, g \rangle = \int_a^b \rho(x) f(x) \overline{g(x)} \, dx$.

Simplicity of Eigenspaces: If a separated boundary condition is used in defining Dom (L), then all eigenspaces of L are one-dimensional.

☞ The maximal dimension is two because $Lv = \lambda v$ is a second-order ODE.

The orthogonality of eigenspaces follows immediately from the symmetry relation $\langle Lu, v \rangle = \langle u, Lv \rangle$ for all u, v in Dom (L). The only choices for the dimension of any eigenspace are one or two. Now if the dimension of the eigenspace corresponding to an eigenvalue λ is two, *all* solutions of the homogeneous equation $(p(x)y')' + q(x)y = \lambda \rho(x)y$ must satisfy the endpoint conditions. This is impossible if Dom (L) is defined with a separated condition at either $x = a$ or $x = b$. So the dimension of the eigenspace is one, as asserted.

THEOREM 10.6.3

Nonpositivity of Eigenvalues. Suppose that L is any differential operator of the type defined in the Symmetric Operator Theorem. Suppose that $q(x)$ is nonpositive on $[a, b]$, and $\alpha\beta \leq 0$, and $\gamma\delta \geq 0$ [if either of the separated conditions in (5) is used to define Dom (L)]. Then the eigenvalues of L are all nonpositive.

To show this use integration by parts to show that $\langle Lu, u \rangle \leq 0$ for all u in Dom (L). The assertion follows from this inequality. Here are two examples.

EXAMPLE 10.6.1

Sturm-Liouville Problem: Separated Conditions
Suppose that the operator L has action $Lu = u''$ and domain Dom $(L) = \{u$ in $\mathbf{C}^2[0, T] : u(0) = u(T) = 0\}$. Then L is a symmetric differential operator with $a = 0$, $b = T$, $\rho(x) = p(x) = 1$, $q(x) = 0$ on $[0, T]$, and the separated boundary conditions in (5) with $\alpha = 0$, $\beta = 1$, $\gamma = 0$, and $\delta = 1$. So all the eigenvalues of L are nonpositive, and all eigenspaces are one-dimensional and mutually orthogonal with the scalar product $\langle f, g \rangle = \int_0^T f(x)g(x) \, dx$.

To find the eigenvalues and eigenspaces, first try $\lambda = 0$ in the eigenvalue equation $u'' = \lambda u$. The general solution of $u'' = 0$ is $u = Ax + B$ for arbitrary constants A, B. The condition $u(0) = u(T) = 0$ in this case yields that $A = B = 0$, so $\lambda = 0$ can't be an eigenvalue of L.

Now try $\lambda = -k^2$ for some positive constant k. The general solution of the eigenvalue equation $u'' = -k^2 u$ is $u = A \cos kx + B \sin kx$. The conditions $u(0) = u(T) = 0$ imply that $A = 0$ and $k = n\pi/T$. So L has the eigenvalues $\lambda_n = -(n\pi/T)^2$, with corresponding eigenspaces spanned by the respective eigenfunctions $u_n = \sin n\pi x/T$, $n = 1, 2, \ldots$.

EXAMPLE 10.6.2

Sturm-Liouville Problem: Periodic Boundary Conditions
As we saw in Section 10.4,

$$\Phi_T = \{1, \cos \pi x/T, \sin \pi x/T, \ldots, \cos n\pi x/T, \sin n\pi x/T, \ldots\}$$

is an orthogonal set in the Euclidean space $\mathbf{C}^0[-\pi, \pi]$ under the scalar product $\langle f, g \rangle = \int_{-\pi}^{\pi} fg\, dx$. The elements of this set Φ_T turn up as eigenfunctions for the Sturm-Liouville problem with periodic boundary conditions

$$y'' = \lambda y, \qquad y(-T) = y(T), \quad y'(-T) = y'(T) \tag{7}$$

Indeed, associated with (7) is the symmetric operator $Ly = y''$ with domain defined by Dom $(L) = \{y$ in $\mathbf{C}^2[-T, T] : y(-T) = y(T), \ y'(-T) = y'(T)\}$. From Theorem 10.6.3 we see that the eigenvalues of L are nonpositive. Note that $\lambda = 0$ is an eigenvalue and the corresponding eigenspace V_0 is spanned by the constant function 1. The other eigenvalues are $\lambda_n = -(n\pi/T)^2$, $n = 1, 2, \ldots$; each corresponding eigenspace V_{λ_n} is two-dimensional and is spanned by the orthogonal set $\{\cos(n\pi x/T),$ $\sin(n\pi x/T)\}$.

We end now with a fundamental property of Sturm-Liouville systems.

THEOREM 10.6.4

Sturm-Liouville Theorem (in the Regular Case). **Suppose that the operator L has action given by M in (1) and (2) and domain in $\mathbf{C}^2[a, b]$ characterized by the separated conditions in (5) at both $x = a$ and $x = b$. Then:**

- The eigenvalues of L form a sequence λ_n, $n = 1, 2, \ldots$, with $|\lambda_n| \to \infty$.
- The corresponding eigenspaces V_{λ_n} are one-dimensional and mutually orthogonal under the scalar product $\langle f, g \rangle = \int_a^b \rho fg\, dx$.
- If Φ is any set consisting of precisely one eigenfunction from each V_{λ_n}, then Φ is a basis for PC$[a, b]$.
- The Fourier series over Φ of any function u in Dom (L) converges uniformly to u on $[a, b]$.

EXAMPLE 10.6.3

Basis for Piecewise Continuous Functions
The Sturm-Liouville system in Example 10.6.1 is in the regular case. So the Sturm-Liouville Theorem (Theorem 10.6.4) says that the set $\Phi = \{\sin n\pi x/T : n = 1, 2, \ldots\}$ is a basis for PC$[0, T]$ under the standard scalar product. For example, look at the quadratic function $v = -(x - T/2)^2 + T^2/4$. Since v is in the domain of the associated operator L, Theorem 10.6.4 implies that the FSS$[v]$ converges uniformly to v on $[0, T]$.

We treat singular (i.e., the operator L is *not* regular) Sturm-Liouville problems in Sections 10.9 and 11.8, where they arise in separating the variables for PDEs in non-Cartesian coordinates.

PROBLEMS

www **1.** In each of the regular Sturm-Liouville systems below identify the operator L associated with it. Find the eigenvalues and eigenspaces of L, and state the orthogonality and basis properties of the eigenfunctions.

(a) $y'' = \lambda y$; $y(0) = 0$, $y(\pi/2) = 0$

(b) $y'' = \lambda y$; $y(0) = 0$, $y'(T) = 0$

(c) $y'' = \lambda y$; $y'(0) = 0$, $y'(T) = 0$

(d) $y'' = \lambda y$; $y(-T) = y(T)$, $y'(-T) = y'(T)$

(e) $y'' = \lambda y$; $y(0) = 0$, $y(\pi) + y'(\pi) = 0$

(f) $y'' - 4y' + 4y = \lambda y$; $y(0) = 0$, $y(\pi) = 0$

10.7 Separation of Variables

The Method of Separation of Variables is used to construct solution formulas for a wide class of boundary/initial value problems for linear PDEs that occur surprisingly often in the applications. We will build on the theory and examples of earlier sections to outline this general method. We will also extend the method to solve boundary/initial value problems involving driven linear PDEs.

The Method of Separation of Variables

Although we will work out the method only in the context of a specific problem, the method itself applies to many other problems (see Sections 10.8, 10.9, and 11.8).

Consider the transverse motion of a taut, flexible string of length L introduced in Section 10.1. Assuming that the endpoints are held fixed, we saw that the deflection $u(x, t)$ of the string at the point x and time t is the unique solution of the boundary/initial value problem given below.

Guitar String Problem

☞ Recall that G^* consists of G and its boundary.

For the region $G = \{(x, t) : 0 < x < L, t > 0\}$ we seek a function u that is both in $\mathbf{C}^2(G)$ and in $\mathbf{C}^1(G^*)$, and such that

$$
\begin{array}{lll}
\text{(PDE)} & u_{tt} - c^2 u_{xx} = 0 & \text{in } G \\
\text{(BC)} & u(0, t) = 0, \quad u(L, t) = 0, \quad t \geq 0 & \quad\quad (1) \\
\text{(IC)} & u(x, 0) = f(x), \quad u_t(x, 0) = g(x), \quad 0 \leq x \leq L
\end{array}
$$

where c is a positive constant and the functions $f(x)$ and $g(x)$ represent the initial deflection and initial velocity, respectively, of the string. A solution of (1) that satisfies the stated smoothness conditions is called a *classical solution*.

We solve the boundary/initial value problem (1) by completing a series of steps.

I. Set Up Mathematical Model as Boundary/Initial Value Problem: We already did this in (1). If (PDE) is driven or if the boundary conditions are driven, the techniques at the end of this section must be applied.

II. Choose Independent Variables Appropriate to Shape of Domain: In general, solutions $u = u(x, y, z, t)$ of PDEs of physical interest are defined on regions in $xyzt$-space of the form $G = H \times \{t > 0\}$, where H is a region in xyz-space. The Method of Separation of Variables will only work when, in the coordinates used to describe H, each part of the boundary of H is a set on which one of these coordinates is constant (i.e., a coordinate level set). In the case of problem (1), H is the interval $0 < x < L$ whose endpoints $x = 0$ and $x = L$ are coordinate level "sets" for the Cartesian coordinate on the real line. If H were, for example, a circular disk, then polar coordinates should be used to express the space derivatives in the PDE (see Section 10.9).

III. Separate the Variables: Look for all separated solutions. Suppose that $u = X(x)T(t)$ satisfies (PDE). Substituting into (PDE), we have

$$X(x)T''(t) - c^2 X''(x)T(t) = 0 \tag{2}$$

or, after dividing through by $X(x)T(t)$,

$$\frac{X''(x)}{X(x)} = \frac{1}{c^2}\frac{T''(t)}{T(t)} \quad \text{in } G \tag{3}$$

The name *Method of Separation of Variables* derives from the fact that equation (2) can be put into the variables-separated form (3). For $t_0 > 0$, where $T(t_0) \neq 0$, we have

$$\frac{X''(x)}{X(x)} = \frac{1}{c^2}\frac{T''(t_0)}{T(t_0)}, \qquad 0 \leq x \leq L$$

so the ratio $X''(x)/X(x)$ is constant for $0 \leq x \leq L$; call the constant λ. Then, repeating the argument for some $0 < x_0 < L$ with $X(x_0) \neq 0$, we see that the ratio $T''(t)/c^2 T(t)$ is also equal to that same constant λ for $t \geq 0$. So there is a *separation constant* λ such that

$$\frac{X''(x)}{X(x)} = \frac{1}{c^2}\frac{T''(t)}{T(t)} = \lambda \tag{4}$$

for all (x, t) in G^*. The functions $X(x)$ and $T(t)$ that solve (2) must necessarily satisfy the pair of equations

$$\begin{aligned} X'' - \lambda X = 0, \qquad & 0 < x < L \\ T'' - \lambda c^2 T = 0, \qquad & t > 0 \end{aligned} \tag{5}$$

for some constant λ. Stated the other way around, any solution of the ODEs in (5) for any real constant λ must be such that $X(x)T(t)$ is a solution of (PDE) in G. This separation argument is the same every time the Method of Separation of Variables is used.

IV. Set Up a Sturm-Liouville Problem: Now we require, in addition, that the solution $X(x)T(t)$ of (PDE) generated by (5) also satisfy (BC). This amounts to a restriction on the choice of λ. For $X(x)T(t)$ to satisfy the (BC), we must have

$$X(0)T(t) = 0, \qquad X(L)T(t) = 0, \quad \text{all } t \geq 0$$

Now if either $X(0) \neq 0$ or $X(L) \neq 0$, we would have $T(t) = 0$ for all t, so $v(x, t) = X(x)T(t)$ would be the trivial solution. To ensure a nontrivial solution, we must take

$X(0) = X(L) = 0$. But recall that $X(x)$ satisfies the ODE $X'' - \lambda X = 0$ on $0 < x < L$, so we have the Sturm-Liouville problem

$$X'' = \lambda X, \qquad X(0) = X(L) = 0 \tag{6}$$

Recall from Section 10.6 that (6) is an eigenvalue problem for the operator d^2/dx^2 and domain $\{X \text{ in } \mathbf{C}^2[0, L] : X(0) = X(L) = 0\}$.

V. Solve the Sturm-Liouville Problem: We looked at (6) in Section 10.6 and found that it has a nontrivial solution if and only if $\lambda = \lambda_n = -(n\pi/L)^2$, $n = 1, 2, \ldots$; the λ_n are the eigenvalues of the operator associated with the Sturm-Liouville problem in (6). The general solution of (6) corresponding to $\lambda = \lambda_n$ is given by

$$X_n(x) = C_n \sin \frac{n\pi x}{L}, \qquad n = 1, 2, \ldots \tag{7}$$

where C_n is an arbitrary real constant. The X_n are the eigenfunctions of the operator associated with (6). Now inserting $\lambda = \lambda_n$ in the second equation in (5) gives the equation $T'' + (n\pi c/L)^2 T = 0$, $n = 1, 2, \ldots$, whose general solution is given by

$$T_n(t) = \alpha_n \cos \frac{n\pi ct}{L} + \beta_n \sin \frac{n\pi ct}{L}, \qquad n = 1, 2, \ldots$$

where the constants α_n and β_n are arbitrary real numbers for each n. The numbers $\lambda_n = n\pi c/L$ are the *natural frequencies* for the problem. We have found all the standing waves $v_n(x, t) = X_n(x)T_n(t)$ supported by the string, that is, all separated solutions of (PDE) that satisfy (BC):

$$v_n(x, t) = \sin \frac{n\pi x}{L} \left(A_n \cos \frac{n\pi ct}{L} + B_n \sin \frac{n\pi ct}{L} \right), \qquad n = 1, 2, \ldots$$

where A_n and B_n are arbitrary constants.

VI. Construct the Formal Solution: Let's look for a solution of (1) as a superposition of the standing waves $\{v_n\}$, that is, the solution of (1) has the form

$$u(x, t) = \sum_{n=1}^{\infty} \sin \frac{n\pi x}{L} \left(A_n \cos \frac{n\pi ct}{L} + B_n \sin \frac{n\pi ct}{L} \right) \tag{8}$$

for some choice of the constants A_n and B_n. Let's now restrict ourselves to only those choices for A_n and B_n for which all the series derived by termwise differentiation of (8) up to two times with respect to t and x converge uniformly in G^*. (It would be very difficult to determine in advance the conditions on A_n and B_n to ensure the uniform convergence of the series mentioned above, but fortunately as we will see, there is no need to do so.) So for *any* such choice of the A_n and B_n, $u(x, t)$ satisfies the (PDE) and both boundary conditions.

It remains only to determine values for A_n and B_n in (8) such that $u(x, t)$ satisfies the initial conditions; $u(x, t)$ will satisfy (IC) in problem (1) if for $0 \le x \le L$,

$$f(x) = \sum_{n=1}^{\infty} A_n \sin \frac{n\pi x}{L} \qquad \text{[Set } t = 0 \text{ in (8).]}$$

$$g(x) = \sum_{n=1}^{\infty} \frac{n\pi c}{L} B_n \sin \frac{n\pi x}{L} \qquad \text{[Differentiate (8) termwise, then set } t = 0.\text{]}$$

Recall from Example 10.6.3 that the set of functions $\Phi_S = \{\sin n\pi x/L : n = 1, 2, \ldots\}$ is a basis for PC[0, L]. It follows from the Fourier-Euler formulas that

$$A_n = \frac{\langle f, \sin(n\pi x/L)\rangle}{\|\sin(n\pi x/L)\|^2} = \frac{2}{L}\int_0^L f(x) \sin\frac{n\pi x}{L}\,dx, \qquad n = 1, 2, \ldots$$

and

$$\frac{n\pi c}{L} B_n = \frac{\langle g, \sin(n\pi x/L)\rangle}{\|\sin(n\pi x/L)\|^2} = \frac{2}{L}\int_0^L g(x) \sin\frac{n\pi x}{L}\,dx, \qquad n = 1, 2, \ldots$$

So the *formal solution* of (1) by the Method of Separation of Variables is

$$u(x, t) = \sum_1^\infty \sin\frac{n\pi x}{L}\left(A_n \cos\frac{n\pi ct}{L} + B_n \sin\frac{n\pi ct}{L}\right)$$

$$A_n = \frac{2}{L}\int_0^L f(x)\sin\frac{n\pi x}{L}\,dx, \qquad n = 1, 2, \ldots \qquad (9)$$

$$B_n = \frac{2}{n\pi c}\int_0^L g(x)\sin\frac{n\pi x}{L}\,dx, \qquad n = 1, 2, \ldots$$

where we have simply *assumed* that the series in (9) and its derived series up to second order converge uniformly. Although $u(x, t)$ satisfies (BC), we need the stated uniform convergence properties to verify (PDE) and (IC).

VII. Determine Whether the Formal Solution is a Classical Solution: To see that the formal solution $u(x, t)$ as defined by (9) is a *classical solution* [i.e., that u satisfies (PDE) and (IC) in (1) as well as (BC)], we must first show that u belongs to $\mathbf{C}^2(G)$ and to $\mathbf{C}^1(G^*)$. To do this, we must impose smoothness conditions on the initial data f and g. Suppose that $\tilde{f}$ and $\tilde{g}$ are the odd extensions of f and g into the interval $[-L, L]$, which are then extended periodically to $\mathbb{R}$.

We will use the extensions in item VII above to show the following result.

THEOREM 10.7.1

> Classical Solution. Suppose that f and g satisfy the following conditions.
> (a) f is in $\mathbf{C}^2[0, L]$, f''' is in PC[0, L], $f(0) = f(L) = f''(0) = f''(L) = 0$;
> (b) g is in $\mathbf{C}^1[0, L]$, g'' is in PC[0, L], $g(0) = g(L) = 0$.
> Then $u(x, t)$ defined by (9) is a classical solution of problem (1).

To show this, let's calculate the Fourier sine series of f and g on [0, L]:

$$\text{FSS}[f](x) = \sum_{n=1}^\infty \alpha_n \sin\frac{n\pi x}{L}, \qquad \text{FSS}[g](x) = \sum_{n=1}^\infty \beta_n \sin\frac{n\pi x}{L}$$

The hypotheses imply that $\tilde{f}$ lies in $\mathbf{C}^2(\mathbb{R})$, $\tilde{g}$ in $\mathbf{C}^1(\mathbb{R})$, while $\tilde{f}'''$ and $\tilde{g}''$ are in PC(I) for every interval I. Apply the Decay of Coefficients Theorem 10.4.7 to FSS[f] and FSS[g] on [0, L]; we have $\alpha_n = O(1/n^4)$, $\beta_n = O(1/n^3)$. So

$$\sum_{n=1}^\infty n^2|\alpha_n| \text{ converges}, \qquad \sum_{n=1}^\infty n|\beta_n| \text{ converges} \qquad (10)$$

Since $A_n = \alpha_n$ and $n\pi c B_n/L = \beta_n$ are Fourier sine series coefficients of f and g, respectively, we have from (10) that

$$\sum_1^\infty n^2 |A_n| \text{ converges}, \qquad \sum_1^\infty n^2 |B_n| \text{ converges} \qquad (11)$$

Using (11) and the Weierstrass M-Test (Theorem 10.4.4), we see that the series obtained by differentiating the series in (9) term by term up to second order in x and t all converge uniformly on G^*. So $u(x, t)$ as defined by (9) belongs to $C^2(G^*)$, and all derivatives of u up to second order can be computed by term by term differentiation. Formulas (9) define a classical solution of problem (1).

Often it is enough to find a formal solution such as (9) without bothering to check that the initial data are smooth enough to produce a classical solution. Indeed, some of the most interesting behavior, at least for the wave equation, occurs precisely in problems where the initial data are not smooth but have "corners" that propagate into G. In this case the formal solution can't possibly be a classical solution since at the interior points where $u(x, t)$ has "corners," u is not even differentiable.

The Method of Eigenfunction Expansions

The Method of Separation of Variables applies if the PDE and boundary conditions involved are undriven [as in (1)], but what do we do if this is not the case? We saw what to do in Section 10.5 for the ODE modeling the charge on the capacitor in a driven electrical circuit. A solution technique was employed that involved the expansion of the driving force in an eigenfunction series of an associated Sturm-Liouville problem. First we describe the process for an ODE, using our experience with the circuit problem as a guide. Then we adapt the method to PDEs.

Let's look at an ordinary differential operator given by

$$M[y] = \frac{1}{\rho(x)} \left\{ \frac{d}{dx}\left[p(x)\frac{dy}{dx} \right] + q(x)y \right\}$$

where the coefficients satisfy the conditions in (2) of Section 10.6. The regular Sturm-Liouville problem with separated boundary conditions

$$\begin{cases} M[y] = \lambda y \\ B_a[y] = \alpha y(a) + \beta y'(a) = 0, & \alpha^2 + \beta^2 \neq 0 \\ B_b[y] = \gamma y(b) + \delta y'(b) = 0, & \gamma^2 + \delta^2 \neq 0 \end{cases} \qquad (12)$$

has eigenvalues λ_j, $j = 1, 2, \ldots$, and a corresponding collection of eigenfunctions $\Phi = \{y_j(x)\}$ that is a basis of PC$[a, b]$. Now consider the boundary value problem

$$M[y] = f, \qquad B_a y = 0, \qquad B_b y = 0 \qquad (13)$$

where f is a given function belonging to PC$[a, b]$. We look for a solution to this problem in the form $y(x) = \sum_{k=1}^\infty c_k y_k$. The function $y(x)$ always satisfies the boundary conditions for any choice of the constants c_k, so it remains just to choose the c_k such that $M[y] = f$ is satisfied (if we can). Expanding f in a Fourier series with respect to the basis $\Phi = \{y_k\}$ of eigenfunctions of the Sturm-Liouville system (12), we have $f = \sum a_k y_k$. Substitution into ODE (13) yields

$$M[y] = M \sum_{k=1}^{\infty} c_k y_k = \sum_{k=1}^{\infty} c_k M[y_k] = \sum_{k=1}^{\infty} c_k \lambda_k y_k = \sum_{k=1}^{\infty} a_k y_k$$

where we have *assumed* that we can interchange M and $\sum$. Now since $\Phi = \{y_n(x)\}$ is a basis for PC[a, b], it follows by matching coefficients with the same index that $c_j \lambda_j = a_j$ $j = 1, 2, \ldots$. If no $\lambda_j = 0$, we only have to take $c_j = a_j / \lambda_j$, $j = 1, 2, \ldots$, and we obtain the solution of (13) in the form

$$y(x) = \sum_{k=1}^{\infty} \frac{a_k}{\lambda_k} y_k \tag{14}$$

This is an example of the *Method of Eigenfunction Expansions*.

Driven Partial Differential Equations

Now we will adapt the Method of Eigenfunction Expansions to boundary/initial value problems for driven PDEs. We begin with the following problem:

$$\begin{cases} u_{tt} - c^2 u_{xx} = F(x, t), & 0 < x < L, \quad t > 0 \\ u(0, t) = u(L, t) = 0, & t \geq 0 \\ u(x, 0) = f(x), & 0 \leq x \leq L \\ u_t(x, 0) = g(x), & 0 \leq x \leq L \end{cases} \tag{15}$$

where $F(x, t)$ is in $C^0(G)$, $G = \{(x, t) : 0 < x < L, t > 0\}$ and f, g belong to PC[0, L]. Now we know that the solution of the undriven version of (15) with $F = 0$ [i.e., the boundary/initial value problem (1)] is a superposition of functions from the basis $\{\sin(n\pi x/L)\}$ with time-varying coefficients. In the Method of Eigenfunction Expansions for (15) with F not identically zero, the solution $u(x, t)$ is expressed as a superposition $\sum U_n(t) \sin(n\pi x/L)$ for some functions $\{U_n(t)\}$ that will be found by the method of undetermined coefficients. We carry out the calculations below.

All the data functions F, f, g may be expanded into Fourier sine series in x (assuming throughout that the functions are sufficiently smooth):

$$\text{FSS}[f](x) = \sum_{n=1}^{\infty} A_n \sin \frac{n\pi x}{L}, \qquad A_n = \frac{2}{L} \int_0^L f(x) \sin \frac{n\pi x}{L} dx \tag{16a}$$

$$\text{FSS}[g](x) = \sum_{n=1}^{\infty} B_n \sin \frac{n\pi x}{L}, \qquad B_n = \frac{2}{L} \int_0^L g(x) \sin \frac{n\pi x}{L} dx \tag{16b}$$

$$\text{FSS}[F](x, t) = \sum_{n=1}^{\infty} C_n(t) \sin \frac{n\pi x}{L}, \qquad C_n(t) = \frac{2}{L} \int_0^L F(x, t) \sin \frac{n\pi x}{L} dx \tag{16c}$$

Now expand the solution $u(x, t)$ of (15) in a Fourier sine series in x:

$$\text{FSS}[u](x, t) = \sum_{n=1}^{\infty} U_n(t) \sin \frac{n\pi x}{L}, \qquad (x, t) \text{ in } G \tag{17}$$

☞ The U_n's are the "undetermined coefficients."

where $U_n(t) = (2/L) \int_0^L u(x, t) \sin(n\pi x/L)\, dx$. Proceeding formally, we assume that $u(x, t)$ is the sum of its Fourier Sine Series FSS$[u](x, t)$, insert (17) into the partial differential equation of (15), and use (16) to calculate the set of coefficient functions $\{U_n(t)\}$. We have [using equation (16c)]

$$\sum_{n=1}^{\infty} \left\{ U_n''(t) + c^2 \left(\frac{n\pi}{L}\right)^2 U_n(t) - C_n(t) \right\} \sin\frac{n\pi x}{L} = 0, \qquad (x, t) \text{ in } G$$

Since $\Phi = \{\sin n\pi x/L : n = 1, 2, \ldots\}$ is a basis for PC$[0, L]$, we have

$$U_n''(t) + \left(\frac{n\pi c}{L}\right)^2 U_n(t) = C_n(t), \qquad t > 0, \quad n = 1, 2, \ldots \qquad (18)$$

Using (16a), (16b) to determine initial conditions for each $U_n(t)$, we have

$$u(x, 0) = f(x) = \sum_{n=1}^{\infty} U_n(0) \sin\frac{n\pi x}{L} = \sum_{n=1}^{\infty} A_n \sin\frac{n\pi x}{L}, \qquad 0 \le x \le L$$

$$u_t(x, 0) = g(x) = \sum_{n=1}^{\infty} U_n'(0) \sin\frac{n\pi x}{L} = \sum_{n=1}^{\infty} B_n \sin\frac{n\pi x}{L}, \qquad 0 \le x \le L$$

The initial conditions on U_n are

$$U_n(0) = A_n, \qquad U_n'(0) = B_n \qquad (19)$$

☞ Problem 8 in Section 3.7 describes the Variation of Parameters Method.

Using the Variation of Parameters Method to solve the IVP (18), (19), we see that

$$U_n(t) = A_n \cos\frac{n\pi c t}{L} + \frac{L B_n}{\pi n c} \sin\frac{n\pi c t}{L} + \frac{L}{\pi n c} \int_0^t \sin\left[\frac{n\pi c}{L}(t - s)\right] C_n(s)\, ds \qquad (20)$$

The *formal solution* of (15) is given by (17) and (20) with A_n, B_n, and C_n given by (16a), (16b), and (16c). Observe that this solution can be written as

$$u(x, t) = u_1(x, t) + u_2(x, t) \qquad (21)$$

where

$$u_1(x, t) = \sum_{n=1}^{\infty} \sin\frac{n\pi x}{L} \left(A_n \cos\frac{n\pi c t}{L} + \frac{L B_n}{\pi n c} \sin\frac{n\pi c t}{L} \right)$$

$$u_2(x, t) = \sum_{n=1}^{\infty} \sin\frac{n\pi x}{L} \left(\frac{L}{\pi n c} \int_0^t \sin\left[\frac{n\pi c}{L}(t - s)\right] C_n(s)\, ds \right) \qquad (22)$$

So $u_1(x, t)$ represents the response of the string to the initial conditions, while $u_2(x, t)$ is the response to the external force $F(x, t)$.

Remark. If $F(x, t)$ is periodic with a frequency near a natural frequency, one might expect resonance to occur. This is indeed the case. See Problem 7 for a related problem.

Shifting the Data

The Method of Separation of Variables handles initial conditions together with un-driven boundary conditions and an undriven linear PDE. The Method of Eigenfunction Expansions extends the solution process to a driven PDE. We now show how to reduce a problem where *all* the conditions have driving terms to one or the other of the problems above. As always, the discussion is in terms of the wave equation on a finite interval.

Consider the vibrating string moving under the influence of a vertically acting external force $F(x, t)$ subject to the usual initial conditions and with time-varying boundary conditions:

$$
\begin{aligned}
&\text{(PDE)} \quad u_{tt} - c^2 u_{xx} = F(x, t), \qquad 0 < x < L, \quad t > 0 \\
&\text{(BC)} \quad u(0, t) = \alpha(t), \qquad\qquad u(L, t) = \beta(t), \quad t \geq 0 \\
&\text{(IC)} \quad u(x, 0) = f(x), \qquad\qquad u_t(x, 0) = g(x), \quad 0 \leq x \leq L
\end{aligned} \tag{23}
$$

Observe that the function $v(x, t) = \alpha(t) + (x/L)(\beta(t) - \alpha(t))$ satisfies the boundary conditions $v(0, t) = \alpha(t)$, $v(L, t) = \beta(t)$. Suppose that $w(x, t)$ is a solution of

$$
\begin{aligned}
w_{tt} - c^2 w_{xx} &= F(x, t) - (v_{tt} - c^2 v_{xx}) \\
w(0, t) &= w(L, t) = 0 \\
w(x, 0) &= f(x) - v(x, 0) \\
w_t(x, 0) &= g(x) - v_t(x, 0)
\end{aligned} \tag{24}
$$

We see that $u(x, t) = w(x, t) + v(x, t)$ is a solution of (23).

What we have done is to introduce the *boundary function* $v(x, t)$, which satisfies both boundary conditions of (23) but none of the other conditions. The effect of letting $w = u - v$ is to introduce a boundary/initial value problem for w in which the boundary data have been *shifted* away from the endpoints $x = 0$ and $x = L$ and attached in altered form to the right-hand sides of the PDE and the initial conditions. But a problem such as (24) for w can be solved by the Method of Eigenfunction Expansions outlined earlier in this section.

In many special cases of practical interest, nonzero data can be shifted in such a way as to make the modified problem have an undriven PDE *and* undriven boundary conditions (but at the expense of altered initial data), which can then be treated directly by the Method of Separation of Variables. For illustrations of this technique, see the problem set.

Comments

The Method of Separation of Variables and the Method of Eigenfunction Expansions are based on the same idea—that of expanding functions in a series of eigenfunctions of an appropriate Sturm-Liouville problem. The Sturm-Liouville problem itself is inherent in the geometry of the boundary/initial value problem being solved.

Different operators and geometries lead to different Sturm-Liouville problems, to different eigenfunctions and eigenvalues, and to different bases of different spaces. The possibilities seem unlimited. In Sections 10.8 and 11.8 we explore some other types of Sturm-Liouville problems.

PROBLEMS

www **1.** Use the Method of Separation of Variables to solve (1) under the following conditions.

(a) $f(x) = 0$, $g(x) = 3\sin(\pi x/L)$, $0 \le x \le L$

(b) $f(x) = g(x) = \begin{cases} x, & 0 \le x \le L/2 \\ L - x, & L/2 \le x \le L \end{cases}$

(c) $f(x) = x(L - x) = -g(x)$, $0 \le x \le L$

2. Look at the problem $u_{tt} - c^2 u_{xx} = 0$, $0 < x < L$, $t > 0$; $u(0, t) = 0$, $u_x(L, t) = -hu(L, t)$, $t \ge 0$, h a positive constant; $u(x, 0) = f(x)$, $u_t(x, 0) = g(x)$, $0 \le x \le L$.

(a) Construct the Sturm-Liouville problem associated with the boundary/initial value problem.

(b) Show that the eigenvalues of the Sturm-Liouville operator of part (a) are $\lambda_n = -s_n^2/L^2$, $n = 1, 2, \ldots$, where s_n is the n-th consecutive positive zero of the function $s + hL\tan s$. Show that there are infinitely many zeros s_n, but do not evaluate them. Find a corresponding basis of eigenfunctions for PC[0, L].

3. (*Damped Wave Equation*). The equation $u_{tt} + b^2 u_t - a^2 u_{xx} = 0$, $0 < x < L$, $t > 0$, models a vibrating string, taking into account air resistance. Find the formal solution $u(x, t)$ of the boundary problem of the damped wave equation

$$u_{tt} + b^2 u_t - a^2 u_{xx} = 0, \qquad 0 < x < L, \quad t > 0$$
$$u(0, t) = u(L, t) = 0, \qquad t \ge 0$$
$$u(x, 0) = f(x), \qquad u_t(x, 0) = 0, \quad 0 \le x \le L$$

where $b^2 < 2\pi a/L$ and a are positive constants and f and g belong to PC[0, L].

4. (*Shifting Data*). The boundary/initial value problem

$$u_{tt} - c^2 u_{xx} = g, \qquad 0 < x < L, \quad t > 0$$
$$u(0, t) = u(L, t) = 0, \qquad t \ge 0$$
$$u(x, 0) = u_t(x, 0) = 0, \qquad 0 \le x \le L$$

models the vertical displacement $u(x, t)$ of a taut flexible string tied at both ends with vanishing initial data and acted on by gravity (g is the constant gravitational acceleration). The outline below shows how to shift the nonhomogeneity in the partial differential equation onto an initial condition.

(a) Find a function $v(x)$ such that $-c^2 v_{xx} = g$, $0 < x < L$, $t > 0$; $v(0) = v(L) = 0$ $t > 0$. [*Hint*: Try $v(x) = Ax + Bx^2$, where A, B are constants.] Note that $v(x)$ is the steady-state sag of the string under the force of gravity.

(b) Let $w = u - v$ and show that w satisfies the same equations as u, but with g replaced by 0 and the condition $u(x, 0) = 0$ replaced by $w(x, 0) = -gx(L - x)/2c^2$.

(c) Find $u(x, t)$.

5. (*Shifting Boundary Data Onto Initial Data*). A string of unit length with $c^2 = 1$ clamped at one end, driven by $\sin \pi t/2$ at the other end, and given an initial velocity is modeled by $u_{tt} - u_{xx} = 0$ $0 < x < 1$, $t > 0$; $u(0, t) = \sin \pi t/2$, $u(1, t) = 0$, $t \ge 0$; $u(x, 0) = f(x)$, $u_t(x, 0) = g(x)$, $0 \le x \le 1$. The following steps show how to shift the boundary data $\sin \pi t/2$ onto an initial condition.

(a) Suppose that v is any solution of $v_{tt} - v_{xx} = 0$, $v(0, t) = \sin \pi t/2$, $v(1, t) = 0$, $t \ge 0$. Set $w = u - v$. Show that w is a solution of the same problem as u except the boundary condition $u(0, t) = \sin \pi t/2$ is replaced by $w(0, t) = 0$ and the data $f(x)$ and $g(x)$ are replaced by $f(x) - v(x, 0)$ and $g(x) - v_t(x, 0)$, respectively.

(b) Find $v(x, t)$ in the form $X(x)T(t)$. [*Hint*: Let $v = X(x)\sin(\pi t/2)$.]

(c) Solve the problem if $g(x) = 0$ for all x.

6. Solve the following problems.

(a) Use the Method of Eigenfunction Expansions to solve

$$u_{tt} - u_{xx} = 6x, \qquad\qquad 0 < x < 1, \quad t > 0$$
$$u(0, t) = u(1, t) = 0, \qquad t \geq 0$$
$$u(x, 0) = u_t(x, 0) = 0, \qquad 0 \leq x \leq 1$$

(b) First solve the problem $v_{tt} - v_{xx} = 0$, $v(0, t) = \sin(3\pi t/2L)$, $v(L, t) = 0$. Then use v to shift the boundary data onto the initial conditions and solve the following problem:

$$u_{tt} - u_{xx} = 0, \qquad\qquad 0 < x < L, \quad t > 0$$
$$u(0, t) = \sin(3\pi t/2L), \quad u(L, t) = 0, \qquad t \geq 0$$
$$u(x, 0) = u_t(x, 0) = 0, \qquad\qquad 0 \leq x \leq L$$

7. (*Breaking a String by Shaking It*). Consider the problem of a string of length L with one end fastened and the other driven:

$$u_{tt} - c^2 u_{xx} = 0, \qquad 0 < x < L, \quad t > 0$$
$$u(0, t) = 0, \qquad u(L, t) = \mu(t), \quad t \geq 0$$
$$u(x, 0) = 0, \qquad u_t(x, 0) = 0, \quad 0 \leq x \leq L$$

Show that there is a periodic function $\mu(t)$ such that the string will eventually break. [*Hint:* What if $\mu(t) = A \cos \omega t$ for ω near a natural frequency $n\pi c/L$?]

10.8 The Heat Equation: Optimal Depth for a Wine Cellar

The heat equation is a linear PDE that models both the flow of heat and a host of other physical, chemical, and biological phenomena involving diffusion processes. We derive the heat equation as a mathematical model of thermal conduction and then solve some heat flow problems. Finally, we discuss the mathematical properties of solutions of the heat equation and the physical meaning of these properties.

The solution technique for solving the heat equation is the Method of Separation of Variables, the same method we have used to solve boundary/initial value problems for the wave equation. So we will find solutions of the heat equation in series form, just as before. Although the method for finding the solution and the form of the solution are nothing new, the properties of the solution of a boundary/initial value problem for the heat equation are completely different from those for the wave equation. We will end the section with a discussion of those differences.

Heat Conduction

Molecular vibrations of a material body generate energy that we feel as heat. Heat flows from warm to cool parts of the body by *conduction*, a process in which the collision of neighboring molecules transfers thermal energy from one molecule to another. It is conduction that is modeled below. Flow of heat also takes place through *convection*, molecules moving from region to region carrying their thermal energies along with them, but we will not model convection here.

The thermal state of a material body at each of its points P at time t is measured by the *temperature* $u(P, t)$. The units of temperature are degrees centigrade (or Celsius). Heat itself is measured in *calories* (or *joules*), 1 calorie being the energy needed to raise the temperature of 1 gram of water by 1 degree (1 joule $= 0.239$ calorie). Each substance has a characteristic *specific heat* c, which is the energy required to raise the temperature of 1 gram of the substance 1 degree (so the specific heat of water is 1). The *heat* in a portion D of a material body at time t is given by

$$H(t) = \int_D c\rho u(P, t)\, dP \tag{1}$$

where ρ is the density of the material and the integration is over D. The integral in (1) is single or multiple according to the spatial dimension of D. The quantity H as defined by (1) has the units of energy. In the derivations below we take D to be three-dimensional. The one- and two-dimensional cases are treated in the same way.

The conductive flow of heat through D is governed by a form of the balance law:

☞ We first ran across the Balance Law in Section 1.4. Doesn't it have a long reach?

$$
\begin{array}{ccccc}
\text{Rate of} & & \text{Heat generated or} & & \text{Heat moving across} \\
\text{change of} & = & \text{consumed inside} & + & \text{the boundary of } D \\
\text{heat in } D & & D \text{ per unit time} & & \text{per unit time}
\end{array} \tag{2}
$$

We will show how (2) can be represented mathematically by the partial differential equation of heat flow. From (1) we see that the left side of the balance equation has the form

$$\frac{dH(t)}{dt} = \frac{d}{dt}\int_D c\rho u(P, t)\, dP = \int_D \frac{\partial}{\partial t}[c\rho u(P, t)]\, dP \tag{3}$$

where c, ρ, and u are assumed to be continuously differentiable functions of t and continuous in P.

The first term on the right of the balance equation in (2) may be expressed as

$$\int_D F(P, t)\, dP \tag{4}$$

where $F(P, t)$ is the heat generated or consumed per unit volume per unit time at the point P at time t. The point P is a *source* if $F(P, t)$ is positive, a *sink* if $F(P, t)$ is negative.

The last term on the right-hand side of (2) can be written as an integral over the boundary ∂D of D. The integrand is the amount of heat moving across the boundary per unit of surface area per unit of time at the point P on the boundary at time t,

$$\int_{\partial D} k\nabla u(P, t) \cdot \mathbf{n}\, dS \tag{5}$$

where ∇u is the spatial gradient of u, $\mathbf{n}$ is the unit outward normal to the boundary ∂D at P, and dS denotes integration over the boundary surface. The coefficient k in (5) is called the *thermal conductivity* and measures the ability of the body to conduct heat. Specifically, k is the time rate of change of heat through unit thickness per unit of surface area per unit of temperature. The integral in (5) is the mathematical representation of the *Euler-Fourier Law of Heat Conduction* (*Fick's Law* in the case of gas or liquid diffusion): Heat flows in the direction of maximal temperature drop per unit of distance at a rate proportional to the magnitude of that drop. Since ∇u points in the direction of maximal rise in temperature, heat flow is actually in the direction of $-\nabla u$. Note that if $\nabla u \cdot \mathbf{n}$ is positive at a point P on the boundary of D, the temperature outside D and near P is higher than that at P, so thermal energy flows into D through P. Using the Divergence Theorem, we replace the surface integral in (5) by a volume integral over D:

☞ The Divergence
Theorem is
Theorem B.5.13.

$$\int_{\partial D} k\nabla u \cdot \mathbf{n}\, dS = \int_D \mathrm{div}(k\nabla u)\, dP \tag{6}$$

In Cartesian x-, y-, z-coordinates,

$$\nabla u = \frac{\partial u}{\partial x}\mathbf{i} + \frac{\partial u}{\partial y}\mathbf{j} + \frac{\partial u}{\partial z}\mathbf{k}$$

$$\mathrm{div}(k\nabla u) = \frac{\partial}{\partial x}\left(k\frac{\partial u}{\partial x}\right) + \frac{\partial}{\partial y}\left(k\frac{\partial u}{\partial y}\right) + \frac{\partial}{\partial z}\left(k\frac{\partial u}{\partial z}\right)$$

Inserting the expressions in (3)–(6) into the balance equation in (2), we have, after some rearranging,

$$\int_D \left[\frac{\partial}{\partial t}(c\rho u) - F - \mathrm{div}(k\nabla u)\right] dP = 0 \tag{7}$$

If the integrand is continuous in P and in t, then (7) is valid for all regions D if and only if the integrand vanishes. This gives the *heat* (or *diffusion*) *equation*

$$\frac{\partial}{\partial t}(c\rho u) - \mathrm{div}(k\nabla u) = F \tag{8}$$

at each point P inside the body for all time t for which the model is valid. The *temperature equation* would be a more precise name for (8). Note that (8) is linear in the unknown u and its derivatives.

If there are no internal sources or sinks, F vanishes and we have the *undriven* (or *source-free*) *heat equation*

$$\frac{\partial}{\partial t}(c\rho u) - \mathrm{div}(k\nabla u) = 0 \tag{9}$$

If the material coefficients c, ρ, and k are constants, the *diffusivity* $K = k/c\rho$ can be defined. The units of diffusivity typically are cm^2/s and its values range from 0.0014 for water (a good insulator, but a poor conductor) up to 1.71 for silver (a good conductor). In this case (9) can be written in terms of the Laplacian operator ∇^2 as

$$u_t - K\nabla^2 u = u_t - K(u_{xx} + u_{yy} + u_{zz}) = 0 \tag{10}$$

Boundary and Initial Conditions

On physical grounds one would not expect the heat equation to be enough to determine uniquely the temperatures $u(P, t)$ within a material body B. Initial and boundary conditions are also needed. The *initial condition* can be written as

$$u(P, 0) = f(P), \quad \text{all } P \text{ in } B \tag{11}$$

where f is a given function. Unlike the situation with the wave equation, it is *not* necessary to prescribe $u_t(P, 0)$ since the heat equation itself does that.

There are three common types of thermal conditions imposed on the boundary of B. The first has to do with the *prescribed boundary temperatures*:

$$u(P, t) = g(P, t), \qquad P \text{ in } \partial B, \quad t \geq 0 \tag{12}$$

where g is a given function. For example, the body may be submerged in an ice bath, so $u(P, t) = 0$ on the boundary for all $t \geq 0$.

Alternatively, the body may be wrapped with *thermal insulation*, which affects the flow of heat across the boundary:

$$\frac{\partial u(P, t)}{\partial n} = \nabla u \cdot \mathbf{n} = h(P, t), \qquad P \text{ in } \partial B, \quad t \geq 0 \tag{13}$$

where h is a prescribed function. If $h = 0$, the insulation is said to be *perfect*, and there is no heat flow through the boundary.

A third type of boundary condition is given by *Newton's Law of Cooling*,

$$\frac{\partial u(P, t)}{\partial n} = r[U(t) - u(P, t)], \qquad P \text{ in } \partial B, \quad t \geq 0 \tag{14}$$

where $U(t)$ is the given *ambient temperature* outside B and r is a given *heat transfer coefficient*. If $r > 0$ and if the ambient temperature is higher than the boundary temperature, $\partial u/\partial n$ is positive, indicating a flow of heat into B across the boundary.

There are other types of boundary conditions, but these three cover most cases. Each boundary condition may be written in terms of a linear boundary operator acting on u and $\partial u/\partial n$. For example, (14) can be written as $B[u] = rU$, where B is the linear operator $(\partial/\partial n) + r$ acting on the linear space of sufficiently smooth functions $u(P, t)$ defined for P in ∂B and $t \geq 0$. In a given problem, part of the boundary can be subject to one condition, another part to a different condition. For example, one end of an iron bar can be immersed in an ice bath, the other end in boiling water, while the middle is covered with perfect insulation.

With these remarks in mind, we can formulate a typical *boundary/initial value problem* for the heat equation in a material body B of constant diffusivity K and with no internal sources and sinks: Find $u(P, t)$ such that

$$
\begin{array}{llll}
\text{(PDE)} & u_t - K\nabla^2 u = 0, & P \text{ inside } B, & t > 0 \\
\text{(BC)} & \alpha \partial u/\partial n + \beta u = f, & P \text{ in } \partial B, & t \geq 0 \\
\text{(IC)} & u = g, & P \text{ in } B, & t = 0
\end{array} \tag{15}
$$

where α, β, and f are prescribed functions of P and t, and g is a function of P. Continuity and smoothness conditions may also be imposed on α, β, f, g, and on ∂B so that (15) is well-posed (see the end of this section). Rather than consider general problems, we take up and solve two simple and illustrative special cases of problem (15).

Temperature in a Rod

Suppose that a straight rod of constant diffusivity has uniform cross sections. Suppose, also, that the lateral surface of the rod is wrapped in perfect insulation, while ice packs are held against the two ends, which then are maintained at a temperature of $0°$. Suppose that at some initial time the temperature distribution is known. The problem is to determine the temperature distribution within the rod at later times.

We can take the central axis of the rod to be the x-axis. Because of the cross-sectional symmetry and the insulation on the lateral walls, we will assume there is no temperature variation in the y- and z-directions. Denoting the diffusivity by K and the length of the rod by L, we see that we must solve the following boundary/initial value problem for the temperature $u(x, t)$:

$$
\begin{aligned}
&\text{(PDE)} &u_t - Ku_{xx} &= 0, &0 < x < L, \quad t &> 0 \\
&\text{(BC)} &u(0, t) = 0, \quad u(L, t) &= 0, &t &\geq 0 \\
&\text{(IC)} &u(x, 0) &= f(x), &0 \leq x &\leq L
\end{aligned}
\tag{16}
$$

where the initial temperature distribution f is given. So the solution of problem (16), $u(x, t)$, lives on the region $R : 0 < x < L, t > 0$. The shape of R and the nature of the problem itself suggest that the Method of Separation of Variables might be used to construct the solution.

First we look for separated solutions $u = X(x)T(t)$ of (PDE) and of (BC) in (16), leaving the satisfaction of (IC) to a subsequent superposition of separated solutions. Inserting $X(x)T(t)$ into (PDE), we now have $X(x)T'(t) = KX''(x)T(t)$. Separating variables in the usual way, we have a function of x equaling a function of time t, which can happen only if both equal a *separation constant* λ:

$$
\frac{X''(x)}{X(x)} = \frac{1}{K}\frac{T'(t)}{T(t)} = \lambda
\tag{17}
$$

Conditions (BC) impose additional restrictions on $X(x)$, but not on $T(t)$. Combining these restrictions with the equations of (17), we see that $X(x)$ and $T(t)$ must be solutions of

$$
X''(x) - \lambda X(x) = 0, \qquad X(0) = 0, \quad X(L) = 0
\tag{18a}
$$

$$
T'(t) - \lambda K T(t) = 0
\tag{18b}
$$

The values of λ for which the Sturm-Liouville problem (18a) is solvable and the solutions $X(x)$ were determined in Section 10.1:

$$
\lambda_n = -(n\pi/L)^2, \qquad X_n(x) = A_n \sin(n\pi x/L), \qquad n = 1, 2, 3, \dots
\tag{19}
$$

where A_n is any constant. Corresponding solutions of (18b) are

$$
T_n(t) = B_n \exp[-K(n\pi/L)^2 t]
$$

where B_n is any constant. Solutions of (PDE) and (BC) in (16) are given by

$$
u_n = X_n(x)T_n(t) = C_n \sin\left(\frac{n\pi x}{L}\right)\exp\left[-K\left(\frac{n\pi}{L}\right)^2 t\right], \qquad n = 1, 2, \dots
\tag{20}
$$

where $C_n = A_n B_n$ is an arbitrary constant.

Since (PDE) and (BC) are undriven in this problem, any superposition of functions of the form given in (20) is again a solution of (PDE) and (BC). We will determine constants C_n so that the superposition

$$u = \sum_{n=1}^{\infty} C_n \sin\left(\frac{n\pi x}{L}\right) \exp\left[-K\left(\frac{n\pi}{L}\right)^2 t\right] \tag{21}$$

is also a solution of the initial condition (IC). That is, we must choose C_n so that

$$u(x, 0) = f(x) = \sum_{n=1}^{\infty} C_n \sin\left(\frac{n\pi x}{L}\right), \qquad 0 \le x \le L$$

This suggests a Fourier Sine Series. The coefficients C_n can then be found:

$$C_n = \frac{\langle f, \sin(n\pi x/L)\rangle}{\|\sin(n\pi x/L)\|^2} = \frac{2}{L}\int_0^L f(x) \sin\frac{n\pi}{L}x \, dx, \qquad n = 1, 2, \ldots \tag{22}$$

Ignoring questions of convergence, the formal solution of (16) is given by (21) and (22). Although we won't show it, the series defines a classical solution $u(x, t)$ that belongs to $\mathbf{C}^2(R)$ and to $\mathbf{C}^0(R^*)$ and satisfies (PDE) in R and (BC) and (IC) on ∂R if f is continuous, piecewise smooth, and $f(0) = f(L) = 0$.

The calculation of (22) may be carried out explicitly in the particular case of a rod of length 2, diffusivity 1, and initial temperature

$$f(x) = \begin{cases} x, & 0 \le x \le 1 \\ 2 - x, & 1 \le x \le 2 \end{cases} \tag{23}$$

In fact, $C_n = (-1)^{(n-1)/2}8/(n\pi)^2$ for n odd, $C_n = 0$ for n even. The temperature function in this case is

$$u(x, t) = \frac{8}{\pi^2}\sum_{\text{odd } n}(-1)^{(n-1)/2}\frac{1}{n^2}\sin\left(\frac{n\pi x}{2}\right)\exp\left[\frac{-n^2\pi^2 t^2}{4}\right] \tag{24}$$

Several time profiles of this temperature function are sketched in Figure 10.8.1. Note the "smoothing property" of the heat operator, and note also the decay of the temperature down from the initial angular profile as the heat in the rod "leaks" out through the ends.

Optimal Depth for a Wine Cellar

One of the first applications of Fourier series was to model heat flow through soil and rock, Fourier himself taking up this question. We will consider a simple case here. First, however, we solve a heat problem different from (16):

(PDE)	$u_t - Ku_{xx} = 0,$	$0 < x < \infty,$	$-\infty < t < \infty$
(BC)	$u(0, t) = A_0 e^{i\omega t},$		$-\infty < t < \infty$
(Boundedness)	$\|u(x, t)\| < C,$	$0 \le x < \infty,$	$-\infty < t < \infty$

$\qquad(25)$

where K, A_0, ω, and C are assumed to be positive constants. The xt-region R is defined by $0 < x < \infty$, $-\infty < t < \infty$, and the data of the problem are sketched in the

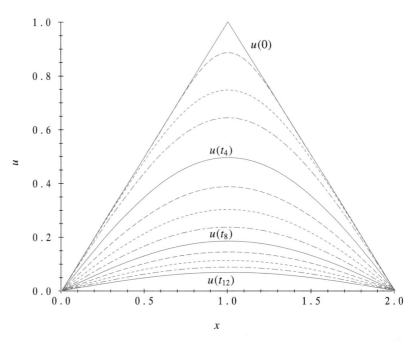

FIGURE 10.8.1 Decaying temperature profiles $u(t)$ in a rod at equally spaced times $0 < t_1 < t_2 < \cdots < t_{12}$ (calculated from solution formula (24)).

$$u(0, t) = A_0 e^{i\omega t}$$
$$\underline{\qquad\qquad} t$$
$$u_t = K u_{xx}$$
$$|u| < C$$
$$x \qquad\qquad R$$

margin. The use of the complex exponential in the boundary condition of (25) is for convenience in calculation. It can be shown that (25) has no more than one solution that belongs to $\mathbf{C}^0(R^*)$ and satisfies (PDE) throughout R. We will find a solution, which then must be the only one.

The exponential "input" $A_0 e^{i\omega t}$ along the boundary suggests that the solution of (25) might have the separated form $u = A(x)e^{i\omega t}$. The amplitude $A(x)$ can be found by inserting u into the equations of (25):

$$i\omega A(x)e^{i\omega t} = K A''(x)e^{i\omega t}, \qquad A(0) = A_0, \quad |A(x)| < C \qquad (26)$$

☞ Use De Moivre's formula (Theorem B.3.2)

Canceling $e^{i\omega t}$ from the first equation of (26) and solving for $A(x)$ [using the fact that $(i\omega/K)^{1/2} = (1+i)(\omega/2K)^{1/2}$], we have

$$A(x) = C_1 e^{\alpha(1+i)x} + C_2 e^{-\alpha(1+i)x}, \qquad \alpha = \left(\frac{\omega}{2K}\right)^{1/2} > 0$$

where C_1 and C_2 are arbitrary constants. The condition $|A(x)| < C$ implies that $C_1 = 0$ since $e^{\alpha x}$ becomes unbounded as $x \to \infty$. The condition $A(0) = A_0$ implies that $C_2 = A_0$. The solution of (25) is

$$u = A_0 e^{-\alpha(1+i)x} e^{i\omega t} = A_0 e^{-\alpha x} e^{i(\omega t - \alpha x)}, \qquad \alpha = \left(\frac{\omega}{2K}\right)^{1/2}, \qquad (x, t) \text{ in } R^* \qquad (27)$$

Problem (25) and its solution (27) can be interpreted in terms of finding the optimal depth for locating a storage cellar. In this setting x is the depth below the surface

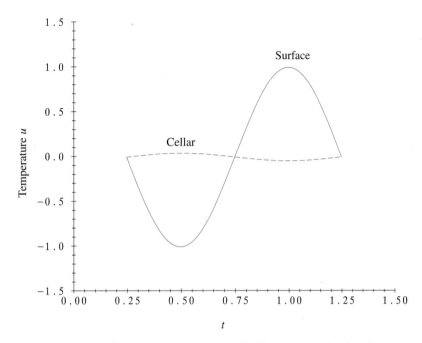

FIGURE 10.8.2 Surface temperature (solid line) and optimal cellar temperature (dashed line) from October ($t = 0.25$) through the following September ($t = 1.25$).

of the earth, while $A_0 \cos \omega t$ (the real part of $A_0 e^{i\omega t}$) is a crude model of surface temperature normalized about a mean of $0°$. Suppose that the period of the surface wave is 1 year (i.e., $2\pi/\omega = 1$ yr $= 3.25 \times 10^7$ s). Then, taking only the real part of $u(x, t)$ from (27), the temperature at depth x at time t is

$$u = A_0 e^{-\alpha x} \cos(\omega t - \alpha x), \qquad \omega = \frac{2\pi}{3.15 \times 10^7}, \qquad \alpha = \left(\frac{\omega}{2K}\right)^{1/2}$$

The *optimal depth* for the storage cellar is defined to be the smallest positive x at which the cellar's "seasons" are 6 months out of phase with the surface seasons. At this depth thermal convection currents from the surface tend to move the cellar temperature even closer to the mean. So the optimal depth satisfies $\alpha x = \pi$. Now the diffusivity K of average dry soil is 0.002 cm²/s. The optimal depth is

$$x = \frac{\pi}{\alpha} = \pi \left(\frac{2K}{\omega}\right)^{1/2} = \pi(0.004)^{1/2} \left(\frac{3.15 \times 10^7}{2\pi}\right)^{1/2} \approx 445 \text{ cm or } 4.45 \text{ m}$$

The amplitude of the surface wave has dropped from A_0 to $A_0 e^{-x\alpha} = A_0 e^{-\pi} \approx A_0/25$ at the optimal depth. This 25-fold reduction in the amplitude coupled with the reversal of the seasons implies a nearly constant temperature in the cellar (see Figure 10.8.2).

Note that there is no need for an initial condition in this problem (indeed, there is no initial time). Note also that the model is only valid near the surface of the earth in a region where there are no subsurface thermal sources or sinks.

Properties of Solutions of the Heat Equation

Boundary/initial value problems for the heat equation have physical significance only if they are *well-posed* in the sense that each problem has a solution, exactly one solution, and the solution changes continuously with the data. Separation of variables can often be used to construct a formal series solution, as we saw above in the problem of the rod with zero boundary data. However, to show that this series solution has enough convergence properties to be a classical solution requires a detailed study of the series (see below). The other two aspects of a well-posed problem (uniqueness and continuity) are somewhat easier to resolve. We will consider the following problem in this regard:

$$
\begin{array}{llll}
\text{(PDE)} & u_t - K u_{xx} = 0, & 0 < x < L, & t > 0 \\
\text{(BC)} & u(0, t) = g_1(t), & u(L, t) = g_2(t), & t \geq 0 \\
\text{(IC)} & u(x, 0) = f(x), & 0 \leq x \leq L &
\end{array}
\qquad (28)
$$

where g_1 and g_2 belong to $\mathbf{C}^0[0, \infty)$ and are bounded for $0 \leq t < \infty$, f belongs to $\mathbf{C}^0[0, L]$, $g_1(0) = f(0)$, and $g_2(0) = f(L)$. The equations of (28) model the temperature in a uniform rod with perfect insulation on the lateral boundary, a rod whose endpoints have prescribed (but varying) temperatures. The conditions on f, g_1, and g_2 ensure the joint continuity of the boundary and initial data.

The first result refers to certain space-time regions shown in the margin. These are defined as follows: $R: 0 < x < L, \ t > 0$; $\quad R_T: 0 < x < L, \ 0 < t \leq T$; $\quad \Gamma_T$: the boundary of R_T with the upper edge deleted. The following basic principle holds for each region R_T.

THEOREM 10.8.1

> Maximal Principle for the Heat Equation. Suppose that $u(x, t)$ is any solution of (PDE) in (28) for which u is in $\mathbf{C}^2(R)$ and also in $\mathbf{C}^0(R^*)$. Then
>
> $$ |u(x, t)| \leq \max_{(y, s) \text{ in } \Gamma_T} |u(y, s)|, \quad \text{for all } (x, t) \text{ in } R_T^* $$

The principle has a clear physical interpretation: In the absence of internal sources and sinks, the magnitudes of the temperatures in the rod at time t do not exceed the extreme magnitudes of the end temperatures up to time t and the extreme magnitudes of the initial temperatures.

The Maximal Principle implies uniqueness and continuity in the data.

THEOREM 10.8.2

> Uniqueness. The initial boundary problem (28) has no more than one solution $u(x, t)$.

Suppose that u_1 and u_2 are both solutions of (28). Then $w = u_1 - u_2$ satisfies (PDE) and the homogeneous conditions, $w(0, t) = w(L, t) = 0$, $w(x, 0) = 0$, $t \geq 0$, $0 \leq x \leq L$. Since $w = 0$ on the three segments of Γ_t (see margin sketch), the Maximal Principle implies that $|w(x,t)| \leq \max |w(\bar{x},\bar{t})| = 0$, where the maximum is taken over all $(\bar{x},\bar{t})$ in Γ_t. So $w(x,t) = 0$ and $u_1(x,t) = u_2(x,t)$. If (28) has a solution at all with the required smoothness, the solution is unique.

THEOREM 10.8.3

Continuity in the Data. Every solution of (28) changes continuously with respect to changes in the data f, g_1, and g_2.

Suppose that u is a solution of (28), and $\bar{u}$ is a solution of (28) with $\bar{f}$, $\bar{g}_1$, $\bar{g}_2$ replacing f, g_1, g_2 throughout. Then $u - \bar{u}$ is a solution of (PDE) with respective initial and boundary data $f - \bar{f}$, $g_1 - \bar{g}_1$, $g_2 - \bar{g}_2$. By the Maximal Principle, for all $t \geq 0$ and all $0 \leq x \leq L$,

$$|u(x,t) - \bar{u}(x,t)| \leq \max_{\substack{0 \leq \bar{x} \leq L \\ 0 \leq \bar{t} \leq t}} \left\{ |f(x) - \bar{f}(x)|, \ |g_1(\bar{t}) - \bar{g}_1(\bar{t})|, \ |g_2(\bar{t}) - \bar{g}_2(\bar{t})| \right\}$$

So small changes in boundary and initial data mean at most small changes in the temperature; the above inequality is the mathematical version of this assertion.

For simplicity we will now set the boundary data g_1 and g_2 equal to 0, reducing boundary/initial value problem (28) to the boundary/initial value problem (16). The series given in (21) with coefficients defined by (22) provides a formal solution to problem (16). The following results give additional properties of that solution.

THEOREM 10.8.4

Smoothing Properties. Suppose that the initial data belong to PC[0, L]. Then the formal solution $u(x,t)$ of (16) defined by (21), (22) belongs to $\mathbf{C}^\infty(R)$ and satisfies (PDE) in R.

The verification of this theorem rests on the Weierstrass M-Test. First observe from (22) that there is a positive constant A such that $|C_n| \leq A$, $n = 1, 2, \ldots$. For $n = 1, 2, \ldots$ and all (x,t) in the set S_{t_0} described by $0 \leq x \leq L, 0 < t_0 \leq t$,

$$C_n \sin\left(\frac{n\pi}{L}x\right)\exp\left[-K\left(\frac{n\pi}{L}\right)^2 t\right] \leq A\exp\left[-K\left(\frac{n\pi}{L}\right)^2 t_0\right]$$

Now the series $\sum \exp[-K(n\pi/L)^2 t_0]$ converges by the Ratio Test. By the Weierstrass M-Test, the series in (21) converges uniformly in S_{t_0} to a function that we call $u(x,t)$. If each term of the series in (21) is differentiated k times (r times in x and $k-r$ times in t, say), the n-th term of the derived series is no larger in magnitude than a term of the form $Bn^{k-r}\exp[-K(n\pi/L)^2 t_0]$ for all (x,t) in S_{t_0} and for some constant B (which may depend on k, but not on n). But the series $\sum_n n^{k-r}\exp[-K(n\pi/L)^2 t_0]$ also converges by the Ratio Test. By the Weierstrass M-Test and a basic theorem on uniform convergence, the derived series converges uniformly to a k-th derivative of $u(x,t)$ (i.e., to $\partial^k u/\partial x^r \partial t^{k-r}$) on the closed region S_{t_0}. Since all of the arguments above hold for every $t_0 > 0$ and for

☞ This basic theorem asserts that if $\sum f_n$ and $\sum f_n'$ converge uniformly on a set A, then $\sum f_n' = (\sum f_n)'$.

every $k = 1, 2, \ldots$, it follows that $u(x, t)$ belongs to $\mathbf{C}^\infty(R)$. A direct calculation involving the term by term differentiation of the series for u shows that (21) satisfies (PDE) in R.

That $u(x, t)$ possesses all derivatives of all orders is quite remarkable since the initial data are only required to be piecewise continuous. Visual evidence of the smoothing properties can be seen in Figure 10.8.1 where the sharp corner in the initial data is immediately rounded off for $t > 0$. Under additional assumptions on the initial data, the series in (21) defines a function $u(x, t)$ which is a classical solution of (16), that is, satisfies the initial conditions of (16) [as well as (PDE)].

THEOREM 10.8.5

> Classical Solution. Suppose that $f(0) = f(L) = 0$ and that $f(x)$ is continuous and piecewise smooth, $0 \le x \le L$. Then the function $u(x, t)$ defined by (21) and (22) belongs to $\mathbf{C}^\infty(R)$ and to $\mathbf{C}^0(R^*)$. Moreover, $u(x, t)$ is a solution of (16).

Comments

The "smoothing property" of the *heat operator* $(\partial/\partial t) - K(\partial^2/\partial x^2)$ is quite unlike any property of the wave operator $(\partial^2/\partial t^2) - c^2(\partial^2/\partial x^2)$. In fact, smoothing gives a direction to time that the wave operator can't. We can argue as follows in the context of (23) and (24) corresponding to piecewise linear initial data with a "corner." By the smoothing character of the heat operator, the temperature function lies in $\mathbf{C}^\infty(R)$ for all $t > 0$. So the "corner singularity" in the initial data does not propagate from ∂R into R, and the process $f(x) \to u(x, t)$ is irreversible. The sequence of temperature profiles in Figure 10.8.1 can't be read backward, and *time has an arrow for heat conduction.*

Although we haven't shown it here, corner singularities in initial data *do* propagate in time for boundary/initial value problems for the wave equation. With a little work it can be shown that the profiles for a plucked guitar string can be read forward or backward in time without distinction. So *time has no arrow for the wave equation in one spatial dimension.*

There is more. It can be shown that initial disturbances propagate with speed c under the influence of the wave equation. It may also be shown, but we will not do so here, that the *speed of propagation of an initial temperature disturbance is infinity.* In fact, suppose that the initial temperature of a uniform rod is $0°$ except for a temperature $T_0 > 0$ in some segment of small length at the middle of the rod. Then (with end temperatures maintained at $0°$) for any positive t, no matter how small, the solution $u(x, t)$ of the corresponding boundary/initial value problem is positive for every point x inside the rod. The thermal disturbance at the center of the rod has traveled infinitely fast and raised the temperature everywhere inside the rod.

The infinite speed of propagation for the heat equation and the periodic repetition of "corners" on initial displacements for the plucked guitar string and the wave operator show the defects of the respective mathematical models. Thermal diffusion and wave motion can't be perfectly modeled by the boundary/initial value problems constructed in this chapter. However, the mathematical models presented here are sufficiently accurate in most regards that they continue to be the models of first choice in the treatment of simple heat and wave phenomena.

PROBLEMS

1. (*Temperature in a Rod*). Use the Method of Separation of Variables to construct a series solution of the one-dimensional heat equation $u_t - Ku_{xx} = 0$, where $0 < x < L$, $t > 0$ and the following initial and boundary conditions are imposed. Describe how the temperature function $u(x, t)$ behaves as $t \to +\infty$. [*Hint*: See the subsection on temperature in a rod.]

   (a) $u(0, t) = u(L, t) = 0$, $\quad u(x, 0) = \sin(2\pi/L)x$

   (b) $u(0, t) = u(L, t) = 0$, $\quad u(x, 0) = x$

   (c) $u(0, t) = u(L, t) = 0$, $\quad u(x, 0) = u_0 > 0$, $\quad u_0$ a constant

   (d) $u(0, t) = u(L, t) = 0$, $\quad u(x, 0) = \begin{cases} u_0, & 0 \leq x \leq L/2 \\ 0, & L/2 < x \leq L \end{cases}$

   (e) $u(0, t) = 0$, $\quad u_x(L, t) = 0$, $\quad u(x, 0) = \sin(\pi x/(2L))$

   (f) $u(0, t) = 0$, $\quad u_x(L, t) = 0$, $\quad u(x, 0) = x$

   (g) $u_x(0, t) = u_x(L, t) = 0$, $\quad u(x, 0) = x$

2. (*Constant End Temperatures*).

   (a) Find the series solution of the problem

   $$u_t - Ku_{xx} = 0, \quad 0 < x < 1, \quad t > 0$$
   $$u(0, t) = 10, \quad u(1, t) = 20, \quad t \geq 0$$
   $$u(x, 0) = 0, \quad 0 < x < 1$$

   [*Hint*: Write $u(x, t) = A(x) + v(x, t)$, where $A''(x) = 0$, $A(0) = 10$, $A(1) = 20$, and

   $$v_t - Kv_{xx} = 0, \quad 0 < x < 1, \quad t > 0$$
   $$v(0, t) = 0, \quad v(1, t) = 0, \quad t \geq 0$$
   $$v(x, 0) = -A(x), \quad 0 \leq x \leq 1$$

   Then find v in the usual way once $A(x)$ has been found.]

   (b) Show that the *steady-state solution* in part (a) is $u = A(x)$ because $v(x, t) \to 0$ as $t \to \infty$.

3. (*Variable End Temperatures*). Suppose that u is a solution to the problem

   $$u_t - Ku_{xx} = 0, \quad 0 < x < 1, \quad t > 0$$
   $$u(0, t) = g_1(t), \quad u(1, t) = g_2(t), \quad t \geq 0$$
   $$u(x, 0) = f(x), \quad 0 \leq x \leq 1$$

   (a) If $V = g_1(t) + [g_2(t) - g_1(t)]x$ and if $U(x, t)$ satisfies the equations

   $$U_t - KU_{xx} = -KV_t, \quad 0 < x < 1, \quad t > 0$$
   $$U(0, t) = U(1, t) = 0, \quad t \geq 0$$
   $$U(x, 0) = f(x) - V(x, 0), \quad 0 \leq x \leq 1$$

   show that $u = V(x, t) + U(x, t)$.

(b) Find $U(x, t)$ if $g_1(t) = \sin t$ and $g_2(t) = 0$ [*Hint*: Use the Eigenfunction Expansion Method of Section 10.7.]

(c) Find $u(x, t)$. [*Hint*: Use your answers to parts **(a)** and **(b)**.]

4. (*Internal Sources/Sinks*). Use the Eigenfunction Expansion Method of Section 10.7 to solve the problem

$$u_t - Ku_{xx} = 3e^{-2t} + x, \qquad 0 < x < 1, \quad t > 0$$
$$u(0, t) = u(1, t) = 0, \qquad t \geq 0$$
$$u(x, 0) = 0, \qquad 0 \leq x \leq 1$$

www **5.** (*Wine Cellars*). Read the material in this section concerning the optimal depth of a wine cellar.

(a) Suppose that the surface temperature wave is $T_0 + A_0 \cos \omega t$, where A_0, T_0, and ω are positive constants. Find the optimal depth of the wine cellar. What is that depth if ω corresponds to 1 day instead of 1 year as in the example given in the text?

(b) Find the temperature function $u(x_0, t)$ at the optimal depth x_0.

(c) Formulate and solve the wine cellar problem where the surface wave has the form

$$T_0 + A_1 \cos \omega_1 t + A_2 \cos \omega_2 t$$

where ω_1 corresponds to 1 year and $\omega_2 = 365\omega_1$ corresponds to 1 day.

6. (*Time's Arrow*).

(a) Prove that the series (24) diverges for every x, $0 < x < 2$, if $t = t_0 < 0$.

(b) Explain this in terms of time's arrow.

(c) Find an initial data function $f(x) \neq 0$ so that (21) and (22) do give solutions defined for all time, even for $t < 0$. [*Hint*: Consider $f(x) = \sin(\pi x/L)$.]

7. (*Variable Boundary Temperatures*). Suppose that the temperature $u(x, t)$ in a rod of length 1 and diffusivity 1 satisfies the problem

$$u_t - u_{xx} = 0, \qquad\qquad 0 < x < 1, \quad t > 0$$
$$u(0, t) = te^{-t}, \qquad\qquad t \geq 0, \qquad u(1, t) = 0, \quad t \geq 0$$
$$u(x, 0) = 0.01x(1 - x), \qquad 0 \leq x \leq 1$$

Show that $|u(x, t)| \leq 1/e$ for $0 \leq x \leq 1$, $t \geq 0$.

10.9 Laplace's Equation

Laplace's equation is the homogeneous, linear, second-order PDE

$$\nabla^2 u = 0 \tag{1}$$

where ∇^2 is the Laplacian operator. Laplace's equation models steady-state temperatures in a body of constant material diffusivity. By "steady state" we mean that the temperature function u does not change with time, although it may vary from point to point within the body. Laplace's equation also models the gravitation and magnetic potentials in empty space, electric potential, and the velocity potential of ideal fluids. For these reasons, PDE (1) is also called the *potential equation*.

The operator has the following forms in various coordinate systems:

Rectangular $\nabla^2 = \partial^2/\partial x^2 + \partial^2/\partial y^2 + \partial^2/\partial z^2$ (2)

Polar $\nabla^2 u = u_{rr} + \dfrac{1}{r}u_r + \dfrac{1}{r^2}u_{\theta\theta}$ (3)

Cylindrical $\nabla^2 u = u_{rr} + \dfrac{1}{r}u_r + \dfrac{1}{r^2}u_{\theta\theta} + u_{zz}$ (4)

Spherical $\nabla^2 u = \dfrac{1}{\rho^2}(\rho^2 u_\rho)_\rho + \dfrac{1}{\rho^2 \sin\phi}(\sin\phi\, u_\phi)_\phi + \dfrac{1}{\rho^2 \sin^2\phi}u_{\theta\theta}$ (5)

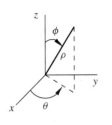

Recall that rectangular coordinates (x, y, z) in $\mathbb{R}^3$ are related to spherical coordinates (ρ, θ, ϕ) as follows: ϕ measures the angle from the vector $\mathbf{k}$ to the vector $\rho = x\mathbf{i} + y\mathbf{j} + z\mathbf{k}$, and θ measures the angle from $\mathbf{i}$ to the vector $\mathbf{r} = x\mathbf{i} + y\mathbf{j}$. So we have that

$$x = \rho\sin\phi\cos\theta, \qquad y = \rho\sin\phi\sin\theta, \qquad z = \rho\cos\phi$$

☞ *Warning*: Some people interchange θ and ϕ.

Laplace's equation has many solutions; for example,

$$u = c_1 e^{-x}\cos y + c_2 z + c_3 e^{-4z}\cos 4x$$

gives solutions in rectangular coordinates for all constants c_1, c_2, c_3, while

$$u = c_1 r\cos\theta + c_2 r^2\sin 2\theta$$

gives solutions of the two-dimensional Laplace's equation in polar coordinates for all c_1 and c_2. Boundary or boundedness conditions are needed to select a unique solution. We will solve representative problems in various coordinate systems and derive the fundamental properties of these solutions.

The Dirichlet Problem: Harmonic Functions

Suppose that G is a region in $\mathbb{R}^2$ or $\mathbb{R}^3$ and that h is a piecewise continuous function defined on ∂G, the boundary of G. Then the *Dirichlet problem* for G with *boundary data h* is defined as the following boundary value problem:

Dirichlet Problem

Find a function u in $\mathbf{C}^2(G)$ such that $\nabla^2 u = 0$ in G with the additional property that if P is a point of ∂G where h is continuous, then $\lim_{x\to P} u(x) = h(P)$, where x is in G.

With this understanding we will briefly write the Dirichlet problem for G with boundary data h as

$$\begin{array}{lll} \text{(PDE)} & \nabla^2 u = 0 & \text{in } G \\ \text{(BC)} & u = h & \text{on } \partial G \end{array} \qquad (6)$$

A solution of Laplace's equation is called a *harmonic function* or a *potential function*. Note that when h is in $\mathbf{C}^0(\partial G)$, a solution u of problem (6) belongs to $\mathbf{C}^0(G^*)$ and $u(P) = h(P)$ for all P in ∂G. In this case, problem (6) is called the *classical* Dirichlet problem and u is called a *classical solution* for the Dirichlet problem.

Steady-State Temperatures in the Unit Disk: Circular Harmonics

Suppose that $G = \{(x, y) : x^2 + y^2 < 1\}$ and that h is a given piecewise continuous function of ∂G. We will construct a solution of the Dirichlet problem in (6) for this region G and the given function h by using the Method of Separation of Variables. If the region G and the conditions (PDE) and (BC) are expressed in rectangular coordinates, separation of variables will *not* provide us with a formal solution to this problem because ∂G is not composed of level curves in the rectangular coordinate system. From this point of view, if we find a problem that is equivalent to (6) but expressed in terms of polar coordinates, separation of variables would have some chance for success. Using identity (3) we see that problem (6) is equivalent to the problem

$$
\begin{array}{lll}
\text{(PDE)} & w_{rr} + (1/r)w_r + (1/r^2)w_{\theta\theta} = 0 & \text{for } \theta \text{ in } \mathbb{R}, \quad 0 < r < 1 \\
\text{(BC)} & w(1, \theta) = f(\theta) & \text{for } \theta \text{ in } \mathbb{R}
\end{array}
\tag{7}
$$

where f is the function h expressed in terms of the polar angle θ and $w(r, \theta) = u(r\cos\theta, r\sin\theta)$. Observe that f belongs to $PC[-\pi, \pi]$ and that w is required to be periodic in θ with period 2π and twice continuously differentiable in the "strip," $0 < r < 1$, $-\infty < \theta < \infty$. Moreover, to ensure that u is twice continuously differentiable in a region containing the origin, we must impose the condition that for any θ_0 in $\mathbb{R}$, the limits of all derivatives of w up to second order exist as $(r, \theta) \to (0, \theta_0)$ and are independent of θ_0. We may interpret problem (7) as the model for steady-state temperatures in a thin homogeneous disk whose top and bottom faces are perfectly insulated and for which there are prescribed edge temperatures.

Following the Method of Separation of Variables to construct a formal solution of problem (7), we first look for all twice continuously differentiable solutions of (PDE) that have the form $R(r)\Theta(\theta)$. After substituting $R(r)\Theta(\theta)$ into (PDE), the variables can be separated and we are led to consider the separated ODEs

$$
\Theta'' - \lambda\Theta = 0, \qquad r^2 R'' + rR' + \lambda R = 0
\tag{8}
$$

where λ is the separation constant. We must demand that the solution $R(r)\Theta(\theta)$ be periodic in θ and smooth across $\theta = \pi$, so we must have the conditions $\Theta(-\pi) = \Theta(\pi)$, $\Theta'(-\pi) = \Theta'(\pi)$. We are led to consider the Sturm-Liouville problem with periodic boundary conditions

$$
\Theta'' - \lambda\Theta = 0, \qquad \Theta(-\pi) = \Theta(\pi), \qquad \Theta'(-\pi) = \Theta(\pi)
$$

But we have considered a similar problem in Example 10.6.2 and found that it has a nontrivial solution if and only if $\lambda = \lambda_n = -n^2$, $n = 0, 1, 2, \ldots$. The solutions of this Sturm-Liouville problem are given by

$$
\Theta_0(\theta) = 1, \qquad \Theta_n(\theta) = A_n \cos n\theta + B_n \sin n\theta, \qquad n = 1, 2, \ldots
$$

where A_n and B_n are arbitrary real numbers. Replacing λ by λ_n in the other separated equation in (8), we obtain the differential equations

$$
r^2 R'' + rR' - n^2 R = 0, \quad n = 0, 1, 2, \ldots
$$

☞ Euler's equation
appears in Problem 4 in
Section 3.7.
For each n this is an Euler equation that has the general solution

$$R(r) = Ar^n + Br^{-n} \quad \text{when } n = 1, 2, \ldots \tag{9a}$$

and

$$R(r) = A + B \ln r \quad \text{when } n = 0 \tag{9b}$$

But since R must be well behaved as $r \to 0^+$, we must take $B = 0$ and so

$$R_n = r^n, \qquad n = 0, 1, 2, \ldots$$

Let's look for a formal solution to problem (7) in the form

$$w(r, \theta) = \frac{A_0}{2} + \sum_{n=1}^{\infty} r^n (A_n \cos n\theta + B_n \sin n\theta) \tag{10}$$

We compute the A_n and B_n now by imposing the boundary condition

$$f(\theta) = \frac{A_0}{2} + \sum_{n=1}^{\infty} (A_n \cos n\theta + B_n \sin n\theta) \quad \text{for } -\pi \leq \theta \leq \pi$$

Recalling that $\{1, \cos x, \sin x, \ldots\}$ is a basis for $PC[-\pi, \pi]$, we have

$$A_n = \frac{1}{\pi} \int_{\pi}^{\pi} f(\theta) \cos n\theta \, d\theta, \qquad n = 0, 1, 2, \ldots$$
$$B_n = \frac{1}{\pi} \int_{\pi}^{\pi} f(\theta) \sin n\theta \, d\theta, \qquad n = 1, 2, \ldots \tag{11}$$

Solution (10) of the Dirichlet problem in (7) is an expansion in *circular harmonics*.

Solution (10) with coefficients defined by (11) is only a formal solution to the Dirichlet problem, since nothing has been said as yet about convergence properties. Although we will not show it, if the boundary function h is piecewise continuous on ∂G, then the function w defined by formulas (10) and (11) belongs to $\mathbf{C}^{\infty}(G)$ and satisfies the PDE in (8). Steady-state temperature functions have the same strong smoothness properties as the time-dependent temperature functions of the diffusion equation in Section 10.8. If, in addition, h is continuous on ∂G and also piecewise smooth, then it can be shown that formulas (10) and (11) define a classical solution of problem (7).

If the disk has radius $r_0 > 0$ rather than radius 1, then (10) and (11) define the solution of the corresponding Dirichlet problem if r in (10) is replaced by r/r_0.

Properties of Harmonic Functions

Harmonic functions have a number of distinctive properties. For simplicity all of the results below are stated only for planar regions.

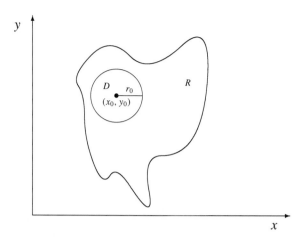

FIGURE 10.9.1 Geometry for the Mean Value Property.

THEOREM 10.9.1

Mean Value Property. Suppose that $w(x, y)$ is a solution of Laplace's equation in a planar region R. Suppose that (x_0, y_0) is a point of R, and D a disk of radius r_0 centered at $P(x_0, y_0)$ and lying entirely in R. Then

$$w(x_0, y_0) = \frac{1}{2\pi} \int_0^{2\pi} w(x_0 + r_0 \cos\theta, y_0 + r_0 \sin\theta) \, d\theta \qquad (12)$$

That is, the value of a harmonic function at a point is the average of its values around any circle centered at the point.

Figure 10.9.1 illustrates the geometrical setup in the statement of the theorem. To show that (12) is valid, first suppose that the polar coordinates r, θ are relative to the point P. Then using solution formula (10) with r replaced by r/r_0, we have

$$w(0, \theta) = \frac{A_0}{2}$$

where by (11) the coefficient A_0 is given by

$$\frac{1}{\pi} \int_{-\pi}^{\pi} u(x_0 + r_0 \cos\theta, y_0 + r_0 \sin\theta) \, d\theta$$

and we are done.

Harmonic functions also satisfy a Maximum Principle.

THEOREM 10.9.2

Maximum Principle for Harmonic Functions. Unless the harmonic function u is a constant function, u can't attain either its maximum or its minimum value inside a region R of $\mathbb{R}^2$.

To see this, suppose that u attains its maximum in R at some interior point P of R. Then this maximum value is the average of the values of u around the edge of any disk centered at P and lying in R (Theorem 10.9.1). But u is continuous, so $u(P)$ cannot be both an average and a maximum unless the values of u are constant. A similar argument applies to the minimum.

In other words, if R is a region and u is continuous on R^*, then the extreme values of a nonconstant harmonic function u are attained only on the boundary of R.

One of the central questions is whether a problem has exactly one solution and whether the solution changes continuously with the data. If so, the problem is said to be *well-posed*. The existence of a solution for an arbitrary Dirichlet problem is not easily shown, although for regions of simple shape, solutions may be constructed by the method of Separation of Variables (e.g., the Dirichlet problem solved above). However, the Maximum Principle may be applied to derive the other two aspects of a well-posed Dirichlet problem (i.e., no more than one solution and continuity in the data).

THEOREM 10.9.3

☞ A region is *bounded* if it can be contained in a ball.

> **Uniqueness and Continuity.** The Dirichlet Problem (6) for a bounded region G in $\mathbb{R}^2$ has no more than one continuous solution u in G^* if the boundary data h are continuous on ∂G. Moreover, the solution (if it exists) varies continuously with respect to the data.

To see this, suppose that $\nabla^2 u = \nabla^2 v = 0$ in G while $u = h$ and $v = h + \epsilon$ on ∂G, where h and ϵ are continuous on ∂G. We will show that

$$\min_{P \text{ on } \partial G} \epsilon(P) \le v - u \le \max_{P \text{ on } \partial G} \epsilon(P) \tag{13}$$

for all values of $v(Q) - u(Q)$, Q in G^*. If $|\epsilon(P)|$ is small, v is close to u and we have continuity in the data. Now to show inequality (13), let $w = v - u$. Then $\nabla^2 w = \nabla^2 v - \nabla^2 u = 0$ in G. By the Maximum Principle (and the corresponding Minimum Principle),

$$\min_{P \text{ on } \partial G} w(P) \le w(Q) \le \max_{P \text{ on } \partial G} w(P)$$

for all $w(Q)$, Q in G^*. But $w = v - u = h + \epsilon - h = \epsilon$ on ∂G, and we are done.

Next, we use the inequality (13) to show that problem (6) has no more than one solution. Suppose that u_1 and u_2 are solutions of problem (6). Then $U = u_1 - u_2$ satisfies (PDE) in G but with $U = 0$ on ∂G. By the same argument as above, $0 \le U \le 0$ for all values of $U(Q)$, Q in G^*. So $U = 0$, and $u_1 = u_2$: the Dirichlet problem in (6) has no more than one solution.

Comments

From the results above we see that harmonic functions have strong and distinctive properties. These properties may be interpreted in terms of steady-state temperatures within a thin plate R whose top and bottom faces are perfectly insulated and whose edge temperatures are prescribed.

PROBLEMS

1. (*Plate Temperatures*).

(a) Show that the formal solution of the steady-state temperature problem in the rectangular plate $G: 0 < x < L, \ 0 < y < M$, modeled by

$$
\begin{array}{llll}
\text{(PDE)} & u_{xx} + u_{yy} = 0 & \text{in } G \\
\text{(BC)}_1 & u(0, y) = \alpha(y), & 0 \le y \le M \\
\text{(BC)}_2 & u(L, y) = 0, & 0 \le y \le M \\
\text{(BC)}_3 & u(x, 0) = 0, & 0 \le x \le L \\
\text{(BC)}_4 & u(x, M) = 0, & 0 \le x \le L
\end{array}
$$

is given by

$$
u_\alpha(x, y) = \sum_{n=1}^{\infty} A_n \sin \frac{n\pi y}{M} \sinh \frac{n\pi}{M}(L - x)
$$

where

$$
A_n \sinh(n\pi L/M) = \frac{\langle \alpha, \sin(n\pi y/M) \rangle}{\| \sin(n\pi y/M) \|^2} = \frac{2}{M} \int_0^M \alpha(y) \sin(n\pi y/M) \, dy
$$

(b) Show that the formal solution of the problem in G with general boundary temperatures

$$
\begin{array}{llll}
\text{(PDE)} & u_{xx} + u_{yy} = 0 & \text{in } G \\
\text{(BC)}_1 & u(0, y) = \alpha(y), & 0 \le y \le M \\
\text{(BC)}_2 & u(L, y) = \beta(y), & 0 \le y \le M \\
\text{(BC)}_3 & u(x, 0) = \gamma(x), & 0 \le x \le L \\
\text{(BC)}_4 & u(x, M) = \delta(x), & 0 \le x \le L
\end{array}
$$

is $u = u_\alpha + u_\beta + u_\gamma + u_\delta$, where u_α is given in part **(a)** and $u_\beta, u_\gamma, u_\delta$ are defined analogously.

(c) Repeat part **(a)** with an insulation condition $u_x(0, y) = \tilde{\alpha}(y), \ 0 \le y \le M$ replacing (BC)$_1$.

(d) Repeat part **(a)** with (BC)$_3$ and (BC)$_4$ replaced by the "perfect insulation" conditions $u_y(x, 0) = 0 = u_y(x, M), \ 0 \le x \le L$.

2. (*Temperatures in a Ring*). A thin ring $G = \{(r, \theta) : \rho < r < R, \ -\pi \le \theta \le \pi\}$ has insulated faces. Find the steady temperature in G if the boundaries are kept at the temperatures $f(\theta)$ and $g(\theta)$ at $r = \rho$ and $r = R$, respectively. [*Hint:* Proceed as in the text for a disk, but keep both terms in formulas (9a), (9b). So in determining the superposition constants we have

$$
f(\theta) = A_0 + B_0 \ln \rho + \sum_n \rho^n [A_n \cos n\theta + B_n \sin n\theta]
$$

$$
+ \sum_n \rho^{-n} [C_n \cos n\theta + D_n \sin n\theta]
$$

and a similar expression with f replaced by g, and ρ by R.]

www 3. (*Temperatures in a Disk*). Solve the Dirichlet problem for steady-state temperatures in the unit disk. Find the maximum and minimum temperatures in the disk:

(a) If $f(\theta) = 3 \sin \theta$ on the edge.

(b) If $f(\theta) = \theta + \pi$ for $-\pi \le \theta < 0$; $f(\theta) = -\theta + \pi$ for $0 \le \theta \le \pi$.

(c) If $f(\theta) = 0$ for $-\pi < \theta < 0$; $f(\theta) = 1$ for $0 < \theta \le \pi$.

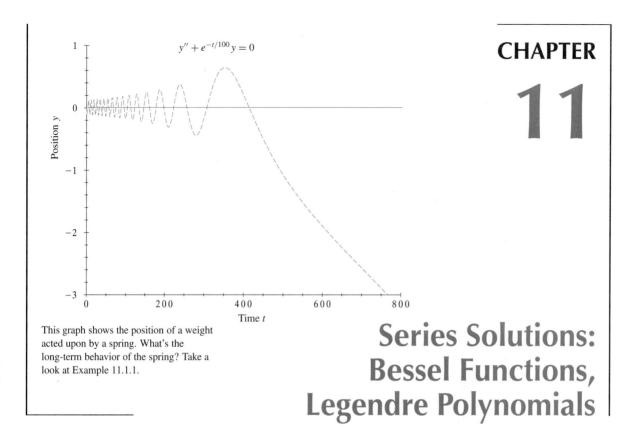

$$y'' + e^{-t/100}y = 0$$

This graph shows the position of a weight acted upon by a spring. What's the long-term behavior of the spring? Take a look at Example 11.1.1.

Series Solutions: Bessel Functions, Legendre Polynomials

Linear ordinary differential equations that model natural processes often have nonconstant coefficients, and the techniques of earlier chapters for finding solution formulas may not apply. In this section we will show how formulas for the solutions of second-order nonconstant coefficient linear ODEs can be expressed in the form of series, if the coefficients themselves can be expressed as power series.

11.1 Aging Springs and Steady Temperatures

The vibrations of an aging spring, the isotherms in a solid cylinder or sphere partly covered with ice, and many other natural processes lead to second-order, nonconstant coefficient linear ODEs. We learned how to find a solution formula for *first*-order linear ODEs with nonconstant coefficients, but so far we haven't done the same for *second*-order linear ODEs. This chapter will help fill that gap.

☞ It might be a good idea to look over Section 3.7 before starting this chapter.

We begin with two examples, and then introduce the method of series to express the solutions of these model ODEs. Let's look first at an aging spring.

EXAMPLE 11.1.1

☞ Springs were modeled in Section 3.1. You might check it out.

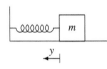

☞ Take a look at Section 3.5 for a discussion of these sinusoidal solutions.

Aging Springs

The spring force of real springs gradually diminishes with age until the spring stretches beyond its elastic limit. Springs are the core elements in openers, closers, dampers, and controllers, and their failure can be catastrophic. So good models for the motion of a spring are of considerable importance. In Section 3.1, we saw that the model of the motion of a body of mass m moving on a smooth table without friction and attached to a wall by an "ageless" Hooke's Law spring is given by

$$my''(t) + ky(t) = 0$$

where k is the spring constant, and the displacement $y(t)$ is measured from the spring's unstretched (and uncompressed) state. The solutions of this ODE are sinusoids that oscillate with period $2\pi\sqrt{m/k}$.

Now let's model the gradual loss of elasticity by replacing the constant k by a decaying function of time $ke^{-\varepsilon t}$, for $\varepsilon > 0$:

$$my''(t) + ke^{-\varepsilon t}y(t) = 0 \qquad (1)$$

What are solutions $y(t)$ of ODE (1) like? The chapter cover figure tells you what to expect. We might expect $y(t)$ to oscillate as long as the elastic coefficient $ke^{-\varepsilon t}$ remains significant. Eventually, however, the restoring force will be so weak that the energy of motion will overwhelm everything else, and the model is no longer valid.

The graph in the chapter cover figure was produced by a numerical solver, but sometimes it's useful to have a solution formula instead. Although ODE (1) is a linear ODE, the solution formula methods of Chapters 3, 4, and 6 do not directly apply because the coefficients are not constant. In Example 11.2.9 we use a method of undetermined coefficients to find solutions of ODE (1) as power series

$$y(t) = a_0 + a_1 t + a_2 t^2 + \cdots + a_n t^n + \cdots$$

where $a_0 = y(0)$, $a_1 = y'(0)$. In Section 11.7 we solve ODE (1) using Bessel functions that are themselves defined in terms of series.

The following example starts with a partial differential equation (PDE) and leads to a second-order linear ODE with nonconstant coefficients.

EXAMPLE 11.1.2

☞ If you haven't covered Section 10.9 you might want to skip this example.

Steady Temperatures in a Cylinder

A solid iron cylinder of radius 1 and height a is enclosed in a blanket of ice that keeps the top and the sidewall at temperature zero. A heat source is fixed to the bottom and keeps the temperatures there at levels that depend only on the radial distance from the cylinder's central axis. Let's find the steady-state temperatures inside the cylinder.

The cylinder is given in cylindrical coordinates by $0 \leq r \leq 1$, $-\pi \leq \theta < \pi$, $0 \leq z \leq a$. If the temperature at the point (r, θ, z) is denoted by $u(r, \theta, z)$, then the temperature in the cylinder is modeled by the boundary problem

$$\nabla^2 u = u_{rr} + \frac{1}{r}u_r + \frac{1}{r^2}u_{\theta\theta} + u_{zz} = 0, \qquad 0 < r < 1, \quad 0 < z < a, \quad -\pi < \theta < \pi$$

$$u(1, \theta, z) = 0, \qquad 0 \le z \le a, \quad -\pi \le \theta < \pi$$

$$u(r, \theta, a) = 0, \qquad 0 \le r \le 1, \quad -\pi \le \theta < \pi \tag{2}$$

$$u(r, \theta, 0) = f(r), \qquad 0 \le r \le 1, \quad -\pi \le \theta < \pi$$

Since the boundary data do not depend on θ, we suspect that the solution u of the boundary value problem (2) is a function of r and z only, that is, $u = u(r, z)$.

Following the Method of Separation of Variables described in Section 10.7, we look for separated solutions of $\nabla^2 u = 0$, that is, solutions of the form $u = R(r)Z(z)$. Inserting $R(r)Z(z)$ for u into $\nabla^2 u = 0$ and recognizing that $u_{\theta\theta} = 0$ [since $u(r, z)$ is assumed to be independent of θ], we have

$$R''(r)Z(z) + \frac{1}{r}R'(r)Z(z) + R(r)Z''(z) = 0$$

Separating the variables, and introducing the separation constant $-\lambda$, we see that

$$\frac{1}{R}R'' + \frac{1}{rR}R' = -\frac{Z''}{Z} = -\lambda$$

In particular, we have a second-order, nonconstant coefficient linear ODE for $R(r)$:

$$rR'' + R' + \lambda r R = 0 \tag{3}$$

Some solutions of ODE (3) may be written as power series

$$R(r) = b_0 + b_1 r + b_2 r^2 + \cdots + b_n r^n + \cdots$$

Section 11.6 gives a method for finding the "undetermined" coefficients b_n.

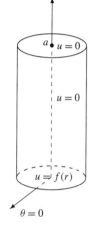

z

$a \bullet u = 0$

$u = 0$

$u = f(r)$

$\theta = 0$

Series Approach to Solving ODEs

☞ It might be a good idea to check out Appendix B.2 on power series at this time.

Examples 11.1.1 and 11.1.2 show how second-order linear ODEs with nonconstant coefficients may arise in modeling natural processes. The following example uses a constant coefficient ODE that we already know how to solve and shows how these solutions can be constructed in the form of series.

EXAMPLE 11.1.3

☞ Notice how cleverly we snuck in the variable x to replace t; we'll use x as the independent variable through much of the chapter.

Power Series Approach to Solving ODEs

We saw in Section 3.4 initial value problem

$$y''(x) + y(x) = 0, \qquad y(0) = 1, \quad y'(0) = 0 \tag{4}$$

has the unique solution $y(x) = \cos x$. Let's see how we can use the ODE itself to construct the solution of IVP (4) in the form of a power series.

Suppose that IVP (4) has a solution in the form of a power series

$$y = a_0 + a_1 x + a_2 x^2 + \cdots + a_n x^n + \cdots = \sum_{n=0}^{\infty} a_n x^n \tag{5}$$

that converges on an interval containing the initial value $x_0 = 0$. The problem is to determine the coefficients a_n such that the sum of the series solves IVP (4).

☞ Differentiating a power series term by term is covered in Appendix B.2, item 4.

Now we know that a power series can be differentiated term by term inside its convergence interval. So differentiating $y = \sum_{n=0}^{\infty} a_n x^n$, we have

$$y' = \sum_{n=0}^{\infty} n a_n x^{n-1} = \sum_{n=1}^{\infty} n a_n x^{n-1}$$

$$y'' = \sum_{n=1}^{\infty} n(n-1) a_n x^{n-2} = \sum_{n=2}^{\infty} n(n-1) a_n x^{n-2} \qquad (6)$$

☞ Zero terms are often dropped from series.

where terms with a coefficient of zero have been dropped from the sums. The first two coefficients a_0 and a_1 are determined by formulas (5), (6), and the initial data:

$$1 = y(0) = a_0 + a_1 \cdot 0 + a_2 \cdot 0^2 + \cdots = a_0$$

$$0 = y'(0) = a_1 + 2a_2 \cdot 0 + 3a_3 \cdot 0^2 + \cdots = a_1$$

Inserting the sums (5), (6) into the ODE of IVP (4) we have

$$y'' + y = \sum_{n=2}^{\infty} n(n-1) a_n x^{n-2} + \sum_{n=0}^{\infty} a_n x^n = 0$$

☞ So lower the range by 2 and increase n by 2 inside the sum to compensate.

To write the sum of these two power series as a single power series, we need to reindex the first summation to make the summand involve x^n instead of x^{n-2}:

$$\sum_{n=0}^{\infty} (n+2)(n+1) a_{n+2} x^n + \sum_{n=0}^{\infty} a_n x^n = 0$$

$$\sum_{n=0}^{\infty} [(n+2)(n+1) a_{n+2} + a_n] x^n = 0 \qquad (7)$$

☞ We use the Identity Theorem a lot in this chapter.

Since zero is represented by the power series $\sum [0] x^n$, the Identity Theorem (see Appendix B.2, item 5) tells us that if the identity (7) holds for all x near $x_0 = 0$, then

$$(n+2)(n+1) a_{n+2} + a_n = 0, \qquad n = 0, 1, 2, \ldots$$

So we have a *recursion formula* for the coefficients:

$$a_{n+2} = -\frac{a_n}{(n+2)(n+1)}, \qquad n = 0, 1, 2, \ldots \qquad (8)$$

Using the recursion formula (8) and the fact $a_0 = 1$, we see that

$$a_2 = -\frac{a_0}{2 \cdot 1} = -\frac{1}{2}, \quad a_4 = -\frac{a_2}{4 \cdot 3} = \frac{1}{24}, \quad a_6 = -\frac{1}{720}, \quad \ldots$$

So, in general, for $k = 1, 2, \ldots$ we have

$$a_{2k} = -\frac{a_{2k-2}}{2k(2k-1)} = \frac{a_{2k-4}}{2k(2k-1)(2k-2)(2k-3)} = \cdots = \frac{(-1)^k}{(2k)!} a_0 = \frac{(-1)^k}{(2k)!}$$

which determines coefficients with even subscripts. For the coefficients with odd subscripts the recursion formula (8) and the fact that $a_1 = 0$ show that

$$a_1 = 0, \quad a_3 = 0, \quad a_5 = 0, \quad \ldots, \quad a_{2k+1} = 0, \qquad \text{for all } k = 1, 2, \ldots$$

Using the values of $a_0, a_2, a_4, a_6, \ldots, a_{2k}, \ldots$ determined above, we have

☞ The Taylor
series for $\cos x$ is in
Appendix B.2, item 9.

$$y(x) = 1 - \frac{1}{2}x^2 + \frac{1}{24}x^4 - \frac{1}{720}x^6 + \cdots + \frac{(-1)^k}{(2k)!}x^{2k} + \cdots = \sum_{k=0}^{\infty} \frac{(-1)^k}{(2k)!}x^{2k} \qquad (9)$$

This is the Taylor series of $\cos x$ about the point $x_0 = 0$, as expected.

The Ratio Test can be applied to determine the convergence interval:

☞ The Ratio Test
is given in
Appendix B.2, item 3.

$$\lim_{k \to \infty} \left| \frac{(-1)^{k+1} x^{2k+2}/(2k+2)!}{(-1)^k x^{2k}/(2k)!} \right| = \lim_{k \to \infty} \frac{x^2}{(2k+2)(2k+1)} = 0$$

for all x. So the series converges to $\cos x$ for all x (but we knew that already).

Solutions in power series form have several uses; for example, they often help us to estimate the values of a solution, as we now show.

EXAMPLE 11.1.4

How Good is a Polynomial Approximation?

Let's use the sum of the first three nonzero terms $p_4(x) = 1 - x^2/2 + x^4/24$ of the alternating series (9) to approximate the solution $y(x)$ of IVP (4) over the interval $|x| \leq 0.1$. On that interval, each term of the alternating series is smaller in magnitude than the preceding term, and the limiting value as $k \to \infty$ of the magnitude of the k-th term $x^{2k}/(2k)!$ is zero for each fixed value of x. So the Alternating Series Test (Appendix B.2, item 10) implies that

$$|y(x) - p_4(x)| \leq \left| \frac{-x^6}{720} \right| \leq 1.4 \cdot 10^{-9}, \quad \text{for } |x| \leq 0.1$$

where $-x^6/720$ is the next term in the series after the three terms of $p_4(x)$. The polynomial $p_4(x)$ is indeed a good approximation to $y(x)$ if $|x| \leq 0.1$.

Because of the convergence properties of power series, we would expect the sum of the first several terms in the series to be a useful approximation near the base point x_0 of the expansion ($x_0 = 0$ in Examples 11.1.3 and 11.1.4), and less useful at some distance from that point. See Figure 11.1.1 for a graph of the solution $y(x)$ of IVP (4) and overlays of some of the polynomial approximations:

$$p_{2k}(x) = 1 - \frac{1}{2}x^2 + \frac{1}{24}x^4 + \cdots + \frac{(-1)^{2k}}{(2k)!}x^{2k} \qquad (10)$$

As k increases, the curves "follow" the exact solution over a longer and longer x-interval before turning away. Observe that the degree $2k$ of p_{2k} must be quite large to obtain a good approximation on the full interval $0 \leq x \leq 8$.

The series approach may be used to solve linear ODEs where, unlike in Example 11.1.3, there are no known closed-form solution formulas. In the nonconstant coefficient case the details change slightly, but the basic process is the same. Let's summarize the steps in this series approach when the coefficients are polynomials.

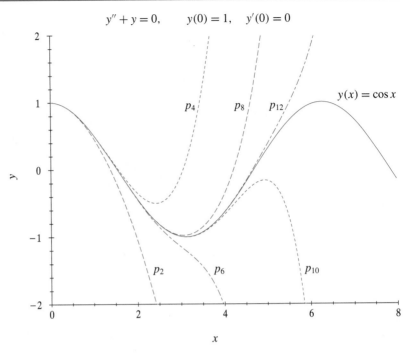

$$y'' + y = 0, \qquad y(0) = 1, \quad y'(0) = 0$$

FIGURE 11.1.1 The solution curve $y(x)$ of an IVP and some of its truncated series approximations (10).

Method of Power Series for $y'' + P(x)y' + Q(x)y = 0$

☞ More on this in the
next section (*much* more).

Here are the key steps in the *Method of Power Series* when the coefficients $P(x)$ and $Q(x)$ are *polynomials* and the base point of the series solutions is $x_0 = 0$:

- Insert $y = \sum_{n=0}^{\infty} a_n x^n$ with undetermined coefficients a_n into the ODE.
- Write the resulting equation as a power series (reindexing if necessary).

$$\sum_{n=0}^{\infty} [\text{terms involving various } a_n] x^n = 0$$

- Use the Identity Theorem to equate each bracketed term $[\cdots]$ to 0; solve these equations recursively for a_n, $n \geq 2$, using the known values $a_0 = y(0)$, $a_1 = y'(0)$.

Comments

Another reason that power series solutions are important is that numerical solvers based on power series can be used to approximate solutions of diffential equations. Most digital computations are based on the four arithmetical operations, and a polynomial approximation like $p_{2k}(x)$ in (10) is well-suited for such calculations. Power series techniques are some of the most widely used methods for constructing solutions of linear ODEs.

PROBLEMS

1. Determine the interval of convergence for each of the following series by using the Ratio Test. [*Hint*: The Ratio Test is given in Appendix B.2, item 3.]

   (a) $\sum_{n=0}^{\infty} \dfrac{x^n}{n!}$ **(b)** $\sum_{n=0}^{\infty} \dfrac{nx^n}{2^n}$ **(c)** $\sum_{n=1}^{\infty} \dfrac{x^{2n+1}}{(2n+1)!}$

2. Expand each function in a Taylor series about the given point, and find the interval of convergence. Over the given interval, plot the graphs of the function and its polynomial approximations using the first 2, 3, and 4 nonzero terms of the series.

   (a) e^x, $x_0 = 0$, $|x| \le 2$ **(b)** $\sin x$, $x_0 = 0$, $|x| \le \pi$

   (c) $(1+x)^{-1}$, $x_0 = 0$, $|x| < 1$ **(d)** $\sqrt{x}$, $x_0 = 1$, $|x - 1| < 1$

3. Reindex each of the following as a series of form $\sum_n [\cdots] x^n$.

   (a) $\sum_{n=0}^{\infty} \dfrac{2(n+1)}{n!} x^{n+1}$ **(b)** $\sum_{n=2}^{\infty} n(n-1)a_n x^{n-2}$ **(c)** $\sum_{n=1}^{\infty} (-1)^{n-1} \dfrac{x^{n+1}}{n(n+1)}$

4. Use the Taylor series of e^x, $\sin x$, $\cos x$, and $1/(1+x)$ about the base point $x_0 = 0$ to obtain the Taylor series of each of the following functions. [*Hint*: See Appendix B.2, item 9.]

   (a) $e^x + e^{-x}$ **(b)** $\sin x - \cos x$ **(c)** $\sin x \cos x$ **(d)** $e^x/(1+x)$

5. Use the Identity Theorem [Appendix B.2, item 5] to determine a recursion formula for a_n in terms of a_k, $k < n$. Then "solve" each recursion formula to express each a_n in terms of a_0 or a_1.

   (a) $\sum_{n=1}^{\infty} (na_n - n + 2)x^n = 0$ **(b)** $\sum_{n=1}^{\infty} [(n+1)a_n - a_{n-1}]x^n = 0$

   (c) $\sum_{n=0}^{\infty} [(n+2)(n+1)a_{n+2} - a_n]x^n = 0$

6. Show that $y = \sum_0^{\infty} x^n/n!$ is the solution of $y'' - y = 0$, $y(0) = 1$, $y'(0) = 1$ by direct substitution. How is this solution expressed in terms of elementary functions?

7. Use the Method of Power Series to perform the tasks below:

   (a) Solve the IVP $y'' - 4y = 0$, $y(0) = 2$, $y'(0) = 2$. Express the solution in terms of exponentials, and describe the connection between this form of the solution and the series.

    (b) Plot the solution and the graphs of the polynomial approximations that use the first 2, 3, and 4 nonzero terms of the series over the interval $|x| \le 1$.

8. Use the Identity Theorem [Appendix B.2, item 5] to solve the IVP. [*Hint*: Try $y = \sum_2^{\infty} a_n x^n$.]

$$y'' - y = \sum_{n=0}^{\infty} x^n, \qquad y(0) = 0, \quad y'(0) = 0, \quad \text{where } |x| < 1$$

www 9. Find the coefficient a_{100} in the series $\sum_{n=0}^{\infty} a_n x^n$ if it is known that $a_0 = a_1 = 1$ and that

$$\sum_{n=0}^{\infty} [(n+1)^2 a_{n+2} - n^2 a_{n+1} + (n-1)a_n]x^n = 0$$

10. Find a series solution of ODE (3) with $\lambda = 1$ such that $R(0) = 1$, $R'(0) = 0$. Find a recursion formula and the first four nonzero terms of the series expansion.

11. (*Aging Spring*). Create a model of an aging spring for which the spring constant is zero after a finite time. Include the following in the analysis of your model.

    - What would you expect to happen to the spring in the long run? Compare your result with what happens in a model ODE with zero spring constant.

    - What is the effect of including damping in the aging spring model?

    - Plot graphs of solutions of each model created, and explain solution behavior.

See Appendix B.2, item 8, for Taylor series.

11.2 Series Solutions Near an Ordinary Point

If the coefficients in a second-order linear ODE are not constant but functions of the independent variable, then sometimes power series can be used to construct solutions. As we saw in Example 11.1.3, a solution in the form of a power series is assumed, and the coefficients of the series are determined after substituting the series into the ODE.

The ODE $a_2(x)y'' + a_1(x)y' + a_0(x)y = 0$ can be written in *normalized form* as

$$y'' + P(x)y' + Q(x)y = 0 \tag{1}$$

after dividing by the leading coefficient $a_2(x)$. ODE (1) is a more convenient form for expressing the results of this section. We present a method for finding power series solutions of the form $\sum a_n(x - x_0)^n$ for ODE (1) when P and Q are *real analytic* at x_0, that is, when the coefficients $P(x)$ and $Q(x)$ can be written as convergent power series in powers of $x - x_0$ (see Appendix B.2, item 8). At the end of the section we use the method to find a series solution for the equation of the aging spring.

☞ Solutions of the form $\sum a_n(x - x_0)^n$ are appropriate when initial data is given at $x = x_0$.

Ordinary Points, Singular Points

We will find that all the solutions of ODE (1) near a point x_0 are power series in powers of $x - x_0$, where x_0 is an *ordinary point* of the ODE. An ordinary point is defined as follows:

❖ **Ordinary Point.** The point x_0 is an *ordinary point* of ODE (1) if both $P(x)$ and $Q(x)$ are real analytic at x_0.

The opposite of "ordinary" is "singular":

❖ **Singular Point.** The point x_0 is a *singular point* of ODE (1) if $P(x)$ and $Q(x)$ are *not* both real analytic at x_0.

If x_0 is an ordinary point, $P(x)$ and $Q(x)$ have Taylor series

$$P(x) = \sum_{n=0}^{\infty} p_n(x - x_0)^n, \qquad Q(x) = \sum_{n=0}^{\infty} q_n(x - x_0)^n \tag{2}$$

that converge inside a common interval centered at x_0. The coefficients p_n and q_n of the Taylor series (2) are

$$p_n = \frac{P^{(n)}(x_0)}{n!}, \qquad q_n = \frac{Q^{(n)}(x_0)}{n!}$$

The identification of the ordinary points of (1) is the first step in using the Method of Power Series. Let's look at some examples that show how to do this.

EXAMPLE 11.2.1

Polynomial Coefficients

Every point x_0 is an ordinary point for ODE (1) if the coefficients $P(x)$ and $Q(x)$ are polynomials in x. For instance, the polynomial $P(x) = 1 - x^2$ can be written as a finite Taylor series about x_0:

$$P(x) = P(x_0) + P'(x_0)(x - x_0) + \frac{1}{2}P''(x_0)(x - x_0)^2$$
$$= (1 - x_0^2) - 2x_0(x - x_0) - (x - x_0)^2$$

So, for example, every point is an ordinary point for $y'' + (1 - x^2)y' + y = 0$.

EXAMPLE 11.2.2

More Ordinary Points

Every point x_0 is an ordinary point of $y'' + e^{-x}y = 0$ since $Q(x) = e^{-x}$ has a convergent Taylor series in powers of $x - x_0$ given by

$$e^{-x} = \sum_{n=0}^{\infty} \frac{1}{n!}(-1)^n e^{-x_0}(x - x_0)^n \qquad (3)$$

which is a series that converges to e^{-x} for all x.

EXAMPLE 11.2.3

☞ Appendix B.2, item 9, has geometric series.

An Ordinary Point and a Singular Point

The point $x_0 = 0$ is an ordinary point of the ODE $y'' + [1/(1 - x)]y = 0$ since $1/(1 - x)$ has a convergent geometric series expansion about $x_0 = 0$:

$$\frac{1}{1 - x} = \sum_{n=0}^{\infty} x^n, \quad \text{where} \quad |x| < 1$$

However, $x = 1$ is a singular point of the ODE since $(1 - x)^{-1}$ is not defined at $x = 1$.

EXAMPLE 11.2.4

Legendre's Equation

The ODE $(1 - x^2)y'' - 2xy' + p(p + 1)y = 0$, where p is a nonnegative constant, is called *Legendre's equation of order p*. In normalized form (1) the coefficients are

$$P(x) = -2x(1 - x^2)^{-1}, \qquad Q(x) = p(p + 1)(1 - x^2)^{-1}$$

The ODE has two singularities, $x = 1$ and $x = -1$. All other points are ordinary because a quotient of polynomials is real analytic everywhere except at the roots of the polynomial in the denominator.

EXAMPLE 11.2.5

A Singular Point

The ODE $y'' + |x|y = 0$ has a singularity at 0 because $|x|$ is not differentiable at $x_0 = 0$, so $|x|$ has no Taylor series expansion about $x_0 = 0$. All other points x_0 are ordinary because $|x|$ is real analytic at any point $x_0 \neq 0$.

Sometimes an apparent singularity is actually an ordinary point. For example, the ODE $y'' + [(x - 1)/(x^2 - 1)]y = 0$ has an apparent singularity at $x = 1$, but this is easily removed by factoring $x^2 - 1$ and canceling the common factor $x - 1$ to obtain $y'' + y/(x+1) = 0$. So 1 is a *removable singularity* (actually, an ordinary point) for this ODE. Removable singularities are treated as ordinary points. The point -1 is a singularity however the given ODE is written, so -1 is *not* a removable singularity.

Series Solutions Near an Ordinary Point

Suppose x_0 is an ordinary point of the ODE $y'' + P(x)y' + Q(x)y = 0$. Theorem 3.7.5 guarantees that all solutions of the ODE that live in a neighborhood of x_0 are generated by a basic solution set $\{y_1(x), y_2(x)\}$. We will actually show that all solutions of the ODE near x_0 have the form of the power series $\sum_0^\infty a_n(x - x_0)^n$ and are real analytic at x_0. For the moment let's assume that solutions of this form do exist and find conditions on the coefficients a_n that turn the series into a solution. Here's an example.

EXAMPLE 11.2.6

☞ Problem 6 of Section 11.3 discusses the general form of Hermite's equation.

Hermite's Equation

The second-order ODE

$$y'' - 2xy' + 2y = 0 \tag{4}$$

is a particular form of *Hermite's equation*. In this ODE we see that $P(x) = -2x$ and $Q(x) = 2$. All points x_0 are ordinary and we look for series solutions of the form $\sum a_n(x - x_0)^n$. If initial data $y(x_0) = y_0$, $y'(x_0) = y_0'$ are given, then $a_0 = y_0$ and $a_1 = y_0'$. For simplicity we set $x_0 = 0$ and find solutions of the form $y = \sum_{n=0}^\infty a_n x^n$. We will find the coefficients a_n, $n = 2, 3, 4, \ldots$, in terms of a_0 and a_1. Differentiating the series term by term, we see that inside the interval of convergence,

$$y' = \sum_{n=1}^\infty n a_n x^{n-1} \quad \text{and} \quad y'' = \sum_{n=2}^\infty n(n-1) a_n x^{n-2}$$

Substituting the series for y, y', and y'' into ODE (4), we have

$$\sum_{n=2}^\infty n(n-1) a_n x^{n-2} - 2x \sum_{n=1}^\infty n a_n x^{n-1} + 2 \sum_{n=0}^\infty a_n x^n = 0 \tag{5}$$

☞ Keys to the success of this method are reindexing so that all of the sums are in terms of the same power of x (usually x^n) and so that the sums start at the same index value (usually $n = 0$).

Reindexing the first two sums, we obtain

$$\sum_{n=2}^\infty n(n-1) a_n\ x^{n-2} = \sum_{n=0}^\infty (n+2)(n+1) a_{n+2} x^n$$

$$-2x \sum_{n=1}^\infty n a_n x^{n-1} = \sum_{n=1}^\infty 2(-n) a_n x^n = \sum_{n=0}^\infty 2(-n) a_n x^n \tag{6}$$

Using (6), we combine all of the series in (5) into a single sum as

$$\sum_{n=0}^\infty [(n+2)(n+1) a_{n+2} - 2n a_n + 2 a_n] x^n = 0 \tag{7}$$

By the Identity Theorem (Appendix B.2, item 2),

$$(n+2)(n+1) a_{n+2} - 2n a_n + 2 a_n = 0, \qquad n = 0, 1, 2, \ldots \tag{8}$$

☞ Steps (7), (8), and (9) are key steps.

and so we have the *recursion formula* for the coefficients:

$$a_{n+2} = \frac{2(n-1)a_n}{(n+2)(n+1)}, \qquad n = 0, 1, 2, \ldots \tag{9}$$

Given a_0 and a_1, we can use recursion formula (9) to determine all the other coefficients a_n. For example,

$$a_2 = \frac{-2a_0}{2 \cdot 1} = -a_0, \quad a_4 = \frac{2a_2}{4 \cdot 3} = -\frac{1}{6}a_0, \quad a_6 = \frac{6a_4}{6 \cdot 5} = -\frac{1}{30}a_0, \quad \ldots$$

$$a_3 = \frac{0a_1}{3 \cdot 2} = 0, \quad a_5 = \frac{4a_3}{5 \cdot 4} = 0, \quad \ldots, \quad a_{2k+1} = 0, \quad n \geq 1$$

The vanishing of all but the first of the odd-indexed coefficients (i.e., a_1) is a fortunate occurrence here. Solutions of ODE (4) can be written as

$$y = a_0 \left[1 - x^2 - \frac{1}{6}x^4 - \frac{1}{30}x^6 - \cdots \right] + a_1 x \tag{10}$$

where a_0 and a_1 are arbitrary constants. In particular, if we let $a_0 = 1$, $a_1 = 0$ and then $a_0 = 0$, $a_1 = 1$, we have two solutions of ODE (4):

$$y_1 = 1 - x^2 - \frac{1}{6}x^4 - \frac{1}{30}x^6 - \cdots, \qquad y_2 = x \tag{11}$$

Observe that $y_1(x)$ and $y_2(x)$ are a basic solution set for ODE (4) because their Wronskian $W[y_1, y_2](0) \neq 0$ (Theorems 3.7.3 and 3.7.4). So every solution of ODE (4) is a linear combination of the solutions $y_1(x)$ and $y_2(x)$.

There are two questions about the solutions of ODE (4) of the form $\sum_{n=0}^{\infty} a_n x^n$, where the a_n are determined by recursion formula (9). What is the interval of convergence of the series solution? What is the specific value of the n-th coefficient a_n? Both questions can be answered quite simply for ODE (4).

EXAMPLE 11.2.7

Interval of Convergence: Continuing Example 11.2.6
We find the interval of convergence of the series solution (11) that has the form $\sum_{k=0}^{\infty} a_{2k} x^{2k}$, where

$$a_{2k+2} = \frac{2(2k-1)a_{2k}}{(2k+2)(2k+1)}, \qquad k = 0, 1, \ldots \tag{12}$$

which we obtain from recursion formula (9) by setting $n = 2k$. Applying the Ratio Test, we have

$$\lim_{k \to \infty} \left| \frac{a_{2k+2}x^{2k+2}}{a_{2k}x^{2k}} \right| = x^2 \lim_{k \to \infty} \left| \frac{2(2k-1)}{(2k+2)(2k+1)} \right| = 0, \quad \text{all } x$$

So the series solution (11) converges for all x.

Occasionally, it is feasible to "solve" a recursion formula for the coefficients of the power series expansion of a solution.

EXAMPLE 11.2.8

Solving the Recursion Identity: Continuing Example 11.2.7

We have from (12) that

$$a_{2k+2} = \frac{2(2k-1)a_{2k}}{(2k+2)(2k+1)} = \frac{2(2k-1)}{(2k+2)(2k+1)} \frac{2(2k-3)a_{2k-2}}{2k(2k-1)} = \cdots$$

$$= \frac{2^{k+1}(2k-1)(2k-3)\cdots 5 \cdot 3 \cdot 1 \cdot (-1)a_0}{(2k+2)(2k+1)\cdots 4 \cdot 3 \cdot 2 \cdot 1}$$

$$= -\frac{2^{k+1}(2k-1)(2k-3)\cdots 5 \cdot 3 \cdot 1}{(2k+2)!}a_0$$

$$= -\frac{2(2k)!}{k!(2k+2)!}a_0 = -\frac{2a_0}{k!(2k+2)(2k+1)}, \qquad k = 0, \ 1, \ 2, \ldots \quad (13)$$

where we have used the identity

$$2^k(2k-1)(2k-3)\cdots 3 \cdot 1 = \frac{(2k)(2k-1)(2k-2)(2k-3)\cdots 3 \cdot 2 \cdot 1}{k(k-1)\cdots 4 \cdot 2}$$

$$= \frac{(2k)!}{k!}$$

So, using (13), the general solution of Hermite's ODE (4) can be written as

$$y = a_0\left\{1 + \sum_{k=0}^{\infty} a_{2k+2}x^{2k+2}\right\} + a_1 x$$

$$= a_0\left\{1 - \sum_{k=0}^{\infty} \frac{2}{k!(2k+2)(2k+1)}x^{2k+2}\right\} + a_1 x$$

$$(14)$$

Although we have "solved" the recursion formula in this case, it is not always useful to do so, even when easy. The "unsolved" recursion formula (9) is already in good form for a computer program to evaluate as many coefficients as desired. The computer must be provided with a_0 and a_1, the recursion formula itself, and a range for n (e.g., $n = 2$ to 100). The computer will do all the work for ODE (4) by iterating (9) for successive even values of n, beginning with $n = 0$.

☞ Series solutions of ODEs are one of the richest sources of useful functions for engineering and the sciences.

Each set of values of a_0 and a_1 in (14) yields a new solution of ODE (4). Each of these solutions is real analytic for all x; moreover, the series defining a solution is easy to differentiate or integrate—just differentiate or integrate the series term by term.

Convergence

The steps used above to solve ODE (4) is called the *Method of Power Series* We have carried out the details of the example above to show how the method works in practice. All the work would be in vain if the constructed series failed to converge to a solution, or perhaps did not converge at all. The following result assures that the method always succeeds.

THEOREM 11.2.1

☞ This assumption just says that $P(x)$ and $Q(x)$ are real analytic at $x_0 = 0$.

Convergence Theorem. Suppose that the series $\sum_{n=0}^{\infty} p_n x^n$, $\sum_{n=0}^{\infty} q_n x^n$ converge to $P(x)$, $Q(x)$, respectively, on a common interval $J = (-\rho, \rho)$, $\rho > 0$. Then the IVP

$$y'' + P(x)y' + Q(x)y = 0, \qquad y(0) = a_0, \quad y'(0) = a_1 \qquad (15)$$

has the unique solution $y = \sum_{n=0}^{\infty} a_n x^n$, where the coefficients a_n, $n \geq 2$, are determined by the *recursion formula*

$$(n+2)(n+1)a_{n+2} + \sum_{k=0}^{n}[p_{n-k}a_{k+1}(k+1) + q_{n-k}a_k] = 0 \qquad (16)$$

The series $\sum_{n=0}^{\infty} a_n x^n$ converges on J.

The recursion formula expresses a_{n+2} in terms of a_k, $k < n + 2$, so (16) can be used to find all coefficients once a_0 and a_1 are given. In practice, (16) is rarely used for a specific equation. It is much easier to insert the unknown series $\sum a_n x^n$ and its respective derivative series into the differential equation, group the coefficients of like powers of x as we did above, and derive the recursion formula directly.

Power series have been around since the time of Euler, but it was Cauchy[1] who made their use into a rigorous tool for differential equations.

If $y_1(x)$ and $y_2(x)$ are solutions of IVP (15), which, respectively, satisfy the initial data $y_1(0) = 1$, $y_1'(0) = 0$ and $y_2(0) = 0$, $y_2'(0) = 1$, then $\{y_1(x), y_2(x)\}$ is a basic

☞ The Wronskian test for a basic solution set is given in Section 3.7.

solution set for the ODE in (15). This follows because the Wronskian of the solution pair $\{y_1, y_2\}$ is

$$W[y_1, y_2](0) = y_1(0)y_2'(0) - y_1'(0)y_2(0) = 1$$

Since the Wronskian is nonzero, it follows that $\{y_1, y_2\}$ is a basic solution set for the ODE in IVP (15). So all solutions of IVP (15) are given by the formula $y = a_0 y_1(x) + a_1 y_2(x)$, where a_0 and a_1 are arbitrary real numbers.

Aging Springs

Let's apply the technique of power series to find solutions of the aging spring ODE that was modeled in Example 11.1.1.

Augustin Louis Cauchy

[1] Augustin Louis Cauchy (1790–1857) was a prolific mathematician, writing over 700 mathematical works. He worked in many areas of mathematics, and among other things, he set the foundation for rigorous mathematical analysis and proved the existence theorem for ODEs in real and complex variables. Graduating from the Ecole Polytechnique in 1807, he became a civil engineer working for Napoleon's army at Cherbourg. In 1830, he refused to swear an oath to the new king, which cost him his job at the Academy of Sciences in Paris and forced him to leave the country. He returned to France after the 1848 revolution, and stayed until his death.

EXAMPLE 11.2.9   **Series Solutions of the Equation of an Aging Spring**

An ODE for the vibrations of an undamped but aging spring (Section 11.1) is

$$my'' + ke^{-\varepsilon t}y = 0$$

👉 Recall that
$e^{-\varepsilon t} = \sum_{n=0}^{\infty} \frac{(-\varepsilon t)^n}{n!}.$

where m, k, and ε are positive constants. Suppose that the value of $k/m = 1$. Since the exponential coefficient is real analytic on the entire t-axis, we can replace $e^{-\varepsilon t}$ by its Taylor series. The ODE meets all conditions of the Convergence Theorem 11.2.1 with, say, $t_0 = 0$ as the ordinary point about which solutions will be expanded in power series. Suppose that the value of k/m is 1 and that the spring is stretched 1 unit from its equilibrium position and then released:

$$y''(t) + \left[\sum_{n=0}^{\infty} \frac{(-\varepsilon t)^n}{n!}\right]y(t) = 0, \qquad y(0) = 1, \quad y'(0) = 0 \qquad (17)$$

The solution of IVP (17) is a power series $\sum_0^\infty a_n t^n$ that converges for all t. So, replacing $y(t)$ by $\sum a_n t_n$ and y'' by $\sum n(n-1)a_n t^{n-2}$, we have

$$\sum_{n=2}^{\infty} n(n-1)a_n t^{n-2} + \left[\sum_{n=0}^{\infty} \frac{(-\varepsilon)^n}{n!}t^n\right]\left[\sum_{n=0}^{\infty} a_n t^n\right] = 0 \qquad (18)$$

Reindexing the first series in (18) and applying the product formula for series (see Appendix B.2, item 6) we see that

$$\sum_{n=0}^{\infty}(n+2)(n+1)a_{n+2}t^n + \sum_{n=0}^{\infty}\left[\sum_{k=0}^{n}\frac{(-\varepsilon)^k}{k!}a_{n-k}\right]t^n = 0$$

$$\sum_{n=0}^{\infty}\left[(n+2)(n+1)a_{n+2} + \sum_{k=0}^{n}\frac{(-\varepsilon)^k}{k!}a_{n-k}\right]t^n = 0$$

From this equation and the Identity Theorem, we have

$$a_{n+2} = -\frac{1}{(n+2)(n+1)}\sum_{k=0}^{n}\frac{(-\varepsilon^k)}{k!}a_{n-k} \qquad (19)$$

This recursion formula can't easily be solved for a_{n+2} in terms of a_0 or a_1, but, as we noted before, this does not restrict its usefulness.

To find the solution of IVP (17), we set $a_0 = 1$ and $a_1 = 0$ since we require that $y(0) = 1$ and $y'(0) = 0$. We can use (19) to find $a_2, a_3, \ldots$:

$$a_0 = 1, \qquad a_1 = 0, \qquad a_2 = -\frac{1}{2}\frac{(-\varepsilon)^0}{0!}a_0 = -\frac{1}{2},$$

$$a_3 = -\frac{1}{3\cdot 2}\sum_{k=0}^{1}\frac{(-\varepsilon)^k}{k!}a_{1-k} = -\frac{1}{6}[a_1 - \varepsilon a_0] = \frac{\varepsilon}{6}$$

Similarly,

$$a_4 = \frac{1}{24}(1-\varepsilon^2), \qquad a_5 = \frac{\varepsilon}{120}(\varepsilon^2 - 4), \qquad a_6 = -\frac{1}{720}(1 - 11\epsilon^2 + \epsilon^4)$$

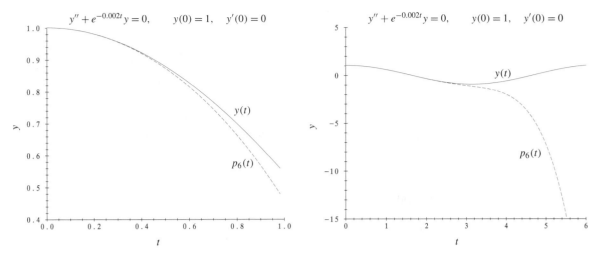

FIGURE 11.2.1 Computed solution $y(t)$ (solid) and a polynomial approximation p_6 (dashed); approximation is very good for $0 \leq t \leq 1$ (Example 11.2.9).

FIGURE 11.2.2 The polynomial approximation p_6 (dashed) is a bad fit to computed solution $y(t)$ (solid) at a distance from $t = 0$ (Example 11.2.9).

So the solution of IVP (17) is

$$y(t) = 1 - \frac{1}{2}t^2 + \frac{\varepsilon}{6}t^3 + \frac{1}{24}(1 - \varepsilon^2)t^4 + \frac{\varepsilon}{120}(\varepsilon^2 - 4)t^5$$
$$- \frac{1}{720}(1 - 11\varepsilon^2 + \varepsilon^4)t^6 + \cdots \tag{20}$$

☞ A good way to check that no mistake was made in the algebra.

If we set $\varepsilon = 0$ in formula (20), then we obtain the Taylor series for $\cos t$ about $t_0 = 0$, which is the solution of IVP (17) if $\varepsilon = 0$.

Figures 11.2.1 and 11.2.2 overlay the computed solution of IVP (17) if $\varepsilon = 0.002$ with the graph of the polynomial approximation $p_6(t)$ consisting of the terms in the series expansion of $y(t)$ through t^6. The approximation is good for $0 \leq t \leq 3$, but not good over the longer time interval. The solution $y(t)$ of IVP (17) computed by our numerical solver shows the spring stretching and then compressing, while the polynomial approximation $p_6(t)$ vastly exaggerates the amount of stretching and delays the onset of compression. Many more terms in series (20) must be used even to begin to correct these exaggerations. As noted previously, a polynomial approximation derived from a power series can be accurate only near the base point of the series expansion.

We will come back to the aging spring model in Section 11.7.

Comments

The examples of this section illustrate both the advantages and the drawbacks of solving IVPs by series expansions. On the one hand, we do have solution formulas in the form of series that can be used to evaluate the solutions (at least approximately) for specific values of the independent variable. On the other hand, although a partial sum of the series can approximate the solution quite well near the ordinary point at which the expansions are based, the approximation can be terrible farther away. Second, solutions may not even be expandable as power series if one of the coefficient functions $P(x)$ or $Q(x)$ is not real analytic. For example, as noted in Example 11.2.5 the method fails to produce a solution of the equation $y''(x) + |x|y(x) = 0$ on any interval containing $x = 0$ since the coefficient function $|x|$ is not real analytic at 0. Nevertheless, power series solutions are useful and appear frequently in both theory and applications.

PROBLEMS

1. Determine whether x_0 is an ordinary or a singular point for each ODE.

   (a) $y'' + k^2 y = 0$, x_0 arbitrary [k is a constant] **(b)** $y'' + \dfrac{1}{1+x} y = 0$, $x_0 = -1$

   (c) $y'' + \dfrac{1}{x} y' - y = 0$, $x_0 = 0$

2. (*Singular Points*). Find all singular points (if any) of the following ODEs.

   (a) $y'' + (\sin x)y' + (\cos x)y = 0$ **(b)** $y'' - (\ln|x|)y = 0$

   (c) $y'' + |x|y' + e^{-x}y = 0$ **(d)** $(1-x)y'' + y' + (1-x^2)y = 0$

   (e) $(1-x^2)y'' + xy' + y = 0$ **(f)** $x^2 y'' + xy' - y = 0$

3. (*Series Solutions About Ordinary Points*). Verify that $x_0 = 0$ is an ordinary point and find all solutions as power series in x. Determine the interval of convergence of the series. Plot the solution satisfying $y(0) = y'(0) = 1$. [*Hint*: Use the Taylor series about $x_0 = 0$ for the familiar solutions of the ODEs in parts **(a)** and **(b)**.]

   (a) $y'' + 4y = 0$ **(b)** $y'' - 4y = 0$

   (c) $y'' + xy' + 3y = 0$ **(d)** $(x^2 + 1)y'' - 6y = 0$

4. (*Airy's Equation*). *Airy's equation* is $y'' - xy = 0$. Certain solutions of the equation are called Airy functions and are used to model the diffraction of light.

   (a) Find all the solutions of the equation $y'' - xy = 0$ as power series in x.

   (b) Plot the solution satisfying $y(0) = 1$, $y'(0) = 0$ over the interval $-20 \le x \le 2$.

   (c) Plot the sums of the first 2 and 4 nonvanishing terms of the series expansion of the solution of $y'' - xy = 0$, $y(0) = 1$, $y'(0) = 0$.

   (d) Why would you expect the solution to oscillate for $x < 0$ but not for $x > 0$? [*Hint*: Look at solutions of the IVPs $y'' + \alpha y = 0$ and $y'' - \alpha y = 0$, $y(0) = 1$, $y'(0) = 0$, where α is a positive constant.]

www 5. (*Mathieu Equation*). The *Mathieu equation* is $y'' + (a + b\cos \omega x)y = 0$. Calculate the first four nonvanishing coefficients in the power series expansion of the solution of the Mathieu equation; use $a = 1$, $b = 2$, $\omega = 1$, and $y(0) = 1$, $y'(0) = 0$.

6. (*Series Solutions of Nonlinear IVPs*). Assume a solution of the form $y = \sum_0^\infty a_n x^n$ and find a_0, $a_1, \ldots, a_7$ for the following IVPs that involve nonlinear ODEs.

   (a) $y' = 1 + xy^2$, $y(0) = 0$ **(b)** $y' = x^2 + y^2$, $y(0) = 0$

7. (*Calculating the Leading Coefficients*). Calculate the first five nonzero terms in the power series expansion of the general solution of the equation $y'' + (1 - e^{-x^2})y = 0$.

8. (*Estimating the Value of a Solution*). Estimate $y(1)$ to three decimal places if $y(x)$ is the solution of the IVP $y'' + x^2 y' + 2xy = 0$, $y(0) = 1$, $y'(0) = 0$.

11.3 Legendre Polynomials

The linear ODE

$$(1 - x^2)y'' - 2xy' + p(p+1)y = 0 \tag{1}$$

where p is a nonnegative constant, is called *Legendre's equation of order p*. ODE (1) is actually a family of differential equations, one ODE for each value of p. The term "order p" refers to a factor in the coefficient of the y-term in ODE (1), not to the "order of the differential equation," which is, of course, 2. Legendre's equation arises in an amazing variety of physical phenomena ranging from the calculation of potential functions to the determination of steady-state temperatures in a solid ball, given the surface temperatures.

The ODE is of particular interest if p is a nonnegative integer, say n, since in that case ODE (1) has a polynomial solution $P_n(x)$ of degree n. Initially, however, we do not restrict p to integer values.

Solutions of Legendre's Equation

The Legendre equation in normalized form is

$$y'' - \frac{2x}{1-x^2}y' + \frac{p(p+1)}{1-x^2}y = 0$$

Because the coefficients $-2x(1-x^2)^{-1}$ and $p(p+1)(1-x^2)^{-1}$ are real analytic in the interval $I = (-1, 1)$, the point $x_0 = 0$ is an ordinary point. To show this use a geometric series (see Appendix B.2, item 9) to expand $(1 - x^2)^{-1}$ in a convergent power series:

$$-2x(1-x^2)^{-1} = -2x(1 + x^2 + x^4 + \cdots + x^{2k} + \cdots), \qquad |x| < 1$$

According to the Convergence Theorem of Section 11.2, solutions of Legendre's equation can be written as convergent power series $y = \sum_{j=0}^{\infty} a_j x^j$. The index j is used rather than n, because n is reserved for later use in this section. In the steps below we will see that the coefficients a_j depend on the value of p.

Substituting $\sum_{j=0}^{\infty} a_j x^j$ and its term-by-term derivative series into (1), we see that $\sum_{j=0}^{\infty} a_j x^j$ is a solution of Legendre's equation (1) for $|x| < 1$ if and only if

$$(1 - x^2)\sum_{j=0}^{\infty} j(j-1)a_j x^{j-2} - 2x\sum_{j=0}^{\infty} ja_j x^{j-1} + p(p+1)\sum_{j=0}^{\infty} a_j x^j = 0$$

or, bringing the factors inside the summations,

$$\sum_{j=0}^{\infty} j(j-1)a_j x^{j-2} - \sum_{j=0}^{\infty} j(j-1)a_j x^j - \sum_{j=0}^{\infty} 2ja_j x^j + \sum_{j=0}^{\infty} p(p+1)a_j x^j = 0$$

Reindexing the first sum after noting that the terms corresponding to $j = 0, 1$ can be dropped, we have

$$\sum_0^\infty \left\{ (j+2)(j+1)a_{j+2} + [-j(j-1) - 2j + p(p+1)]a_j \right\} x^j = 0$$

Since $-j(j-1) - 2j + p(p+1) = (p+j+1)(p-j)$, we have the *Legendre recursion formula*

$$a_{j+2} = -\frac{(p+j+1)(p-j)}{(j+2)(j+1)}a_j, \qquad j = 0, 1, 2, \ldots \tag{2}$$

☞ So if $a_0 = 0$, $a_2 = a_4 = \cdots = 0$ as well. If $a_1 = 0$, then $a_3 = a_5 = \cdots = 0$.

Observe that the value of a_0 determines $a_2, a_4, \ldots$, while a_1 determines $a_3, a_5, \ldots$. Using (2) and an argument by induction, we have

$$a_{2k} = (-1)^k \frac{(p+2k-1)\cdots(p+1)p\cdots(p-2k+2)}{(2k)!}a_0 \tag{3a}$$

$$a_{2k+1} = (-1)^k \frac{(p+2k)\cdots(p+2)(p-1)\cdots(p-2k+1)}{(2k+1)!}a_1 \tag{3b}$$

The convergence for $|x| < 1$ of any infinite series solution of ODE (1) follows from (2) and the Ratio Test because

$$\left| \frac{a_{j+2}x^{j+2}}{a_j x^j} \right| = \left| \frac{(p+j+1)(p-j)}{(j+2)(j+1)} \right| x^2$$

which approaches x^2 as $j \to \infty$.

For each value of p, a basic solution set $\{y_1(x), y_2(x)\}$ for the Legendre equation can be constructed by first setting $a_0 = 1$ and $a_1 = 0$ to find a solution y_1, and then reversing the values to $a_0 = 0$, $a_1 = 1$ to find a second solution y_2:

$$y_1(x) = 1 - \frac{(p+1)p}{2 \cdot 1}x^2 + \cdots + (-1)^k a_{2k}x^{2k} + \cdots \tag{4}$$

$$y_2(x) = x - \frac{(p+2)(p-1)}{3 \cdot 2}x^3 + \cdots + (-1)^k a_{2k+1}x^{2k+1} + \cdots \tag{5}$$

where a_{2k} and a_{2k+1} are respectively given in (3a) and (3b). Observe that $y_1(x)$ is an even function, while $y_2(x)$ is odd. The pair $\{y_1, y_2\}$ is a basic solution set since the Wronskian $W[y_1(0), y_2(0)] = y_1(0)y_2'(0) - y_1'(0)y_2(0) = 1 \cdot 1 - 0 \cdot 0 = 1$. The general solution of Legendre's equation for $|x| < 1$ is

$$y = c_1 y_1(x) + c_2 y_2(x)$$

Note that $c_1 = y(0)$ and $c_2 = y'(0)$ and that $\{y_1, y_2\}$ is a family of basic solution sets [just as ODE (1) is a family of ODEs], one set for each value of p. Figure 11.3.1 displays the graphs of $y_1(x)$ and $y_2(x)$ in the case $p = 4.5$; the series for $y_1(x)$ and $y_2(x)$ diverge for $|x| = 1$, and this shows up in the graphs.

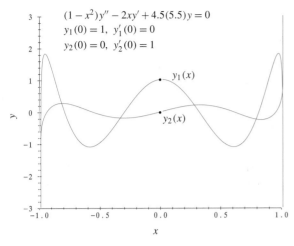

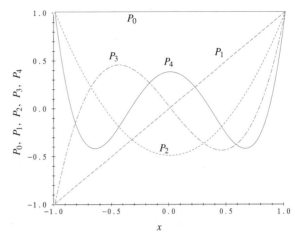

FIGURE 11.3.1 The graphs of $y_1(x)$ and $y_2(x)$ for $p = 4.5$ [formulas (4), (5)]: unbounded at $x = \pm 1$.

FIGURE 11.3.2 Graphs of the first five Legendre polynomials.

Legendre Polynomials

Suppose now that $p = n$, where n is a nonnegative integer. From (2) with p and j replaced by n, we see that $a_{n+2} = 0$, so $a_{n+4} = a_{n+6} = \cdots = 0$. The Legendre equations for $n = 2k$ and $n = 2k + 1$ have the respective polynomial solutions q_{2k} and q_{2k+1} of degree $2k$ and $2k + 1$ [take $a_0 = 1$, $a_1 = 0$ for q_{2k}, and $a_0 = 0$, $a_1 = 1$ for q_{2k+1}]:

$$y = q_{2k}(x) = 1 - \frac{(2k+1)2k}{2}x^2 + \cdots + (-1)^k a_{2k}x^{2k} \tag{6}$$

$$y = q_{2k+1}(x) = x - \frac{(2k+3)2k}{6}x^3 + \cdots + (-1)^k a_{2k+1}x^{2k+1} \tag{7}$$

where a_{2k} and a_{2k+1} are given by (3a) and (3b) with $p = 2k$ and $p = 2k + 1$, respectively. Any nonzero constant multiple of a polynomial solution is another polynomial solution of the same degree. To normalize the polynomials, it is customary to divide each by its value at $x = 1$. The *Legendre polynomials* are given by

☞ This works because for all n, $q_n(1) \neq 0$.

$$P_{2k}(x) = \frac{q_{2k}(x)}{q_{2k}(1)}, \qquad P_{2k+1}(x) = \frac{q_{2k+1}(x)}{q_{2k+1}(1)}, \qquad k = 0, 1, 2, \ldots \tag{8}$$

There is a single (complicated!) formula for these polynomials,

$$P_n(x) = \frac{1}{2^n} \sum_{j=0}^{[n/2]} (-1)^j \frac{(2n-2j)!}{j!(n-j)!(n-2j)!} x^{n-2j} \tag{9}$$

where $[n/2]$ denotes the largest integer not larger than $n/2$, that is, the integer part of the number $n/2$. We have from (6)–(8) or from (9) that

$$P_0 = 1, \quad P_1 = x, \quad P_2 = \frac{1}{2}(3x^2 - 1), \quad P_3 = \frac{1}{2}(5x^3 - 3x)$$

$$P_4 = \frac{1}{8}(35x^4 - 30x^2 + 3), \quad P_5 = \frac{1}{8}(63x^5 - 70x^3 + 15x)$$

See Figure 11.3.2. Equations (6)–(8) or (9) are not efficient ways to construct the Legendre polynomials. Fortunately, there is a recursive process that will do the trick.

THEOREM 11.3.1

Legendre Recursion Relation. For $n = 2, 3, \ldots,$

$$P_n(x) = \left(2 - \frac{1}{n}\right)x P_{n-1}(x) - \left(1 - \frac{1}{n}\right)P_{n-2}(x) \tag{10}$$

The recursion relation (10) can be verified by using (9) to express each Legendre polynomial and then observing that the terms on the left and right sides of (10) (after combining terms of like powers of x) are identical. So, given $P_0 = 1$ and $P_1 = x$, we can use (10) to find $P_2, P_3, \ldots$.

Properties of Legendre Polynomials

Recall from Section 10.4 that the family of sine functions $\{\sin nx : n = 1, 2, \ldots\}$ have the "orthogonality property" over the interval $0 \le x \le \pi$:

$$\int_0^\pi \sin nx \sin mx\, dx = \begin{cases} 0, & n \neq m \\ \pi/2, & n = m \end{cases}$$

The family of Legendre polynomials has a similar property for $-1 \le x \le 1$.

THEOREM 11.3.2

Orthogonality of Legendre Polynomials. **Legendre polynomials** $P_n(x), n = 0, 1, 2, \ldots$ have the property that

$$\int_{-1}^1 P_n(x) P_m(x)\, dx = \begin{cases} 0, & n \neq m \\ \dfrac{2}{2n+1}, & n = m \end{cases} \tag{11}$$

Here's why this is so. $P_n(x)$ and $P_m(x)$ are solutions of the Legendre ODEs of orders n and m, respectively:

$$(1 - x^2)P_n'' - 2x P_n' + n(n+1)P_n = 0$$
$$(1 - x^2)P_m'' - 2x P_m' + m(m+1)P_m = 0$$

If the first equation is multiplied by P_m and the second by P_n, subtraction yields

$$(1 - x^2)[P_n'' P_m - P_m'' P_n] - 2x[P_n' P_m - P_m' P_n]$$
$$= [m(m+1) - n(n+1)]P_n P_m \tag{12}$$

The left side of (12) can be expressed as the derivative

$$\frac{d}{dx}[(1 - x^2)(P_n' P_m - P_m' P_n)] \tag{13}$$

Integrating both sides of (12) from -1 to 1 and using (13), we have

$$[(1-x^2)(P'_n P_m - P'_m P_n)]\Big|_{x=-1}^{x=1} = [m(m+1) - n(n+1)] \int_{-1}^{1} P_n(x) P_m(x)\, dx$$

The left-hand side of this equality vanishes, so for $n \neq m$ we must have

$$\int_{-1}^{1} P_n(x) P_m(x)\, dx = 0$$

We skip showing the result in (11) for $n = m$.

The orthogonality property (11) for Legendre polynomials has the same importance in solving boundary problems for Laplace's equation in spherical coordinates (Section 11.8) that the orthogonality of the family of sine functions has in solving boundary problems for the diffusion of heat in a rod (Section 10.8).

Legendre polynomials have many other properties, and Table 11.3.1 shows the most important ones.

Let's derive part of the "Roots and Bounds" property.

THEOREM 11.3.3

Roots and Bounds. The Legendre polynomials have the following properties:
(a) $P_n(x)$, $n > 0$, has n distinct real roots and all lie in the interval $(-1, 1)$.
(b) $|P_n(x)| \leq 1$ if $|x| \leq 1$ and $n = 0, 1, 2, \ldots$.

☞ Consult the Student Resource Manual for the verification of part **(b)** under Proof of Bounds for Legendre Polynomials.

(a) Since $P_0(x) = 1$ and $\int_{-1}^{1} P_n(x)\, dx = \int_{-1}^{1} P_n(x) P_0(x)\, dx = 0$ if $n > 0$ [by (11)], $P_n(x)$ must take on both positive and negative values on the interval $-1 \leq x \leq 1$. Since $P_n(x)$ is continuous and changes sign on the interval, it must have at least one real root in the interval. Let $x_1, \ldots, x_k$ be the real roots of $P_n(x)$ in $(-1, 1)$. These roots are simple; that is, if x_j were a multiple root, then

$$P_n(x_j) = P'_n(x_j) = 0$$

By the Fundamental Theorem (Theorem 3.2.1), these conditions determine a unique solution of Legendre's equation of order n. Since the trivial solution is one solution of this initial value problem and P_n is another, we must have $P_n(x) = 0$ for all x. This contradiction shows that the roots are simple.

In order to show that $k = n$, suppose that $Q_k(x)$ is the polynomial of degree k with simple roots at $x_1, \ldots, x_k$ defined by

$$Q_k(x) = (x - x_1) \cdots (x - x_k)$$

Since $P_n(x)$ and $Q_k(x)$ each change sign at each x_j, we see that $P_n(x) Q_k(x)$ has a fixed sign on $[-1, 1]$, so $\int_{-1}^{1} P_n(x) Q_k(x)\, dx \neq 0$. However, if $Q_k(x)$ has degree less than n, then using the result of Problem 3 in the problem set we see that $\int_{-1}^{1} P_n(x) Q_k(x)\, dx = 0$. This contradiction shows that $Q_k(x)$ has degree at least n. Consequently, P_n has exactly n distinct roots in $(-1, 1)$ since, as a polynomial of degree n, P_n can't have more than n distinct roots.

Comments

The family of Legendre polynomials is not the only set of orthogonal polynomials that can arise when finding the series solutions of a family of ODEs. Hermite polynomials and Chebyshev polynomials arise in much the same way, but from different ODEs and with a different interpretation of orthogonality. See the Problems for some of their properties. As is true for the Legendre polynomials, Hermite and Chebyshev polynomials are often used in solving boundary problems for PDEs, and they also appear in many other applications.

We will see Legendre polynomials again in Section 11.8 when we find a solution formula for steady temperatures in a sphere. The table below summarizes often-used properties of Legendre polynomials.

TABLE 11.3.1 Legendre Polynomials $P_n(x)$

Differential equation	$(1 - x^2)y'' - 2xy' + n(n+1)y = 0$
Expansion	$P_n(x) = \dfrac{1}{2^n} \displaystyle\sum_{j=0}^{[n/2]} \dfrac{(-1)^j(2n - 2j)!}{j!(n - j)!(n - 2j)!} x^{n-2j}$
Rodrigues's formula	$P_n(x) = \dfrac{1}{2^n n!} \dfrac{d^n}{dx^n}\left[(x^2 - 1)^n\right]$
$P_n(x), \ n = 0, \ldots, 5$	$\begin{cases} P_0(x) = 1 & P_1(x) = x \\ P_2(x) = (3x^2 - 1)/2 & P_3(x) = (5x^3 - 3x)/2 \\ P_4(x) = (35x^4 - 30x^2 + 3)/8 & P_5(x) = (63x^5 - 70x^3 + 15x)/8 \end{cases}$
Recursion formula	$P_n = (2 - 1/n)xP_{n-1} - (1 - 1/n)P_{n-2}$
Orthogonality	$\displaystyle\int_{-1}^{1} P_n(x)P_m(x)\,dx = \begin{cases} 0, & n \neq m \\ 2/(2n + 1), & n = m \end{cases}$
Roots and bounds	$\begin{cases} P_n \text{ has } n \text{ real roots, all in } (-1, 1); \\ |P_n(x)| \leq 1, \ x \text{ in } [-1, 1] \end{cases}$
Values	$\begin{cases} P_n(1) = 1, & P_n(-1) = (-1)^n, \\ P_{2n+1}(0) = 0, & P_{2n}(0) = (-1)^n \dfrac{(2n)!}{2^{2n}(n!)^2} \end{cases}$
Identities	$\begin{cases} P_n' = nP_{n-1} + xP_{n-1}', & xP_n' - nP_n = P_{n-1}' \\ nP_n = nxP_{n-1} + (x^2 - 1)P_{n-1}', & (1 - x^2)P_n' + nxP_n = nP_{n-1} \end{cases}$

PROBLEMS

www **1.** (*Legendre's ODE and the Wronskian Method*). Legendre's equation of orders 0 and 1 have respective solutions $P_0 = 1$ and $P_1 = x$. Use the Wronskian Reduction Method (Problem 6, Section 3.7) to find a second independent solution for each case in terms of elementary functions. Plot the second solution for $|x| < 1$.

2. (*Values of Legendre Polynomials*). Show that

$$P_{2n+1}(0) = 0, \qquad P_{2n}(0) = (-1)^n \frac{(2n)!}{2^{2n}(n!)^2}, \qquad P_n(-1) = (-1)^n$$

[*Hint*: Use induction on n and an identity in Table 11.3.1 to prove the third formula.]

3. (*An Orthogonality Property*). Show that $\int_{-1}^{1} P_n(x)Q(x)\,dx = 0$ if $n > 0$ and $Q(x)$ is any polynomial of degree $k < n$. [*Hint*: Use induction on k to show that there are constants $c_0, \ldots, c_k$ such that $Q(x) = c_0 P_0 + c_1 P_1(x) + \cdots + c_k P_k(x)$. Then use Theorem 11.3.2.]

4. (*Legendre Polynomials and Zonal Harmonics*). The boundary value problem

$$\sin\phi(\rho^2 u_\rho)_\rho + (\sin\phi u_\phi)_\phi = 0, \qquad 0 < \phi < \pi, \quad \rho < 1$$
$$u(1, \phi) = f(\phi), \qquad 0 \le \phi \le \pi$$

models the steady-state temperatures $u(\rho, \phi)$ in the unit ball $\rho \le 1$ if the temperature at the point with spherical coordinates (ρ, ϕ, θ) depends only on ρ and ϕ, but not on θ. See Sections 10.9 and 11.8 for the derivation of these boundary value problems.

(a) Show that the Method of Separation of Variables (see Section 10.7) implies that separated solutions $u = R(\rho)\Phi(\phi)$ of the PDE in the boundary value problem satisfy

$$\rho^2 R'' + 2\rho R' - \lambda R = 0, \qquad 0 < \rho < 1$$

$$\Phi'' + \cot\phi\,\Phi' + \lambda\Phi = 0, \qquad 0 < \phi < \pi$$

where λ is a separation constant, which is assumed to be nonnegative.

(b) Assuming that the separation constant λ in part **(a)** is nonnegative, show that the change of variables $x = \cos\phi$ changes the ODE for $\Phi(\phi)$ in part **(a)** to Legendre's equation

$$(1 - x^2)y'' - 2xy' + p(p+1)y = 0$$

where $y(x) = \Phi(\phi)$ and $p = (1/2)(-1 + \sqrt{1 + 4\lambda})$. Explain why, for physical reasons, p must be a nonnegative integer. [*Hint*: If p is not an integer, show that $\Phi(\phi)$ becomes unbounded as $\phi \to 0$ or π.]

5. (*Orthogonal Polynomials*). Suppose $\{R_n(x) : n = 0, 1, 2, \ldots\}$ is a family of polynomials indexed by degree (i.e., deg $R_n = n$). Let $\rho(x)$ be positive and continuous on an open interval $I = (a, b)$, and assume that $\lim_{x \to a^+} \rho(x)$ and $\lim_{x \to b^-} \rho(x)$ exist (although the limiting values may be infinite). The family of polynomials is said to be *orthogonal on I with respect to the density ρ* (or *weight*, or *scale factor*) if the *scalar product*

$$\langle R_n, R_m \rangle = \int_a^b R_n(x)R_m(x)\rho(x)\,dx \quad \text{is zero for all } n, m, n \ne m$$

(a) What are the values of a, b, and ρ for the orthogonal family of Legendre polynomials?

(b) Show that if $\{R_n : n = 0, 1, 2, \ldots\}$ is a family of orthogonal polynomials on I with respect to ρ, then there are constants a_n, b_n, c_n such that the *three-term recursion relation*

$$R_n = (a_n x + b_n)R_{n-1} + c_n R_{n-2}, \qquad n = 2, 3, \ldots$$

is satisfied. [*Hint*: $R_n = \alpha_n x R_{n-1} + \alpha_{n-1} R_{n-1} + \alpha_{n-2} R_{n-2} + \cdots + \alpha_0 R_0$ for some constants α_i. Then calculate the scalar products $\langle R_n, R_j \rangle$, $j \le n - 2$.]

(c) (*Recursion Formula for Legendre Polynomials*). Use (10) to find the values of a_n, b_n, and c_n in the three-term recursion formula for Legendre polynomials given in Table 11.3.1.

6. (*Hermite's Equation and Polynomials*). *Hermite's equation* of order n, where n is a nonnegative integer, is $y'' - 2xy' + 2ny = 0$.

☞ Take a look at Examples 11.2.6–11.2.8 for Hermite's equation of order 1 and its solutions.

(a) Show that the equation has solutions that are polynomials of degree n, and these polynomials are even functions if n is even, odd functions if n is odd.

(b) The *Hermite polynomials* H_n are determined by selecting (for each n) the polynomial solution with leading coefficient 2^n : $H_n(x) = 2^n x^n + (\cdot)x^{n-2} + \cdots$. Using the notation of Problem 5, show that the family of Hermite polynomials is orthogonal on the real line with respect to the density $\rho = e^{-x^2}$.

(c) It can be shown that the three-term recursion formula for the Hermite polynomials is

$$H_n = 2xH_{n-1} - 2(n-1)H_{n-2}$$

Using $H_0 = 1$, $H_1 = 2x$ and this formula, construct $H_2, \ldots, H_5$. Plot $H_2, \ldots, H_5$ for $|x| \le 3$, and make a conjecture about the number and the nature of the roots of $H_n(x)$.

7. (*Chebyshev's Equation and Polynomials*). *Chebyshev's equation* of order n, n a nonnegative integer, is $(1 - x^2)y'' - xy' + n^2 y = 0$.

(a) Show that the equation has solutions that are polynomials of degree n and that these polynomials are even functions if n is even, odd functions if n is odd.

(b) The *Chebyshev polynomials* T_n, $n > 0$, are determined by selecting the polynomial solution with 2^{n-1} as the coefficient of x^n. Let $T_0 = 1$. Show that the family of Chebyshev polynomials is orthogonal on $I = [-1, 1]$ with respect to the density $(1 - x^2)^{-1/2}$.

(c) It is known that the three-term recursion formula for the Chebyshev polynomials is $T_n = 2xT_{n-1} - T_{n-2}$. Use the formula and $T_0 = 1$, $T_1 = x$ to construct $T_2, \ldots, T_5$. Plot $T_0, \ldots, T_5$. Make a conjecture about the number, location, and nature of the real roots of $T_n(x)$. [In fact, $T_n(x) = \cos(n \arccos x)$, but we don't show that here.]

11.4 Regular Singular Points

The Method of Power Series is successful in constructing a convergent power series solution of a second-order linear ODE on an interval centered at an ordinary point. If the point is not ordinary, but singular, the method may fail. Since a number of ODEs that come up in connection with modeling phenomena of wave motion, heat flow, and electric potential are of interest primarily in the neighborhoods of singular points, new methods of finding solution formulas are needed. In the 150 years or so since the need was recognized, methods have been found that will work with most of the singularities likely to occur in practice.

Some ODEs with singularities can be solved explicitly without series, and we will do so in this section. You might think that in general, solutions will be difficult to describe near a singularity, but that isn't always the case, as we will now see.

Regular Singular Points

The ODE

$$a_2(x)y'' + a_1(x)y' + a_0(x)y = 0 \tag{1}$$

can be normalized by dividing by $a_2(x)$ to obtain the ODE

$$y'' + P(x)y' + Q(x)y = 0 \qquad (2)$$

where $P = a_1/a_2$ and $Q = a_0/a_2$. Our goal in the remainder of this chapter is to modify the series and undetermined coefficient techniques of Section 11.2 so that we can construct solution formulas for ODE (2) in the neighborhood of a "regular" singular point. Recall from Section 11.2 that x_0 is a singular point of ODE (2) if P or Q fails to be real analytic at x_0. "Regularity" is defined as follows.

> ❖ **Regular Singular Points.** A singular point x_0 of ODE (2) is *regular* if the functions
>
> $$p(x) = (x - x_0)P(x) \quad \text{and} \quad q(x) = (x - x_0)^2 Q(x)$$
>
> are real analytic at x_0.

The (finite) values of $p(x_0)$ and $q(x_0)$ are given, respectively, by $\lim_{x \to x_0}(x - x_0)P(x)$ and $\lim_{x \to x_0}(x - x_0)^2 Q(x)$. A singular point of ODE (2) that is not regular is said to be *irregular*

If x_0 is a regular singular point of ODE (2), then multiplying by $(x - x_0)^2$ we see that the ODE can be written in the *standard form*

$$(x - x_0)^2 y'' + (x - x_0)p(x)y' + q(x)y = 0 \qquad (3)$$

where p and q are real analytic at x_0. We sometimes use the linear operator L, which, applied to $y(t)$, produces

$$L[y] = (x - x_0)^2 D^2 + (x - x_0)p(x)D + q(x)$$

The following examples show how to classify the singularities of an ODE as regular or irregular and how to write an ODE in standard form with respect to each regular singularity.

EXAMPLE 11.4.1

☞ Notice that this is an Euler equation (check out Problem 4 in Section 3.7).

A Regular Singular Point
The point $x_0 = 0$ is a singular point of the normalized ODE

$$y'' - 2x^{-1}y' + 2x^{-2}y = 0$$

because $P(x) = -2x^{-1}$ and $Q(x) = 2x^{-2}$ are infinite at $x = 0$. However, the singularity is regular because the (constant) functions $p(x) = x \cdot P(x) = -2$ and $q(x) = x^2 \cdot Q(x) = 2$ are real analytic at $x_0 = 0$. The standard form is

$$x^2 y'' - 2xy' + 2y = 0$$

There are no singular points other than $x_0 = 0$.

EXAMPLE 11.4.2

Legendre's Equation and Regular Singular Points
Legendre's equation of order r,

$$(1 - x^2)y'' - 2xy' + r(r + 1)y = 0$$

has singularities at $x = \pm 1$. Dividing by $1 - x^2$, we obtain the ODE

$$y'' - \frac{2x}{1 - x^2}y' + \frac{r(r+1)}{1 - x^2}y = 0$$

So $P(x) = -2x(1 - x^2)^{-1}$ and $Q(x) = r(r+1)(1 - x^2)^{-1}$. With $x_0 = 1$, we have

$$p(x) = (x-1)P(x) = 2x(1+x)^{-1},$$
$$q(x) = (x-1)^2 Q(x) = -r(r+1)(x-1)(1+x)^{-1} \tag{4}$$

To show that p and q are real analytic at 1, expand $(1 + x)^{-1}$ in a geometric series:

☞ This is a geometric series in powers of $(x - 1)/2$ instead of x.

$$\frac{1}{1+x} = \frac{1}{2} \cdot \frac{1}{1 + (x-1)/2} = \frac{1}{2}\sum_{n=0}^{\infty}(-1)^n \frac{1}{2^n}(x-1)^n$$

which converges for $|x - 1| < 2$. The standard form at $x = 1$ is

$$(x-1)^2 y'' + (x-1)p(x)y' + q(x)y = 0$$

where $p(x)$ and $q(x)$ are given in (4). So Legendre's equation of order r has a regular singularity at $x = 1$. Similarly, $x = -1$ is a regular singularity.

EXAMPLE 11.4.3

An Irregular Singular Point
The ODE

$$y'' + 2x^{-3}y = 0$$

has an irregular singularity at $x_0 = 0$ because $Q(x) = 2x^{-3}$ becomes infinite at 0 and $q(x) = x^2 \cdot 2x^{-3} = 2x^{-1}$ still becomes infinite at 0 (so q is *not* real analytic at 0).

EXAMPLE 11.4.4

Regular and Irregular Singular Points
The ODE $x^3(x - 1)y'' + y = 0$ is written in normal form as

$$y'' + x^{-3}(x-1)^{-1}y = 0$$

Because $Q(x) = x^{-3}(x-1)^{-1}$ becomes infinite at $x = 0$ and at $x = 1$, these points are singular points. Because $q(x) = x^2 \cdot x^3(x-1)^{-1} = x^{-1}(x-1)^{-1}$ still becomes infinite at $x = 0$, the origin is an irregular singular point. However, $q(x) = (x-1)^2 \cdot x^{-3}(x-1)^{-1} = x^{-3}(x-1)$ is real analytic at $x = 1$ (it has the value 0 at that point), so $x = 1$ is a regular singular point. The standard form of the ODE $y'' + x^{-3}(x-1)^{-1}y = 0$ is

$$(x-1)^2 y'' + x^{-3}(x-1)y = 0$$

for the regular singular point $x = 1$.

EXAMPLE 11.4.5

Another Irregular Singular Point
The point $x_0 = 0$ is an irregular singularity of

$$y'' + |x|y' + y = 0$$

because $p(x) = x^2|x|$ is *not* real analytic at $x_0 = 0$.

All other points in the examples above are ordinary points, and the methods of Section 11.2 can be used to find real analytic solutions in the form of power series convergent in the neighborhood of such a point. Our goal here and in the rest of the chapter is to solve (2) near a regular singular point x_0. We will suppose from now on that $x_0 = 0$ is a regular singular point of (2). If $x_0 \neq 0$ is a regular singular point, it can be translated to 0 by the change of variable $\bar{x} = x - x_0$, and the resulting ODE in $\bar{x}$ has 0 as a regular singular point. That's why from this point on we solve only ODEs with a regular singular point at $x_0 = 0$.

Euler ODEs

☞ Series are used in the next section to solve more complicated ODEs with regular singularities.

The simplest kind of ODE with a regular singular point at 0 is the *Euler* (or *Cauchy*, or *Cauchy-Euler*, or *equidimensional*) *equation* first encountered in Problem 4 of Section 3.7,

$$x^2 y'' + p_0 x y' + q_0 y = 0 \tag{5}$$

where p_0 and q_0 are real constants. Euler ODEs can be solved completely in terms of elementary functions.

The form of ODE (5) suggests that $y = x^r$ may be a solution for some value of r. So substituting x^r for y in ODE (5) we have that

$$x^2 r(r-1) x^{r-2} + p_0 x r x^{r-1} + q_0 x^r = 0$$

$$[r(r-1) + p_0 r + q_0] x^r = 0$$

So $y = x^r$ is a solution of ODE (5), if and only if r is a root of the *indicial polynomial* $f(r) = r^2 + (p_0 - 1)r + q_0$ associated with ODE (5). These roots, r_1 and r_2, are

$$r_1, r_2 = -\frac{1}{2}(p_0 - 1) \pm \frac{1}{2}\sqrt{(p_0 - 1)^2 - 4q_0} \tag{6}$$

and are called the *indicial roots* of ODE (5). We consider separately the three cases where r_1 and r_2 are real and distinct, real and equal, or complex conjugates.

Real and Distinct Roots

Let's look for all solutions defined on the interval $x > 0$. If the roots r_1 and r_2 are real and distinct, then a basic solution set for ODE (5) is $\{x^{r_1}, x^{r_2}\}$. The reason for this is that each of the functions is a solution and the Wronskian

$$W[x^{r_1}, x^{r_2}] = x^{r_1} r_2 x^{r_2-1} - r_1 x^{r_1-1} x^{r_2} = (r_2 - r_1) x^{r_1+r_2-1} \neq 0$$

which is evidently nonzero for $x \neq 0$. The general solution of ODE (5) in this case is

$$y = c_1 x^{r_1} + c_2 x^{r_2}, \qquad x > 0 \quad \text{or} \quad x < 0 \tag{7}$$

where c_1 and c_2 are arbitrary reals. It turns out that $y = c_1 |x|^{r_1} + c_2 |x|^{r_2}$, for arbitrary reals c_1 and c_2, is the general solution over the interval $x < 0$.

EXAMPLE 11.4.6   **Euler ODE: Distinct Real Roots**

The ODE $x^2 y'' - 2xy' + 2y = 0$ is an Euler ODE with indicial polynomial $f(r) = r^2 - 3r + 2 = (r-1)(r-2)$, and indicial roots $r_1 = 1$, $r_2 = 2$. Observe that in this case the restriction to $x > 0$ is unnecessary, and the general solution $y = c_1 x + c_2 x^2$ is defined for all x. The general solution consists of a family of lines ($c_2 = 0$) and parabolas ($c_2 \neq 0$), all of which go through the origin. The "singular behavior" in this case lies in the fact that the IVP

$$x^2 y'' - 2xy' + 2y = 0, \qquad y(0) = a, \quad y'(0) = b \qquad (8)$$

has no solution at all if $a \neq 0$, and infinitely many if $a = 0$. This does not violate the Existence and Uniqueness Theorem for IVPs (Theorem 3.2.1), because the normalized ODE $y'' - 2x^{-1} y' + 2x^{-2} y = 0$ does not satisfy the hypotheses of that theorem in a rectangle in the xy-plane that includes the point $(0, b)$, for any b. See Figure 11.4.1 for solutions of IVP (8) in the case $a = 0$, $b = 2$. In this case solutions have the form $y = 2x + c_2 x^2$, where c_2 is an arbitrary constant; solutions are plotted for three negative and three positive values of c_2 (the dashed parabolas) as well as for $c_2 = 0$ (the solid line).

Equal Roots

If the roots r_1 and r_2 of the indicial polynomial are equal, then $y_1 = x^{r_1}$ is one solution of ODE (5), but we must do some work to find a second, independent solution. The easiest way to proceed in this case is to change the independent variable x in ODE (5) to s via the formula $x = e^s$. The transformed ODE for $y(s)$ is

$$\frac{d^2 y}{ds^2} + (p_0 - 1)\frac{dy}{ds} + q_0 y = 0 \qquad (9)$$

Note that the characteristic polynomial of ODE (9) is exactly the indicial polynomial of ODE (5), so if the indices r_1 and r_2 are equal, then from Theorem 3.4.1, the general solution of ODE (9) is $y(s) = C_1 e^{r_1 s} + C_2 s e^{r_1 s}$, where C_1 and C_2 are arbitrary reals. Changing back to the x-variables, we see that $s = \ln x$ and that $e^{r_1 s} = (e^s)^{r_1} = x^{r_1}$, so the general solution of ODE (5) is

$$y = C_1 x^{r_1} + C_2 x^{r_1} \ln x, \qquad x > 0 \qquad (10)$$

where C_1 and C_2 are any real constants. As expected, $\{x^{r_1}, x^{r_1} \ln x\}$ is a basic solution set for $x > 0$ since the Wronskian is not zero. For $x < 0$ use $|x|$ instead of x in the above formula.

EXAMPLE 11.4.7   **Euler ODE: Equal Roots**

The indicial polynomial of the Euler ODE $x^2 y'' + y/4 = 0$ is $f(r) = r^2 - r + 1/4 = (r - 1/2)^2$. The functions $y_1 = x^{1/2}$ and $y_2 = x^{1/2} \ln x$ form a pair of independent solutions on the interval $x > 0$. See Figure 11.4.2 for graphs of y_1, y_2 (solid curves), and some linear combinations of y_1 and y_2 (dashed curves).

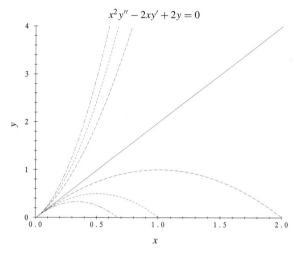

$$x^2 y'' - 2xy' + 2y = 0$$

$$x^2 y'' - y/4 = 0$$

FIGURE 11.4.1 Solution curves of this Euler equation pass through the origin (Example 11.4.6).

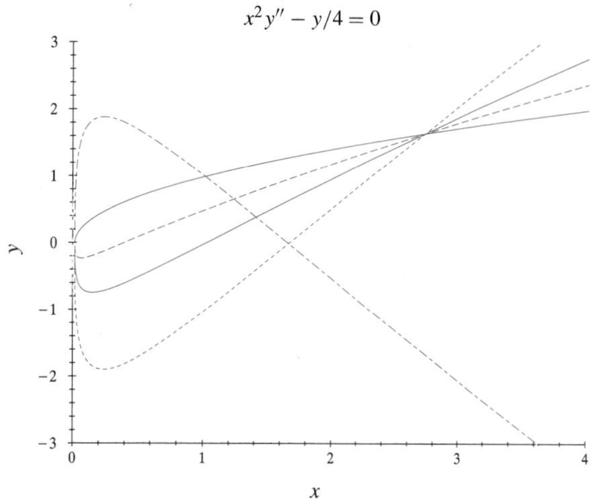

FIGURE 11.4.2 Solution curves of this Euler equation also go through the origin (Example 11.4.7).

Complex Conjugate Roots

Suppose that the roots of the indicial polynomial are $r_1 = \alpha + i\beta = \bar{r}_2$, $\beta \neq 0$. Since $L[x^{\alpha+i\beta}] = 0$, we see that $x^{\alpha+i\beta}$ is a complex-valued solution of ODE (5). We have from Appendix B.3 that

$$x^{\alpha+i\beta} = e^{(\alpha+i\beta)\ln x} = e^{\alpha\ln x}[\cos(\beta\ln x) + i\sin(\beta\ln x)]$$

$$= x^{\alpha}\cos(\beta\ln x) + ix^{\alpha}\sin(\beta\ln x)$$

So according to Theorem 3.6.3, the functions

$$y_1 = x^{\alpha}\cos(\beta\ln x), \qquad y_2 = x^{\alpha}\sin(\beta\ln x), \qquad x > 0 \tag{11}$$

form a basic solution set for ODE (5). Use $|x|$ instead of x if $x < 0$.

EXAMPLE 11.4.8   **Euler ODE: Complex Conjugate Roots**
The Euler ODE

$$x^2 y'' + (1 - 2\alpha)xy' + (\alpha^2 + 400)y = 0 \tag{12}$$

where α is a real constant, has indicial polynomial

$$r^2 - 2\alpha r + \alpha^2 + 400$$

whose roots are $r_1, r_2 = \alpha \pm 20i$. So

$$y_1 = x^{\alpha}\cos(20\ln x), \qquad y_2 = x^{\alpha}\sin(20\ln x)$$

generate the solution space. Figure 11.4.3 shows the graphs of $y_1(x)$ for $\alpha = -0.5, 0$, and 0.5. The rapidly oscillating behavior near the origin corresponds to the fact that $\ln x \to -\infty$ as x decreases toward 0, which implies that the frequency of the cosine term tends to infinity. The factor x^{α} controls the amplitude of the oscillations.

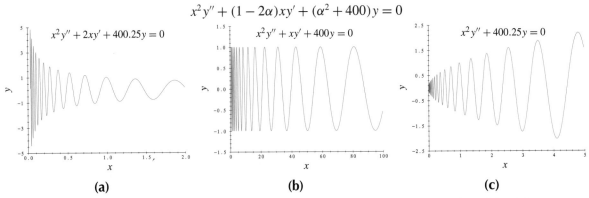

FIGURE 11.4.3 Solution curves $y = x^\alpha \cos(20 \ln x)$ of the Euler equation (12): **(a)** $\alpha = -0.5$, **(b)** $\alpha = 0$, **(c)** $\alpha = 0.5$ (Example 11.4.8).

Comments

The examples and figures of this section show that solutions of an Euler equation for $x > 0$ may, or may not, behave in unusual ways near the regular singularity at the origin. The behavior is completely determined by the nature of the roots of the indicial polynomial. The solutions make sense for negative values of x by replacing x by $|x|$ in all of the solution formulas. Observe that in some cases, solutions are defined at $x = 0$ (see, e.g., the ODEs of Examples 11.4.6, 11.4.7).

Summarizing, we have the following formulas for the general solution of the Euler ODE (5) with real coefficients in terms of the roots r_1 and r_2 of the indicial polynomial $r^2 + (p_0 - 1)r + q_0$:

$$y = \begin{cases} C_1|x|^{r_1} + C_2|x|^{r_2} & \text{if } r_1 \neq r_2, \quad r_1, r_2 \text{ real} \\ C_1|x|^{r_1} + C_2|x|^{r_1}\ln|x| & \text{if } r_1 = r_2 \\ |x|^\alpha[C_1\cos(\beta\ln|x|) + C_2\sin(\beta\ln|x|)] & \text{if } r_1 = \bar{r}_2 = \alpha + i\beta \end{cases} \quad (13)$$

where $x \neq 0$.

The extension of these ideas to finding solutions near a regular singularity of an ODE *not* of Euler form is taken up in the remaining sections of this chapter.

PROBLEMS

1. (*Regular and Irregular Singularities*). Determine if the given point is a regular or an irregular singular point for the corresponding ODE.

   (a) $x^2y'' + xy' + y = 0$, $x_0 = 0$ (b) $xy'' + (1-x)y' + xy = 0$, $x_0 = 0$

   (c) $x(1-x)y'' + (1-2x)y' - 4y = 0$, $x_0 = 1$ (d) $x^2y'' + 2y'/x + 4y = 0$, $x_0 = 0$

2. (*Classifying Singular Points*). Find and classify all singular points of the following equations.

   (a) $(1 - x^2)y'' - xy' + 2y = 0$

   (b) $y'' + \left(\dfrac{x}{1-x}\right)^2 y' + (1+x)^2 y = 0$

   (c) $x(1-x^2)^3 y'' + (1-x^2)^2 y' + 2(1+x)y = 0$

(d) $x^3(1-x^2)y'' - x(x+1)y' + (1-x)y = 0$

www **3.** (*Euler ODEs*). Find the general, real-valued solution (for $x > 0$) of each of the following equations. Plot some solutions of each equation.

(a) $x^2y'' - 6y = 0$ **(b)** $x^2y'' + xy' - 4y = 0$ **(c)** $x^2y'' + xy' + 9y = 0$

(d) $x^2y'' + xy'/2 - y/2 = 0$ **(e)** $xy'' - y' + (5/x)y = 0$ **(f)** $x^2y'' + 7xy' + 9y = 0$

4. (*Nonhomogeneous Euler ODEs*). Find the general solution for $x > 0$ of each of the following ODEs. [*Hint*: Use Undetermined Coefficients to find a particular polynomial solution.]

     (a) $x^2y'' - 4xy' + 6y = 2x + 5$ **(b)** $x^2y'' - 6y = 2x^2$

5. (*Decaying Solutions of Euler ODEs*). Show that every solution $y(x)$ of the ODE $x^2y'' + p_0xy' + q_0y = 0$, p_0, q_0 real, has the property that $|y(x)| \to 0$ as $x \to \infty$ if and only if $p_0 > 1$ and $q_0 > 0$.

6. (*The n-th-order Euler ODE*). The *n-th-order Euler ODE* is

$$x^n y^{(n)} + a_{n-1}x^{n-1}y^{(n-1)} + \cdots + a_1 xy' + a_0 y = 0, \quad x > 0 \tag{14}$$

where $a_0, a_1, \ldots, a_{n-1}$ are real constants. The *indicial polynomial* is $p(r) = r(r-1)\cdots(r - n + 1) + r(r-1)\cdots(r - n + 2)a_{n-1} + \cdots + ra_1 + a_0$.

(a) Show that $y = x^{r_1}$, $x > 0$, is a solution of (14) if and only if r_1 is a root of $p(r)$.

(b) Use part **(a)** to find three independent solutions of $x^3y''' + 4x^2y'' - 2y = 0$, $x > 0$.

11.5   Series Solutions Near Regular Singular Points, I

ODEs with regular points are pervasive in the mathematical modeling of physical phenomena. If the ODE has Euler form, it may be solved explicitly in terms of powers, logarithms, or trigonometric functions of logarithms of the independent variable. If the ODE has a regular singularity but is not of Euler form, we must look elsewhere in our search for solution formulas. As it turns out, a combination of the power series of Section 11.2 and the Euler solutions of the preceding section gives us just what we want, solutions such as

$$x^r \sum_{n=0}^{\infty} a_n x^n, \quad x^r \ln x \sum_{n=0}^{\infty} a_n x^n, \quad \text{or} \quad x^{\alpha} \cos(\beta \ln x) \sum_{n=0}^{\infty} a_n x^n$$

In this section we focus our attention on solutions of the first type. The approach we will use is a special case of techniques developed during the nineteenth century by Sturm, Fuchs, Frobenius, and many others. Later the theory was extended by Weyl to irregular singular points.

Indicial Polynomials and Frobenius Series

Suppose that $x = 0$ is a regular singular point of the ODE

$$a_2(x)y'' + a_1(x)y' + a_0(x)y = 0$$

☞ Looks like an Euler equation, but p and q aren't necessarily constants.

Multiplying the equation by x^2/a_2, we obtain the ODE in standard form,

$$x^2 y'' + xp(x)y' + q(x)y = 0 \tag{1}$$

where $p = xa_1/a_2$ and $q = x^2a_0/a_2$. The functions p and q are assumed to be real analytic at 0 and to have power series expansions of the form

$$p(x) = \sum_{n=0}^{\infty} p_n x^n, \qquad q(x) = \sum_{n=0}^{\infty} q_n x^n \qquad (2)$$

which converge on a common interval J centered at 0. The following definition extends a similar definition given in the last section.

> ❖ **Indicial Polynomial.** If p_0, q_0 are the constant terms in power series (2), then the *indicial polynomial* of ODE (1) is given by $f(r) = r^2 + (p_0 - 1)r + q_0$. Its roots r_1 and r_2 are the *indicial roots* of ODE (1) at $x_0 = 0$.

If all the coefficients of the series expansions of $p(x)$ and $q(x)$ vanish (except, possibly, p_0 and q_0), then ODE (1) is an Euler equation, and the definition of the indicial polynomial agrees with that given before.

Our goal is to find solutions of ODE (1) in the form of a *Frobenius series*,

$$y = x^r \sum_{n=0}^{\infty} a_n x^n \qquad (3)$$

As we will see, the exponent r will turn out to be an indicial root r_1 or r_2, and the coefficients $a_0, a_1, \ldots$ can be found by an extension of the Method of Power Series used in Section 11.2. The following example shows how solutions in the form of a Frobenius series can be determined.

EXAMPLE 11.5.1

Finding the Recursion Relation for a Frobenius Series
Consider the equation

$$4xy'' + 2y' + y = 0 \qquad (4)$$

Multiplying by $x/4$ to get the equation into the standard form (1), we have

$$x^2 y'' + \frac{x}{2} y' + \frac{x}{4} y = 0 \qquad (5)$$

We see that $p(x) = 1/2$, $q(x) = x/4$, so 0 is a regular singular point. The indicial polynomial is

$$f(r) = r^2 - r/2$$

since $p_0 = 1/2$, and $q_0 = 0$ [since $q(x) = x/4$ has constant term 0 in its Taylor series]. The indicial roots are 1/2 and 0. In what follows it will be more convenient to use (4) rather than the standard form (5). Take x to be positive in the analysis below.

Ignoring convergence questions for the time being, we assume that (4) has a solution of the form $y = \sum_0^{\infty} a_n x^{n+r}$. We will insert this series and its first two term-by-term derivatives in ODE (4) for y, y', y'':

$$4x \sum_{n=0}^{\infty} (n+r)(n+r-1)a_n x^{n+r-2} + 2\sum_{n=0}^{\infty} (n+r)a_n x^{n+r-1} + \sum_{n=0}^{\infty} a_n x^{n+r} = 0$$

Bringing the factors $4x$ and 2 inside the summations, we have

$$\sum_{n=0}^{\infty} 4(n+r)(n+r-1)a_n x^{n+r-1} + \sum_{n=0}^{\infty} 2(n+r)a_n x^{n+r-1} + \sum_{n=0}^{\infty} a_n x^{n+r} = 0 \quad (6)$$

Reindexing the last sum, we have $\sum_0^{\infty} a_n x^{n+r} = \sum_1^{\infty} a_{n-1} x^{n+r-1}$. Isolating the first term ($n=0$) in each of the other two sums so that all three begin at $n=1$, we can replace (6) by

$$4(0+r)(0+r-1)a_0 x^{r-1} + 2(0+r)a_0 x^{r-1}$$

$$+ \sum_{n=1}^{\infty} [4(n+r)(n+r-1)a_n + 2(n+r)a_n + a_{n-1}]x^{n+r-1} = 0$$

or, after some algebraic simplification,

$$x^{r-1}\left\{4r\left(r-\frac{1}{2}\right)a_0 + \sum_{n=1}^{\infty}[2(n+r)(2n+2r-1)a_n + a_{n-1}]x^n\right\} = 0 \quad (7)$$

The left side of (7) is 0 for all x in some interval centered at $x=0$ if and only if the parenthetical expression is 0. In turn, this can happen if and only if the coefficient of each power of x in that expression is 0 (the Identity Theorem for power series). We have the following recursion relations:

$$n=0: \quad 4r\left(r-\frac{1}{2}\right)a_0 = 0 \quad (8a)$$

$$n \geq 1: \quad 2(n+r)(2n+2r-1)a_n + a_{n-1} = 0 \quad (8b)$$

From (8a) we conclude that either $r=1/2$ or 0, or $a_0=0$. Recall from Example 11.5.1 that

$$f(r) = r(r-1/2)$$

is the indicial polynomial and $r_1 = 1/2$ and $r_2 = 0$ are the indicial roots. If we set $r=1/2$ or 0, we can take a_0 to be an arbitrary constant. The remaining coefficients are functions of r and a_0. They may be determined from the general recursion formula

$$a_n(r) = \frac{-a_{n-1}(r)}{2(n+r)(2n+2r-1)}, \quad n = 1, 2, \ldots \quad (9)$$

which is a rearrangement of (8b).

Finding the recursion relation for a Frobenius series solution is the critical step, and that has now been accomplished with the derivation of formula (9). Now we can construct a pair of independent solutions of ODE (4).

EXAMPLE 11.5.2

Constructing a Frobenius Series
Returning to ODE (4), its indicial polynomial, and its recursion formula (9), we note that if r is either of the roots $1/2$, 0 of the indicial polynomial, then the denominator

in (9) is *not* zero for $n = 1, 2, \ldots$. Using the recursion relation (9) to obtain a_n in terms of a_0, we have for $n \geq 1$

$$
\begin{aligned}
a_n(r) &= \frac{-a_{n-1}(r)}{2(n+r)(2n+2r-1)} \\
&= \frac{a_{n-2}(r)}{2(n+r)(2n+2r-1)2(n-1+r)(2n-2+2r-1)} = \cdots \\
&= \frac{(-1)^n a_0}{2^n(n+r)(n+r-1)\cdots(r+1)(2n+2r-1)(2n+2r-3)\cdots(2r+1)}
\end{aligned}
\tag{10}
$$

Observe from (10) that $a_0 = 0$ produces either the trivial solution $y = 0$ for all x (since every coefficient $a_n = 0$), or else the recursion process breaks down if a factor of the denominator in (10) vanishes. For $r = 1/2$ or 0, this difficulty can't occur. Let's evaluate $a_n(r)$ first for $r = 0$, then for $r = 1/2$:

$$
a_n(0) = \frac{(-1)^n a_0}{2^n n!(2n-1)(2n-3)\cdots 1} = \frac{(-1)^n a_0}{(2n)!}
\tag{11}
$$

$$
\begin{aligned}
a_n\left(\frac{1}{2}\right) &= \frac{(-1)^n a_0}{2^n(n+1/2)(n-1/2)\cdots(3/2)(2n)(2n-2)\cdots 2} \\
&= \frac{(-1)^n a_0}{(2n+1)(2n-1)\cdots 3(2n)(2n-2)\cdots 2} \\
&= \frac{(-1)^n a_0}{(2n+1)!}
\end{aligned}
\tag{12}
$$

We can use (11) and (12) to construct a basic solution pair for ODE (4):

$$
y_1(x) = x^{1/2}\sum_{n=0}^{\infty}\frac{(-1)^n x^n}{(2n+1)!}, \qquad y_2(x) = \sum_{n=0}^{\infty}\frac{(-1)^n x^n}{(2n)!}
\tag{13}
$$

More precisely, the solution $y_1(x)$ is defined for $x > 0$, or for $x < 0$ if we replace $x^{1/2}$ by $|x|^{1/2}$. But $y_1(x)$ is *not* a solution at $x = 0$, since the term $x^{1/2}$ is not differentiable there. The second solution is defined and twice continuously differentiable for all x.

The power series in (13) converge everywhere (use the Ratio Test). It is straightforward now to justify all our earlier steps and check that the two functions defined in (13) actually are solutions of (4). This can be done by direct insertion of y_1 and y_2 into (4). Of course, the recursion relations were chosen specifically so that y_1 and y_2 are solutions. Since the power series parts of y_1 and y_2 converge for all x, it is all right to differentiate them term by term. The functions y_1 and y_2 are independent since $y_2(0) = 1$, while $|y_1(x)| \to 0$ as $x \to 0$. So $\{y_1, y_2\}$ is indeed a basic solution pair for ODE (4).

Observe that even if we had not determined the indicial polynomial and indicial roots before beginning the process of determining the coefficients, the polynomial (or a multiple of it) appears as the coefficient of a_0 in (8a), and we could at that point have chosen r to be a root of the polynomial. We will find this useful fact to hold for the general ODE (1) as well.

Method of Frobenius

The Method of Power Series used in the example above is called the *Method of Frobenius*, and the series solutions obtained are Frobenius series. A Frobenius series may be a standard power series [see $y_2(x)$ in (13)], or involve x to a power that is not an integer [see $y_1(x)$ in (13)]. The technique will always generate at least one nontrivial solution defined near (but perhaps not at) the regular singular point. Many times the process will also produce a second independent solution as in the example above—but not always. We have the following basic theorem.

THEOREM 11.5.1

Frobenius's Theorem I. Suppose that $x_0 = 0$ is a regular singular point for

$$x^2 y'' + x p(x) y' + q(x) y = 0 \tag{14}$$

and that p_0 and q_0 are the constant terms in the Taylor series expansion $\sum p_n x^n$ and $\sum q_n x^n$ of $p(x)$ and $q(x)$ about $x_0 = 0$. Suppose that the indicial polynomial

$$f(r) = r^2 + (p_0 - 1)r + q_0$$

has real roots, r_1 and r_2, with $r_2 \leq r_1$. Then ODE (14) has a solution of the form

$$y_1 = |x|^{r_1} \sum_{n=0}^{\infty} a_n(r_1) x^n, \qquad x > 0 \quad \text{or} \quad x < 0 \tag{15}$$

with $a_0 \neq 0$, where the coefficients are determined by the *recursion relation*

☞ This relation is not easy to derive.

$$f(r_1 + n) a_n(r_1) = -\sum_{k=0}^{n-1} [(k + r_1) p_{n-k} + q_{n-k}] a_k(r_1), \qquad n \geq 1 \tag{16}$$

where $a_0 = 1$. The series in (15) converges to a real analytic function on the interval J. Moreover, if $r_1 - r_2$ is *not* an integer, a second independent solution is given by the series

☞ Take a look at the discussion below for the reason we assume that $r_1 - r_2$ is not an integer.

$$y_2 = |x|^{r_2} \sum_{n=0}^{\infty} a_n(r_2) x^n, \qquad x > 0 \quad \text{or} \quad x < 0 \tag{17}$$

where the coefficients are determined by (16) with r_2 replacing r_1 and $a_0 \neq 0$. The series in (17) converges on J to a real analytic function.

Usually the most efficient way to solve a specific ODE is to proceed by direct substitution and the Method of Power Series as in the earlier example, rather than to use the recursion relations of (16) directly.

When $r_1 - r_2$ is an integer, some ingenuity must be used to construct a solution that is independent of y_1. One can see from (16) just where the recursion formula breaks down in this case. For example, suppose that $r_1 - r_2 = 3$. Then the indicial polynomial $f(r) = r^2 + (p_0 - 1)r + q_0$ factors into $f(r) = (r - r_2)(r - r_2 - 3)$, and we see that $f(r_2 + 3) = 3 \cdot 0 = 0$. The recursion formula for $r = r_2$ and $n = 3$ becomes $0 \cdot a_3 = -\sum_{k=0}^{2}[\cdots]a_k$, and we can't in general determine a_3.

We have seen that the construction of a Frobenius series solution $y = x^r \sum_0^\infty a_n x^n$ of the ODE consists of the following steps in the case where the indicial roots r_1 and r_2 are real, $r_2 \leq r_1$.

- Check that the differential equation can be written in the form

$$x^2 y'' + xp(x)y' + q(x)y = 0$$

  where $p = \sum_0^\infty p_n x^n$ and $q = \sum_0^\infty q_n x^n$, each series converging on an open interval containing 0. The point $x = 0$ is then a regular singularity and the Method of Frobenius will apply. This step must not be omitted since you will need the numbers $p_0 = p(0)$ and $q_0 = q(0)$.

- Write out the indicial polynomial $f(r) = r^2 + (p_0 - 1)r + q_0 = 0$ and find its roots, that is, the indicial roots r_1 and $r_2 \leq r_1$, which we assume are real.

- Clear any denominators from the differential equation and cancel common factors. For example, the first ODE below should be written as the second:

$$x^2 y'' + xy' + [x^3/(1+x)]y = 0$$

$$x(1+x)y'' + (1+x)y' + x^2 y = 0$$

  The resulting equation need not have the form $x^2 y'' + xp(x)y' + q(x)y = 0$.

- Insert the series $\sum_0^\infty a_n x^{n+r}$ and its term-by-term derivatives into the ODE found above.

- Reindex and rearrange the resulting series until you have

$$f(r)a_0 x^r + \sum_{n=1}^\infty [\cdots]x^{n+r} = 0$$

  In reindexing and rearranging, be particularly careful not to "lose" terms at the head of the series (e.g., terms corresponding to $n = 0$ or 1). Equating the bracketed expression to zero gives the recursion formula for the coefficients.

- Solve the recursion formula if it is simple enough to do so, and find the coefficients a_n in terms of a_0. Otherwise, use the recursion formula to find the first few coefficients in terms of a_0.

- Set $r = r_1$ in the recursion formula to find a solution $y_1 = x^{r_1} \sum_0^\infty a_n x^n$ with $a_0 \neq 0$.

- If $r_1 - r_2$ is *not* an integer, repeat the process with $r = r_2$ to find a second, independent solution $y_2 = x^{r_2} \sum_0^\infty b_n x^n$ with $b_0 \neq 0$.

- The numbers a_0 and b_0 are arbitrary, but to obtain specific solutions set a_0 and b_0 equal to specific constants.

- If $r_1 - r_2$ *is* an integer, find a second independent solution corresponding to the indicial root r_2. (We give examples of how to do this in Section 11.8.)

Comments

We have assumed that 0 is a regular singular point of the equation. If it is an ordinary point, the series methods of Section 11.2 should be used instead. If 0 is an irregular singular point, series methods may or may not work, but are worth a try. If $x_0 \neq 0$ is a regular singularity of (1), then replace x by $x = \bar{x} + x_0$ in (1) and use the steps described above for the ODE in $\bar{x}$.

PROBLEMS

1. (*Frobenius Series*). Find solutions $y_1 = x^{r_1} \sum_0^\infty a_n x^n$ and $y_2 = x^{r_2} \sum_0^\infty b_n x^n$, where r_1 and r_2 are the indicial roots, $r_2 < r_1$, and $a_0 = 1$. Find the recursion formula for the coefficients.

(a) $9x^2 y'' + 3x(x+3)y' - (4x+1)y = 0$ (b) $4x^2 y'' + x(2x+9)y' + y = 0$

(c) $x^2 y'' + x(1-x)y' - 2y = 0$ (d) $x^2(1-x^2)y'' + x(x-1)y' + 8y/9 = 0$

2. (*Nonhomogeneous ODEs*). Find the first four nonzero terms of the power series expansion of a particular solution of the ODE

$$x^2 y'' + x(1-x)y' - 2y = \ln(1+x)$$

Then find the general solution. [*Hint*: First find the general solution of the undriven ODE $x^2 y'' + x(1 - xy' - 2y = 0$ [see Problem 1(c)]. Next expand $\ln(1+x)$ in a Taylor series about $x_0 = 0$ (see Appendix B.2, item 8). Then assume a particular solution of the form $y_d = \sum_0^\infty a_n x^n$ and use the Method of Power Series to find the a_n's.]

www 3. (*Frobenius Series*). Solve the ODE $3x^2 y'' + 5xy' - e^x y = 0$ by expanding e^x in a Taylor series about $x_0 = 0$ and recalling the formula for the product of two series (Appendix B.2). You only need to find the first four terms in the Frobenius series explicitly.

4. (*Frobenius Series and Nonuniqueness*). Find a solution of $xy'' - y' - 4x^3 y = 0$ for which $y(0) = y'(0) = 0$, but $y(x)$ is not identically zero. Why does this not contradict the uniqueness part of the Fundamental Theorem 3.2.1?

5. Show that the ODE $x^3 y'' + y = 0$ has no nontrivial solutions of the form $y = x^r \sum_0^\infty a_n x^n$. Why does this not contradict the results of this section?

6. (*Laguerre's Equation and Polynomials*). The ODE $xy'' + (1-x)y' + py = 0$ is *Laguerre's equation of order* p.

(a) Show that 0 is a regular singular point.

(b) Find the recursion formula for the coefficients of a Frobenius series solution. Solve to find the coefficient a_{n+1} of the series as a multiple of a_0. Set $a_0 = 1$ and write out the corresponding Frobenius series solution of Laguerre's equation of order p.

(c) Show that if p is a nonnegative integer, then there are polynomial solutions of degree p.

(d) The *Laguerre polynomials* are chosen to be those polynomial solutions $y = L_p(x)$ for which $L_p(0) = 1$. Show that the family of Laguerre polynomials $\{L_p : p = 0, 1, 2, \ldots\}$ is orthogonal on the interval $[0, \infty)$ with density $\rho(x) = e^{-x}$. Find the first five Laguerre polynomials.

7. (*Frobenius Series and Legendre's Equation*). The Legendre equation of order p is

$$(1 - x^2)y'' - 2xy' + p(p+1)y = 0$$

(a) Find the indicial polynomial and the indicial roots at the regular singularity $x = 1$.

(b) Find a series solution in powers of $x - 1$. [*Hint*: Change the variable x in the ODE to $s = x - 1$.]

11.6 Bessel Functions

The most important differential equation with a regular singularity is *Bessel's equation*[2] *of order* p,

$$x^2 y'' + xy' + (x^2 - p^2)y = 0 \tag{1}$$

where p is a nonnegative real constant. The term "order p" refers to a parameter in Bessel's equation, not to the order of (1) as a differential equation. Actually, ODE (1) is not one differential equation, but a family of equations, one for each value of p. Solutions of ODE (1) are called *Bessel functions of order* p. After the trigonometric, exponential, and logarithmic functions, Bessel functions are the most widely used transcendental functions in applied mathematics. Arising in studies of vibrational, thermal, elastic, and gravitational phenomena, Bessel functions have been widely studied and their values tabulated. [3] Many software packages have Bessel functions in their lists of predefined functions. Bessel functions of assorted kinds and orders are studied in this section and the next.

Solving Bessel's Equation

Bessel's equation of order $p \geq 0$ has a regular singularity at $x = 0$; all other points are ordinary. Here $p(x) = 1$ (so $p_0 = 1$) while $q(x) = x^2 - p^2$ (so $q_0 = -p^2$). From now on we won't use "$p(x)$" because of possible confusion with the order of a Bessel function. The indicial polynomial at the singularity is $f(r) = r^2 - p^2$, whose roots are p and $-p$. Frobenius's Theorem I in Section 11.5 guarantees the existence of a series solution $y_1 = x^p \sum_0^\infty a_n x^n$, $x > 0$, with $a_0 \neq 0$. To determine the coefficients, insert $y_1 = \sum_0^\infty a_n x^{n+p}$ and the corresponding derivative series for y_1' and y_1'' into Bessel's equation, obtaining

$$x^2 \sum_{n=0}^{\infty} (n+p)(n+p-1)a_n x^{n+p-2} + x \sum_{n=0}^{\infty} (n+p)a_n x^{n+p-1}$$

$$+ x^2 \sum_{n=0}^{\infty} a_n x^{n+p} - p^2 \sum_{n=0}^{\infty} a_n x^{n+p} = 0$$

Friedrich Wilhelm Bessel

[2]Friedrich Wilhelm Bessel (1784–1846) was a German mathematician, astronomer, and celestial mechanist who introduced the equation to model perturbations of planetary orbits. Bessel was the first person to calculate accurately the distance to a fixed star and thereby set the scale for all stellar distances. Like John Couch Adams and Urbain Jean Leverrier, Bessel tried to calculate the orbit of a suspected (but at that time undetected) planet by its gravitational effect on the orbit of the planet Uranus. But Bessel did not finish his calculations, and Adams and Leverrier made the mathematical discovery of the planet Neptune, a discovery that was soon verified by telescope.

[3]See M. Abramowitz and I. A. Stegun, *Handbook of Mathematical Functions* (Washington, D.C.: National Bureau of Standards, 1964, and New York: Dover Publications, 1965); J. Spanier and K. B. Oldham, *An Atlas of Functions*, (New York: Hemisphere, 1987); and D. Zillinger (Ed.), *CRC Standard Mathematical Tables and Formulae*, 30th ed. (Boca Raton, Fla.: CRC Press, 1996).

Bring the factors on the left of the summation symbols to the inside of the summations, take the factor x^p outside, and combine the first, second, and fourth summations into a single sum to obtain

$$x^p \left\{ \sum_{n=0}^{\infty} [(n+p)(n+p-1)+(n+p)-p^2]a_n x^n + \sum_{n=0}^{\infty} a_n x^{n+2} \right\} = 0$$

Reindex the final summation to the form $\sum_2^{\infty} a_{n-2} x^n$ and separate out the terms corresponding to $n = 0$ and $n = 1$ in the first summation to obtain

$$x^p \left\{ [p(p-1)+p-p^2]a_0 + [(1+p)(1+p-1)+(1+p)-p^2]a_1 x \right\}$$
$$+ x^p \sum_{n=2}^{\infty} \left\{ [(n+p)(n+p-1)+(n+p)-p^2]a_n + a_{n-2} \right\} x^n = 0 \qquad (2)$$

or

$$(1+2p)a_1 x + \sum_{n=2}^{\infty} [n(n+2p)a_n + a_{n-2}]x^n = 0 \qquad (3)$$

The coefficient of a_0 vanishes in (2) because its value is $f(p) = p^2 - p^2 = 0$, so a_0 is an arbitrary constant. The *recursion formulas* for the coefficients a_n in the sum $\sum_0^{\infty} a_n x^{n+p}$ follow from (3) by equating the coefficients of the powers of x to zero:

$$a_1 = 0$$
$$a_n = \frac{-a_{n-2}}{n(n+2p)}, \qquad n = 2, 3, \ldots \qquad (4)$$

We must set $a_1 = 0$ because its coefficient in (2) is always nonzero (recall that $p \geq 0$). It follows from (4) that $a_3 = a_5 = a_7 = \cdots = 0$, while the even-indexed coefficients are given by

$$a_{2k} = \frac{-a_{2k-2}}{2k(2k+2p)} = \frac{a_{2k-4}}{2k(2k+2p)(2k-2)(2k+2p-2)}$$

$$= \cdots = \frac{(-1)^k a_0}{2^{2k}k!(k+p)(k-1+p)\cdots(1+p)}$$

The corresponding series solutions of ODE (1) are given by

$$y_1 = a_0 x^p \left\{ 1 + \sum_{k=1}^{\infty} \frac{(-1)^k x^{2k}}{2^{2k}k!(1+p)\cdots(k+p)} \right\}, \qquad x > 0 \qquad (5)$$

where a_0 is any constant. If p is *not* an integer and $a_0 \neq 0$, then a second solution independent of y_1 is defined by replacing p in (5) by $-p$, the other indicial root. Note that if p *is* an integer, say 10, then when we replace 10 by -10 in (5) we divide by zero if the summation index is $k = 10$. That's why we require for now that p *not* be an integer when we construct a second solution.

Frobenius's Theorem I (Theorem 11.5.1) implies that the power series in (5) converges for all x; this can also be shown directly by using the Ratio Test. Formula (5) can be extended to $x \leq 0$ by replacing the factor x^p by $|x|^p$. The "normalizing" constant a_0 plays a curiously important role; a judicious choice of a_0 will result in some remarkable formulas, as we will now see.

Bessel Functions of the First Kind of Integer Order

Suppose that p is a nonnegative integer n. Then define the coefficient a_0 in (5) by

$$a_0 = \frac{1}{2^n n!}$$

After writing $n!(1+n)(2+n)\cdots(k+n)$ as $(k+n)!$ in (5), we obtain a Bessel function with a compact form that is given a special name.

❖ **Bessel Functions of the First Kind of Integer Order.** The *Bessel function of the first kind of integer order n* is

$$J_n(x) = \left(\frac{x}{2}\right)^n \sum_{k=0}^{\infty} \frac{(-1)^k}{k!(k+n)!}\left(\frac{x}{2}\right)^{2k} \tag{6}$$

$J_n(x)$ is a solution of Bessel's equation $x^2 y'' + x y' + (x^2 - n^2)y = 0$ of order n.

Since n is a nonnegative integer, (6) defines $J_n(x)$ for all x. It follows from (6) that

$$\begin{array}{ll} J_0(0) = 1, & J_0'(0) = 0 \\ J_1(0) = 0, & J_1'(0) = 1/2 \\ J_2(0) = 0, & J_2'(0) = 0 \\ \quad\vdots & \quad\vdots \\ J_n(0) = 0, & J_n'(0) = 0 \end{array}$$

EXAMPLE 11.6.1 **Three Bessel Functions**
The first three Bessel functions of the first kind are

$$J_0(x) = 1 - \frac{x^2}{4} + \frac{x^4}{64} - \frac{x^6}{2304} + \cdots + (-1)^k \frac{x^{2k}}{(k!)^2 2^{2k}} + \cdots \tag{7}$$

$$J_1(x) = \frac{x}{2}\left\{1 - \frac{x^2}{8} + \frac{x^4}{192} - \frac{x^6}{9216} + \cdots + (-1)^k \frac{x^{2k}}{k!(k+1)! 2^{2k}} + \cdots\right\} \tag{8}$$

$$J_2(x) = \frac{x^2}{4}\left\{\frac{1}{2} - \frac{x^2}{24} + \frac{x^4}{768} - \frac{x^6}{46080} + \cdots + (-1)^k \frac{x^{2k}}{k!(k+2)! 2^{2k}} + \cdots\right\} \tag{9}$$

See Figure 11.6.1 for the graphs, which resemble decaying sinusoids. This is not surprising because the normalized form of Bessel's equation

$$y'' + \frac{1}{x}y' + \left(1 - \frac{p^2}{x^2}\right)y = 0$$

resembles the equation with constant coefficients

$$y'' + \frac{1}{a}y' + \left(1 - \frac{p^2}{a^2}\right)y = 0 \tag{10}$$

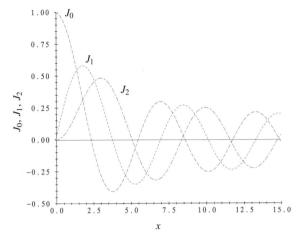

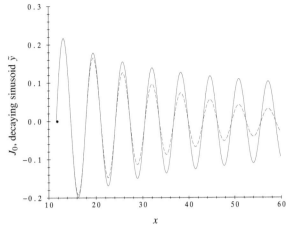

FIGURE 11.6.1 Graphs of the Bessel functions J_0, J_1, and J_2 (Example 11.6.1).

FIGURE 11.6.2 J_0 (solid) and an exponentially decaying sinusoid (dashed). See Example 11.6.1.

whose solutions (for sufficiently large values of a) are linear combinations of $e^{\alpha x}\cos\beta x$ and $e^{\alpha x}\sin\beta x$, where $\alpha = -1/(2a)$ and $\beta = [1 - p^2/a^2 - 1/(4a^2)]^{1/2}$. See Figure 11.6.2 for a comparison of $J_0(x)$ with a solution $\tilde{y}(x)$ of $y'' + y'/10 + y = 0$ [take $p = 0$, $a = 10$ in ODE (10)]. Both J_0 and $\tilde{y}$ pass through the point shown with a dot in Figure 11.6.2 with a common slope. As x increases, J_0 and $\tilde{y}$ appear to oscillate together, but $\tilde{y}$ decays faster.

Truncations of a convergent, alternating series of terms whose magnitudes decrease as n increases are useful in calculating approximate values for the sum of the series. The following example shows how this may be done for the values of a Bessel function of the first kind of integer order.

EXAMPLE 11.6.2

Approximating $J_0(x)$

The terms in the alternating series (7) for $J_0(x)$ diminish in magnitude as k increases if $|x| < 1$. In this case, the magnitude of the difference between the value of $J_0(x)$ and a partial sum of the series is no more than the magnitude of the first term omitted from the partial sum. For example, if $x = 1/10$, then from (7)

$$\left| J_0\left(\frac{1}{10}\right) - \left[1 - \frac{1}{4}\left(\frac{1}{10}\right)^2\right]\right| \le \frac{1}{64}\left(\frac{1}{10}\right)^4 \approx 1.6 \times 10^{-6}$$

So $1 - 1/400 = 399/400$ is a good approximation to the value of $J_0(1/10)$. If x is large, then k may have to be very large before the magnitudes of the terms in series (7) begin to diminish, so other methods are used to approximate $J_n(x)$ for large x.

Before we can derive a solution formula for Bessel's equation of noninteger order, we will need to define a new function.

The Gamma Function

Formula (6) might make sense when n is replaced by a noninteger number p if we knew how to caclulate the "factorial" of a noninteger number. Such a calculation involves introducing the gamma function $\Gamma(z)$, which was invented by Euler to solve the problem of constructing an infinitely differentiable factorial function defined for all real $z \geq 0$ that interpolates the integer factorials. The new function $\Gamma(z)$ is defined for all $z \geq 0$, and $n\Gamma(n)$ has value $n!$ for any positive integer n.

❖ **Gamma Function.** The *gamma function* is defined for $z > 0$ by

$$\Gamma(z) = \int_0^\infty x^{z-1} e^{-x} \, dx \tag{11}$$

Although the integral in (11) is improper, it can be shown that $\Gamma(z)$ and its derivatives with respect to z,

$$\Gamma'(z) = \int_0^\infty (x^{z-1} \ln x) e^{-x} \, dx, \quad \Gamma''(z) = \int_0^\infty [x^{z-1} (\ln x)^2] e^{-x} \, dx, \quad \ldots$$

are convergent improper integrals for $z > 0$. This means that $\Gamma(z)$ is continuous and so are all of its derivatives for $z > 0$. It can also be shown that $\Gamma(z) \to +\infty$ as $z \to 0^+$. The gamma function has other properties that justify the name of "the interpolated factorial."

THEOREM 11.6.1

Properties of the Gamma Function. The infinitely differentiable function $\Gamma(z)$ satisfies the formulas

$$\Gamma(z+1) = z\Gamma(z), \qquad z > 0 \tag{12}$$

$$\Gamma(n+1) = n!, \qquad n = 1, 2, \ldots \tag{13}$$

$$\Gamma\left(n + \frac{1}{2}\right) = \frac{(2n)!\sqrt{\pi}}{2^{2n}n!}, \qquad n = 0, 1, 2, \ldots \tag{14}$$

Here is why formula (12) is true. Replace z by $z + 1$ in (11) and integrate by parts; we have

$$\Gamma(z+1) = \int_0^\infty x^z e^{-x} \, dx = -x^z e^{-x} \Big|_{x=0}^{x=\infty} + \int_0^\infty z x^{z-1} e^{-x} \, dx$$

$$= 0 + z \int_0^\infty x^{z-1} e^{-x} \, dx = z\Gamma(z), \qquad z > 0$$

because $-x^z e^{-x}$ is zero at $x = 0$ and approaches 0 as $x \to +\infty$. This verifies formula (12).

To verify formula (13) we use induction on n. First note that

$$\Gamma(1) = \int_0^\infty e^{-x} \, dx = 1$$

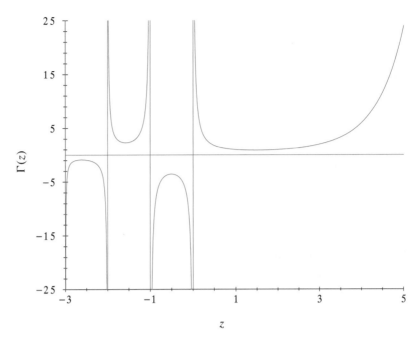

FIGURE 11.6.3 The graph of the gamma function $\Gamma(z)$.

Then suppose that $\Gamma(n+1) = n!$ and show that $\Gamma(n+2) = (n+1)!$; we have

$$\Gamma(n+2) = \int_0^\infty x^{n+1}e^{-x}\,dx = -x^{n+1}e^{-x}\Big|_{x=0}^{x=\infty} + (n+1)\int_0^\infty x^n e^{-x}\,dx$$

$$= (n+1)\Gamma(n+1) = (n+1)n! = (n+1)!$$

where integration by parts and the supposition that $\Gamma(n+1) = n!$ have been used. By induction, (13) holds for all positive integers. For $n = 0$, equation (13) would suggest $\Gamma(1) = 0!$. Since $\Gamma(1) = 1$, you can now see why it is conventional to take $0! = 1$.

Problem 7 outlines one way to show (14).

Property (12) written as $\Gamma(z) = \Gamma(z+1)/z$ is used to define the gamma function on $-1 < z < 0$ by means of its values for $0 < z < 1$. For example, $\Gamma(-1/2) = \Gamma(1/2)/(-1/2) = -2\sqrt{\pi}$ since $\Gamma(1/2) = \sqrt{\pi}$ by (14). Then the values of $\Gamma(z)$ for $-1 < z < 0$ are used to define $\Gamma(z)$ for $-2 < z < -1$, and so on. In this way, the gamma function is defined for all z except the negative integers [the above process breaks down at negative integers since $\Gamma(0)$ is infinite]. The extended gamma function is differentiable to all orders wherever it is defined. See Figure 11.6.3 for its graph.

We can use the gamma function to simplify the formula for a Bessel function of arbitrary order.

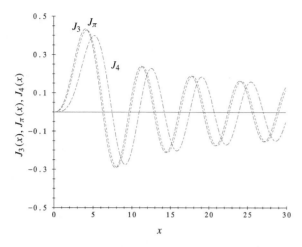
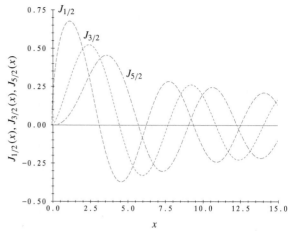

FIGURE 11.6.4 Graphs of the Bessel functions J_3, J_π, and J_4 stay close together.

FIGURE 11.6.5 Graphs of the half-integer Bessel functions $J_{1/2}$, $J_{3/2}$, $J_{5/2}$ (Example 11.6.3).

Bessel Functions of the First Kind of Any Order

The gamma function is used to extend the definition of a_0 in (5) to any choice of $p \geq 0$:

$$a_0 = \frac{1}{2^p \Gamma(p+1)}$$

where we have used the fact that $\Gamma(p+1) = p!$. Use this value of a_0 in (5) and the fact that $\Gamma(p+1)(1+p) \cdots (k+p) = \Gamma(k+p+1)$, which follows from repeated application of identity (12) to $\Gamma(k+p+1)$, to define $J_p(x)$ for any real $p \geq 0$:

❖ **Bessel Functions of the First Kind of Order** p. The *Bessel function of the first kind of order $p \geq 0$* is defined by

$$J_p(x) = \left(\frac{x}{2}\right)^p \sum_{k=0}^{\infty} \frac{(-1)^k}{k! \Gamma(k+p+1)} \left(\frac{x}{2}\right)^{2k}, \qquad x > 0 \qquad (15)$$

If $x < 0$, replace $(x/2)^p$ by $(|x|/2)^p$. Formula (15) reduces to (6) if p is an integer n.

Formula (15) looks exotic because of the appearance of the gamma function, but the behavior of $J_p(x)$ and its graph resembles that of $J_n(x)$ for n a positive integer "near" p. See Figure 11.6.4 for the graphs of J_3, J_π, and J_4. Finally, note that if p is replaced by $-p$ in (15), then a second solution, J_{-p}, of Bessel's equation (1) is defined as long as the positive number p is not an integer [remember that $\Gamma(z)$ is undefined when z is 0 or a negative integer]. So for nonintegral p, it can be shown that the pair $\{J_p, J_{-p}\}$ is a basic solution set.

Bessel functions of the first kind satisfy several recursion formulas that are somewhat similar to the Legendre recursion identities of Section 11.3. Formula (16) below is particularly important because it can be used to generate J_{p+1} from J_p and J_{p-1}.

THEOREM 11.6.2

Recursion Formulas. Bessel functions of the first kind satisfy the identities

$$J_{p+1} = \frac{2p}{x} J_p - J_{p-1}, \qquad p \geq 1 \tag{16}$$

$$J_{p+1} = -2J'_p + J_{p-1}, \qquad p \geq 1 \tag{17}$$

$$(x^p J_p)' = x^p J_{p-1}, \qquad p \geq 1 \tag{18}$$

$$(x^{-p} J_p)' = -x^{-p} J_{p+1}, \qquad p \geq 0 \tag{19}$$

EXAMPLE 11.6.3

Half-Integer Bessel Functions of the First Kind

The function $J_{1/2}(x)$ is not nearly as formidable as its definition via (15) makes it appear. In fact, $J_{1/2}$ is a sine function with a decaying amplitude:

$$J_{1/2}(x) = \sqrt{\frac{2}{\pi x}} \sin x, \qquad x > 0 \tag{20}$$

This is shown by letting $p = 1/2$ in (15):

$$J_{1/2}(x) = \left(\frac{x}{2}\right)^{1/2} \sum_{k=0}^{\infty} \frac{(-1)^k}{k! \Gamma(k+3/2)} \left(\frac{x}{2}\right)^{2k} = \left(\frac{x}{2}\right)^{1/2} \sum_{k=0}^{\infty} \frac{(-1)^k 2^{2k+2}(k+1)!}{k!(2k+2)! \sqrt{\pi}} \left(\frac{x}{2}\right)^{2k}$$

$$= \sqrt{\frac{2}{\pi x}} \sum_{k=0}^{\infty} \frac{(-1)^k}{(2k+1)!} x^{2k+1} = \sqrt{\frac{2}{\pi x}} \sin x \tag{21}$$

where the last series in (21) is the Taylor series for $\sin x$ about $x_0 = 0$. In the second step of the derivation, (14) is used with $n = k + 1$:

$$\Gamma(k+3/2) = \frac{(2k+2)! \sqrt{\pi}}{2^{2k+2}(k+1)!}$$

The third summation in (21) is obtained from the second by rearranging and canceling some terms.

By the same kind of argument it can be shown that

$$J_{3/2}(x) = \sqrt{\frac{2}{\pi x}} \left(-\cos x + \frac{\sin x}{x}\right) \tag{22}$$

Recursion formula (16) can be used to define the remaining *half-integer Bessel functions of the first kind*, $J_{n+1/2}$, $n = 2, 3, \ldots$. For example, by (16), we have

$$J_{5/2} = \frac{3}{x} J_{3/2} - J_{1/2} = \sqrt{\frac{2}{\pi x}} \left[\left(\frac{3}{x^2} - 1\right) \sin x - \frac{3}{x} \cos x\right]$$

The graphs of $J_{1/2}$, $J_{3/2}$, and $J_{5/2}$ are displayed in Figure 11.6.5.

Properties of Bessel Functions

The functions $J_p(x)$ seem to behave like decaying sinusoids such as $f(x) = e^{-x} \sin x$. The aim now is to describe the similarities more precisely. The nonnegative zeros of

$f(x) = e^{-x} \sin x$ are at $\pi, 2\pi, 3\pi, \ldots$, each zero is simple [i.e., $f'(n\pi) \neq 0$], the zeros are isolated from one another, successive zeros are π units apart, and the increasing sequence of zeros diverges to $+\infty$. The following theorems show that each nontrivial Bessel function [not just $J_p(x)$] has properties that are similar to those of $f(x)$, but not precisely the same.

THEOREM 11.6.3

> Zeros of Bessel Functions. Suppose that $y(x)$ is a nontrivial solution of Bessel's equation (1) for $x > 0$. Then the positive zeros λ_n of $y(x)$ are simple and isolated from each other. They form an increasing sequence, $0 < \lambda_1 < \lambda_2 < \cdots < \lambda_n < \cdots$, that diverges to $+\infty$, and $(\lambda_{n+1} - \lambda_n) \to \pi$ as $n \to \infty$.

EXAMPLE 11.6.4

The Zeros of J_0
The list below gives several pairs of zeros λ_n and λ_{n+1} of J_0; note how the difference between successive zeros is close to π if n is large. The zeros of J_0 through J_{20} have been calculated to several decimal places because of their importance in analysis.

λ_n	λ_1	λ_2	λ_5	λ_6	λ_{19}	λ_{20}
Value	2.4048	5.5201	14.9309	18.0711	58.9070	62.0485
$\lambda_{n+1} - \lambda_n$		3.1153		3.1402		3.1415

The oscillatory character of Bessel functions follows from the fact that each function is continuous and has a sequence of zeros as described in Theorem 11.6.3. The zeros are simple and this implies that the graph of a Bessel function crosses the x-axis at nonzero angles at the zeros, rising at one zero, falling at the next, just as the graphs indicate. It is known that between its successive zeros, a Bessel function takes on exactly one local extreme value, and this is suggested by the visual evidence of the graphs in this section.

The zeros of $J_p(x)$ play an important role in the orthogonality of Bessel functions on the interval $[0, 1]$ with weight function x.

THEOREM 11.6.4

> Orthogonality. Let $p \geq 0$, and let λ and μ be positive zeros of $J_p(x)$. Then
>
> $$\int_0^1 x J_p(\lambda x) J_p(\mu x) \, dx = \begin{cases} 0, & \lambda \neq \mu \\ \frac{1}{2}(J_p'(\lambda))^2, & \lambda = \mu \end{cases} \qquad (23)$$

Here is why $J_p(\lambda x)$ and $J_p(\mu x)$ are orthogonal if $\lambda = \mu$. Suppose first that λ is any positive number. The Chain Rule shows that if L is the *Bessel operator* which takes $y(x)$ into $L[y] = x^2 y'' + xy' + (x^2 - p^2)y$, then

$$L[J_p(\lambda x)] = x^2 \frac{d^2}{dx^2} J_p(\lambda x) + x \frac{d}{dx} J_p(\lambda x) + (\lambda^2 x^2 - p^2) J_p(\lambda x) = 0 \quad (24)$$

A straightforward (but lengthy) calculation shows that for any nonnegative λ and

μ we have the identity

$$0 = \frac{1}{x}[J_p(\mu x)L[J_p(\lambda x)] - J_p(\lambda x)L[J_p(\mu x)]]$$

$$= \frac{d}{dx}\left\{x[J_p(\mu x)J'_p(\lambda x) - J'_p(\mu x)J_p(\lambda x)]\right\} + x(\lambda^2 - \mu^2)J_p(\lambda x)J_p(\mu x)$$

which is written in abbreviated form as

$$\frac{d}{dx}\{\cdot\} + x(\lambda^2 - \mu^2)J_p(\lambda x)J_p(\mu x) = 0$$

☞ The · stands for the long ugly expression in the curly braces in the equation above.

If, in addition, λ and μ are zeroes of $J_p(x)$, then integrating each side of the above equation from $x = 0$ to $x = 1$ gives:

$$\left\{x[J_p(\mu x)J'_p(\lambda x) - J'_p(\mu x)J_p(\lambda x)]\right\}\Big|_{x=0}^{x=1} + (\lambda^2 - \mu^2)\int_0^1 xJ_p(\lambda x)J_p(\mu x)dx = 0$$

The term $\{\cdot\}|_{x=0}^{x=1}$ is zero since λ and μ are zeros of J_p. The desired result in formula (23) for $\lambda \neq \mu$ follows. We will not derive Formula (23) for $\lambda = \mu$.

☞ The notion of orthogonality is discussed in Section 10.2.

The orthogonality property looks strange, but it is analogous to the orthogonality property of the functions $\sin n\pi x$ and $\sin m\pi x$ on the interval $[0, 1]$:

$$\int_0^1 \sin(n\pi x)\sin(m\pi x)\,dx = \begin{cases} 0, & n \neq m \\ 1/2, & n = m \end{cases}$$

Just as $n\pi$ and $m\pi$ are zeros of $\sin x$, λ and μ are zeros of $J_p(x)$. The difference is that the orthogonality condition for a Bessel function uses a weight factor x.

The next theorem shows that nontrivial solutions of Bessel's equation do indeed decay, but like $1/\sqrt{x}$, not like $e^{-\alpha x}$.

THEOREM 11.6.5

Decay. For every nontrivial solution $y(x)$ of Bessel's equation there are constants A and φ and a bounded function $h(x)$ such that

$$y(x) = \sqrt{\frac{2}{\pi x}}\left[A\sin(x + \varphi) + \frac{1}{x}h(x)\right], \qquad x \geq 1 \qquad (25)$$

EXAMPLE 11.6.5

Decay of $J_{1/2}$ and $J_{3/2}$

Using the formulas of Example 11.6.3 for $J_{1/2}$ and $J_{3/2}$, we see that $J_{1/2}$ already is in the form (25) with $A = 1$, $\varphi = 0$, and $h = 0$. From (22) we see that

$$J_{3/2}(x) = \sqrt{\frac{2}{\pi x}}\left[\sin(x + \pi/2) + \frac{1}{x}\sin x\right]$$

with $A = 1$, $\varphi = \pi/2$, and $h = \sin x$.

The form of (25) indicates that all nontrivial solutions of Bessel's equation decay as $x \to +\infty$ in much the same way as the half-integer Bessel functions decay.

Comments

The formula (15) that defines each Bessel function of the first kind is formidable, but the properties of the functions themselves are not very different from those of the familiar decaying sinusoids that appear as solutions of second-order differential equations with constant coefficients. The Recursion Formulas of Theorem 11.6.2 and the formulas given in Problem 9 show that a differential and integral calculus of the functions $J_p(x)$ can be constructed. Bessel functions of the second kind are defined in the next section. They have many of the features of the Bessel functions of the first kind.

PROBLEMS

1. Find the first four terms in the series for $J_3(x)$.

2. (*Properties of $J_3(x)$*).The first few terms of the series for $J_3(x)$ can be used to determine some of its properties.

(a) Let $f(x)$ be the ninth-degree polynomial of the first four terms of the series for $J_3(x)$. Plot $f(x)$ for $0 \le x \le 2$. Explain why $|J_3(x) - f(x)| < 0.02$ for $0 \le x \le 4$.

(b) Explain why the first positive zero of $J_3(x)$ is larger than 4.

(c) Explain why the graph of $J_3(x)$ is concave upward at $x = 0$.

3. (*General Solution of Bessel's Equation*).Use the Wronskian Reduction of Order Method (Problem 6, Section 3.7) to show that the general solution of Bessel's equation of order p can be written in the form

$$y = c_1 J_p(x) + c_2 J_p(x) \int^x \frac{ds}{s[J_p(s)]^2}, \qquad c_1 \text{ and } c_2 \text{ any constants}$$

4. (*Convergence*). Use the Ratio Test to show that the series for $J_p(x)$ converges for all x.

www 5. Plot J_0, J_1, J_2, and J_3. Then plot $J_{1/3}$, $J_{4/3}$, $J_{7/3}$, $0 \le x \le 20$.

6. (*Zeros of Bessel Functions*).

(a) Locate the zeros of J_0, J_1, J_2 and $J_{1/2}$, $J_{3/2}$, $J_{5/2}$ in Figures 11.6.1 and 11.6.5, and make a conjecture about the relative positions of the positive zeros of J_p and of J_{p+1}.

(b) Verify your conjecture.

7. (*Values of $\Gamma(n + 1/2)$*). Formula (14) for $\Gamma(n + 1/2)$ can be verified by completing parts **(a)** and **(b)** below.

(a) Verify that $\Gamma(1/2) = \sqrt{\pi}$ by carrying out the following steps. First make a change of variable in the integral that defines $\Gamma(1/2)$ and show that $\Gamma(1/2) = 2\int_0^\infty e^{-u^2}\, du$. Then show that $(\Gamma(1/2))^2 = 4\int_0^\infty \int_0^\infty e^{-u^2-v^2}\, du\, dv$, and evaluate that double integral by switching to polar coordinates and integrating over the region $r \ge 0, 0 \le \phi \le \pi/2$.

(b) Use **(a)**, the property $\Gamma(z + 1) = z\Gamma(z)$, and induction to show that

$$\Gamma\left(n + \frac{1}{2}\right) = \frac{(2n)!\sqrt{\pi}}{2^{2n}n!} \quad \text{for } n = 0, 1, 2, \ldots$$

(c) Use the result in **(a)** to find $\Gamma(3/2)$ and $\Gamma(5/2)$. Then find $\Gamma(-3/2)$ and $\Gamma(-5/2)$.

8. Express J_2, J_3, and J_4 in terms of J_0 and J_1 by using recursion formula (16).

9. (*Integrating $J_p(x)$*). The recursion formulas (16)–(19) and integration by parts can be used to derive integration formulas for Bessel functions of the first kind. The symbol C in the formulas below denotes an arbitrary constant of integration.

(a) Show that $\int^x J_1(t)\,dt = -J_0(x) + C$ and that $\int^x t J_0(t)\,dt = x J_1(x) + C$.

(b) Show that $\int^x J_0(t)\sin t\,dt = x J_0(x)\sin x - x J_1(x)\cos x + C$ and

$$\int^x J_0(t)\cos t\,dt = x J_0(x)\cos x + x J_1(x) + x J_1(x)\sin x + C$$

[*Hint*: Show that the derivative of the right side is the given integrand.]

(c) Show that

$$\int^x t^2 J_0(t)\,dt = x^2 J_1(x) + x J_0(x) - \int^x J_0(t)\,dt$$

[*Hint*: Use part **(a)**.]

(d) Show that

$$\int^x t^5 J_2(t)\,dt = x^5 J_3(x) - 2x^4 J_4(x) + C$$

[*Hint*: $\int^x t^5 J_2\,dt = \int^x t^2 (t^3 J_2)\,dt = \int^x t^2 (t^3 J_3)'\,dt$.]

10. Show that

$$J_{-1/2} = \sqrt{\frac{2}{\pi x}}\cos x$$

is a solution of Bessel's equation of order $1/2$ and that $\{J_{-1/2}, J_{1/2}\}$ is a basic solution set for that equation for $x > 0$.

 11. Plot $J_1(x)$ and find (graphically) the first pair of its successive zeros λ_n, λ_{n+1} for which $|(\lambda_{n+1} - \lambda_n) - \pi| < 0.0001$.

12. (*Modified Bessel Equations*). ODEs related to Bessel's equation often appear in the applications, but the properties of their solutions can be quite different from those of Bessel functions.

(a) Replace the independent variable x in Bessel's equation of order p by the complex quantity $-it$, and obtain the *modified Bessel equation of order p*: $t^2 y'' + t y' - (t^2 + p^2)y = 0$.

(b) Show that one solution to the modified Bessel equation of order 0, $t \geq 0$, is

$$I_0(t) = \sum_{n=0}^{\infty} \frac{1}{(n!)^2}\left(\frac{t}{2}\right)^{2n}$$

(c) (*Modified Bessel Function of the First Kind*). Show that

$$I_p(t) = \left(\frac{t}{2}\right)^p \sum_{n=0}^{\infty} \frac{1}{n!\,\Gamma(n+p+1)}\left(\frac{t}{2}\right)^{2n}$$

is a solution of the modified Bessel equation of order $p \geq 0$. I_p is called the *modified Bessel function of the first kind of order p*.

 13. (*Recursion Properties of Bessel Functions*). Show that the recursion formulas (16)–(19) hold for the Bessel functions J_p.

11.7 Series Solutions Near Regular Singular Points, II

Bessel functions are used to describe the swinging motion of a lengthening pendulum, the longitudinal vibrations of a tapered bar, the oscillations of a hanging chain, the buckling of a steel disk under tension, tidal motion in a canal open to the sea, and a host of other natural processes. Bessel functions were constructed early in the nineteenth century by the method of generalized power series, but, as noted in Section 11.6, the method does not always produce a second, independent solution of Bessel's equation. The puzzle of finding a second solution remained unresolved until 1867, when C. G. Neumann (1832–1925) published his pioneering work on the matter. Frobenius extended Neumann's methods to a complete treatment of all solutions near any regular singular point of any second-order linear ODE. The extended theory of Frobenius always allows us to find a second solution even when the roots of the indicial polynomial differ by an integer.

In this section we will outline the extended Frobenius theory, define Bessel functions of the second kind, and solve the problem of the aging spring modeled in Example 11.1.1 using Bessel functions of both the first and second kinds.

The Extended Method of Frobenius

To review briefly, the goal is to find a pair of independent solutions near the regular singular point $x = 0$ of the second-order linear ODE

$$x^2 y'' + x p(x) y' + q(x) y = 0 \tag{1}$$

where $p(x)$ and $q(x)$ are real analytic on an interval J centered at the origin,

$$p(x) = \sum_0^\infty p_n x^n \quad \text{and} \quad q(x) = \sum_0^\infty q_n x^n$$

Associated with (1) are the quadratic indicial polynomial

$$f(r) = r^2 + (p_0 - 1)r + q_0$$

and its indicial roots r_1 and r_2. We will assume that the indicial roots r_1 and r_2 are real and that $r_2 \le r_1$. The Method of Frobenius outlined in Section 11.5 always yields a nontrivial solution of ODE (1) of the form

$$y_1 = |x|^{r_1} \sum_0^\infty a_n x^n$$

However, the method may not always yield a second solution when $r_1 = r_2$, or even when $r_1 - r_2$ is a positive integer. In the former case, experience with Euler equations suggests that the second solution should involve a logarithm, but the case where $r_1 - r_2$ is a positive integer is somewhat mysterious. What form do the second solutions have in that situation, and how can they be found? The following complete version of Frobenius's Theorem clears up all of these questions.

THEOREM 11.7.1

Frobenius's Theorem II. Assume that the indicial roots r_1 and r_2 for the ODE $x^2 y'' + xp(x)y' + q(x)y = 0$ are real and that $r_2 \leq r_1$. Then the Method of Power Series produces a nontrivial solution y_1 of the form

$$y_1(x) = |x|^{r_1} \sum_0^\infty a_n x^n, \qquad x > 0 \quad \text{or} \quad x < 0 \tag{2}$$

where a_0 is arbitrary. The a_n's are determined in terms of a_0 by the recursion formula that makes use of the indicial polynomial $f(r) = r^2 + (p_0 - 1)r + q_0$:

$$f(r_1 + n)a_n = -\sum_{k=0}^{n-1} [(k + r_1)p_{n-k} + q_{n-k}]a_k, \qquad n = 1, 2, \dots \tag{3}$$

A second independent solution y_2 of the ODE can also be determined by the Method of Power Series. It has one of the forms described below:

- **Case I:** If $r_1 - r_2$ is not an integer, then

$$y_2 = |x|^{r_2} \left(1 + \sum_1^\infty b_n x^n \right), \qquad x > 0 \quad \text{or} \quad x < 0 \tag{4}$$

  where the b_n's are determined by the same recursion formula (3) as the a_n's, with r_2 replacing r_1.

- **Case II:** If $r_1 = r_2$, then

$$y_2 = y_1 \ln|x| + |x|^{r_1} \sum_0^\infty c_n x^n, \qquad x > 0 \quad \text{or} \quad x < 0 \tag{5}$$

- **Case III:** If $r_1 - r_2$ is a positive integer, then

$$y_2 = \alpha y_1 \ln|x| + |x|^{r_2} \left(1 + \sum_1^\infty d_n x^n \right), \qquad x > 0 \quad \text{or} \quad x < 0 \tag{6}$$

  where α is a constant (possibly 0).

The power series in formulas (4)–(6) converge in an interval J centered at 0.

The proof of Frobenius's Theorem II is omitted. Recursion formula (3) shows just where the difficulty lies. In Case II, if r_1 in formula (3) is replaced by r_2, no new solution independent of y_1 results. In Case III, if $r_1 - r_2$ is the positive integer m, then (3) may not produce a second solution independent of y_1 when r_1 is replaced by r_2 because $f(r_2 + m) = f(r_1) = 0$, and (3) may not be solvable for a_m.

The coefficients c_n, d_n, and α in (5) and (6) can be determined by substituting the appropriate form for y_2 into ODE (1) and matching coefficients of like powers of x to obtain recursion relations. The first example illustrates Case II.

EXAMPLE 11.7.1

☞ The standard
form of this ODE is
$x^2 y'' + xy' + 2xy = 0$,
so $p_0 = 1$ and $q_0 = 0$.
The indicial
polynomial is
$f(r) = r^2$.

Case II: Equal Indicial Roots
The ODE

$$xy'' + y' + 2y = 0$$

has a regular singular point at 0. The indicial roots are $r_1 = r_2 = 0$ (Case II of Frobenius's Theorem II). The solution space is spanned by a Frobenius series

$$y_1 = \sum_{n=0}^{\infty} a_n x^n$$

and a function with a logarithmic term,

$$y_2 = y_1 \ln|x| + \sum_{n=0}^{\infty} c_n x^n$$

The Method of Power Series can be used to find the values of a_n in y_1:

$$x \sum_{n=0}^{\infty} n(n-1)a_n x^{n-2} + \sum_{n=0}^{\infty} na_n x^{n-1} + 2\sum_{n=0}^{\infty} a_n x^n = 0$$

$$\sum_{n=1}^{\infty} [n(n-1)a_n + na_n + 2a_{n-1}]x^{n-1} = 0$$

This leads to the recursion relation $n^2 a_n + 2a_{n-1} = 0$, so

$$a_n = \frac{-2a_{n-1}}{n^2} = \frac{4a_{n-2}}{n^2(n-1)^2} = \cdots = \frac{(-1)^n 2^n a_0}{(n!)^2}$$

Setting $a_0 = 1$, we obtain

$$y_1 = 1 + \sum_{n=1}^{\infty} \frac{(-1)^n 2^n x^n}{(n!)^2}, \quad \text{all } x \tag{7}$$

For convenience, we put $L[y] = xy'' + y' + 2y$, and turn to the calculation of y_2 for $x > 0$, where y_2 has the form given in formula (5). Using the fact that L is a linear operator, we have

$$0 = L[y_2] = L[y_1 \ln x] + L\left[\sum_{n=0}^{\infty} c_n x^n\right]$$

$$= x\left(y_1'' \ln x + \frac{2y_1'}{x} - \frac{y_1}{x^2}\right) + \left(y_1' \ln x + \frac{y_1}{x}\right)$$

$$+ 2y_1 \ln x + \sum_{n=1}^{\infty} (n^2 c_n + 2c_{n-1})x^{n-1}$$

After collecting the terms that include $\ln x$, the above expression may be rearranged:

$$0 = L[y_2] = L[y_1]\ln x + 2y_1' + \sum_{n=1}^{\infty} (n^2 c_n + 2c_{n-1})x^{n-1}$$

$$= 2y_1' + \sum_{1=1}^{\infty}(n^2c_n + 2c_{n-1})x^{n-1} \quad \text{(since } L[y_1] = 0\text{)}$$

$$= \sum_{n=1}^{\infty}\frac{(-1)^n 2^{n+1}x^{n-1}}{(n-1)!n!} + \sum_{n=1}^{\infty}(n^2c_n + 2c_{n-1})x^{n-1}$$

where it is only in the final step that the actual series (7) for y_1 is used. This leads to the recursion relation

$$n^2c_n + 2c_{n-1} + \frac{(-1)^n 2^{n+1}}{(n-1)!n!} = 0, \qquad n = 1, 2, \ldots \tag{8}$$

It is hard to solve this set of recursion relations to find c_n in terms of c_0. Instead, we will find the coefficients c_1, c_2, and c_3 in terms of c_0, using the recursion (8) with $n = 1, 2, 3$:

$$\begin{aligned}
c_1 + 2c_0 - 4 = 0, &\quad \text{so } c_1 = 2(2 - c_0)\\
4c_2 + 2c_1 + 4 = 0, &\quad \text{so } c_2 = -3 + c_0\\
9c_3 + 2c_2 - 4/3 = 0, &\quad \text{so } c_3 = 22/27 - 2c_0/9
\end{aligned}$$

For simplicity, set $c_0 = 0$ and obtain $c_1 = 4$, $c_2 = -3$, and $c_3 = 22/27$. Then

$$y_2 = y_1 \ln x + \sum_{n=1}^{\infty} c_n x^n = y_1 \ln x + 4x - 3x^2 + \frac{22}{27}x^3 + \cdots, \qquad x > 0 \tag{9}$$

where the coefficients c_n are defined recursively by (8) with $c_0 = 0$. For $x < 0$, replace x by $|x|$ in the logarithm.

Note once more that for all practical purposes an unsolved recursion relation is just as useful as a solved relation. If a computer is being used to calculate the coefficients, the unsolved form is to be preferred. Initializing with c_0 and then iterating with (8) did the trick in Example 11.7.1.

The following example illustrates Case III of Frobenius' Theorem II.

EXAMPLE 11.7.2

☞ The standard form of this ODE is $x^2 y'' - xy = 0$, so $p_0 = 0$ and $q_0 = 0$. The indicial polynomial is $f(r) = r^2 - r$.

Case III: Indicial Roots Differ by a Positive Integer
The ODE

$$xy'' - y = 0$$

has a regular singular point at 0, and its indicial polynomial has roots $r_1 = 1$ and $r_2 = 0$. The equation has a Frobenius series solution of the form

$$y_1 = x\sum_{n=0}^{\infty} a_n x^n = \sum_{n=0}^{\infty} a_n x^{n+1}, \qquad a_0 \neq 0$$

where the a_n's can be found in the usual way. The Method of Power Series with $a_0 = 1$ can be applied to show that

$$y_1 = x\sum_{n=0}^{\infty} \frac{x^n}{(n+1)!n!} = \sum_{n=0}^{\infty} \frac{x^{n+1}}{(n+1)!n!}$$

From Frobenius's Theorem II, a second solution has the form

$$y_2 = \alpha y_1 \ln|x| + \sum_{n=0}^{\infty} d_n x^n, \qquad d_0 = 1, \qquad x > 0 \quad \text{or} \quad x < 0 \tag{10}$$

Let x be positive and apply to y_2 the linear operator L defined by $L[y] = xy'' - y$:

$$0 = L[y_2] = \alpha L[y_1 \ln x] + L\left[\sum_{n=0}^{\infty} d_n x^n\right]$$

$$= \alpha \ln x L[y_1] + \alpha\left(2y_1' - \frac{y_1}{x}\right) + \sum_{n=0}^{\infty}[n(n+1)d_{n+1} - d_n]x^n$$

$$= \alpha \sum_{n=0}^{\infty} \frac{2n+1}{(n+1)!n!}x^n + \sum_{n=0}^{\infty}[n(n+1)d_{n+1} - d_n]x^n$$

where we have used the fact that $L[y_1] = xy_1'' - y_1 = 0$. This gives the recursion relation

$$n(n+1)d_{n+1} - d_n = \frac{-\alpha(2n+1)}{(n+1)!n!}, \qquad n = 0, 1, 2, \dots \tag{11}$$

The first three equations are

$$-d_0 = -\alpha, \qquad 2d_2 - d_1 = \frac{-3\alpha}{2}, \qquad 6d_3 - d_2 = \frac{-5\alpha}{12}$$

Since $d_0 = 1$ is already built into (6), we see that $\alpha = 1$. Thus,

$$\alpha = d_0 = 1, \qquad d_2 = d_1/2 - 3/4, \qquad d_3 = d_1/12 - 7\alpha/36$$

Letting $d_1 = 0$ (or any other convenient value), the recursion formula (11) can be used to determine d_n, $n = 2, 3, \dots$. So we can use formula (10) for $y_2(x)$ if we set $\alpha = 1$, $d_0 = 1$, $d_1 = 0$, $d_2 = -3/4$, $d_3 = -7/36$, ... and use recursion formula (11) for d_n, $n \geq 4$. Then $\{y_1, y_2\}$ is a basic solution set for the ODE $xy'' - y = 0$.

Although $\alpha \neq 0$ in Example 11.7.2, it may happen that α vanishes. In such a case, the second solution would be a generalized power series like the first, and there would be no logarithmic term.

Bessel Functions of the Second Kind of Integer Order

We now apply Frobenius's Theorem II to the problem of finding a second solution of Bessel's equation of order n,

$$x^2 y'' + xy' + (x^2 - n^2)y = 0 \tag{12}$$

where n is a nonnegative integer. One solution, $J_n(x)$, of the Bessel equation (12) was found in Section 11.6.

Because the difference of the two roots n and $-n$ of the indicial polynomial $f(r) = r^2 - n^2$ is an integer, Frobenius's Theorem I does not apply. There are two cases.

Case 1: If $n = 0$, then Case II of Frobenius's Theorem II applies. A second solution of Bessel's equation of order 0 independent of J_0 has the form

$$y_2 = y_1 \ln |x| + \sum_{n=0}^{\infty} c_n x^n \qquad (13)$$

and the c_n's can be determined by the Method of Power Series. This can be done, but it requires ingenuity and patience. In practice, the second solution is taken to be a certain linear combination of $J_0(x)$ and y_2 as given by (13). This is the *Bessel function of the second kind* (or *Weber function*) *of order* 0 and is defined by

$$Y_0(x) = \frac{2}{\pi} \left(\gamma + \ln \left| \frac{x}{2} \right| \right) J_0(x) - \frac{2}{\pi} \sum_{k=0}^{\infty} \frac{(-1)^k h(k)}{(k!)^2} \left(\frac{x}{2} \right)^{2k}$$

where $h(k) = 1 + 1/2 + \cdots + 1/k$ and γ is *Euler's constant*,[4]

$$\gamma = 1 + \sum_{n=2}^{\infty} \left(\frac{1}{n} + \ln \frac{n-1}{n} \right) = \lim_{k \to \infty} [h(k) - \ln k]$$

Case 2: If $n > 0$, then Case III of Frobenius's Theorem II applies. In practice, the second solution is taken to be

$$
\begin{aligned}
Y_n(x) = {} & \frac{2}{\pi} \left(\gamma + \ln \left| \frac{x}{2} \right| \right) J_n(x) - \frac{1}{\pi} \left| \frac{x}{2} \right|^{-n} \sum_{k=0}^{n-1} \frac{(n-k-1)!}{k!} \left(\frac{x}{2} \right)^{2k} \\
& - \frac{1}{\pi} \left| \frac{x}{2} \right|^{n} \sum_{k=0}^{\infty} \frac{(-1)^k [h(k) + h(k+n)]}{k!(n+k)!} \left(\frac{x}{2} \right)^{2k}
\end{aligned}
\qquad (14)
$$

It is called the *Bessel function of the second kind* (or a *Weber function*) *of order* n.

Although the first and third terms in the expression for Y_n remain bounded near 0, the second does not, and $Y_n(x) \to -\infty$ as $x \to 0^+$. See Figure 11.7.1 for the graphs of the first three functions of integer order Y_0, Y_1, and Y_2. The zeros and the oscillatory character of Y_n, and indeed of any nontrivial solution of Bessel's equation, have been treated in Section 11.6.

Bessel Functions of the Second Kind of Noninteger Order

Now suppose that the order p of Bessel's equation (12) is nonnegative, but not an integer. Then one solution is given by

$$J_p(x) = \left(\frac{x}{2} \right)^p \sum_{k=0}^{\infty} \frac{(-1)^k}{k!\Gamma(k+p+1)} \left(\frac{x}{2} \right)^{2k} \qquad (15)$$

As noted in Section 11.6, a second independent solution, $J_{-p}(x)$, is easily constructed by replacing p by $-p$ in (15). However, it is customary not to use J_{-p}, but a certain linear combination of J_p and J_{-p} as the second independent solution of (12).

---

[4]Euler's constant is 0.5772156649...; it is not known whether the number is rational, algebraic, or transcendental.

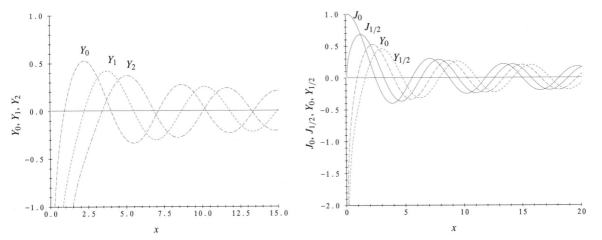

FIGURE 11.7.1 Graphs of Bessel functions of the second kind Y_0, Y_1, and Y_2 [formula (14)].

FIGURE 11.7.2 Graphs of J_0 and $J_{1/2}$, Y_0 and $Y_{1/2}$.

❖ **The Bessel Function of the Second Kind of Order** p. For p nonnegative and not an integer,

$$Y_p(x) = \frac{\cos(p\pi)\, J_p(x) - J_{-p}(x)}{\sin p\pi} \tag{16}$$

is the *Bessel function of the second kind of order p*.

Since it can be shown from formula (16) that

$$\lim_{p \to n} Y_p(x) = Y_n(x)$$

where $Y_n(x)$ is defined by formula (14), formula (16) can be considered to define Y_p for all values of p. See Figure 11.7.2 for the graphs of J_0 and Y_0, $J_{1/2}$ and $Y_{1/2}$.

The functions $Y_p(x)$ satisfy the same Recursion Formulas given in Theorem 11.6.2 for $J_p(x)$ and have the same oscillatory behavior and structure of the set of zeros as J_p. The most notable visual difference between the graphs of $J_p(x)$ and $Y_p(x)$ is the behavior at the singularity $x = 0$: $J_p(0)$ is 1 (if $p = 0$) or 0 (if $p > 0$), but $\lim_{x \to 0^+} Y_p(x) = -\infty$ for every nonnegative p. That unbounded behavior of $Y_p(x)$ at $x = 0$ plays a central role in the problem of the aging spring, as we will now see.

The Aging Spring and Bessel Functions

Let's examine the ODE of the aging spring

$$my''(t) + ke^{-\varepsilon t} y(t) = 0 \tag{17}$$

that was first introduced in Section 11.1. Let's introduce a new measure of time, $s = \alpha e^{\beta t}$, where α and β are constants to be chosen later, and see what happens when we change from ty-variables to sy-variables in equation (17). An exponential measure of

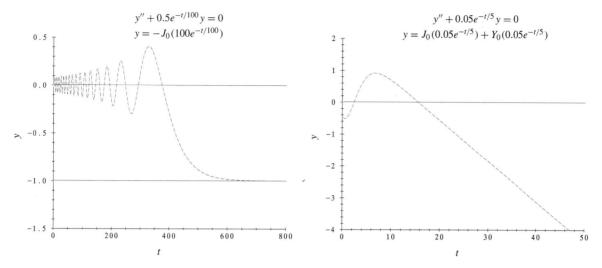

FIGURE 11.7.3 Spring stretches to a finite length.   **FIGURE 11.7.4** A rapidly aging spring stretches.

time is suggested by the form of the elastic coefficient $ke^{-\varepsilon t}$ in (17). Since $ds/dt = \beta\alpha e^{\beta t} = \beta s$, and $(s/\alpha)^{-\varepsilon/\beta} = e^{-\varepsilon t}$, we have by the Chain Rule that

$$\frac{dy}{dt} = \frac{dy}{ds}\frac{ds}{dt} = \frac{dy}{ds}\beta s$$

$$\frac{d^2 y}{dt^2} = \frac{d}{dt}\left(\frac{dy}{dt}\right) = \frac{d}{ds}\left(\frac{dy}{dt}\right)\frac{ds}{dt} = \frac{d}{ds}\left(\frac{dy}{ds}\beta s\right)\frac{ds}{dt} = \frac{d^2 y}{ds^2}(\beta s)^2 + \frac{dy}{ds}\beta^2 s$$

Substituting into ODE (17), we have

$$\frac{d^2 y}{dt^2} + \frac{k}{m}e^{-\varepsilon t}y = \beta^2 s^2 \frac{d^2 y}{ds^2} + \beta^2 s\frac{dy}{ds} + \frac{k}{m}\left(\frac{s}{\alpha}\right)^{-\varepsilon/\beta}y = 0$$

Divide the above ODE in y and s by β^2, choose α and β as follows

$$\beta = -\frac{\varepsilon}{2}, \qquad \alpha = \left(\frac{2}{\varepsilon}\right)\sqrt{\frac{k}{m}}$$

and obtain the Bessel equation of order 0:

$$s^2\frac{d^2 y}{ds^2} + s\frac{dy}{ds} + s^2 y = 0 \tag{18}$$

The general solution of (18) is

$$y = c_1 J_0(s) + c_2 Y_0(s) \tag{19}$$

where c_1 and c_2 are arbitrary constants. In terms of the time variable t,

$$y = c_1 J_0\left(2\alpha e^{-\varepsilon t/2}\right) + c_2 Y_0\left(2\alpha e^{-\varepsilon t/2}\right), \quad \text{where } \alpha = \sqrt{\frac{k}{m}} \tag{20}$$

As time t increases, the argument $2\alpha e^{-\varepsilon t/2}$ of J_0 and of Y_0 in (20) tends to 0, so the asymptotic behavior of the aging spring is modeled by the behavior of solutions of

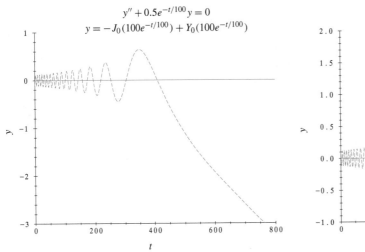

$$y'' + 0.5e^{-t/100}y = 0$$
$$y = -J_0(100e^{-t/100}) + Y_0(100e^{-t/100})$$

$$y'' + 0.5e^{-t/100}y = 0$$
$$y = J_0(100e^{-t/100}) - Y_0(100e^{-t/100})$$

FIGURE 11.7.5 A slowly aging spring stretches.   **FIGURE 11.7.6** A slowly aging spring compresses.

Bessel's equation of order 0 near the regular singularity at the origin. There are two distinct possibilities. First, if $c_2 = 0$, then as $t \to +\infty$ the displacement y from the rest position [as modeled by (20)] asymptotically approaches the value c_1 since $J_0(0) = 1$.

☞ Recall that $Y_0(s) \to -\infty$ as $s \to 0^+$.

It is more likely however, that c_2 is nonzero and that the displacement tends to $c_1 J_0(0) + c_2 Y_0(0)$, that is, to $\mp\infty$ depending on whether c_2 is positive or negative. If c_2 is negative, then the spring oscillates with increasing amplitude and eventually begins to stretch (or compress) without bound. In reality, the spring will stretch beyond its elastic limit and snap, or else will behave in a completely inelastic matter that is not modeled by ODE (17) at all. Figures 11.7.3–11.7.6 illustrate some of the possibilities. The case $c_2 < 0$ is illustrated in Figure 11.7.6 and corresponds to the unreal situation of approach to infinite compression (i.e., $y \to +\infty$). In fact, for any real spring there is a limit to the amount the spring can be compressed, and the model ODE (17) is not valid when $y(t)$ approaches that limit.

Steady Temperatures and Bessel Functions

Finally, we close this section with a reconsideration of the equation connected with the model for the steady temperatures in a solid cylinder (see Example 11.1.2).

EXAMPLE 11.7.3   **Steady Temperatures in a Cylinder**
A solution $R(r)$ of the equation

☞ Skip this example if you haven't covered Section 10.7.

$$rR''(r) + R'(r) + \lambda r R(r) = 0, \qquad 0 \le r \le 1 \tag{21}$$

is one factor of the temperature function $u(r, z) = R(r)Z(z)$ constructed in Example 11.1.2 as part of a model for steady temperatures in a solid cylinder. The full set of model equations is given by boundary value problem (3) in Section 11.1. The parameter λ in ODE (21) is a separation constant and is assumed to be positive. The conditions given in Example 11.1.2 imply that $R(r)$ must be bounded for $0 \leq r \leq 1$, while the second and third boundary conditions of boundary value problem (3) in Section 11.1 imply that the separation constant λ in (21) must be chosen so that $R(1) = 0$. The problem, then, is to find all positive values of λ such that

$$rR''(r) + R'(r) + \lambda r R(r) = 0, \qquad 0 \leq r \leq 1$$
$$R(1) = 0, \qquad |R(r)| \text{ bounded for } 0 \leq r \leq 1 \tag{22}$$

First, the ODE in (21) can be transformed to the Bessel equation of order 0 by multiplying the equation by r and introducing the variable changes

$$r = x\lambda^{-1/2}, \qquad y(x) = R(x\lambda^{-1/2}) \tag{23}$$

Using the Chain Rule, we have

$$\frac{dy}{dx} = \frac{dR}{dr}\frac{dr}{dx} = R'(r)\lambda^{-1/2}, \qquad \frac{d^2y}{dx^2} = \frac{d}{dr}\left(R'(r)\lambda^{-1/2}\right)\frac{dr}{dx} = R''(r)\lambda^{-1}$$

So ODE (21) becomes

$$x^2 y'' + xy' + x^2 y = 0 \tag{24}$$

which is Bessel's equation of order 0. The general solution of ODE (24) is

$$y = c_1 J_0(x) + c_2 Y_0(x)$$

where c_1 and c_2 are constants.

In xy-variables, problem (22) becomes [using the formulas of (23)]

$$x^2 y'' + xy' + x^2 y = 0, \qquad 0 \leq x \leq \lambda^{1/2}$$
$$y(\lambda^{1/2}) = 0, \qquad |y(x)| \text{ bounded for } 0 \leq x \leq \lambda^{1/2} \tag{25}$$

Because the general solution of ODE (24) involves the function $Y_0(x)$ that is infinite at $x = 0$, that function must be excluded on physical grounds, and the desired solution of (25) has the form

$$y = c_1 J_0(x)$$

The boundary condition $y(\lambda^{1/2}) = 0$ implies that $\lambda^{1/2}$ must be any one of the positive zeros λ_n of J_0 studied in Section 11.6. The acceptable solutions of (22) are (since $x = r\lambda^{1/2}$)

$$R_n(r) = c_1 J_0(r\lambda_n), \qquad \lambda_n \text{ a positive zero of } J_0 \tag{26}$$

where c_1 is any constant. Solutions (26) of the Sturm-Liouville Problem (22) will be used in Section 11.8 in the solution of some boundary value problems.

Comments

It often happens that an equation superficially quite unlike a Bessel equation can be transformed to a Bessel equation by a judiciously chosen change of variable. Problems 3 and 4 list two families of ODEs that can be transformed to Bessel equations.

PROBLEMS

www **1.** (*Frobenius's Theorem II: Cases II, III*). Check that 0 is a regular singular point of each equation and find a basis for the solution space on the interval $(0, \infty)$. [*Hint*: In parts **(a)**, **(b)**, **(c)**, the solution y_1 is easily found in closed form. Then use the Wronskian Reduction Method of Problem 6 of Section 3.7 to find a second independent solution.]

    (a) $xy'' + (1+x)y' + y = 0$ **(b)** $x^2 y'' + x(x-1)y' + (1-x)y = 0$

    (c) $xy'' - xy' + y = 0$ **(d)** $xy'' - x^2 y' + y = 0$

2. (*Equations Reducible to Bessel's Equation*). Many second-order, nonconstant coefficient linear ODEs are equivalent to Bessel's equation. Here is one.

    (a) If $w(s)$ is the general solution of Bessel's equation of order p, $s^2 w'' + s w' + (s^2 - p^2)w = 0$, show that $y(x) = e^{ax} w(bx)$ is the general solution of the equation

$$x^2 y'' + x(1 - 2ax)y' + [(a^2 + b^2)x^2 - ax - p^2]y = 0$$

    where a and b are real constants, $b \neq 0$.

    (b) Find the general solution of $x^2 y'' + x(1-2x)y' + (2x^2 - x - 1)y = 0$.

3. (*Equations Reducible to Bessel's Equation*). Here are more ODEs equivalent to Bessel's equation.

    (a) Show that if $y(x)$ is a solution of Bessel's equation of order p, then $w(z) = z^{-c} y(az^b)$ is a solution of

$$z^2 w'' + (2c+1)zw' + [a^2 b^2 z^{2b} + (c^2 - p^2 b^2)]w = 0$$

    (b) (*Airy's equation*). Use part **(a)** to show that the general solution of *Airy's equation* (see Section 11.2, Problem 4), $y'' - xy = 0$, is

$$y = |x|^{1/2} \left[c_1 J_{1/3}\left(\frac{2|x|^{3/2}}{3} \right) + c_2 J_{-1/3}\left(\frac{2|x|^{3/2}}{3} \right) \right]$$

    The *Airy functions* $Ai(x)$ and $Bi(x)$ are respectively defined for $x < 0$ by setting $c_1 = c_2 = 1/3$, and $c_1 = -c_2 = -1/\sqrt{3}$. Show that $\{Ai(x), Bi(x)\}$ is an independent set. [*Hint*: First introduce a new variable $x = -z$.]

    (c) Find the general solution of $x^2 y'' + \left(\frac{1}{8} + x^4 \right) y = 0$.

    (d) Find the general solution of $y'' + x^4 y = 0$.

4. (*Aging Spring*). Suppose that $y = J_0(6e^{-t})$ is the solution of a model of an aging spring. Plot y as a function of t, $0 \le t \le 6$. What happens as $t \to \infty$?

5. (*Bessel Functions of the Second Kind*). If p is not an integer, use the formula

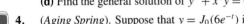

$$Y_p = \frac{\cos(p\pi)J_p - J_{-p}}{\sin p\pi}$$

to show that Bessel functions of the second kind of noninteger order satisfy the recursion formulas (16)–(19) in Section 11.6.

6. (*Aging Spring*). Summarize what has been shown in this chapter about the problem of the aging spring. Critique the model's validity. Propose and analyze other models. In particular, consider the model

$$y'' + (b^2 + a^2 e^{-\varepsilon t})y = 0$$

where a, b, and ε are constants. Show that solutions are given by

$$y = c_1 J_p(s) + c_2 Y_p(s)$$

where $p = 2ib/\varepsilon$ and $s = 2a\varepsilon^{-1}e^{-\varepsilon t/2}$. Critique this model. Plot solutions for various values of a, b, ε, and interpret these solutions in terms of motions of the aging spring.

11.8 Steady Temperatures in Spheres and Cylinders

Now we return to the Method of Separation of Variables developed in Section 10.7. The main topic is now Laplace's equation $\nabla^2 u = 0$, but over spheres and cylinders where Legendre polynomials and Bessel functions turn out to play a critical role. It was pointed out in Section 10.7 that the Method of Separation of Variables only succeeds when the physical boundaries of the region where the solution u is defined consist of level sets of the coordinate system in which $\nabla^2 u$ is expressed. We experienced this in Section 10.9 when the region was a thin circular plate and we needed to express $\nabla^2 u$ in terms of polar coordinates. As we will see in this section, when the region in which u is defined is a sphere or a cylinder, then the PDE $\nabla^2 u = 0$ will have to be expressed in spherical coordinates or cylindrical coordinates, respectively, before the Method of Separation of Variables can be applied. In these new coordinate systems the Sturm-Liouville problem encountered will no longer be a regular one, so we will take care of this difficulty first.

Singular Sturm-Liouville Problems

Suppose that the functions $\rho(x)$, $p(x)$, and $q(x)$ are such that

$$\rho, q \text{ are in } \mathbf{C}^0[a, b], \qquad p \text{ is in } \mathbf{C}^1[a, b], \qquad \rho, p \neq 0 \text{ on } (a, b) \qquad (1)$$

Then a regular Sturm-Liouville problem for the operator L with action $L[y] = (py')' + xy$ asks us to find all values of λ (called eigenvalues of L) for which the ODE

$$(py')' + qy = \lambda \rho y \qquad (2)$$

has nontrivial solutions (called eigenfunctions of L) in $\mathbf{C}^2[a, b]$ that satisfy separated boundary conditions at the endpoints $x = a$, $x = b$ where $p \neq 0$. For example, a separated condition at $x = a$ has the form $\alpha u'(a) + \beta u(a) = 0$, where the constants α and β are not both zero. It was shown in Theorem 10.6.2 that eigenfunctions corresponding to distinct eigenvalues are orthogonal with respect to the scalar product

$$\langle f, g \rangle = \int_a^b \rho f g \, dx, \qquad f, g \text{ in } \mathbf{C}^2[a, b]$$

If $p(a) = 0$ or $p(b) = 0$, then ODE (2) has a singularity at one or both endpoints of the interval $a \le x \le b$, so the Sturm-Liouville problem is not regular. The two examples that follow show what happens in this case.

EXAMPLE 11.8.1

Bessel Functions

Consider the eigenvalue equation with operator $L[u]$ and action $L[u] = (1/x)(xu')'$:

$$\frac{1}{x}(xu')' = \lambda u$$

where u is in $\mathbf{C}^2[0, R_0]$, $R_0 > 0$, and $u(R_0) = 0$. In this case we have $a = 0$, $b = R_0$, $\rho = x$, $p = x$, $q = 0$ in (1) and (2). Notice that $p(0) = 0$ and that we have a separated

☞ It would be helpful to review Sections 11.3 (Legendre Polynomials) and 11.6 (Bessel Functions) before plunging into this material .

condition at $x = R_0$. L is a symmetric differential operator, and since a separated condition was used, the eigenspaces are all one-dimensional and mutually orthogonal (see Theorem 10.6.2), with the weighted scalar product

$$\langle f, g \rangle = \int_0^{R_0} x f(x) g(x)\, dx \tag{3}$$

and the eigenvalues are nonpositive.

Let's find the eigenvalues of L. First try $\lambda = 0$; the equation $(xu')' = 0$ has the general solution $u = A \ln x + B$. But the conditions u in $\mathbf{C}^2[0, R_0]$ and $u(R_0) = 0$ imply that $A = B = 0$, so $\lambda = 0$ is *not* an eigenvalue. Now try $\lambda = -k^2$, for $k > 0$. The eigenvalue equation $(xu')' + k^2 xu = 0$ has only one solution (up to constant multiples) that belongs to the class $\mathbf{C}^2[0, R_0]$; it is $u = J_0(kx)$, where J_0 is the Bessel function of order zero. The condition $u(R_0) = 0$ implies that $J_0(kR_0) = 0$. But we know from Section 11.6 that J_0 has infinitely many positive zeros: $x_1, x_2, \ldots$. This implies that $k_n = x_n/R_0$, $n = 1, 2, \ldots$, so the eigenvalues of L are

$$\lambda_n = -(x_n/R_0)^2, \qquad n = 1, 2, \ldots$$

The corresponding eigenspaces are spanned by the eigenfunctions

$$u_n = J_0(x_n x/R_0), \qquad n = 1, 2, \ldots$$

It turns out that $\{u_n\}$ is a basis of PC$[0, R_0]$, a so-called *Bessel basis*, with the weighted scalar product given by (3).

EXAMPLE 11.8.2

Legendre Polynomials

Consider the eigenvalue equation with operator L and action $L[u] = [(1 - x^2)u']'$:

$$[(1 - x^2)u']' = \lambda u$$

for u in $\mathbf{C}^2[-1, 1]$. In this case $a = -1$, $b = 1$, $\rho = 1$, $p(x) = (1 - x^2)$, $q(x) = 0$, and since $p(-1) = p(1) = 0$, Dom $(L) = \mathbf{C}^2[-1, 1]$ in (1) and (2). So L is a symmetric operator under the scalar product

$$\langle f, g \rangle = \int_{-1}^{+1} f(x) g(x)\, dx \tag{4}$$

The eigenvalues are nonpositive and the eigenspaces are mutually orthogonal. Although Theorem 10.6.2 does not guarantee it, we will see that the eigenspaces are all one-dimensional.

For $\lambda = -n(n + 1)$, $n = 0, 1, 2, \ldots$, the eigenvalue equation

$$L[u] = ((1 - x^2)u')' = -n(n + 1)u$$

is Legendre's equation (Section 11.3) whose only solution (up to constant multiples) in $\mathbf{C}^2[-1, 1]$ is $P_n(x)$, the Legendre polynomial $P_n(x)$ of order n, $n = 0, 1, 2, \ldots$. It is a fact that $\{P_n(x)\}$ is a basis for PC$[-1, 1]$, a so-called *Legendre basis*, with the scalar product given by (4).

We use this fact to show that L has no eigenvalues, other than $\lambda_n = -n(n + 1)$, $n = 0, 1, 2, \ldots$. Assume, to the contrary, that $\mu \neq -n(n + 1)$, $n = 0, 1, 2, \ldots$, is an eigenvalue of L with corresponding eigenfunction $v(x)$. Now L is a symmetric operator, so

$$\mu \langle v, P_n \rangle = \langle \mu v, P_n \rangle = \langle Lv, P_n \rangle = \langle v, LP_n \rangle = -n(n+1) \langle v, P_n \rangle$$

So $\langle v, P_n \rangle = 0$, for all $n = 0, 1, 2, \ldots$. By Theorem 10.3.5, $v = 0$, so v can't have been an eigenfunction, nor μ an eigenvalue. The eigenvalues and corresponding eigenfunctions of L are

$$\lambda_n = -n(n+1), \qquad u_n = P_n(x), \qquad n = 0, 1, 2, \ldots$$

If f is in PC$[-1, 1]$, then f has the orthogonal expansion

$$f(x) = \sum_0^\infty A_n P_n(x), \qquad A_n = \frac{\langle f, P_n \rangle}{||P_n||^2} = \frac{2n+1}{2} \int_{-1}^1 f(x) P_n(x)\, dx$$

where we have used the formula $||P_n||^2 = 2/(2n+1)$ from Table 11.3.1.

The Sturm-Liouville systems in Examples 11.8.1 and 11.8.2 are not in the regular case. Such systems are said to be *singular Sturm-Liouville systems*. Nevertheless, it is true that conclusions similar to those in the regular Sturm-Liouville Theorem (Theorem 10.6.4) hold. For example, $\{J_0(x_n x / R_0) : n = 1, 2, \ldots\}$, where x_n is the n-th positive zero of J_0, is a basis for PC$[0, R_0]$ under the scalar product $\langle f, g \rangle = \int_0^{R_0} xfg\, dx$. Also, as mentioned earlier, $\{P_n(x) : n = 1, 2, \ldots\}$, is a basis for PC$[-1, 1]$ under the standard scalar product, where P_n is the Legendre polynomial of order n.

Steady Temperatures in a Ball: Zonal Harmonics

Now let G be the interior of the unit ball in $\mathbb{R}^3 : x^2 + y^2 + z^2 < 1$, and consider the Dirichlet problem

$$
\begin{array}{lll}
\text{(PDE)} & \nabla^2 u = 0 & \text{in } G \\
\text{(BC)} & u = f(\theta, \phi) & \text{on } \partial G
\end{array}
\tag{5}
$$

where the boundary function $f(\theta, \phi)$ is expressed in spherical coordinates and is assumed to be a continuous function. Since the physical boundary of G is a sphere, we would first need to express the PDE in (5) in spherical coordinates if we want to use the Method of Separation of Variables. Using (5) from Section 10.9 to transform boundary problem (5) into spherical coordinates, we obtain

$$\text{(PDE)} \qquad \frac{1}{\rho^2}(\rho^2 u_\rho)_\rho + \frac{1}{\rho^2 \sin \phi}(\sin \phi u_\phi)_\phi + \frac{1}{\rho^2 \sin^2 \phi} u_{\theta\theta} = 0 \tag{6}$$

$$\text{(BC)} \qquad u(1, \theta, \phi) = f(\theta, \phi)$$

We will assume from here on that f is independent of θ. We will find a solution u that is also independent of θ. As usual, it is assumed that u belongs to $\mathbf{C}^2(G)$ and to $\mathbf{C}^0(G^*)$.

In using the Method of Separation of Variables we need only look for solutions of (PDE) of the form $R(\rho)\Phi(\phi)$ since the boundary data f depend only on ϕ. The variables are separated to obtain the ODEs

☞ We have left out the calculations that lead to these two ODEs

$$\rho^2 R'' + 2\rho R' + \lambda R = 0, \qquad 0 < \rho < 1 \tag{7}$$

$$(\sin\phi)\Phi'' + (\cos\phi)\Phi' - (\lambda\sin\phi)\Phi = 0, \qquad 0 < \phi < \pi \tag{8}$$

where λ is the separation constant. Now since we assumed that $R(\rho)\Phi(\phi)$ belongs to $\mathbf{C}^2(G)$, we will look for those constants λ such that ODEs (7) and (8) have solutions R and Φ with

$$R \text{ in } \mathbf{C}^2[0, 1] \tag{9a}$$

$$\Phi \text{ in } \mathbf{C}^2[0, \pi] \tag{9b}$$

Observe that ODEs (7) and (8) are *nonnormal* because the leading coefficients vanish when $\rho = 0$ and when $\phi = 0$ or π, so conditions (9a) and (9b) are not redundant. The Method of Separation of Variables provides us with a singular Sturm-Liouville problem to solve.

If we change the independent variable from s to ϕ with the function $s = \cos\phi$, we see that the interval $0 \le \phi \le \pi$ is transformed one-to-one onto the interval $-1 \le s \le 1$ and (using the Chain Rule) that (8) and (9b) become the singular Sturm-Liouville system

$$(1 - s^2)\frac{d^2\Phi}{ds^2} - 2s\frac{d\Phi}{ds} - \lambda\Phi = 0, \qquad -1 < s < 1, \quad \Phi \text{ in } \mathbf{C}^2[-1, 1] \tag{10}$$

Now, we have seen the ODE in (10) before; indeed, if λ is replaced by $-n(n + 1)$, then ODE (10) is precisely Legendre's equation and has as one of its solutions the Legendre polynomial $P_n(s)$. According to Example 11.8.2, we have $\lambda_n = -n(n + 1)$, $n = 0, 1, \ldots$ and $\Phi_n(\phi) = P_n(\cos\phi)$, $n = 0, 1, \ldots$.

If we substitute λ_n for λ in (7), we arrive at the problem

$$\rho^2 R'' + 2\rho R' - n(n + 1)R = 0, \qquad R \text{ in } \mathbf{C}^2[0, 1], \quad n = 0, 1, 2, \ldots \tag{11}$$

Now ODE (11) is an Euler equation and has the general solution

$$R(\rho) = A\rho^n + B\rho^{-n-1}, \qquad n = 0, 1, 2, \ldots$$

Since R must be in $\mathbf{C}^2[0, 1]$ [see (9b)], we must take $B = 0$, so we have

$$R_n(\rho) = A\rho^n, \qquad n = 0, 1, 2, \ldots$$

Now we take the functions $P_n(x) = P_n(\cos\phi)$ and ρ^n and look for a solution for (6) in the form

$$u(\rho, \phi) = \sum_{n=0}^{\infty} A_n \rho^n P_n(\cos\phi) \tag{12}$$

Inserting $\rho = 1$ into (12), we have the condition

$$f(\phi) = \sum_{n=0}^{\infty} A_n P_n(\cos\phi), \qquad 0 < \phi < \pi \tag{13}$$

If we make the change of variables $x = \cos\phi$ in (13) and let g be the function on $|x| \le 1$ such that $g(\cos\phi) = f(\phi)$ for all $0 \le \phi \le \pi$, then (13) becomes

$$g(x) = \sum_{n=0}^{\infty} A_n P_n(x), \qquad |x| \le 1 \qquad (14)$$

Referring to Example 11.8.2, we see that (14) holds if we choose A_n by using the Fourier-Euler formula appropriate to a Legendre basis of $PC[-1, 1]$:

$$A_n = \frac{\langle g, P_n \rangle}{\| P_n \|^2} = \frac{2n+1}{2} \int_{-1}^{1} g(x) P_n(x)\, dx, \qquad n = 0, 1, 2, \ldots \qquad (15)$$

where $g(x) = f(\arccos x)$. The series (12) with coefficients given by (15) gives a *formal* solution to (6).

Those values of ϕ for which $\cos\phi$ is a root of $P_n(\cos\phi)$ [recall that $P_n(x)$ has n distinct roots, all of which lie in the interval $(-1, 1)$] determine *nodal latitudes* on the unit sphere. The regions between consecutive nodal latitudes are called *zones* and the functions $\rho^n P_n(\cos\phi)$ are called *zonal harmonics*. The solution (12), (15) is an *expansion in zonal harmonics*.

The results above may be interpreted in terms of steady-state temperatures in a ball, given boundary temperatures independent of longitude θ.

Steady Temperatures in a Cylinder: Cylindrical Harmonics

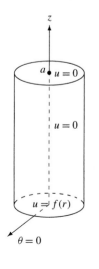

☞ We first saw this problem in Example 11.1.2.

Suppose that C is the cylinder described in cylindrical coordinates by $0 \le r \le 1$, $0 \le z \le a$, and that $u(r, \theta, z)$ is the steady temperature in C at the point whose cylindrical coordinates are given by (r, θ, z). We consider the problem of finding the steady temperature in C when the base of the cylinder ($z = 0$) is maintained at a given temperature $f(r)$ degrees, independent of θ, and the rest of ∂C is maintained at $0°$. So u is the solution of the boundary problem

$$
\begin{aligned}
\nabla^2 u &= 0, & 0 \le r &< 1, & -\pi \le \theta &< \pi, & 0 &< z < a \\
u(1, \theta, z) &= 0, & & & -\pi \le \theta &< \pi, & 0 &\le z \le a \\
u(r, \theta, a) &= 0, & 0 \le r &\le 1, & -\pi \le \theta &< \pi \\
u(r, \theta, 0) &= f(r), & 0 \le r &\le 1, & -\pi \le \theta &< \pi
\end{aligned}
\qquad (16)
$$

where $\nabla^2 u$ is expressed in cylindrical coordinates [see (4) in Section 10.9]. Since the data are a function of the variable r only, we are led to suspect that the solution of (16) is a function of r and z only. Assuming this and separating the variables in the usual way, we obtain the ODEs

$$R'' + \frac{1}{r} R' - \lambda R = 0, \qquad |R(0+)| < \infty, \quad R(1) = 0 \qquad (17)$$

$$Z'' + \lambda Z = 0, \qquad Z(a) = 0 \qquad (18)$$

The singular Sturm-Liouville problem (17) was treated in Example 11.8.1. The differential equation in (17) is an eigenvalue equation for Bessel's operator of order $p = 0$. The separation constant λ can assume only the values $\lambda = \lambda_n = -x_n^2$, where the x_n are the consecutive positive zeros of $J_0(x)$, $n = 1, 2, 3, \ldots$. The corresponding eigenfunctions for (17) are given by

$$R_n(r) = J_0(x_n r), \qquad n = 1, 2, \ldots$$

So (18), with λ replaced by $\lambda = -x_n^2$, yields

$$Z_n(z) = \sinh x_n(a - z), \qquad n = 1, 2, \ldots$$

☞ Recall that
$\sinh x = \frac{1}{2}(e^x - e^{-x})$.

We are led to expect a solution of boundary problem (16) in the form

$$u(r, z) = \sum_{n=1}^{\infty} A_n \sinh x_n(a - z) J_0(x_n r) \tag{19}$$

The remaining condition in (16) that must be satisfied now implies that

$$f(r) = u(r, 0) = \sum_{n=1}^{\infty} A_n \sinh(x_n a) J_0(x_n r)$$

This suggests an orthogonal expansion of f using the basis $\{J_0(x_n r), \ n = 1, 2, \ldots\}$ (see Example 11.8.1). Using the Fourier-Euler formula for the coefficients with the inner product given by (3), we see that

$$A_n = \frac{2}{[J_1(x_n)]^2 \sinh(x_n a)} \int_0^1 r f(r) J_0(x_n r)\, dr \tag{20}$$

where we have used the formula $\int_0^1 r J_0^2(x_n r)\, dr = \frac{1}{2} J_1^2(x_n)$ (see Problem 4). The series (19) with coefficients given by (20) defines the formal solution to (16). Although we will not do it here, it can be shown that this formal solution is indeed a classical solution if $f(r)$ is smooth enough and $f(1) = 0$. The functions $\sinh x_n(a - z) J_0(x_n r)$ are called *cylindrical harmonics*, and (19) is an *expansion in cylindrical harmonics*.

Comments

We have analyzed a sample of Dirichlet problems that can be solved by the Method of Separation of Variables. As always, the boundary of the region involved must be composed of level sets of the coordinate variables. In practice, this restricts the method to rectangular or boxlike regions, balls, cylinders, disks, or to simple combinations of these regions. We have also showed by two examples that nonrectangular geometry may lead to singular Sturm-Liouville problems, which lead in turn to special orthogonal series.

    Given the complexities of actually constructing solutions of the Dirichlet problem, it is remarkable that properties of solutions (e.g., the Maximal Principle) can be proved independently of the particular shape of the region involved. Although we proved the properties in Section 10.9 only for planar regions, they remain true, after appropriate reformulations, in any number of dimensions.

    Other boundary conditions can be used in the problems considered in this section, so we have just scratched the surface of this subject.

PROBLEMS

1. (*Temperatures in a Solid Ball*). Find the steady-state temperatures in a ball of unit radius if the surface temperatures f in spherical coordinates are as given. [*Hint*: Use (12), (15), and properties of Legendre polynomials given in Table 11.3.1.]

   (a) $f = \cos 2\phi - \sin^2 \phi$

   (b) $f(\phi) = \begin{cases} 1, & 0 < \phi < \pi/2 \\ 0, & \pi/2 < \phi < \pi \end{cases}$ [*Hint*: Use the identity $xP_n' - nP_n = P_{n-1}'$ in Table 11.3.1 and integration by parts to evaluate $\int_{-1}^{1} P_n(x)\,dx$.]

   (c) $f(\phi) = |\cos \phi|$

2. (*Temperatures in a Hollow Ball*). Find the steady-state temperatures in a spherical shell, $0 < \rho_1 < \rho < \rho_2$, if the temperatures on the inner sphere are given by $f(\phi)$, on the outer sphere by $g(\phi)$. [*Hint*: Proceed as in the text, but keep both terms in $R(\rho) = A\rho^b + B\rho^{-n-1}$.]

www 3. Solve the boundary value problems below:

   (a) Solve boundary problem (16) if $f(r) = 1$, $0 \le r \le 1$.

   (b) Solve the boundary value problem that arises from (16) if the side-wall temperature condition $u(1, \theta, z) = 0$ is replaced by the perfect-insulation condition $u_r(1, \theta, z) = 0$, $0 \le z \le a$, $-\pi \le \theta \le \pi$. [*Hint*: Use (19) in Theorem 11.6.2 with $p = 0$, and use the identity below in Problem 4 with $a = x_n^*$, where x_n^* is a positive zero of J_0' (and so of J_1).]

4. Show that

   $$\int_0^1 r J_0^2(ar)\,dr = [J_0^2(a) + J_1^2(a)]/2$$

   where a is any real number. [*Hint*: Multiply Bessel's equation of order 0 by $2J_0'(r)$ and rewrite to obtain $[r^2(J_0')^2]' + r^2(J_0^2)' = 0$. Integrate from 0 to a using an integration by parts, and then use the recursion formula (19) of Section 11.6 with $p = 0$: $J_0' = -J_1$.]

5. (*Neumann Problems*). Let G be a bounded planar region whose boundary is a simple smooth closed curve. The problem $\nabla^2 u = 0$ in G, $\partial u/\partial n = f$ on ∂G, where u belongs to $\mathbf{C}^2(G)$ and to $\mathbf{C}^0(G^*)$ and f is continuous on ∂G, is called a *Neumann problem* for G.

   (a) Show that any two solutions of the Neumann problem differ by a constant. [*Hint*: Let $w = u_1 - u_2$, where u_1 and u_2 solve the Neumann problem. Apply the Divergence Theorem (Theorem B.5.13) to the vector $\nabla(w^2)$, and use the facts that $\nabla^2 w = 0$ in G and $\partial w/\partial n = 0$ on ∂G to show that ∇w is the zero vector on G^*.]

   (b) Show that if the Neumann problem has a solution at all, then $\int_{\partial G} f(s)\,ds = 0$.

   (c) Solve the Neumann problem above if G is the unit disk in the plane and if $f(\theta) = \sin\theta$, while $u = 0$ at $\theta = 0$, $r = 1$. [*Hint*: Use formula (10), Section 10.9.]

6. (*Singular Sturm-Liouville Problems*). In each of the Sturm-Liouville problems below identify the operator L, find the eigenvalues and eigenspaces of L, and state the orthogonality and basis properties of the eigenfunctions.

   (a) $x^4 y'' + 2x^3 y' = \lambda y$; $y(1) = 0$, $y(2) = 0$. [*Hint*: To solve the ODE, try the substitution $s = 1/x$.]

   (b) $xy'' - y' = \lambda x^3 y$; $y(0) = 0$, $y(a) = 0$. [*Hint*: To solve the ODE, try the substitution $s = x^2$.]

   (c) $\dfrac{1}{x}\left[(xy')' - \dfrac{p^2}{x}y\right] = \lambda y$, y in $\mathbf{C}^2[0, R_0]$, $y(R_0) = 0$, where p is a positive number. [*Hint*: $L[y] = [(xy')' - p^2 y/x]/x$ is a Bessel operator of order p.]

APPENDIX A

Basic Theory of Initial Value Problems

The aim of this Appendix is to present the basic theory for the initial value problem for a single state variable, and to give some of the proofs omitted in Chapter 2. The basic questions of uniqueness, existence, extension, and sensitivity are each taken up in turn.

A.1 Uniqueness

The uniqueness question for the initial value problem (IVP)

$$y' = f(t, y), \qquad y(t_0) = y_0 \tag{1}$$

is easier to handle than the existence question, so we will treat it first. For convenience we first define a *region* in the plane.

❖ **Interior Point, Region.** A point P is an *interior point* of a set R in the ty-plane if for some $r > 0$ all points less than distance r away from P are also in R. A *region* is a connected set (i.e., not in disjoint pieces) whose points are all interior points. For example, the points inside a rectangle form a region.

THEOREM A.1.1

> **Uniqueness Theorem.** In IVP (1) suppose that f and $\partial f/\partial y$ are continuous on some region R containing the interior point (t_0, y_0). Then on any t-interval I containing t_0 there is at most one solution of IVP (1).

Proof. We will prove the theorem only in the case where I is the interval $t_0 \leq t \leq t_0 + a$ and R contains the rectangle S: $t_0 \leq t \leq t_0 + a$, $y_0 - b \leq y \leq y_0 + b$, for some positive constants a, b. This special case can be used to establish the more general claim, but we omit the details. Now suppose that $y_1(t)$ and $y_2(t)$ are two solutions of IVP (1) that remain in S over the interval I: $t_0 \leq t \leq t_0 + a$ (see Figure A.1.1). Then

$$[y_1(t) - y_2(t)]' = f(t, y_1(t)) - f(t, y_2(t)), \qquad t_0 < t < t_0 + a \tag{2}$$

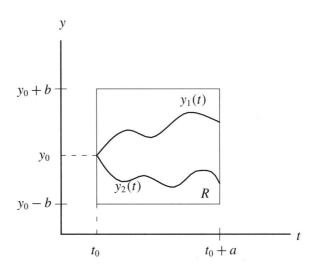

FIGURE A.1.1 Specialized geometry for IVP (1).

Notice that $y_1(t) - y_2(t)$ is continuous on I and has value 0 at $t = t_0$. Integrating each side of equation (2) from t_0 to some t in I, we have

$$y_1(t) - y_2(t) = \int_{t_0}^{t} [f(s, y_1(s)) - f(s, y_2(s))]\,ds, \quad t \text{ in } I \tag{3}$$

☞ These two theorems are listed as Theorems B.5.4 and B.5.1 in Appendix B.

Now since $\partial f/\partial y$ is continuous on S, it follows from the Mean Value Theorem and the Maximum-Minimum Value Theorem that there is a positive constant L such that[1]

$$|f(t, y_1) - f(t, y_2)| \le L|y_1 - y_2|, \quad \text{for any points } (t, y_1), (t, y_2) \text{ in } S \tag{4}$$

Functions $f(t, y)$ that satisfy inequality (4) for some constant $L > 0$ are said to satisfy a *Lipschitz condition* in y on S; the constant L is called a *Lipschitz constant* Taking absolute values of each side of (3), denoting $|y_1(t) - y_2(t)|$ by $w(t)$, estimating the integral, and using (4), we obtain the integral inequality

☞ These integral estimates are given in Theorem B.5.9 in Appendix B.

$$0 \le w(t) \le \int_{t_0}^{t} |f(s, y_1(s)) - f(s, y_2(s))|\,ds \le L \int_{t_0}^{t} w(s)\,ds, \quad \text{all } t \text{ in } I \tag{5}$$

We will show that the only nonnegative solution $w(t)$ of inequality (5) is the function $w(t) = 0$ for all t.

Suppose that

$$v(t) = \int_{t_0}^{t} w(s)\,ds, \quad \text{all } t \text{ in } I \tag{6}$$

---

[1] L may be taken to be max $|\partial f/\partial y|$ for (t, y) in S.

Since $v'(t) = w(t)$ on I, we can rewrite (5) as the inequality

$$v'(t) - Lv(t) \leq 0, \quad \text{all } t \text{ in } I \tag{7}$$

Multiplying inequality (7) through by the "integrating" factor e^{-Lt}, we have the following inequality (recall that I is the interval $t_0 \leq t \leq t_0 + a$)

$$[v(t)e^{-Lt}]' \leq 0, \quad t_0 \leq t \leq t_0 + a$$

So $v(t)e^{-Lt}$ is a nonincreasing function of t for $t_0 \leq t \leq t_0 + a$. This implies that

$$v(t)e^{-Lt} \leq v(t_0)e^{-Lt_0}, \quad t_0 \leq t \leq t_0 + a$$

But since $v(t_0) = 0$, it follows that $v(t) \leq 0$ for all t in I. On the other hand, since $v(t) = \int_{t_0}^{t} w(s)\, ds$, $t > t_0$, and $w \geq 0$, we know that $v(t) \geq 0$ for all t in I. Since $v(t) \leq 0 \leq v(t)$, we see that $v(t) = 0$ for all t, so $v'(t) = w(t) = 0$, all t, as well, and we have shown that $y_1(t) = y_2(t)$ for all t in I. So IVP (1) can't have more than one solution on I, finishing our proof.

PROBLEMS

1. Let's look at the ODE $y' = f(t, y)$, where $f = |y|$.

    (a) Show that the function f does *not* satisfy the hypotheses of the Uniqueness Principle in any region containing all or part of the t-axis in the ty-plane.

    (b) Show that the IVP $y' = |y|$, $y(t_0) = 0$, has a unique solution even so. Is there a contradiction?

    (c) Show that the hypothesis "$\partial f/\partial y$ is continuous" in the Uniqueness Principle can be replaced by "$f(t, y)$ satisfies a *Lipschitz condition* in y." Use this conclusion to prove again that $y' = |y|$, $y(t_0) = 0$, has a unique solution.

2. Suppose that m and n are positive integers without common factors (i.e., m and n are *relatively prime*). Here's an IVP that has a unique solution for some values of m and n and infinitely many solutions for other values:

    $$y' = |y|^{m/n}, \quad y(0) = 0$$

    (a) Show that the IVP has the unique solution $y(t) = 0$ if $m \geq n$.

    (b) Show that the IVP has infinitely many solutions if $m < n$.

A.2 The Picard Process for Solving an Initial Value Problem

Let's establish the existence of a solution for the IVP

$$y' = f(t, y), \quad y(t_0) = y_0 \tag{1}$$

where f and $\partial f/\partial y$ are both continuous functions on a region R of the ty-plane containing the interior point (t_0, y_0). The method we will use is to construct a sequence of iterate functions—called *Picard iterates*—that converges to the unique solution of IVP (1).

An Equivalent Integral Equation

To approach the task of finding a solution for IVP (1), we will first convert it into an "equivalent" integral equation. Suppose that $y(t)$ is a solution of IVP (1) on the interval I containing t_0 such that the graph of $y(t)$ lies in R. Then integrating each side of the relation $y'(t) = f(t, y(t))$ from t_0 to some t in the interval I and using the initial condition, we have

$$y(t) - y_0 = \int_{t_0}^t y'(s)\, ds = \int_{t_0}^t f(s, y(s))\, ds$$

If $y(t)$ is a solution of IVP (1) on the interval I containing t_0, then $y(t)$ is a solution of the integral equation

$$y(t) = y_0 + \int_{t_0}^t f(s, y(s))\, ds, \qquad t \text{ in } I \tag{2}$$

☞ Theorem B.5.5 in Appendix B is the Fundamental Theorem of Calculus.

The other way around, suppose that $y(t)$ is continuous on I, its graph lies in R, and $y(t)$ satisfies the integral equation (2). Then $y(t_0) = y_0$. From the Fundamental Theorem of Calculus, $y(t)$ is differentiable at each interior point of I, and [using (2)]

$$y'(t) = \frac{d(y(t))}{dt} = \frac{d}{dt}\left(y_0 + \int_{t_0}^t f(s, y(s))\, ds\right) = f(t, y(t))$$

It follows that $y(t)$ is a solution of IVP (1) on the interval I. This is the sense in which IVP (1) and integral equation (2) are equivalent. So the solvability of IVP (1) is equivalent to the solvability of integral equation (2). We shall exploit this equivalence.

Existence of a Solution

After this detour, we are ready to prove the existence of a solution to integral equation (2), and so of a solution to IVP (1). Under the condition that f and $\partial f/\partial y$ are continuous on R, we showed in Appendix A.1 that IVP (1) cannot have more than one solution on any interval containing t_0. Now let's show that there is a solution.

THEOREM A.2.1

> Existence Theorem. If the functions f and $\partial f/\partial y$ are continuous on a region R of the ty-plane and if (t_0, y_0) is a point of R, then IVP (1) has a solution $y(t)$ on an interval I containing t_0 in its interior.

Let's explain why this fundamental result is true. The first step is to construct in R a rectangle S that has the form $t_0 \le t \le t_0 + a$, $y_0 - b \le y \le y_0 + b$, where a and b are positive numbers (Figure A.2.1). Such a construction is possible since (t_0, y_0) is an interior point of R. Now the assumptions of the theorem imply that f and $\partial f/\partial y$ are continuous on the rectangle S. There are positive constants M and L such that

☞ The Maximum/Minimum Value Theorem (Theorem B.5.1) and inequality (4) in Appendix A.1 give us the inequalities in (3).

$$|f(t, y)| \le M, \qquad\qquad \text{all } (t, y) \text{ in } S$$
$$|f(t, y_1) - f(t, y_2)| \le L|y_1 - y_2|, \quad \text{all } (t, y_1), (t, y_2) \text{ in } S \tag{3}$$

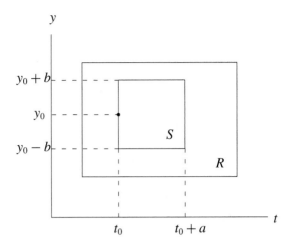

FIGURE A.2.1 Geometry for a forward initial value problem.

Our analysis will eventually produce the unique solution $y(t)$ of the differential equation $y' = f(t, y)$ on the interval $t_0 \leq t \leq t_0 + c$, where $c = \min\{a, b/M\}$, with $y(t_0) = y_0$. The corresponding solution curve remains in the rectangle S. It may be regarded as a solution of the forward initial value problem associated with IVP (1). The same technique can be used to show the existence of a solution for IVP (1) on an interval $-d \leq t \leq t_0$ for some $d > 0$. Putting this backward solution together with the forward solution, we would then have a two-sided solution to IVP (1) on the interval $t_0 - d \leq t \leq t_0 + c$, as asserted in the statement of the Existence Theorem (Theorem A.2.1). We turn now to the construction of the forward solution.

Construction of the Picard Iterates

Let's begin with the equivalent formulation of the forward problem for IVP (1) as integral equation (2). Recall that $y(t)$ is a solution of IVP (1) over an interval $t_0 \leq t \leq t_0 + c$ if and only if $y(t)$ satisfies the integral equation

$$y(t) = y_0 + \int_{t_0}^{t} f(s, y(s))\, ds, \qquad t_0 \leq t \leq t_0 + c \qquad (4)$$

We now describe a procedure that uses (4) for generating a sequence of functions $y_0(t), y_1(t), y_2(t), \ldots$, where all of these functions are defined and continuous on $t_0 \leq t \leq t_0 + c$, $c = \min\{a, b/M\}$. These functions will be used to generate a solution $y(t)$ of (4). The functions $y_n(t)$, $n = 0, 1, 2, \ldots$, constructed in this proof arise in a distinctive way and are called *Picard iterates*. They are generated recursively [i.e., $y_{n+1}(t)$ is computed directly from $y_n(t)$ for each $n = 0, 1, 2, \ldots$]. Taking $y_0(t) = y_0$, $t_0 \leq t \leq t_0 + c$, we define $y_1(t), y_2(t), \ldots$ by the recursion relation

$$y_{n+1}(t) = y_0 + \int_{t_0}^{t} f(s, y_n(s))\, ds, \qquad n = 0, 1, 2, \ldots \qquad (5)$$

with $t_0 \leq t \leq t_0 + c$.

In constructing the Picard iterates using (5), we must take care at each stage that the graph of $y = y_n(t)$ remain in S for all $t_0 \le t \le t_0 + c$ before we can go on to compute $y_{n+1}(t)$. Recalling that $c = \min\{a,\ b/M\}$, we will now show by induction that the graph of $y_n(t)$, $n = 0, 1, 2, \ldots$ also lies in (5). The graph of $y = y_0(t) = y_0$ belongs to S, so the first step of an inductive proof is complete. Assume that for some integer $n \ge 0$, the graph of $y_n(t)$ lies in S for all $t_0 \le t \le t_0 + c$. Using (5), (3), and integral estimates, we have the estimate

☞ The integral estimates are given in Theorem B.5.9

$$|y_{n+1}(t) - y_0| \le |t - t_0|M \le cM \le b, \qquad t_0 \le t \le t_0 + c \qquad (6)$$

since $c = \min\{a, b/M\}$. So the graph of $y = y_{n+1}(t)$ remains in S for all $t_0 \le t \le t_0 + c$. The induction is complete, and there is no difficulty in constructing the Picard iterates using (5) if we restrict t to the interval $[t_0, t_0 + c]$.

Uniform Convergence of a Sequence of Functions

Let's interrupt our verification of the Existence Theorem to describe the sense in which the Picard iterates $y_0(t)$, $y_1(t)$, ... are used to generate the solution of integral equation (4). Our remarks actually apply to any sequence of functions defined over a common interval, so we will not restrict ourselves to Picard iterates at the moment.

❖ **Uniform Convergence of a Sequence of Functions.** A sequence of functions $\{y_n(t)\}$, $n = 0, 1, 2, \ldots$, all defined on a common interval I, is said to converge *uniformly* to the function $y(t)$ on I if given any error tolerance $E > 0$, there exists a positive integer N such that

$$|y_n(t) - y(t)| < E, \quad \text{for all } t \text{ in } I \text{ and all } n \ge N \qquad (7)$$

As we will see, uniform convergence is the way we will generate the solution of integral equation (4) from the Picard iterates.

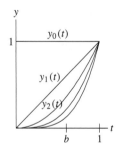

Here's an example of convergence that *is* uniform on one interval, but *not* uniform on another. The sequence $y_n(t) = t^n$, $n = 0, 1, 2, \ldots$, converges uniformly to the zero function on any closed interval $[0, b]$, where $0 < b < 1$. It does *not* converge uniformly on the closed interval $[0, 1]$ (see margin sketch). It is sometimes useful to view uniform convergence graphically as follows. The sequence $\{y_n(t)\}$ converges to $y(t)$ uniformly over I if for every $E > 0$ the graphs of $y_n(t)$ eventually all remain inside a "tube" of radius E about the graph of $y(t)$. Draw your own tubes of radius E less than $1/2$ around the limit function $y(t)$ in the example just given. This should convince you that there is uniform convergence on $[0, b]$, $0 < b < 1$, but not on $[0, 1]$.

Inequality (7) is not a very useful test for uniform convergence of a sequence $y_n(t)$ because one must "guess" the limit function $y(t)$ in advance before the test can be applied. As with sequences of numbers, there is a *Cauchy Test for uniform convergence*. It has the advantage that one need not know the limit function in advance before the test can be applied; on the other hand, the Cauchy Test has the disadvantage that when it succeeds one does not necessarily know the limit function.

THEOREM A.2.2

> **Cauchy Test for Uniform Convergence.**   The sequence $\{y_n(t)\}$ of functions defined over a common interval I is uniformly convergent to some function $y(t)$ if, given any $E > 0$, there is a positive integer N such that
>
> $$|y_n(t) - y_m(t)| < E, \quad \text{all } t \text{ in } I, \quad \text{all } n, m \geq N \qquad (8)$$

Although the limit is not known from the Cauchy Test, the limit function is unique and can be uniformly approximated by the sequence element $y_n(t)$ as closely as desired by choosing n large enough. Another fact we accept without proof is that the uniform limit of a sequence of continuous functions on a common interval I is also continuous on that same interval. The margin sketch on the previous page provides an example of this phenomenon of a continuous limit function ($y = 0$ in this case) on the interval $[0, b]$, as well as an example of what goes wrong when the convergence is not uniform (look at what happens over the closed interval $[0, 1]$).

Convergence of the Picard Iterates

Now we show that the sequence of Picard iterates $\{y_n(t)\}$ converges uniformly to a solution of integral equation (4) on the interval $[t_0, t_0 + c]$. First recall from (3) that

$$|f(t, y_1) - f(t, y_2)| \leq L|y_1 - y_2|, \quad \text{all } (t, y_1), (t, y_2) \text{ in } I \qquad (9)$$

Using (5) and (9), we have the estimate

$$|y_{n+1}(t) - y_n(t)| \leq L \int_{t_0}^{t} |y_n(s) - y_{n-1}(s)| \, ds, \quad t_0 \leq t \leq t_0 + c, \quad n = 1, 2, \dots \quad (10)$$

Now observe from (6) that

$$|y_1(t) - y_0| \leq b, \quad t_0 \leq t \leq t_0 + c \qquad (11)$$

So, using (10) with $n = 1$, we have

$$|y_2(t) - y_1(t)| \leq Lb(t - t_0), \quad \text{for } t_0 \leq t \leq t_0 + c \qquad (12)$$

Using (12) and (10) with $n = 2$, we see that

$$|y_3(t) - y_2(t)| \leq L^2 b \frac{(t - t_0)^2}{2}, \quad \text{for } t_0 \leq t \leq t_0 + c$$

Proceeding along in this manner, we can use (10) to establish the estimate

$$|y_{n+1}(t) - y_n(t)| \leq L^n b \frac{(t - t_0)^n}{n!}, \quad t_0 \leq t \leq t_0 + c, \quad n = 0, 1, 2, \dots \quad (13)$$

So for any $n > m$, and using (13) and the triangle inequality, $[|a + b| \leq |a| + |b|]$, we have the estimate

$$
\begin{aligned}
|y_n(t) - y_m(t)| &= |y_n(t) - y_{n-1}(t) + y_{n-1}(t) + \cdots + y_{m+1}(t) - y_m(t)| \\
&\leq |y_n(t) - y_{n-1}(t)| + \cdots + |y_{m+1}(t) - y_m(t)| \\
&\leq b \sum_{k=m}^{n-1} \frac{L^k (t - t_0)^k}{k!}
\end{aligned}
\qquad (14)
$$

☞ The Taylor series
for e^x is given in
Appendix B.2, item 9.

If we denote by $S_n(t)$ the n-th partial sum of the Taylor series for $e^{L(t-t_0)}$, that is,

$$S_n(t) = \sum_{k=0}^{n-1} \frac{L^k (t-t_0)^k}{k!}$$

then (14) can be written as

$$|y_n(t) - y_m(t)| \leq b[S_n(t) - S_m(t)], \qquad t_0 \leq t \leq t_0 + c \qquad (15)$$

Now since the partial sums $\{S_n(t)\}$ converge uniformly to $e^{L(t-t_0)}$ on any interval of the t-axis, from what was said earlier it follows that the sequence $\{S_n(t)\}$ of partial sums satisfies the Cauchy Test (8) and, from (15), so does the sequence of Picard iterates $\{y_n(t)\}$ over the interval $[t_0, t_0 + c]$. It follows that $y_n(t)$ converges uniformly to a continuous function $y(t)$ over $[t_0, t_0 + c]$.

We need the following result.

THEOREM A.2.3

> Integral Convergence Theorem. Suppose that the sequence $\{y_n(t)\}$ of continuous functions converges uniformly to $y(t)$ on $t_0 \leq t \leq t_0 + c$, and suppose that $|y_n(t) - y_0| \leq b$ for all $t_0 \leq t \leq t_0 + c$, where b, c are defined as above. Suppose that $f(t, y)$ satisfies the conditions of IVP (1). Then, as $n \to \infty$, we have
>
> $$\int_{t_0}^{t} f(s, y_n(s)) \, ds \to \int_{t_0}^{t} f(s, y(s)) \, ds, \qquad t_0 \leq t \leq t_0 + c \qquad (16)$$

Finally, we can show that the Picard iterates converge to a solution of IVP (1). Observe that the sequence of Picard iterates $\{y_n(t)\}$ satisfies the conditions leading to (16), and also satisfy relation (5) over the interval $[t_0, t_0 + c]$. Taking limits of each side of (5) for each fixed t in $[t_0, t_0 + c]$, we see from (16) that $y(t)$ satisfies integral equation (4), and so also satisfies the forward IVP (1).

Here's an example of how the Picard process works.

EXAMPLE A.2.1

Picard Iterates and Their Convergence

Look at the initial value problem

$$y' = ty + 1, \qquad y(0) = 1$$

Comparing this IVP with IVP (1), we see that $f(t, y) = ty + 1$, $t_0 = 0$ and $y_0 = 1$. We may as well take the region R to be the whole ty-plane. Let's take the rectangle S to be $0 \leq t \leq a$, $1 - b \leq y \leq 1 + b$, where a, b are arbitrary positive numbers. Now

$$|ty + 1| \leq |ty| + 1 \leq a(b+1) + 1$$

on S, so we can take $M = a(b+1) + 1$. Consequently,

$$c = \min\left\{a, \frac{b}{a(b+1) + 1}\right\}$$

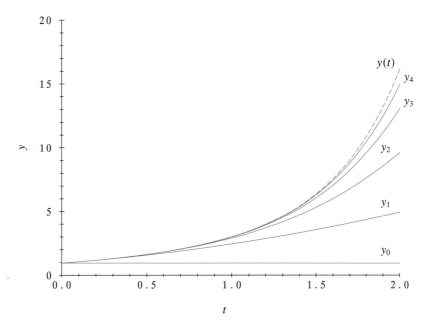

FIGURE A.2.2 First five Picard iterates (solid) and the "true" solution (dashed) of the IVP $y' = ty + 1$, $y(0) = 1$ (Example A.2.1).

Because $f(t, y)$ is defined and well behaved over the entire ty-plane, the Picard algorithm (5) defines the iterates for all t, but our proof only guarantees convergence of iterates over the interval $0 \leq t \leq c$. We have

$$y_0(t) = 1$$

$$y_1(t) = 1 + \int_0^t f(s, y(s))\, ds = 1 + \int_0^t [sy_0(s) + 1]\, ds$$

$$= 1 + \int_0^t (s + 1)\, ds = 1 + t + t^2/2$$

$$y_2(t) = 1 + \int_0^t [sy_1(s) + 1]\, ds$$

$$= 1 + \int_0^t \left[s\left(1 + s + \frac{s^2}{2}\right) + 1 \right] ds$$

$$= 1 + t + \frac{t^2}{2} + \frac{t^3}{3} + \frac{t^4}{8}$$

and so on. Taking $a = 1$, $b = 8$, we have $c = \min\{1, 0.8\} = 0.8$, and the Picard iterates converge in the prescribed sense at least over the interval $0 \leq t \leq 0.8$ (see Figure A.2.2 where the first five Picard iterates are shown).

Summarizing the Picard approach, we have the

❖ **Picard Iteration Scheme.** The *Picard Iteration Scheme* for the initial value problem, $y' = f(t, y)$, $y(t_0) = y_0$, consists of constructing the sequence

$$y_0(t) = y_0, \quad \ldots, \quad y_{n+1}(t) = y_0 + \int_{t_0}^t f(s, y_n(s)) \, ds, \quad n = 0, 1, 2, \ldots$$

According to the Existence and Uniqueness Theorem, as $n \to \infty$, $y_n(t) \to y(t)$ [the solution of the IVP] for each t in some interval containing t_0.

PROBLEMS

1. This problem involves Picard iterates for the IVP $y' = -y$, $y(0) = 1$.

   (a) Find the Picard iterates y_1, y_2, y_3.

   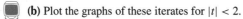 (b) Plot the graphs of these iterates for $|t| < 2$.

   (c) Find a formula for the n-th iterate y_n by induction.

   (d) Find the exact solution $y(t)$ and sketch its graph along with those of y_0, y_1, y_2, and y_3.

   (e) Show that $y_n(t)$ is a partial sum of the Taylor series for the solution $y(t)$ of the IVP.

2. Repeat Problem 1 for the IVP $y' = 2ty$, $y(0) = 2$.

3. This problem involves Picard iterates for the IVP $y' = y^2$, $y(0) = 1$.

   (a) Find the Picard iterates y_1, y_2, and y_3.

   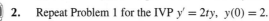 (b) Find the solution $y(t)$ directly and plot y_0, y_1, y_2, y_3, and y on their maximal domains.

   (c) Show by induction that the n-th Picard iterate $y_n(t)$ is a polynomial of degree $2^n - 1$.

   (d) Show that every Picard iterate is defined for all t, but that the exact solution is defined only for $t < 1$. Then show that the Taylor series of the exact solution converges only for $|t| < 1$. So Picard iterates, the exact solution, and the Taylor series of the exact solution may be defined on different intervals.

4. The Picard iteration scheme is sturdy enough to survive a bad choice for the first approximation $y_0(t)$. Show that the Picard iterates for the IVP $y' = -y$, $y(0) = 1$, converge to the exact solution e^{-t} even if the wrong starting "point" $y_0(t) = 0$ is used to start the iteration scheme. [*Hint*: The Picard iteration scheme becomes in this case $y_{n+1}(t) = 1 + \int_0^t (-y_n(s)) \, ds$, for $n = 0, 1, 2, \ldots$.]

5. Repeat Problem 4 but with the absurd first approximation $y_0(t) = \sin t$. Show that the sequence of iterates still converges to e^{-t}.

6. This problem involves Picard iterates for the IVP $y' = 1 + y^2$, $y(0) = 0$.

   (a) Find the Picard iterates y_1, y_2, and y_3.

   (b) Prove by induction that $y_n(t)$ is a polynomial of degree $2^n - 1$.

   (c) Find the exact solution $y(t)$ and identify the maximal interval on which it is defined.

   (d) Does $y_n(t) \to y(t)$ for all t? Explain.

7. How many Picard iterates can you calculate for $y' = \sin(t^3 + ty^5)$, $y(0) = 1$? What is the practical difficulty in finding the iterates?

8. Let $y_0(t) = 3$, $y_1(t) = 3 - 27t$, $y_n(t) = 3 - \int_0^t y_{n-1}^3(s) \, ds$, $n = 0, 1, 2, \ldots$. Find $\lim_{n\to\infty} y_n(t)$. [*Hint*: Consider a certain IVP related to the Picard iteration scheme.]

9. *(Uniform Convergence of a Sequence of Functions).*

(a) Show that the sequence $\{(1/n)\sin(nt)\}_{n=1}^{\infty}$ converges uniformly to the zero function on any interval I. [*Hint*: $|(1/n)\sin(nt) - 0| \leq 1/n$.]

(b) Show that the sequence $\{t^n\}_{n=1}^{\infty}$ converges uniformly to the zero function on the interval $0 \leq t \leq 0.5$, while it converges, but not uniformly, on the interval $0 \leq t \leq 1$ to the function $f(t) = 0,\ 0 \leq t < 1,\ f(1) = 1$.

A.3 Extension of Solutions

A solution of the IVP

$$y' = f(t, y), \qquad y(t_0) = y_0 \tag{1}$$

defined on an interval can be extended outside that interval when the conditions on the data allow it. This brings up the question of the relationship between the maximally extended solutions of $y' = f(t, y)$ and the region where f is defined. In the proof of the Existence Theorem (Theorem A.2.1), we showed that for any rectangle S in the ty-plane where f and $\partial f/\partial y$ are continuous, the solution curve that "enters" S through the left side of S (S includes its edges) will "exit" through one of the other sides. The result below shows that this behavior of solution curves is typical.

THEOREM A.3.1

> Extension Principle. Suppose that R is a region in the ty-plane and suppose that f and $\partial f/\partial y$ are continuous on R. Suppose that D is a bounded region that includes its edges and that D is contained in R. Consider IVP (1) with (t_0, y_0) a point in the interior of D. Then the solution of IVP (1) can be extended forward and backward in t until its solution curve exits through the boundary of D.

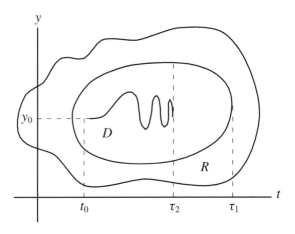

FIGURE A.3.1 Geometry for the Extension Principle.

To verify this, let's consider only the forward extension assertion since the backward assertion is proved in the same way. Since D is bounded, there is a *smallest* value τ_1 such that no point of D lies in the half-plane $t > \tau_1$. If the forward solution curve exits D through its boundary, then there is nothing to prove. To have something to prove, let's suppose that the forward maximally extended solution of IVP (1) remains in the interior of D but that it is defined for all t in the interval $t_0 \leq t \leq \tau_2$, where $\tau_2 < \tau_1$. We will show that this assumption leads to a contradiction, so the forward solution curve can't both lie inside D and also be to the left of some line, $t = \tau_2 < \tau_1$ (see Figure A.3.1). So the solution curve must indeed exit D. To get at this question recall that IVP (1) is equivalent to the integral equation

$$y(t) = y_0 + \int_{t_0}^{t} f(s, y(s))\, ds$$

So

$$y(t_2) - y(t_1) = \int_{t_1}^{t_2} f(s, y(s))\, ds \qquad t_0 \leq t_1, \quad t_2 \leq \tau_2 \tag{2}$$

Since f is continuous on a closed bounded region there is a constant $M > 0$ such that $|f(t, y)| \leq M$ for all (t, y) in D. So (2) implies by an estimate for integrals that

☞ The integral estimates are given in Theorem B.5.9, Appendix B.

$$|y(t_1) - y(t_2)| \leq M|t_1 - t_2|, \qquad t_0 \leq t_1, \quad t_2 < \tau_2 \tag{3}$$

From (3) and the Cauchy Test for convergence (Theorem A.2.2), it follows that the limit $y(\tau_2^-)$ exists. Now the point $(\tau_2, y(\tau_2^-))$ can't lie in the interior of D, for then the method described in the proof of the Existence Theorem would allow us to extend $y(t)$ beyond $t = \tau_2$. So the point $(\tau_2, y(\tau_2^-))$ must lie on the boundary of D, and we are done.

Comments

Just because a rate function $f(t, y)$ is continuously differentiable for all t and y does *not* mean that the solution of $y' = f$, $y(t_0) = y_0$ is defined for all t. For example, the solution $y = (1 - t)^{-1}$, $t < 1$, of the IVP $y' = y^2$, $y(0) = 1$, can't be extended to $t = 1$, even though the rate function y^2 is continuously differentiable everywhere. Nevertheless, for *linear* problems the situation is much simpler (Problem 3).

PROBLEMS

1. Find a solution formula for each IVP below, and determine the largest t-interval on which the solution is defined. What happens to the solution as t approaches each endpoint of the interval?

    (a) $y' = -y^3$, $y(0) = 1$ **(b)** $y' = -te^{-y}$, $y(0) = 2$

2. Find a first-order ODE $y' = f(t, y)$, each solution of which is defined only on a finite t-interval.

3. Show that the solution of the IVP $y' + p(t)y = q(t)$, $y(t_0) = y_0$, where $p(t)$ and $q(t)$ are continuous for all t in an open interval I containing t_0, is defined for all t in I. [*Hint*: Use the solution formula (3) of Section 1.4.]

A.4 Sensitivity of Solutions to the Data

There remains a last question concerning the solution of IVP

$$y' = f(t, y), \qquad y(t_0) = y_0 \tag{1}$$

Loosely phrased, the question amounts to this: Is it always possible to find bounds on the determination of the data $f(t, y)$ and y_0 in IVP (1) that will guarantee that the corresponding solution will be within prescribed error bounds over a given t-interval? If this question can be answered in the affirmative, one consequence is that any small enough change in the data of an initial value problem produces only a small change in the solution (*continuity in the data*). Or, stated another way: In spite of the fact that the initial value problem arising from a model has empirically determined components (and therefore can never be known precisely), the model is still relevant and useful.

In our study of IVP (1), we have been guided by the questions posed in Section 2.1: Does IVP (1) have any solution? How many? How far in time can a solution be extended? How do solutions respond to changes in the data (i.e., how sensitive is the solution to data changes)? If the functions f and $\partial f / \partial y$ are continuous in some region R in the ty-plane and (t_0, y_0) is a point of R, we gave satisfactory answers to the first, second, and third questions in Sections 2.1 and 2.2. The question on sensitivity was discussed in Sections 2.3 and 2.4, but not completely answered. In this section of the appendix we show that simple conditions on the data lead to a satisfactory answer to the sensitivity question.

In addressing the last question, it would be extremely helpful to have a formula for the solution of IVP (1) in which the data appear explicity. But for general nonlinear differential equations, this rarely happens. We will have to find some other way to answer the question.

To estimate the change in the solution to IVP (1) as the data $f(t, y)$ and y_0 are modified, it will be helpful to have an estimate of the type given below.

THEOREM A.4.1

Basic Estimate. For given constants A, B, and L with $L > 0$, suppose that $z(t)$ is a continuous function on the interval $[a, b]$ that satisfies the integral inequality

$$z(t) \leq A + B(t - a) + L \int_a^t z(s)\, ds \quad \text{for all } a \leq t \leq b \tag{2}$$

Then $z(t)$ satisfies the estimate

$$z(t) \leq Ae^{L(t-a)} + \frac{B}{L}\left[e^{L(t-a)} - 1\right] \quad \text{for all } a \leq t \leq b \tag{3}$$

☞ The Basic Estimate is also called *Gronwall's inequality* after the American mathematician T.H. Gronwall.

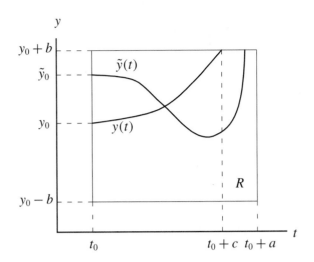

FIGURE A.4.1 Geometry for perturbation estimate.

To show the validity of the Basic Estimate, we first define $Q(t)$:

$$Q(t) = \int_a^t z(s)\,ds \quad \text{for } a \le t \le b$$

☞ Theorem B.5.5 is the Fundamental Theorem of Calculus.

Then Q is continuous on $[a, b]$, and the Fundamental Theorem of Calculus implies that $Q'(t) = z(t)$ for $a < t < b$; so (2) can be written as

$$Q'(t) - LQ(t) \le A + B(t-a), \qquad a < t < b \tag{4}$$

Multiply each side of (4) by e^{-Lt}, and use the fact that $\left[e^{-Lt}Q\right]' = e^{-Lt}Q' - e^{-Lt}LQ$ to obtain

$$\left[Q(t)e^{-Lt}\right]' \le [A + B(t-a)]e^{-Lt}, \qquad a < t < b \tag{5}$$

Since $F(t) = -L^{-2}[B + LA + LB(t-a)]e^{-Lt}$ is an antiderivative to the right-hand side of (5), we see that (5) can be written as

$$\left[Q(t)e^{-Lt} - F(t)\right]' \le 0, \qquad a \le t \le b \tag{6}$$

The function $G(t) = Q(t)e^{-Lt} - F(t)$ is continuous on the interval $[a, b]$, and so because of inequality (6), we see that $G(t) \le G(a)$ for $a \le t \le b$, or after simplification,

$$A + B(t-a) + LQ(t) \le Ae^{L(t-a)} + \frac{B}{L}\left[e^{L(t-a)} - 1\right], \qquad a \le t \le b \tag{7}$$

where we have used the fact that $Q(a) = 0$. From (2) we see that

$$z(t) \le A + B(t-a) + LQ(t), \qquad a \le t \le b$$

which together with (7) yields the inequality (3), and we are done.

There is also a backward form of the Basic Estimate, which under the same hypotheses states that the inequality

$$z(t) \le A + B(b - t) + L \int_t^b z(s)\,ds, \quad a \le t \le b \tag{8}$$

implies that

$$z(t) \le Ae^{L(b-t)} + \frac{B}{L}\left[e^{L(b-t)} - 1\right], \quad a \le t \le b \tag{9}$$

The derivation of (9) from (8) is very similar to the foregoing derivation of (3) from (2), and so is omitted.

We need some estimate of the change in the solution of IVP (1) occurs when the data $f(t, y)$ and y_0 are modified (or "perturbed").

THEOREM A.4.2

Perturbation Estimate. Suppose that the function f in IVP (1) is continuous along with $\partial f/\partial y$ in a rectangle R described by the inequalities $t_0 \le t \le t_0 + a$ and $|y - y_0| \le b$. Suppose that $g(t, y)$ and $\partial g(t, y)/\partial y$ are also continuous functions on R. Suppose that on some common interval $t_0 \le t \le t_0 + c$, with $c \le a$, the solution $y(t)$ of IVP (1) and the solution $\tilde{y}(t)$ of the "perturbed" IVP

$$y' = f(t, y) + g(t, y), \quad y(t_0) = \tilde{y}_0 \tag{10}$$

both have solution curves which lie in R (see Figure A.4.1).

Then we have the estimate

$$|y(t) - \tilde{y}(t)| \le |y_0 - \tilde{y}_0|e^{L(t-t_0)} + \frac{M}{L}\left[e^{L(t-t_0)} - 1\right], \quad t_0 \le t \le t_0 + c \tag{11}$$

where L and M are any numbers such that

$$|g(t, y)| \le M, \quad \left|\frac{\partial f}{\partial y}\right| \le L, \quad \text{all } (t, y) \text{ in } R$$

The Perturbation Estimate follows from the Basic Estimate. Using the fact proven in Section A.2 that the solution of an IVP also solves an equivalent integral equation, we see that $y(t)$ and $\tilde{y}(t)$ must satisfy the integral equations (for $t_0 \le t \le t_0 + c$)

$$y(t) = y_0 + \int_{t_0}^t f(s, y(s))\,ds$$

$$\tilde{y}(t) = \tilde{y}_0 + \int_{t_0}^t [f(s, \tilde{y}(s)) + g(s, \tilde{y}(s))]\,ds$$

Subtracting the second equation from the first, for any $t_0 \le t \le t_0 + c$ we have

$$y(t) - \tilde{y}(t) = y_0 - \tilde{y}_0 + \int_{t_0}^t [f(s, y(s)) - f(s, \tilde{y}(s))]\,ds - \int_{t_0}^t g(s, \tilde{y}(s))\,ds \tag{12}$$

Putting $z(t) = |y(t) - \tilde{y}(t)|$ and using inequality (4) in Section A.1 to estimate the

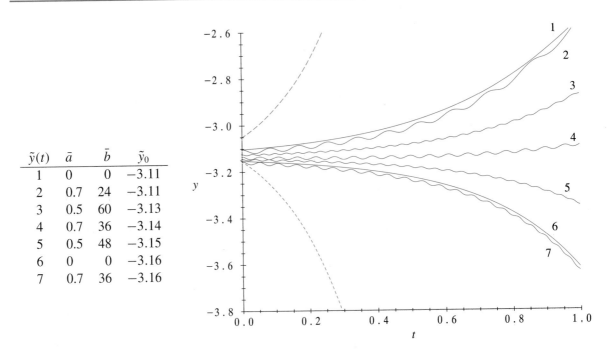

$\tilde{y}(t)$	$\bar{a}$	$\bar{b}$	$\tilde{y}_0$
1	0	0	-3.11
2	0.7	24	-3.11
3	0.5	60	-3.13
4	0.7	36	-3.14
5	0.5	48	-3.15
6	0	0	-3.16
7	0.7	36	-3.16

FIGURE A.4.2 Solution curves for (14).

integrals in (12), we have the inequality

$$z(t) \le z(t_0) + M(t - t_0) + L \int_{t_0}^{t} z(s)\, ds, \qquad t_0 \le t \le t_0 + c \tag{13}$$

Comparing (13) with (12) and applying the Basic Estimate with A and B replaced by $|y_0 - \tilde{y}_0|$ and M, respectively, we obtain the desired estimate (11), concluding our proof.

EXAMPLE A.4.1

As an illustration of the Perturbation Estimate, we graph in Figure A.4.2 the solutions of the initial value problems

$$\begin{cases} y' = y \sin y \\ y(0) = y_0 \end{cases} \qquad \begin{cases} y' = y \sin y + \bar{a} \sin(\bar{b} t y) \\ y(0) = \tilde{y}_0 \end{cases} \tag{14}$$

for various values of the constants $\tilde{y}_0$, $\bar{a}$, and $\bar{b}$. The solid lines are the solution curves of $y(t)$ that correspond to the data in the table at the left of Figure A.4.2.

The dashed lines come from the error bounds given by the Perturbation Estimate (11). Notice that the error bound is not very sharp as we move away from $t_0 = 0$.

Now we are in a position to answer the fourth of the basic questions.

THEOREM A.4.3

Continuity in the Data. Suppose that f, $\partial f/\partial y$, g, and $\partial g/\partial y$ are continuous functions of t and y on the rectangle R defined by $t_0 \le t \le t_0 + a$, $|y - y_0| \le b$. Suppose that $E > 0$ is a given error tolerance. Then there exist positive constants $H < b$ and $c \le a$ such that the respective solutions $y(t)$ and $\tilde{y}(t)$ of the IVPs

$$(a) \begin{cases} y' = f(t, y) \\ y(t_0) = y_0 \end{cases} \quad (b) \begin{cases} y' = f(t, y) + g(t, y) \\ y(t_0) = \tilde{y}_0 \end{cases} \tag{15}$$

satisfy the inequality

$$|y(t) - \tilde{y}(t)| \le E, \qquad t_0 \le t \le t_0 + c \tag{16}$$

for any choice of $\tilde{y}_0$ for which $|y_0 - \tilde{y}_0| \le H$.

Here's how to derive inequality (16). Suppose that H is any positive constant $0 < H < b$, and that $\tilde{M}$ is such that $\max |g(t, y)| \le \tilde{M}$ for all (t, y) in R. Then according to the Existence and Uniqueness Theorems (Theorems A.1.1 and A.2.1), the solution $\tilde{y}(t)$ of the second IVP in (15) must be defined at least over the interval: $t_0 \le t \le t_0 + c$, with

$$c = \min\left\{a, \frac{b - H}{M + \tilde{M}}\right\} \tag{17}$$

where M is a positive constant such that $|f(t, y)| \le M$ for all (t, y) in R. Now by the Perturbation Estimate (Theorem A.4.2), we have

$$|y(t) - \tilde{y}(t)| \le He^{Lc} + \frac{\tilde{M}}{L}(e^{Lc} - 1), \qquad t_0 \le t \le t_0 + c \tag{18}$$

where $L > 0$ is any number such that $|\partial f/\partial y| \le L$ for all (t, y) in R. Take the positive constants H and c so small that $He^{Lc} < E/2$ and $\tilde{M}L^{-1}(e^{Lc} - 1) < E/2$, so the value of the right-hand side of (18) is no more than E. We see that $|y(t) - \tilde{y}(t)| \le E$ holds, and we are done.

Properties of Solutions

Putting together the results of this section with those of previous sections, we now have a comprehensive view of how solutions of IVPs such as IVP (1) behave. If $f(t, y)$ and $\partial f/\partial y$ are continuous on a closed region R of the ty-plane, then:

1. Through each point in the interior of R, there is one and only one solution curve of the equation $y' = f(t, y)$, and this solution curve can be extended up to the boundary of R (from the Existence Theorem, the Uniqueness Principle, and the Extension Principle).

2. The change in the solution of IVP (1) is small if the changes in the data $f(t, y)$ and y_0 are sufficiently small (Continuity in the Data Theorem).

The first property above shows that an IVP is an effective tool in solving problems that involve dynamical systems. The second property indicates that models involving IVPs are not invalidated even though empirically determined elements of the model do not have their "true" values.

PROBLEMS

1. Use the Perturbation Estimate to approximate $|\bar{y}(t) - y(t)|$ if $\bar{y}(t)$ solves the IVP $y'(t) = (e^{-y^2} + 0.1)y$, $y(0) = \tilde{a}$, and $y(t)$ solves the IVP $y'(t) = 0.1y$, $y(0) = a$.

2. Let $y(t)$ be a continuously differentiable function on the real line such that $|y'(t)| \le |y(t)|$ for all t. Show that either $y(t)$ never vanishes or else that $y(t) = 0$ for all t.

3. *(Perturbation Estimate When Initial Data Changes)*. Suppose that the rate function $f(t, y)$ is defined in some rectangle $R: a \le t \le b$, $c \le y \le d$ and that f, f_y are continuous on R. If K and L are constants such that $K \le f_y(t, y) \le L$, and if $y(t)$ and $\bar{y}(t)$ are any solutions on $a \le t \le b$ of the ODE $y' = f(t, y)$ whose solution curves lie in R, then follow the outline below to show that

$$|y(a) - \bar{y}(a)|e^{K(t-a)} \le |y(t) - \bar{y}(t)| \le |y(a) - \bar{y}(a)|e^{L(t-a)}, \quad \text{for } a \le t \le b$$

- If $K = 0$, show that $|y(a) - \bar{y}(a)| \le |y(t) - \bar{y}(t)|$ for all $a \le t \le b$. If $L = 0$ show that $|y(t) - \bar{y}(t)| \le |y(a) - \bar{y}(a)|$ for all $a \le t \le b$. [*Hint*: First convert the ODE $y' = f(t, y)$ to an equivalent integral equation.]

- Let $0 < K < L$ and follow the proof of the Basic Estimate (Theorem A.4.1) to show the asserted estimate. What goes wrong with this approach when either K or L is negative? [*Hint*: Note that if $y(a) > \bar{y}(a)$, then $y(t) > \bar{y}(t)$ for all $t \ge a$.]

- If either $K < 0$ or $L < 0$, show that the appropriate part of the asserted estimate is valid. [*Hint*: Use iteration on (2) in the Basic Estimate (Theorem A.4.1).]

- Consider the function $f(t, y) = (0.1)(-y + \ln(1 + y))$ defined on the rectangle $R: 0 < t < \infty$, $y \ge 0$. Find the sharpest possible values for the constants K and L such that $K \le f_y(t, y) \le L$ for all (t, y) in R. Let $y(t)$ be the solution of the IVP $y' = f(t, y)$, $y(0) = 4$, and let $\bar{y}(t)$ be the solution of the IVP $y' = f(t, y)$, $y(0) = 1$. Show by direct measurement at $t = 10, 20, 30$, and 40 that the asserted estimate holds.

- Repeat this analysis for the solution $y(t)$ of the IVP

$$y' = \sin t - (0.1)(0.5\cos y + y), \qquad y(0) = 5$$

and the solution $\bar{y}(t)$ of the IVP $y' = \sin t - (0.1)(0.5\cos y + y)$, $y(0) = -5$.

APPENDIX B

Background Information

B.1   Engineering Functions
Description of the piecewise-continuous functions (square waves, step functions, etc.) that are commonly used in science and engineering.

B.2   Power Series
Summary of the properties of power series. Includes convergence properties, tests for convergence, Taylor's formula with remainder, real analytic functions, and the calculus of power series.

B.3   Complex Numbers and Complex-Valued Functions
Review of the arithmetic of complex numbers and the calculus of complex-valued functions of a real variable.

B.4   Useful Formulas
Review of trigonometric identities, hyperbolic functions, polynomials, logarithms, and exponentials.

B.5   Theorems from Calculus
Tabulation of useful results from single and multi-variable calculus.

B.6   Scaling and Units
Description of commonly used unit systems and conversion factors, scaling and nondimensionalizing ODEs.

B.1   Engineering Functions

The eight functions defined and graphed below are frequently used in science and engineering applications. These functions are divided into three groups: *step* or *stair functions*, *pulses*, and *wavetrains*. The "on-time" (where the function is nonzero) is described as follows: The step and stair functions and the square pulses are first turned on at $t = 0$. The step and stair functions remain on for $t \geq 0$. The square pulse turns on at $t = 0$, switches off at $t = a$, and remains off. The on-time for sawtooth and triangular pulses is $0 \leq t < a$. The on-time of a square wave, triangular wave, and sawtooth wave over the period $0 \leq t \leq T$ is $0 \leq t < (d/100)T$, where the *duty cycle* d is the percentage of on-time in a cycle. For example, if $d = 50$ and $T = 4$, then the wave is on for $0 \leq t < 2$ (since $2 = 50\%$ of $T = 4$) and off for $2 \leq t < 4$.

$\mathrm{sqw}(t, d, T)$:	periodic square wave, duty cycle $d\%$, period T
$\mathrm{trw}(t, d, T)$:	periodic triangular wave, duty cycle $d\%$, period T
$\mathrm{sww}(t, d, T)$:	periodic sawtooth wave, duty cycle $d\%$, period T
$\mathrm{step}(t)$:	unit step at $t = 0$
$\mathrm{sqp}(t, a)$:	unit square pulse over interval $0 \leq t < a$
$\mathrm{trp}(t, a)$:	unit triangular pulse over interval $0 \leq t \leq a$
$\mathrm{swp}(t, a)$:	unit sawtooth pulse over interval $0 \leq t \leq a$
$\mathrm{stair}(t, a)$:	stair, rise of 1 (starting at $t = 0$), run of a

If the value of an engineering function is constant over a time interval of length L, it may be a good idea to change the maximum internal step size of your numerical solver to no more than $L/10$. If this isn't done, an adaptive solver may overlook the points where the function values suddenly change.

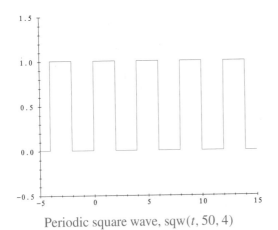

Periodic square wave, $\mathrm{sqw}(t, 50, 4)$

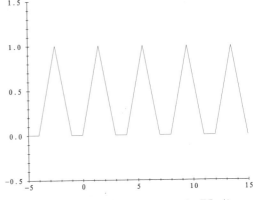

Periodic triangular wave, $\mathrm{trw}(t, 75, 4)$

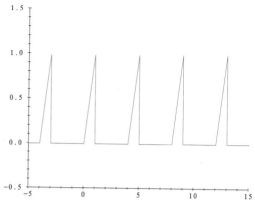

Periodic sawtooth wave, sww(t, 25, 4)

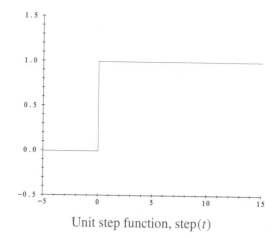

Unit step function, step(t)

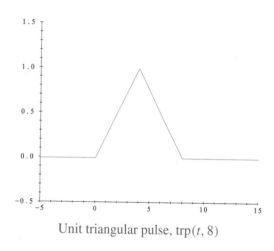

Unit square pulse, sqp(t, 5)

Unit triangular pulse, trp(t, 8)

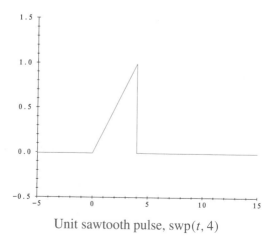

Unit sawtooth pulse, swp(t, 4)

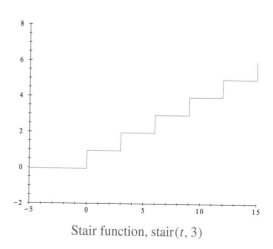

Stair function, stair(t, 3)

B.2 Power Series

Power series have some remarkable properties that make them especially well suited for applications. We will list these properties below in a form that allows easy reference.

1. *Power Series.* The series

$$a_0 + a_1(x - x_0) + a_2(x - x_0)^2 + \cdots + a_n(x - x_0)^n + \cdots = \sum_{n=0}^{\infty} a_n(x - x_0)^n$$

where x_0 and the coefficients a_n are real, is a *power series* based at x_0. We treat only series of the form $\sum_{n=0}^{\infty} a_n x^n$ since $\sum_{n=0}^{\infty} a_n(x - x_0)^n$ may be converted to that form by replacing $x - x_0$ by a new variable. We write $\sum_0^{\infty} a_n x^n$ or $\sum a_n x^n$ for $\sum_{n=0}^{\infty} a_n x^n$ if there is no danger of confusion.

2. *Intervals of Convergence.* $\sum a_n x^n$ converges at a point x if $\lim_{N \to \infty} \sum_{n=0}^{N} a_n x^n$ exists. Unless $\sum a_n x^n$ converges only at $x = 0$, there is a largest positive number R (possibly infinity) such that $\sum a_n x^n$ converges for all $|x| < R$. R is the *radius of convergence* and $J = (-R, R)$ is the *open interval of convergence.* The series diverges (i.e., does not converge) if $|x| > R$, and may or may not converge if $x = \pm R$. We will only work with a series inside its open interval of convergence.

3. *Ratio Test.* The *Ratio Test* is a useful way to find the radius of convergence of a power series. For example, suppose that $\lim_{n \to \infty} |a_{n+1}/a_n| = a$. Then

$$\lim_{n \to \infty} \left| \frac{a_{n+1} x^{n+1}}{a_n x^n} \right| = a|x|$$

According to the Ratio Test, we have convergence if $a|x| < 1$ and divergence if $a|x| > 1$. The radius of convergence R is $1/a$ and $\sum a_n x^n$ converges for all $|x| < 1/a$. Note that $R = \infty$ if $a = 0$.

4. *Calculus of Power Series.* The function $f(x) = \sum_{n=0}^{\infty} a_n x^n$, x in the interval of convergence J, is continuous and possesses derivatives of all orders. Moreover, $f'(x) = \sum_{n=1}^{\infty} n a_n x^{n-1}$, and this series also has J as its open interval of convergence. In general, $f^{(k)}(x) = \sum_{n=k}^{\infty} n(n - 1) \cdots (n - k + 1) a_n x^{n-k}$, where the series converges for x in J. In addition, the series

$$\sum_0^{\infty} \frac{a_n}{n + 1} (x^{n+1} - b^{n+1})$$

converges to $\int_b^x \sum_0^{\infty} (a_n s^n) \, ds$ for all b and x in J.

5. *Identity Theorem.* If $\sum_{n=0}^{\infty} b_n x^n = \sum_{n=0}^{\infty} a_n x^n$ for all x in an open interval, then $b_n = a_n$ for $n = 0, 1, 2, \ldots$. In particular, if $\sum_{n=0}^{\infty} a_n x^n = 0$ for all x near 0, then $a_n = 0$, $n = 0, 1, 2, \ldots$. This is called the *Identity Theorem* for power series.

6. *Algebra of Power Series.* Convergent power series may be added or multiplied for all x in a common interval J of convergence. The resulting *sum* or *product* series converges on that interval. Specifically:

(i) $\displaystyle\sum_{n=0}^{\infty} a_n x^n + \sum_{n=0}^{\infty} b_n x^n = \sum_{n=0}^{\infty} (a_n + b_n) x^n$

(ii) $\displaystyle\left(\sum_{n=0}^{\infty} a_n x^n\right)\left(\sum_{n=0}^{\infty} b_n x^n\right) = \sum_{n=0}^{\infty} c_n x^n$, where $c_n = \displaystyle\sum_{k=0}^{n} a_k b_{n-k}$

(iii) If $\sum_{n=0}^{\infty} b_n x^n \neq 0$ on the interval J, the quotient series,

$$\left(\sum_{n=0}^{\infty} a_n x^n\right)\bigg/\left(\sum_{n=0}^{\infty} b_n x^n\right) = \sum_{n=0}^{\infty} d_n x^n$$

converges on that interval. Using (ii), coefficients d_n can be found by successively solving for the coefficients d_n, $n = 0, 1, 2, \ldots$, from the formulas $a_n = d_0 b_n + d_1 b_{n-1} + \cdots + d_{n-1} b_1 + d_n b_0$.

7. *Reindexing.* A series may be reindexed in any convenient way. For example, the following three summations are equivalent:

$$\sum_{n=0}^{\infty} a_n x^n, \qquad \sum_{n=2}^{\infty} a_{n-2} x^{n-2}, \qquad \sum_{n=-3}^{\infty} a_{n+3} x^{n+3}$$

Writing out the first few terms of a series explicitly helps to determine the equivalence of indexing schemes. Sometimes it is convenient to introduce negative indexed coefficients, but it is understood that these coefficients have value zero. For example, we could rewrite $\sum_{n=0}^{\infty} a_n x^n + \sum_{n=2}^{\infty} b_{n-2} x^n$ either as $a_0 + a_1 x + \sum_{n=2}^{\infty} (a_n + b_{n-2}) x^n$ or as $\sum_{n=0}^{\infty} (a_n + b_{n-2}) x^n$, where it is understood that $b_{-2} = 0$, and $b_{-1} = 0$.

8. *Taylor Series and Real Analytic Functions.* A function $f(x)$ that has a *Taylor series expansion* about the *base point* x_0

$$\sum_{n=0}^{\infty} \frac{f^{(n)}(x_0)}{n!} (x - x_0)^n$$

that converges to $f(x)$ on an interval $(x_0 - R, x_0 + R)$ is said to be *real analytic* on that interval, or, alternatively, at x_0. If $x_0 = 0$, the Taylor series is sometimes called the *Maclaurin series* of f. A real analytic function f at x_0 has a unique power series in powers of $x - x_0$; this must be the Taylor series of f.

The definitions above extend to functions of several variables. For example, the Taylor series of $f(x, y)$ about the base point $x_0 = 0$, $y_0 = 0$ is

$$\sum_{n=0}^{\infty} \sum_{i=0}^{n} \binom{n}{i} \frac{1}{n!} \frac{\partial^n f(0, 0)}{\partial x^{n-i} \partial y^i} x^{n-i} y^i =$$

$$f(0, 0) + \frac{\partial f(0, 0)}{\partial x} x + \frac{\partial f(0, 0)}{\partial y} y$$

$$+ \frac{1}{2} \frac{\partial^2 f(0, 0)}{\partial x^2} x^2 + \frac{\partial^2 f(0, 0)}{\partial x \partial y} xy + \frac{1}{2} \frac{\partial^2 f(0, 0)}{\partial y^2} y^2 + \cdots$$

where $\binom{n}{i}$ is the binomial coefficient, $n!/[i!(n-i)!]$, and $\binom{n}{0} = 1 = \binom{n}{n}$.

9. Here are some widely used Taylor series about the base point $x_0 = 0$:

(*Sine*) $\qquad \sin x = x - \dfrac{x^3}{3!} + \dfrac{x^5}{5!} - \cdots + (-1)^n \dfrac{x^{2n+1}}{(2n+1)!} + \cdots,$ all x

(*Cosine*) $\qquad \cos x = 1 - \dfrac{x^2}{2!} + \dfrac{x^4}{4!} - \cdots + (-1)^n \dfrac{x^{2n}}{(2n)!} + \cdots,$ all x

(*Exponential*) $\quad e^x = 1 + x + \dfrac{x^2}{2!} + \cdots + \dfrac{x^n}{n!} + \cdots,$ all x

(*Geometric*) $\qquad \dfrac{1}{1-x} = 1 - x + x^2 - \cdots + (-1)^n x^n + \cdots,$ $|x| < 1$

(*Binomial*) $\qquad (1+x)^\alpha = 1 + \alpha + \dfrac{\alpha(\alpha-1)}{2}x^2 + \cdots$

$\qquad\qquad\qquad + \dfrac{\alpha(\alpha-1)\cdots(\alpha-n+1)}{n!}x^n + \cdots,$ $|x| < 1$

10. *Error Estimate for Alternating Series.* Suppose that $\alpha_n > 0$, $n = 0, 1, 2, \ldots$. The *alternating series* $\sum_{n=0}^{\infty}(-1)^n \alpha_n$ converges if $\lim_{n \to \infty} \alpha_n = 0$ and $\alpha_n \geq \alpha_{n+1}$ for every n. The *error* in using the partial sum $\sum_{n=0}^{N}(-1)^n \alpha_n$ to approximate the true sum $\sum_{n=0}^{\infty}(-1)^n \alpha_n$ has magnitude no greater than α_{n+1}:

$$\left| \sum_0^\infty (-1)^n \alpha_n - \sum_0^N (-1)^n \alpha_n \right| \leq \alpha_{N+1}$$

11. *Taylor's Formula with Remainder.* Let $f(x)$ be at least $N + 1$ times continuously differentiable in the interval $x_0 - R < x < x_0 + R$. Then

$$f(x) = \sum_{n=0}^{N} \frac{f^{(n)}(x_0)}{n!}(x - x_0)^n + \frac{f^{(N+1)}(c)}{(N+1)!}(x - x_0)^{N+1}$$

where c is some point between x and x_0; the last term on the right is the *Lagrange form of the remainder*. The remainder has order at least $N + 1$ at x_0.

The following examples illustrate the properties given above.

EXAMPLE B.2.1

Ratio Test

$\sum_{n=0}^{\infty} n^5 (x^n/2^n)$ converges for $|x| < 2$ by the Ratio Test since

$$\lim_{n \to \infty} \left| \frac{(n+1)^5 x^{n+1}/2^{n+1}}{n^5 x^n/2^n} \right| = |x| \lim_{n \to \infty} \frac{(n+1)^5}{2n^5} = \frac{|x|}{2} \lim_{n \to \infty} \left(1 + \frac{1}{n}\right)^5 = \frac{|x|}{2}$$

and $|x|/2 < 1$ for all $|x| < 2$.

EXAMPLE B.2.2

Differentiating a Power Series

$$-\frac{1}{(1+x)^2} = \frac{d}{dx}\left(\frac{1}{1+x}\right) = \frac{d}{dx}(1 - x + x^2 + \cdots + (-1)^n x^n + \cdots)$$

$$= -1 + 2x + \cdots + (-1)^n n x^{n-1} + \cdots$$

if $|x| < 1$ (differentiating a geometric series within the interval of convergence).

EXAMPLE B.2.3   **Product of Two Power Series**

$$\left(\sum_{n=0}^{\infty} nx^n\right)\left(\sum_{n=0}^{\infty} \frac{x^n}{n+1}\right) = \sum_{n=0}^{\infty}\left(\sum_{k=0}^{n} \frac{k}{n-k+1}\right)x^n$$

$$= 0 + \left(\sum_{k=0}^{1} \frac{k}{1-k+1}\right)x + \left(\sum_{k=0}^{2} \frac{k}{2-k+1}\right)x^2 + \cdots$$

$$= x + \frac{5}{2}x^2 + \cdots$$

The product series converges to the product of the two factor series for $|x| < 1$ since each factor series converges for $|x| < 1$.

EXAMPLE B.2.4   **Reindexing**

$$\sum_{n=0}^{\infty} \frac{x^n}{n+3} + \sum_{n=2}^{\infty} x^n = \frac{1}{3} + \frac{1}{4}x + \sum_{n=2}^{\infty}\left(\frac{1}{n+3}+1\right)x^n = \frac{1}{3} + \frac{1}{4} + \sum_{n=2}^{\infty} \frac{n+4}{n+3}x^n$$

B.3   Complex Numbers and Complex-Valued Functions

It is customary to write *complex numbers* in the form $a + ib$, where a and b are real numbers. Complex numbers are introduced to facilitate calculation of roots of polynomials of degree 2 or higher. We review below the basic features of the complex number system. Our interest, however, goes beyond the calculation of roots of polynomials. To develop a simple solution technique for linear differential equations with constant coefficients, we will need to develop the notion of differentiability of complex-valued functions of a single real variable. At first glance this approach may seem like a far cry from our goal of studying real-world dynamical systems, but complex-valued functions are so useful that most mathematicians, scientists, and engineers use them.

☞ Sums and products of complex numbers.

    If $z = a + ib$, the real number a is called the *real part* of z and is denoted by $\text{Re}[z]$; the real number b is called the *imaginary part* of z and is denoted by $\text{Im}[z]$. Real numbers are identified with those complex numbers whose imaginary parts vanish. The complex number $a - ib$ is called the *complex conjugate* of $z = a + ib$ and is denoted by $\bar{z}$. Complex numbers are added and multiplied as linear polynomials in the variable i with real coefficients, remembering to replace i^2 by -1:

$$(a + ib) + (c + id) = (a + c) + i(b + d)$$

$$(a + ib)(c + id) = (ac - bd) + i(ad + bc)$$

This definition extends to complex numbers the corresponding operations for real numbers. Sometimes the representation $z = a + ib$ is called the *Cartesian* form of the

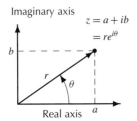

Imaginary axis

Real axis

complex number z since we can associate with z the point (a, b) in the Cartesian plane (margin sketch). For this reason we also sometimes refer to the collection of all complex numbers as the *complex plane*. Observe that $\overline{z + w} = \bar{z} + \bar{w}$ and $\overline{zw} = \bar{z}\bar{w}$.

Using polar coordinates for $z = a + ib$, we see that $a = r\cos\theta$ and $b = r\sin\theta$, where $r = (a^2 + b^2)^{1/2}$, $0 \le \theta < 2\pi$; we have the alternative representation $z = r(\cos\theta + i\sin\theta)$, called the *polar* form for z. For any real number θ we shall write

THEOREM B.3.1

> **Euler's Formula.** For any real number θ,
>
> $$e^{i\theta} = \cos\theta + i\sin\theta \tag{1}$$

The polar form of z can be written as $z = re^{i\theta}$; note that $\bar{z} = re^{-i\theta}$ since $\cos(-\theta) = \cos\theta$ and $\sin(-\theta) = -\sin\theta$. The nonnegative real number r is called the *absolute value* or the *modulus* of z, denoted also by $|z|$, and θ is called the *argument* or *angle* of z. Note that $|z|$ is the absolute value if z is a real number. The complex number zero (denoted simply by 0) is given by $0 + i0$; also note that $z = 0$ if and only if $|z| = 0$. Observe that $z\bar{z} = |z|^2 = r^2 = a^2 + b^2$. From this formula we can write $1/z$ in Cartesian form:

$$\frac{1}{z} = \frac{\bar{z}}{z\bar{z}} = \frac{a}{a^2 + b^2} - i\frac{b}{a^2 + b^2} = \frac{a - ib}{a^2 + b^2}$$

It is a consequence of the addition laws for the sine and cosine functions that for any complex numbers in polar form $z = re^{i\theta}$, $w = \rho e^{i\phi}$,

$$\frac{1}{z} = \frac{1}{re^{i\theta}} = \frac{1}{r}e^{-i\theta}, \quad \text{if } z \ne 0$$

$$zw = r\rho e^{i(\theta + \phi)}$$

We can use these identities to prove the following useful result.

THEOREM B.3.2

> **De Moivre's Formula.** For any integer n, positive, negative, or zero,
>
> $$z^n = r^n e^{in\theta} = r^n[\cos n\theta + i\sin n\theta]$$

One advantage of the polar form for complex numbers is the simple geometric characterization of the product and quotient of complex numbers.

Just as for real numbers, absolute value can be used to measure the distance between complex numbers; indeed, the *distance* between the complex numbers z and w is defined to be $|z - w|$. The sequence $\{z_n\}$ is said to *converge* if and only if there exists a z_0 in the complex plane such that $|z_n - z_0| \to 0$ as $n \to \infty$.

Now suppose that f is a complex-valued function defined over an interval I on the real t-axis; that is, $w = f(t)$ is a complex number for each t in the interval I. We can define the derivative for such a function in a manner consistent with the usual notion of derivative for real-valued functions.

❖ **Derivative of a Complex-Valued Function**. We say that the complex-valued function $f(t)$ defined on the real t-interval I is differentiable at some t_0 in I if and only if there exists a complex number z_0 such that

$$\left| \frac{f(t_0 + h) - f(t_0)}{h} - z_0 \right| \to 0 \quad \text{as } h \to 0$$

The number z_0 is said to be the *derivative* of f and is denoted by $f'(t_0)$.

If f is real-valued, the definition above reduces to the usual notion of differentiation for real-valued functions. No confusion can arise if we do not know in advance whether a function is real- or complex-valued, since differentiation of complex-valued functions of a real variable is defined in precisely the same formal manner as for real-valued functions of a real variable. It is not surprising that all the usual differentiation formulas (such as the Chain Rule, differentiation of sums, products, etc.) hold for these more general functions.

We now present a computational device for computing derivatives of complex-valued functions of a real variable. Suppose that $\alpha(t)$ and $\beta(t)$ are the respective real and imaginary parts of $f(t)$. Then f can be uniquely written as $f(t) = \alpha(t) + i\beta(t)$, where $\alpha(t)$ and $\beta(t)$ are real-valued functions defined over an interval I.

THEOREM B.3.3

> Differentiation Principle for Complex-Valued Functions. The complex-valued function of a real variable $f(t) = \alpha(t) + i\beta(t)$ is differentiable at t_0 if and only if $\alpha(t)$ and $\beta(t)$ are both differentiable at t_0. If $f'(t_0)$ exists, then
>
> $$f'(t_0) = \alpha'(t_0) + i\beta'(t_0) \qquad (2)$$

The verification of this result is a straightforward result of applying the definitions. Antidifferentiation (i.e., integration) of $f(t) = \alpha(t) + i\beta(t)$ is equivalent to performing the same operation on both $\alpha(t)$ and $\beta(t)$, and conversely.

The function $f(t) = e^{ikt} = \cos kt + i \sin kt$, where k is a real constant, is differentiable for all t. Using (2) and the definition of e^{ikt}, we see that

$$(e^{ikt})' = (\cos kt + i \sin kt)' = -k \sin kt + ik \cos kt = ik(\cos kt + i \sin kt) = ike^{ikt} \quad (3)$$

We can define and then differentiate the *complex exponential function* e^{zt} for any complex constant $z = a + ib$ in the following way:

$$e^{zt} = e^{(a+ib)t} = e^{at}e^{ibt}$$
$$(e^{zt})' = (e^{at}e^{ibt})' = ae^{at}e^{ibt} + e^{at}ibe^{ibt} = (a + ib)e^{at}e^{ibt} = ze^{zt} \qquad (4)$$

This result is easy to remember because it is just like the formula for differentiating e^{at}, where a is a real number.

Finally, everything done here directly extends to vectors and matrices whose entries are complex constants or complex-valued functions.

B.4 Useful Algebra and Trigonometry Formulas

Trigonometric Identities

Sum and Difference

$$\sin(\alpha \pm \beta) = \sin\alpha\cos\beta \pm \cos\alpha\sin\beta$$
$$\cos(\alpha \pm \beta) = \cos\alpha\cos\beta \mp \sin\alpha\sin\beta$$

Double Angle

$$\sin(2\alpha) = 2\sin\alpha\cos\alpha$$
$$\cos(2\alpha) = \cos^2\alpha - \sin^2\alpha$$

Product

$$\sin\alpha\sin\beta = \tfrac{1}{2}\{\cos(\alpha-\beta) - \cos(\alpha+\beta)\}$$
$$\sin\alpha\cos\beta = \tfrac{1}{2}\{\sin(\alpha-\beta) + \sin(\alpha+\beta)\}$$
$$\cos\alpha\cos\beta = \tfrac{1}{2}\{\cos(\alpha-\beta) + \cos(\alpha+\beta)\}$$

Phase-shift

$$A\sin\theta + B\cos\theta = \sqrt{A^2+B^2}\cos(\theta-\delta),$$
where $\sin\delta = A/\sqrt{A^2+B^2}$, $\cos\delta = B/\sqrt{A^2+B^2}$
and $-\pi/2 \le \delta < \pi/2$

Lagrange Identity

$$\frac{1}{2} + \sum_{k=1}^{n}\cos(ks) = \frac{\sin((n+1/2)s)}{2\sin(s/2)}$$

Law of Cosines

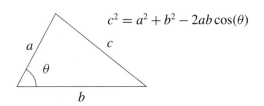

$$c^2 = a^2 + b^2 - 2ab\cos(\theta)$$

Euler's Formulas (a, θ real)

$$e^{i\theta} = \cos\theta + i\sin\theta \qquad \cos\theta = (e^{i\theta} + e^{-i\theta})/2$$
$$e^{a+i\theta} = e^a(\cos\theta + i\sin\theta) \quad \sin\theta = i(e^{-i\theta} - e^{i\theta})/2$$

Logarithms and Exponentials

(for all $A \ne 0$, $B \ne 0$, and all a, b)

$$\ln|A| + \ln|B| = \ln|AB| \qquad e^{a+b} = e^a e^b$$
$$\ln|A| - \ln|B| = \ln|A/B| \qquad e^{a-b} = e^a/e^b$$
$$a\ln|B| = \ln(|B|^a) \qquad\qquad e^{a\ln|B|} = |B|^a$$

Hyperbolic Trigonometric Functions

$$e^x = \cosh x + \sinh x$$
$$e^{-x} = \cosh x - \sinh x$$
$$\cosh x = (e^x + e^{-x})/2$$
$$\sinh x = (e^x - e^{-x})/2$$
$$\tanh x = (e^x + e^{-x})/(e^x - e^{-x})$$

Polynomials

Fundamental Theorem of Algebra (Factorization Formula). Suppose that $P(x) = x^n + a_{n-1}x^{n-1} + \cdots + a_0$, where n is a positive integer and $a_{n-1}, \ldots, a_0$ are constants. Then there are distinct numbers $\{r_1, \ldots, r_k\}$, some of which may be complex, and positive integers $\{m_1, \ldots, m_k\}$ such that

$$P(x) = (x - r_1)^{m_1} \cdots (x - r_k)^{m_k}$$

where $m_1 + \cdots + m_k = n$. The unique numbers $r_1, \ldots, r_k$ are the *zeros* (or *roots*) of $P(x)$ and the unique positive integers $m_1, \ldots, m_k$ are the corresponding *multiplicities*.

Coefficients and Roots. Suppose that $s_1, \ldots, s_n$ are the n roots (not necessarily distinct) of the polynomial $x^n + a_{n-1}x^{n-1} + \cdots + a_0$, with each root listed as often as its multiplicity. Then

$$s_1 + s_2 + \cdots + s_n = -a_{n-1}$$
$$s_1 \cdot s_2 \cdots s_n = (-1)^n a_0$$

B.5   Useful Results from Calculus

In this section, we state the theorems of single- and multivariable calculus most frequently encountered in the study of differential equations.

Results from Single-Variable Calculus

In the theorems that follow f is a function of a single variable. We represent the closed interval from a to b by $[a, b]$, and the open interval by (a, b).

THEOREM B.5.1

Maximum/Minimum Value Theorem. Suppose that f is continuous on $[a, b]$. Then f is bounded on $[a, b]$ and attains its maximum and its minimum values at points in $[a, b]$. If one of these points is in the open interval (a, b) and if f is differentiable on (a, b), then $f' = 0$ at the point.

THEOREM B.5.2

L'Hôpital's Rule. Suppose that $f(x)$ and $g(x)$ are differentiable functions and that $\lim_{x \to a} f(x) = \lim_{x \to a} g(x) = 0$, or else $\lim_{x \to a} f(x) = \lim_{x \to a} g(x) = \pm\infty$, where a may be $\pm\infty$. If $\lim_{x \to a} f'(x)/g'(x)$ exists, then so does $\lim_{x \to a} f(x)/g(x)$, and

$$\lim_{x \to a} \frac{f(x)}{g(x)} = \lim_{x \to a} \frac{f'(x)}{g'(x)}$$

THEOREM B.5.3

Intermediate Value Theorem. Suppose that f is continuous on $[a, b]$ and that c is any real number between $f(a)$ and $f(b)$ inclusive. Then there exists a point t^* in $[a, b]$ such that $f(t^*) = c$.

THEOREM B.5.4

Mean Value Theorem. **Integral Form:** Suppose that f is continuous on $[a, b]$. Then there exists a point t^* in $[a, b]$ such that $\int_a^b f(t)\, dt = f(t^*)(b - a)$. **Differential Form:** If f is differentiable on (a, b) and continuous on $[a, b]$, then there exists a point t^* in (a, b) such that $f(b) - f(a) = f'(t^*)(b - a)$.

THEOREM B.5.5

Fundamental Theorem of Calculus. **Form 1:** Suppose that f is continuous on $[a, b]$ and that F is an antiderivative of f. Then $\int_a^b f(t)\, dt = F(b) - F(a)$. **Form 2:** Suppose that f is continuous on $[a, b]$ and that c is any point in $[a, b]$. If $F(t) = \int_c^t f(s)\, ds$, then $F'(t) = f(t)$ at each point t in (a, b).

THEOREM B.5.6

Leibniz's Rule. Suppose that R is the rectangle in the st-plane given by $a \le t \le b$, $c \le s \le d$. Suppose that $\alpha(t)$, $\beta(t)$ are continuously differentiable functions on $[a, b]$, whose graphs lie in R, and that $f(s, t)$ and $\partial f/\partial t$ are continuously differentiable on R. Then $\int_{\alpha(t)}^{\beta(t)} f(s, t)\, ds$ is a continuously differentiable function of t and

$$\frac{d}{dt} \int_{\alpha(t)}^{\beta(t)} f(s, t)\, ds = \beta'(t) f(\beta(t), t) - \alpha'(t) f(\alpha(t), t) + \int_{\alpha(t)}^{\beta(t)} \frac{\partial}{\partial t} f(s, t)\, ds$$

THEOREM B.5.7

Chain Rule. Suppose that $g(t)$ is differentiable at the point t and that $f(s)$ is differentiable at the point $s = g(t)$. Then the composition $f(g(t))$ is differentiable at the point t, and $f(g(t))' = f'(g(t)) \cdot g'(t)$. Stated another way: Put $s = g(t)$ and $y = f(s)$, then the composition $y = f(g(t))$ is differentiable and

$$\frac{dy}{dt} = \left[\frac{dy}{ds} \cdot \frac{ds}{dt}\right]_{s=g(t)}$$

THEOREM B.5.8

Integration by Parts. Suppose that $f(x)$ and $g(x)$ are differentiable functions on $[a, b]$. Then

$$\int_a^b f(x) g'(x)\, dx = f(x) g(x) \Big|_{x=a}^{x=b} - \int_a^b f'(x) g(x)\, dx$$

THEOREM B.5.9

Integral Estimate. Suppose that f is continuous on $[a, b]$. Suppose that M is any real number such that $|f(t)| \le M$ for all $a \le t \le b$. Then

$$\left|\int_a^b f(t)\, dt\right| \le \int_a^b |f(t)|\, dt \le (b-a)M$$

Results from Multivariable Calculus

In the theorems that follow, the real-valued function f has two or more real variables.

THEOREM B.5.10

Maximum/Minimum Value Theorem. Suppose that $f(x, y)$ is continuous on a bounded region R that contains its edges. Then f is bounded on R and attains its maximum and its minimum values at some points in R. If one of these points is in the interior of R and if f is differentiable, then $\partial f/\partial x$ and $\partial f/\partial y$ are both 0 at the point.

THEOREM B.5.11

Fubini's Theorem: Interchanging the Order of Integration. Suppose that $f(x, y)$ is piecewise continuous on the rectangle R: $a \leq x \leq b, c \leq y \leq d$. Then

$$\iint_R f \, dA = \int_a^b \left(\int_c^d f(x, y) \, dy \right) dx = \int_c^d \left(\int_a^b f(x, y) \, dx \right) dy$$

THEOREM B.5.12

Green's Theorem. Suppose that R is a simply connected planar region whose boundary is a closed, piecewise-smooth curve C traversed counterclockwise. If $P(x, y)$ and $Q(x, y)$ have continuous first partial derivatives on some region containing R, then

$$\int_C P(x, y) \, dx + Q(x, y) \, dy = \iint_R \left(\frac{\partial Q}{\partial x} - \frac{\partial P}{\partial y} \right) dA$$

If $f(x, y, z)$ is a continuous function, its *gradient* is the vector

$$\nabla f = \frac{\partial f}{\partial x} \mathbf{i} + \frac{\partial f}{\partial y} \mathbf{j} + \frac{\partial f}{\partial z} \mathbf{k}$$

At each point of a level set $f(x, y, z) = C$, the gradient vector is perpendicular to the tangent to the level set and points in the direction of the maximal increase in the values of f. If $\mathbf{F}(x, y, z) = f(x, y, z)\mathbf{i} + g(x, y, z)\mathbf{j} + h(x, y, z)\mathbf{k}$, then the *divergence* of $\mathbf{F}$ is

$$\nabla \cdot \mathbf{F} = \frac{\partial f}{\partial x} + \frac{\partial g}{\partial y} + \frac{\partial h}{\partial z}$$

THEOREM B.5.13

Divergence Theorem. Suppose that G is a solid with boundary surface A oriented by outward unit normals $\mathbf{n} = n_1\mathbf{i} + n_2\mathbf{j} + n_3\mathbf{k}$. If $\mathbf{F}(x, y, z) = f(x, y, z)\mathbf{i} + g(x, y, z)\mathbf{j} + h(x, y, z)\mathbf{k}$, where f, g, and h have continuous first partial derivatives on some region containing G, then

$$\iint_A \mathbf{F} \cdot \mathbf{n} \, dS = \iiint_G \nabla \cdot \mathbf{F} \, dV$$

That is,

$$\iint_A (fn_1 + gn_2 + hn_3) \, dS = \iiint_G \left(\frac{\partial f}{\partial x} + \frac{\partial g}{\partial y} + \frac{\partial h}{\partial z} \right) dV$$

THEOREM B.5.14

Chain Rule. Suppose that $x = x(t)$ and $y = y(t)$ are differentiable at t and that $z = f(x, y)$ is differentiable at the point $(x(t), y(t))$. Then

$$\frac{dz}{dt} = \frac{\partial z}{\partial x}\frac{dx}{dt} + \frac{\partial z}{\partial y}\frac{dy}{dt}$$

THEOREM B.5.15

> Taylor's Theorem for a Function of Two Variables. Suppose that the function $f(x, y)$ is twice continuously differentiable in a region S containing a point (x_0, y_0). Then there is a rectangle R, $|x - x_0| < a$, $|y - y_0| < b$, such that for all (x, y) in R
>
> $$f(x, y) = f(x_0, y_0) + f_x(x_0, y_0)(x - x_0) + f_y(x_0, y_0)(y - y_0) + P(x, y)$$
>
> where P has order at least 2 at (x_0, y_0).

THEOREM B.5.16

> Taylor's Theorem for a Vector Function of a Vector Variable. Suppose that $F(x)$ is a function of the vector x in $\mathbb{R}^n$, where the values of F are in $\mathbb{R}^m$. Suppose that F is twice continuously differentiable in an open ball $B_r(x^0)$ of $\mathbb{R}^n$ of radius r and centered at x^0. Then for all x in R
>
> $$F(x) = F(x^0) + [\partial F_i / \partial x_j]_{x=x^0} (x - x^0) + P(x)$$
>
> where $[\partial F_i / \partial x_j]_{x=x^0}$ is the $m \times n$ *Jacobian matrix* of the first partial derivatives of the components of F with respect to the components of x, all evaluated at x^0, and $P(x)$ has order at least 2 at x^0.

Rectangular, Cylindrical, Spherical Coordinates

Suppose that G is a three-dimensional region. If G and its boundary surfaces are described in rectangular coordinates x, y, z, then we will denote G by G_1; in cylindrical coordinates, by G_2; and in spherical coordinates, by G_3.

Suppose that G is a simple solid and that $f(x, y, z)$ is continuous on G. Then

$$\text{Rectangular:} \quad \int_G f \, dV = \iiint_{G_1} f(x, y, z) \, dz \, dy \, dx$$

$$\text{Cylindrical:} \quad \int_G f \, dV = \iiint_{G_2} f(r, \theta, z) r \, dz \, dr \, d\theta$$

$$\text{Spherical:} \quad \int_G f \, dV = \iiint_{G_3} f(\rho, \theta, \phi) \rho^2 \sin\phi \, d\rho \, d\phi \, d\theta$$

The Laplacian Operator

Laplace's equation is the linear-second order PDE $\nabla^2 u = 0$, where ∇^2 is the *Laplacian operator*. The Laplacian operator in the rectangular coordinates of 3-space is

$$\text{Rectangular:} \quad \nabla^2 = \frac{\partial^2}{\partial x^2} + \frac{\partial^2}{\partial y^2} + \frac{\partial^2}{\partial z^2}$$

The operator ∇^2 has the following form in other coordinate systems:

$$\text{Polar:} \quad \nabla^2 = \frac{\partial^2}{\partial r^2} + \frac{1}{r}\frac{\partial}{\partial r} + \frac{1}{r^2}\frac{\partial^2}{\partial \theta^2}$$

Cylindrical: $\nabla^2 = \dfrac{\partial^2}{\partial r^2} + \dfrac{1}{r}\dfrac{\partial}{\partial r} + \dfrac{1}{r^2}\dfrac{\partial^2}{\partial \theta^2} + \dfrac{\partial^2}{\partial z^2}$

Spherical: $\nabla^2 = \dfrac{1}{\rho^2}\dfrac{\partial}{\partial \rho}\left(\rho^2 \dfrac{\partial}{\partial \rho}\right) + \dfrac{1}{\rho^2 \sin\phi}\dfrac{\partial}{\partial \phi}\left(\sin\phi \dfrac{\partial}{\partial \phi}\right) + \dfrac{1}{\rho^2 \sin^2\phi}\dfrac{\partial^2}{\partial \theta^2}$

B.6 Scaling and Units

Units

Three systems of units are used throughout the text—SI units [the International System of units, also known as MKS (meter-kilogram-second) units], CGS (centimeter-gram-second) units, and British Engineering units. Although the SI system is most frequently used in the scientific community, the text also uses the British Engineering system (BE) due to its continued widespread use in the United States. Table B.6.1 summarizes the units of length, force, mass, acceleration, and energy in these three systems. In CGS and SI units, the three fundamental physical units are length (L), mass (M), and time (T). In BE units, the three fundamental physical dimensions are length, force (ML/T^2), and time.

When working with physical systems involving these units of measure, the following points are useful to keep in mind.

- Although any system of units may be used, it is crucial to be consistent in the use of a single system for any given problem.

- Time is measured in seconds (s), hours (h), days (d), or years (y).

- The dimensions of force are the same as the product of mass and acceleration. For example, 1 N = 1 kg·m/s^2, and 1 lb = 1 slug·ft/s^2. Also, 1 N = 0.2248 lb.

- The acceleration due to gravity is denoted by g and has a magnitude near the earth's surface of approximately 9.8 m/s^2 in SI units, 980 cm/s^2 in CGS units, and 32 ft/s^2 in BE units. The *weight* of an object is the gravitational force exerted on it by the earth, and is equal to the product of the mass m of the object and the acceleration g due to gravity.

- Although the SI unit of energy is the joule (1 J = 1 N·m), the calorie is still occasionally used [1 cal = 4.184 J = 3.968 × 10^{-3} BTU (British thermal units)].

- A useful method of ensuring that you are using units correctly is to check that the dimensions are the same on both sides of any equation. For example, the undamped Hooke's Law spring equation is $my'' = -ky$. Both sides of the equation should have the dimensions of force. Using the BE system, m is measured in slugs and y'' is in ft/s^2. So my'' is measured in slug·ft/s^2 (that is, in lbs). Similarly, k is measured in slug/s^2 and y is measured in ft, so ky is also measured in slug·ft/s^2. In the CGS system, m is measured in grams and y'' has units of cm/s^2; my'' has units of gm·cm/s^2. Since y has units of cm, k must have units of gm/s^2.

TABLE B.6.1 Comparison of Basic Units in Various Systems

System	Length (L)	Force (ML/T^2)	Mass (M)	Acceleration (L/T^2)	Energy (ML^2/T^2)
SI	meter (m)	newton (N)	kilogram (kg)	m/s^2	joule (J)
CGS	centimeter (cm)	dyne	gram (gm)	cm/s^2	erg
BE	foot (ft)	pound (lb)	slug	ft/s^2	BTU

The problem of units can be avoided by rescaling equations and using dimensionless variables.

Scaling: Dimensionless Variables and Parameters

An initial value problem may involve many variables and parameters. For example, the IVPs that model radioactive decay, the motion of an oscillating spring, and the changing concentrations in a chemical reactor include a host of variables, physical parameters, and initial data. Even an abstract IVP that does not model a specific natural system may be written in terms of several variables and parameters. Variables and parameters can be measured in any one of several systems of units. Is there a way to scale the state variables and time so that the variables of the rescaled IVP are dimensionless? Any study of the solutions of the dimensionless system would then give information about the behavior of solutions regardless of the units used. Is there a way to study the sensitivity of an IVP to changes in some of the parameters or initial data so that the results apply independently of the system of units used? These two questions are clearly related, and the three examples given below show how a suitable rescaling of the state variables and time resolves the questions. In the process, we see that rescaling reduces the number of parameters that we have to work with—always an advantage. Finally, scaling a problem before computing is essential to avoid awkward scales (e.g., an interval 10^5 units long on the x-axis and 2 units long on the y-axis). One important consequence of rescaling to dimensionless quantities is that the results of a sensitivity study of the dimensionless IVP are inherent in the IVP and are not just a consequence of using particular systems of units, scales, and dimensions.

Radiocarbon Dating (Section 1.5)

The model IVP for a radioactive carbon-14 (^{14}C) dating process is

$$q'(t) = -kq(t), \qquad q(0) = q_0, \qquad T \le t \le 0 \qquad (1)$$

where $q(t)$ is the (dimensionless) fraction of the mass of ^{14}C divided by the total mass of all forms of carbon in a sample of once-living material, k is a positive decay constant (measured in (yr)$^{-1}$), $q(0)$ is the current fraction of ^{14}C (at $t = 0$), T is the time in years in the past when the material was last alive, and q_T is the fraction of ^{14}C in the sample at time T. Observe that $q(t)$ is already dimensionless since it is a ratio of masses. The four parameters here are k, q_0, T, and q_T.

The fraction $q(t)$ is often very small; to avoid working with small numbers it is convenient to rescale $q(t)$ by dividing it by its current value q_0. In dating problems the time span may be very large (thousands of years). Time can be rescaled (and nondimensionalized) by dividing by the *half-life* of ^{14}C, $\tau = 5568$ years. Denote the rescaled time by the dimensionless time variable $s = t/\tau$. In terms of the rescaled state variable, $y = q/q_0$, and of the rescaled time s, we have

$$y(s) = \frac{q(t)}{q_0} = \frac{q(\tau s)}{q_0} \tag{2}$$

From (1), (2), and the Chain Rule (Theorem B.5.7) we have (since $dt/ds = \tau$)

$$\frac{dy(s)}{ds} = \frac{1}{q_0} \frac{dq(t)}{dt} \frac{dt}{ds} = \frac{\tau}{q_0} \frac{dq(t)}{dt}$$

$$= \frac{\tau}{q_0}(-kq(t)) = \frac{\tau}{q_0}(-kq_0 y(s)) = -k\tau y(s) = -(\ln 2)y(s)$$

where we have made use of the half-life relationship $k\tau = \ln 2$. After setting $S = T/\tau$, IVP (1) becomes

$$\frac{dy(s)}{ds} = -(\ln 2)y(s), \qquad y(0) = 1, \qquad S \le s \le 0 \tag{3}$$

Observe that by rescaling the state variable and time we have reduced the number of parameters from four to two, S and y_S ($= q_T/q_0$).

The radioactive dating problem now reduces to finding the value $S < 0$ for which $y(S) = q_T/q_0$. Observe also that by rescaling time by dividing by the half-life, the time scale has been reduced from thousands of years (possibly) to a dimensionless time range (usually) of 4 or 5.

Soft Spring (Section 3.1)

The IVP

$$m\frac{d^2 y(t)}{dt^2} = -c\frac{dy(t)}{dt} - ky(t) + jy^3(t) + A_0 \cos \omega t, \quad y(0) = 0, \quad y'(0) = 0 \tag{4}$$

models the changing displacement $y(t)$ from equilibrium of a body of mass m attached to a soft spring whose action on the mass is modeled by the spring force $-ky(t) + jy^3(t)$ and which is subject to the driving force $A_0 \cos \omega t$. We assume that the spring is attached to a wall and that the body slides horizontally so that we can ignore the force of gravity. The term $-c\,dy(t)/dt$ represents a frictional damping force. The variables y and t have respective units of length and time. The six parameters m, c, k, j, A_0, and ω also have physical units. By rescaling y and t we can simultaneously introduce dimensionless variables and parameters and reduce the number of parameters from six to three. Let's set $u = y/A$ and $s = t/B$, where the scale factors A and B are to be determined. By the Chain Rule (Theorem B.5.7) we have

$$\frac{dy(t)}{dt} = \frac{dy}{du}\frac{du}{ds}\frac{ds}{dt} = A\frac{du}{ds}\frac{ds}{dt} = \frac{A}{B}\frac{du}{ds}$$

$$\frac{d^2y(t)}{dt^2} = \frac{d}{dt}\left(\frac{dy(t)}{dt}\right) = \frac{d}{ds}\left(\frac{A}{B}\frac{du}{ds}\right)\frac{ds}{dt} = \frac{A}{B^2}\frac{d^2u}{ds^2}$$

The differential equation in IVP (4) transforms to

$$m\frac{A}{B^2}\frac{d^2u}{ds^2} = -c\frac{A}{B}\frac{du}{ds} - kAu + jA^3u^3 + A_0\cos(\omega Bs)$$

which is normalized by dividing by mA/B^2 to yield the IVP

$$\frac{d^2u}{ds^2} = -\frac{c}{m}B\frac{du}{ds} - \frac{k}{m}B^2u + \frac{j}{m}(AB)^2u^3 + \frac{A_0}{m}\frac{B^2}{A}\cos(\omega Bs), \quad u(0) = u'(0) = 0 \quad (5)$$

Recalling that the scale factors A and B are yet to be determined, we decide which effects we want to focus on in a computer study of the solution of IVP (4). For example, we may want to see how the frictional force, the soft spring force, and the frequency of the driving force affect the solution. In that case, the coefficients of the terms involving du/ds (friction), $-u^3$ (soft spring), and ω (the frequency) will be the new parameters, while A and B will be chosen to set the remaining coefficients at some convenient value, say 1. So choose A and B so that

$$\frac{k}{m}B^2 = 1, \qquad \frac{A_0}{m}\frac{B^2}{A} = 1$$

that is,

$$B = \sqrt{\frac{m}{k}}, \qquad A = \frac{A_0}{k}$$

Renaming the other coefficients in the ODE in (5) as $c_1 = cB/m$, $j_1 = j(AB)^2/m$, and $\omega_1 = \omega B$, we have the IVP

$$\frac{d^2u}{ds^2} = -c_1\frac{du}{ds} - u + j_1u^3 + \cos(\omega_1 s), \qquad u(0) = 0, \quad u'(0) = 0 \quad (6)$$

with three parameters c_1, j_1, and ω_1 (instead of the original six).

In fact, u, s, c_1, j_1, and ω_1 are all dimensionless quantities. To show this, we begin by identifying the physical dimensions of the parameters and variables in IVP (4), using M for mass, L for length, and T for time dimensions. All the summands in the ODE of IVP (4) must have the force dimensions ML/T^2 of the first term on the left. This establishes the dimensions of the parameters and so of A and B. The appropriate dimensions for all parameters and variables are tabulated in Table B.6.2.

We can use Table B.6.2 to show the nondimensionality of the variables and parameters of IVP (6). For example, $u = y/A$ has the dimension of y divided by that of A, in other words L/L, which is dimensionless. The parameter $c_1 = cB/m$ has the dimensions of c, times the dimension of B, divided by that of m [i.e., $(M/T) \cdot T/M$], which is again dimensionless.

TABLE B.6.2 Dimensions of Soft Spring Parameters

Parameter	Dimension		Parameter	Dimension
y	L		t	T
dy/dt	L/T		d^2y/dt^2	L/T^2
m	M		c	M/T
k	M/T^2		j	$M/(LT)^2$
A_0	ML/T^2		ω	$1/T$
$A = A_0/k$	L		$B = \sqrt{m/k}$	T

A computer study of the effect of the damping force, the soft spring force, and the driving frequency on solution behavior should use IVP (6), varying the dimensionless parameters c_1, j_1, and ω_1 from some standard values and computing the dimensionless displacement u and velocity du/ds over some dimensionless time interval. The results of such a study then apply to the behavior of the solutions of any damped and driven soft spring equation in any units.

The Autocatalator (Section 5.1)

The IVP for the changing concentrations $x_1, \ldots, x_4$ of four chemical species in an autocatalytic reaction are

$$
\begin{aligned}
x_1' &= -k_1 x_1, & x_1(0) &= A > 0 \\
x_2' &= k_1 x_1 - k_2 x_2 - k_3 x_2 x_3^2, & x_2(0) &= 0 \\
x_3' &= k_2 x_2 - k_4 x_3 + k_3 x_2 x_3^2, & x_3(0) &= 0 \\
x_4' &= k_4 x_3, & x_4(0) &= 0
\end{aligned}
\tag{7}
$$

where $k_1, \ldots, k_4$ are positive rate constants. The total concentration remains fixed at the value $x_1(0)$ since $(x_1 + x_2 + x_3 + x_4)' = 0$. The units of $x_1, \ldots, x_4$ are concentrations C (e.g., moles/liter), and the unit T of time is (usually) seconds. The five parameters of the problem have units:

$$
k_1, \ k_2, \ k_4 \text{ in } T^{-1}
\tag{8}
$$

$$
k_3 \text{ in } T^{-1}C^{-2}
\tag{9}
$$

$$
x_1(0) \text{ in } C
\tag{10}
$$

Rescaling the four concentrations and time will simultaneously nondimensionalize the concentrations and time and reduce the number of parameters from five to any three of our choice.

Suppose that we introduce new variables y_i and s:

$$
y_i = x_i/a_i, \quad i = 1, 2, 3, 4; \quad s = t/a_5
\tag{11}
$$

where the to-be-chosen scale constants $a_1, \ldots, a_4$ have the dimension C of concentration and a_5 has the dimension T of time. Changing the variables of IVP (7) accordingly and using the Chain Rule (Theorem B.5.7) to calculate the derivatives, we have

$$\frac{dy_1}{ds} = -a_5 k_1 y_1, \qquad\qquad\qquad y_1(0) = A/a_1$$

$$\frac{dy_2}{ds} = \frac{a_1 a_5}{a_2} k_1 y_1 - a_5 k_2 y_2 - a_3^2 a_5 k_3 y_2 y_3^2, \qquad y_2(0) = 0$$

$$\frac{dy_3}{ds} = \frac{a_2 a_5}{a_3} k_2 y_2 - a_5 k_4 y_3 + a_2 a_3 a_5 k_3 y_2 y_3^2, \qquad y_3(0) = 0$$

$$\frac{dy_4}{ds} = a_5 k_4 y_3, \qquad\qquad\qquad y_4(0) = 0$$

For convenience, set

$$a_1 = a_2 = a_3 = a_4$$

Then we have $d(y_1 + y_2 + y_3 + y_4)/ds = 0$, and the total scaled concentration $y_1(s) + y_2(s) + y_3(s) + y_4(s)$ remains fixed at $y_1(0) = A/a_1$. Suppose that we want to study the sensitivity of the system to changes in the initial concentration of the first chemical species, as well as to changes in the first and second rate constants. Then we should take as the parameters

$$\alpha = y_1(0) = A/a_5, \qquad \beta = a_5 k_1, \qquad \gamma = a_5 k_2$$

For convenience set $a_5 k_4$ and $a_3^2 a_5 k_3$ equal to 1; then we have

$$a_5 = 1/k_4, \qquad a_1 = a_2 = a_3 = a_4 = (k_4/k_3)^{1/2}$$

Observe that a_5 has the units T of time, while a_1, a_2, a_3, and a_4 have the units of $(k_4/k_3)^{1/2}$, that is, of $(T^{-1}/T^{-1}C^{-2})^{1/2} = C$. From these observations and (11), we see that the scaled variables s, y_1, y_2, y_3, y_4 are dimensionless.

In the dimensionless variables y_i and s and the new parameters α, β, γ, the IVP becomes

☞ Although the names of the variables and parameters are changed, IVP (12) is the same as IVP (17) in Section 5.1.

$$\frac{dy_1}{ds} = -\beta y_1, \qquad\qquad y_1(0) = \alpha$$

$$\frac{dy_2}{ds} = \beta y_1 - \gamma y_2 - y_2 y_3^2, \qquad y_2(0) = 0$$

$$\frac{dy_3}{ds} = \gamma y_2 - y_3 + y_2 y_3^2, \qquad y_3(0) = 0 \qquad\qquad (12)$$

$$\frac{dy_4}{ds} = y_3, \qquad\qquad\qquad y_4(0) = 0$$

There is nothing unique about this rescaling or choice of parameters. The particular factors a_i used here were chosen to facilitate sensitivity studies in $x_1(0)$, k_1, and k_2 [rescaled as $y_1(0)$, β, and γ]. It would be equally straightforward to choose, say, k_1, k_3, and k_4 for a sensitivity study.

Answers to Selected Problems

Section 1.1   Page 8.

1(a). $y = y_0 e^{at}$.
2(a). The overcrowding coefficient is $1/9$ (ton·year)$^{-1}$; the harvesting rate is $8/9$ tons/year.

Section 1.2   Page 14.

2(a). $y_1 = 1$, $y_2 = 10$. Solution curves fall in the horizontal band between y_1 and y_2 and rise outside the band.
3(a). $y < 1$.   **3(c)**. $y < 1 + t$.

Section 1.3   Page 22.

1(a). First order, linear, $y' + t^3 y = \sin t$.
1(c). First order, nonlinear.
1(e). Second order, nonlinear.
1(g). Second order, linear, $y'' + e^{-t}(\sin t)y' + 3e^{-t}y = 5$.
2(a). $r = -3$.   **2(c)**. $r = 0, \pm\sqrt{2}, \pm 1$.
3(b). $r = 2/29$.   **4(a)**. $r = (-3 \pm \sqrt{5})/2$.
5(a). $y = 5t + \sin t + C$.   **5(d)**. $y = C_1 t + C_2$.
5(f). $y = t^3/3 + C_1 t^2 + C_2 t + C_3$.
6(a). $y = 1 + Ce^{-t}$, C is any constant.
6(c). $y = t + Ce^{-t}$, C is any constant.
6(e). $y = \pm\sqrt{(t^2/2) + C}$, C is any constant, $t^2/2 > -C$.
7(a). $y = Ce^{t^2} - 1/2$, C is any constant and $-\infty < t < \infty$.
7(c). $y = Ce^{\cos t} + 1$, $-\infty < t < \infty$.
7(e). $y = Ce^{-t^2/2} + 1/2$, $-\infty < t < \infty$.
8(a). $y = te^{-t} + Ce^{-t}$, $y(t) = (t+1)e^{-t}$, $-\infty < t < \infty$. As $t \to +\infty$, $y(t) \to 0$.
8(c). $y(t) = 1 + Ce^{-t^2}$, $y(t) = 1$, $-\infty < t < \infty$. As $t \to +\infty$, $y(t) \to 1$.
10(a). $y = t^2/4 + Ct^{-2}$, $t > 0$. As $t \to 0^+$, $y(t) \to \pm\infty$ or 0, depending on C. As $t \to +\infty$, $y(t) \to +\infty$.
11(a). $y = (1/t^2)\sin t - (1/t)\cos t + C$, $y(t) = (1/t^2)\sin t - (1/t)\cos t$, for $t > 0$. As $t \to +\infty$, $y(t) \to 0$.
11(c). $y = \sin t + C \csc t$, $y(t) = \sin t + 2\csc t$, $0 < t < \pi$. As $t \to 0^+$, $y \to +\infty$.

Section 1.4   Page 32.

3(a). 12 lb salt when $t = 2$; 85.3 lb salt when $t = 25$.
4(a). $y_d = t^2 - 2t + 2$, $y = Ce^{-t} + t^2 - 2t + 2$.
4(c). $y_d = te^{-2t}$, $y = Ce^{-2t} + te^{-2t}$.

4(e). $y_d = -\cos(2t) + 2\sin(2t)$, $y = Ce^{-t} - \cos(2t) + 2\sin(2t)$.
14(a). $y' = k[y(t) - m(t)]$, k negative.
15(a). 9.42%.

Section 1.5   Page 45.

3. $t = 278.5$ days.   **5(a)**. 2041 cm at $t = 2.04$ sec.
7(a). $H = 62.8$ meters.
12. 2028 B.C., approximately.
14(a). $K(t) = K_0 e^{-kt}$, $A(t) = (K_0 k_1/k)(1 - e^{-kt})$, $C(t) = (K_0 k_2/k)(1 - e^{-kt})$.

Section 1.6   Page 54.

3(a). $y = x$, $-1 < x < \infty$.   **3(c)**. $y = \exp(2 - e^{-x})$, for all x.
3(e). $y = (5 - x^2)^{1/2}$, $|x| \le 5^{1/2}$.
5(e). 1.016×10^8 individuals.   **8(a)**. 3.29 sec.

Section 1.7   Page 65.

5(a). $y = (-3t/2 + 1)^{2/3}$.
7. $y = -t^{-2}\sin t + C_1 t^{-1} + C_2 t^{-2}$, $t > 0$, C_1 and C_2 any constants.

Section 1.8   Page 74.

1(a). $x'(t) = -3x$, $y'(t) = x$ and $z'(t) = 2x$. The solution is $x(t) = e^{-3t}$, $y(t) = (1 - e^{-3t})/3$, and $z(t) = 2(1 - e^{-3t})/3$. As $t \to +\infty$, $x \to 0$, $y \to 1/3$, and $z \to 2/3$.
1(c). $x' = 1 + \sin t - 3x$, $y' = 3x - y$. The solution is $x = 1/3 + (3\sin t - \cos t)/10 - 7e^{-3t}/30$, $y = 1 + 3(\sin t - 2\cos t)/10 + 7e^{-3t}/20 - 3e^{-t}/4$. As $t \to +\infty$, x oscillates about $1/3$ and y oscillates about 1.
3(a). $x = Ae^{-k_1 t}$, $y = Ak_1(e^{-k_1 t} - e^{-k_2 t})/(k_2 - k_1) + Be^{-k_2 t}$.
7(a). $x' = I - k_1 t$, $x(0) = 0$, $y' = k_1 x - k_2 y$, $y(0) = 0$.

Section 1.9   Page 82.

1(a). $y = x\ln|x| + Cx$, where $x > 0$ or $x < 0$, and C is any constant.
1(c). $x^3(x - 2y)(x + y)^2 = C$, where C is a constant.
3(a). $y = -x + 2\arctan(x + C) + n\pi$, where C is any constant and n is any integer.

3(c). $3(x + 2y - 2\ln|x + 2y + 2|) = x + C$, where C is any constant and $|x + 2y + 2| \neq 0$.

10(c). $y = e^t + [Ce^{-3t} - e^{-t}/2]^{-1}$, where C is any constant and $2Ce^{-3t} \neq e^{-t}$.

11(a). $[\sqrt{2}(x^2 + y^2)^{1/2} + x - y]^2 = a^{\sqrt{2}}(3 + 2\sqrt{2})|x + y|^{2 - \sqrt{2}}$.

Section 2.1　Page 95.

3. $y(t) = t^{n+1}/n + Ct$, for any constant C.

5(c). Infinitely many solutions; e.g., $y(t) = -1$ for $t < (2n - 1)\pi$, $y(t) = \cos t$ for $(2n - 1)\pi \leq t \leq 2n\pi$, $y(t) = 1$ for $t > 2n\pi$, n a nonpositive integer.

Section 2.2　Page 105.

3. Escape time is $t = -(1/2)\ln[(1 + y_0)/(y_0 - 1)]$.

6(a). Curves rise if $-8 < y < 0$, and either $-6 < t < -3\pi/2$, or $-\pi/2 < t < \pi/2$, or $3\pi/2 < t < 6$; or if $0 < y < 8$, and either $-3\pi/2 < t < -\pi/2$, or $\pi/2 < t < 3\pi/2$. Curves fall elsewhere in the rectangle R.

Section 2.3　Page 114.

2. $P(t) \leq P_0 + R_0/r_0$, for $t \geq 0$.

5(b). $|y_c(t) - y_5(t)| \leq 0.1(1 - e^{-t})$, $\quad t \geq 0$.

Section 2.4　Page 121.

7(a). No more than 12 licenses.

Section 2.5　Page 130.

1(a). $y(1) \approx y_{10} = 0.348678$.

3(a). $y(1) \approx 0.283786, 0.325058, 0.329236$.

Section 2.6　Page 135.

1(a). $y = (1 - 2t)^{-1/2}$, $-\infty < t < 1/2$.

3(a). $y = Ce^{-2t} + 0.4\cos t + 0.2\sin t$, where C is an arbitrary constant. As $t \to +\infty$, $y(t) \to y_p(t) = 0.4\cos t + 0.2\sin t$.

5(a). $y = y_0[y_0 + (1 - y_0)e^{-t}]^{-1}$, $t_0 > 0$. As $t \to +\infty$, $y \to 1$.

Section 2.7　Page 145.

2. $h = 0.023$: 2-cycle about the equilibrium.
$h = 0.025$: 4-cycle about the equilibrium.

Section 3.1　Page 155.

4. $mz'' = -kz - \alpha/(b + z)^2$, where α is a positive constant and $z > -b$; nonlinear.

Section 3.2　Page 164.

1(a). Single equilibrium solution, $y(t) = 0$, all t. Periodic nonconstant orbits.

1(c). $y(t) = 0$, all t. Solutions oscillate and decay towards 0 as time goes backward.

1(e). $y(t) = 0$, all t; periodic solutions as $t \to +\infty$.

2(a). All solution curves are concave down everywhere.

5. $y(t) = 1/(1 - t)$, $t < 1$.

Section 3.3　Page 176.

1(a). $y = C_1 + C_2 t$, C_1 and C_2 arbitrary constants.

8(a). $y = -(1/5)\sin 2t + C_1 e^{-t} + C_2 e^t$, C_1 and C_2 arbitrary constants.

Section 3.4　Page 184.

1(a). $y(t) = c_1 e^{-t} + c_2 e^{2t}$, where c_1 and c_2 are real constants.

2(a). $y'' + 7y' + 10y = 0$.

2(c). $y'' + 8y' + 17y = 0$.

7(a). $y(t) = k_1 e^{-2it} + k_2 e^{it}$, where k_1 and k_2 are any complex constants.

8(a). $y(t) = ce^{-t}$, where c is any real constant.

Section 3.5　Page 189.

3. Yes; $y(t) = C_1 \cos 3t + C_2 \sin 3t - (3/5)\cos 2t$.

Section 3.6　Page 200.

1(a). $[i(e^{-it} - e^{it}) + (e^{2it} - e^{-2it})]/2$.

1(c). $-(1 + t)/2 - (1 + t + 2it^2)e^{2it}/4 + (2it^2 - t - 1)e^{-2it}/4$.

1(e). $(1 - t)(e^{4it} + e^{-2it})/2$.

3(a). $(D^2 + 1)^3$.　　　**3(c).** $D^2(D^2 + 1)$.

5(a). $y(t) = -e^{-2t}/2 + e^{2t} + 2t - 1/2$.

5(c). $y(t) = -2\cos t - 4\sin t + 2e^{2t}$.

6(a). $y = c_1 e^{-t} + c_2 e^{2t} + (\cos 2t - 3\sin 2t)/10$, c_1, c_2 any constants.

6(c). $y = (c_1 + c_2 t)e^t - t^3 e^t/6$, c_1, c_2 any constants.

6(e). $y = (c_1 + c_2 t)e^{-t} + e^t(3\cos t + 4\sin t)/25$, c_1, c_2 any constants.

6(g). $y = e^{-2t}(c_1 \cos t + c_2 \sin t) + e^{-t}/2 - 12/5 + 3t$, c_1, c_2 any constants.

7(a). $y = e^{2t} - e^t$.

Section 3.7　Page 209.

1(a). $y = c_1 \ln t + c_2$, $0 < t < \infty$, c_1 and c_2 arbitrary constants.

4(c). $y(t) = t - t^2$, for all t.

Section 4.1   Page 221.

4(a). $T = 2\pi\sqrt{L/g}$.

Section 4.2   Page 230.

1(a). $y(t) = \cos 2t - \cos 3t$; beats.

Section 4.3   Page 238.

1. $y_p = F_0 M(\omega)\cos(\omega t + \varphi(\omega))$, where
$M(\omega) = [(k^2 - \omega^2)^2 + 4c^2\omega^2]^{-1/2}$ and
$\varphi(\omega) = \cot^{-1}[(\omega^2 - k^2)/(2c\omega)]$ with
$-\pi \le \varphi(\omega) \le 0$.

5(a). $2c = 0.613$/sec.

Section 4.4   Page 249.

1(a). $I(t) = \exp(-t)[\cos t - \sin t]$.

2(a). $I(t) = k_1 e^{r_1 t} + k_2 e^{r_2 t}$,
$q(t) = 10^{-3} + k_1[(e^{r_1 t} - 1)/r_1 - (e^{r_2 t} - 1)/r_2]$,
where $r_1 \approx -2.33$, $r_2 \approx -47.7$,
$k_1 = -k_2 \approx -0.0024$.

4. $q(t) = (1/26)[-e^{-t}(9\cos 3t + 7\sin 3t) + 6\sin 2t + 9\cos 2t]$.

Section 5.1   Page 261.

1(a). $y = C_1 e^{-2t} + C_2 e^{2t}$; $x_1' = x_2$, $x_2' = 4x_1$ with general
solution $x_1 = C_1 e^{-2t} + C_2 e^{2t}$,
$x_2 = -2C_1 e^{-2t} + 2C_2 e^{2t}$.

1(c). $y = C_1 e^{-4t} + C_2 e^{-t}$; $x_1' = x_2$, $x_2' = -4x_1 - 5x_2$ with
general solution $x_1 = C_1 e^{-4t} + C_2 e^{-t}$,
$x_2 = -4C_1 e^{-4t} - C_2 e^{-t}$.

1(e). $y'' + 16y = 0$ with general solution
$y = C_1 \cos 4t + C_2 \sin 4t$; $x_1 = C_1 \cos 4t + C_2 \sin 4t$,
$x_2 = -4C_1 \sin 4t + 4C_2 \cos 4t$.

2(a). $x(t) = e^{-t/2}\left[\cos(\sqrt{3}t/2) + (1/\sqrt{3})\sin(\sqrt{3}t/2)\right]$,
$y(t) = -(2/\sqrt{3})e^{-t/2}\sin(\sqrt{3}t/2)$.

2(b). $x(t) = [(\sqrt{5}+3)/(2\sqrt{5})]e^{(-3+\sqrt{5})t/2} + [(\sqrt{5} - 3)/(2\sqrt{5})]e^{(-3-\sqrt{5})t/2}$,
$y(t) = -(\sqrt{5}/5)e^{(-3+\sqrt{5})t/2} + (\sqrt{5}/5)e^{(-3-\sqrt{5})t/2}$.

3(a). $x_1(t) = 10e^{-3t}$, $x_2(t) = 10(1 - e^{-3t})/3$,
$x_3 = 20(1 - e^{-3t})/3$; the limiting values as
$t \to +\infty$ are $x_1 = 0$, $x_2 = 10/3$, $x_3 = 20/3$.

3(b). $x_1(t) = 10e^{-t}$, $x_2(t) = 5e^{-t} + 15e^{-3t}$; the limiting
values as $t \to +\infty$ are $x_1 = x_2 = 0$.

5(a). $x_2 = Cx_1^{-3}$ and $x_1 = 0$.

6(a). $x_1 = C_1 e^{-2t} + 3C_2 e^{2t}$, $x_2 = -C_1 e^{-2t} + C_2 e^{2t}$.

7(a). $x' = -k_1 xy$, $y' = -k_1 xy$, $z' = k_1 xy - k_2 z$,
$w' = k_2 z$.

Section 5.2   Page 275.

2(a). $(0, 0)$ and $(1, 1)$; cycles enclosing $(1, 1)$.

2(c). No equilibrium points or cycles.

3(a). $x(t) = C_1 e^{3t}$, $y(t) = C_2 e^{-t}$, $-\infty < t < \infty$. The
solution $x(t - T)$, $y(t - T)$ is defined on
$-\infty < t < \infty$ as well.

3(c). $x(t) = C_1(1 + 2C_1^2 t)^{-1/2}$, $y(t) = t + C_2$,
$t > -1/2C_1^2$. The solution $x(t - T)$, $y(t - T)$ is
defined on $t > -1/2C_1^2 + T$ as well.

4(a). $(1, 1)$, $(1, -1)$, $(-1, 1)$, $(-1, -1)$.

Section 5.3   Page 284.

1(d). x-species changes logistically and is prey of
predator y-species, which is subject to overcrowding
effects and to periodic restocking.

2(a). x-nullclines: $x = 0$, $y = x - 5$; y-nullclines: $y = 0$,
$y = x/5 + 2$; equilibrium points: $(0, 0)$, $(0, 2)$,
$(5, 0)$, $(35/4, 15/4)$. The species cooperate, and
approach the equilibrium point $(35/4, 15/4)$ as
$t \to \infty$.

4(a). $2a$ measures the effectiveness of the y-species in
using the x-species to promote growth in y.

Section 5.4   Page 292.

1(a). $x(t)$ is the predator populatiom, $y(t)$ is the prey
population. The average populations are both 1. The
orbits turn clockwise about the equilibrium point
$(1, 1)$ as time increases.

Section 5.5   Page 298.

3. $dz/ds = z(c - y) - (1 - c - r)y$, $dy/ds = (z - r)y$.
The equilibrium points are $(0, 0)$, $(r, cr/(1 - c))$.

Section 6.1   Page 306.

1(a). $3/s^2 - 5/s$, $s > 0$.   **1(c).** $n!/s^{n+1}$, $s > 0$.

1(e). $1/(s - a)^2$, $s > a$.

2(a). $(1 + e^{-s\pi})/(1 + s^2)$, $-\infty < s < \infty$.

2(c). $[1 - (1 + s)e^{-s}]/s^2$, $s > 0$.   **3(a).** $y = e^{-2t}$.

Section 6.2   Page 315.

1(a). $a/(s^2 - a^2)$.   **1(c).** $2/(s - a)^3$.

1(e). $-\varphi'(s - 2)$, where $L[f](s) = \varphi(s)$.

1(g). $e^{-s}(1/s^2 + 2/s)$.

1(i). $e^{-(s-a)}/(s - a) - e^{-2(s-a)}/(s - a)$.

2(a). III.7: $\mathcal{L}[f] = 3(e^{-2s} - e^{-5s})/s$.

2(c). II.3: $\mathcal{L}[f] = 12!/(s - 5)^{13}$.

2(f). II.17: $f = (e^{2t} - e^{-5t})/t$.

3(a). $y(t) = (e^{3t} + 4e^{-2t})/5$.   **3(c).** $y(t) = e^t \sin t$.

3(e). $y(t) = e^t(1 - t) + \text{step}(t - 1)[1 + (t - 2)e^{t-1}]$.

4(a). $[(s - 1)^{-2} - (s - 1)^{-1}]s^{-1}$.

4(c). $(\sqrt{2}s)^{-1}[((s - 1)^2 + 1)^{-1} - (s - 1)((s - 1)^2 + 1)^{-1}]$.

7(c). $q_a(t) = \text{step}(t-a)CE_0 \times$
$\left[1 + r_2(r_1 - r_2)^{-1}e^{r_1(t-a)} + r_1(r_2 - r_1)^{-1}e^{r_2(t-a)}\right]$,
where $q_a(t)$ is the charge at time $t \geq 0$ if the switch is turned on at time $t = a$.

9(a). $x_1(t) = -9 + 3t + 9\cos t - 3\sin t$,
$x_2(t) = -15 + 5t + 15\cos t$.

Section 6.3 Page 324.

1(a). $[1 + (s-1)^2]^{-1}$.   **1(c).** $(s^2 + 2)s^{-1}(s^2 + 4)^{-1}$.
1(e). $-2a(3s^2 - a^2)(a^2 + s^2)^{-3}$.
1(g). $(1/\sqrt{2})(s+1)[(s+3)^2 + 4]^{-1}$.
2(a). $1 + \text{step}(t-1)$.
2(c). $t^{n-1}e^{at}/(n-1)!$.   **2(e).** $t + (2-t)\text{step}(t-1)$.
3(a). $1 - e^{-t}$.   **3(c).** $(1/(a-b))[e^{at} - e^{bt}]$.
3(e). $-(4/5)e^{-t}\cos t + (3/5)e^{-t}\sin t + (4/5)e^t$.

Section 6.4 Page 330.

1(a). $1/s^4$.   **1(c).** $2/s^4$.
2(a). $\int_0^t \cos(t-u)\sin u\, du$.   **2(c).** $\int_0^t (t-u)e^{-u}du$.
2(e). $\int_0^t [2e^{-2(t-u)} - e^{-(t-u)}]ue^{-2u}\, du$.
2(g). $\int_0^t f(t-u)\sin u\, du$.
3(a). $\int_0^t \sin(t-u)[u\,\text{step}(u-1)]\, du$.
3(c). $(1/3)\int_0^t f(t-u)[e^{u/2} - e^{-u}]\, du$.
4(a). $g(t) = (1/2)e^{-3t}\sin 2t$.
5(a). $y = [(e^{-3t}\sin 2t)/2] * f(t)$.

Section 6.5 Page 335.

1(a). $y(t) = -\text{step}(t-\pi)e^{-(t-\pi)}\sin t$.
1(c). $y(t) = [-3\cos t + \sin t + 5e^{-t} - 2e^{-2t}]/5 +$
$\text{step}(t-\pi)[e^{-(t-\pi)} - e^{-2(t-\pi)}]$.
1(e). $y = [e^t - \cos t - \sin t]/2 + \text{step}(t-1)\sin(t-1)$.

Section 7.1 Page 344.

1(a). $x_1 = 1800$, $x_2 = 699$, $x_3 = 200583$.
4(a). $x_1'(t) = -20x_1 + 4x_2 + x_3$, $x_2'(t) = x_1 - 4x_2 + 2x_3$,
$x_3'(t) = 3x_1 - 8x_3 + 1$.
5(b). $x_1 = I_1/k_{01}$, $x_2 = (k_{21}/k_{12})(I_1/k_{01})$.

Section 7.2 Page 351.

1. $\begin{bmatrix} 1 & -3 \\ -1 & 4 \end{bmatrix}$

7(a). Linear with standard form

$$\begin{bmatrix} x_1 \\ x_2 \\ x_3 \end{bmatrix}' = \begin{bmatrix} 1 & 1 & -2 \\ -5 & -1 & 2 \\ -1 & -2 & 1 \end{bmatrix}\begin{bmatrix} x_1 \\ x_2 \\ x_3 \end{bmatrix} + \begin{bmatrix} 0 \\ e^{-t} \\ \sin t \end{bmatrix}$$

Section 7.3 Page 361.

2(c). Points (a, b, c) on the plane through the origin
$c - 3a - 2b = 0$.

8(a). 156.   **9(a).** Dependent.
9(c). Independent.

Section 7.4 Page 369.

1(a). 1 is a double eigenvalue; $[0 \quad 1]^T$ is a basis of V_1.
1(c). The eigenvalues are 3 and -3 and are simple. V_3 is
spanned by $[1 \quad 1]^T$; V_{-3} is spanned by $[1 \quad -1]^T$.
1(e). The eigenvalues are $3 + 5i$ and $3 - 5i$ and are
simple. V_{3+5i} is spanned by $[1 \quad -i]^T$; V_{3-5i} is
spanned by $[1 \quad i]^T$.
1(g). 1 is a simple eigenvalue and 2 is a double
eigenvalue. V_1 is spanned by $[3 \quad -1 \quad 3]^T$ and V_2
has basis $\{[2 \quad 1 \quad 0]^T, [2 \quad 0 \quad 1]^T\}$.
1(i). -1 is a double eigenvalue and 5 is a simple
eigenvalue. V_{-1} is spanned by $[1 \quad 0 \quad 0]^T$ and V_5 is
spanned by $[262 \quad 27 \quad 6]^T$.

Section 7.5 Page 379.

1(a). $x(t) = c_1 e^{2t}\begin{bmatrix} 1 & -1 \end{bmatrix}^T + c_2 e^{4t}\begin{bmatrix} 3 & -1 \end{bmatrix}^T$, c_1, c_2
arbitrary reals.
1(b). $x(t) = c_1 e^{4t}\begin{bmatrix} 2 & 1 \end{bmatrix}^T + c_2 e^{9t}\begin{bmatrix} 1 & -2 \end{bmatrix}^T$, c_1, c_2
arbitrary reals.
3(a). $x = c_1 e^{-4t}[1 \quad -4]^T + c_2 e^{2t}[1 \quad 2]^T$, where c_1 and
c_2 are arbitrary reals.
3(b). $x = c_1 e^{2t}[2 \quad 1]^T + c_2 e^{-t}[1 \quad 2]^T$, where c_1 and c_2
are arbitrary reals.
3(c). $x = c_1 e^{5t}[1 \quad 3]^T + c_2 e^{3t}[1 \quad 1]^T$, where c_1 and c_2
are arbitrary reals.
6(a). $x = c_1 e^t[1 \quad -1 \quad 1]^T + c_2 e^{-t}[0 \quad 1 \quad -1]^T +$
$c_3 e^{-t}([1 \quad 1 \quad 0]^T + t[0 \quad 1 \quad -1]^T)$, where c_1, c_2,
and c_3 are arbitrary reals.

Section 7.6 Page 388.

1. $x = c_1 e^{2t}[\cos t \quad \sin t]^T + c_2 e^{2t}[\sin t \quad -\cos t]^T$, c_1
and c_2 arbitrary reals.
3(a). $x = c_1[\cos 8t \quad -8\sin 8t]^T + c_2[\sin 8t \quad 8\cos 8t]^T$,
where c_1 and c_2 are arbitrary reals.

Section 7.7 Page 398.

2(a). $p(\lambda) = \lambda^2 + \lambda - 6$; $\lambda_1 = 2$, $\lambda_2 = -3$; $v^1 = [1 \quad 1]^T$,
$v^2 = [1 \quad -4]^T$. $[x \quad y]^T = c_1 v^1 e^{2t} + c_2 v^2 e^{-3t}$, for
arbitrary reals c_1, c_2. Saddle.
2(c). $p(\lambda) = \lambda^2 - 13\lambda + 30$; $\lambda_1 = 10$, $\lambda_2 = 3$;
$v^1 = [2 \quad 1]^T$, $v^2 = [-3 \quad 2]^T$.
$[x \quad y]^T = c_1 v^1 e^{10t} + c_2 v^2 e^{3t}$, for arbitrary reals c_1,
c_2. Improper node.
2(e). $p(\lambda) = \lambda^2 + 2\lambda + 5$; $\lambda_1 = -1 + 2i$, $\lambda_2 = -1 - 2i$;
$v^1 = [2 \quad -i]^T$, $v^2 = [2 \quad i]^T$.
$[x \quad y]^T = c_1 e^{-t}[2\cos 2t \quad \sin 2t]^T +$
$c_2 e^{-t}[2\sin 2t \quad -\cos 2t]^T$, for arbitrary reals c_1, c_2.
Spiral point.

6(c). $a = 0$, $b > 0$.

7(a). $[x \quad y]^T =$
$c_1[2\cos 2t \quad -\sin 2t]^T + c_2[2\sin 2t \quad \cos 2t]^T$, for
arbitrary reals c_1, c_2.

Section 7.8   Page 408.

1(a). $e^{tA} = (1/4)\begin{bmatrix} e^{-2t} + 3e^{2t} & -3e^{-2t} + 3e^{2t} \\ -e^{-2t} + e^{2t} & 3e^{-2t} + e^{2t} \end{bmatrix}$;
$x(t) = (1/4)[-5e^{-2t} + 9e^{2t} \quad 5e^{-2t} + 3e^{2t}]^T$.

1(c). $e^{tA} = \begin{bmatrix} e^{-t} & te^{-t} \\ 0 & e^{-t} \end{bmatrix}$; $x(t) = e^{-t}[1 + 2t \quad 2]^T$.

2(a). $e^{tA} = \begin{bmatrix} 1 & t & t+t^2/2 \\ 0 & 1 & t \\ 0 & 0 & 1 \end{bmatrix}$;
$x(t) = [1 + 5t + 3t^2/2 \quad 2 + 3t \quad 3]^T$.

4(a). $e^{tA}x^0 = \begin{bmatrix} a\cos 2t + b\sin 2t \\ b\cos 2t - a\sin 2t \end{bmatrix}$; $e^{-sA}F(s) = \begin{bmatrix} \cos 2s \\ \sin 2s \end{bmatrix}$.

4(c). $e^{tA}x^0 = \begin{bmatrix} a\cos t + (2a - 5b)\sin t \\ (a - 2b)\sin t + b\cos t \end{bmatrix}$;
$e^{-sA}F(s) = \begin{bmatrix} \cos^2 s + 2\sin s \cos s \\ \sin s \cos s \end{bmatrix}$.

5(a). $x = \begin{bmatrix} e^{it} & e^{-it} \\ (2-i)e^{it} & (2+i)e^{-it} \end{bmatrix} c + \begin{bmatrix} 3e^t/2 - 1 \\ 5e^t/2 - 2 \end{bmatrix}$,
where c is a constant column vector.

Section 7.9   Page 417.

1(a). Not a steady state.   **2(a)**. $\alpha < -3/2$.
7(b). At least one root has nonnegative real part.
7(d). All roots have negative real parts.

Section 7.10   Page 424.

5(c). $a \approx 0.0102$.

Section 8.1   Page 440.

1(c). Equilibrium points at $(\pm 1, 0)$ are (locally)
asymptotically stable; point at $(0, 0)$ is unstable.

Section 8.2   Page 449.

1(a). Asymptotically stable.   **1(c)**. Unstable.
2(a). Asymptotically stable at $(0, 0)$, unstable at $(-2, 0)$.
2(c). Unstable at $(0, 0)$, asymptotically stable at $(0, -1)$.
2(e). Unstable at each point on the y-axis.
5(a). $(0, 0)$, $(1, 1)$, $(-1, 1)$.

Section 8.3   Page 461.

1(a). $K(x, y) = y^3 x$. The origin is an unstable saddle
point.
1(c). $K(x, y) = (x^4 + y^4)/4$. The origin is neutrally
stable.

Section 8.4   Page 470.

1(a). Asymptotically stable.
1(c). Asymptotically stable.
1(e). Asymptotically stable.   **1(g)**. Unstable.
3(a). Asymptotically stable.   **3(b)**. Unstable.

Section 9.1   Page 478.

1(a). $(0, 0)$, unstable; attracting limit cycle $r = 4$ and
repelling limit cycle $r = 5$.
1(c). The equilibrium points are $(0, 0)$ (unstable) and all
points on the circle $r = 1$ (neutrally stable);
repelling limit cycle $r = 2$.   **1(e)**. $(0, 0)$,
unstable; attracting limit cycles $r = 2n + 1/2$ and
repelling limit cycles $r = 2n + 3/2$.
2(a). Asymptotically stable equilibrium point $(0, 0)$.

Section 9.2   Page 486.

1(a). For the orbit through the equilibrium point $(0, 0)$,
the negative and positive limit sets are the point
$(0, 0)$ itself. The negative and positive limit sets of
the orbit through $(1, 1)$ are the orbit itself, which is a
cycle.
1(e). The negative and positive limit sets of the orbit
through $r = 1/2$, $\theta = 0$ are, respectively, the origin
and the cycle $r = 1$, where the origin is an
equilibrium point. The negative and positive limit
sets of the orbit through $r = 3/2$, $\theta = 0$ are,
respectively, the cycles $r = 2$ and $r = 1$. The
negative and positive limit sets of the orbit through
$r = 5/2$, $\theta = 0$ are, respectively, the empty set and
the cycle $r = 2$.
5(a). No cycles because x' is always positive.

Section 9.3   Page 497.

1(a). Saddle-node bifurcation at $c = 0$.
1(c). Transcritical bifurcation at $c = 0$.
1(e). Pitchfork bifurcation at $c = 0$.
5(a). $u' = cu + v - u(u^2 + v^2)$ and
$v' = -u + cv - v(u^2 + v^2)$, where $u = x - 5c$ and
$v = y - 5c$; $r' = cr - r^3$, $\theta' = -1$.

Section 9.4   Page 511.

1(b). $\begin{bmatrix} -\sigma & \sigma & 0 \\ r - z & -1 & -x \\ y & x & -b \end{bmatrix}$ is the Jacobian matrix at
(x, y, z).

Section 10.1   Page 524.

1(a). $u_n(x, t) =$
$\cos(n\pi x/L)(A_n\cos(n\pi ct/L) + B_n\sin(n\pi ct/L))$.

Section 10.2   Page 530.

  3(a). $\sin x \cos ct$, $c = \sqrt{T/\rho}$.
  5(b). 0 for the first pair of functions; $(e^2 - 1)/4$ for the second.

Section 10.3   Page 538.

  1. $\sum_{k=1}^{N} (-1)^{k+1}/k \sin kx$.
  2(a). $2 \sum_{k=1}^{\infty} (-1)^{k+1} (\sin kx)/k$.
  3(a). $(2A/\pi) \sum_{k=1}^{\infty} (\sin kx)/k$.

Section 10.4   Page 546.

  1(a). $A_0 = 5$, $B_1 = -1$, $B_3 = -7$, $A_6 = -4$, all other coefficients are zero.
  2(a). $-2 + 4 \sum_{k=1}^{\infty} (-1)^{k+1} k^{-1} \sin kx$.
  2(c). $(a + \pi^2 c/3) + \sum_{k=1}^{\infty} \left(4c(-1)^k k^{-2} \cos kx + 2b(-1)^{k+1} k^{-1} \sin kx\right)$.
  2(e). $\pi/2 - (4/\pi) \sum_{k=1}^{\infty} (2k+1)^{-2} \cos(2k+1)x$.
  2(g). $AB/\pi + (2A/\pi) \sum_{k=1}^{\infty} k^{-1} \sin kB \cos kx$.
  3(a). $\sin x$.

Section 10.5   Page 552.

  1(a). $(4/\pi) \sum_{odd\ k} k^{-1} \sin kx$; 1.
    2(a). $(e^{ix} + e^{-ix})/2$.
  3(b). $B_{2m} = 0$;
    $B_{2m-1} = (4/\pi)[(4m-3)^{-1}\sin(m-3/4)\pi - (4m-1)^{-1}\sin(m-1/4)\pi]$, for $m = 1, 2, \dots$.

Section 10.6   Page 557.

  1(a). $L[y] = y''$;
    Dom$(L) = \{y$ in $\mathbf{C}^2[0, \pi/2] : \ y(0) = y(\pi/2) = 0\}$.
    $\lambda_n = -4n^2$ for $n = 1, 2, \dots$;
    basis $\Phi = \{\sin 2nx : \ n = 1, 2, \dots\}$.
  1(c). $L[y] = y''$;
    Dom$(L) = \{y$ in $\mathbf{C}^2[0, T] : \ y'(0) = y'(T) = 0\}$.
    $\lambda_n = -(n\pi/T)^2$, $n = 1, 2, \dots$;
    basis $\Phi = \{\cos(n\pi x/T) : \ n = 0, 1, 2, \dots\}$.
  1(e). $L[y] = y''$, Dom$(L) = \{y$ in $\mathbf{C}^2[0, \pi] : \ y(0) = y(\pi) + y'(\pi) = 0\}$. $\lambda_n = -(r_n/\pi)^2$, where the r_n satisfy $\tan r_n = -\pi r_n$;
    basis $\Phi = \{\sin(r_n x/\pi) : \ n = 1, 2, \dots\}$.

Section 10.7   Page 565.

  1(a). $u(x, t) = (3L/\pi c) \sin(\pi ct/L) \sin(\pi x/L)$.
  2(a). The SL problem is as follows: find all λ such that $X'' = \lambda X$, $X(0) = 0$, $X'(L) + hX(L) = 0$ has a nontrivial solution in $\mathbf{C}^2[0, L]$.
  4(a). $v = gx(L - x)/(2c^2)$.
  6(a). $u(x, t) = \sum_{n=1}^{\infty} (-1)^{n+1} \left[12/(n^3\pi^3 c^2)\right] \sin(n\pi x)(1 - \cos n\pi ct)$.

Section 10.8   Page 577.

  1(a). $u(x, t) = \sin(2\pi x/L) \exp\left[-K(2\pi/L)^2 t\right]$.
  1(c). $u(x, t) = (4u_0/\pi) \sum_{odd\ n} n^{-1} \sin(n\pi x/L) \exp\left[-K(k\pi/L)^2 t\right]$.
  1(e). $u(x, t) = \sin(\pi x/2L) \exp\left[-K(\pi/2L)^2 t\right]$.
  2(a). $u(x, t) = 10(1 + x) - \sum_{n=1}^{\infty} (20/(n\pi))[1 - 2(-1)^n] \sin n\pi x \exp[-K(n\pi)^2 t]$.
  3(b). $U(x, t) = \sum_{n=1}^{\infty} e^{-K(n\pi)^2 t}[2\int_0^1 f(x)\sin n\pi x\, dx - (2K/n\pi)\int_0^t \cos \tau e^{K(n\pi)^2 \tau}\, d\tau] \sin n\pi x$.
  4. $U(x, t) = 2\sum_{n=1}^{\infty} \Big\{ (-1)^n (K(n\pi)^3)^{-1}[e^{-K(n\pi)^2 t} - 1]$
    $+ [3(-1)^n - 3][n\pi(Kn^2\pi^2 - 2)]^{-1}[e^{-K(n\pi)^2 t} - e^{-2t}]\Big\} \sin n\pi x$.
  5(b). $u = T_0 + A_0 e^{-\pi} \cos(\omega t - \pi)$.

Section 10.9   Page 584.

  3(a). $u(r, \theta) = 3r \sin \theta$. The maximum and minimum temperatures in the disk are ± 3, respectively.

Section 11.1   Page 591.

  1(a). $-\infty < x < \infty$.   **3(a).** $\sum_{n=1}^{\infty} 2nx^n/(n-1)!$.
  4(a). $\sum_{k=0}^{\infty} 2x^{2k}/(2k)!$.
  4(c). $(1/2) \sum_{k=0}^{\infty} (-1)^k (2x)^{2k+1}/(2k+1)!$.
  5(a). $na_n - n + 2 = 0$, for $n = 1, 2, \dots$; $a_n = 1 - 2/n$.
  8. $y(x) = x^2/2 + x^3/6 + \sum_4^{\infty} a_n x^n$, where $a_n = \sum_{k=1}^{[n/2]} (n - 2k)!/n!$.
  10. $R(r) = 1 - r^2/4 + r^4/(2 \cdot 4)^2 - r^6/(2 \cdot 4 \cdot 6)^2 + \cdots$.

Section 11.2   Page 600.

  1(a). Ordinary.   **2(a).** No singular points.
  2(c). $x = 0$.   **2(e).** $x = \pm 1$.
  4(a). $y = a_0\left[1 + \sum_{k=1}^{\infty} x^{3k}/[(3k)(3k-1)(3k-3)(3k-4)\cdots 3 \cdot 2]\right]$
    $a_1\left[x + \sum_{k=1}^{\infty} x^{3k+1}/[(3k+1)(3k)(3k-2)(3k-3)\cdots 4 \cdot 3$
    where a_0 and a_1 are arbitrary.
  5. $y = 1 - (3/2)x^2 + (13/48)x^4 - (23/288)x^6 + \dots$.
  6(a). $y = x + (1/4)x^4 + (1/14)x^7 + \cdots$.
  8. $y(1) \approx 1 - (1/3) + (1/18) - (1/162) + (1/1944) \approx 0.7166$.

Section 11.3   Page 607.

  1. $v_0 = (C_1/2) \ln[(1 + x)/(1 - x)] + C_2$;
    $v_1 = x[C_3 + C_1(-1/x + (1/2) \ln[(1 + x)/(1 - x)])]$.
  5(a). $a = -1$, $b = 1$, $\rho(x) = 1$.

Section 11.4   Page 614.

  1(a). Regular.   **1(c).** Regular.
  2(a). $x = \pm 1$ are regular singular points.

2(c). $x = 0, -1$ are regular singular points; $x = 1$ is an irregular singular point.

4(a). $y(x) = c_1 x^3 + c_2 x^2 + x + 5/6$, where c_1 and c_2 are any constants.

6(b). $y_1 = x^{-1}$, $y_2 = x^{-\sqrt{2}}$, $y_3 = x^{\sqrt{2}}$.

Section 11.5   Page 621.

1(a). $y_1 = x^{1/3} + x^{4/3}/5$, $y_2 = x^{-1/3} \sum_{n=0}^{\infty} a_n x^n$ where $a_n = (-1)^n 10 / (3^n n! (3n-2)(3n-5)) a_0$, $n \geq 2$.

1(c). $y_1 = x^{\sqrt{2}} \sum_{n=0}^{\infty} a_n(\sqrt{2}) x^n$,
$y_2 = x^{-\sqrt{2}} \sum_{n=0}^{\infty} a_n(-\sqrt{2}) x^n$ where $a_0 = 1$ and
$a_n(\pm\sqrt{2}) =$
$[(n \pm \sqrt{2} - 1)(n \pm \sqrt{2} - 2) \cdots (\pm\sqrt{2}) a_0] / [n! (n \pm 2\sqrt{2})(n - 1 \pm 2\sqrt{2}) \cdots (1 \pm 2\sqrt{2})]$.

3. $y_1 =$
$x^{1/3} + x^{4/3}/7 + 9x^{7/3}/280 + 227x^{10/3}/32760 + \cdots$,
$y_2 = x^{-1} - 1 - x/8 - 11x^2/360 + \cdots$.

7(a). The indicial polynomial is r^2 with roots $r_1 = r_2 = 0$.

Section 11.6   Page 632.

1. $J_3(x) =$
$x^3/48 - x^5/768 + x^7/30720 - x^9/2211840 + \cdots$.

Section 11.7   Page 644.

1(a). $y_1 = e^{-x}$,
$y_2 = e^{-x}[\ln x + x + x^2/4 + \cdots + x^n/(n!n) + \cdots]$.

1(c). $y_1 = x$,
$y_2 = x[-1/x + \ln x + \cdots + x^{n-1}/(n!(n-1)) + \cdots]$.

2(b). $y(x) = e^x[c_1 J_1(x) + c_2 Y_1(x)]$.

3(c). $y = x^{1/2}[c_1 J_p(x^2/2) + c_2 J_{-p}(x^2/2)]$,
$p = 1/(4\sqrt{2})$, $x > 0$.

Section 11.8   Page 651.

1(a). $u = -1 + 2P_2(\cos\phi)\rho^2$.

3(a). $u(r, z) = \sum_{n=1}^{\infty} (2/(x_n)^2) \cdot (\sinh x_n(a - z)/\sinh(x_n a)) \cdot (J_0(x_n r)/J_1(x_n))$.

6(a). $L[y] = (x^2 y')'/x^{-2}$;
$\text{Dom}(L) = \{y \text{ in } \mathbf{C}^2[1, 2] : y(1) = y(2) = 0\}$.
$\lambda_n = -4n^2\pi^2, n = 1, 2, \ldots$.
$\Phi = \{\sin(2n\pi/x) : n = 1, 2, \ldots\}$ is a basis.

Section A.1   Page 655.

2(b). One infinite set of solutions is
$$y(t) = \begin{cases} (at - aC)^{1/a}, & t \geq C \\ 0, & t < C. \end{cases}$$

Section A.2   Page 662.

1(a). $y_0(t) = 1$, $y_1(t) = 1 - t$, $y_2(t) = 1 - t + t^2/2$,
$y_3(t) = 1 - t + t^2/2 - t^3/6$.

6(a). $y_0(t) = 0$, $y_1(t) = t$, $y_2(t) = t + t^3/3$,
$y_3(t) = t + t^3/3 + 2t^5/15 + t^7/63$.

Section A.3   Page 664.

2. The ODE $y' = 2ty^2$ has solutions $y = (C - t^2)^{1/2}$ that are defined only for $|t| < C^{1/2}$ if $C > 0$.

Section A.4   Page 670.

1. $|y(t) - \tilde{y}(t)| \leq |a - \tilde{a}|e^{t/10} + 5(e^{t/10} - 1)$, $0 \leq t$, where we have chosen M to be $1/2$.

Photo Credits

Chapter 1

Page 37, The New York Public Library. Page 39, The New York Public Library. Page 66, The New York Times. Page 68, Courtesy Edward Spitznagel.

Chapter 2

Page 89, History of Mathematics Archive, University of St. Andrew. Page 123, The New York Public Library. Page 128, AIP Emilio Serge Visual Archives/Lande Collection. Page 144, Courtesy James A. Yorke.

Chapter 4

Page 243, AIP Niels Bohr Library W. F. Meggers Collection.

Chapter 5

Page 276, Courtesy Leah Edelstein-Keshet. Page 288, AIP Emilio Segre Visual Archives.

Chapter 6

Page 300, The New York Public Library. Page 310, The New York Public Library. Page 331, Corbis–Bettmann.

Chapter 9

Page 477, David Eugene Smith Collection Rare Book & Manuscript Library Columbia University. Page 480, AIP Emilio Serge Visual Archives/Bridgeman Collection. Page 500, Courtesy Edward N. Lorenz/MIT. Page 511, Courtesy of the Mistress and Fellows of Girton College.

Chapter 10

Page 540, Sketch by Biolly/Corbis–Bettmann. Page 554, The New York Public Library. Page 554, History of Mathematics Archive, University of St. Andrew.

Chapter 11

Page 597, Courtesy Columbia University. Page 622, Corbis–Bettmann.

Index

Abel, Niels Henrik, 205
Abel's Formula, 206, 210
Abel's Theorem, 205
Absolute value
    of a complex number, 678
Aging spring, 148, 586, 591, 597,
    640, 644
Airy's equation, 600, 644
Aliasing, 187
Alternating series, 676
Amplitude, 185, 186
Angular velocity, 458
Antiderivative, 16
Approximate slope function, 125
Archimedes' Principle, 231
Aspect ratio, 11
Attractor, 434, 472
    and conservative systems, 457
    basin of attraction, 219, 438
    global, 504
    strange, 508
Autocatalator, 258, 260, 499, 689
Autonomous ODEs, 56, 98, 150
    equilibrium solutions, 98
    translation property of
        solutions, 99, 162
Autonomous systems of ODEs
    planar, 56, 266
        equilibrium solution, 265
        polar form, 272
    properties of, 162
    separation of orbits, 265
    stability of, 434

Backward IVP, 44
Balance Law, 24, 567
Basic Questions for IVPs, 88
Basic solution matrix, 401, 428
Basic solution set, 205, 386, 426
Basin of attraction, 219, 438
Basis, 359, 535, 536
Beats, 228–230
Belousov, Boris, 258
Bendixson's Criterion, 485
Bernoulli brothers, 83

Bernoulli's equation, 83
Bessel function of the first kind
    Bessel basis, 646
    decay of, 631
    half integer, 629
    integration of, 633
    modified, 633
    of any order, 628
    of integer order, 624
    orthogonality of, 630
    oscillation of, 630
    recursion formulas for, 629
    solving a PDE, 649
    zeros of, 630
Bessel function of the second kind
    of any order, 640
    of integer order, 639
Bessel Inequality, 537
Bessel's equation, 622
    modified, 633
    recursion formula, 623
    solution of, 622
Bessel, Friedrich Wilhelm, 622
BIBO, 109, 413
Bifurcation, 116, 489
    analysis, 115
    diagram, 119, 490, 502
    Hopf, 490, 492
        subcritical, 492, 498
        supercritical, 492
    period-doubling, 140, 506
    pitchfork, 120, 121, 497, 502
    saddle-node, 116, 118, 121,
        489
    tangent, 118
    transcritical, 122, 497
Bilinearity, 527
Binomial series, 676
Bodé plots, 235
Body axes, 458
Boundary conditions, 519–520,
    553–554
Boundary function, 564
Boundary operators, 519

Boundedness
    of Legendre polynomials, 605
    of orbits, 483
    positive, 184

Capacitance, 243
Cartwright, Mary Lucy, 511
Cascade, 73, 261
Cauchy Uniform Convergence
    Test, 659
Cauchy, Augustin Louis, 597
Cauchy-Schwarz Inequality, 221,
    349, 529
Center-spiral, 479
Chain Rule, 682, 683
Chaos, 141, 142, 508
    routes to, 506
Chaotic wandering, 141
Characteristic polynomial
    of a linear ODE, 167, 172
    of a matrix, 365
Characteristic roots
    of a linear ODE, 167
Chebyshev polynomials, 608
Chemical Law of Mass Action, 259
Chemical rate law
    first-order, 109, 340
    second-order, 109
Chemical reaction, 258
Circuits, electrical, 241
    charge, 241
    Coulomb's Law, 243
    current, 241
    elements of, 242
    Faraday's Law, 242
    Kirchhoff's Laws
        current, 244
        voltage, 243
    LC, 316, 318, 324
    multiloop, 247
    Ohm's Law, 242
    RC, 103
    RLC, 244, 474, 475, 551
    tuning, 247
    two-loop, 248, 420

voltage, 241
Circular frequency, 224, 227
Circular harmonics, 581
Classical solution, 557, 560, 579
Clearance coefficient, 70
Closed box, 263
Closed rectangle, 10
Closed set, 263
Closure property, 172, 371
Cobweb diagrams, 146
Coefficient Test, 414
Cofactor Expansion, 357
Combat model, 62
Communications channel, 103
Compartment model, 24, 67, 72, 339, 419
Complex exponential function, 179, 679
Complex numbers, 381–382, 677–679
  complex-valued functions
    derivative of, 178, 679
    imaginary part, 178
    real part, 178
  conjugate, 381, 677
  De Moivre's Formula, 678
  Euler's formula, 678
  polar form, 678
Complex plane, 678
Component curve, 57, 159, 256
Conduction, 566
Conservative systems, 452, 455
  and attractors, 457
Constant steady state, 409
Continuity in the data, 112, 113, 158, 665
Continuity set, 202, 359, 516
Contour, 49, 452
Convergence
  Integral Convergence
    Theorem, 660
  interval of, 674
  mean, 531, 533
  of a complex sequence, 678
  pointwise, 534
  radius of, 674
  ratio test for, 674
  uniform, 534, 658
    Cauchy test for, 659
Convergence Theorem, 597
Convolution, 326
  solving an IVP, 329
Convolution Theorem, 327

Coordinate systems
  cylindrical, 684
  polar, 217
  rectangular, 684
  spherical, 684
Coulomb's Law, 243
Critically damped, 225
Current, 241
Curves of pursuit, 78
Cycle-graph, 481
Cycles, 265
  k-cycle, 139
  attracting, 276, 472
  Bendixson's Criterion, 485
  limit, 472
  repelling, 276, 472
  semistable, 479
  van der Pol, 475
Cylindrical harmonics, 650

Damping constant, 148
Darwin, Charles, 285
Data, 88, 112, 158, 263
Decay estimate, 412
Decay of Coefficients, 537
Definiteness, 463, 466
Delta function, 331
  solving an IVP, 332
Dense orbit, 508
Derivative
  complex-valued function, 679
  following the motion, 455
Determinant, 356, 357
De Moivre's Formula, 678
Difference equation, 315, 316
Differential equation, *see* ODE
Differential operator D, *see* Operators
Differential systems, 34
Differential-delay equation, 301, 321
Diffusion equation, 523, 568
Dimension
  Hausdorff, 510
  of a linear space, 359
Dimensionless variables, 82, 259, 686
Direction field, 11, 266
Dirichlet Problem, 579
  boundary data for, 579
  classical, 579
  expansion in circular
    harmonics, 581

well-posed, 583
Discretization scheme, 125
Distance, 434, 528
Divergence Theorem, 683
Dot product, 214
  bilinearity, 221
  positive definiteness, 221
  symmetry, 221
Driving term, 35, 42, 190, 202, 371
Duffing's equation, 513
Dulac's Criterion, 488
Duty cycle, 672
Dynamical system, 42

Edelstein-Keshet, Leah, 276
Eigenbasis, 364
Eigenfunction, 554
Eigenline, 389
Eigenspace, 362, 554
  deficient, 367
  nondeficient, 367
Eigenspace Dimension, 367
Eigenspace Property, 364
Eigenvalue, 362, 554
  multiplicity, 366
  simple, 366
Eigenvector, 362, 554
  generalized, 367
Elastic coefficient, 586
Elementary row operations, 353
Elliptic integral, 223
Energy
  kinetic, 453
    angular rotation, 458
  potential, 453
  total, 453
Engineering functions, 28, 672
Equilibrium point, 265, 434
  center (or vortex), 389
  deficient node, 389
  elementary, 449
  improper node, 389
  saddle, 389
  spiral (or focus), 389
  star (or proper) node, 389
Equilibrium solution, 4, 12, 98, 150, 265, 434
Errors, 126
  global discretization, 126
  in the mean, 532
  local discretization, 126
  mean-square, 533
  round-off, 132

total, 132
Escape time, finite, 98, 105
Escape velocity, 62
Estimating the Deviation, 111
Euclidean space, 527
   complex, 529
Euler, Leonhard, 123
Euler ODE, 165, 209, 581, 611
Euler solution, 123
Euler's constant, 639
Euler's formulas, 678, 680
Euler's Method, 122, 512
Euler-Fourier Law of Heat
   Conduction, 567
Exact equation, 55, 461
Existence and Uniqueness
   Theorem, 22, 88, 113,
   158, 202, 263
Existence Theorem, 656
Exponential matrix, 402
Exponential order, 305
Extension Principle, 96, 113, 158,
   263, 663

Faraday's Law, 242
Fick's Law, 567
First-order linear ODE, 18
   coefficient $p(t)$, 18
   driven, 18
   driving term, 18
   input $q(t)$, 18
   integrating factor for, 18
   IVP
     solution of, 25
   undriven, 18
First-order system, 56, 254
Forced oscillation, 103–104, 199,
   226, 232, 240, 397, 410,
   552
Formal solution, 560
Forward IVP, 44
Fourier, Joseph, 540
Fourier coefficient, 535
Fourier series, 536, 540
   decay estimates, 544
   decay of coefficients, 542
   exponential, 550
   odd/even extensions, 548
   trigonometric, 539
Fourier-Euler formulas, 535, 539
Fredholm Alternative, 362, 370
Frequency, 185
   circular, 185

Frequency response modeling, 232
Frobenius Method, 635
Frobenius series, 616
Frobenius's Theorem, 619, 635
Fubini's Theorem, 683
Function vector space, 350
Functions
   continuity set of, 202
   definite, 463, 466
   even, 540
   indefinite, 466
   odd, 540
   of exponential order, 305
   order of, 442, 516
   order zero homogeneous, 76
   piecewise continuous, 93
   polynomial-exponential, 194
   quadratic form, 466
   semidefinite, 466
Fundamental matrix, 401
Fundamental Theorem, 263
   for first-order ODEs, 113
   for linear ODEs, 202
   for linear systems, 425
   for second-order ODEs, 158
   for systems, 263

Gain, 235
Galilei, Galileo, 37
Gamma function, 626
General solution, 48, 49, 372
General solution theorem
   for $P(D)[y] = f(t)$, 208
   for system $x' = Ax$, 402
   second-order
     complex-valued, 182
     real-valued, 181
Generalized Eigenspace Property,
   370
Generalized function, 334
Geometric series, 317, 676
Geometric vectors, 214
Geometry of orbits
   boundedness, 483
   Dulac's Criterion, 488
   Ragozin's Negative Criterion,
     488
   separation, 265
   unboundedness, 483
Gerschgorin disks, 416
Gibbs phenomenon, 541
Global attractor, 504
Gradient, 683

Green's kernel, 211
Green's Theorem, 683
Gronwall's inequality, 665

Half-life, 36, 687
Hamiltonian system, 461
Harmonic equation
   Maximum Principle, 582
   Mean Value Property, 582
   uniqueness of solutions, 583
Harmonic function, 579
Harmonic motion, 186, 224
   circular frequency of, 224
   resonance, 226
   simple, 186, 224
Harmonic oscillator, 165, 186
Harvesting
   constant effort, 279, 290
   constant rate, 2
   Law of, 290
   logistic population, 116
Hausdorff dimension, 510
Heat equation, 523, 568, 574–576
   Maximal Principle, 574
   undriven, 568
Heat operator, 576
Heaviside, Oliver, 310
Heaviside function, 310
Hermite's equation, 594, 608
Hertz, 104
Heun's Method, 127
Heun, Karl, 127
Hilbert, David, 477
Homogeneity, 529
Homogeneous of order zero, 76
Hooke's Law, 148
Hopf bifurcation, *see* Bifurcation,
   Hopf
Hopf Bifurcation Theorem, 492

Ideal Gas Law, 222
Identity Theorem, 674
Impedance, 246
Implicit solution, 49
Impulse, 334
Indicial polynomial, 611, 616
Inductance, 242
Inertial ellipsoid, 458
Initial condition, 44, 88, 158, 263,
   518
Initial value problem, *see* IVP
Input, 18, 73
Insulation, 569

Integral, 49, 452, 455
Integral curve, 49
Integral equation, 89
Integral estimate, 682
Integrating factor, 18
Integration by parts, 682
Interior
    point, 516, 653
    region, 516
Intermediate Value Theorem, 681
IVP, 3, 10, 44, 254
    approximation of solutions, 122
        Euler's Method, 122, 137
        Heun's Method, 127
        implementation problems, 132
        Runge–Kutta Methods, 128
    backward, 44
    basic questions for, 88
    Existence and Uniqueness
        Theorem, 88, 113, 158, 263
    for first-order linear ODEs
        continuity in data, 669
        Sensitivity Estimate, 665
    forward, 44
    Picard iteration scheme, 89, 655
    properties of solutions, 669
    two-sided, 44

Jacobian matrix, 446, 684
Jump discontinuity, 93

Kepler, Johannes, 61
Kernel function, 211
Kirchhoff's Laws, 243, 244
Kirchhoff, Gustav, 243
Kutta, M. W., 128

L'Hôpital's Rule, 681
l^2, 531
Lagrange form, 676
Lagrange's Identity, 547, 680
Laguerre polynomials, 621
Laguerre's equation, 621
Lanchester, Frederick William, 62
Laplace transform
    Convolution Theorem, 327
    definition of, 301
    for systems, 409
    inverse of, 303

linearity of, 303
of f', 302
of $f^{(n)}$, 307
of $t^n f(t)$, 307
of a periodic function, 317
of an integral, 309
of ramp function, 312
of square pulse, 305
of square wave, 318, 324
of step function, 312
Shifting Theorem, 311
smoothness and decay, 306
solving IVP by, 304
triangular wave, 325
Laplace's equation, 523, 578, 684
Laplace, Pierre-Simon de, 300
Laplacian, 524, 684
Law of Averages, 289
Legendre basis, 646
Legendre polynomials, 603–606
    recursion formula, 602, 604
    table of properties for, 605
Legendre's equation, 593, 601, 609, 621
Leibniz's rule, 682
Leibniz, Gottfried, 83
Level set, 49, 452, 455
Libby, William, 37
Liénard xy-plane, 479
Limit cycle, 472
Limit sets, 482
Limiting velocity, 40, 53
Linear approximation, 154
Linear combination, 173
Linear differential equation
    first-order, 18
        driven, 18
        general solution, 25
        undriven, 18
    second-order, 202
Linear differential system, 34, 350
    second order, 255, 394
Linear equations
    solvability, 358
Linear independence, 358
Linear operator, 173, 303
Linear space, 526
    basis for, 359
    dimension of, 359
    linear independence in, 358
    subspace of, 530
Linear system, 339–432
    driving term, 351

input, 351
state vector, 351
system matrix, 351
undriven, 256, 351
unforced, 351
Linearity property, 172
Linearization
    hard spring, 154
    nearly linear system, 444
    pendulum, 220
    planar system, 292
Lipschitz condition, 654, 655
Lissajous curve, 257, 397
Logarithmic identities, 680
Logistic curves, 53
Logistic ODE, 51
Long-term behavior
    Euler solutions, 134, 137
    first-order linear, 101, 106
    general systems, 434, 439, 445, 480–488, 501
    linear systems, 389–393, 411, 418
    second-order linear, 235
Lorenz system, 501
    chaos, 507
    limit cycles, 505
    Lorenz attractor, 503
    Lorenz squeeze, 503
    orbit map of, 503
    shadowing property, 509
    strange attractor, 508
Lorenz, E.N., 500
Lotka-Volterra system, 286
Low-pass filter, 104, 420, 422
Lyapunov, Aleksandr
        Mikhailovich, 462
Lyapunov functions
    strong, 463
    strong local, 466
    weak, 468
Lyapunov's First Theorem, 463

Maclaurin series, 675
Manifold
    unstable, 504
Mathieu equation, 600
Matrices, 345
    addition, 346
    augmented, 353
    block, 348
    characteristic polynomial, 365

derivative of, 350
determinant, 356, 357
diagonal, 346
eigenvalue of transpose, 370
eigenvalues of, 362
elementary row operations, 353
identity, 348
inverse, 355
invertibility of, 355
main diagonal, 346
minor, 356
operations, 346
product, 346
singular, 355
skew-symmetric, 432, 461
square, 345
symmetric, 348
transpose, 348
triangular, 346
zero, 348
Matrix estimate, 349
Matrix exponential, 402
Maximal Principle, 574
Maximally extended solutions, 97, 158, 263
Mean Approximation Theorem, 532
Method of
    Eigenfunction Expansions, 562
    Eigenvectors, 372
    Frobenius, 619
    Integrating Factors, 19
    Power Series, 590, 596
    Separation of Variables, 47, 557
    Undetermined Coefficients, 194
    Variation of Parameters, 210
    Varied Parameters, 66
Model, 34
    autocatalator, 258, 260, 499, 689
    breaking a string, 566
    car following, 300, 320
        separation control model, 326
    cascade, 73
    circuit
        LC, 318
        RC, 103
        RLC, 244

scroll, 494
cold pills, 67, 325
combat, 62
communications channel, 103
compartment, 72, 339
competition, 282
cooperation, 280
    stability, 448
double pendulum, 388
elements of, 42
falling body
    Newtonian damping, 53
    sky diver, 55
frequency response, 103, 199, 232
    curves of, 235
guitar string, 521, 525, 557
harvesting a population, 1, 116, 313, 316
heat conduction, 566
lead in human body, 339, 418
low-pass filter, 104, 325
optimal depth for a wine cellar, 572
pendulum, 216, 271, 446, 514
pollutant accumulation, 24, 26
population, 1, 116, 278
    computer simulations, 280–283
    harvesting, 3, 116, 279, 290
    interacting, 276
    Principle of Competitive Exclusion, 283
possum plague, 294–296
potassium-argon dating, 46
predator-prey, 286
pursuit, 78
radioactivity, 35, 316
range of validity of, 43
SIR disease, 298
sky diver, 53
springs, 148
    aging, 148
    coupled, 253, 394
    hard, 148
    Hooke's Law, 149
    soft, 148
temperature in a ball, 647
temperature in a cylinder, 586, 642
temperature in a rod, 569

temperature of a horse, 33
tennis racket, 458, 469
validation of, 43
vertical motion, 37, 53, 61
vibrating string, 516
viscous damping, 40
whiffle ball, 40
Modeling principles
    Balance Law, 24
    Chemical Law of Mass Action, 259
    Coulomb's Law, 243
    Euler-Fourier Law of Heat Conduction, 567
    Faraday's Law, 242
    Fick's Law, 567
    Ideal Gas Law, 222
    Kirchhoff's Laws, 243, 244
    Newton's Law of Cooling, 33, 569
    Newton's Law of Universal Gravitation, 61
    Newton's Laws of Motion, 39, 215, 216
    Ohm's Law, 242
    Population Law of Mass Action, 278
    Radioactive Decay Law, 35
Modulus, 678
Multiplicity of eigenvalue, 366
Multiplicity of roots, 680

Natural frequencies, 522, 559
Natural laws, 42
Natural parameter, 42
Natural variables, 42
Nearly linear system, 444
Neumann problems, 651
Newton, Isaac, 39
Newton's Law of Cooling, 33, 569
Newton's Laws of Motion, 39, 215, 216
Newtonian damping, 53
Newtonian mechanics
    inertial frames, 215
    momentum, 216
Nilpotent, 405
Nodal latitude, 649
Node, 522
    linear, 389, 449
    nonlinear, 449
Nonlinear
    ODE, 16

system, 35
Norm, 528
Normal form, 10, 34, 157, 254, 592
Normal linear form, 16, 149, 158,
        202, 255
Normal mode, 559
        circular frequencies, 397
        oscillations, 397
        vectors, 397
Null space, 204, 362, 372, 519
Nullcline, 11, 266
Numerical methods
        Euler's method, 122
        Heun's method, 127
        one-step methods, 125
        Runge–Kutta method, 128
Numerical solvers, 130, 131
        generating solution curves, 11

ODE
        Airy's equation, 600
        Bernoulli's equation, 83
        Bessel's equation, 622
        Chebyshev's equation, 608
        Duffing's equation, 513
        Euler equation, 209, 581, 611
        Exact, 55
        Hermite's equation, 594, 608
        Laguerre's equation, 621
        Legendre's equation, 601, 609
        Liénard equation, 479
        Linear, 18, 167, 202
        Mathieu equation, 600
        Painlevé transcendent, 165
        Rayleigh equation, 479, 498
        Riccati's equation, 83
Ohm's Law, 242
One-step method, 125
Open ball, 442
Open interval of convergence, 674
Operators
        D, 171
        P(D), 172, 204
        action of, 204
        codomain of, 204
        domain of, 204
        Fourier series, 536
        Laplace transform operator,
                see Laplace transform
        Laplacian, 524
        linear, 173
        Linearity property, 190
        range of, 204

symmetric, 553
Orbit, 57, 152, 159, 256
        bounded, 482
        simple pendulum, 218
Orbit map, 503
Order
        of a function, 443, 516
        of a one-step method, 126
        of a system, 34
        of an ODE, 15
Ordinary differential equation, see
        ODE
Ordinary point, 592
Orthogonal projection, 531
Orthogonal set, 529
Orthogonality, 529
        of Bessel functions, 630
        of Legendre polynomials, 604
        relation, 526
        with respect to a density, 607
Oscillation
        critically damped, 225
        forced, 103, 227, 410
        free periodic, 224, 227
        overdamped, 225
        underdamped, 225
Output, 43
Overdamped, 225
Overflow, 135

Painlevé transcendent, 165
Parallelogram Law, 214
Parameter identification, 237
Parameters, 686
Parseval Relation, 537
Partial differential equation, see
        PDE
Partial fractions, 314
Particular solution, 208, 372
PDE
        diffusion equation, 523, 568
        driven, 524
        heat equation, 523, 568
                undriven, 568
        Laplace's equation, 523, 578
        linear, 524
        order, 524
        separated solution, 521
        undriven, 524
        wave equation, 518, 523
Pendulum, 216–221
        damped simple, 218, 486
        double, 388

driven, 514
        effects of changing mass, 271
        energy of, 453
        orbits of, 218, 222
        stability, 446
        viscous damping, 222
Period
        fundamental, 185
Period doubling, 137, 506
Period three implies chaos, 142
Periodic Cycle, Law of, 288
Periodic extension, 542
Periodic function, 185
Periodic steady state, 409
Perturbation
        higher order, 444
Phase angle, 186
Phase shift, 235
Picard iterate, 89, 655, 657, 662
Piecewise continuity, 93, 528
Piecewise smooth, 541
Pigs
        three little, see Teddy bears
Poincaré, Jules Henri, 480
Poincaré time section, 508
Point-orbit, 159
Polar coordinates, 272, 487
Polycycle, 481
Polynomial operators, 172
Polynomial-exponential functions,
        387
Population Law of Mass Action,
        278
Population quadrant, 5, 277
Portrait, planar systems, 389
Positive definite, 463, 527, 529
Potential difference, 241
Potential equation, 523, 578
Potential function, 241, 579
Power series, 674
        algebra of, 674
        calculus of, 674
        Method of, 590, 596
Predator-prey interaction, 278,
        285–293
        growth coefficients, 278
        Law of Averages, 289
        natural decay, 278
        satiable predation, 280, 493
Pulse function, 310, 672

$\mathbb{R}^3$
        subspaces of, 350